T0180521

Dilip Singh Sisodia · Lalit Garg ·
Ram Bilas Pachori · M. Tanveer
Editors

Machine Intelligence Techniques for Data Analysis and Signal Processing

Proceedings of the 4th International
Conference MISP 2022, Volume 1

Editors
Dilip Singh Sisodia
Department of Computer Science
and Engineering
National Institute of Technology Raipur
Raipur, Chhattisgarh, India

Lalit Garg
Department of Information
and Communication Technology
University of Malta
Msida, Malta

Ram Bilas Pachori
Department of Electrical Engineering
Indian Institute of Technology Indore
Indore, India

M. Tanveer
Department of Mathematics
Indian Institute of Technology Indore
Indore, India

ISSN 1876-1100   ISSN 1876-1119 (electronic)
Lecture Notes in Electrical Engineering
ISBN 978-981-99-0087-9   ISBN 978-981-99-0085-5 (eBook)
https://doi.org/10.1007/978-981-99-0085-5

This Springer imprint is published by the registered company Springer Nature Singapore Pte Ltd.
The registered company address is: 152 Beach Road, #21-01/04 Gateway East, Singapore 189721,
Singapore

Contents

About the Editors

Dilip Singh Sisodia is an Associate Professor and Head at the Department of Computer Science Engineering of National Institute of Technology Raipur. Dr. Sisodia contributed over 100 high-impact articles in reputed journals, conference proceedings, and edited volumes. He also edited Scopus indexed research book volumes published by Springer Nature and IGI Global. Dr. Sisodia was included in the World Ranking of Top 2% Scientists/researchers in 2022 by a study of Scientists from Stanford University USA and Published by Elsevier B.V. So far he has supervised four Ph.D. thesis, eight M.Tech. dissertations and more than 50 B.Tech. projects. He is a Senior Member of IEEE and ACM. He received a Ph.D. degree in computer science and engineering from the National Institute of Technology Raipur, India. He earned his M.Tech. and B.E. degrees, respectively, in information technology (with specialization in artificial intelligence) and computer science and engineering from the Rajiv Gandhi Technological University, Bhopal, India. His research interests include the applications of machine learning/soft computing techniques, artificial intelligence, biomedical signal, and image processing.

Lalit Garg is a Senior Lecturer in the Department of Computer Information Systems at the University of Malta, Malta, and an Honorary Lecturer at the University of Liverpool, UK. He has been a Researcher at the Nanyang Technological University, Singapore, and Ulster University, UK, has supervised 200+ Master's dissertations, two DBA, and two Ph.D. theses, and published 120+ high-impact publications in refereed journals/conferences/books, five edited books, and 20 patents. His research interests are business intelligence, machine learning, data science, deep learning, cloud computing, mobile computing, the Internet of Things (IoT), information systems, management science, and their applications mainly in healthcare and medical domains.

Ram Bilas Pachori received a B.E. degree with honors in Electronics and Communication Engineering from Rajiv Gandhi Technological University, Bhopal, India, in 2001, the M.Tech. and Ph.D. degrees in Electrical Engineering from Indian Institute of Technology (IIT) Kanpur, Kanpur, India, in 2003 and 2008, respectively. He was Postdoctoral Fellowship Holder at Charles Delaunay Institute, the University of Technology of Troyes, Troyes, France, from 2007 to 2008. He is a Professor in the Department of Electrical Engineering at IIT Indore, India. His research interests are in the areas of signal and image processing, biomedical signal processing, non-stationary signal processing, speech signal processing, brain-computer interfacing, machine learning, and artificial intelligence in health care. He has supervised 14 Ph.D., 20 M.Tech., and 41 B.Tech. students for their theses and projects. He has over 235 publications which include journal papers, conference papers, books, and chapters. He has worked on various research projects with funding support from SERB, DST, DBT, and CSIR.

M. Tanveer is an Associate Professor and Ramanujan Fellow at the Discipline of Mathematics of the Indian Institute of Technology Indore. Prior to that, he worked as Postdoctoral Research Fellow at the Rolls-Royce@NTU Corporate Lab of the Nanyang Technological University, Singapore. He received a Ph.D. degree in Computer Science from the Jawaharlal Nehru University, New Delhi, India. His research interests include support vector machines, optimization, machine learning, deep learning, and applications to Alzheimer's disease and dementia. He has published over 80 research papers in journals of international repute.

On Diverse and Serendipitous Item Recommendation: A Reinforced Similarity and Multi-objective Optimization-Based Composite Recommendation Framework

Rahul Shrivastava, Dilip Singh Sisodia, and Naresh Kumar Nagwani

1 Introduction

Recommender system is an effective business strategy for dealing with an overwhelming volume of content available on web applications such as e-commerce websites, social media news, and other applications [1]. The recommender system can infer customer interest through different information sources and provide personalized information access [2, 3]. Statistical analysis techniques and knowledge data discovery techniques help the recommender system achieve personalization [4]. However, most of the research is monotonous towards considering a single performance indicator as accuracy [5]. Recommending popular items and achieving better accuracy may not always be a suggested way of recommendation [6]. Improved precision-based top-n personalized recommendation algorithm may lack in generating a diversified novel list of items and which may, in turn, leads to the problem of a user losing interest in the recommendation [7]. Nowadays, the recommender system should have the capability to produce an unexpected novel yet relevant recommendation [8]. Recent studies have proposed numerous works that consider beyond-accuracy objectives such as diversity, serendipity, and novelty [9, 10]. However, the

R. Shrivastava (✉) · D. S. Sisodia · N. K. Nagwani
Department of Computer Science and Engineering, National Institute of Technology Raipur, GE Road Raipur, Chhattisgarh 492010, India
e-mail: rshrivastava.phd2018.cs@nitrr.ac.in

D. S. Sisodia
e-mail: dssisodia.cs@nitrr.ac.in

N. K. Nagwani
e-mail: nknagwani.cs@nitrr.ac.in

© The Author(s), under exclusive license to Springer Nature Singapore Pte Ltd. 2023
D. S. Sisodia et al. (eds.), *Machine Intelligence Techniques for Data Analysis and Signal Processing*, Lecture Notes in Electrical Engineering 997,
https://doi.org/10.1007/978-981-99-0085-5_1

memory-based collaborative filtering approach lacks handling data sparsity problems [11] and is incapable of producing better fitness results for objective functions. Additionally, these methods only consider the intuitive definition of beyond-accuracy objectives and adapt the traditional rating evaluation approach, with poor performance with sparse data.

Hence, this research proposes an evolutionary optimization-based multi-objective recommendation framework (MORF). MORF establishes the trade-off solution between the three conflicting objectives, i.e. accuracy, diversity, and serendipity. It employs a reinforced similarity-based rating prediction model [11] to mitigate the data sparsity problems and evaluate the unknown rating of an item. The reinforced similarity model identifies implicit user–user and item–item similarity and enhances the capability of user–item rating metrics. The objective accuracy function incorporating popular items in a recommendation list diversity improves by incorporating novel items of different categories in the recommendation list (RL). The third objective function, serendipity, brings the surprise factor in the RL and includes unexpected yet relevant items. Finally, joint optimization of three objective functions is performed using NSGA-II [12] to generate a Pareto-optimal solution. The following are the key contributions of this article:

- This study proposes a multi-objective recommendation framework (MORF) to recommend diverse and serendipitous items by simultaneous optimization of conflicting objectives that includes accuracy, diversity, and serendipity.
- We designed objective functions for diversity and serendipity by incorporating their associated component metrics strength for improving the capability of recommending unexpected, novel, yet relevant items.
- This study employs a reinforced similarity computation-based rating prediction model to improve the fitness of objective functions.
- The performance of MORF is compared with baseline using Pareto front and evaluated with accuracy and non-accuracy-based measures.

The rest of the article, organized as follows in Sect. 2, describes the related work and background details like rating evaluation strategy and multi-objective evolutionary algorithm. Section 3 describes the necessary steps, objective functions, and workflow of the MORF. Further, experimental results and discussion are presented in Sect. 4. Lastly, Sect. 5 concludes this paper with a possible future direction.

2   Related Work

This section presents the recent literature related to traditional recommendation systems, traditional rating evaluation, and multi-objective recommendation models. This section also briefly describes the background details of the rating evaluation strategy and multi-objective evolutionary algorithm.

2.1 End-User Personalized Traditional Recommendation System

Neighbourhood or memory-based [13] collaborative filtering technique constructs user–user and item–item similarity matrix to predict the user's unknown preferences [14]. Model-based collaborative filtering [15] techniques accommodate; accuracy improves personalized top-n recommendation. The model-based algorithm explores the semantic relationship and builds a user–item relationship. Bayesian model [16] and matrix factorization [17] are popular approaches that improve collaborative filtering accuracy. The rating prediction task in traditional neighbourhood-based collaborative filtering (CF) techniques like user-based CF [18] and item-based CF [13] is entirely based on practical similarity computation between user and item. The item-based and user-based CF model uses the historical rating information available for users and items for similarity computation. These approaches suffer from the data sparsity problem. In most real-world recommender system applications, the sparsity level is above 90% [11]. Effective preference prediction for target users becomes difficult due to data sparsity.

2.2 Multi-Objective Recommendation Model

Several studies such as [19, 20] discuss that only improving the accuracy of generated recommendations may not generate diversified novel item suggestions. It also leads to a problem of the user losing interest in the recommendation framework. In the same context of user interest, numerous works presented in [21] describe users who are surprised by receiving unexpected [22] yet valuable recommendations. Several efforts have been made in [23] to keep user interest high in the Pareto efficient multi-objective optimization-based recommendation system and improve the item suggestion in a recommendation list. A long-tail items recommendation using a multi-objective optimization approach has been proposed in [24] to address the problem of recommending popular and unpopular items. A multi-objective evolutionary algorithm for recommender systems is proposed in [25] to produce a set of non-dominated solutions for the interested user. A multi-objective optimization using an evolutionary algorithm [26] was found to be a majorly adapted approach for generating a trade-off recommendation solution. Recently, a topic popularity-based item diversification approach was proposed [27]. A probabilistic multi-objective evolutionary algorithm proposed in [9] incorporates the probabilistic crossover operator and considers accuracy and diversity objective functions to optimize. But in the algorithms mentioned above, the objective functions are intuitive and lack in capturing broader insight of diversified novel item recommendations and producing unexpected yet relevant solutions.

2.3 *Motivation*

The recent studies consider only one non-accuracy indicator, like only diversity or coverage. Still, there is a need to develop an approach to incorporate more non-accuracy indicator objective functions to enhance the recommendation list's capability in real-world application scenarios. Additionally, the above algorithms consider only a traditional user-based or item-based collaborative filtering approach for unknown item rating evaluation and suffer from data sparsity [11]. The generated user–item rating matrix and similarity matrix are adequately used to calculate the respective values of objective functions. However, the user–item rating matrix values may not be efficient enough to produce an accurate, objective function value due to data sparsity. Therefore, the mentioned aspects need to be considered when multiple conflicting objectives participate in a recommendation system.

3   Proposed Multi-stakeholder Recommendation Model

In this section, the different phases involved in the MORF have been presented. The discussion starts with a brief overview of the MORF. Next, the discussion regarding reinforced similarity-based rating evaluation, objective functions, and evaluation metrics is presented.

3.1 *The Proposed Multi-objective Recommendation Framework (MORF)*

The MORF is shown in Fig. 1. An enhanced reinforced similarity-based rating evaluation model [11] performs the rating prediction and generates user and item similarity matrix and user–item rating matrix, which further helps in evaluating the fitness value of each respective fitness or objective function of accuracy [9, 28], diversity [29], and serendipity [30, 31]. In Fig. 2, we have shown that the three objectives are extensively evaluated by considering their sub-component. In the MORF, we believe separate sub-components of the different objective functions and integrate them into a single multi-objective recommendation framework. Further, MORF performs the evolutionary optimization of the fitness functions and generates the Pareto-optimal trade-off recommendation solution. Conventional genetic operators such as selection, crossover, and mutation are being used to create new child recommendation solutions.

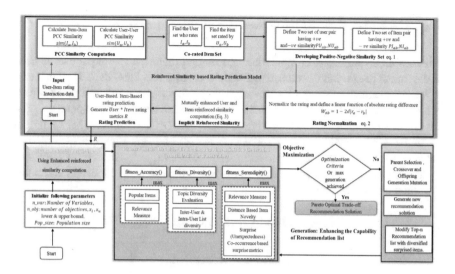

Fig. 1 Proposed multi-objective recommendation framework

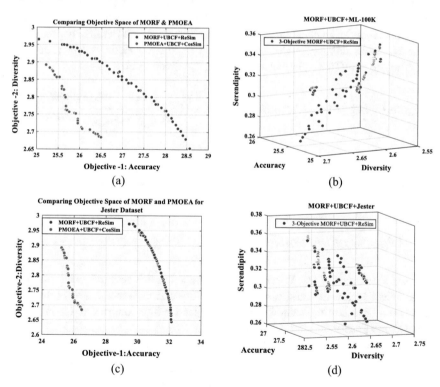

Fig. 2 Pareto front describing the trade-off recommendation solution for ML-100 K data set **a** MOR + UBCF + ReSim **b** PMOEA + UBCF + CosSim **c** Comparing Pareto front (a) and (b) **d** 3-Objectives (accuracy–diversity–serendipity) Pareto front of MORF + UBCF + ReSim

3.2 Reinforced Similarity-Based Rating Prediction

The reinforced similarity computation and rating prediction technique can be used to determine the implicit similarity between user–user and item–item. It contributes to the achievement of more accurate user–item rating metrics and optimized fitness values. Equations (1–3) briefly describe the reinforced similarity computation. The $\text{sim}\left(I_{uj}, I_{uk}\right)$ denotes Pearson correlation coefficient (PCC) similarity between item I_{uj} and I_{uk}. The $\text{Positive}_{\text{Similarity}}$ represents the set of items where PCC similarity between items is greater than zero. Similarly, $\text{Negative}_{\text{Similarity}}$ represents the set of items where PCC similarity between items is less than zero. Equation (1) computes the union of both similarities.

$$\text{PN}_{\text{similarity}} : \text{Positive}_{\text{Similarity}} \cup \text{Negative}_{\text{Similarity}} \tag{1}$$

$$W_{MN} = 1 - 2d\left|r_{uj} - r_{uk}\right| \tag{2}$$

In Eq. (2), the W_{MN} measures the implicit contribution of item pair and the similarity between ratings defined as an absolute rating difference. α is a damping factor that controls the degree of contribution of the reinforced similarity procedure. r_{uj} and r_{uk} is the rating given by user u to item j and k, respectively. The reinforced similarity $\text{Re\_sim}(U_j, U_k)$ between two users is defined in Eq. (3).

$$\text{Re\_sim}(U_j, U_k) = (1 - \alpha).\text{sim}(U_j, U_k) + \alpha. \frac{\Sigma_{(I_{u_j}, I_{u_k}) \in PN_{\text{similarity}}} W_{MN}.\text{sim}(I_{uj}, I_{uk})}{\Sigma_{(I_{u_j}, I_{u_k}) \in PN_{\text{similarity}}} |W_{MN}|} \tag{3}$$

where $\text{Reinforced\_sim}(U_j, U_k)$ represents reinforced similarity between the user U_j and U_k, $\text{sim}(U_j, U_k)$ represents PCC similarity between both the users.

3.3 Objective Functions

This section describes the objective functions used in this research. The accuracy-based objective function addresses the popularity and relevance-based preferences of the user. Diversity and serendipity-based objective functions ensure the inclusion of novel and unexpected items in the RL.

The objective function of accuracy is described as Eq. (4).

$$\text{Accuracy} = \frac{\Sigma_{u \in U} \Sigma_{l \in T} \left(\widehat{P\_R}_{u,l}\right)_{\text{RE\_Sim}}}{|U|.|T|} \tag{4}$$

where U is the set of all users, T is the top-n recommendation list, $\widehat{P\_R}_{u,l}$ is the predicted rating between the user and item l in the recommendation list, $|U|$ is the total number of users, $|T|$ denotes the length of the recommendation list, and RE_Sim represents reinforced similarity. The diversity objective function may promote unpopular yet useful items that may differ in features like various movies termed as diversity in topics [9]. The objective function of the topic diversity described is Eq. (5).

$$\text{Topic}_{\text{Diversity}} = \frac{\sum_{U \in \text{User}} \sum_{L \in \text{Top}-n_{\text{List}}} \left(\text{Topic}_{\text{div}}\right)}{|U|} \tag{5}$$

where $\text{Top} - n_{\text{List}}$ represents the top-n recommendation list generated for every user. The $\text{Topic}_{\text{div}}$ represents the topic distribution diversity in the recommendation list. Further, we explored another dimension of diversity. The Hamming distance measurement [32] was used to measure the difference between the two user's top-n recommendation list; the greater the value of the more diverse recommendation is, the intra-user diversity $\text{IU}_{\text{Diversity}}$ is described in Eq. (6).

$$\text{IU}_{\text{Diversity}} = \text{Hamming Distance}_{i,j} = 1 - \frac{\left|\text{Common}_{\text{Item}}(i, j)_{\text{Top}-n\text{List}}\right|}{\left|\text{Top} - n\text{List}\right|} \tag{6}$$

where $\text{Common}_{\text{Item}}(i, j)$ is the common item in the top-n recommendation list of users i and j. If the recommendation list is the same for both, the user's hamming distance will be zero. If the list is entirely different, the hamming distance will be one. Mean hamming distance can be obtained by averaging all user pairs. A higher value signifies better diversity in the recommendation. Equations (5) and (6) of topic diversity and inter-user diversity are objective maximization problems. The maximum among the topic and inter-user diversity in the proposed framework can be selected as mentioned in Eq. (7) to gain the recommendation list's overall diversity.

$$\text{Diversity} = \max(\text{Topic}_{\text{Diversity}}, \text{Inter}_{\text{User}_{\text{Diversity}}}) \tag{7}$$

Pointwise mutual information (PMI) formulated in Eq. (8) for the pair of item i and j:

$$\text{PMI}(i, j) = -\log_2 \frac{p(i, j)}{p(i)p(j)} / \log_2 p(i, j), \tag{8}$$

where $p(i)$ and $p(j)$ are the probabilities for the item i, j to be rated by any user from the set of all users. $P(i, j)$ is the likelihood for both items to be rated by users from the group of all users. Higher pointwise mutual information value signifies higher co-occurrence between items and lowers the chances of being surprised by the recommended item. The objective function for the serendipity is described in Eq. (9).

$$\text{Serendipity} = \max_{J \in \text{Item}_r} \text{PMI}(\text{Recommended}_{\text{item}}, j) \tag{9}$$

3.4 Evaluation Metrics

The performance of MORF is validated over precision [32]-, diversity-, and novelty [31]-based evaluation metrics. Equations (10), and (11) present the definition for the precision and novelty evaluation metrics. And diversity metrics is already defined in Eqs. (5–7).

$$\text{Precision} = \frac{1}{|U|} \sum_{u \in U} \frac{\text{Relevant}_{\text{item}}}{|L_R|} \tag{10}$$

$$\text{Novelty} = \frac{1}{|L_R|} \sum_{J \in \text{Item}_r} \text{distance}(i, j) \tag{11}$$

where $\text{distance}(i, j) = 1 - \text{RE}_{\text{sim}}(i, j)$, Relevant$_{\text{item}}$ represents the set of relevant items (highly rated items in the RL), and L_R denotes the length of the RL.

4 Experiments

This section describes the experimental parameter setting and data set used to evaluate the proposed model. Further, the comparison of obtained Pareto front and evaluation results with competing baseline algorithms is discussed.

4.1 Experimental Data

We use two data sets Movie Lens data set [33] and Jester data set [34], for evaluating the performance of the MORF. The Movie Lens data set consists of 100 K rating ranges from 1 to 5 from 943 users and 1682 movies. Each user has rated at least 20 movies—necessary contextual information like age, gender occupation, and zip contained by the data set. The data set was collected through the grouplens website, which made available the rating data set (https://grouplens.org/datasets/movielens/). The Jester data set contains a jester joke recommender system and includes 4.1 million continuous ratings (-10.00 to $+10.00$) of 100 jokes from 73,496 users. All the ratings from the Jester data set are mapped in the range 1–5. The large Jester data

Table 1 Evaluating model performance based on objective strength with cosine and reinforced similarity

Models	Precision	Topic diversity	Intra-user diversity	Serendipity
UBCF + Cos_Sim	0.528	2.072	1.091	1.163
UBCF + Re_Sim	**0.557**	2.181	1.271	1.227
PMOEA + UBCF + Cos_Sim	0.479	2.791	2.361	2.061
PMOEA + UBCF + Re_Sim	0.488	2.837	2.539	2.251
MORF + UBCF + Cos_Sim	**0.533**	**3.371**	**3.371**	**2.931**
MORF + UBCF + Re_Sim	**0.541**	**3.859**	**3.815**	**3.185**

Bold values represent the superior performance of the model for different evaluation metrics

set [24] was used in this paper. Table 1 represents the statistics of the data set used in the proposed work.

4.2 Competing Algorithm

The performance of MORF is compared with the following baseline algorithm:

- **UBCF** [35]: The traditional single rating user-based collaborative filtering recommendation model learns user expectations with a co-rated item set.
- **IBCF** [13: The traditional single rating item-based collaborative filtering recommendation model learns user expectations with a co-rated item set.
- **PMOEA + UBCF** [9]: Probabilistic multi-objective evolutionary algorithm with UBCF as rating evaluation approach.
- **PMOEA + IBCF** [9]: Probabilistic multi-objective evolutionary algorithm with IBCF as rating evaluation approach.

4.3 Experimental Results Analysis and Discussion

To analyse the Pareto front of the proposed multi-objective recommendation framework with a reinforced similarity-based rating evaluation model with two and three recommendation objectives, we conducted 30 independent executions of each competing algorithm over the Movie Lens and Jester data set on a 64-bit machine with i7 10th generation @2.3Ghz with 16 GB RAM in the windows environment.

Figure 2a represents the comparative Pareto front obtained by MORF and PMOEA over ML-100 K data set with reinforced similarity (ReSim) and cosine similarity-based rating evaluation using UBCF. Figure 2c compares both the Pareto fronts obtained by MORF and PMOEA. And Fig. 2b and d present the three-objective-based Pareto front obtained by MORF with accuracy, diversity, and serendipity as the objective functions, respectively, over Movie Lens and Jester data set. The Pareto front results conclusively describe that the Pareto front obtained by MORF outperforms PMOEA in terms of objective space and fitness value coverage. The balanced trade-off obtained by MORF presents the non-dominated Pareto-optimal recommendation solution. The major observations which justify the performance of MORF are first is enhancing the strength of objective functions by considering the sub-components metrics. The second is incorporating the reinforced similarity in rating prediction that improves the fitness of the objective functions. The precision-, diversity-, and novelty-based evaluation results are presented in Fig. 3. Interestingly, in Fig. 3a, UBCF obtains the highest precision due to exclusive consideration of accuracy metrics. However, MORF outperforms baseline algorithms except for UBCF in terms of precision metrics. MORF simultaneously considers diversity and serendipity metrics that cause MORF to achieve slightly less precision than UBCF. But, Fig. 3b, c indicate that MORF achieves balanced trade-off growth in diversity and novelty metrics over the baseline models.

The results presented in Table 1 demonstrate that MORF, when incorporated with reinforced similarity, substantially improves the strength of objective function when considering sub-components metrics. Precision, topic diversity, intra-user diversity, and serendipity are the metrics utilized to design the objective functions. Table 1 indicates that the precision of the UBCF + Re_Sim outperforms all baseline models due to the inclusion of popular and highly rated items in the RL. In contrast, MORF + UBCF + Re_Sim outperforms all baseline models regarding topic diversity, intra-user diversity, and serendipity. The performance of all the models are evaluated using cosine similarity and reinforced similarity, and a significant improvement has been observed when models are executed with reinforced similarity.

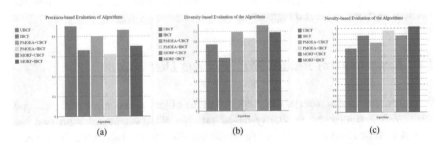

Fig. 3 Precision, diversity, and novelty based evaluation of different algorithms over ML-100 K data set

5 Conclusion

This study was undertaken to design the evolutionary optimization-based recommendation framework and establish the trade-off relationship between the conflicting objectives, including accuracy, diversity, and serendipity. The three key strengths of this study are first is incorporating a reinforced similarity-based rating prediction model. The second is to design an objective function for diversity and serendipity by also considering its component metrics. Finally, establishing a trade-off recommendation solution by employing NSGA-II-based evolutionary optimization approach. The extensive experimental and evaluation results indicate the capabilities of MORF in delivering diverse, unexpected, novel yet relevant recommendation solutions. A natural progression of this work is to analyse the performance of MORF in a multi-stakeholder recommendation environment.

References

1. Resnick P, Varian HR (1997) Recommender systems. Commun ACM 40:56–58. https://doi.org/10.1145/245108.245121
2. Aggarwal CC (2016) An introduction to recommender systems. Recomm Syst Textb 1–28. Springer International Publishing, Cham. https://doi.org/10.1007/978-3-319-29659-3_1
3. Pujahari A, Sisodia DS (2020) Aggregation of preference relations to enhance the ranking quality of collaborative filtering based group recommender system. Expert Syst Appl 156:113476. https://doi.org/10.1016/j.eswa.2020.113476
4. Bobadilla J, Ortega F, Hernando A, Gutiérrez A (2013) Recommender systems survey. Knowledge-Based Syst 46:109–132. https://doi.org/10.1016/j.knosys.2013.03.012
5. Boratto L, Fenu G, Marras M (2021) Connecting user and item perspectives in popularity debiasing for collaborative recommendation. Inf Process Manag 58:102387. https://doi.org/10.1016/j.ipm.2020.102387
6. Liu H, He J, Wang T, Song W, Du X (2013) Combining user preferences and user opinions for accurate recommendation. Electron Commer Res Appl 12:14–23 (2013). https://doi.org/10.1016/j.elerap.2012.05.002
7. Fortes RS, Lacerda A, Freitas A, Bruckner C, Coelho D, Goncalves M (2018) User-oriented objective prioritization for meta-featured multi-objective recommender systems. UMAP 2018—Adjun Publ 26th Conf User Model Adapt Pers 311–316 (2018). https://doi.org/10.1145/3213586.3225243
8. Pujahari A, Sisodia DS (2019) Modeling side information in preference relation based restricted Boltzmann machine for recommender systems. Inf Sci (Ny) 490:126–145. https://doi.org/10.1016/j.ins.2019.03.064
9. Cui L, Ou P, Fu X, Wen Z, Lu N (2017) A novel multi-objective evolutionary algorithm for recommendation systems. J Parallel Distrib Comput 103:53–63. https://doi.org/10.1016/j.jpdc.2016.10.014
10. Lee Y (2020) Serendipity adjustable application recommendation via joint disentangled recurrent variational auto-encoder. Electron Commer Res Appl 44:101017. https://doi.org/10.1016/j.elerap.2020.101017
11. Hu YAN, Shi W, Li H, Hu X (2017) Mitigating data sparsity using similarity reinforcement-enhanced 17:1–20
12. Deb K, Pratap A, Agarwal S, Meyarivan T (2002) A fast and elitist multiobjective genetic algorithm: NSGA-II. IEEE Trans Evol Comput 6:182–197. https://doi.org/10.1109/4235.996017

13. Sarwar B, Karypis G, Konstan J, Riedl J (2001) Item-based collaborative filtering recommendation algorithms. In: Proceeding 10th international conference world wide web, WWW 2001, Association for Computing Machinery, Inc, New York, New York, USA, 2001: pp 285–295. https://doi.org/10.1145/371920.372071

14. Zhang F, Lee VE, Jin R, Garg S, Choo KKR, Maasberg M, Dong L, Cheng C (2019) Privacy-aware smart city: a case study in collaborative filtering recommender systems. J Parallel Distrib Comput 127:145–159. https://doi.org/10.1016/j.jpdc.2017.12.015

15. Pujahari A, Sisodia DS (2020) Pair-wise preference relation based probabilistic matrix factorization for collaborative filtering in recommender system. Knowledge-Based Syst 196:105798. https://doi.org/10.1016/j.knosys.2020.105798

16. Hernando A, Bobadilla J, Ortega F (2016) A non negative matrix factorization for collaborative filtering recommender systems based on a Bayesian probabilistic model. Knowl-Based Syst 97:188–202. https://doi.org/10.1016/j.knosys.2015.12.018

17. Luo X, Xia Y, Zhu Q (2012) Incremental collaborative filtering recommender based on regularized matrix factorization. Knowl-Based Syst 27:271–280. https://doi.org/10.1016/j.knosys.2011.09.006

18. Breese JS, Heckerman D, Kadie C (2013) Empirical analysis of predictive algorithms for collaborative filtering. http://arxiv.org/abs/1301.7363

19. Jungkyu HAN, Yamana H (2017) A survey on recommendation methods beyond accuracy. In: IEICE Trans Inf Syst 2931–2944. https://doi.org/10.1587/transinf.2017EDR0003

20. Kotkov D, Wang S, Veijalainen J (2016) A survey of serendipity in recommender systems. Knowl-Based Syst 111:180–192. https://doi.org/10.1016/j.knosys.2016.08.014

21. Kotkov D, Veijalainen J, Wang S (2016) Challenges of serendipity in recommender systems. WEBIST 2016—Proceeding 12th international conference web information systems technol vol 2, pp 251–256. https://doi.org/10.5220/0005879802510256

22. Adamopoulos P, Tuzhilin A (2011) On unexpectedness in recommender systems: or how to expect the unexpected. CEUR Workshop Proc 816:11–18

23. Lin X, Chen H, Pei C, Sun F, Xiao X, Sun H, Zhang Y, Ou W, Jiang P (2019) A pareto-efficient algorithm for multiple objective optimization in e-commerce recommendation. In: RecSys 2019—ACM conference on recommender systems, 20–28. https://doi.org/10.1145/3298689.3346998

24. Wang S, Gong M, Li H, Yang J (2016) Multi-objective optimization for long tail recommendation. Knowl-Based Syst 104:145–155. https://doi.org/10.1016/j.knosys.2016.04.018

25. Geng B, Li L, Jiao L, Gong M, Cai Q, Wu Y (2015) NNIA-RS: a multi-objective optimization based recommender system. Phys A Stat Mech Its Appl 424:383–397. https://doi.org/10.1016/j.physa.2015.01.007

26. Deb K (2011) Multi-objective optimisation using evolutionary algorithms: an introduction. In: Multi-objective Evolutionary Optimisation for Product. Design and Manufacturing Springer London, pp 3–34. https://doi.org/10.1007/978-0-85729-652-8_1

27. Huang H, Shen H, Meng Z (2019) Item diversified recommendation based on influence diffusion. Inf Process Manag 56:939–954. https://doi.org/10.1016/j.ipm.2019.01.006

28. Zuo Y, Gong M, Zeng J, Ma L, Jiao L (2015) Personalized recommendation based on evolutionary multi-objective optimization [Research frontier]. IEEE Comput Intell Mag 10:52–62. https://doi.org/10.1109/MCI.2014.2369894

29. Karabadji NEI, Beldjoudi S, Seridi H, Aridhi S, Dhifli W (2018) Improving memory-based user collaborative filtering with evolutionary multi-objective optimization. Expert Syst Appl 98:153–165. https://doi.org/10.1016/j.eswa.2018.01.015

30. Kaminskas M, Bridge D (2014) Measuring surprise in recommender systems. In: RecSys REDD 2014 International workshop on recommender systems evaluation: dimensions and design, vol 69, pp 2–7. https://doi.org/10.1007/978-0-387-85820-3_4

31. Vargas S, Castells P (2011) Rank and relevance in novelty and diversity metrics for recommender systems. In: RecSys'11—Proceeding 5th ACM Conference Recomm Syst pp 109–116. https://doi.org/10.1145/2043932.2043955

32. Lü L, Medo M, Yeung CH, Zhang YC, Zhang ZK, Zhou T (2012) Recommender systems. Phys Rep 519:1–49. https://doi.org/10.1016/j.physrep.2012.02.006
33. Miller BN, Albert I, Lam SK, Konstan JA, Riedl J (2003) MovieLens unplugged: experiences with an occasionally connected recommender system. In: Proceedings of the 4th international conference on Intelligent user interfaces, pp 263–266
34. Goldberg KY, Roeder TM (2014) Eigentaste: a constant time collaborative filtering algorithm. CEUR Workshop Proc 1225:41–42. https://doi.org/10.1023/A
35. Herlocker JL, Konstan JA, Borchers A, Riedl J (1999) An algorithmic framework for performing collaborative filtering. In: Proceedings of the 29th annual international ACM SIGIR conference on research and development in information retrieval, pp 230–237

Comparative Analysis of Node-Dependent and Node-Independent Graph Matrices for Brain Connectivity Network

Mangesh Ramaji Kose, Mitul Kumar Ahirwal, and Mithilesh Atulkar

1 Introduction

The human brain is made up of complex networks of interconnected neurons. It controls all the physical as well as cognitive activities performed by a person [1]. Various disorders affect the functionality of the brain. Schizophrenia is one of the disorders causing disability to the person. It is characterized by hallucination, illusions, disorganized thinking, etc. [2]. Various neuroimaging techniques like electroencephalogram (EEG), magnetic resonance imaging (MRI), diffusion tensor imaging (DTI), etc., provide the means for analysis of brain functionality. EEG signals are the noninvasive, less costly, and readily available means of recording electrical activity that occurs inside the brain. EEG electrodes are placed over the specific region of the scalp to record the EEG signal [3]. To detect brain-related disorders, various automated techniques have been proposed by researchers. For example, Ahmed I. Sharaf et al. proposed an approach for epilepsy detection using EEG signals. The authors have segmented the input EEG signal into sub-bands and extracted features from each segment of the signal. The extracted features are then classified with a random forest classifier [4]. In [5], the authors have extracted multiple features from the EEG dataset and applied an independent component analysis (ICA)-based feature selection approach for selecting the best features. A fuzzy classifier is applied to classify the features to detect epileptic EEG signals.

M. R. Kose (✉) · M. Atulkar
Department of Computer Application, NIT, Raipur, C.G. 492010, India
e-mail: mangeshkose@gmail.com

M. Atulkar
e-mail: matulkar.mca@nitrr.ac.in

M. K. Ahirwal
Department of Computer Science and Engineering, MANIT, Bhopal, M.P. 462003, India
e-mail: ahirwalmitul@gmail.com

© The Author(s), under exclusive license to Springer Nature Singapore Pte Ltd. 2023
D. S. Sisodia et al. (eds.), *Machine Intelligence Techniques for Data Analysis and Signal Processing*, Lecture Notes in Electrical Engineering 997,
https://doi.org/10.1007/978-981-99-0085-5_2

Brain-related disorders are complex, having multi-factorial causes which makes it more complicated to diagnose the abnormality than the conventional methodology of the training model. The human brain consists of billions of neurons and different regions that work together to generate particular brain conditions. Hence, to perform diagnosis, the whole brain needs to be considered in addition to functional interconnection among different regions [6]. The conventional approach does not consider this fact. Functional connectivity network forms a connectivity pattern among functionally correlated discrete brain regions [1].

Correlation-based connectivity assessment is considered for the evaluation of connectivity among discrete cortical regions [7]. Graph theory-based approach for analyzing brain connectivity networks is an important means to compare brain connectivity networks for different brain conditions [8]. The comparison among different brain connectivity networks can be performed using quantitative measures. Various graph matrices can be calculated from each network to characterize the brain connectivity network [1]. The selection of graph matrices depends on whether the numbers of nodes in the network is the same or different. Some graph matrices are dependent on the numbers of nodes can be abbreviated as DNN graph matrices while some of the graph matrices are independent of the numbers of nodes in the network can be abbreviated as INN graph matrices.

The present study constructs the brain connectivity network for the EEG signals having a different number of channels corresponding to schizophrenia diseased patients as well as the healthy subject. The DNN and INN graph matrices are calculated from the brain connectivity network. Finally, the comparison between DNN and INN graph matrices has been performed and analyzed.

The paper is arranged as follows: the current section provides a brief introduction to various techniques of brain analysis. Section 2 describes in detail various databases used and the methodology applied in this study. In Sect. 3, the comparison of results obtained from different databases is performed. Section 4 provides the concluding remarks on the study with future scope.

2   Methodology

The main focus of this study is the graph matrices-based feature extraction from EEG signals having a different number of channels and identifying the effect of change in the number of nodes on the graph matrices. The EEG signals corresponding to the schizophrenia diseased patient and healthy subjects are used. The process flows the study are presented in Fig. 1.

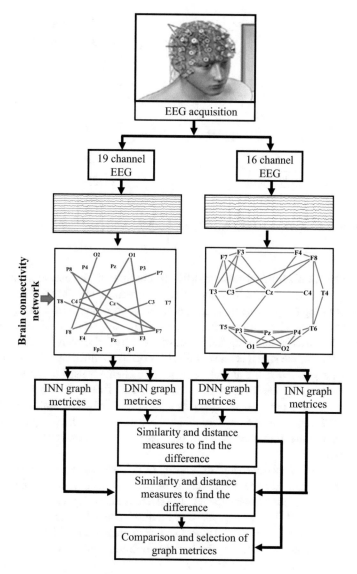

Fig. 1 Process flow of the proposed method

2.1 Dataset

The study validated the experimentation using two EEG datasets corresponding to schizophrenia diseased subjects having a different number of electrodes. The details of the dataset used are given below.

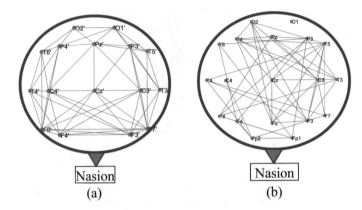

Fig. 2 Brain connectivity network from **a** 16 channel EEG **b** 19 channel EEG with 0.4 threshold value

Dataset 1: Schizophrenia vs. healthy (16 channels): The dataset is taken from Laboratory for Neurophysiology and Neuro-Computer Interfaces, containing 16 channel EEG signals related to schizophrenia diseased and healthy subjects. The dataset contains EEG records related to 39 healthy subjects and 45 schizophrenia diseased subjects. The sampling frequency of the recorded signals is 128 Hz, and the recording time for the signal is one minute; hence, the obtained signal has a length of 7680 samples [9]. The name and placement of 16 channels used for EEG recording are presented in Fig. 2a. To have the same number of records in both classes, 14 records from each group are considered in this study. The EEG signal corresponding to mental activities ranges from 1 to 45 Hz [10]. Hence, the EEG signals from the obtained dataset are filtered using a FIR bandpass filter to allow the frequency range of 1–45 Hz.

Dataset 2: Schizophrenia versus healthy (19 channels): This database contains schizophrenia diseased EEG signals from 14 participants and normal EEG signals from the same number of participants. The database is freely available and can be downloaded from [11]. EEG signals were recorded using 19 channels with 250 Hz sampling frequency. The name and placement of 19 channels used for recording EEG signals are presented in Fig. 2b. The location between 'Fz' and 'Cz' electrodes is considered as a reference location [12].

2.2 Brain Connectivity Network

The connectivity or relation between different brain regions can be modeled as a brain connectivity network/graph using equation $G = \{V, E\}$, where the nodes V interconnected through edges E [13]. To generate the brain connectivity network graph, it is necessary first to identify the vertices and the edges of the graph. In this study, the electrode locations are considered as vertices. Edges are used to show the connection between the pair of vertices [14]. To show the connection between different regions of the brain, we need to calculate the similarity between EEG signals corresponding to those respective locations. This study considered the multichannel EEG signals for constructing brain connectivity graphs. If there is a similarity between EEG signals from two different electrodes, a link can be established between vertices corresponding to those electrodes' location. There are various measures available in the literature to calculate the similarity among EEG signals from different electrodes/channels [15]. The time domain measures for estimating similarity include correlation coefficient, cross-correlation, etc., whereas the frequency domain measures include coherency, partial coherence, and spectral density function [16].

The study uses a correlation coefficient as a similarity measure between EEG signals from different electrodes. Correlation ($\rho_{x,y}$) is calculated using (1) [15]. The threshold value of correlation must be selected carefully, to generate an optimal brain connectivity network. The large value of the threshold will create a sparser network, and the small value of the threshold will result in a denser network [3]. Depending on this threshold value, the adjacency matrix is created. Brain connectivity network is constructed using the adjacency matrix. Following (2) interprets the logic behind the construction of the adjacency matrix by considering the threshold value. Here, the correlation value 0.4 is considered as the threshold based on the trial-and-error approach for link establishment.

$$\rho_{x,y} = \frac{1}{N} \sum_{n=1}^{N} \frac{(x(n) - \varphi_x)(y(n) - \varphi_y)}{\tau_x \tau_y}. \tag{1}$$

Here, N represents the signal length; $x(n)$ is a sample value of signal x at time n. φ_x and τ_x are the mean and standard deviation of signal x, respectively. Similarly, $y(n)$ is a sample value of signal y at time n. φ_y and τ_y are the mean and standard deviation of signal y.

$$\mathrm{Adj}_{A,B} = \begin{cases} \mathrm{corr}(A, B) \geq \text{Threshold then, } \mathrm{Adj}(A, B) = 1 \\ \mathrm{corr}(A, B) < \text{Threshold then, } \mathrm{Adj}(A, B) = 0 \end{cases} \tag{2}$$

where Adj is the adjacency matrix, $\mathrm{corr}(A, B)$ is a correlation value between EEG signals corresponding to channel number A and B. After preparing the adjacency matrix, a functional brain connectivity network is constructed such that each channel location in a 10–20 electrode placement arrangement is considered as a vertex. The link between two vertices is established, if the value at adjacency matrix corresponding to channel numbers representing those vertices is equal to 1, otherwise the link is not be established between those vertices.

Various graph matrices can be extracted from the brain connectivity network/graph. The graph matrices are used for quantitative analysis of brain connectivity networks. In this study, the brain connectivity networks are characterized by calculating network-level measures, as they contain a different number of nodes. Following are the network level/global measures calculated from the brain connectivity network.

INN graph matrices: The graph matrices which are independent of the number of nodes are as follows:

Characteristic Path Length (CPL): The measure is responsible for presenting the ability of the network connectivity and exchange of information in the network. It is an average of the least length paths between nodes [17]. It is calculated using (3).

$$L = \frac{1}{N} \sum_i l_i, \tag{3}$$

$$l_i = \frac{1}{(N-1)} \sum_{i \neq j} l_{ij}, \tag{4}$$

Here, l_{ij} is the least length path between node i and j, and l_i is the average least length paths between node i and remaining nodes.

Assortativity Coefficient: Assortativity is an essential measure of the resilience of the network; it is denoted by r. It represents the ability of a highly connected node to connect other nodes with large connectivity in the network. Assortativity value ranges from -1 to $+1$. The network with an assortativity value near to -1 means that it is a disassortative network, and the network having an assortativity value near to $+1$ means it is more assortative [18]. Equation (5) is used to calculate the assortativity coefficient.

$$r = \frac{l^{-1} \sum_{(i,j) \in L} k_i k_j - \left[l^{-1} \sum_{(i,j) \in L} \frac{1}{2}(k_i + k_j) \right]^2}{l^{-1} \sum_{(i,j) \in L} \frac{1}{2}(k_i^2 + k_j^2) - \left[l^{-1} \sum_{(i,j) \in L} \frac{1}{2}(k_i + k_j) \right]^2} \tag{5}$$

Here, k_i represents degree of ith node; L is the CPL; l is the count of edges with at least 'k' degree.

Modularity: In a complex network, modularity is an important measure that represents the ability of the network to get separated into different communities. Each community represents the group of nodes having denser internal connectivity but are sparsely connected to the external vertices [19]. The modularity for the brain connectivity network is calculated using (6).

$$Q = \frac{1}{2m} \sum_{vw} \left[A_{vw} - \frac{k_v k_w}{2m} \right] \delta(C_v C_w) \tag{6}$$

Here, m represents the total number of network edges; A_{vw} is an element of adjacency matrix, i.e.,

$$A_{vw} = \begin{cases} 1, \text{ if vertices } V \text{ and } W \text{ are connected} \\ 0, \text{ Otherwise} \end{cases},$$

k_v is the degree of vertex v, k_w is the degree of vertex w. C_v and C_w are the separate modules or communities where vertex v and vertex w belong, respectively.

Link density: Link density is the measure of connectivity strength of a network, i.e., the ability to resist link failure. Link density is the ratio of the total number of edges that exist in the connectivity graph to the number of edges possible in the network [20]. Link density is calculated using (7).

$$LD = \frac{E\_Total}{N(N-1)/2} \tag{7}$$

where E_Total is the total number of edges present in the connectivity network. N represents the count of nodes in the connectivity network.

Network transitivity: Network transitivity represents the ability of a specialized region in the network to get connected with its neighbors. In other words, the ability of the network to get separated into independent local neighborhoods [21]. The network transitivity is calculated using (8)

$$NT = \frac{3 \times \text{number of triangles in netwok}}{\text{Number of paths of length 2}} \tag{8}$$

DNN graph matrices: The DNN graph matrices are extracted when the number of nodes in all the networks is equal. The DNN graph matrices from the networks with the different numbers of nodes are biased. The DNN matrices used in this study are explained below.

Node degree: It measures the connectedness of a given node, i.e., how many other nodes are connected to the current node. It represents the strength of the node in the network [1]. The node having a high degree is considered more central to the network. Equation (9) calculates the degree for ith node.

$$K_i = \sum_{i \neq j} A_{ij} \tag{9}$$

Here, the value of A_{ij} *is equal to 1 if node i and j are connected otherwise the value will* be 0.

Betweenness centrality: It represents the ability of the node to perform data transfer in the network. The betweenness centrality for node 'i' is defined as the total number of least length paths that pass through node i divided by the total number of minimum length paths [1]. Equation (10) presents the mathematical expression for calculating the betweenness centrality for any node i.

$$C_b(i) = \frac{2}{(N-1)(N-2)} \sum_{j \neq h \neq i} \frac{n_{hj}(i)}{n_{hj}} \tag{10}$$

Here, $n_{hj}(i)$ is the count of minimum length paths between hth and jth a node that uses node i as an intermediate node. n_{hj} is the count of least length connections between node h and j.

Closeness centrality: It calculates the nearness of a node with all the remaining nodes. It is the ratio of, the number of network vertices to the sum of the least path length from the current node to all the other nodes in the connectivity network [22]. The mathematical expression for calculating closeness centrality is given in (11).

$$C_c(i) = \frac{N-1}{\sum_{i \neq j} l_{ij}} \tag{11}$$

Here, l_{ij} is the minimum length path between ith and jth node. N represents the count of vertices.

Participation coefficient: The measure used to find the ability of a node to connect the different modules in the network. The node having connectivity with the maximum number of modules will have a maximum participation coefficient value [23]. Equation (12) is used to calculate the participation coefficient for ith node.

$$PC_i = 1 - \sum_{\mu=1}^{C} \left(\frac{K_{i,\mu}}{K_i} \right) \tag{12}$$

Here, $K_{i,\mu}$ represents the degree of ith node belonging to the module μ. C is the total number of modules in the network.

Within module degree(Z-score): It is an important measure of the node used to identify the strength of connectivity of a node within the same module. The node connecting the maximum number of other nodes within the module will have the maximum value of Z-score [23]. Equation (13) is used to calculate the Z-score value for ith node.

$$Z_i = \frac{K_{i,\mu} - K_\mu}{\Gamma_{K_\mu}} \tag{13}$$

Here, K_μ is the mean degree of all the nodes belonging to module μ, and Γ_{K_μ} represents the standard deviation of K.

2.3 Distance and Similarity Measures Calculation

Consider a situation where we want to classify the schizophrenia disease patient and healthy person with brain connectivity network approach as mentioned in Sect. 2.2. The dataset may contain multichannel EEG signals with a different number of electrodes. The graph matrices extracted from networks with unequal number nodes result in biased classification results. To remove this limitation, this study proposes an approach to extract and validate the graph matrices unbiased on a number of nodes. Here, two categories of graph matrices, i.e., DNN and INN graph matrices are extracted from 16 and 19 channel EEG-based BCN. The similarity and difference measures are calculated for each category of features corresponding to schizophrenia diseased EEG and healthy EEG BCN. If the difference value and similarity value is maximum and minimum, respectively, for the particular type of features then that category of features will be suitable for the classification of brain networks with a different number of nodes. Various similarity and distance measures like correlation coefficient, Euclidean distance, city block distance, mean square error, maximum error, etc., are calculated to represent the variability of features from both 16 and 19 channel datasets.

Correlation coefficient: It is an important time domain measure used to find out the similarity between two entities. It is calculated as the linear correlation of two entities divided by the product of their standard deviation [24]. The mathematical expression for calculating the correlation coefficient is given in Eq. (14).

$$r_{XY} = \frac{C_{X,Y}}{\emptyset_X \emptyset_Y} \tag{14}$$

$$C_{X,Y} = \frac{1}{N-1} \sum_{i=1}^{N} (x(i) - \varphi_X)(Y(i) - \varphi_Y) \tag{15}$$

φ_X and φ_Y represent mean of variable X and Y, and the standard deviation is represented by $\emptyset$.

Euclidean distance: It is the widely used measure for calculating the distance between the variables on Euclidean space [25]. The mathematical expression for calculating the Euclidean distance is given in Eq. (16).

$$EU = \sqrt{\sum_{i=1}^{n} (X_i - Y_i)^2} \tag{16}$$

City block distance: It is also called taxi-cab distance or Manhattan distance. The city block distance between pair of points is obtained by calculating a sum of the absolute value of their coordinate difference [26]. The mathematical expression for calculating city block distance is given in Eq. (17).

$$CB_d(X, Y) = \sum_{i=1}^{N} |X_i - Y_i| \tag{17}$$

Mean squared error: It is the most common distance measure, represents the average squared distance between the two variables [27]. The mathematical expression for calculating MSE is given in Eq. (18).

$$\text{MSE}(X, Y) = \frac{1}{N} \sum_{i=1}^{N} [X_i - Y_i]^2 \tag{18}$$

Maximum error: It is the measure of the maximum possible difference between two variables [28]. The mathematical expression for calculating maximum error is given in Eq. (19).

$$\text{ME} = \max_{i=1..N} (|X_i - Y_i|) \tag{19}$$

3 Results

The study focused on the comparison of the DNN and INN features extracted from EEG-based BCN by calculating the similarity and difference measures. The idea behind the calculation of similarity and difference measures is to identify the variation in the INN and DNN matrices due to changes in the number of nodes. If the difference and similarity value is minimum and maximum, respectively, for the particular type of metric, then that category of matrices will be suitable for the classification of brain networks with a different number of nodes.

From the obtained value of similarity measures, it is observed that the INN graph matrices from 16 channel EEG and 19 channel EEG are more similar. Hence, INN graph matrices are considered to be useful for network comparison when the number of nodes in the network is different. Various similarity and distance measures like correlation coefficient, Euclidean distance, city block distance, mean square error, maximum error, etc., are calculated to represent the variability of features from both 16 and 19 channel datasets.

Table 1 presents the similarity and distance measures calculated between the graph-based matrices from EEG signals with 16 channels and EEG signals with 19 channels. The first row of the table presents the similarity measure correlation; hence, the category of matrices that has maximum value is suitable. Rows 2–5 of Table 1 present the distance measures, i.e., Euclidean distance, city block distance, mean square error, and maximum error. The category of graph matrices having minimum value for the distance measure is considered suitable for the graph matrices-based classification from BCN with a different number of nodes. From Table 1, it can be observed that the INN features represent less variability for 16 channels and 19 channels signals whereas the DNN features from 16 channels EEG and 19 channels EEG have a significant difference. Hence, the INN features are more suitable for comparing the BCN with a different number of channels.

Table 1 Similarity and distance measure for graph matrices

Similarity/difference	INN metrics	DNN metrics
Correlation	0.83	0.81
Euclidean distance	137.21	453.04
City block distance	937.01	2569.95
Mean square error	134.48	1466.01
Maximum error	43.62	180.20

4 Conclusion and Future Scope

The presented study provides a comparative analysis of the graph matrices-based features from the brain connectivity network. The network is constructed using 16 channels and 19 channels EEG signals of schizophrenia diseased and healthy subjects. The variability of the features from 19 and 16 channels EEG is analyzed using distance and similarity measures. It has been observed that the DNN graph matrices from the networks having a different number of nodes show biased results, whereas the INN graph matrices are not much affected by the variable nodes in the networks. Hence, for comparing the networks with the different number of nodes, the INN graph matrices must be extracted. The presented study can be further extended by performing classification on the obtained graph-based matrices using various classification algorithms. Further, the use of a weighted brain connectivity network can provide more information about the network.

References

1. Liu J, Li M, Pan Y, Lan W, Zheng R, Wu FX, Wang J (2017) Complex brain network analysis and its applications to brain disorders: a survey. Complexity 2017:1–27. https://doi.org/10.1155/2017/8362741
2. Das K, Pachori RB (2021) Schizophrenia detection technique using multivariate iterative filtering and multichannel EEG signals. Biomed Sig Proc Cont 67:102525. https://doi.org/10.1016/j.bspc.2021.102525
3. Van Diessen E, Zweiphenning WJEM, Jansen FE, Stam CJ, Braun KPJ, Otte WM (2014) Brain network organization in focal epilepsy: a systematic review and meta-analysis. PLoS ONE 9:1–21. https://doi.org/10.1371/journal.pone.0114606
4. Sharaf AI, El-Soud MA, El-Henawy IM (2018) An automated approach for epilepsy detection based on tunable Q -Wavelet and firefly feature selection algorithm. Int J Biomed Imag 2018 (2018). https://doi.org/10.1155/2018/5812872
5. Harpale V, Bairagi V (2018) An adaptive method for feature selection and extraction for classification of epileptic EEG signal in significant states. J King Saud Univ Comput Inf Sci. https://doi.org/10.1016/j.jksuci.2018.04.014
6. Sargolzaei S, Cabrerizo M, Goryawala M, Eddin AS, Adjouadi M (2015) Scalp EEG brain functional connectivity networks in pediatric epilepsy. Comput Biol Med 56:158–166. https://doi.org/10.1016/j.compbiomed.2014.10.018
7. Ahirwal MK, Kumar A, Londhe ND, Bikrol H (2016) Scalp connectivity networks for analysis of EEG signal during emotional stimulation. In: International conference on communication

and signal processing, ICCSP 2016, IEEE, 2016: pp 592–596. https://doi.org/10.1109/ICCSP. 2016.7754208

8. Brier MR, Thomas JB, Fagan AM, Hassenstab J, Holtzman DM, Benzinger TL, Morris JC, Ances BM (2014) Functional connectivity and graph theory in preclinical Alzheimer's disease. Neurobiol Aging 35:757–768. https://doi.org/10.1016/j.neurobiolaging.2013.10.081

9. Laboratory for Neurophysiology and Neuro-Computer Interfaces, M. V. Lomonosov Moscow State University, Faculty of Biology, EEG Database—Schizophrenia (2016) http://brain.bio. msu.ru/eeg_schizophrenia.htm (Accessed 27 Dec 2019)

10. Newson JJ, Thiagarajan TC (2019) EEG frequency bands in psychiatric disorders: a review of resting state studies. Front Human Neurosc 12. https://doi.org/10.3389/fnhum.2018.00521

11. EEG in schizophrenia - IBIB PAN - Department of Methods of Brain Imaging and Functional Research of Nervous System, (n.d.). https://repod.icm.edu.pl/dataset.xhtml?persistentId=doi: 10.18150/repod.0107441 (Accessed 16 Oct 2020)

12. Olejarczyk E, Jernajczyk W (2017) Graph-based analysis of brain connectivity in schizophrenia. PLoS ONE 12:1–28. https://doi.org/10.1371/journal.pone.0188629

13. Kumar S, Sharma VK, Kumari R (2013) A novel hybrid crossover based artificial bee colony algorithm for optimization problem. Int J Comput Appl 82:18–25. https://doi.org/10.5120/14136-2266

14. Eldeeb S, Akcakaya M, Sybeldon M, Foldes S, Santarnecchi E, Pascual-Leone A, Sethi A (2019) EEG-based functional connectivity to analyze motor recovery after stroke: a pilot study. Biomed Sig Proc Control 49:419–426. https://doi.org/10.1016/j.bspc.2018.12.022

15. van Mierlo P, Papadopoulou M, Carrette E, Boon P, Vandenberghe S, Vonck K, Marinazzo D (2014) Functional brain connectivity from EEG in epilepsy: seizure prediction and epileptogenic focus localization. Prog Neurobiol 121:19–35. https://doi.org/10.1016/j.pneurobio.2014.06.004

16. Jalili M (2016) Functional brain networks: does the choice of dependency estimator and binarization method matter? Sci Rep 6:29780. https://doi.org/10.1038/srep29780

17. Qi S, Meesters S, Nicolay K, ter Haar Romeny BM, Ossenblok P (2016) Structural brain network: what is the effect of life optimization of whole brain tractography? Front Comput Neurosc 10:1–18. https://doi.org/10.3389/fncom.2016.00012

18. Newman ME (2002) Assortative mixing in networks. Phys Rev Lett 89:1–5

19. Newman MEJ (2006) Modularity and community structure in networks. Proc Nat Acad Sci 103:8577–8582

20. Liu J, Li M, Pan Y, Lan W, Zheng R, Wu FX, Wang J (2017) Complex brain network analysis and its applications to brain disorders: a survey. Complexity. https://doi.org/10.1155/2017/8362741

21. Humphries MD, Gurney K (2008) Network "small-world-ness": a quantitative method for determining canonical network equivalence. PLoS ONE 3. https://doi.org/10.1371/journal.pone.0002051

22. Opsahl T, Agneessens F, Skvoretz J (2010) Node centrality in weighted networks: generalizing degree and shortest paths. Soc Netw 32:245–251. https://doi.org/10.1016/j.socnet.2010.03.006

23. Aggarwal P, Gupta A (2019) Multivariate graph learning for detecting aberrant connectivity of dynamic brain networks in autism. Med Image Anal 56:11–25. https://doi.org/10.1016/j.media.2019.05.007

24. Ahirwal MK, Kumar A, Singh GK, Londhe ND, Suri JS (2016) Scaled correlation analysis of electroencephalography: a new measure of signal influence. IET Sci Meas Technol 10:585–596. https://doi.org/10.1049/iet-smt.2015.0299

25. Panda A, Pachori RB, Sinnappah-Kang ND (2021) Classification of chronic myeloid leukemia neutrophils by hyperspectral imaging using Euclidean and Mahalanobis distances. Biomed Sig Proc Cont 70. https://doi.org/10.1016/j.bspc.2021.103025

26. Prabhakar SK, Rajaguru H (2016) Code converters with city block distance measures for classifying epilepsy from EEG signals. Proced Comput Sci 87:5–11. https://doi.org/10.1016/j.procs.2016.05.118

27. Chachlakis DG, Zhou T, Ahmad F, Markopoulos PP (2021) Minimum mean-squared-error autocorrelation processing in coprime arrays. Dig Sig Proc: A Rev J 114. https://doi.org/10.1016/j.dsp.2021.103034
28. Dong H, Ying W, Zhang J (2020) Maximum error estimates of a MAC scheme for Stokes equations with Dirichlet boundary conditions. Appl Numer Math 150:149–163. https://doi.org/10.1016/j.apnum.2019.09.017

Facial Expression Recognition from Low Resolution Facial Segments Using Pre-trained Networks

Dhananjay Theckedath and R. R. Sedamkar

1 Introduction

Facial expressions have been an area of interest from time immemorial as they are considered to be the richest source of nonverbal channel of communication [1]. Great philosophers like Aristotle, Stewart and Darwin took to studying facial expressions. Facial expression recognition (FER) refers to detection and classification of facial features into universal expressions, viz. happiness, sadness, fear, disgust, surprise and anger. These universal features were presented by Ekman in 1971 [2]. Human beings are capable of comprehending various facial expressions with ease; however, for computer-based systems, it is a daunting task.

Deep learning is one of the most exciting areas in machine learning, and with advancements in Graphics Processing Unit (GPU) technology, it is now possible to implement real-time deep learning algorithms. Deep learning models are excellent in discovering complex arrangements and have dramatically improved accuracies in detection and classification in speech recognition and visual object recognition [3]. Due to the efficiency of deep learning algorithms coupled with powerful GPUs, facial expression recognition has become a widely explored topic [4].

FER depends on accurately identifying various facial features. Image occlusion and low resolution images are conspicuous issues while identifying various facial expressions. The research uses deep learning networks to study the effect of reduction in spatial resolution on the performance metrics for full as well as occluded facial images.

D. Theckedath (✉)
Thadomal Shahani Engineering College, Mumbai, India
e-mail: dhananjay.theckedath@thadomal.org

R. R. Sedamkar
Thakur College of Engineering and Technology, Mumbai, India

D. S. Sisodia et al. (eds.), *Machine Intelligence Techniques for Data Analysis and Signal Processing*, Lecture Notes in Electrical Engineering 997,
https://doi.org/10.1007/978-981-99-0085-5_3

2   Related Work

Extensive literature survey was carried out in the domain of FER. Stress was given to papers that dealt with detection and classification of facial expressions from low resolution and occluded images.

Tian [5] is considered as a every important work in the area of low resolution facial expression recognition. Work is carried out for five different resolutions of the head region (288 × 384, 144 × 192, 72 × 96, 36 × 48, and 18 × 24). Gabor wavelets are employed to extract facial appearances changes. These are obtained in the form of multi-scale and multi-resolution coefficients. They obtain validation accuracies of 93.2%, 93%, 92.8%, 89% and 68.2%, respectively.

Khan et al. [6] this study presents a unique framework which recognises various facial expressions with high accuracy for low resolution facial images. A novel descriptor called Pyramid of Local Binary Pattern (PLBP) is proposed. Validation accuracies of 96%, 95%, 94%, 93% and 90% are reported for the different spatial resolutions. The results obtained outperform state-of-the-art methods for low resolution FER. The proposed network achieves an FER accuracy of 95.3% on Extended Cohn-Kanade data set (CK+).

In the paper [7], various machine learning methods are studied on several databases. The authors use Boosted LBP to extract the discriminant LBP features. It is noted that Support Vector Machine classifiers give the best FER performance. Work is carried out on slow spatial resolution images. They demonstrate that the LBP features have an advantage of being low in computations and yet perform better than most of the reported techniques.

The authors in the paper [8] investigate the use of different frame rates and resolutions on bench mark data sets like SMIC and CASME. They use three feature types, viz. LBP in Three Orthogonal Planes, 3D Histograms of Oriented Gradient and Histogram of Oriented Optical Flow. The results show that the performance of micro-expression recognition is dependent on frame rate and resolution.

In the paper [9], a new deep CNN configuration known as SPLITFACE is introduced. This configuration accomplishes attribute detection in partially occluded faces. Different segments of the face are taken and have been able to identify attributes associated with different sections of the face. They have validated that the proposed method outdoes other recent methods.

The paper [10] is an important contribution in the area of spatial patches. The authors propose an innovative framework for FER which uses features of selected facial patches. A few prominent facial patches are extracted, and LBP are applied on the selected patches. Experiments are performed on CK+ and JAFFE data sets. They also study the effect of low spatial resolution.

In the paper [11], the authors select a new feature called LBP map. Their method extracts texture information for sparse representation-based classification (SRC). The extracted features have been shown to be robust to grey scale variation. Experiments are performed on the Cohn-Kanade database, and it is shown that the LBP map along with SRC gives a high accuracy and is also fast compared to techniques which use different features.

Mahmood et al. [12] partial facial occlusions make the process of extracting features extremely complex. Here, both, texture features as well as geometric features are obtained effective FER. Using facial landmark detection, the authors introduce the concept of effective geometric features.

3 Database and Pre-processing

In this research, we have used the Extended Cohn-Kanade data set (CK+) [13] and the Facial Expression Research Group Database (FERGB-DB) [14]. For the CK+ data set, we not only use the target frame but also a few more preceding frames as well. This gives us an increased number of images to carry out our experiments. The FERG-DB database exhibits an animated model of facial expressions. The database provides important insights into expressions demonstrated by six animated stylized characters. The animated characters were created using MAYA software. Images from the CK+ and FERG-DB database are shown in Figs. 1 and 2.

The distribution of labelled images used in our research is given in Table 1.

For training the network, we need to separate the data into training set and validation set. For each of the data sets, 70% images are randomly chosen to train the network while 30% images are used for validation. Hence for the CK+ data set which has 2502 images, 1751 images were used for training the network while 751 images were used for validation. For FERG-DB, of the 3974 images, 2781 images were used

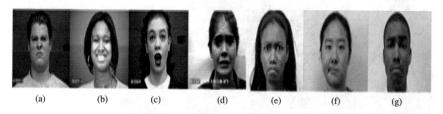

Fig. 1 Peak expressions in CK+ database **a** disgust, **b** happy, **c** surprise, **d** fear, **e** angry, **f** contempt and **g** sad

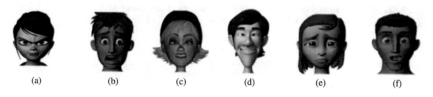

Fig. 2 Peak expressions in FERG-DB database **a** anger, **b** fear, **c** disgust, **d** joy, **e** sad and **f** surprise

Table 1 Distribution of labelled images used for experimentation

Data set	Angry	Contempt	Disgust	Fear	Happy	Sad	Surprise	Total
CK+	412	105	325	220	604	313	523	2502
FERG	672	–	845	899	489	392	677	3974

for training while 1193 images were used for validation. Face extraction was carried out on each of the images of the data sets using the Viola Jones algorithm.

3.1 Reduction in Spatial Resolution

This paper studies the effect of reduction in spatial resolution of full face as well as facial segments on the FER results. The spatial resolution of each of the images from the two data sets were reduced by down sampling to obtain the following resolutions, 256×256, 150×150, 100×100, 50×50, 25×25 and 10×10. Figure 3 shows us the down sampled images of full face images of the two data sets.

Along with down sampling the full face images, we also down sample half face segments as well as quarter face segments. This is shown in Figs. 4 and 5.

This process of partitioning and down sampling was done on FERG-DB image data set as well.

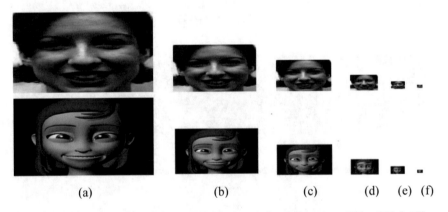

(a)   (b)   (c)   (d)   (e) (f)

Fig. 3 Reduction in spatial resolution on full face images for all data sets. **a** 256×256, **b** 150×150, **c** 100×100, **d** 50×50, **e** 25×25 and **f** 10×10

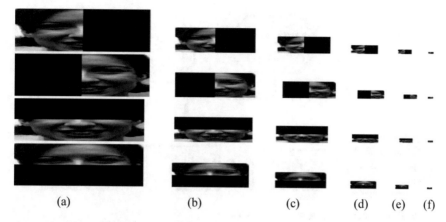

Fig. 4 Reduction in spatial resolution for half face segments of the CK+ data set. **a** 256 × 256, **b** 150 × 150, **c** 100 × 100, **d** 50 × 50, **e** 25 × 25 and **f** 10 × 10

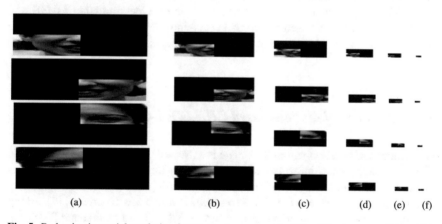

Fig. 5 Reduction in spatial resolution for quarter face segments of the CK+ data set. **a** 256 × 256, **b** 150 × 150, **c** 100 × 100, **d** 50 × 50, **e** 25 × 25 and **f** 10 × 10

4 Proposed Methodology

Convolutional neural networks (CNN) consist of a large number of parameters: number of layers, weights biases filter size, max pooling, activation function, stride, optimizers, learning rate, optimizers, etc.

In this work, we use three networks, viz. spatial exploitation-based CNN (SEB-CNN), depth-based CNN (DB-CNN) and channel exploitation-based CNN (CEB-CNN). The SEB-CNN uses the VGG-16 network while the DB-CNN uses the ResNet-50 network. The CEB-CNN uses a hybrid network which incorporates ResNet-50 along with a squeeze and excite network. The basic block diagram of the proposed methodology is shown in Fig. 6. We will briefly discuss the three

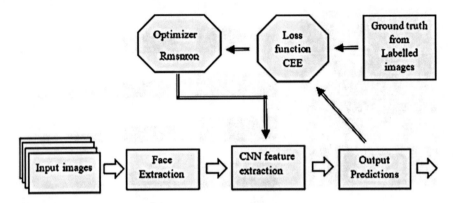

Fig. 6 Block diagram of proposed methodology

networks in the next section. The data was shuffled in order to train the network better. The image batch size was set to 32. We used a learning rate of 0.001 and a Softmax classifier. We used categorical cross-entropy loss function in all the tree networks along with RMSprop and Adam as optimizers. A total of epochs run were 25.

4.1 Spatial Exploitation-Based CNN (SEB-CNN)

Fine feature information within the image is extracted using small size filters while large size coarse feature information is captured using large sized filters. Different sets of features within an image can thus be extracted using this convolution operation. In this framework, we use the VGG-16 network which has 16 layers [12]. Along with this, we use the Categorical Cross-Entropy (CEE) loss function and RMSprop optimizer. This network is called spatial exploitation-based convolutional neural network (SEB-CNN). To accommodate the network to our necessities, the last layer the architecture was modified and changed to the size of our data class. A Softmax classifier was used. A detailed explanation of this network is given in [15] by the same author.

4.2 Depth-Based CNN (DB-CNN)

Deep CNNs have layers that run deep because of which back propagation of error results in small gradient values at lower layers, thus making the network difficult to train. Thus, the main problem with depth-based CNNs is slow training as well as low speed of convergence.

In the proposed method, we use the ResNet-50 architecture [16]. Adaptive moment estimation (Adam) optimizer and the CEE loss function are used in the proposed architecture. This network is called depth-based CNN (DB-CNN). A detailed explanation of this network is given in [17] by the same author. The CNN used comprises 50 layers with each layer comprising of a bypass as shown in Fig. 7a.

4.3 Channel Exploitation-Based CNN (CEB-CNN)

As we are aware, in CNNs, features are dynamically selected by the network itself by tuning the weights of the kernel. Because of the deep layers of the network, diverse features are extracted. This enormous amount of feature maps could act as an impediment, creating noise, and hence leading to over fitting of the network. This leads to a concept of suppressing the less significant feature maps and giving high weightage to the important feature maps.

Squeeze and excitation (SE) network is a recent innovative architectural block having an objective of improving the quality of representations. The proposed new block selects only specific feature maps, also called channels, which are relevant to object detection and classification. In the proposed method, SE block is assimilated into the ResNet-50 architecture to achieve a channel wise squeeze operation. It is inserted after the nonlinearity as shown in Fig. 7b. A detailed explanation of this network is given in [18–20].

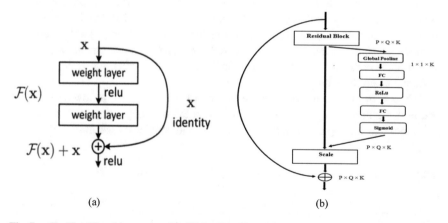

(a)   (b)

Fig. 7 **a** ResNet-50 architecture, and **b** SE-ResNet-50 architecture

5   Results

In this section, we discuss the results obtained for FER using full face as well as half and quarter face segments under the effect of low spatial resolution using the proposed networks. All the three networks were trained and validated using the training set and validation set, respectively. In order to train the networks better, the data was randomly shuffled. The image batch size was set to 32. We used a learning rate of 0.001 and a Softmax classifier. A total of epochs run were 25.

5.1   Results of SEB-CNN, DB-CNN and CEB-CNN on Full Face Low Resolution Images

We have used validation accuracy and confusion matrix as performance metrics. Precision and recall values can be calculated using the confusion matrix as well. Validation accuracies obtained for the three networks for various spatial resolutions is listed in Table 2.

From Table 2, we observe that even for resolutions as low as 10×10, for the CK+ data set, DB-CNN and CEB-CNN networks give us validation accuracies of 93%, and 98%, respectively, with CEB-CNN performing marginally better. For the FERG-DB data set, the validation accuracies for SEB-CNN, DB-CNN and CEB-CNN are 76%, 99% and 99%, respectively, for a resolution of 10×10. From the values obtained we conclude that even for very low resolutions, the proposed networks perform exceedingly well with DB-CNN and CEB-CNN performing better. We can thus state that deep networks perform better in the event of low resolution images. Limited research has been carried out in this area, and we now compare our results with already published works. Comparison is based on the validation accuracy results obtained using the CK+ data set.

We have been able to reduce the resolution of our data set to as low as 10×10. We notice that the proposed work outperforms the existing results. From Table 3, it is clear that the top performing networks are the ones proposed in this work. We

Table 2 Validation accuracy for full face images under different resolutions using the three networks

		256×256	150×150	100×100	50×50	25×25	10×10
CK+	SEB	97	98	96	93	88	60
	DB	99	99	99	99	99	93
	CEB	98	98	97	96	98	98
FERG	SEB	100	100	100	99	94	76
	DB	100	100	100	100	100	99
	CEB	100	100	100	100	100	99

Table 3 Comparison of results obtained for full face with existing literature for CK+ data set

Algorithm	288×384	144×192	72×96	36×48	18×24	10×10
PLBP-SVM [21]	96	95	94	93	90	–
LBP-JIVC [7]	93	92.6	89.9	87.3	79.6	–
Gabor-JIVC [7]	90	89.8	89.2	86.4	75.1	–
Gabor-CVPRW [5]	93.2	93	92.8	89	68.2	–
PHOG-ICIP [22]	95	90	82	75	72	–
Proposed networks	256×256	150×150	100×100	50×50	25×25	10×10
SEB-CNN	97	98	93	93	88	60
DB-CNN	99	99	99	99	99	93
CEB-CNN	98	98	97	96	98	98

observe that CEB-CNN performs slightly better than the DB-CNN for low resolution images. The accuracy for the SEB-CNN network drops for a resolution of 10×10.

5.2   Results of SEB-CNN, DB-CNN and CEB-CNN on Half Face Low Resolution Images

We now compare the results obtained for low resolution images from half segments of the face. Half segments of the face were down sampled and given to the three proposed networks (Table 4).

From the validation accuracy values, we note that for the DB-CNN and CEB-CNN give high accuracies even for 10×10 resolutions with the CK+ data set. The validation values marginally drop for the upper segment. The SEB-CNN cannot handle a low resolution of 10×10, and there is a significant drop for all segments of the face.

Even for the FERG-DB data set, DB-CNN and CEB-CNN outperform the SEB-CNN. We can thus conclude that accurate facial expression recognition can be achieved form half face segments even for very low resolutions. The proposed networks perform exceedingly well with DB-CNN and CEB-CNN performing better, thus indicating that deep networks perform better in the event of low resolution images.

5.3   Results of SEB-CNN, DB-CNN and CEB-CNN on Quarter Face Low Resolution Images

In this subsection, we discuss the results obtained for low resolution images from quarter segments of the face under the effect of low resolution.

Table 4 Validation accuracy for half face segments under different resolutions using the three networks

		256 × 256		150 × 150		100 × 100		50 × 50		25 × 25		10 × 10	
		CK	FER	CK	FER	CK	FER	CK	FER	CK	FER	CK	FER
RIGHT	SEB	97	100	98	100	98	100	97	99	90	98	68	93
	DB	99	100	99	100	99	100	99	100	99	100	98	100
	CEB	98	100	96	100	97	100	96	100	94	100	94	99
LEFT	SEB	97	100	96	100	97	100	95	98	87	98	68	88
	DB	100	100	99	100	97	100	96	100	97	100	97	100
	CEB	97	100	97	100	97	100	96	100	94	99	93	100
LOWER	SEB	97	100	97	100	97	100	96	99	88	97	68	86
	DB	99	100	99	100	99	100	98	100	99	100	96	100
	CE	97	100	97	100	97	100	96	100	96	100	96	100
UPPER	SEB	85	100	93	100	96	100	89	99	77	98	59	88
	DB	94	100	98	100	98	100	95	100	97	100	93	99
	CEB	97	100	96	100	96	100	94	100	95	99	92	99

From Table 5, we note that for the DB-CNN and CEB-CNN give high accuracies even for 10 × 10 resolutions with the CK + data set for all four quarter segments of the face, with segment QUL being a little lower at 89% (original value for 256 × 256 is 93%) for CEB-CNN. The validation values for the FERG-DB data set are close to100% for DB-CNN and CEB-CNN. We thus conclude that the proposed networks perform very well even for low resolution quarter face segments. Along with validation accuracies, we were also able to show that precision, recall and f1-score are also high for the various facial segments.

6 Conclusion and Future Scope

In this paper, we aim to understand the relationship between facial segments and spatial resolution reduction on recognition rates in FER. We use three proposed networks, viz. SEB-CNN, DB-CNN and CEB-CNN. The results of three methods are compared on full face un-occluded images, facial segments, low resolution full face un-occluded images and low resolution facial segments.

We have reduced the resolution of all the segments of the faces down to 10 × 10 and observed very little deterioration in the results compared to the original segments. Hence, the results obtained hold true even when there is large reduction in spatial resolution in facial segment images. This kind of detailed analysis on facial segments under the effect of reduction in spatial resolution has not been found in recent literature.

Table 5 Validation accuracies for low resolution quarter face segments

		256 × 256		150 × 150		100 × 100		50 × 50		25 × 25		10 × 10	
		CK	FER	CK	FER	CK	FER	CK	FER	CK	FER	CK	FER
QLR	SEB	95	100	97	100	97	100	92	99	80	95	63	80
	DB	99	100	99	100	99	100	98	100	98	100	94	100
	CEB	97	100	97	100	97	100	98	100	97	100	95	99
QLL	SEB	100	100	97	100	97	100	92	100	84	95	70	82
	DB	100	100	99	100	99	100	99	100	99	100	96	99
	CEB	97	100	97	100	97	100	98	100	97	100	95	99
QUR	SEB	81	100	90	100	89	99	84	99	74	95	53	85
	DB	99	100	96	100	97	100	96	100	94	100	95	100
	CEB	97	100	97	100	96	100	96	99	94	100	91	100
QUL	SEB	98	100	98	100	95	100	97	100	94	100	90	99
	DB	93	100	97	100	96	100	95	100	94	99	89	98
	CEB	98	100	98	100	95	100	97	100	94	100	90	99

Our evaluation shows that all the three networks, viz. SEB-CNN, DB-CNN and CEB-CNN are successful in detecting and classifying the universal affect state from all eight low resolution segments of the face, viz. right half, left half, upper half, lower half, quarter lower right, quarter lower left, quarter upper right and quarter upper left segment of the face under low spatial resolution. In this work, we have shown that CEB-CNN performs marginally better than the DB-CNN for resolutions as low as 10 × 10 for the CK+ as well as the FERG-DB data set. The SEB-CNN recognition rate drops for resolutions below 25 × 25. From the experiments carried out, it is clear that the top performing networks are the ones proposed in this work, viz. DB-CNN and CEB-CNN. DB-CNN and CEB-CNN are both deep networks and have 50 layers as opposed to only 16 layers in the SEB-CNN and hence are capable of extracting finer details. The major contribution of this work is in successfully demonstrating that even with extreme spatial resolution reduction in both, full face images as well as facial segment images, accurate FER can be achieved using the proposed frameworks, and there is negligible effect on any the performance metrics used. From the results discussed, we conclude that even when the human computer interface system has access to entire fontal face images, only a small segment of the image is required for accurate FER, and the results obtained are not affected by reduction in spatial resolution. This work deals only with posed facial images. We plan to extend this work for facial images captured from natural settings.

Acknowledgements This work is supported in part by NVIDIA GPU grant programme. We thank NVIDIA for giving us Titan XP GPU as a grant to carry out our work in deep learning.

References

1. Ambady N, Rosenthal R (1992) Thin slices of expressive behavior as predictors of interpersonal consequences: a meta-analysis. Psychol Bull. https://doi.org/10.1037/0033-2909.111.2.256
2. Ekman P (1971) Universals and cultural differences in facial expressions of emotion. Nebraska Symp Motiv
3. Woolf B, Burleson W, Arroyo I, Dragon T, Cooper D, Picard R (2009) Affect-aware tutors: recognizing and responding to student affect. Int J Learn Technol. https://doi.org/10.1504/ijlt.2009.028804
4. Gong C (2009) Human-computer interaction: the usability test methods and design principles in the human-computer interface design. https://doi.org/10.1109/ICCSIT.2009.5234724
5. Tian YL (2004) Evaluation of face resolution for expression analysis. In: Conference on computer vision and pattern recognition workshop 2004 Jun 27 IEEE, pp 82–82
6. Khan RA, Meyer A, Konik H, Bouakaz S (2013) Framework for reliable, real-time facial expression recognition for low resolution images. Pattern Recognit Lett. https://doi.org/10.1016/j.patrec.2013.03.022
7. Shan C, Gong S, PM-I (2009) Vision computing, and undefined 2009. Facial expression recognition based on local binary patterns: a comprehensive study. Elsevier, Accessed: Aug. 01, 2020. [Online]. Available: https://www.sciencedirect.com/science/article/pii/S0262885608001844
8. Merghani W, Davison AK, Yap MH (2019) The implication of spatial temporal changes on facial micro-expression analysis. Multimed Tools Appl 78(15):21613–21628. https://doi.org/10.1007/s11042-019-7434-6
9. Mahbub U, Sarkar S, Chellappa R (2020) Segment-based methods for facial attribute detection from partial faces. Accessed: 01 Aug 2020. [Online]. Available: https://ieeexplore.ieee.org/abstract/document/8326549/
10. Happy SL, Routray A (2020) Automatic facial expression recognition using features of salient facial patches. Accessed: 01 Aug 2020. [Online]. Available: https://ieeexplore.ieee.org/abstract/document/6998925/
11. Ouyang Y, Sang N, Optik RH (2013) Robust automatic facial expression detection method based on sparse representation plus LBP map. Elsevier, Accessed: 01 Aug 2020. [Online]. Available: https://www.sciencedirect.com/science/article/pii/S0030402613007468
12. Mahmood SH, Iqbal K, Elkilani WS, Perez-Cisneros M (2019) Recognition of facial expressions under varying conditions using dual-feature fusion. hindawi.com. https://doi.org/10.1155/2019/9185481
13. Lucey P, Cohn JF, Kanade T, Saragih J, Ambadar Z, Matthews I (2020) The extended Cohn-Kanade dataset (CK+): a complete dataset for action unit and emotion-specified expression. Accessed: 01 Aug 2020. [Online]. Available: https://ieeexplore.ieee.org/abstract/document/5543262/
14. Aneja D, Colburn A, Faigin G, Shapiro L, Mones B (2017) Modeling stylized character expressions via deep learning. Lect Notes Comput Sci (including Subser Lect Notes Artif Intell Lect Notes Bioinformatics) 10112(LNCS):136–153. https://doi.org/10.1007/978-3-319-54184-6_9
15. Theckedath D, Sedamkar S (2020) Classification of affect states from facial segments using transfer learning. Intell Dec Technol 1–11. https://doi.org/10.3233/IDT-190077
16. He K, Zhang X, Ren S, Sun J (2016) Deep residual learning for image recognition. In: Proceedings of the IEEE conference on computer vision and pattern recognition, pp 770–778
17. Theckedath D, Sedamkar RR (2020) Detecting affect states using VGG16, ResNet50 and SE-ResNet50 Networks. SN Comput Sci 1(2):1–7. https://doi.org/10.1007/s42979-020-0114-9
18. Khan A, Zahoora SU, Qureshi AS (2020) A survey of the recent architectures of deep convolutional neural networks. Artif Intell Rev 53(8):5455–5516
19. Hu J, Shen L, Sun G (2018) Squeeze-and-excitation networks. In: Proceedings of the IEEE conference on computer vision and pattern recognition, 7132–7141
20. Theckedath D, Sedamkar RR (2020) Affect state classification from face segments using resnet-50 and SE-Resnet-50. Int J Innovat Technol Exp Eng 9(3). ISSN: 2278-3075. https://doi.org/10.35940/ijitee.B6196.019320

21. Khan RA, Meyer A, Konik H, Bouakaz S (2013) Framework for reliable, real-time facial expression recognition for low resolution images. Pattern Recogn Lett. https://doi.org/10.1016/j.patrec.2013.03.022
22. Khan RA, Meyer A, Konik H, Bouakaz S (2012) Human vision inspired framework for facial expressions recognition. In 2012 19th IEEE international conference on image processing, pp 2593–2596

Design and Analysis of Quad Element UWB MIMO Antenna with Mutual Coupling Reduction Techniques

B. Ananda Venkatesan, K. Kalimuthu, and Palaniswany Sandeep Kumar

1 Introduction

With the continuous evolution of the wireless communication, there was steady increase in the need for high-data rate communication. Consequently, systems with larger data capacity were needed. The required high-data rate can be obtained by increasing the bandwidth of the communication signal. This proved to be the starting point of the ultra-wideband (UWB) communications.

As stated by the Federal Communications Commission (FCC), ultra-wide-band is a wireless communication technology which uses a bandwidth greater than 500 MHz or 20 percentage of the central frequency. It was in the year 2002 that FCC issued the first set of standards for the utilization of UWB technology. A bandwidth of 7.5 GHz for frequencies that range from 3.1 to 10.6 GHz was allotted for UWB. This allotted bandwidth was larger than the bandwidth of the previously existing terrestrial systems as noted by Schantz [1]. UWB devices are commonly used as an integral part of imaging systems and radar systems. These devices also find their utility as a part of communications and measurement systems.

Multiple technical standards have been issued to make UWB as a part of the physical layer technology. IEEE 802.15.3a has been issued for small range (10 m) with high rate wireless personal area networks (WPANs). IEEE 802.15.4a has been set up for short-range low-rate WPANs. The standard IEEE 802.15 study community for body area networks (SG-BANs) has been set up for the regulation of wireless

B. Ananda Venkatesan (✉) · K. Kalimuthu · P. Sandeep Kumar
Department of ECE, SRM Institute of Science and Technology, Chennai 603203, India
e-mail: anandavb@srmist.edu.in

K. Kalimuthu
e-mail: kalimutk@srmist.edu.in

P. Sandeep Kumar
e-mail: sandeepp@srmist.edu.in

© The Author(s), under exclusive license to Springer Nature Singapore Pte Ltd. 2023
D. S. Sisodia et al. (eds.), *Machine Intelligence Techniques for Data Analysis and Signal Processing*, Lecture Notes in Electrical Engineering 997,
https://doi.org/10.1007/978-981-99-0085-5_4

43

communications in and around human body at a distance of 2–3 m. However, IEEE 802.15.3a has now become obsolete due to uncertainties in market and regulations.

The large bandwidth of ultra-wideband can provide a channel with high capacity without increasing the transmission power. This is in accordance with Shannon's channel capacity theorem. Therefore, the relatively new UWB technology can make use of the existing spectrum without causing degenerative interference. According to the Cramer-Rao bound (CRB) analysis, better timing resolution and highly precise ranging and position determining capability can be achieved with the use of UWB. These properties can be credited to UWB's large bandwidth. This makes UWB technology an attractive choice for short-range very high-data rate communication systems, radar and imaging systems, wireless sensing, highly precise localization and tracking systems [2].

The advantages of using UWB are many with the primary one being the high-data rate communications. The other upsides of using UWB as pointed out by the authors in [3] include, small size, more affordable equipment and utilization of less power. The penetration capability of UWB is high due to energy distribution at varied frequencies. These unique features of Ultra-wideband make it an attractive candidate for wireless communication applications. However, UWB poses a major challenges such as, short-range communication as a result of its low transmission power and high rate of attenuation. But, it is possible to overcome this with the use of the multiple input-multiple output (MIMO) methodology, thereby increasing range of the UWB signals [4].

Electromagnetic interactions between the adjacent elements in the MIMO array system generate problem of mutual coupling (MC). The effect of mutual coupling changes the current distribution resulting of the electric field caused by an element, in addition to the deformations that are exposed to the radiation pattern of other elements. Relatively low price, easy to fabricate, and low profile are the advantages of micro-strip antennas (MSAs), in addition to the multiplicity of applications that are compatible with it. Meanwhile, low impedance bandwidth, low gain are the main drawbacks. To enhance the antenna and improve the bandwidth, many techniques have studied, and experiments have implemented, such as use patches made of stacked parasitic [5–7], create slots on the ground plane or patch [8, 9], make changes in the electromagnetic feeding structures [10–12].

MIMO is a technique which improves the throughput of the antenna system significantly in multipath environment. In MIMO antenna system, more than one antenna has been used for the transmission and reception of signals. When many antennas are placed in a single FR4 substrate, the issue of mutual coupling arises. This coupling effect may degrade the antenna system performance. The coupling between the adjacent antennas can be efficiently reduced by deploying anyone of following techniques such as decoupling structure, neutralization line technique, and use of stub in the ground structure [13].

This work presents the UWB MIMO antenna and different techniques to minimize the mutual coupling such as decoupling structure, neutralization line, and stub in the ground plane. Then, the comparison is made among the various mutual coupling techniques in terms of frequency band, relative bandwidth percentage, ECC, diversity gain (DG).

2 Antenna Design

The proposed UWB planar antenna has three layers, namely a dielectric substrate layer, a rectangular-shaped radiator in the top, and a partial ground in the bottom side of the substrate as shown in Fig. 1. The FR4 dielectric material is used as substrate with 0.8 mm thick; it has a relative permittivity of 4.3, and thermal conductivity of FR4 is 0.3 [W/K/m]. The conductive nature of the rectangular patch and ground plane leads to the formation of an EM resonant cavity. This cavity radiates EM waves at a particular resonant frequency. The proposed antenna consists of simple rectangular radiator with square slits in the bottom edges to increase the electrical length and tapered partial ground in the rear plane. A squared shaped slit of dimension 1×1 mm^2 is introduced just behind the feed line in the ground plane, to match the impedance for wide range of frequencies, and it produces bandwidth of 7.46 GHz. S_{11} curves are used to depict the radiation characteristics of the planar antenna. The proposed antenna provides a usable operating frequency of 3.15–10.61 GHz as shown in Fig. 2.

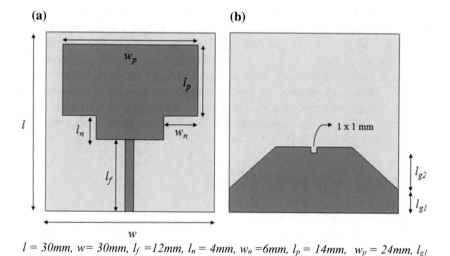

$l = 30mm$, $w = 30mm$, $l_f = 12mm$, $l_n = 4mm$, $w_n = 6mm$, $l_p = 14mm$, $w_p = 24mm$, $l_{g1} = 4mm$, $l_{g2} = 7mm$

Fig. 1 Schematic diagram of UWB antenna **a** Front view **b** Rear view

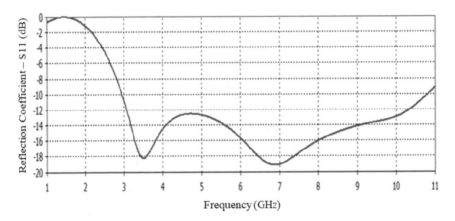

Fig. 2 Reflection coefficient (S_{11}) characteristics of unit element

3 Mimo System Design

Throughput is the important requirement in wireless communication applications. MIMO technique is a great method to enhance the throughput of the system. MIMO system has the ability to transmit and receive a big amount of streaming data at the same time. This property is very precious in the communication systems. The MIMO antenna array can be constructed by placing multiple antennas in the single substrate. When the multiple antennas are placed in the close proximity, then the problem of mutual coupling arises. So, it is necessary to enhance the isolation characteristics between the adjacent radiating elements without affecting the other performance metrics of the antenna system. The isolation characteristics of the MIMO array have been analyzed after employing the different mutual coupling reduction techniques such as decoupling structure, neutralization line, and stub in the ground plane.

3.1 Double-Sided MIMO Antenna

The quad element MIMO antenna has been constructed by placing the unit elements orthogonal to each other. Two elements are placed on the one side of the substrate, and remaining two elements are placed on the other side of the substrate [14] as shown in Fig. 3. This method provides good isolation characteristics without employing any of the mutual coupling reduction methods.

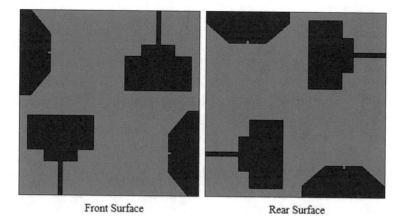

Front Surface   Rear Surface

Fig. 3 Double-sided MIMO antenna

3.2 Decoupling Structure Technique

Decoupling structure (Fig. 4) minimizes the coupling of space wave and surface wave between the antenna elements [15]. Further additional power dissipation occurs due to parasitic resistance which may cause unavoidable coupling of EM energy. As the frequency increases, coupling due to parasitic resistance also increases, and this phenomena may limit the effectiveness of the decoupling structure in the high frequencies.

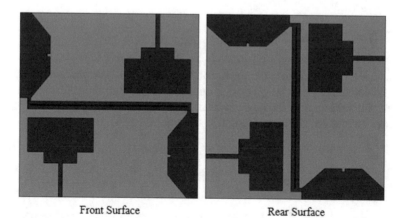

Front Surface   Rear Surface

Fig. 4 MIMO antenna with decoupling structure

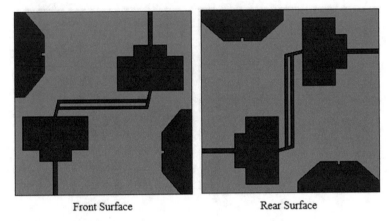

Front Surface   Rear Surface

Fig. 5 MIMO antenna with neutralization line technique

3.3 Neutralization Line Technique

In neutralization line technique (Fig. 5), the current from one radiator is taken and phase inverted by employing the neutralization line [16] having appropriate length. Then, the phase inverted current is utilized for cancelling out the coupling due to another adjacent element. This technique provides effective isolation without modifying the ground structure, but there is no specific standard procedure for fixing the positon and length of the line.

3.4 Ground Plane Stub

To enhance the isolation between the radiating elements, "L"-shaped stub is included in the ground plane as shown in Fig. 6. Introduction of the stub will change the surface current distribution and thereby considerable improvement [17] in the isolation characteristics of MIMO array (Fig. 7).

4 Results Ad Discussion

The proposed double-sided MIMO antenna's bandwidth has been considerable widened compared to single element antenna. The reflection coefficient (S_{11}) characteristics are lower than −10 dB from 2.7 to 10.5 GHz with some undesirable effects at the 4.73 GHz where the curve (black line) then rises above −10 dB because of mutual coupling effects. So, the double-sided MIMO antenna needs an improvement in its reflection coefficient by reducing the mutual coupling effects.

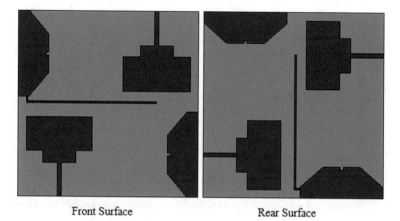

Front Surface   Rear Surface

Fig. 6 MIMO antenna with stub in the ground plane

Fig. 7 S-parameter
characteristics of
double-sided MIMO antenna

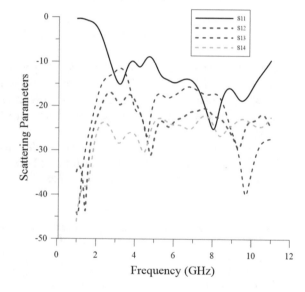

The different types of mutual coupling reduction techniques are incorporated in the double-sided MIMO array, and the performance metrics are analyzed

Figure 8 depicts the S-parameter characteristics of MIMO array with decoupling structure. The MIMO antenna array with decoupling structure in the ground plane produces resonant from 2.54 and 10.8 GHz with relative bandwidth 124%. Thus, this method improves the isolation, but the level of ECC and DG gets affected.

The envelope correlation coefficient (ECC) is calculated using (1) and found to be less than 0.015, and diversity gain (DG) is evaluated using (2), and it is higher than 9.93 (Figs. 9 and 10).

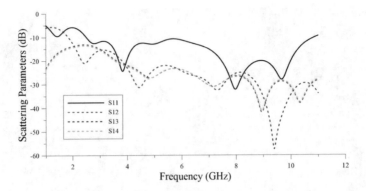

Fig. 8 S-parameter characteristics of double-sided MIMO antenna with decoupling structure

$$\rho_e = \frac{|S_{11}{}^*S_{12} + S_{21}{}^*S_{22}|^2}{(1 - |S_{11}|^2 - |S_{21}|)^2(1 - |S_{22}|^2 - |S_{12}|^2)} \tag{1}$$

$$DG = 10 * \sqrt{\left(1 - ECC^2\right)} \tag{2}$$

The S-parameter characteristics of MIMO array with of neutralization line technique are plotted in Fig. 11. The simulated -10 dB antenna bandwidth spreads between 2.6 and 10.58 GHz where the proposed antenna with this technique achieves good isolation between its element without affecting ECC and DG values (Figs. 12 and 13).

Based on Eq. (1) mentioned above, the ECC values have been calculated.

The diversity gain can be calculated using Eq. (2) mentioned above. The resulting diversity gain (DG) for the maximum value of ECC is ≥ 9.96 dB at all the frequency band.

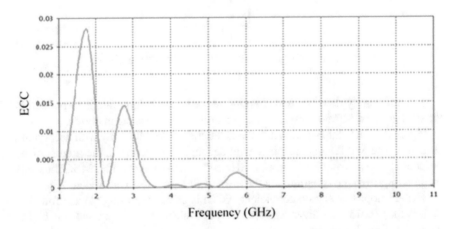

Fig. 9 ECC of MIMO antenna with decoupling structure

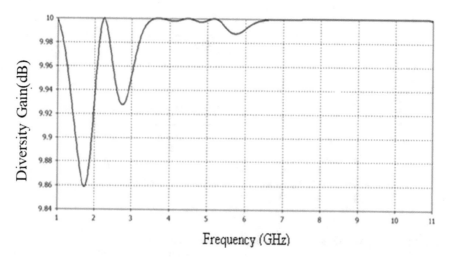

Fig. 10 Diversity gain (DG) of MIMO antenna with decoupling structure

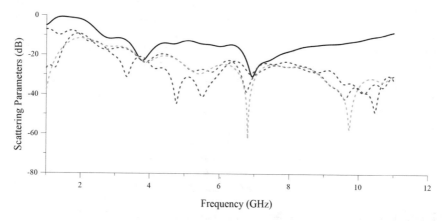

Fig. 11 S-parameter characteristics of double-sided MIMO antenna with neutralization line technique

Figure 14 shows the scattering parameter characteristics of the MIMO antenna after including the stub in the ground plane. This MIMO array produces resonance coefficient between 2.5 and 10.46 GHz with relative bandwidth of 127%. It is noted that the ECC values are lower than 0.01, and diversity gain is greater than 9.9 dB across the bandwidth (Figs. 15 and 16).

After implementing the various mutual coupling reduction techniques, the performance characteristics of MIMO antenna such as bandwidth, envelope correlation coefficient, and diversity gain have been calculated and listed in Table 1.

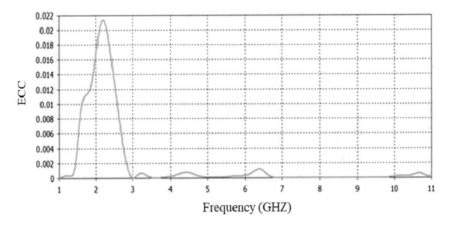

Fig. 12 ECC of MIMO antenna with neutralization line technique

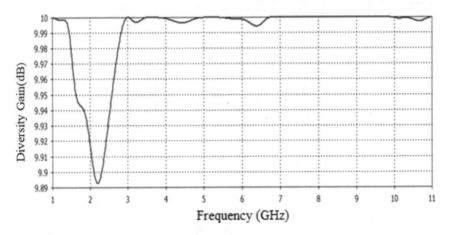

Fig. 13 Diversity gain (DG) of MIMO antenna with neutralization line technique

5 Conclusion

The double-sided quad element MIMO array with different mutual coupling techniques has been proposed. An ultra-wide band antenna which resonates from 3.15 to 10.61 GHz has been designed on FR4 substrate with the dimension of $30 \times 30 \times 0.8$ mm^3. The quad element MIMO antenna has been constructed by placing unit elements orthogonal to each other in the double-sided configuration. The proposed MIMO antenna produces a 10 dB of 7.8 GHz between 2.7 and 10.5 GHz with relative bandwidth 118%. Then, the different types of mutual coupling reduction techniques such as decoupling structure, neutralization line technique, and stub in the ground plane have been implemented, and performance of the MIMO antenna array is analyzed in terms of isolation, ECC, and diversity gain. The MIMO antenna's

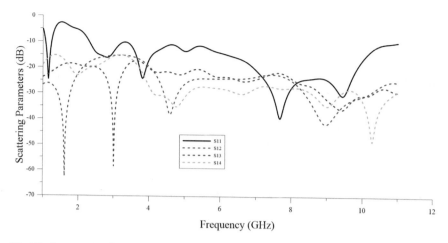

Fig. 14 S-parameter characteristics of double-sided MIMO antenna with stub in the ground plane

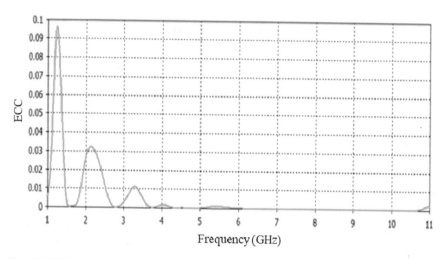

Fig. 15 ECC of MIMO antenna with stub in the ground plane

bandwidth and isolation has been significantly enhanced after incorporating mutual coupling reduction techniques.

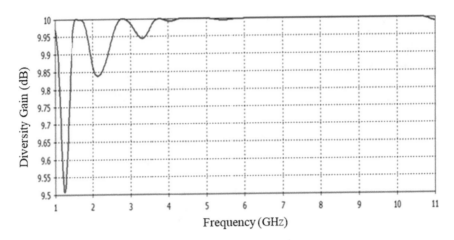

Fig. 16 Diversity gain (DG) of MIMO antenna with stub in the ground plane

Table 1 Comparison between various mutual coupling reduction techniques

Parameter	Type of technique			
	Double-sided MIMO antenna	Decoupling structure	Neutralization line technique	Stub in the ground plane
Bandwidth	2.7–10.5 GHz	2.54–10.8 GHz	2.6–10.58 GHz	2.5–10.46 GHz
% Bandwidth	118%	124%	121%	127%
ECC values	≤ 0.02	≤ 0.015	≤ 0.01	≤ 0.01
DG values	≥ 9.7	≥ 9.93	≥ 9.96 dB	≥ 9.9

References

1. Wang C, Wang C, Wan GC, Tong MS, Guan S, Xie LY (2019) RFID antenna sensor for quantitatively monitoring surface crack growth. In: IEEE International Conference on Computational Electromagnetics (ICCEM), pp 1–3
2. Ojaroudi M, Ghobadi C, Nourin J (2009) Small square monopole antenna with inverted T-shaped notch in the ground plane for UWB application. IEEE Antennas Wirel Propag Lett 8:728–731
3. Schantz HG (2004) A brief history of UWB antennas. IEEE Aerosp Electron Syst Mag 19(4), 22–26
4. Tata U, Huang H, Carter RL, Chiao JCM (2008) Exploiting a patch antenna for strain measurements. Meas Sci Technol IOP 20:1–6
5. Kumar G, Gupta K (1985) Nonradiating edges and four edges gap-coupled multiple resonator broad-band microstrip antennas. IEEE Trans Antennas Propag 33(2):173–178
6. Yang D, Zhai H, Guo C, Li H (2020) A compact single-layer wideband Microstrip antenna with filtering performance. IEEE Antennas Wirel Propag Lett 19(5):801–805
7. Targonski SD, Waterhouse RB, Pozar DM (2018) Design of wide-band aperture-stacked patch microstrip antennas. IEEE Trans Antennas Propag 46(9):1245–1251
8. Huynh T, Lee K (1995) Single-layer single-patch wideband microstrip antenna. In: Electron Lett 31(16):1310–1312

9. Yang F, Zhang X-X, Ye X, Rahmat-Samii Y (2001)Wide-band E-shaped patch antennas for wireless communications. IEEE Trans Antennas Propag 49(7), 1094–1100
10. Mak CL, Luk KM, Lee KF, Chow YL (2000) Experimental study of a microstrip patch antenna with an L-shaped probe. IEEE Trans Antennas Propag 48(5):777–783
11. Kim S, Nam S (2020) Compact ultrawideband antenna on folded ground plane. IEEE Trans Antennas Propag 68(10):7179–7183
12. Liu Z-X, Zhu L, Liu N-W (2020) A compact omnidirectional patch antenna with Ultrawideband harmonic suppression. IEEE Trans Antennas Propag 68(11):7640–7645
13. Boologam AV, Krishnan K, Palaniswamy SK, Manimegalai CT, Gauni S (2020) On the design and analysis of compact super-wideband quad element chiral MIMO array for high data rate applications. Electronics 9(12)
14. Wael AE, Ali AAI (2017) A compact double-sided MIMO antenna with an mproved isolation for UWB applications. AEU Int J Electron Commun 82:7–13
15. Saeed Khan M, Capobianco A-D, Najam AI, Shoaib I, Autizi E, Shafique MF (2014) Compact ultra-wideband diversity antenna with a floating parasitic digitated decoupling structure. IET Microw Antennas Propag 747–753
16. Huang H-F, Xiao S-G (2015) Compact triple band-notched UWB MIMO antenna with simple stepped stub to enhance wideband isolation. Prog Electromagnet Rese Lett 56:59–65
17. Yu Y, Liu X, Gu Z, Yi L (2016)A compact printed monopole array with neutralization line for UWB applications. In: 2016 IEEE international symposium on antennas and propagation (APSURSI), pp 1779–1780

Enhancing Agricultural Outcome with Multiple Crop Recommendations Using Sequential Forward Feature Selection

Souvik Mondal, Bibek Upadhyaya, Kamal Nayan, Vinay Singh, and Jyoti Singh Kirar

1 Introduction

The population of the world is increasing at a steady rate. From 7.9 billion in 2020, it is expected to reach 10 billion at the end of this century [1]. As the population is growing, so does the demand for food. Speaking of the population, our country India is worked to be the most populous country over the next two decades, surpassing our neighbor nation China [2]. So, it is necessary to have a good strategy and technique for crop production to meet the demand of the increasing population.

In India, each year, we receive various reports of crop failures, less and less production and immature crops that waste the resources required to produce the crops [3]. Farmers are unable to produce crops to fulfill the demand of the population. It is necessary to consider the soil composition, texture and various topographic factors for healthy and sufficient production. But, due to lack of technological advances and various other factors like less awareness and knowledge, people tend to grow their crops on some hypothetical cropping probability [4]. So, it is necessary to know the composition of the soil and conditions required for various crops' production. For this purpose, we are making a recommendation system for the crops to the farmers keeping in mind a few of the factors required and are essential for crop production.

In this system, we are going to use data to predict the crop which can be grown in an area based on the given chemical and topographical factor of the soil. Or, in order to grow a specific crop in what amount we are going to use a particular fertilizer in the soil. This would help farmers to avoid the use of unnecessary fertilizer in the soil and particularly know exactly what do their soil need and what is the soil capable of growing, if the recommended amount of fertilizer is given.

S. Mondal · B. Upadhyaya · K. Nayan · V. Singh · J. S. Kirar (✉)
Banaras Hindu University, Varanasi, India
e-mail: kirarjyoti@gmail.com

© The Author(s), under exclusive license to Springer Nature Singapore Pte Ltd. 2023
D. S. Sisodia et al. (eds.), *Machine Intelligence Techniques for Data Analysis and Signal Processing*, Lecture Notes in Electrical Engineering 997,
https://doi.org/10.1007/978-981-99-0085-5_5

2   Literature Review

There have been lots of research in the field of agriculture, especially mentioning crop recommendations in precision agriculture. Precision agriculture is in trend nowadays. It helps the farmers to get informed decisions about the farming strategy. There have been plenty of machine learning models used to recommend crops such as Random Forest, Decision Tree, Support Vector Machine, Bayes, Neural Network for making the precision agriculture recommended to everyone in the country with ease [5, 6].

Our effort has been to build a recommendation system for precision agriculture to help our country's farmers grow crops available to them with ease. One more problem that is faced by farmers is that by using recommendations, a particular region, if recommended a particular crop, grows the same crops in abundance, thus reducing the cost of crops. So, here we have added one more feature in this crop recommendation: recommending second and further best crops. For this, we have used various techniques: Principal Component Analysis (PCA), Forward Feature Selection, Feature Selection using Correlation, Feature Selection using Mutual Information, Use of accuracy table in feature selection.

In [7], the authors have prepared a crop recommendation system with the help of soil characteristics and environmental characteristics. They also built rainfall prediction models and visual mapping. They have obtained 90.20% accuracy in the Decision Tree Classifier, 89.78% in KNN Classifier, 90.43% accuracy in Random Forest Classifier and 91% accuracy in Neural Network. Our approach is to improve the accuracy with minimum features and use Naive Bayes Classifier for multiple crop recommendations.

3   Methodology

Our whole analysis procedure is presented below in the form of a flowchart (Fig. 1).

The main objective of our work is to predict the crop to be cultivated with the help of different features which affect the growth of the crop. While predicting the crop to be cultivated, we have also considered the factor of demand and supply. If the same type of crop is cultivated in an area, then the farmer may suffer a great loss due to an increase in crop supply and decrease in demand. So, we will do multiple crop predictions in order to give options to farmers to grow the crop by considering the factor of demand and supply.

3.1   Dimensionality Reduction

Principal Component Analysis (PCA): PCA is an unsupervised machine learning technique that transforms the feature from the original feature space to a new space to

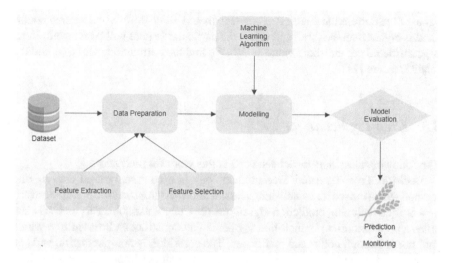

Fig. 1 Flowchart describing complete analysis procedure

increase the distance between data. The direction of the maximum separation of data or maximum variance is determined by the maximum magnitude of the eigenvalues and their corresponding eigenvector [8].

3.2   Feature Subset Identification

Forward Feature Selection: It is the feature selection technique that uses an iterative method in selecting the features [9]. Here, we begin with the null model—a model that contains an intercept but no predictors. We then fit p simple linear regressions and add to the null model the variable that results in the lowest Residual Sum of Squares (RSS). We then add to that model the variable that results RSS for the new two-variable model. This approach is continued until some stopping rule is satisfied [10].

    Mutual Information: Mutual Information is the technique that calculates the statistical similarities between two different variables. It is the extended version of information gain which is applied to variable selection. It computes the amount of information obtained about one random variable by observing the other random variable. It is based on the concept of entropy of a random variable which means the amount of information contained in a random variable [11].

    Correlation method: Correlation is the method that measures the degree of similarity between two random variables. If the correlation coefficient between them is + 1, then the variables are positively correlated, whereas if the correlation coefficient is −1, the variables are negatively correlated. If the variables are uncorrelated, the value of the correlation coefficient between them will be 0. If the correlation coefficient is higher than the threshold value, the particular feature will be selected. Here, we considered the threshold value as 0.7. We find the correlation matrix, to find the related feature [11].

3.3 Crop Prediction

The following classifiers have been used in this work for prediction:

Decision Tree: Decision Tree methodology is a commonly used data mining method which is used for establishing classification systems based on multiple covariates or for developing prediction algorithms for a target variable. This method classifies a population into branch-like segments that construct an inverted tree with a root node, internal nodes and leaf nodes. The algorithm is non-parametric and can efficiently deal with large, complicated datasets without imposing a complicated parametric structure [12].

Support Vector Machine (SVM): Support Vector Machines (SVMs) are classifier which create a maximum margin hyperplane to classify the training datasets. The SVM learning problem can be formulated as a convex optimization problem, in which efficient algorithms are available to find the global minimum of the objective function [8, 13].

Naive Bayes: The Naïve Bayes (NB) Classifier is a family of simple probabilistic classifiers which is based on a general assumption that all features are independent of each other, given the category variable, and it is often used as the baseline in text classification [14].

K-Nearest Neighbor (KNN): KNN Classifier is used to classify unlabeled observations by assigning them to the class of the most similar labeled examples. Characteristics of observations are collected for both training and test datasets [15].

4   Experimental Results and Discussion

In this work, we have utilized the publicly available dataset on Kaggle collected from Indian territories [16]. We have approached the work in a step-by-step manner as mentioned in Sect. 3. We checked the shape of our data, which were found to have 2200 entries and eight fields (seven features and one target variable). The seven features here are N, P, K, temperature, humidity, pH and rainfall, and the target variable is "label", which determines the crop to be grown. Then, we check if there are any null values in the data using a heatmap (Fig. 2).

It is clear from the heatmap that the dataset does not contain any missing value.

Next, we checked unique crop labels from the data. We found that we have 22 unique crop labels and each with 100 samples. The crops are: "apple", "banana", "blackgram", "chickpea", "coconut", "coffee", "cotton", "grapes", "jute", "kidney

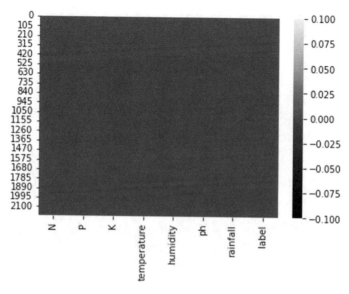

Fig. 2 Heatmap of different features

beans", "lentil", "maize", "mango", "mothbeans", "mungbean", "muskmelon", "orange", "papaya", "pigeon peas", "pomegranate", "rice", and "watermelon". Our dataset is balanced as there are an equal number of samples for each unique crop label. Next, we explored the dataset visually plotting graphs.

First, we checked the correlation between all the features (Fig. 3).

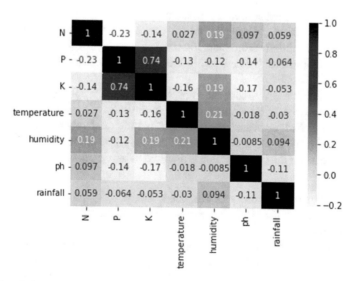

Fig. 3 Correlation matrix of different features

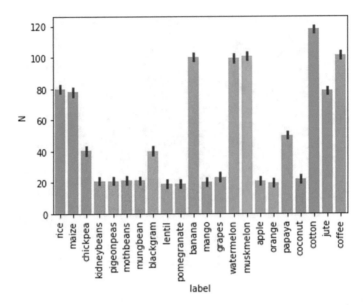

Fig. 4 Different levels of N

Here, from the heatmap, we concluded that the features "P" and "K" are highly linearly correlated. The other feature combinations are not much correlated.

We used bar plots to observe the required levels of the different features (Figs. 4, 5, 6, 7, 8 and 9).

From the given charts for the features, we observed that:

- The soil pH is almost similar for all the crops.
- For the crops "Rice", "Coffee", "Jute", "Coconut", "Papaya", etc., it is required to have heavy rainfall (more than 100 mm).
- A high proportion of "Potassium" in the soil is required for the fruits "Grapes", "Apple".
- For "Lentils", "Peas", "Beans the Phosphorus Lana", "Apple" and "Grapes", it is required to have a high proportion of "P" in the soil.
- "Chickpea", "Kidney beans" and "Mango" require relatively less humid weather.
- The lentils, peas, beans and some fruits required less amount of nitrogen in the soil, whereas, for the other crops, it is required to have a relatively large proportion of N.

Next, we performed Principal Component Analysis.

We got Table 1, which indicates explained variation corresponding to a number of components.

Therefore, the first four components explained 96.8% of the variation of the dataset. Our objective here is to get an idea about the minimum number of features which can predict our target variable with maximum accuracy. We will now try to select a maximum of four-feature combination for further process. Next, we split our

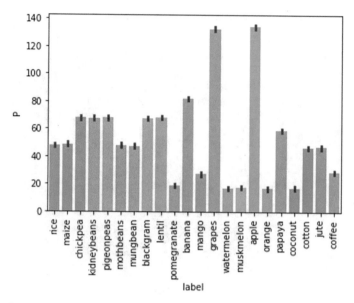

Fig. 5 Different levels of *P*

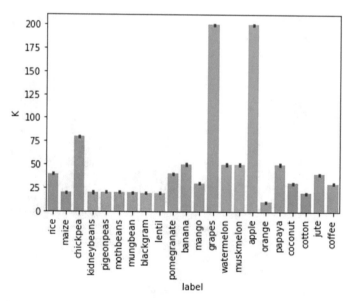

Fig. 6 Different levels of *K*

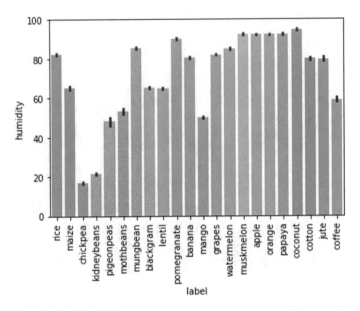

Fig. 7 Different levels of humidity

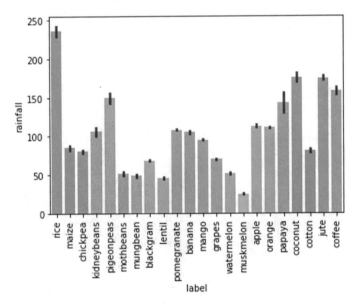

Fig. 8 Different levels of rainfall

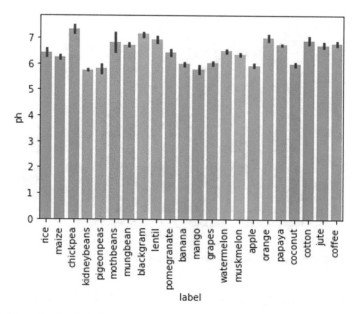

Fig. 9 Different levels of rainfall

Table 1 Explained *variance*

Number of components	Explained variation ratio
1	0.401
2	0.343
3	0.158
4	0.066
5	0.029
6	0.002
7	0.001

dataset into train and test datasets with a ratio of 8:2 and also stratified with respect to unique crop labels. Next, we performed some feature selection techniques on the training set to select the features for final model training. For this, we trained our classifiers and calculated accuracies for each classifier for each feature combination sequentially [12]. For almost all classifiers, the accuracy maximized for the feature combined with all features. But, there exist some feature combinations for which we are having a bit lesser accuracy score. We can either use all the features or use some lesser number of features that have higher accuracy. We also used the mutual information corresponding to each feature. We got the following info from Fig. 10.

Here, the feature "Humidity" has the highest magnitude of the mutual information and "pH" has the lowest magnitude. We found the accuracy for [humidity, K, rainfall, P] is 0.970.

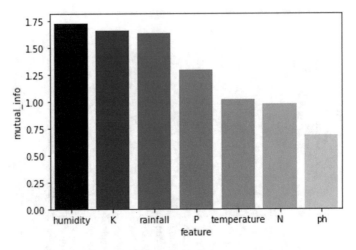

Fig. 10 Mutual information for different features' info

Now, we want to check the best four correlated features with the target variable using correlation matrix (Fig. 11).

The required four features obtained based on the magnitude of correlation with the target variable are P, K, humidity and rainfall and their accuracy is 0.929.

So, considering the forward feature selection technique, we are taking the following feature combination for training the Naive Bayes Classifiers: N, K, humidity, rainfall as their accuracy is 0.977 which is the highest among the previous accuracies obtained.

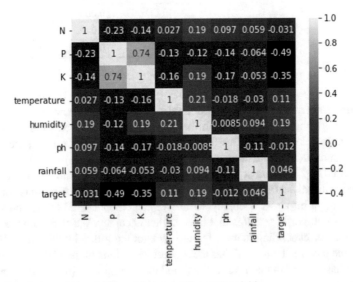

Fig. 11 Correlation matrix of different features with target variable

We used four different classifiers, namely, Decision Tree, Naive Bayes, k-Nearest Neighbors and Support Vector Classifiers.

To find a good choice of "k" in the case of the nearest neighbor classifier, we have taken an iterative method. In our method, we have started from $k = 1$ and run to 50, and in each iteration, we trained the classifier and calculated error values for each k. The following graph shows the error value versus k value graph (Fig. 12).

We got the least error at $k = 3$. We will use $k = 3$ for our nearest neighbor classifier. Also, as a distance measure, we are using "Minkowski distance".

We trained the classifiers on the training dataset and validated or tested with the test dataset. We have measured the value of different performance metrics for each classifier. Table 2 contains the average value of each metric.

We have also performed k-fold cross-validation with $k = 5$. We got the following cross-validation score for the different algorithms (Table 3).

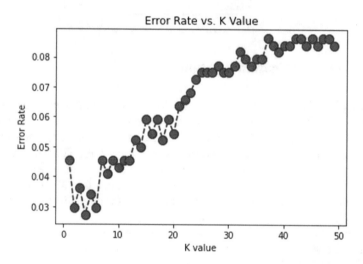

Fig. 12 K value versus error rate

Table 2 Average *metrics score*

Metrics	Decision tree	SVC	KNN	Naïve Bayes
Accuracy	0.977	0.954	0.979	0.990
Precision	0.980	0.960	0.980	0.990
Recall	0.980	0.950	0.980	0.990
F1-score	0.980	0.950	0.980	0.990

Table 3 K-fold cross-validation

Models	Fold 1	Fold 2	Fold 3	Fold 4	Fold 5
Decision tree	0.989	0.964	0.975	0.982	0.973
SVC	0.952	0.959	0.954	0.954	0.952
KNN	0.959	0.968	0.964	0.975	0.970
Naïve Bayes	0.995	0.993	0.990	0.989	0.989

Table 4 Prediction of multiple crops for a terrain

Crops	Probability
Kidney beans	0.759
Pigeon peas	0.240
Coffee	$1.36 * 10^{-17}$
Maize	$6.17 * 10^{-22}$
Jute	$1.25 * 10^{-34}$

5 Multi Crop Recommendation

Among all the used classifiers, the "Naive Bayes" Classifier has approximately 99% accuracy. Also, if we look at the other evaluation metrics and the cross-validation scores corresponding to the Naive Bayes Classifier, we can conclude that the trained classifier performs very well. So, we can use our trained NB Classifier to get our prediction. We have also used the same model to get probabilities for each target class, by which we can predict more than one crop name, that will also be helpful for farmers. We present below one of the examples of the output in Table 4 when the input for different features were [30, 23, 18, 174].

It clearly shows that the features with input as given in this example favor the growth of kidney beans with a probability equal to 0.759 and pigeon peas with a probability of 0.240. The other crops are not favored to be grown in these conditions.

6 Comparison

We have obtained 97.6% accuracy in the Decision Tree Classifier and 96.7% accuracy in KNN Classifier which is higher than the accuracy obtained in [2]. The feature selection techniques like PCA, Forward Feature Selection, Mutual Information and Correlation Matrix have possibly improved our accuracy. Also, the iterative method used for finding the value of k has improved the accuracy. Further, we have obtained 95.4% accuracy in SVC. In the Naive Bayes Classifier, we have obtained 99.1% accuracy. With the help of the Naive Bayes Classifier, we can obtain the probability of different crops. This will give farmers the option to cultivate different crops based on their requirements. Accuracies obtained are compared in Table 5.

Table 5 Accuracy comparison

Classifiers	Proposed model	Existing model
Decision tree	0.9766	0.902
SVC	0.9542	–
KNN	0.9672	0.8978
Naïve Bayes	0.9912	–
Neural network	–	0.91
Random forest	–	0.9043

7 Conclusion and Future Enhancement

In this work, we have used sequential forward feature selection to identify multiple crops for agriculture on different terrains of India on the basis of different features.

The proposed work has provided significant results on publicly available datasets. Though we have reached our primary goal, there are some enhancements we should make. First of all, we have to implement an application that can be directly used by the farmers. Also, the whole dataset is secondary data. If it is possible, we can collect real-time data and implement our model to validate the performance. And lastly, we shall find a way to predict which fertilizer a farmer should use for a specific crop. Furthermore, we should look for different issues related to this sector and come up with an application with solutions to the issues.

References

1. Gates B, Pitron G, Mann ME, Gates BBDCATHA 978–0385546133, Pitron TRMWBG, 978–1950354313 (2022) How to avoid a climate disaster, the new climate war, The Rare Metals War 3 Books Collection Set. Knopf/Scribe US Ltd
2. https://www.thehindu.com/news/national/india-to-be-most-populous-by-2027-un/article28067167.ece
3. https://agricoop.nic.in/sites/default/files/Web%20copy%20of%20AR%20%28Eng%29_7.pdf
4. https://economictimes.indiatimes.com/news/economy/agriculture/view-empowering-the-indian-farmer/articleshow/77904360.cms?from=mdr
5. Pande SM, Ramesh PK, Anmol A, Aishwarya BR, Rohilla K, Shaurya K (2021) Crop recommender system using machine learning approach. In: 2021 5th International Conference on Computing Methodologies and Communication (ICCMC). IEEE, pp 1066–1071
6. Raja SKS, Rishi R, Sundaresan E, Srijit V (2017) Demand based crop recommender system for farmers. In: 2017 IEEE technological innovations in ICT for agriculture and rural development (TIAR). IEEE, pp 194–199
7. Doshi Z, Nadkarni S, Agrawal R, Shah N (2018) AgroConsultant: intelligent crop recommendation system using machine learning algorithms. In: 2018 fourth international conference on computing communication control and automation (ICCUBEA). IEEE, pp 1–6
8. Kumar PTMSAKV (2022) Introduction to data mining, 2nd edn. Pearson India
9. Kirar JS, Agrawal RK (2018) Relevant frequency band selection using sequential forward feature selection for motor imagery brain-computer interfaces. IEEE Sympos Ser Comput Intell (SSCI) 2018:52–59. https://doi.org/10.1109/SSCI.2018.8628719

10. James G, Witten D, Hastie T, Tibshirani R (2021) An introduction to statistical learning: with applications in R (Springer Texts in Statistics) (2nd ed. 2021 ed.). Springer
11. Kirar JS, Agrawal RK (2018) Relevant feature selection from a combination of spectral-temporal and spatial features for classification of motor imagery EEG. J Med Syst 42(5):1–15
12. Yan Y-S (2015) Decision tree methods: applications for classification and prediction. Shanghai, Shanghai Municipal Bureau of Publishing.ncbi.nlm.nih.gov
13. Armin S (2005) "SVM." Data mining and knowledge discovery handbook I:2
14. Shou X (2016) Bayesian Naïve Bayes classifiers to text classification. J Inf Sci 44(1):59. https://doi.org/10.1177/0165551516677946
15. Zhang Z (2016) Introduction to machine learning: k-nearest neighbours. Ncbi 1(1):1
16. https://www.kaggle.com/atharvaingle/crop-recommendation-dataset

Kernel-Level Pruning for CNN

Pragnesh Thaker and Biju R. Mohan

1 Introduction

Many real-life applications use deep learning, such as self-driving cars, natural language processing, and many more with acceptable accuracy. Designing and learning a neural network from scratch face two challenges. The first one is that deep learning requires a vast amount of dataset, and for many applications, the dataset is relatively small, which leads to overfitting. The second one is that training from scratch is computationally expensive even when we have a vast dataset. Thus, most AI developers use transfer learning to choose a standard neural model trained on similar problems on a vast dataset. For example, we choose a standard model such as VGGNet, ResNet, and GoogLeNet [1–3] for image classification problems trained on ImageNet dataset.

As all these standard models are trained for a complex problem with a vast dataset, they are enormous with millions of parameters and huge computational costs that make deploying on resource-constrained devices difficult. There are two approaches to building an optimal neural network. The first is a constructive approach, which starts with a small neural network and adds complexity by adding more neurons at each layer and more layers. This approach has limited success. Second, the destructive approach, where we start with a large model and keep compressing the model, destructive approach has shown promising results using several techniques, and pruning is one of them [4].

The l-th layer of convolution layer has N_l number of channel and it generate N_l output tensor of size $W \times H$. Each channel contains N_{l-1} number of kernel where N_{l-1} is number of output generated by previous layer, so there are total $N_{l-1} \times N_l$ number of kernel. We can perform structured pruning at different granularities. We perform

P. Thaker (✉) · B. R. Mohan
National Institute of Technology, Karnataka, India
e-mail: pragnesh.187it001@nitk.edu.in
URL: http://nitk.ac.in

© The Author(s), under exclusive license to Springer Nature Singapore Pte Ltd. 2023
D. S. Sisodia et al. (eds.), *Machine Intelligence Techniques for Data Analysis and Signal Processing*, Lecture Notes in Electrical Engineering 997,
https://doi.org/10.1007/978-981-99-0085-5_6

pruning at channel level and kernel-level granularity in structured pruning for the convolution layer. We performed pruning at the neuron level for the dense layer, where we removed neurons along with all incoming and outgoing edges.

When we want to reduce the memory requirement, we should perform the pruning on the dense layer, and when we want to reduce the computational cost, we should perform the pruning on the convolution layer. Consider that we have $K_h \times K_w$ kernel where K_h is the height of the kernel and K_w is the kernel's width. The input size is $W \times H$, where W is the width of the input tensor and H is the height of the input tensor. There are two important types of convolution operation (i) *valid convolution* where no padding is applied and (ii) the *same convolution* where padding is applied in such a way size of the output is the same as the input. For each value of output, we need to perform $K_h \times K_w$, and we have a total $W \times H$ values, and thus we need $K_h \times K_w \times W \times H$ to generate one output channel using one input channel. For N_l number of channel in current layer and N_{l-1} number of channel in previous layer then total number of operation in one convolution layer is $K_h \times K_w \times W \times H \times N_l \times N_{l-1}$. The convolution layer has 10% of the parameter but contributes to 99% of computation, whereas the dense layer has 90% of the parameter but only contributes 1% of computation cost [1].

$$Y_{i,j}^l = \sum_{p=i}^{i+k_h} \sum_{q=j}^{j+k_w} F_{p-i,q-j} \times Y_{p,q}^{l-1} \quad \forall i, j \in (1, W), (1, H) \tag{1}$$

The following paper will focus on structured pruning as compression directly transfers to reduce computational cost and memory requirement. In unstructured pruning, we require special hardware and software library written for specific hardware to exploit the advantage of the compression. A kernel is the smallest structural unit for structured pruning in the convolution layer. Consider there are N_{l-1} input channel and N_l output channel then there will be $N_{l-1} \times N_l$ number of kernel where N_l number of channel in working on each input channel. We will perform iterative pruning in the following method to prune the k kernels from each output channel. We will compute similarities between each kernel working on the same input channel to find redundant kernels in each channel [5–7].

Generally four step iterative method is followed for the pruning. (i) Fine-tuned model, (ii) Select parameter to be prune, (iii) Remove selected parameter and fine-tune, (iv) Evaluate prune model, and repeat from step 1 if after evaluation we can still prune model. Most important step in pruning is how to select parameter to be pruned [8].

2 Related Work

Simonyan [1] shown that with an increase in the depth of the CNN, we can improve accuracy by training models of different depths in VGG11, VGG13, VGG16,

Table 1 Related work summary

S. No.	Paper	Pros	Cons
1	Pruning CNN for resource efficient inference [5]	Explain computation limitation and explained need of saliency-based pruning	Explored only L_1 norm for saliency-based approach
2	Pruning filters for efficient ConvNets [8]	It prunes the model iteratively till accuracy drop is within threshold	Computationally expensive process
3	Channel pruning for accelerating DNN [9]	Uses layer by layer iterative approach for pruning	Pruning process is computationally expensive
4	Learning to prune neuron in dense layer [10]	Removes neuron from dense layer using group regularization	Helps to reduce memory bandwidth but computation cost remains almost unchanged

VGG19, where the last two-digit number indicates the number of a learnable layer in the CNN. For example, VGG16 has 16 learnable layers (13 convolution layers and 3 dense layers) [1].

Exhaustive search to select parameters to be pruned is infeasible as every subset of the parameter needs to be evaluated. There is 2^n number of subset where n is the number of parameters. Here, we look for the quantitative analysis to predict the effect of parameter on output and prune the parameter, which has minimal effect on output. L_1 norm value is widely used for quantitative analysis [5, 8].

The model is trained with L_1 and L_2 norm regularization. Regularization decreases values of the different parameters at a different rate. Few parameter values reduce to near-zero values selected for the pruning as their effect on the output may be negligible (Table 1).

He et al. [6] used L_2 norm value for iterative pruning where we remove a small portion of channels followed by fine-tuning in every iteration. This method provides better compression with the use of a massive amount of computation for pruning and fine-tuning.

Alvarez et. al. [10] propose a group regularization technique. Regularization is applied to all the incoming and outgoing connection of each neuron individually and help to remove neuron from the dense layer. This method reduces memory requirements for the standard network such as VGGNet.

In many applications, we may have a threshold for accuracy. We want to maximize the compression of the given model until the accuracy drop is within the threshold limit. Li et al. [11] developed an iterative approach where the threshold is given as input for pruning. We evaluate the model after every iteration of pruning. If the model accuracy is within the limit, we continue the next iteration of pruning. When the accuracy drop is above the threshold, we roll back to the last acceptable compressed model.

The smallest structural unit to be pruned is kernels. Consider there are N_{l-1} channel in the previous layer and N_l number of layer in the current layer then there are total $N_{l-1} \times N_l$ number of kernel. We like to maintain a regular shape tensor to facilitate parallel execution with a simpler representation of all parameters. Thus, we prune the kernel in such a way that its effect will be like channel pruning [12].

3   Methodology

VGGNet is very successful model for Image Classification. There are several variant of VGGNet based on the number of learnable layer present in it. We are working on VGG16 with 16 learnable layer (13 convolution layer and 3 dense layer). Here we are working on cifar10 and Intel Image Classification dataset. Cifar10 has 10 classes of object, while Intel Image classification has 6 different class of object. For convolution layer l, weight tensor W_l is a 4 dimension. The l-th layer of convolution layer has N_l number of channel and it generate N_l output tensor of size $W \times H$. Each channel contains N_{l-1} number of kernel where N_{l-1} is number of output generated by previous layer, so there are total $N_{l-1} \times N_l$ number of kernel. We can perform structured pruning at different granularities. We perform pruning at channel level and kernel-level granularity in structured pruning for the convolution layer. We performed pruning at the neuron level for the dense layer, where we removed neurons along with all incoming and outgoing edges.

Tensor. $W^l \in R^{n_l \times n_{l-1} \times w \times h}$ where W^l is four-dimensional set of real numbers R. n_l indicates the number of channels in the current layer. n_{l-1} indicates the number of channels in the previous layer. w and h indicate the width and height of the feature map.

We are going to use transfer learning where we will use VGG16 which is trained on ImageNet dataset and we will fine-tune it for given specific problem. First we will replace output layer which suits given problem and initialize there weight randomly. Then we train the model by freezing weight of all layer except last layer with random initialization. Once we start getting good accuracy, we will unfreeze all parameter and fine-tune for some epoch using state of art techniques such as Adam optimizer, gradient clipping with learning rate finder [13–15]. We will fine-tune the model using $L1$ and L_2 norm. L_1 is absolute sum of each parameter and L_2 is summation of square value of each parameter.

$$r(\theta) = \sum_{i=1}^{L} \|W^l\| + \sum_{i=1}^{L} (W^l)^2 \tag{2}$$

Total loss is summation of original loss and regularization loss.

$$L(\theta) = L(\theta) + r(\theta) \tag{3}$$

Algorithm 1 KernelPruning

Input: PruneLayer, PruneAmount
Output: PrunedWeight
  Initialisation: PrunedWeight = Weight
  $KernelList = []$
  for $l = 0$ to L **do**
    $KernelLayer$ = ComputeKernekScore($PruneLayer[l]$)
    $KernelList$.append($KernelLayer$)
  end for
  for $i = 0$ to L **do**
    sort($KernelLayer[i]$)
    $extract\_K$ = get_K($KernelLayer[i], PruneAmount[i]$)
    mask kernel in $extract\_K$
  end for
  perform deepCopy()
  return $PruneWeight$

Algorithm 2 ComputKernelScore

Input: WeightMatrix of a Layer
Output: Layer_Kernel_Score
  for $j = 0$ to N_{l-1} **do**
    for $i1 = 0$ to N_l **do**
      $dist[j][i1][i2]$ = distance($kernel[i1][j], kernel[i2][j]$)
      ChKernelScore.append($j, i1, i2, dist[j][i1][i2]$)
    end for
    $LayerKernelScore.append(ChKernelScore)$
  end for
  return $LayerKernelScore$

$W^{n_{l-1}, n_l, k_h, k_w}$ is the weight matrix of lth layer. Here, n_{l-1}, n_l index gives n_l-th kernel of l-th layer operating on n_{l-1}-th input tensor. Thus, we have total N_l kernel operating on each input tensor.

For the given algorithms, we specify the amount to be pruned from each layer as input. Then for each layer, we will compute the kernel score. Kernel score is the distance between two kernels of the same filter operating on the same input tensor, as shown in algorithm 2. In algorithm 1, we will append the kernel score for each kernel pair and store required index values to identify the kernel. After that, we will sort the KernelLayer list based on the score value in ascending order. First, k values indicate minimum score values, and thus we will extract the first k values from the KernelLayer list. The value of k is equal to the prune amount for a given layer.

In the ComputeKernelScore algorithm, we want to compute the distance between a pair of kernels belonging to the same channel and working on the same input. Thus, index value j indicates input, and $i1$ and $i2$ form a pair of kernel working on the same input.

4 Result and Analysis

In the following experiment, we train VGG13 and VGG16 on cifar10 dataset and Intel Image Classification dataset and perform pruning over the trained model. We used transfer learning; here, we replace the final layer of the pre-train model for a given problem and initialize it with random weight. We trained the model for 20 epochs and got 84.14% accuracy. We perform fine-tuning where we unfreeze all the weight and train the model for 10 more epochs with a lower learning rate and achieve 88.16%. Here, we perform kernel-level pruning, and we select a different amount of kernel to prune for each layer of VGGNet in an iteration and store them in a list. We perform several iterations of pruning as shown in the Table 2. We can see that the accuracy drop till 20% pruning is less than 1%, but accuracy starts dipping after that.

We follow the same steps for pruning of VGG13 on Intel Image Classification. Here, we get 86.17% accuracy after fine-tuning. We performed several iterations of pruning on the model, and results are given in Table 3. Here, we can see that the accuracy drop is less than 2% till 25% pruning but start dipping after that.

Similarly, we trained VGG16 on the cifar10 dataset and achieved 90.87% accuracy after fine-tuning. Here, we performed several iterations on pruning as given in the Table 4. Here, we can see that the accuracy drop is nearly 2.69% till 30% but start dipping from next iteration

Table 2 Compression of VGG13 model trained on cifar10 dataset

Iteration	Val Acc (%)	Pruning sparsity (%)	Accuracy drop (%)
1	88.12	05	0.04
2	88.04	10	0.12
3	87.61	15	0.55
4	87.23	20	0.93
5	84.73	25	3.14
6	81.48	30	6.76

Table 3 Compression of VGG13 model trained on Intel Image Classification

Iteration	Val Acc (%)	Pruning sparsity (%)	Accuracy drop (%)
1	86.18	05	−0.01
2	85.89	10	0.28
3	85.14	15	1.03
4	84.73	20	1.44
5	84.42	25	1.75
6	82.18	30	3.99
7	79.44	35	6.73

Table 4 Compression of VGG16 model trained on cifar10

Iteration	Val Acc (%)	Pruning sparsity (%)	Accuracy drop (%)
1	90.37	05	0.50
2	90.06	10	0.81
3	89.58	15	1.29
4	89.12	20	1.75
5	88.89	25	1.98
6	88.18	30	2.75
7	85.23	35	5.64
8	83.62	40	7.25

Table 5 Compression of VGG13 model trained on Intel Image Classification

Iteration (%)	Val Acc (%)	Pruning sparsity (%)	Accuracy drop (%)
1	88.38	05	0.18
2	88.12	10	0.44
3	87.73	15	0.83
4	87.06	20	1.5
5	86.12	25	2.44
6	84.08	30	4.48
7	81.51	35	7.05

Similarly, when we trained VGG16 on the Intel Image Classification problem using transfer learning, we achieved an accuracy of 88.38% on the validation dataset. After performing several iterations of pruning, we observe that till 20% pruning, the accuracy drop is slow but increases sharply after it (Table 5).

5 Conclusion and Future Work

Here, we can observe that kernel-level pruning can be useful for pruning in the convolution layer and we can prune up to 20% with minimum drop in accuracy and prune up to 35% with acceptable accuracy drop. Kernel pruning is independent to channel pruning so in future, we can combine channel pruning with kernel-level pruning to achieve better reduction in the computational cost. We can perform neuron pruning at dense layer for memory reduction of the model.

References

1. Simonyan K, Zisserman A (2014) Very deep convolutional networks for large-scale image recognition, arXiv preprint arXiv:1409.1556
2. He K, Zhang X, Ren S, Sun J (2016) Deep residual learning for image recognition. In: Proceedings of the IEEE conference on computer vision and pattern recognition, pp 770–778
3. Szegedy C, Liu W, Jia Y, Sermanet P, Reed S, Anguelov D, Erhan D, Vanhoucke V, Rabinovich A (2015) Going deeper with convolutions. In: Proceedings of the IEEE conference on computer vision and pattern recognition, pp 1–9
4. Cheng Y, Wang D, Zhou P, Zhang T (2017) A survey of model compression and acceleration for deep neural networks, arXiv preprint arXiv:1710.09282
5. Molchanov P, Tyree S, Karras T, Aila T, Kautz J (2016) Pruning convolutional neural networks for resource efficient inference, arXiv preprint arXiv:1611.06440
6. He Y, Zhang X, Sun J (2017) Channel pruning for accelerating very deep neural networks. In: Proceedings of the IEEE international conference on computer vision, pp 1389–1397
7. Li Y, Lin S, Zhang B, Liu J, Doermann D, Wu Y, Huang F, Ji R (2019) Exploiting kernel sparsity and entropy for interpretable CNN compression. In: Proceedings of the IEEE/CVF conference on computer vision and pattern recognition, pp 2800–2809
8. Huang SYQ, Zhou K, Neumann U (2018) Learning to prune filters in convolutional neural networks. IEEE, pp 709–718
9. Zhang Q, Shi Y, Zhang L, Wang Y, Tian Y (2020) Learning compact networks via similarity-aware channel pruning. In: 2020 IEEE conference on multimedia information processing and retrieval (MIPR). IEEE, pp 145–148
10. Alvarez JM, Salzmann M (2016) Learning the number of neurons in deep networks. In: Advances in neural information processing systems, pp 2270–2278
11. Li H, Kadav A, Durdanovic I, Samet H, Graf HP (2016) Pruning filters for efficient convnets, arXiv preprint arXiv:1608.08710
12. Lin C, Zhong Z, Wu W, Yan J (2018) Synaptic strength for convolutional neural network, arXiv preprint arXiv:1811.02454
13. Kingma DP, Ba J (2014) Adam: a method for stochastic optimization, arXiv preprint arXiv:1412.6980
14. Ede JM, Beanland R (2020) Adaptive learning rate clipping stabilizes learning. Mach Learn: Sci Technol 1(1):015011
15. Smith LN (2017) Cyclical learning rates for training neural networks. In: IEEE winter conference on applications of computer vision (WACV). IEEE, pp 464–472

Linear Regression Model for Predicting Virtual Machine Consolidation Within the Cloud Data Centers (LrmP_VMC)

M. Tejaswini, T. Hari Sumanth, and K. Jairam Naik

1 Introduction

Cloud computing is the field of information technology, which is a sort of model that provides on demand accessibility to configure the computing resources. Cloud computing integrates a bunch of tools and technologies in which some are pre-existing such as virtualization. This process of virtualization [1] is the concept where the cloud resources can be given to customers at a lower cost along with high flexibility.

Through the process of live migration [2] of virtual machines, virtual machines are dynamically allocated to respective host physical machines or hosts in an efficient way that can lead to less energy utilization and resource utilization. This method of live migration of virtual machines and determination of destined host machines is known as the virtual machine consolidation process [3]. Although the dynamic virtual machine consolidation can help in decreasing the energy utilization in a best way, the migration of virtual machines can lead to Service Level Agreement Violations (SLAVs) [4].

In this paper, we are using the Genetic Algorithm (a regression-based technique), for the allocation of virtual machines. This is also known as the famous Roulette Wheel Selection Strategy, where we compare the virtual machines and cloudlets with the chromosome instances. The lesser the power/fitness value, the better it is for the consolidation. These in turn result in reducing the SLAVs, which is much more

M. Tejaswini · T. Hari Sumanth · K. Jairam Naik (✉)
Department of Computer Science & Engineering, National Institute of Technology Raipur, Great Eastern Rd, Raipur, Chhattisgarh 492010, India
e-mail: jnaik.cse@nitrr.ac.in

M. Tejaswini
e-mail: mamidipakatejaswini@gmail.com

T. Hari Sumanth
e-mail: harisumanth1637@gmail.com

D. S. Sisodia et al. (eds.), *Machine Intelligence Techniques for Data Analysis and Signal Processing*, Lecture Notes in Electrical Engineering 997,
https://doi.org/10.1007/978-981-99-0085-5_7

beneficial. The objective is to understand how Genetic Algorithms work and to find the best algorithm in terms of power and cost utilization. The power evaluation is done based on the fitness value.

There are many works existing and have been worked upon, but the objective of this paper is to obtain the most optimal model for the virtual machine consolidation and the resources work with them.

The main contribution of our paper is as follows:

(1) Demonstrating all of the supremacies of using this type of prediction models.
(2) Presenting one of the heuristic algorithms of novel type, that considers a virtual machine's memory and CPU when selecting these virtual machines for their migration.
(3) Presenting an algorithm of novel type for the selection of the destination hosts for the selected virtual machines to be migrated.

2 Related Work

Virtual machine consolidation strategies and techniques have been researched upon by many of the researchers in the elapsed years—one of the following is of Beloglazov and Raj Kumar Buyya, where they used the Robust Local Regression and the most common local regression model of algorithms to foretell the physical host's CPU consumption and utilization and then decided to compare it to those of the dynamic brink in order to detect the identification of the overly utilized physical hosts.

Yousefpour puts forward a mathematical predictive model which used the Genetic Algorithm based on heuristic and metaheuristic methodologies, that are energy, power and cost virtual machine consolidation for resolving this problem of reducing and maintaining the power usage and the whole cost by using this type of efficient virtual machine allocation algorithm strategy [5].

Masoumzadeh et al. [6] proposed a model to dwell upon on the concept of virtual machine selection and also validated their unique approach with the simulation platform using the real-world tracing log of cloud resources that is the PlanetLab work log. Portaluri et al. [7] compared a calibrated bundle of virtual machine specific allocators for the respective cloud data centers. Janpan et al. [8] proposed a virtual machine consolidation framework for the cloud stack called the Apache Cloud Stack, which is popularly used for open-source software suite as a cloud platform. Paplinski et al. [9] also contributed to the cloud resource management. Md Anit Khan et al. proposed a better heuristic dynamic virtual machine consolidation algorithm, the so-called RTDVMC that excels in minimizing the energy consumption. Madhusudhan et al. [10] have worked on a rather complex hybrid approach for the consolidation type of resource allocation mostly in the domain of cloud infrastructure. In this paper, we further compare our LrmP_VMC model with this method. Another relatively new method based on the Ant Colony System [11] has been known for its much integrating ability into any type of problem. Additionally, the approaches proposed in

[12–15] have discussed on a type of multi-objective thresholds, Genetic solutions and Deadline-based Elastic approaches in VM consolidation.

3 Design and Implementation

3.1 Overall Design

The whole design and architecture of the project framework consist of three main roles. These include a user, a cloud broker and a cloud provider. The cloud agent is the principal piece of the framework plan, which goes about as a halfway between the clients and suppliers to deal with the utilization of cloud administrations and its conveyance, thinking about the presentation prerequisites. The cloud broker ensures that the cloud suppliers are undetectable to the clients.

3.2 Architecture Diagram

User sends a request in the form of cloudlets to the broker and the broker assigns it to the cloud provider. As we can see from Fig. 1, the user first puts forward the request of his choice, which is carried by the cloud broker to the cloud provider, which allocates all the required cloud resources to the user as per the need.

3.3 Power Model

The power model depicts the whole power utilized or consumed by all the servers in the data centers. The static state describes the state of inactivity or the state where no virtual machine is present or working on the server and the server is running. And the other state, i.e., running state describes the state where there is an interaction between the virtual machines. The power model can be understood easily from Eq. 1.

$$Total\ Power = Power\_idle + Power\_switch \\ + Power\_vm + Power\_placement. \qquad (1)$$

From Eq. 1, we can define the terms as *Power_idle* id the power consumed by the servers when the servers are inactive. *Power_switch* is the power consumed by them when the state changes by switching from on and off simultaneously, and *Power_vm* is the power consumed by the virtual machines running on the particular server. And finally, the *Power_placement* is the utilized power during the placement of the

ARCHITECTURE DIAGRAM

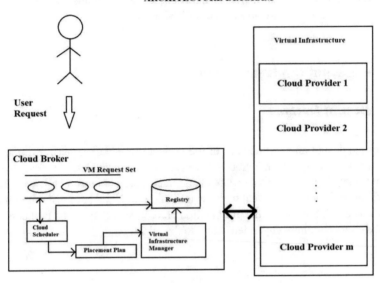

Fig. 1 Architectural diagram of the proposed method

respective virtual machines. The summation of all these powers results in the total power consumed by the machine which contributes to the power model.

3.4 Evaluation and Selection

The Genetic Algorithm constitutes the process of evaluation and selection of the chromosomes as per the requirement. Chromosome is the main component of the proposed Genetic Algorithm. The performance of the algorithm depends mostly on the encoding of this chromosome. In the project, every chromosome constitutes two parts and has its length in the form of $L + 2N + M - 2$.

The starting population is generated by taking the number of servers into account, number of virtual machines and number of virtual machine instance types.

3.4.1 Evaluation Function

The generated chromosome will be evaluated against a power and cost function, and lower boundary value signifies the better performance of the chromosome. We have shown the significance of each term used in the equations below in the following Table 1 for reference.

Table 1 Parameters and their significance

Indices	Notations
v_T	Total power function value
i	Host starting from index, $i = 1$ to M
j	Instance indexing type of virtual machine, $j = 1$ to L
k	Index of type, $k = 1$ to N
Parameters	
X_{ijk}	1, if virtual machine k can be done—instance type j, else 0
Y_j	1, if 1 virtual machine instance type at least runs on server i, else 0
R_j^{CPU}	Request of each virtual machine instance type j by CPU
R_j^{Memory}	Request of each virtual machine instance type j on memory
R_j^{Disk}	Request of each virtual machine instance type j on disk
C_i^{CPU}	Size of each server i (CPU)
C_i^{Memory}	Size of each server i (memory)
C_i^{Disk}	Size of each server i (disk)

$$v_i^{CPU} = \max\left(\frac{\sum_{j=1}^{L} \sum_{k=1}^{N} R_j^{CPU} X_{ijk}}{Y_i * C_i^{CPU}} - 1.0\right), \quad \forall i \int \{1, 2, \ldots, M\}, \quad (2)$$

$$v_i^{Memory} = \max\left(\frac{\sum_{j=1}^{L} \sum_{k=1}^{N} R_j^{Memory} X_{ijk}}{Y_i * C_i^{Memory}} - 1.0\right), \quad \forall i \in \{1, 2, \ldots, M\}, \quad (3)$$

$$v_i^{Disk} = \max\left(\frac{\sum_{j=1}^{L} \sum_{k=1}^{N} R_j^{Disk} X_{ijk}}{Y_i * C_i^{Disk}} - 1.0\right), \quad \forall i \in \{1, 2, \ldots, M\}, \quad (4)$$

$$v_T = \frac{\sum_{i=1}^{M} \left(v_i^{CPU} + v_i^{Memory} + v_i^{Disk}\right)}{M}, \quad \forall i \in \{1, 2, \ldots, M\}. \quad (5)$$

3.4.2 Selection Strategy

In this model that is put forward, we utilized the Genetic Algorithm system that gives the likelihood of picking every chromosome based on their fitness esteem. The stages involved in allocation are described below:

Stage 1: Virtual machine chooses the destined instance type.

Stage 2: Virtual machine chooses the destined physical machine [pm].

In every stage of the process, the fitness value of every single one or being in the population considered is computed, assessed and all the multiple individuals are again selected from each of the current generation and then are modified and redesigned to form a new set of population.

3.4.3 Crossover Operator

One of the prominent operators used in the Genetic Algorithms is the crossover operator. The process of crossover is very vital in the generation of new chromosomes where the combination of two or more parent chromosomes takes place with the small hope that this will create a brand new and a very efficient set of chromosomes. The crossover process transpires after the selection of the pairs of these parent chromosomes is completed and which in result helps in the information exchange between the two parents selected to create the destined offspring or children. The algorithm for the crossover operator is briefly presented below in Algorithm 1.

Algorithm 1: Crossover Operator
Input: $CHb = a1\ ba2 \ldots b$ -- $CHc = a1\ ca2\ c\ldots aqc \rightarrow$ q genes-parent
Output: $CHd = a1\ da2\ d\ldots aq\ d$ -- $CHe = a1\ ea2\ e\ldots aqe \rightarrow$ offspring
1. *For Each*
a. Randomly Integer Encoding $\rightarrow$ size 'r'
b. R1 = Intersect (CHb [1…r], CHc [r+1…Segment Size]);
c. R2 = Intersect (CHc [1…r], CHb [r+1…Segment Size]);
d. Find r1 in CHb [1 … r]
e. Find r2 in CHc [1 … r]
f. CHd [1…Segment Size] = concatenate (CHb [1…r], CHc [r+1…Segment Size]);
g. CHe [1…Segment Size] = concatenate (CHc [1…r], CHb [r+1…Segment Size]);
2. *End for Each*
3. *Return CHe*;

While the crossover process is going on, the parent chromosomes are chosen in pairs and their respective genes have to be exchanged in a certain selected order so as to obtain the destined offspring. This resulting offspring becomes the next-generation parent chromosomes for the further resulting process. This offspring generation is shown in Fig. 2.

3.4.4 Mutation Operator

The process of mutation operator is done to cease the process of getting the solution at a very early stage so that it gets a new solution or an answer. Conflicting to the crossover operator as shown in Fig. 2, the gene value of the chromosome generated is changed in this process of mutation as shown in Fig. 2. The blue ones in Fig. 2 represent the new characteristic or the gene produced in the resulting offspring. The

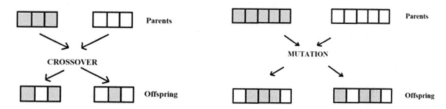

Fig. 2 Crossover operator and mutation operator

mutation operator is explained in the form of algorithm below in Algorithm 2. This mutation operator has the objective of bringing a whole newly generated chromosome into the pre-existing population with a set of new qualities and attributes along with the gene values into the present population.

Algorithm 2: Mutation Operator
Input: CH=$a1a2...aq$
Output: $CH'=a1'\ a2'\ ...aq' \to$ mutated
1. *For Each*
a. $CH' = $ CH;
b. Integer encoding$\to$ segment size called r1, r2;
c. Exchange at indexes r1 and r2
2. *End for Each*
3. *Return CH'*;

3.5　The Overall Proposed Algorithm

The proposed algorithm is the Genetic Algorithm—Regression model, a form of Roulette Wheel Selection Strategy Mechanism. The whole algorithm is stated below in Algorithm 3. The input and output as per the proposed algorithm are also presented in Algorithm 3.

Algorithm 3: Genetic Algorithm for *LrmP_VMC*
Input: VMList, PMList, InstanceTypeList
Output: Allocation of VMs
1. Initially Pop Size, CP;
2. For 2000 iterations ($t>2000$);
nc Pop size * CP; nm Pop size * MP;
3. population InitializePopulation(Pop Size);
4. EvaluatePopulation(population);

(continued)

(continued)

5. While terminate () do;
t = *t*+1;
FOR *i*=1…*nc*/2;
Parents selection(population,2);
Offsprings Crossover(parents);
EvaluatePopulation(offsprings);
Population.add(offsprings);
End FOR;
FOR *i*=1…nm
Parents selection(population,1);
Off springs mutation(parent);
EvaluatePopulation(offsprings);
Population.add(offsprings);
End FOR;
6. End While;
7. Population.sort();
8. Allocationpopulation.get(first);
9. Return allocation;

The diagram Fig. 3 represents the control structure diagram for the proposed Genetic Method for our LrmP_VMC technique.

4 Simulation and Experimental Results

This part deals with analysis of results. First, the metrics to be considered for evaluation of the modules are discussed and then analyzed the performance of the system based on the values obtained.

4.1 Experimental Setup

Simulation Platform used: CloudSim is a type of framework basically written in Java Programming language. This simulation platform was developed by Raj Kumar Buyya and began in 2009 in the laboratory of the Grid Computing and Distributed Systems (GRIDS) University of Melbourne. The characteristics of the VM, host and cloudlet used in the simulation are detailed below (Table 2).

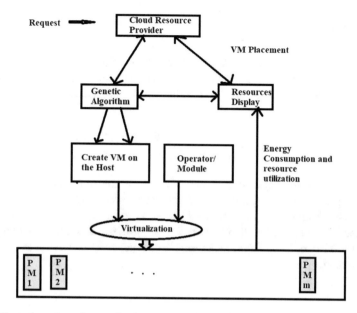

Fig. 3 Control structure diagram for the genetic process after the population generation

Table 2 Specifications of all those involved

Type	Specifications	
Virtual machine (each host has a VM)	Rating of MIPS	250
	Disk image size	10,000 MB
	Random Access Memory [RAM]	512 MB
	Band width	1000
	Pes—number of requirement	1
	Hypervisor	Xen
Host (every data center has one host)	Random Access Memory [RAM]	200,000
	Storage	1,000,000
	Bandwidth	100,000
	Rating of the corresponding processing entity [MIPS]	1000
Cloudlets	Length (in the form of instructions)	40,000
	File size (Input)	300
	File size (Output)	300

4.2 Evaluation Metrics

The fitness function calculates the power utilization of the placement or allocation strategy depending on the data that is obtained from the chromosome. The resulting graph is plotted as the obtained fitness value versus the generations involved in it, where the fitness value is obtained as

$$\text{Fitness Value} = \frac{\text{Total fitness of the population}}{\text{Number of generations (populations)}}. \tag{6}$$

Total fitness of the population = sum of fitness values of all individual chromosomes in a generation, where individual chromosome fitness value is calculated based on formulas: From Eq. 5, we calculate the total power function from the previously calculated individualized, CPU from Eq. 2, disk from Eq. 3 and the whole memory powers from Eq. 4. Their significant terms have been described in Table 1 for clear understanding.

4.3 Performance Evaluation

Firstly, with the help of Integer Encoding, we generate a set of string values as a form of dataset for the proposed methodology. Then, we calculate the fitness values according to the number of cloudlets and VMs assigned and compare by increasing either the generations or the virtual machines. By calculating the fitness value using Eq. 6, we can compare the generations with the calculated fitness value for clear understanding.

4.4 Comparing the Genetic Algorithm with Other Methods

The basic method used to compare different ML models is the MSE value. MSE refers to the root mean squared error, which is very helpful in finding the best model or technique among different existing algorithms. Apart from the total power function as in Eq. 5, we can compare the results with these MSE values.

Mathematically,

$$\text{MSE} = \sum_{i=1}^{N} (t_i - u_i)^2. \tag{7}$$

Here, N is total number of operations, t_i refers to the actual observation from the executed experiment and the term $u_i$refers to the predicted values. After calculating

Fig. 4 Graph showing MSE
versus generations for
LrmP_VMC method

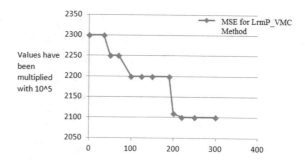

the MSE for a set of values using Eq. 7, we get the following graph as in Fig. 4. The MSE values in the graph are in the form of MSE value $*10^{-5}$.

Figure 4 depicts the graph plotted with the values of MSE with the generations involved as shown in Table 3. The procured MSE values are calculated with the help of Eq. 7.

We have plotted a graph using the values in Table 4. There is an unnatural increase in between the increment of the virtual machines. But, in the case of linear regression, it is evident that the technique performs well when the numbers of virtual machines participating are more in number.

The Random Forest method also showed a great accuracy, but it failed to achieve the least error compared to our LrmP_VMC. As from the graphical representation in Fig. 5, we can clearly predict that the model LrmP_VMC is better than the Random Forest method as when the number of virtual machines working is less, it has shown a high value of MSE than the LrmP_VMC, and with the increase in number of VMS, the MSE decreased, but still there are some fluctuations in between the process.

Table 3 Generations versus
MSE value

Generations	MSE ($*10^{-5}$)
0	2300
35	2300
50	2250
70	2250
100	2200
125	2200
150	2200
190	2200
200	2110
220	2100
250	2100
300	2100

Table 4 Number of VMs versus MSE for LrmP_VMC and Random Forest

No. of VMs	MSE for LR genetic approach (*LrmP_VMC*)	MSE for random forest approach [10]
800	0.01	0.025
700	0.01	0.0201
600	0.01	0.02
500	0.01	0.0201
400	0.01	0.02
300	0.01	0.02001
200	0.02	0.03
100	0.025	0.049
75	0.03	0.053
50	0.04	0.053

Fig. 5 Graph showing cross-validation for the VM's versus MSEs of the two methods (LrmP_VMC–Random Forest)

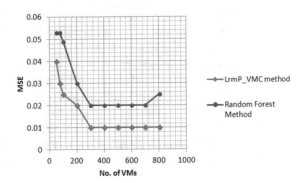

5   Conclusion and Future Scope

This part discusses the results of the project, the limitations and the future work to be carried out. In this particular area of cloud programming and computing, the power utilization and management are researched and designed mostly with the goal of reducing the total power usage done in the respective data centers. To obtain this reduction in the power usage and management, researchers have proposed many unique solutions and approaches related in terms of software as well as hardware such as the DVFS method. In this project report, we particularly focused on software-related optimization techniques using the proposed Genetic Algorithm that is based on the particular better virtual machine's allocation techniques for virtual machine consolidation. Future work is to carry out binding of data centers and VM using Genetic Algorithm for the given data centers, cloudlet and VM. Binding of VM and cloudlets and VM and data centers will provide much efficient results in terms of power consumption when compared to only binding of VM and cloudlets, as data centers also play a major role in the total power consumption; moreover, different selection strategies other than Roulette Wheel Selection Strategy can also be implemented to checkout which would produce good results in terms of power consumption.

References

1. Rashid A, Chaturvedi A (2019) Virtualization and its role in cloud computing environment. Int J Comput Sci Eng
2. Clark C, Fraser K, Hand S, Hansen JG, Jul E, Limpach C, Pratt I, Warfield A (2005) Live migration of virtual machines. In: Proceedings of the 2nd conference on symposium on networked systems design & implementation, vol 2. USENIX Association, pp 273–286
3. Moghaddam SM, Sullivan SF (2019) Machine learning prediction models for energy efficient VM consolidation within Cloud data centers. University of Aukland, Australia
4. Ranabahu AH, Patel P, Sheh A (2009) Service level agreement in cloud computing. Wright State University
5. Korf RE (2016) An improved algorithm for optimal bin packing. Computer Science Department University of California, Los Angeles
6. Yousefipour A, Rahmani AM, Jahanshahi M (2018) Energy and cost-aware virtual machine consolidation in cloud computing 48(10):1758–1774
7. Abdelsamea A, El-Moursy AA, Hemayed EE, Eldeeb H (2017) VM consolidation enhancement using hybrid regression Algo. Egyptian Inf J 18(3):161–170
8. Masoumzadeh SS, Hlavacs H (2013) Integrating VM selection criteria in distributed dynamic VM consolidation using fuzzy Q learning. In: 2013 9th international conference on network and service management (CNSM). IEEE, pp 332–338
9. Portaluri G, Adami D, Gabbrielli A, Giordano S, Pagano M (2017) Power consumption-aware virtual machine placement in cloud data center. IEEE Trans Green Commun Netw 1(4):541–550
10. Madhusudhan HS, Satish Kumar T, Mustapha S, Gupta P (2021) Hybrid approach for resource allocation in cloud infrastructure using random forest and genetic algorithm. Sci Programming 2021(4):1–10. https://doi.org/10.1155/2021/4924708
11. Khan MA, Paplinski AP, Khan AM, Murshed M, Buyya R (2018) Exploiting user provided information in dynamic consolidation of virtual machines to minimize energy consumption of cloud data centers. In: 2018 3rd international conference on fog and mobile edge computing (FMEC)
12. Jairam Naik K (2021) A deadline based elastic approach for balanced task scheduling in computing cloud environment. Int J Cloud Comput (IJCC) 10(5/6):579–602
13. Jairam Naik K, Chandra S, Agarwal P (2021) Dynamic workflow scheduling in the cloud using a neural network-based multi-objective evolutionary algorithm. Int J Commun Netw Distributed Syst 27(4):424–451
14. Naik KJ (2021) Optimized genetic algorithm for efficient allocation of virtualized resource in cloud (OGA_EAVRC). In: International conference on artificial intelligence and smart systems (ICAIS 2021), March-2021, JCT College of Engineering and Technology, Coimbatore (TN), India
15. Jairam Naik K (2021) A cloud-fog computing system for classification and scheduling the information-centric IoT applications. Int J Commun Netw Distributed Syst 27(4):388–423

Detection of Fraudulent Credit Card Transactions Using Advanced LightGBM Approach

Kotireddy Yazna Sai, Repalle Venkata Bhavana, and N. Sudha

1 Introduction

One of the most often utilized financial instruments is the credit card, which is intended to be used to make purchases when funds are unavailable. When one individual uses another person's credit card for personal reasons without the owner's knowledge, this is referred to as credit card fraud. Credit cards are the most beneficial since they offer several rewards in the form of points when used for a variety of purchases. Typically, big hotels and car rental companies need the buyer to have a credit card. By the end of 2005, massive sums of sales were made, amounting to nearly $190.6 billion, simply by circulating approximately 56.4 million credit cards in Canada [1]. Fraud is defined as obtaining money, services, or goods through illegal or unethical ways. Banking fraud is defined as "the unlawful use of a person's sensitive information to make purchases or withdraw money from the user's account".

With the rise of e-commerce and the world's shift toward digitization and cashless transactions, the use of credit cards has expanded fast, as has the number of frauds associated with them. Fraud detection is seen as a major difficulty for machine learning, owing to the fact that only a few of all transactions in credit cards are found to be fraudulent. A method for identifying fraud that makes use of a LightGBM and a Bayesian-based Hyperparameter Optimization technique is introduced in this paper. We compare the performance of this approach with other classical machine learning techniques using two different datasets.

K. Yazna Sai · R. Venkata Bhavana (✉) · N. Sudha
SASTRA Deemed University, Thanjavur, India
e-mail: repallebhavana123@gmail.com

K. Yazna Sai
e-mail: yaznasai.16@gmail.com

N. Sudha
e-mail: sudha@cse.sastra.edu

© The Author(s), under exclusive license to Springer Nature Singapore Pte Ltd. 2023   93
D. S. Sisodia et al. (eds.), *Machine Intelligence Techniques for Data Analysis and Signal Processing*, Lecture Notes in Electrical Engineering 997,
https://doi.org/10.1007/978-981-99-0085-5_8

2 Related Works

A considerable amount of work has been done on the detection of fraudulent credit card transactions. Rai et al. [2] suggested a system for identifying fraud in credit card data that uses an unsupervised learning technique based on Neural Networks (NNs). It outperformed previous techniques such as Local Outlier Factor, Isolation Forest, and K-means clustering in terms of accuracy. The accuracy values for Naive Bayes, k-nearest neighbors, and logistic regression classifiers in [3] are 97.92%, 97.69%, and 54.86%, respectively. The results show that K-nearest neighbors (KNN) surpassed Naïve Bayes and logistic regression methods.

In [4], Bayesian Belief Networks obtained the highest accuracy (90.3%), whereas Decision Trees obtained an accuracy of 73.6% and Neural Networks accuracy of 80%. To tackle class imbalance in datasets without utilizing resampling methods, Priscilla et al. [5] presented an Optimized XGBoost methodology. The Randomized Search CV hyperparameter optimization approach is used to determine the best XGBoost settings. This method had a high level of accuracy in detecting fraudulent transactions. Fiore et al. [6] suggested a strategy for creating synthetic instances based on generative adversarial networks to improve the detection of fraudulent credit card transactions by addressing imbalanced dataset difficulties.

Ge et al. [7] proposed a detection algorithm using LightGBM. This method performed better than other classical algorithms like Random Forest (RF), XGBoost, and Support Vector Machine (SVM). This also demonstrates the significance of feature engineering, which is useful for performance tuning and feature selection. Many algorithms were implemented in [8] to classify transactions as fraudulent or real, including Random Forest, Naive Bayes, Logistic Regression, and Multilayer Perceptron. Since the dataset was severely unbalanced, the SMOTE method was applied for oversampling.

3 Proposed Approach

The proposed approach consists of four major stages: Data Preprocessing, Feature Selection, Performing Advanced LightGBM, and Evaluation Metrics, and these are detailed in the following subsections. Figure 1 presents the entire framework of the proposed approach.

3.1 Data Preprocessing

Two independent real-world datasets are used to assess the applicability of the proposed approach. The dataset-1 [9] comprises credit card transactions made during two days in September 2013 by European cardholders. This dataset contains

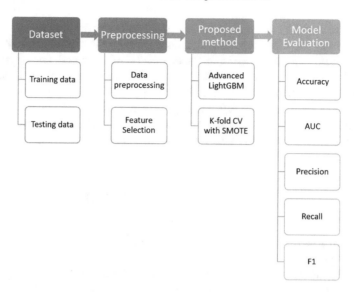

Fig. 1 Flow diagram for the proposed approach

284,807 transactions, 492 of which are fraudulent. Positive transactions account for 0.172 percent of all transactions in the dataset, which is unbalanced. It has numerical input variables that are the result of a PCA transformation. The features are V1 through V28, Time, and Amount. Except for 'Time' and 'Amount', all of the features had been transformed using PCA. The 'Time' feature keeps track of seconds elapsed between each transaction and the first transaction in the dataset. The feature 'Amount' represents the transaction amount, which may be used for example-dependent cost-sensitive learning.

The dataset-2 [10] contains the features amount, hour1, state1, zip1, field1, domain1, field2, hour2, flag1, total, field3, field4, field5, indicator1, indicator2, flag2, flag3, flag4, flag5, Class. It contains 2094 fraudulent transactions and 92,588 non-fraudulent transactions. For both datasets, the target variable is 'Class', which has a value of 1 if there is fraud and 0 otherwise.

An imbalanced dataset tends to give poor accuracy results. As the two datasets are imbalanced, a cross-validation method is used to get accurate scores. While evaluating the proposed method, the Synthetic Minority Oversampling *Technique* (SMOTE) along with cross-validation is performed to avoid overlearning/overfitting.

3.2 Feature Selection

For detecting credit card risk in huge and high-dimensional data, feature selection is critical for improving classification performance and fraud detection. Information gain (IG) is a popular feature selection approach that works well with large datasets. IG works by examining the correlation between the card transactions and then weighting the most important data depending on the classification of fraudulent credit card transactions and non-fraudulent credit card transactions (legitimate). In the proposed methodology, IG is used as a feature selection method due to its computational efficiency and enhanced accuracy.

To define IG, we must first define entropy, a metric often employed in information theory. Entropy essentially informs us how impure a set of data is; now, we can assess the efficacy of an attribute in categorizing the training set. Entropy can be calculated using the following formula:

$$Entropy = -\sum_{i=1}^{c} P(x_i) \log_b P(x_i),$$

where c is the number of classes, $P(x_i)$ is the probability of value occurring in data, and b is the number of bits (here, $b = 2$).

The information gain metric is just the predicted reduction in entropy generated by splitting the dataset based on this property. The information gain $Gain(S, F)$ of a feature F relative to a dataset collection S is defined as:

$$Gain(S, F) = Entropy(S) - \sum_{v \in Values(F)} \frac{|S_v|}{|S|} Entropy(S_v),$$

where $Values(F)$ is all possible values of features F, and S_v is subset of S, for which feature F has value v.

IG has good performance; it tends to discover the most predictive features. In this proposed approach, IG is used to improve the detection of fraudulent credit card transactions. In terms of classification performance on a bigger dataset, the experimental results in this paper suggest that the proposed algorithm beats the classical algorithms.

3.3 Performing Advanced LightGBM

Boosting machine learning algorithms like LightGBM are widely employed because they outperform simple techniques in terms of accuracy. The performance of these algorithms is determined by hyperparameters. The algorithm of LightGBM is shown in Fig. 2 [11]. The parameters of LightGBM are grouped under these categories, namely: parameters for better accuracy, parameters to combat overfitting, parameters which affect the learning and the structure of the decision trees, and parameters which affect training speed. A well-chosen set of parameters among all the parameters of LightGBM can aid in achieving improved accuracy. Tuning hyperparameters by hand is time-consuming and computationally expensive. As a result, automation of hyperparameter tuning is critical. Hyperparameters are usually optimized using

Random Search, GridSearchCV, and Bayesian optimization. When compared to other methods, Bayesian optimization produces better and faster outcomes.

In this paper, Bayesian optimization is used to get the best values for hyperparameters of LightGBM. Python in-built package bayes_opt is adopted to perform Bayesian-based Hyperparameter Optimization. Hyperparameters like max_depth, num_leaves are chosen to control the tree structure and overfitting problem. learning_rate, which controls the learning speed, is used for better accuracy, and max_bin is used to increase the speed of the model. These parameters are tuned within their ideal ranges. When deciding which hyperparameters set to examine next, Bayesian optimization considers previous evaluations in a step-by-step manner. It focuses on those parts of the parameter space that it predicts will give the most

Input: Training data D = $\{(\chi1, y1), (\chi2, y2), ..., (\chi N, yN)\}$, χi χ, $\chi \subseteq$ R, $yi\_\{-1,+1\}$; loss function: $L(y, \_(\chi))$; iterations: M; sampling ratio of large gradient data: a; sampling ratio of small gradient data: b;

1. Merge mutually exclusive features(i.e. features never take nonzero values simultaneously) of χi, $i = \{1, ...,N\}$ by exclusive feature bundling(EFB) method;
2. Initialize $\_0(\chi) = arg\ minc\_Ni\ L(yi, c)$;
3. **for** $m = 1$ toMdo
4. Compute absolute values of gradients:

$$ri = \left| \frac{\partial L(y_i, \theta(x_i))}{\partial \theta(x_i)} \right|_{\partial \theta(x)=\theta_{m-1(x)}}, i = \{1, \ldots, N\}$$

5. Resampled dataset by Gradient-based One-Side Sampling (GOSS) method:
   topN = a × len(D); randN = b × len(D);
   sorted = GetSortedIndices(abs(r));
   A = sorted [1 : topN]; B = RandomPick(sorted[topN : len(D)] , randN);
   D_ = A + B;
6. Compute information gains:

$$Vj(d) = \frac{1}{n} \left(\frac{\left(\Sigma_{x_i \in A_l} r_i + \frac{1-a}{b} \Sigma_{x_i \in B_l} r_i \right)^2}{N_l^j(d)} + \frac{\left(\Sigma_{x_i \in A_r} r_i + \frac{1-a}{b} \Sigma_{x_i \in B_r} r_i \right)^2}{N_r^j(d)} \right)^{nt}$$

7. Get a new decision tree $\theta_m(X)'$ on set D'
8. Update $\theta_m(X) = \theta_{m-1}(X) + \theta_m(X)'$
9. **end for**
10. **return** $\theta_M'(X) = \theta_M(X)$

Fig. 2 Algorithm of LightGBM

promising validation scores by making intelligent parameter combinations. The final values are assigned to the hyperparameters of LightGBM during evaluation.

3.4 Evaluation Metrics

To measure the performance of the suggested approach for detecting fraudulent credit card transactions, a cross-validation test is used. The proposed strategy for identifying fraudulent credit card transactions is thoroughly evaluated in this paper utilizing the k-fold cross-validation (CV) method. K-fold cross-validation is a statistical analysis technique that is commonly used in academic research to evaluate the efficacy of machine learning techniques [12]. When choosing the value of k in K-fold CV, the size of the dataset is considered. As large datasets are used for experimenting, the value of k should be relatively smaller to get good results. During the experiment, k = 5 gave better results than other values of k. So, a fivefold CV test is implemented in this paper to evaluate the performance of the proposed technique. There is a class imbalance in the two datasets examined: there are more legal transactions than fraudulent ones. CV is utilized for training and testing the model in each subset of the two different datasets. Mean of the highlighted measures is computed throughout the whole dataset [13].

Because the fraud-to-non-fraud transaction ratios for dataset-1 and dataset-2 are roughly 1:570 and 1:50, respectively, the SMOTE technique is employed to evaluate the metrics while executing the fivefold CV. SMOTE is one of the oversampling techniques; it generates the synthetic samples for minority class. Several metrics, including Accuracy, Precision, Recall, AUC, and F1-score, are used to evaluate the suggested approaches' performance.

4 Experimental Results

The proposed approach was implemented in Python, evaluated with two datasets and with other machine learning approaches. Jupyter Notebook was used to perform this experiment. Bayesian-based Hyperparameter Optimization was conducted for training the model with the best hyperparameters. For evaluating the metrics of Advanced LightGBM, the SMOTE technique along with a fivefold CV was used for two real-world datasets.

The results of the proposed method for two datasets are listed in Tables 1 and 2. Advanced LightGBM achieved an average accuracy of 99.92% for dataset-1 and 97.59% for dataset-2. This approach is able to outperform existing methods like KNN, Logistic Regression (LR), SVM RBF, SVM Linear, Gaussian Naive Bayes, Decision Tree, and Random Forest (RF) for both datasets.

The ROC-AUC curve is an important metric to compare the efficiency of different approaches. Figures 3 and 5 show the ROC-AUC curves of the proposed method for

Table 1 Metric values of advanced LightGBM based on fivefold CV with SMOTE for dataset-1

Fold no.	Accuracy	AUC	Precision	Recall	F1-score
1	0.99918	0.87642	0.76315	0.75324	0.75816
2	0.99920	0.89588	0.75308	0.79220	0.77215
3	0.99934	0.92243	0.80681	0.84523	0.82558
4	0.99936	0.87794	0.87323	0.75609	0.81045
5	0.99942	0.93225	0.8	0.86486	0.83116
Avg	0.99930	0.90098	0.79926	0.80233	0.79950

Table 2 Metric values of advanced LightGBM based on fivefold CV with SMOTE for dataset-2

Fold no.	Accuracy	AUC	Precision	Recall	F1-score
1	0.97709	0.69411	0.42307	0.39933	0.41086
2	0.97683	0.70432	0.46308	0.41945	0.44019
3	0.97636	0.69324	0.46020	0.39701	0.42628
4	0.97491	0.70503	0.43292	0.42261	0.42771
5	0.97471	0.70402	0.48297	0.41935	0.44892
Avg	0.97598	0.70014	0.45245	0.41155	0.43079

all the five folds in fivefold CV technique for dataset-1 and dataset-2, respectively. The Precision–Recall curves for dataset-1 and dataset-2 are represented in Figs. 4 and 6, respectively.

From Table 3, for dataset-1, Advanced LightGBM has the highest average accuracy, while Decision Tree has the lowest average accuracy of 88.94%. From Table 4, for dataset-2, the Gaussian Naive Bayes approach obtained the least average accuracy of 31.43%. Figures 7 and 8 exhibit comparison of ROC-AUC curves

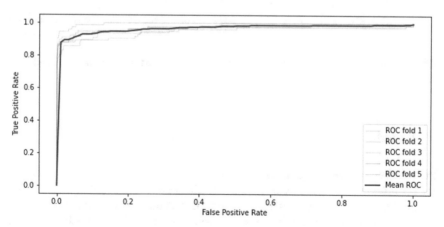

Fig. 3 ROC-AUC curves of advanced LightGBM based on fivefold CV with SMOTE for dataset-1

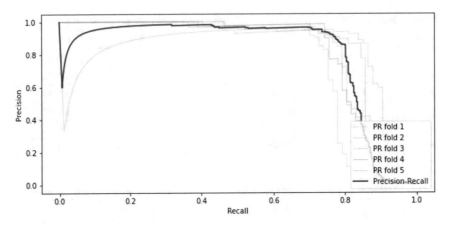

Fig. 4 Precision–Recall curves of advanced LightGBM based on fivefold CV with SMOTE for dataset-1

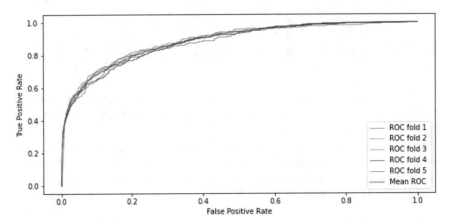

Fig. 5 ROC-AUC curves of advanced LightGBM based on fivefold CV with SMOTE for dataset-2

for Advanced LightGBM and other classical ML algorithms using dataset-1 and dataset-2, respectively.

5 Conclusion

Increased credit card usage needs fraud detection system. More efficient methods for identifying fraudulent credit card transactions are necessary due to the rising complexity of detecting fraudulent credit card transactions, as well as the significant and recurrent financial losses that financial institutions incur. An approach, Advanced

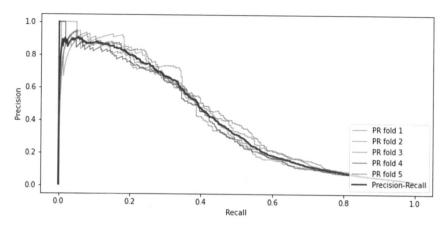

Fig. 6 Precision–Recall curves of advanced LightGBM based on fivefold CV with SMOTE for dataset-2

Table 3 Metric values of Advanced LightGBM and other ML techniques for dataset-1

Approach	Accuracy	AUC	Precision	Recall	F1-score
Logistic regression	0.94666	0.94768	0.98130	0.91081	0.94434
SVM RBF	0.94768	0.94714	0.98064	0.91131	0.94442
SVM linear	0.94333	0.94247	0.96629	0.91678	0.94054
KNN	0.93024	0.92923	0.95642	0.89903	0.92662
Decision trees	0.88947	0.88953	0.88974	0.88802	0.88765
Random forest	0.94187	0.94103	0.97756	0.90226	0.93789
Gaussian Naïve Bayes	0.92444	0.92277	0.98980	0.85395	0.91632
Proposed approach	0.99930	0.90098	0.79926	0.80233	0.79950

Table 4 Metric values of Advanced LightGBM and other ML techniques for dataset-2

Approach	Accuracy	AUC	Precision	Recall	F1-score
Logistic regression	0.55705	0.48563	0.85910	0.44292	0.58426
SVM RBF	0.45690	0.54539	0.85612	0.40424	0.54516
SVM linear	0.55971	0.43701	0.87040	0.36497	0.51348
KNN	0.58342	0.56031	0.84778	0.59638	0.69981
Decision trees	0.69116	0.73462	0.93921	0.66512	0.77827
Random forest	0.84198	0.77363	0.92081	0.88276	0.90099
Gaussian Naïve Bayes	0.31436	0.54750	0.90689	0.17821	0.29776
Proposed approach	0.97598	0.70014	0.45245	0.41155	0.43079

Fig. 7 ROC-AUC curves of advanced LightGBM and other ML techniques for dataset-1

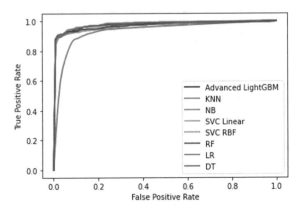

Fig. 8 ROC-AUC curves of advanced LightGBM and other ML techniques for dataset-2

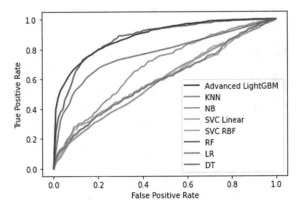

Light Gradient Boosting Machine, is proposed that uses Bayesian-based Hyperparameter Optimization to detect fraudulent credit card transactions. A lot of trials are done using two datasets. Using Random Forest, Logistic Regression, SVM RBF, SVM Linear, KNN, Gaussian Naive Bayes, and Decision Tree, the performance of the suggested strategy is compared to the existing research findings.

The results of the experiments reveal that the proposed method outperforms all other machine learning techniques. The findings further emphasize the relevance and value of using an effective parameter optimization strategy in improving detection performance.

References

1. Tiwari P, Mehta S, Sakhuja N, Kumar J, Singh AK (2021) Credit card fraud detection using machine learning: a study. arXiv preprint arXiv:2108.10005
2. Rai AK, Dwivedi RK (2020) Fraud detection in credit card data using unsupervised machine learning based scheme. In: 2020 international conference on electronics and sustainable communication systems (ICESC), IEEE
3. Awoyemi JO, Adetunmbi AO, Oluwadare SA (2017) Credit card fraud detection using machine learning techniques: a comparative analysis. In: 2017 international conference on computing networking and informatics (ICCNI). IEEE, pp 1–9
4. Kirkos E, Spathis C, Manolopoulos Y (2007) Data mining techniques for the detection of fraudulent financial statements. Expert Syst Appl 32(4):995–1003
5. Priscilla CV, Prabha DP (2020) Influence of optimizing XGBoost to handle class imbalance in credit card fraud detection. In: 2020 third international conference on smart systems and inventive technology (ICSSIT). IEEE, pp 1309–1315
6. Fiore U, De Santis A, Perla F, Zanetti P, Palmieri F (2019) Using generative adversarial networks for improving classification effectiveness in credit card fraud detection. Inf Sci 479:448–455
7. Ge D, Gu J, Chang S, Cai J (2020) Credit card fraud detection using lightgbm model. In: 2020 international conference on e-commerce and internet technology (ECIT). IEEE, pp 232–236
8. Varmedja D, Karanovic M, Sladojevic S, Arsenovic M, Anderla A (2019) Credit card fraud detection-machine learning methods. In: 2019 18th international symposium INFOTEH-JAHORINA (INFOTEH). IEEE, pp 1–5
9. Credit Card Fraud Dataset. Available: https://www.kaggle.com/mlg-ulb/creditcardfraud/data. Last accessed 4 Sept 2019
10. UCSD: University of California, San Diego Data Mining Contest 2009. Available: https://www.cs.purdue.edu/commugrate/data/credit_card/. Last accessed 14 Jan 2019
11. Mohammed MA, Kadhem SM, Maisa'a AA (2021) Insider attacker detection using light gradient boosting machine. Tech-Knowledge 1(1):67–76
12. Kumar RD, Searleman AC, Swamidass SJ, Griffith OL, Bose R (2015) Statistically identifying tumor suppressors and oncogenes from pan-cancer genome-sequencing data. Bioinformatics 31(22):3561–3568
13. Russell S, Norvig P (2002) Artificial intelligence: a modern approach

Approach of Different Classification Algorithms to Compare in N-gram Feature Between Bangla Good and Bad Text Discourses

Abu Kowshir Bitto⊙, Md. Hasan Imam Bijoy⊙, Saima Khan⊙, Imran Mahmud⊙ and Khalid Been Badruzzaman Biplob⊙

1 Introduction

We live in a digital age when everything can be accomplished with the help of smart technology. The amount of time we spend interacting with computers is steadily increasing. Every country has its own native language. We are all contents to converse in our own language. We all speak our own languages, which we use to communicate, comprehend one another, and connect with other people and nations. Even though the human language is a form of communication, humans find it challenging to communicate with machines. NLP is a method of allowing humans and computers to communicate, understand, and operate machines as if they were speaking the same language. Our native language is Bangla, and we are from Bangladesh. To communicate inside our own tongue with a machine or computer, we must first create a link that helps to clarify our language. Bengalis, on the other hand, have just recently begun to utilize computers to teach our language. NLP for computer recognition has gotten a lot of press, but it has not

A. K. Bitto · S. Khan · I. Mahmud · K. B. B. Biplob
Department of Software Engineering, Daffodil International University,
Dhaka 1207, Bangladesh
e-mail: abu.kowshir777@gmail.com

S. Khan
e-mail: saima35-2392@diu.edu.bd

I. Mahmud
e-mail: imranmahmud@daffodilvarsity.edu.bd

K. B. B. Biplob
e-mail: khalid@daffodilvarsity.edu.bd

M. H. I. Bijoy (✉)
Department of Computer Science and Engineering, Daffodil International University,
Dhaka 1207, Bangladesh
e-mail: hasan15-11743@diu.edu.bd

D. S. Sisodia et al. (eds.), *Machine Intelligence Techniques for Data Analysis and Signal Processing*, Lecture Notes in Electrical Engineering 997,
https://doi.org/10.1007/978-981-99-0085-5_9

gotten much attention. However, a growing number of scholars are concentrating on natural language processing in Bengali. Data preparation is the most challenging part of any machine learning algorithm, and presenting the Bengali speech to the computer is also too difficult due to the different challenges we must solve. Exploring Bengali data takes longer, and forecasting the outcomes of any study is challenging. After preprocessing data, tokenize, proceed, and use several machine learning techniques. Facebook, YouTube, and Twitter are examples of social media platforms that have enhanced the twenty-first century. It has sped up modern communication with a limited range of services. In 2018, 4.021 billion individuals accessed the internet, while 3.196 billion used social media, according to estimates. Each year, the number of individuals using the internet and Facebook increases by 7% and 13%, respectively. More than 200 million people speak Bangla as their primary language in Bangladesh and several Indian regions. Large volumes of digital data, such as review comments and Facebook status updates, have lately emerged in Bangla, allowing us to measure sentiment in the language. Everyone's daily routine now includes some form of social media. It provides for rapid communication, easy sharing, and the exchange of ideas and perspectives from all over the world. At the same time, this exhibition of liberty has resulted in a rise in online hate speech and harsh language. In comparison to conventional media, the language employed in social media is substantially different. It has a wide range of linguistic features. As a result, automatically identifying hate speech is a challenging undertaking. Bengali is used by a vast number of people all over the world to interact with one another. This study proposes a novel classification technique in N-gram feature to compare and contrast Bangla's good and bad texts. We are aiming to create a machine learning model that can recognize Bengali text in this environment. Bengalis utilize a variety of phrases to communicate, educate, express emotions, and accomplish other goals. These terms are used to categorize a person. Text categorization entails specific topic content categorization and textual type-based classification, as demonstrated by the order concept. Our main objective is to find the sentence's variation quickly.

In this approach, the remainder of the paper is in order. Section 2 of the paper is a review of the literature. The approach for recognizing Bengali bad and good text discourses is discussed in Sect. 3. The experimental results are demonstrated and discussed in Sect. 4. Section 5 of the document, certainly, brings the whole thing to a conclusion.

2   Literature Review

Many papers, articles, and research projects focus on text categorization, text recognition, and categories, while some focus on particular points. Here are some of the work reviews that have been provided below to make a connection to our work.

Alam et al. [1] presented a methodology for analyzing sentiments in Bangla comments texts. They applied Convolutional Neural Network (CNN) architecture to build a model that can distinguish between positive and negative sentiments, with a 99.87% accuracy. Taher et al. [2] address the sentiment analysis problem largely utilizing SVM for classification and the N-gram technique for vectorization in N-gram Based Sentiment Mining for Bangla Text Using Support Vector Machine. They manually gathered data and discovered that using the N-gram vectorization technique yields the best accuracy (91.581% for linear SVM and 89.271% for nonlinear SVM, respectively). The authors of [3] used support vector machine to do sentiment analysis on Bengali data regarding the Bangladesh cricket squad. Hassan et al. [4] used Deep Recurrent Model and Long Short-Term Memory to study not just standardized Bangla, but also Banglish (Bangla words combined with English words) and Romanized Bangla. However, the accuracy of their suggested system is unsatisfactory (78%). To detect hostile Bengali language, Ishmam et al. [5] compare machine learning and deep learning-based models. Using a gated recurrent neural network (GRNN) approach on a dataset of six classes and 5000 documents collected from several Facebook sites, their method obtained 70.10% accuracy. The lack of training documents in each class is the cause of this low accuracy (approximately 900). Most significantly, they did not properly identify the classes, which is vital to the performance of classifying hostile texts. Farhadloo et al. [6], to study customer happiness, colleagues used the aspect-level sentiment on unstructured review data. The data were collected from 51 different California state parks, and the model was built using the Bayesian approach. Other work has been found in [7], which includes tests that use the transformation-based Brill's method and provide excellent results. On a small corpus of about 5000 Bangla tokens, this work uses Brill's technique along with various other approaches such as HMMs and n-grams, and the findings show that the transformation-based approach performs better for Bangla. However, the corpus used in this study is too small for stochastic methods to produce the frequency distribution tables needed to construct acceptable tag sequences for specific phrases. A recent study [8] described a different machine and deep learning approach for detecting abusive Bengali remarks. Using RNN on 4700 Bengali text documents, the model achieved an accuracy of 82%. Six classifiers were used by researchers to recognize offensive remarks in a manually gathered dataset. The authors preprocessed data from various sites in this study. The findings of their experiments indicated that a deep learning-based model excels other models by a large margin; however, the use of a limited dataset with just 4700 comments is a limiting issue in this research. Bijoy et al. [9] proposed an automated term by applying many machine learning techniques to categorize the sentences. Almost 740 text documents were gathered for this research work, and the data went through a process from a primary cleaning to purified text. For applying the NLP model to the dataset, they had to generate six classification algorithms such as KNN, RF, DT, MNB, SVM, and XGB. They were ensuring the best model by assessing certain statistical performance and accuracy. Among all, Random Forest and Extreme Gradient Boosting showed a high accuracy of 96.39%. Salehin et al. [10] used an LSTM-based Recurrent Neural Network to score words based on their sentiment in a corpus of Bengali Facebook postings, and they obtained a high level of

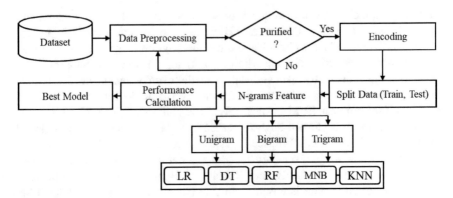

Fig. 1 Step-by-step working procedure to approach different classification algorithms to compare in N-gram feature between Bangla good and bad text discourses

success (at least 72.86% accuracy). Kumar et al. [11] attempts were made to forecast customer happiness based on product reviews. For prediction rating, it used a unique multimodal approach. The global reviews were compiled using several data sources. The worldwide rating score of chemicals is calculated using natural language processing. The Random Forest regression technique was used to create predictions using EEG data. Artificial Bee Colony was utilized to improve the overall performance of both global and local Artificial Bee Colonies.

3   Methodology

Our research's major purpose is to create a model with an N-gram feature that can recognize Bengali good and bad sentences by machine. To achieve our goal, we must go through several steps, including dataset collection, data preprocessing, model building, and so on. The working method is being introduced in Fig. 1.

3.1   Dataset Description

To deliver great performance and accuracy, any expert machine learning system requires a large amount of high-quality data. For this data collection, we gathered Bengali words from different channels like social media, Facebook, YouTube, and Bengali websites where people express their positive and negative sentiments. Figure shows that our dataset could gather 1499 text documents, with a total of 650 good reviews and 849 bad reviews, and storing our data in a two-column format, with the first column containing text or class or remark and second categorizing them as either good or bad. The dataset distribution is depicted in Fig. 2 as a number of 'Good' text documents and 'Bad' text documents are divided into two categories.

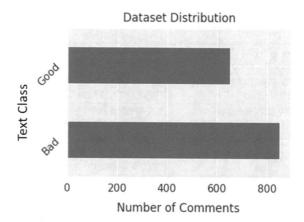

Fig. 2 Number of comments class-wise in the working dataset

3.2 Data Preprocessing

We all know that data preparation is one of the most crucial tasks for every machine learning system. As a result, in NLP, we must go through a data filtering or processing process. To preprocess data, we apply the methods outlined in Fig. 3.

First and foremost, we check our data to see whether it is nicely ordered, if there is any misspelled content, and so on. If there is a major problem or error, we personally remove it. Then, we add some Bengali contraction to get the proper meaning of sentence as like 'ইঞ্জি:' to 'ইঞ্জিনিয়ার', 'মো:' to 'মোহাম্মদ', 'ব্রিগে:' to 'ব্রিগেডিয়ার', etc. After that, we work on regular expressions. We remove any superfluous regular expressions such as punctuation, special characters, numeric, such as '{', '!', '?', '২', '।'. Bengali stop words are looked after by us. The Bengali phrase contains a high amount of stop words. Stop words, on the other hand, make it harder to evaluate data or construct models. So, in this case, we need to remove stop words from our dataset by utilizing the Bangla stop word corpus. Then, we compute the length of a sentence by word counting a sentence. We remove 40 small reviews, then finally select total reviews of 1459. Finally, we are preparing or obtaining cleaned or purified text in order to proceed to the following step of our research investigation. Table 1 shows raw data and preprocess data. We computed data statistics as a category of data in our dataset preprocessing; there are 1499 documents (650 good and 849 bad) in addition to 13,035 words, 4879 unique words in the 'Bad' class and 8328 words, 2936 unique words in the 'Good' class which represent in Fig. 4.

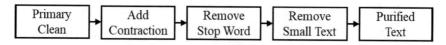

Fig. 3 Step-by-step data preprocessing procedure to extract purified text

Table 1 Sample of dataset before preprocess and after processed data with class

Raw data	Preprocessed data	Class
অনেক পরে হলেও বুজলেন৷ ঠিক কথা বলেছেন ৷ আপনাকে ধন্যবাদ:)	অনেক পরে হলেও বুজলেন ঠিক কথা বলেছেন আপনাকে ধন্যবাদ	Good
'আসলে তোমরা নাস্তিক রা হলা সর্ব হারা, মরার সময়ই বুঝবি শালারা৷ '৷	আসলে তোমরা নাস্তিক রা হলা সর্ব হারা মরার সময়ই বুঝবি শালারা	Bad
ভালোবাসার দিনে ভালোবাসার রঙে সজেছিলি অমর একুশে বইমেলা৷ ☺☺	ভালোবাসার দিনে ভালোবাসার রঙে সেজেছিল অমর একুশে বইমেলা	Good
রেন্ডিয়া নিজের দেশের মানুষের নিরাপত্তা দিতে ব্যর্থ হওয়া সত্ত্বেও বাংলাদেশকে নিয়ে নাক গলায়৷ ৷	রেন্ডিয়া নিজের দেশের মানুষের নিরাপত্তা দিতে ব্যর্থ হওয়া সত্ত্বেও বাংলাদেশকে নিয়ে নাক গলায়	Bad

Fig. 4 Data statistics as per category of total documents, words, unique words in our dataset with class-wise

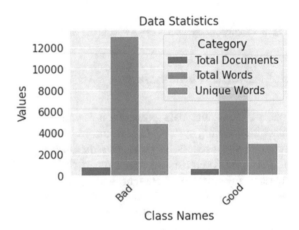

3.3 Data Encoding

The encoding dataset is processed by Label Encoder. Our data are ready to be trained using the machine learning algorithm once the encoding procedure is completed.

3.4 Model Implementation

In our model implementation phase, we first split our dataset into two halves with an 80:20 ratio, with 80% of the data being used to train the model and the remaining 20% data being utilized for testing. We used N-gram features in association with five classification methods to achieve our objectives. We used five methods to be using unigram, bigram, and trigram from the N-gram feature: Logistic Regression (LR), Decision Tree Classifier (DTC), Random Forest (RF),

Multinomial Naive Bayes (MNB), and K-Nearest Neighbors (KNN). The model-relevant theory is given below.

N-Gram: A series of N tokens is known as an N-gram (or words). The likelihood of a given N-gram inside any sequence of words in a language is predicted by an N-gram language model [12]. We can forecast p(w | h)—what is the likelihood of seeing the word 'w' given a history of prior words h—where the history comprises n-1 words—if we have a strong N-gram model. A one-word sequence is known as a unigram. A bigram is a two-word word sequence. A trigram is a three-word word sequence.

For example, considering a good text 'তুমি খুব ভালো' then,
the unigram probability of this sentence is
P(তুমি খুব ভালো) = P(তুমি) × P(খুব) × P(ভালো).
Then, using bigram probability of this sentence is
P(তুমি খুব ভালো) = P(তুমি | খুব) × P(খুব | ভালো).
Now, using trigram probability of this sentence is
P(তুমি খুব ভালো) = P(তুমি | খুব | ভালো).

Logistic Regression (LR): Logistic Regression is a classification algorithm [13]. It predicts discrete values such as 0 or 1 based on a collection of independent variables provided as input. In other words, it estimates the likelihood of any event occurring based on the data. Because it guesses the probability, the results are always incorrect between 0 and 1. The sigmoid function is a Logistic Regression function. The formula is as follows in '(1)':

$$y = \frac{1}{1 + e^{-z}}.$$

$$(1)$$

Decision Tree Classifier (DTC): This approach may be used in a variety of different situations. In most cases, this is accomplished using a machine learning (ML) method that discovers different criteria that may be utilized to split a set of data. For these solutions, Decision Tree analysis [14] might be a common, foresightful presenting technique. This is a non-performed learning model that may be used for both classification and regression. The aim is to develop a demonstration that learns precise decision rules from data and anticipates the dependent variable's value. The formula follows in '(2)':

$$H(X) = -\sum_{i=1}^{n} P(x_i) \log_2 P(x_i).$$

$$(2)$$

Random Forest (RF): This is a strategy that may be used for both classification and regression. A Random Forest set decision tree classifier was used in distinct sub-samples. It is a meta-estimator that uses the average of sub-samples to improve prediction accuracy and control overfitting. It was usually taught using the bagging approach. Bagging is a technique for combining several learning models. It is utilized as a classified as well as a regression model. It introduces unpredictability into the tree-growing paradigm. It selects the best feature from a randomly divided subgroup. The random subset of a feature is considered while splitting a node. Random Forest is used for missing data imputation, grouping, and feature selection. The formula follows in '(3)':

$$RFfi_i = \frac{\sum_j norm\, fi_{ij}}{\sum_{j\,\in\,all\,features,\,k\,\in\,all\,trees} norm\, fi_{jk}}.$$

(3)

Multinomial Naive Bayes (MNB): Multinomial Naïve Bayes (MNB) is a probabilistic technique that is commonly used in Natural Language Processing (NLP) [15]. The Bayes theorem is used to create the algorithm, which guesses the articles. For the following instance, it determines the frequency of each label and then assigns the output to the label with the best frequency. MNB formula is as follows in '(4)':

$$P(X|Y) = P(X) \times \frac{P(Y|X)}{P(Y)}$$

(4)

K-Nearest Neighbors (KNN): This approach may be used in a variety of different situations. In most cases, this is accomplished using a machine learning (ML) method that discovers different criteria that may be utilized to split a set of data. Nearest neighbor (NN) classifiers, particularly the KNN method, are among the most basic and yet effective classification rules, and they are frequently employed in practice. During learning, this rule simply keeps the complete training set and gives a class to each query based on the majority label of its k-nearest neighbors in the training set. When k = 1, the Nearest Neighbor rule (NN) is the simplest form of KNN [16]. The formula follows in '(5)':

$$Dist(X, Y) = \sqrt{\sum_{i=1}^{D} (X_i Y_i)^2}.$$

(5)

3.5 Performance Calculation

After training the models, we utilized test data to estimate their performance. Here are some of the performance evaluation metrics that were calculated. Using these criteria, we found the best model to predict in this case. Many percentage performance metrics have been generated using Eqs. (6–12) based on the confusion matrix provided by the model.

$$Accuracy = \frac{True\ Good\ Text + True\ Bad\ Text}{Total\ Number\ of\ Text\ Documnet} \times 100\%,$$

(6)

$$True\ Positive\ Rate\ (TPR) = \frac{True\ Good\ Text}{True\ Good\ Text + False\ Bad\ Text} \times 100\%,$$

(7)

$$True\ Negative\ Rate\ (TNR) = \frac{True\ Bad\ Text}{False\ Good\ Text + True\ Bad\ Text} \times 100\%,$$

(8)

$$\text{False Positive Rate (FPR)} = \frac{\text{False Good Text}}{\text{False Good Text} + \text{True Bad text}} \times 100\%, \quad (9)$$

$$\text{False Negative Rate (FNR)} = \frac{\text{False Bad Text}}{\text{False Bad text} + \text{True Good Text}} \times 100\%, (10)$$

$$\text{Precision} = \frac{\text{True Good Text}}{\text{True Good Text} + \text{False Good Text}} \times 100\%, \quad (11)$$

$$\text{F1 Score} = 2 \times \frac{\text{Precision} \times \text{Recall}}{\text{Precision} + \text{Recall}} \times 100\%. \quad (12)$$

4 Results and Discussions

NLP approaches' ultimate purpose is to see how significantly higher the given models work. The classification method can undoubtedly produce exact and accurate results based on class. Logistic Regression (LR), Decision Tree Classifier (DTC), Random Forest (RF), Multinomial Naive Bayes (MNB), and K-Nearest Neighbors (KNN) models constructed utilizing unigram, bigram, and trigram features based on two classes as Bangla good text and bad text are used in our dataset. Since this is a binary classification problem, then each model will produce a 2×2 confusion matrix. Table 2 shows the confusion matrix created by each of the classifiers.

To find the best model and compare performance among N-grams feature with the combination of five classifiers to our work and evaluate this work, accuracy,

Table 2 Confusion matrices for all N-gram features (unigram, bigram, trigram)

N-grams	Model	TP	FN	FP	TN
Unigram	LR	127	37	13	115
	DTC	116	33	24	119
	RF	137	31	22	101
	MNB	168	12	19	91
	KNN	156	41	6	89
Bigram	LR	139	52	15	86
	DTC	107	24	36	125
	RF	146	18	39	98
	MNB	165	9	29	88
	KNN	136	30	13	113
Trigram	LR	158	3	84	47
	DTC	134	27	35	96
	RF	157	4	60	71
	MNB	118	43	10	121
	KNN	135	26	20	111

Table 3 Performance evaluation for unigram feature

Unigram	Model	Accuracy (%)	TPR (%)	FNR (%)	FPR (%)	TNR (%)	Precision (%)	F1-score (%)
	LR	82.88	77.44	22.57	10.16	89.84	90.71	83.55
	DTC	80.48	77.85	22.15	16.78	83.22	82.86	80.28
	RF	81.79	81.55	18.45	17.89	82.11	86.16	83.79
	MNB	89.31	93.33	6.67	17.27	82.73	89.84	91.55
	KNN	83.90	79.19	20.81	6.32	93.68	96.30	86.91

Table 4 Performance evaluation for bigram feature

Bigram	Model	Accuracy (%)	TPR (%)	FNR (%)	FPR (%)	TNR (%)	Precision (%)	F1-score (%)
	LR	77.05	72.77	27.23	14.85	85.15	90.26	80.58
	DTC	79.45	81.68	18.32	22.36	77.64	74.83	78.10
	RF	81.06	89.02	10.98	28.47	71.53	78.92	83.67
	MNB	86.94	94.83	5.17	24.79	75.21	85.05	89.67
	KNN	85.27	81.93	18.07	10.32	89.68	91.28	86.35

Table 5 Performance evaluation for trigram feature

Trigram	Model	Accuracy (%)	TPR (%)	FNR (%)	FPR (%)	TNR (%)	Precision (%)	F1-score (%)
	LR	70.02	98.14	1.86	64.12	35.88	65.29	78.41
	DTC	78.77	83.23	16.77	26.72	73.28	79.29	81.21
	RF	78.08	97.52	2.48	45.80	54.20	72.35	83.07
	MNB	81.85	73.29	26.71	7.63	92.36	92.19	81.66
	KNN	84.25	83.85	16.15	15.27	84.73	87.10	85.44

TPR, TNR, FPR, FNR, precision, and F1-score from the above confusion matrix are computed. The result of several performance evaluation metrics for five applied machine learning algorithms with unigram, bigram, and trigram features is presented in Tables 3, 4, and 5, gradually.

Tables 3 and 4 indicate that Multinomial Naive Bayes (MNB) outperformed all other classifiers in the unigram and bigram features, respectively, with the highest accuracy of 89.31% and 86.94 percent. Table 5 shows that K-Nearest Neighbors (KNN) achieve maximum accuracy of 84.25% in the trigram feature. The F1-score of MNB for unigram is 91.55% and 89.67% for the bigram feature. Then, in the trigram feature, KNN reaches 85.44% of the F1-score. Figure 5 depicts the comparison in graph representation for convenient visualization.

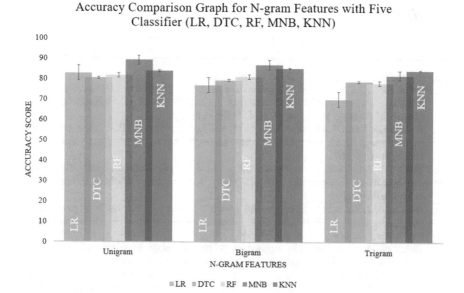

Fig. 5 Accuracy comparison graph for N-gram features with five classifiers of combination (LR, DTC, RF, MNB, and KNN) by accuracy score

5   Conclusion

In Artificial Intelligence, Natural Language Processing in Bangla is a relatively new challenge. Bangla is presently used on a range of platforms, including social media, communication platforms, news media, and so forth, as a result of its rapid expansion. Bangla is the fourth most spoken language in the world, and day by day, it is increasing. We used different classification algorithms with the N-gram feature for classifying Bangla's good and bad texts. Logistic Regression (LR), Decision Tree Classifier (DTC), Random Forest (RF), Multinomial Naive Bayes (MNB), and K-Nearest Neighbors (KNN) models are applied to the dataset using unigram, bigram, and trigram feature approaches. Multinomial Naive Bayes (MNB) surpassed all other classifiers in the unigram and bigram features, with the greatest accuracy of 89.31% and 86.94%, respectively. The K-Nearest Neighbors (KNN) trigram feature yields an accuracy of 84.25%, and the resulting model can categorize the Bangla language document as good or bad discourse. In the future, we can apply more classification algorithms with an added neutral position of data. Also, more data can reduce or improve accuracy in this research. Based on this observation, we may conclude that our tested model functions admirably and provides the computer with the capacity to identify Bengali text properly, making it simpler to engage with humans. Another goal is to create a bigger Bangla domain-wise dataset that can be used to adapt to any machine modernization effort.

References

1. Alam MH, Rahoman M-M, Azad MAK (2017) Sentiment analysis for Bangla sentences using convolutional neural network. In: 2017 20th international conference of computer and information technology (ICCIT). IEEE, pp 1–6
2. Taher SM, Abu KAA, Azharul Hasan KM (2018) N-gram based sentiment mining for Bangla text using support vector machine. In: 2018 international conference on Bangla speech and language processing (ICBSLP). IEEE, pp 1–5
3. Mahtab SA, Islam N, Rahaman MM (2018) Sentiment analysis on Bangladesh cricket with support vector machine. In: 2018 international conference on Bangla speech and language processing (ICBSLP). IEEE, pp 1–4
4. Hassan A, Amin MR, Mohammed N, Azad AKA (2016) Sentiment analysis on Bangla and romanized Bangla text (BRBT) using deep recurrent models. arXiv preprint arXiv:1610.00369
5. Ishmam AM, Sharmin S (2019) Hateful speech detection in public facebook pages for the Bengali language. In: 2019 18th IEEE international conference on machine learning and applications (ICMLA). IEEE, pp 555–560
6. Farhadloo M, Patterson RA, Rolland E (2016) Modeling customer satisfaction from unstructured data using a Bayesian approach. Decis Support Syst 90:1–11
7. Hasan FM, UzZaman N, Khan M (2007) Comparison of different POS tagging techniques (N-Gram, HMM and Brill's tagger) for Bangla. In: Advances and innovations in systems, computing sciences and software engineering. Springer, Dordrecht, pp 121–126
8. Emon EA, Rahman S, Banarjee J, Das AK, Mittra T (2019) A deep learning approach to detect abusive Bengali text. In: 2019 7th international conference on smart computing & communications (ICSCC). IEEE, pp 1–5
9. Imam Bijoy MH, Hasan M, Tusher AN, Rahman MM, Mia MJ, Rabbani M (2021) An automated approach for Bangla sentence classification using supervised algorithms. In: 2021 12th international conference on computing communication and networking technologies (ICCCNT), pp 1–6. https://doi.org/10.1109/ICCCNT51525.2021.9579940
10. Salehin SM, Samiul RM, Islam MS (2020) A comparative sentiment analysis on Bengali Facebook posts. In: Proceedings of the international conference on computing advancements, pp 1–8
11. Kumar S, Yadava M, Roy PP (2019) Fusion of EEG response and sentiment analysis of products review to predict customer satisfaction. Inf Fusion 52:41–52
12. Robertson AM, Willett P (1998) Applications of n-grams in textual information systems. J Documentation
13. Peng C-Y, Joanne KLL, Ingersoll GM (2002) An introduction to logistic regression analysis and reporting. J Educ Res 96(1): 3–14
14. Safavian SR, Landgrebe D (1991) A survey of decision tree classifier methodology. IEEE Trans Syst Man Cybern 21(3): 660–674
15. Raschka S (2014) Naive Bayes and text classification i-introduction and theory. arXiv preprint arXiv:1410.5329
16. Laaksonen J, Oja E (1996) Classification with learning k-nearest neighbors. In: Proceedings of international conference on neural networks (ICNN'96), vol 3. IEEE, pp 1480–1483

Effective Heart Disease Prediction Using Hybrid Ensemble Learning Model

Deepthi Karuturi, Keerthi Pendyala, and N. Sudha

1 Introduction

Heart disease is one of the main ill-health conditions that attacks a huge range of people in intermediate or old age, and in several incidents, this ultimately leads to death [1]. Over the years, several people are succumbing to death because of heart disease. Age, gender, smoking, family background, cholesterol, unhealthy diet, high blood pressure, obesity, physically inactive, and consumption of alcohol are treated as the risk factors for heart disease. Some of these are tractable. Apart from these risk factors, daily routines such as eating, being physically inactive, and obesity are also pondered to be major risk factors [2–4]. Manual examination of the disparity of having heart disease based on the risk factors is very tough [5]. Machine learning (ML) approaches are found convenient for the prediction of the outcome from the originally existing data.

Several models based on K-Nearest Neighbors (KNNs), Decision Trees (DTs), Logistic Regression (LR), Naive Bayes (NB), Deep Learning (DL), Random Forest (RF), Support Vector Machine (SVM), Gradient Boosted Trees (GBTs) are available [6, 7]. In this paper, we propose a hybrid ensemble learning model to improve the accuracy of prediction.

D. Karuturi · K. Pendyala (✉) · N. Sudha
SASTRA Deemed to be University, Thanjavur 613401, India
e-mail: keerthi.pendyala01@gmail.com

D. Karuturi
e-mail: deepthikaruturi1@gmail.com

N. Sudha
e-mail: sudha@cse.sastra.edu

© The Author(s), under exclusive license to Springer Nature Singapore Pte Ltd. 2023
D. S. Sisodia et al. (eds.), *Machine Intelligence Techniques for Data Analysis and Signal Processing*, Lecture Notes in Electrical Engineering 997,
https://doi.org/10.1007/978-981-99-0085-5_10

2 Data Description

Dataset was imported from the UCI ML repository. Out of the four databases, Cleveland, Switzerland, Hungary, and the VA Long Beach, Cleveland database is used in this experiment which is a widely taken database for ML analyzer. It contains 303 records and 14 attributes. Table 1 gives detail information about the attributes. Here, thirteen features are used for forecasting the illness and one is used as a result of forecasting the disease [8, 9].

Here, an attribute/feature named num shows the presence/absence of heart disease on a scale of 0–4. The value 0 depicts the healthy person with no heart disease, and the rest of the values depict the patients having heart disease.

Table 1 Attribute information in the dataset

Attribute/feature	Definition	Attribute type
Age	Person's age	Numeric
Sex	Represents gender 1: male 0: female	Nominal
Cp	Shows the types of chest pains 1. Typical angina 2. Atypical angina 3. Non-angina pain 4. Asymptomatic	Numeric
Trestbps	Level of BP in resting mode	Numeric
Chol	Values of cholesterol in mg/dl	Nominal
FBS	Blood pressure in case of fasting > 120 mg/dl is noted as 1—Yes and 0—No	
Resting	Resting electrocardiographic results 0. normal 1. ST-T wave abnormality 2. Definite left ventricular hypertrophy by Estes' criteria	Nominal
Thali	Maximal heart rate obtained	Numeric
Exangct	It is the chest pain while exercising 1—True 0—False	Nominal
Oldpeak	Exercise relative to rest	Numeric
Slope	Slope while in peak exercise 1—Up sloping 2— Flat 3—Down sloping	Nominal
Ca	Count of major vessels (0–3)	Numeric
Thal	It is the nature of heart through three unique values 3—normal, 6—fixed defect, and 7—reversible defect	Nominal
Num	1–4 denotes presence of disease 0 denotes the absence of disease	Nominal

3 Proposed Model

Figure 1 gives a glimpse into the process workflow. Firstly, the UCI Cleveland dataset was imported. The process begins with the preprocessing and is succeeded by the feature selection which is helpful to choose significant features followed by classification modeling to generate prediction models [10, 11]. The analysis was continued with all the ML models using all the features. The performance analysis of each model is done based on selected attributes which is noted, and the conclusions of the outputs are presented once the overall proceedings are done.

3.1 Hybrid Ensemble Model

The technique ensemble learning has a very good history of presenting better conduct in various machine learning applications [12]. These applications include the domains of classification and regression. In this paper, we used a heterogeneous group of weak learners for creating a hybrid ensemble model. In this task, various machine learning techniques are collected to trial classification problem.

LR+DT+SVM+KNN+NB

Here, five unique machine learning techniques are utilized as weak learners to come up with a hybrid ensemble model (Fig. 2). The five techniques used in this are Logistic Regression, SVM, KNN, Decision Tree, and Naïve Bayes. Ensemble models are

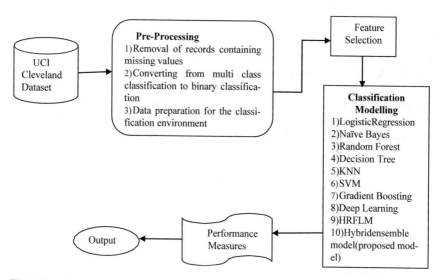

Fig. 1 Experiment workflow

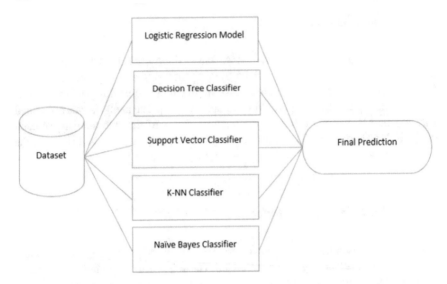

Fig. 2 Workflow of the hybrid ensemble model

a homogeneous collection of weak learners, whereas here a heterogeneous group consisting of weak learners is chosen which in turn is the reason for the tag hybrid.

In this above implementation, we used each of the five ML models five times that providing a blend of 25 weak learners. Lastly, the Max Voting Classifier method is utilized where the class which has been predicted frequently by the weak learners will be taken into consideration. This model produced an accuracy about 80.6%.

RF+LR

These two models were brought together as they have shown better results among other models when implemented individually. In this model, we grouped Random Forest (RF) and Logistic Regression (LR) to build a hybrid ensemble model. This implementation is carried out similar to the above-mentioned model. In this model, we combine a homogeneous ensemble model like Random Forest with Logistic Regression, thus improving the performance of the model (Fig. 3).

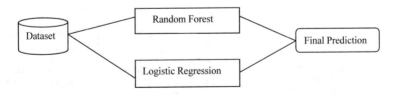

Fig. 3 Workflow of RF+LR

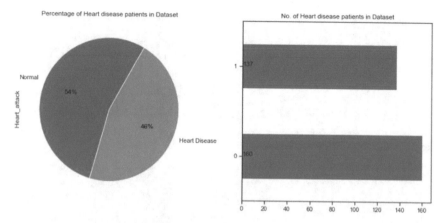

Fig. 4 Count of diseased and normal patients in the dataset

4 Experimental Results

Jupyter Notebook (python) was used to implement the proposed model and evaluate it.

4.1 Data Preprocessing Results

In this step, we handle the missing values. We have 297 records in the dataset. Figure 4 shows the count of heart-diseased and normal patients in the dataset. It shows that there are 137 people with heart disease and 160 normal patients.

4.2 Data Analysis Results

In this process, we accumulate and assemble the data which in return help to draw useful conclusions. Figures 5, 6, 7 and 8 shows univariate and bivariate data analyses, respectively, that are helpful to find relationship between attributes and to gain more understanding about the data.

4.3 Performance Evaluation

Here, various ideal performance metrics such as accuracy, F1-score, precision, recall classification error, sensitivity, specificity have been taken into consideration for

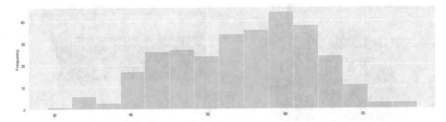

Fig. 5 Distribution of heart attack versus age

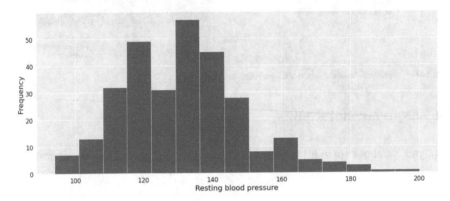

Fig. 6 Distribution of heart attack versus resting blood pressure

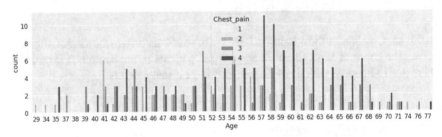

Fig. 7 Dependence of chest pain on age

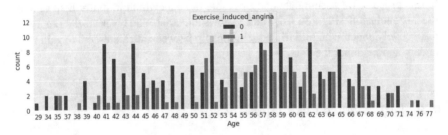

Fig. 8 Dependence of exercise-induced angina on age

Table 2 Performance metrics of various models

Models	Accuracy (in %)	Classification error	Precision	F1-score	Sensitivity	Specificity
Logistic regression	83.33	16.66	81.48	91.6	96.66	56.66
Naive Bayes	81.66	18.33	80.76	79.24	87.87	70.37
Deep learning	83.33	16.66	91.66	81.48	93.33	60.0
Decision tree	85.0	15.0	84.0	82.35	88.23	80.76
Random forest	88.33	11.66	80.76	85.71	94.11	3.84
SVM	86.66	13.33	84.61	84.61	100.0	0.0
Gradient boosted	76.99	23.33	91.66	81.48	76.66	70.0
KNN	83.33	16.66	91.66	81.48	100.00	23.33
HRFLM	85.24	14.75	88.46	83.6	91.00	79.00
Ensemble (LR, DT, SVM, KNN, NB)	80.61	19.38	95.65	86.27	88.00	79.00
Hybrid ensemble (RF, LR) Proposed model	88.88	11.11	91.66	81.48	93.00	73.00

measuring performance efficiency of the model [12, 13]. Accuracy refers to how close a measured value is. Precision is the proximity of measured values to each other. Classification error is the error present in the occurrences. To observe the vital features of heart disease, the above performance measures are used which also help in better awareness of different models [14, 15]. We can see Table 2 for the obtained results. Machine learning technique keeps an eye on the best working model in comparison with already existing models. We introduced a hybrid ensemble model (i.e., a hybrid ML technique with Random Forest and Logistic Regression) which obtains max accuracy and a less amount of error in classification in the heart disease prediction. The performance of all models is calculated separately and is clearly depicted in Figs. 9 and 10, and all observations are noted for any further analysis.

5 Conclusion

As we observe the histograms of performance measures, it is clear that the hybrid ensemble model (RF+LR) has performed absolutely well with an accuracy of 88.88% than rest of the machine learning models. Validating this can be done by observation of classification report, confusion matrices of separate models and the proposed

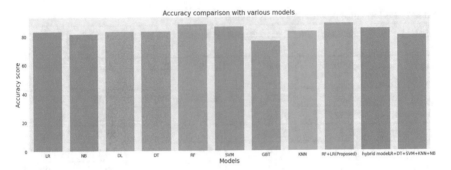

Fig. 9 Accuracy comparison with various models

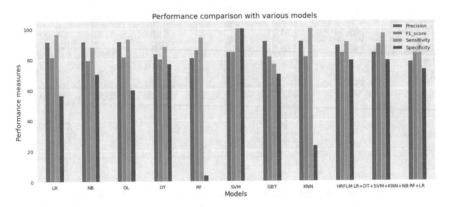

Fig. 10 Performance comparison with various models

models. In future, even more hybrid ensemble models can be performed by bringing together more diverse homogeneous and heterogeneous weak models.

References

1. Alkeshuosh AH, Moghadam MZ, Al Mansoori I, Abdar M (2017) Using PSO algorithm for producing best rules in diagnosis of heart disease. In: Proceedings of the international conference on computer and applications (ICCA), pp 306–311
2. Anooj PK (2012) Clinical decision support system: risk level prediction of heart disease using weighted fuzzy rules. J King Saud Univ Comput Inf Sci 24(1): 27–40. https://doi.org/10.1016/j.jksuci.2011.09.002
3. Esfahani HA, Ghazanfari M (2017) Cardiovascular disease detection using a new ensemble classifier. In: Proceedings of IEEE 4th international conference on knowledge based engineering and innovation (KBEI), pp 1011–1014
4. Banu NKS, Swamy S (2016) Prediction of heart disease at early stage using data mining and big data analytics: a survey. In: Proceedings of the international conference on electrical, electronics, communication, computer and optimization techniques (ICEECCOT), pp 256–261

5. Abdullah AS, Rajalaxmi RR (2012) A data mining model for predicting the coronary heart disease using random forest classifier. In: Proceedings of the international conference on recent trends computational methods, communication and controls, pp 22–25
6. Durairaj M, Revathi V (2015) Prediction of heart disease using back propagation MLP algorithm. Int J Sci Technol Res 4(8):235–239
7. Gavhane A, Kokkula G, Pandya I, Devadkar K (2018) Prediction of heart disease using machine learning. In: Proceedings of the 2nd international conference on electronics, and communication aerospace technology (ICECA), pp 1275–1278
8. Gandhi M., Singh SN (2015) Predictions in heart disease using techniques of data mining. In: Proceedings of the international conference on futuristic trends computational analysis and knowledge management (ABLAZE), pp 520–525
9. Rathnayakc BSS, Ganegoda GU (2018) Heart diseases prediction with data mining and neural network techniques. In: Proceedings of the 3rd international conference on convergence in technology (I2CT), pp 1–6
10. Baccour L (2018) Amended fused TOPSIS-VIKOR for classification (ATOVIC) applied to some UCI data sets. Expert Syst Appl 99:115–125. https://doi.org/10.1016/j.eswa.2018.01.025
11. Dammak F, Baccour L, Alimi AM (2015) The impact of criterion weights techniques in TOPSIS method of multi-criteria decision making in crisp and intuitionistic fuzzy domains. In: Proceedings of the IEEE international conference on fuzzy systems (FUZZ-IEEE), vol 9, pp 1–8
12. Al-milli N (2013) Backpropagation neural network for prediction of heart disease. J Theor Appl Inf Technol 56(1):131–135
13. Devi CA, Rajamhoana SP, Umamaheswari K, Kiruba R, Karunya K, Deepika R (2018) Analysis of neural networks based heart disease prediction system. In: Proceedings of the 11th international conference on human system interaction (HSI), Gdansk, Poland, pp 233–239
14. Das R, Turkoglu I, Sengur A (2009) Effective diagnosis of heart disease through neural networks ensembles. Expert Syst Appl 36(4):7675–7680. https://doi.org/10.1016/j.eswa.2008.09.013
15. Cheng C-A, Chiu H-W (2017) An artificial neural network model for the evaluation of carotid artery stenting prognosis using a national-wide database. In: Proceedings of the 39th annual international conference on IEEE engineering in medicine and biology society (EMBC), pp 2566–2569

A SAR ATR Using a New Convolutional Neural Network Framework

Bibek Kumar, Ranjeet Kumar Ranjan, and Arshad Husain

1 Introduction

SAR imagery systems are very popular nowadays due to their capability of all-weather, day and night working. There are several applications of SAR imagery, including agriculture, oceanography, earth observation, and civil wars. SAR systems and their applications are very effective in the field of surveillance and battlefield operations [1]. In SAR, images, described by grayscale with high-intensity regions, represent targets. In contrast, optical images represent rich colors. In general, the value of each pixel is determined in two ways. The first method is the single reflection from the target surface. The method depends on the roughness, shape, and target material of the surface. The other method is the secondary reflection of the electromagnetic waves, and it has an excellent connection with radar height and shooting angle [2]. The characteristics of SAR images included high resolution, large size, speckle noise, one channel, and scattering electromagnetic. These characteristics are responsible for making large size data information in the SAR images and performing huge manual works in detecting targets in the huge SAR image. The characteristics of SAR images make it difficult to recognize any object, which leads to the selection of this research topic in the remote sensing field. A common architecture of SAR automatic target recognition (ATR) can be divided into the following three phases; (1) Detection, which is used to separate target regions, (2) Discrimination, which is used to identify regions located by targets as per the result of the first phase, and (3) Classification to identify the target category [3] as shown in Fig. 1.

Generally, features of target are obtained by transformation of image, morphological procedure, and scattering center approximation. An atomic norm minimization method is used to approximate the scattering center approximation in which authors transform the problem into a rank minimization problem [4]. A two-dimensional

B. Kumar (✉) · R. K. Ranjan · A. Husain
DIT University, Dehradun, UK 248009, India
e-mail: bkknith@gmail.com

© The Author(s), under exclusive license to Springer Nature Singapore Pte Ltd. 2023
D. S. Sisodia et al. (eds.), *Machine Intelligence Techniques for Data Analysis and Signal Processing*, Lecture Notes in Electrical Engineering 997,
https://doi.org/10.1007/978-981-99-0085-5_11

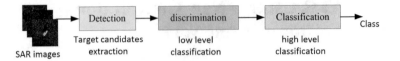

Fig. 1 SAR-ATR architecture

Markov chains-based edge segmentation approach is proposed in which an actual image is transformed into the lab color space, and the second element is used for detection of edges [5]. An adaptive boosting approach used with the radial basis function (RBF) in which the multiclass problem decomposed in a bunch of binary ones with the help of error correcting output code (ECOC) [6]. The conventional classification models of SAR automatic target recognition (ATR) comprise three categories: pattern-based technique [7]; model-based technique [2], and template-based technique [8]. A template-based SAR-ATR framework uses least mean square error (MSE) measures to distinguish the objective kind from a data set of putting away objective reference pictures. The model-based framework dissects each picture exhaustively what's more, distinguishes each piece of a mark commitment to the acknowledgment. As compared to both methods, the pattern-based technique creates a bunch of feature extractors to change the raw image into low dimensional feature vectors, and afterward the output vectors are categorized by a classifier.

In recent years, the convolutional neural network has become very popular and commonly used in the field of computer vision. With the advancement of innovation and computation power, CNNs are being created at a more acute level [9]. The network depth has continuously increased and several pre-trained CNN modes have been proposed, including AlexNet [10], ResNet [11], VGGNet [12], etc.

Deep neural networks (DNNs) given us excellent performance such as; faster region-based CNN (faster RCNN), VGGnet, RetinaNet [13] and single shot multi box detector (SSD) [14]. These models have been extensively used in high resolution SAR imagery and produced excellent results in the case of automatic target recognition. However, currently, there are several variants of DNNs are available which are not performing well in small scale target detection. Many of them are using a large down sampling method. After applying numerous phases of down sampling, the map of features of input images is reduced significantly, and due to this, the knowledge of small scale target reduced lost easily.

The biggest problem with the SAR imagery data is the multiplicative noise known as 'speckle'. The speckle appeared in the SAR images due to its coherent nature of electromagnetic scattering. Due to electromagnetic scattering, high resolution, single channel, and speckle noise presence in this imagery, it becomes more complex to analyze these images. It may lead to pose challenges in classifying objects present in these datasets [15].

In this paper, we used a framework to despeckle SAR imagery dataset and then compare the effect of despeckling in object classification tasks with the proposed neural network model in terms of various performance metrices.

2 Proposed Methodology

2.1 Dataset Used

In this work, we have used the Moving and Stationary Target Acquisition and Recognition (MSTAR) dataset that is freely available [16]. The dataset is created by the Sandila National Laboratory jointly with the Defence Advanced Research Project Agency. The dataset contains SAR images of 10 different armed vehicles used in war. The image data captured with 9.60 GHz functioned X-band radar and bandwidth of 0.591 GHz. The cross-range resolution and range are identical (0.3047 m). The size of images is 128×128 pixels. The sample images of dataset are shown in Fig. 2.

2.2 Data Preprocessing

To remove speckle noise from the images, we have used Multi-Objective Enhanced Fruit Fly Optimization (MO-EFOA) Framework for Despeckling as shown in Fig. 3.

In this model, bivariate shrinkage based dual tree complex wavelet transform (DTCWT) denoising filter has been used for despeckling. The objective of applying DTCWT is to calculate the wavelet coefficient in the manner of neighboring variance estimation. Further, in the second step, an enhanced fruit fly optimization approach has been applied to find the optimal fitness function values in the form of Peak signal-to-noise ratio (PSNR) and mean structured similarity index (MSSIM). PSNR parameter indicates the good quality of despeckling whereas the MSSIM indices indicate high edge preservation in the despeckled images. After applying this model, we have improved the PSNR and MSSIM parameters in the SAR imagery dataset. The overall average of PSNR and MSSIM values we obtained are 34.89 dB and 0.89 respectively, for the MSTAR dataset.

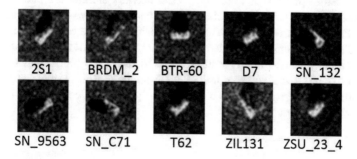

Fig. 2 Sample images from MSTAR dataset

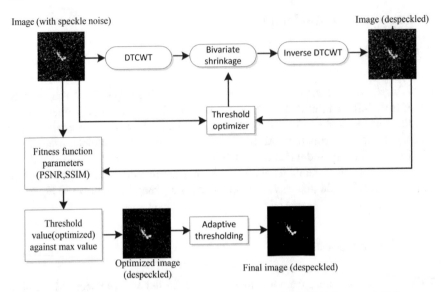

Fig. 3 A Multi-Objective Enhanced Fruit Fly Optimization (MO-EFOA) framework [17]

Next, we have used Data augmentation. Data augmentation is the process to regularize the data to prevent overfitting of a classification model and enhance the imbalance amongst class problems [18]. Different methods are being used in the augmentation step of the images, such as image cropping, image rotation, image translations.

2.3 Proposed Convolutional Neural Network

The advantage of using a neural network is its automatic feature reduction. In CNN, the main challenging task is to design perfect network architecture just because of many blocks (layers). Each convolution layer consists of two-dimensional kernels of the same size. Stride components are used in a CNN model to compress the images. Padding parameters are also used to control boundaries and output effect in CNN. Generally, an activation function relates to each CNN block, such as tangent hyperbolic (tanh), rectified linear unit (ReLU), sigmoid. ReLu is more popular due to its fast-learning capabilities, but it is more costly in term of computation time. The proposed CNN is shown in Fig. 4.

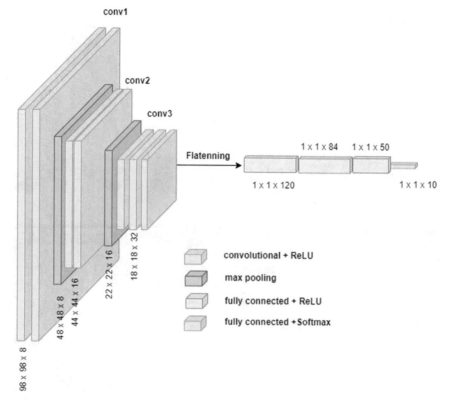

Fig. 4 The proposed CNN architecture

3 Result and Discussion

We have used different training and testing dataset for evaluation of the proposed CNN model. The training dataset is captured on 17°, and the testing dataset is captured on 15° azimuth angle as described in Table 1. In this work, we have performed image augmentation on the dataset and cropped images to 100×100 pixels size from 128×128 pixels. The two-dimensional convolutional neural network has been used with 5×5 kernel size, stride value is (1, 1), zero padding and the ReLu activation function. The network architecture comprises of convolutional layers followed by pooling, flatten, and fully connected layers. The first advantage of using max-pooling is to convolve input in findings of sharpest features and improve computational cost. The second reason to use max-pooling is that it will help in minimizing the scale, rotation, and shift invariance. The flattening layer is being used to flatten the output of convolutional and pooling layer, which is further used as input for the classification by fully connected layers. Flattening is changing over the information into a 1-layered cluster for contributing it to the next following layer. We flatten the result of the convolutional layers to make a individual long feature vector.

Fig. 5 Epoch versus loss

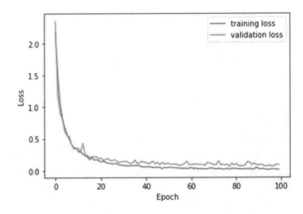

Fig. 6 Epoch versus accuracy

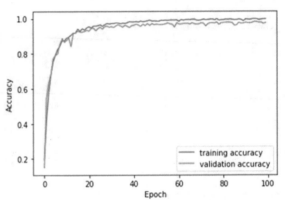

To evaluate the performance of the proposed method we have considered two parameter; accuracy and loss. The accuracy can be calculate with the ratio of correct prediction to the total sample size. It is one of the crucial parameter in the field of image recognition and classification. The second parameter we have considered is loss and error of the predicted value and true value.

We have executed this model for 100 epochs and achieved 99.67% training accuracy and 98.98% validation accuracy. The max-pooling produced higher training accuracy of 99.67% and validation accuracy 98.98% which is better than the average pooling. To handle the problem of vanishing gradient, we have used the ReLu activation function. Figures 5 and 6 present the epoch versus loss and epoch versus accuracy graph of the proposed model training and validation, respectively. A comparison of the proposed model with other existing state-of-arts techniques has been presented in Table 2.

Table 1 Class wise size of training and testing dataset

Class name	Number of training sample (Depression angle : 17°)	Number of testing sample (Depression angle : 15°)
BTR60	256	195
T72	195	233
2S1	299	274
D7	299	274
ZIL131	299	273
BTR70	196	232
T62	299	273
BRDM2	298	274
BMP2	196	233
ZSU234	299	274
Total classes: 10	Total: 2636	Total: 2535

Table 2 Comparison of accuracy results with other techniques

State of art techniques	Accuracy (%)
Borodinov et al. [1]	95.09
Inkawhich et al. [19]	96
Gorovyi et al. [15]	96.80
Yu et al. [20]	96.30
Ozkaya et al. [21]	96.02
Wang et al. [22]	97.10
Proposed	98.98

4   Conclusion and Future Scope

In this work, we have proposed a new methodology to classify SAR images using a CNN model. The key features of the proposed technique are despeckling and augmentation in the preprocessing phase and the proposal of a new CNN architecture. The SAR images used in this work are affected by multiplicative noise which resulted in a low accuracy score in object classification. The despeckling and data augmentation has improved the proposed CNN model accuracy significantly. The experimental results show that the proposed SAR automatic target recognition technique has outperformed existing state-of-arts techniques. The proposed approach has resulted the training accuracy of 99.67% and validation accuracy of 98.98%. In this work, the proposed model is evaluated using only one dataset which has limited numbers of objects. In future, large datasets consisting of more numbers of objects can be used for generalization and the same preprocessing step may be apply with ensemble deep learning.

References

1. Borodinov A, Myasnikov V (2018) Classification of radar images with different methods of image preprocessing. In: CEUR proceedings, vol 2210, pp 6–13
2. Zhai J, Dang X, Chen F, Xie X, Zhu Y, Yin H (2019) SAR image generation using structural Bayesian deep generative adversarial network. In: 2019 photonics and electromagnetics research symposium-fall (PIERS-Fall), pp 1386–1392
3. Qi B, Jing H, Chen H, Zhuang Y, Yue Z, Wang C (2019) Target recognition in synthetic aperture radar image based on PCANet. J Eng 2019:7309–7312
4. Xi Y, Liu H, Wang Y, Li Y (2019) SAR image scattering center estimation via atomic norm minimization. In: 2019 6th Asia-Pacific conference on synthetic aperture radar (APSAR), pp 1–4
5. Kurbatova E, Laylina V (2019) Detection of roads from images based on edge segmentation and morphological operations. In: 2019 8th Mediterranean conference on embedded computing (MECO), pp 1–4
6. Sun Y, Liu Z, Todorovic S, Li J (2007) Adaptive boosting for SAR automatic target recognition. IEEE Trans Aerosp Electronic Syst 43:112–125
7. Prasad S, Bruce L (2008) Limitations of principal components analysis for hyperspectral target recognition. IEEE Geoscience Remote Sens Lett 5:625–629
8. Li Z, Jia Z, Yang J, Kasabov N et al (2020) A method to improve the accuracy of SAR image change detection by using an image enhancement method. ISPRS J Photogrammetry Remote Sens 163:137–151
9. Joshi C, Ranjan R, Bharti V (2021) A fuzzy logic based feature engineering approach for Botnet detection using ANN. J King Saud Univ-Comput Inf Sci
10. Krizhevsky A, Sutskever I, Hinton G (2012) Imagenet classification with deep convolutional neural networks. Adv Neural Inf Process Syst 25
11. He K, Zhang X, Ren S, Sun J (2016) Deep residual learning for image recognition. In: Proceedings of the IEEE conference on computer vision and pattern recognition, pp 770–778
12. Simonyan K, Zisserman A (2014) Very deep convolutional networks for large-scale image recognition. ArXiv Preprint ArXiv:1409.1556
13. Lin T, Goyal P, Girshick R, He K, Dollár P (2017) Focal loss for dense object detection. In: Proceedings of the IEEE international conference on computer vision, pp 2980–2988
14. Liu W, Anguelov D, Erhan D, Szegedy C, Reed S, Fu C, Berg A (2016) Ssd: Single shot multibox detector. In: European conference on computer vision, pp 21–37
15. Gorovyi I, Sharapov D (2017) Comparative analysis of convolutional neural networks and support vector machines for automatic target recognition. In: 2017 IEEE microwaves, radar and remote sensing symposium (MRRS), pp 63–66
16. Keydel E, Lee S, Moore J (1996) MSTAR extended operating conditions: a tutorial. Algorithms Synthetic Aperture Radar Imagery III(2757):228–242
17. Kumar B, Ranjan R, Husain A (2021) A Multi-Objective Enhanced Fruit Fly Optimization (MO-EFOA) framework for despeckling SAR images using DTCWT based local adaptive thresholding. Int J Remote Sens 42:5493–5514
18. Zhu H, Wang W, Leung R (2020) SAR target classification based on radar image luminance analysis by deep learning. IEEE Sens Lett 4:1–4
19. Inkawhich N, Inkawhich M, Davis E, Majumder U, Tripp E, Capraro C, Chen Y (2021) Bridging a gap in SAR-ATR: training on fully synthetic and testing on measured data. IEEE J Selected Topics Appl Earth Observations Remote Sens 14:2942–2955
20. Sun Z, Liu M, Liu P, Li J, Yu T, Gu X, Yang J, Mi X, Cao W, Zhang Z (2021) SAR image classification using fully connected conditional random fields combined with deep learning and superpixel boundary constraint. Remote Sens 13:271
21. Özkaya U (2020) Automatic target recognition (ATR) from SAR imaginary by using machine learning techniques. In: Avrupa Bilim Ve Teknoloji Dergisi, pp 165–169
22. Wang L, Bai X, Zhou F (2019) SAR ATR of ground vehicles based on ESENet. Remote Sens 11:1316

A Distributed Data Fusion Approach to Mitigate Node Redundancies in an IoT Network

Pallavi Joshi and Ajay Singh Raghuvanshi

1 Introduction

1.1 Motivation

The inherent characteristic of being less costly and self-organized makes WSN a most versatile technology in many applications involving the IoT, Edge networks, fog computing, etc. WSN is employed in many applications like surveillance, smart city administration, cyberattacks, and many more due to the pervasive node deployment and self-configuring aspect. The sensor nodes possess limited coverage and communication range. To combat this limitation, clustering provides the best solution which leads to appreciable energy-saving and enhancement in network lifetime. But still, some problems exist like redundancy in data collection, node malfunctioning, and presence of anomaly nodes. So, to identify the redundant data and anomalous node measurement in an application-based WSN which focuses the data collection from only significant nodes, an efficient data fusion technique develops a strong incentive for our proposed work.

1.2 Literature Review

The redundancy at the node level is identified by anomalous nodes, whereas the redundancy at the data level is identified by the outliers deflecting from the normal

P. Joshi (✉) · A. S. Raghuvanshi
National Institute of Technology Raipur, Raipur 492010, India
e-mail: drpallavijoshi1722@gmail.com

A. S. Raghuvanshi
e-mail: asraghuvanshi.etc@nitrr.ac.in

© The Author(s), under exclusive license to Springer Nature Singapore Pte Ltd. 2023
D. S. Sisodia et al. (eds.), *Machine Intelligence Techniques for Data Analysis and Signal Processing*, Lecture Notes in Electrical Engineering 997,
https://doi.org/10.1007/978-981-99-0085-5_12

data pattern of sensor measurements. In WSN, the redundancies may be in the form of varied data patterns, abnormal measurements, or repetitive sensor readings. To identify these, some techniques like divergence, spatial and temporal correlations are adopted.

Cao et al. [1] have proposed a data fusion approach using an optimized extreme learning machine network. It uses the bat optimization to optimize the input parameters of the ELM and trains it to obtain the data without redundancies and sends that data to the sink. To reduce the number of transmissions, a data fusion scheme employing the prediction approach of the gray model along with the implementation of optimally pruned ELM [2]. Gavel et al. [3] proposed a kernel density estimation-based technique to detect a long-span intrusive attack in a WSN. To detect long-term anomalies in a sensor network, a distributed kernel density estimation approach based on KL divergence is proposed in [4]. A novel model for data aggregation aiming to remove noises and redundancies from the sensed data is proposed in [5]. It uses ELM in combination with the radial basis function based on the Mahalanobis distance (MELM).

The above methods focus on enhancing the life and reducing energy consumption, but at the same time, the accuracy of the model is also a major concern. Many times, it is noticed that a node may show redundant behavior for a while and then again behaves normally. Our proposed method acts like a two-level data fusion process that utilizes both divergence and optimized ELM methods to identify abnormalities at the data and node level.

1.3 Contributions and Paper Organization

Following are the major contributions of the paper:

1. A randomly deployed WSN is considered in this paper which is clustered using K-means algorithm. The distance equation in the K-means is used as the fitness function for the salp-swarm optimization algorithm to further optimize the clusters.
2. The redundancies found at the data level are removed using the threshold-based Pearson's divergence method.
3. The input parameters to the ELM are optimized using bacterial foraging optimization algorithm (BFO).
4. The results obtained by the proposed model for the 3 datasets (taken from UCI repository) are validated on the grounds of accuracy, false-positive rates, energy utilization, and the number of alive nodes.

The structure of the paper is as follows: Sect. 2 gives a derivation on threshold-based Pearson's method for identifying the redundancies. Section 3 provides the proposed model along with the algorithm, and Sect. 4 presents simulation results and comparisons. Finally, the paper is concluded in Sect. 5.

2 Threshold-Based Pearson's Divergence to Identify the Redundancies

The information provided by every node has some patterns. The data infusion process is meant to be aggregated based on density, redundancy, etc. Here, we are using Pearson's divergence which is a probabilistic function and works on the probability density functions (PDFs) [3]. The method uses recursive distributed kernel density estimation (KDE) to get locally estimated PDFs, and then, by calculating covariance and bandwidth matrices, the global PDF values are estimated. The below equation defines the PDF for KDE:

$$\text{PDF}_{\text{KDE}}(d) = \frac{1}{n} \sum_{i=1}^{n} \text{KDE}_B(d - d_i) \tag{1}$$

where

$$\text{KDE}_B(d) = |B|^{-1} \text{KDE}(B^{-1}d) \tag{2}$$

'd' is the m-dimensional distribution and $d \in D^m$, 'B' is the bandwidth of size $m \times m$. KDE denotes kernel of second order and KDE: $D^m \rightarrow D$. To track the data values 'n' in the distribution, the below time varying equation is used:

$$\text{PDF}_{\text{KDE}^{t+1}}(d) = \propto \text{PDF}_{\text{KDE}^t}(d) + \frac{1-\propto}{n} \sum_{\text{KDE}=tn+1}^{t(n+1)} \text{KDE}_B(d) \tag{3}$$

$\propto$ is the forgetting factor where $\propto \in (0, 1)$. Let the data samples be evenly distributed in 'p' localizations, and the total estimation is given by:

$$\text{PDF}^t(d) = \frac{1}{p} \sum_{j=1}^{p} \text{PDF}_{\text{KDE}}^t(d) \tag{4}$$

Using Eqs. (1) and (2), the local estimation by cluster members in a cluster can be obtained. Each cluster heads have global values of PDFs given by Eq. (4) obtained by integrating all locally estimated PDFs of the cluster members in a cluster. The 'B' bandwidth matrix gives accuracy.

$$B_i = \left(\frac{4}{m+2}\right)^{\frac{1}{m+4}} \text{cov}^t N^{-\frac{1}{m+4}} \tag{5}$$

where

$$\text{cov} = \begin{bmatrix} \sigma_1 & 0 & \cdots & 0 \\ 0 & \sigma_2 & \cdots & 0 \\ & & \cdots & \\ 0 & 0 & \cdots & \sigma_i \end{bmatrix}$$

where σ_i denotes standard deviation for the required bandwidth. And cov^t is given by:

$$\text{cov}^t = \frac{1}{p} \sum_{j=1}^{p} \text{cov}^t_j \tag{6}$$

Using Eq. (3), the recursive covariance matrix can be written as:

$$\text{cov}^{t+1}_j = \alpha \text{cov}^t_j + \frac{1-\alpha}{n} \text{KDE}_B\left(d_j, d'_j\right)$$

From Eq. (5), the bandwidth matrix is given by:

$$B^t_j = \gamma_{\text{local}} \text{cov}^t_j \tag{7}$$

and $\gamma_{\text{local}} = \frac{4}{m+2}^{\frac{1}{m+4}} N^{-\frac{1}{m+4}}$ is a constant. For global PDF, the global bandwidth matrix is given by:

$$B^t = \gamma_{\text{local}} \text{cov}^t p^{-\frac{1}{m+4}} \tag{8}$$

Finally, by substituting Eqs. (6) and (7) in the Eq. (7), we get

$$B^t = \gamma_{\text{global}} \sum_{j=1}^{p} B^t_j \tag{9}$$

where $\gamma_{\text{global}} = p^{-\frac{m+5}{m+4}}$ is a constant in Eq. (8). So, it is observed that the cluster members handle local bandwidth matrix as given by Eq. (6), and the cluster head handles global bandwidth denoted by Eq. (9). The KDE gives global PDF values, and the difference between two consecutive PDFs determines the redundant nature of node. To obtain the difference between PDFs, Pearson's divergence has been used. Let 'R' and 'S' be two distinct PDFs given by:

$$R = \frac{1}{p} \sum_{j=1}^{p} R_j \quad \text{and} \quad S = \frac{1}{p} \sum_{j=1}^{p} S_j$$

'R' and 'S' have local and global PDFs generated using recursive distributed KDE. Here, $R = \text{PDF}^{t-1}$, $R_j = \text{PDF}_j^{t-1}$, $S = \text{PDF}^{t-1}$, and $S_j = \text{PDF}_j^{t-1}$. 't' is the next value, and '$t-1$' gives previous value.

The Pearson's divergence for 'R' and 'S' is defined as:

$$\text{PE}(R||S) = \sum S\left(\frac{R}{S} - 1\right)^2 \tag{10}$$

Theorem 1 *If $R = \frac{1}{p}\sum_{j=1}^{p} R_j$ and $S = \frac{1}{p}\sum_{j=1}^{p} S_j$, the total PE divergence can be expressed as $\frac{1}{p}\sum_{j=1}^{p} \text{PE}(R_j||S_j)$*

$$\text{PE}(R||S) = \frac{1}{p}\int \sum_{j=1}^{p} S_j \left(\frac{\sum_{j=1}^{p} R_j}{\sum_{j=1}^{p} S_j} - 1\right)^2 \tag{11}$$

Thus, $\text{PE}(R_j||S_j) \leq \frac{1}{p}\sum_{j=1}^{p} S_j\left(\frac{R_j}{S_j} - 1\right)^2 = \frac{1}{p}\sum_{j=1}^{p} \text{PE}(R_j||S_j)$. Placing these values, we obtain,

$$D_{\text{PE}} = \sum_{j=1}^{p}(V^t - V^{t+1}) \tag{12}$$

The above theorem helps to approximate the PE divergence through the log sum inequality (LSI) method. The probability for R with majority is in the V^t, and for S, it is in V^{t+1}. Equation (12) indicates final function D_{PE} with distributed probabilities. At time 't,' the cluster head collects V^t data from member nodes, and once the V^t updates to V^{t+1}, the divergence is computed. $D_{\text{PE}}^t = \{m_1, \ldots, m_{t-1}, m_t\}$ denotes time series for PE method. The threshold value is initialized for making decision on the data values of nodes using normal observations. If an observation is found to have repeated and/or abnormal/unusual value, it is treated as a redundancy. The threshold λ can be expressed as $\lambda = \max\{|D_{\text{PE}}^t|\}$. Depending on D_{PE}^t value, the λ is updated as shown in Eq. (13).

$$\lambda = \eta\lambda + (1 - \eta)|D_{\text{PE}}^{t+1}| \tag{13}$$

where η is the neutralizing factor and $\eta \in (0, 1)$. Using the neutralizing factor, 'λ' is updated.

3 Proposed Model

In this paper, a data fusion model for IoT-enabled WSN is designed which can handle the redundancies at data level as well as node level and can efficiently aggregate the data to the destination. The model consists of three stages: clustering, classifying redundancies using kernel density-based estimation along with Pearson's divergence, and an optimized extreme learning machine for further classifying the abnormal nodes from normal. The node clustering is performed using the K-means clustering technique based on Euclidean distance between nodes, which gives centroids as cluster heads, and the position of the cluster heads is further optimized by applying a swarm intelligence algorithm called salp-swarm algorithm [6]. A model utilizing a divergence-based method described in Sect. 2, and extreme learning machine is designed to detect the redundancies within a dataset produced by various nodes and to eliminate the redundant nodes. The input parameters of the ELM network [7] are optimized by using bacterial foraging optimization (BFO) [8]. The ELM is trained for each cluster head and the associated cluster members. Figure 1 depicts the workflow of the proposed method. Figure 2 shows the ELM network of cluster members, cluster heads at the training phase (Fig. 2a) and testing phase (Fig. 2b). The algorithm for the proposed method shows how each stage in the proposed model is connected and implemented at the node level, data level, and cluster head level.

Usually, the parameters for hidden biases and input weights are chosen arbitrarily, and the Moore–Penrose generalized inverse determines the output weights. So, to adjust the weights and biases in the ELM, an optimized approach is adopted

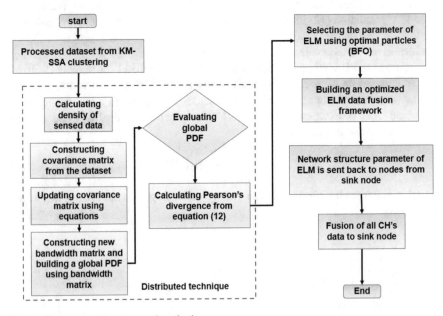

Fig. 1 Flowchart of the proposed method

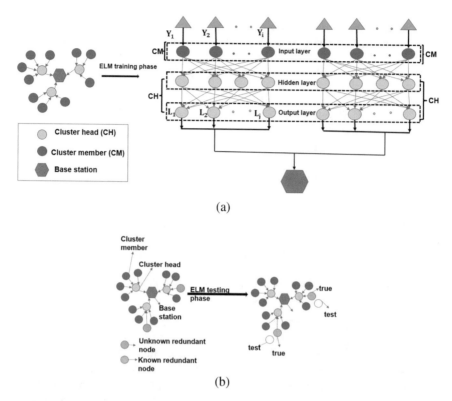

Fig. 2 **a** BFO-ELM training phase. **b** BFO-ELM testing phase

called bacterial foraging optimization in extreme learning machine (BFO-ELM) [9]. Proposed by Passino in the year 2002, the bacterial foraging algorithm is a novel swarm-based optimization approach inspired by the mechanism of searching food through the competitive cooperative approach in E. coli bacterium in the human intestine. The technique works in four phases, namely: swarming, chemotaxis, reproduction, and elimination dispersal. In ELM, we train the SLFNs. SLFNs with 'k' hidden neurons and the activation function '$a(Y)$' to learn 'S' samples (Y_i, L_i) are modeled as:

$$\sum_{i=1}^{k} \gamma_i a(w_i.Y_j + b_i) = L_j, \quad j = 1, 2, \ldots, S \quad (14)$$

$L_i \in \mathbb{R}^m$ is the output against the input Y_i. $Y_i \in \mathbb{R}^s$ denotes the input value matrix for m-dimensional and s-dimensional output and input, respectively. $w_i = [w_{i1}, w_{i2}, \ldots, w_{is}]^T$ is a weight vector which connects the input neurons with ith hidden neuron. $w_i.y_j$ is the inner product of w_i and y_j. $\gamma_i = [\gamma_{i1}, \gamma_{i2}, \ldots, \gamma_{im}]^T$ is a weight vector which connects the ith hidden neuron with output neuron; 'b_i' is the bias applied to hidden layer. '$a(Y)$' is the activation function that can approximate these 's' samples with zero error. The above 's' equations can be written as:

$$H\gamma = T \tag{15}$$

$$H = H(w_1, \ldots, w_k, b_1, \ldots, b_k, Y_1, \ldots, Y_S)$$
$$= \begin{bmatrix} a(w_1.Y_1 + b_1) & \cdots & a(w_k.Y_1 + b_k) \\ \vdots & \ddots & \vdots \\ a(w_1.Y_S + b_1) & \cdots & a(w_k.Y_S + b_k) \end{bmatrix}_{S \times k} \tag{16}$$

$$T = \begin{bmatrix} L_1^T \\ \vdots \\ L_S^T \end{bmatrix}_{S \times m}, \gamma = \begin{bmatrix} \gamma_1^T \\ \vdots \\ \gamma_k^T \end{bmatrix}_{k \times m}$$

The following steps are adopted to train the neurons of ELM using optimized parameters from bacterial foraging algorithm:

Step 1: Initializing all input parameters and loop counters, i.e., elimination dispersion (p), reproduction (q), and chemotaxis loop counter (r). Set the bacterium index i to 0 where $i = 1, 2, \ldots, B$. 'B' is bacterium number (number of matrixes with input weights and hidden biases) in the population; 's' is the search space dimension (hidden biases and input weights); 'C_{chemo}' is chemotactic steps; 'R' is reproduction steps; 'E_d' is number of elimination-dispersal events; 'E_{dp}' is elimination dispersal with probability; '$R(i)$' is size of step in random direction determined by the tumble.

Step 2: Initializing the chemotactic step for 'i' bacterium where $i = 1, 2, \ldots, B$ and finding the fitness function values using Eq. (17):

$$F_{fitness}(i, p, q, r) = \frac{1}{1 + \text{Err}_{ms}} \tag{17}$$

Err_{ms} is the mean square error for training data.

Step 3: Generating a random vector $\Delta(i) = \mathbb{R}^s$, where $\Delta(i) = (\Delta_1(i), \ldots, \Delta_n(i)), n = (1, 2, \ldots, l)$ which is a random number ranging $[-1, 1]$. Assuming Eq. (18) for taking step of size $R(i)$ in tumble's direction.

$$\varphi^i(r + 1, q, p) = \varphi^i(r, q, p) + R(i)\frac{\Delta(i)}{\sqrt{\Delta^T(i)\Delta(i)}} \tag{18}$$

where $\varphi^i(r, q, p)$ symbolizes the ith bacterium at rth chemotactic, qth reproductive, and pth elimination-dispersal steps.

Step 4: Calculate $F(i, p + 1, q, r)$ by $p = p + 1$.

Step 5: If $p < E_d$ then, do $q = q + 1$, otherwise proceed with the pair of optimal parameters and train the ELM by eliminating and dispersing each bacterium with probability E_{dp}. And disperse the eliminated bacterium to arbitrary position in optimization space.

Step 6: If $q < R$, then do $r = r + 1$ otherwise perform elimination dispersal for died and migrated bacterial individuals and jump to step 4.

Step 7: If $r < C_{chemo}$, then train the ELM using the optimal values using Eq. (17) perform steps 3 to 5 and do $r = r + 1$. If $r > C_{chemo}$, perform reproduction and do $q = q + 1$.

Algorithm: Proposed method

1:　Initialize the population, i.e., 'n' data points or nodes
2:　Calculate fitness for each particle using equation

$$D(d, g_k) = \sqrt{\sum_{i=1}^{n} (d_i - g_{ki})^2}$$

'D' is distance function that calculates distance of a datapoint from the cluster center in n-dimensional space; 'd' is datapoint; $G = \{g_1, g_2, \ldots, g_k\}$ is vector for k cluster centers in space where $k > 2$.

3:　Update best solution (salp fitness)
4:　Update c_1 using equation: $c_1 = 2e^{-\left(\frac{4t}{a}\right)}$ where 't' is current iteration and 'a' is maximum iterations.
5:　For $i = 1$ number of data points do
6:　If $j == 1$ (Run K-means algorithm for obtaining clusters and centroids)
7:　Else (Update salp position using equation:

$$d_j^1 = \begin{cases} S_j + c_1\big((u_j - l_j)c_2 + l_j\big), c_3 \geq 0 \\ S_j - c_1\big((u_j - l_j)c_2 + l_j\big), c_3 < 0 \end{cases},$$

where d_j^1, S_j, u_j, and l_j are positions of leader (first) salp, food source, and upper and lower bounds in jth dimension. c_1, c_2 and c_3 define upper and lower bounds.)

8:　Return the best salp (solution).
9:　Initialize clustered dataset
10:　For each member node of cluster 'k' with CH$_i$ do
11:　Initializing ELM parameters $Y_j, L_j, a(Y), k$, and BFO parameters p, q, r
12:　Calculate Pearson's divergence between two data values using Eq. (10)
13:　Using LSI method for approximation find D_{PE} using Eq. (12) and update threshold value depending on D_{PE} using Eq. (13)

14: Redundant nodes with redundancies identified using threshold values are
    classified based on ELM
15: Finding fitness values using Eq. (17)
16: If ($r < C_{chemo}$), then train the ELM according to Eqs. (14)–(16)
17: Else (Perform elimination dispersal as given in step 4 of BFO)
18: ELM network structure sent to nodes for data fusion at sink node

4 Results and Comparative Analysis

This section analyzes the simulation carried out in MATLAB and Python simulation
platform for WSN and ELM to evaluate the overall accuracy, false-positive rates
with respect to the number of iterations, energy consumption, and the count of alive
nodes. The experimentation for the proposed method is implemented on three real-
time datasets (Ds) described in Table 1. The redundant values are separated based on
Pearson's divergence and are further projected to ELM for classification of normal
nodes from redundant nodes measuring the wrong data or repeated data. 75 and 25%
of the data are used for the training and testing phases, respectively. Table 2 gives
the parameter settings for the simulation.

To verify the divergence technique described in Sect. 2, we are taking a swarm
behavior dataset in which the flocking and non-flocking birds are classified using the
3 divergence approaches that are Kernel density estimation with distributed approach
using Pearson's divergence, Kullback–Leibler (KL) divergence using the distributed
technique, and Renyi's centralized approach shown in Fig. 3. Pearson's divergence
and KL divergence show similar performance outperforming Renyi's method. The
black symbols in the graph are threshold points that separate the redundancies (red
squares) from the observed data values (blue symbols). The divergence for every
two consecutive instances is taken, and kernel density estimation is implemented to
find the redundancies. D_{PE} is the deciding factor for the threshold calculation and is
initialized from certain observations in the dataset. The threshold value is calculated
using Eq. (13) and the observations from the sensor measurements. The raw labeling
in the dataset is not considered if more than 10% of measurements are repetitive, they
are labeled as abnormal/redundant. The forgetting factor 'α' is kept constant, and
the neutralizing factor 'η' is calculated depending upon the nature of the dataset. The

Table 1 Description of datasets

S. No	Dataset	Instances	Features	Class	Training	Testing
1	Ds1 (Air quality dataset)	9358	15	Binary	7018	2340
2	Ds2 (Condition monitoring of hydraulic systems dataset)	2205	43,680	Binary	1653	552
3	Ds3 (Swarm behavior dataset)	24,017	2400	Binary	18,012	6005

Table 2 Parameters initialized in the simulation

Parameters	Value
Size of network (N)	$100 \text{ m}^2 \times 100 \text{ m}^2$
Number of nodes (n)	500
Initial energy of nodes (E_0)	0.01 J
Circuit energy consumption (E_{elec})	50 nJ/bit
Free space channel parameter (ε_{fs})	10 pJ/bit/m^2
Multipath channel parameter (ε_{mp})	0.0013 pJ/bit/m^4
Packet length (l)	1000 bits
Distance threshold (d_0)	$\sqrt{\varepsilon_{\text{fs}}/\varepsilon_{\text{mp}}}$
Initial number of bacteria in the population (S)	1
Initial elimination-dispersal steps	1
Initial reproductive steps	1
Initial number of chemotactic steps	8
Initial value of probability-based elimination dispersal	0.5
Initial step size by the tumble	0.01
Activation function	Sigmoid function: $a(Y) = \frac{1}{(1+e^{-Y})}$

threshold value keeps on changing depending on the observations in the datasets for different values of 'n' the threshold is taken depending on $\left(\frac{n}{1-\alpha}\right)$ factor. Table 3 gives the values for the training and testing accuracies and the number of testing errors for ELM [2, 9], MELM [5] and the proposed method, validated on three datasets.

Figure 4a shows the overall accuracy measured in percentage for the Ds 1. The accuracy for the proposed method varies between 81 and 95% and for Renyi's (65%, 87%), and KL divergence (66%, 86%) is observed for 100 rounds. For low false-positive rate values, the proposed approach performs better than the other two approaches. It ranges from 0.02 to 0.76% as depicted in Fig. 4b. The low value of FPR indicates low errors and enhancement in classification accuracy. The decrement in the number of alive nodes with the increase in iterations is shown in Fig. 4c. The nodes reach 300 for the 100th iteration in the case of the proposed method whereas the number of alive nodes decreases to 250, 150, and 100 for MELM [5], perceptron neural network-based data aggregation (PNNDA) [10], and low-energy adaptive clustering hierarchy (LEACH) [11], respectively. In Fig. 4d, we can notice that there is a 33.3% decrease in the energy consumption for the proposed model when compared with MELM.

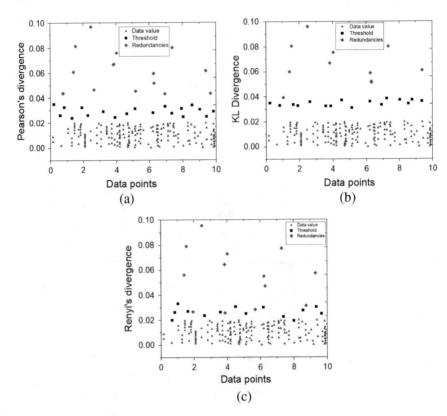

Fig. 3 a Data classification using Pearson's divergence approach for Ds 1, **b** data classification using Kullback–Leibler divergence approach for Ds 1, **c** data classification using Renyi's divergence approach for Ds 1

Table 3 Comparison of training and testing accuracies for ELM, MELM, and proposed model

Datasets	Number of hidden neurons	Training accuracy			Testing accuracy			Number of testing errors		
		[2]	[5]	Proposed	[2]	[5]	Proposed	[2]	[5]	Proposed
Ds1	70	0.90	0.92	0.94	0.87	0.90	0.92	690	347	194
Ds2	40	0.94	0.95	0.97	0.88	0.93	0.95	120	58	37
Ds3	200	0.96	0.98	0.98	0.90	0.95	0.97	1058	980	766

The accuracy in percentage for the proposed method along with the other two methods is depicted in Fig. 5a for Ds 2. The accuracy varies between 88 and 95% for the proposed method and for Renyi's (66%, 86%) and KL divergence (68%, 83%) is observed for 100 iterations. In Fig. 5b, the proposed method records low FPR ranging from 0.1 to 0.8% which is significantly lower as compared to Renyi's (0.14–0.92%) and KL divergence (0.24–0.9%) techniques. Figure 5c depicts that the alive

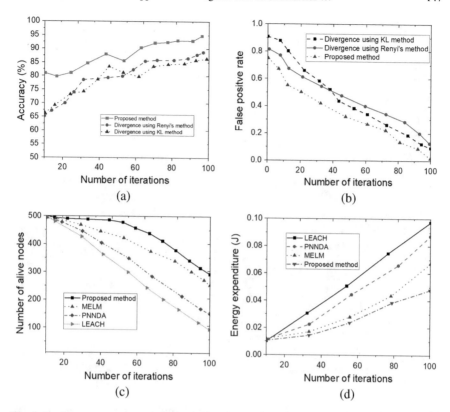

Fig. 4 Performance measures for Ds 1 **a** accuracy, **b** false-positive rate, **c** number of alive nodes, **d** energy consumption

nodes count reaches for 100th iteration in the case of the proposed method whereas the number of alive nodes decreases to 180, 50, and 10 for MELM, PNNDA, and LEACH, respectively. In Fig. 5d, it can be noticed that the energy requirement for the proposed model reduces by 11.1% as compared to MELM.

The accuracy for the proposed method in case of Ds 3 varies between 84 and 97% and for Renyi's (65%, 85%) and KL divergence (70%, 90%) is observed for 100 number of rounds as revealed by Fig. 6a. The proposed method gives FPR values varying from 0.04 to 0.8% as depicted in Fig. 6b thereby performing better than the other two approaches. Similarly, Fig. 6c shows the reduction in alive node numbers with respect to the number of iterations. The energy consumption shows 31.6% decrease for the proposed technique as compared with MELM as shown in Fig. 6d.

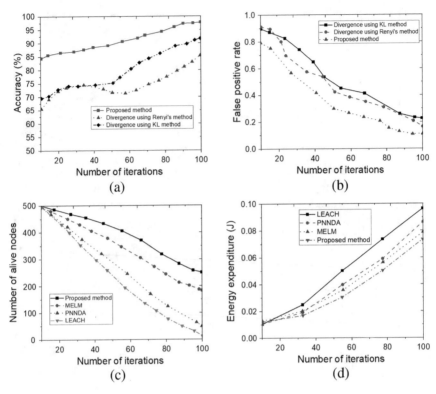

Fig. 5 Performance measures for Ds 2 **a** accuracy, **b** false-positive rate, **c** number of alive nodes, **d** energy consumption

5 Conclusion

In this paper, a novel data fusion method has been proposed which employs extreme learning machine that efficiently collects data at cluster head before fusing it to the sink node. To remove the redundancies in the data measured by the sensors, a kernel density estimation utilizing Pearson's divergence as a distributed approach has been implemented before projecting the data to the ELM network. The clustering of the sensor nodes to mitigate the energy consumption and network lifetime issues has been performed by KM-SSA algorithm. Bacterial foraging optimization has been used with ELM to ensure proposer data processing by efficiently setting the parameters of ELM network. The results show that there has been 10–30% reduction in the utilization of energy and 80–95% increase in accuracy. Also, very low false-positive rates indicate decrease in the errors. The present model can be further extended to process and handle more complex high dimensional datasets by using various machine learning methods.

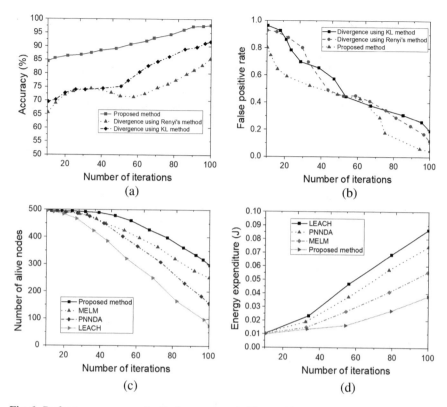

Fig. 6 Performance measures for Ds 3 **a** accuracy, **b** false-positive rate, **c** number of alive nodes, **d** energy consumption

References

1. Cao L, Cai Y, Yue Y, Cai S, Hang B (2020) A novel data fusion strategy based on extreme learning machine optimized by bat algorithm for mobile heterogeneous wireless sensor networks. IEEE Access 8:16057–16072
2. Luo X, Chang X (2015) A novel data fusion scheme using grey model and extreme learning machine in wireless sensor networks. Int J Control Autom Syst 13(3):539–546
3. Gavel S, Raghuvanshi AS, Tiwari S (2021) A novel density estimation-based intrusion detection technique with Pearson's divergence for Wireless Sensor Networks. ISA Trans 111:180–191
4. Xie M, Hu J, Guo S, Zomaya AY (2016) Distributed segment-based anomaly detection with Kullback-Leibler divergence in wireless sensor networks. IEEE Trans Inf Forensics Secur 12(1):101–110
5. Ullah I, Youn HY (2020) Efficient data aggregation with node clustering and extreme learning machine for WSN. J Supercomput 76(12):10009–10035
6. Mirjalili S, Gandomi AH, Mirjalili SZ, Saremi S, Faris H, Mirjalili SM (2017) Salp Swarm algorithm: a bio-inspired optimizer for engineering design problems. Adv Eng Softw 114:163–191
7. Albadra MAA, Tiuna S (2017) Extreme learning machine: a review. Int J Appl Eng Res 12(14):4610–4623
8. Passino KM (2010) Bacterial foraging optimization. Int J Swarm Intell Res (IJSIR) 1(1):1–16

9. Cho JH, Lee DJ, Chun MG (2007) Parameter optimization of extreme learning machine using bacterial foraging algorithm. J Korean Inst Intell Syst 17(6):807–812
10. Hevin Rajesh D, Paramasivan B (2015) Data aggregation framework for clustered sensor networks using multilayer perceptron neural network. Int J Adv Res Comput Eng Technol (IJARCET) 4(4)
11. Jiang S (2018) LEACH protocol analysis and optimization of wireless sensor networks based on PSO and AC. In: 10th international conference on intelligent human-machine systems and cybernetics 2018 (IHMSC), vol 2. IEEE, pp 246–250

ColCompNeT: Deep Learning-Based Colorization-Based Coding Network

Dipti Mishra and Satish Kumar Singh

1 Introduction

Lots of data on the web and transmission channel are available in the form of images and videos, which occupy a lot of storing space. Nowadays, data compression is a critical issue that aims to reduce the size of data that needs to be stored or transmitted via a communication channel, to save the transmission bandwidth and storage space. To minimize the mentioned parameters, one can reduce bandwidth utilization by sending (or storing) only one channel via the channel and retrieving the other two channels at the receiver side with the help of any technique. Any color or RGB image consists of two components: one responsible for brightness (luminance) and the other responsible for providing color (red and blue chrominance). Colorization is the process of providing plausible colors to a gray image by exploiting the luminance channel and producing the other two chrominance channels.

The aim is to make the output image as realistic as the input image, although not necessarily the same as the ground truth version; however, the challenge with compression using colorization is to produce the replica of the original ground truth image. There are various color models available in the literature, like CMYK (cyan, magenta, yellow, black) RGB (red, green, blue), laboratory, YCbCr, YUV, HSL, HSV, etc. Any color model can be chosen for the image compression, but YCbCr is mostly used for image compression. This is due to the maximum de-correlation among Y, Cb, and Cr channels than the other color spaces. Earlier colorization methods [1–3] aim to colorize the entire image; however, the efficiency of each of such techniques

D. Mishra (✉)
Mahindra University, Hyderabad 500043, India
e-mail: dipti.mishra@mahindrauniversity.edu.in
URL: https://www.mahindraecolecentrale.edu.in/

S. K. Singh
Indian Institute of Information Technology Allahabad, Allahabad 211012, India
e-mail: sk.singh@iiita.ac.in

© The Author(s), under exclusive license to Springer Nature Singapore Pte Ltd. 2023
D. S. Sisodia et al. (eds.), *Machine Intelligence Techniques for Data Analysis and Signal Processing*, Lecture Notes in Electrical Engineering 997,
https://doi.org/10.1007/978-981-99-0085-5_13

151

depends upon whether the algorithm is taking care of generating color for each object in the whole image. In the pre and post-deep learning era, a huge amount of work was reported in the direction of developing compression-decompression algorithms [4–11]. But only a few advancements have been done in the domain of the development of compression algorithms using the concept of colorization [12, 13]. In 2007, Cheng et al. [14] presented a graph-based semi-supervised learning-based image compression technique that predicts color to pixels based on some reference pixels. The method is based on storing color labels of some image pixels. The network is trained based on those reference pixels for predicting the colors of other pixels. But storing some side information in the form of labels is itself a problem. Another problem is that the images used for testing are facial images that are not diversified in color. So it is effortless to provide facial skin color to face and black color to hair. The type of images used for training plays an important role in having a good performance in colorizing gray images. Baig et al. [1] have investigated or experimented with colorization for the application of image compression. Then in 2009, He et al. [15] came up with another solution for compressing images using colorization, which is called graph regularized experimental design (GRED). The idea behind this active learning-based GRED algorithm is to select the most representative pixels to compress the image. The selection of the most significant pixels is based on active learning, while colorization is a semi-supervised learning approach. Therefore, in the proposed architecture, the encoding scheme uses active learning while decoding uses semi-supervised learning. This technique outperformed Cheng's approach [14] concerning the compression ratio, quality of the image, and processing time. But, the drawback is that every time some side information is needed to be stored, which increases the storage space and hence becomes a problem for the prospect of image compression. The algorithm reported by Ono et al. [16] automatically extracts representative pixels from an original color image at an encoder and restores a full-color image by using colorization at a decoder. In 2013, Lee et al. [17] proposed a colorization-based compression scheme that is optimized using ℓ-1 minimization problem. The author has used YCbCr color space for compressing data. The method has outperformed JPEG [18] and JPEG 2000 [19], the benchmark algorithms for compression, but at the cost of very high computational cost and memory which makes it non-feasible for real-time applications. Later on, Deshmukh et al. [20] designed a colorization algorithm for compression using smooth ℓ-0 minimization in which the luminance information of the image is segmented using fuzzy c-means to develop colorization matrix.

Motivated through this notion, we have trained two separate networks, i.e., compression and colorization, for the same images. At the time of testing, the gray image is compressed and fed as the input to the colorization network, producing the less storage required compressed images. The objective to use CNN is basically for colorization so that the network can learn plausible color for an object belonging to different classes. Here we have used LAB color space.

Using this mechanism, the algorithm maintains to preserve the semanticity by learning the features of cropped objects, followed by the fusing of object-level and full image features. The colorization method exploited multiple instances and potentially overlapping instances to produce spatially coherent colorization. As far as we know, our approach is good because it combines deep learning-based image colorization for the application of image compression. Our major contributions are as follows.

- We have designed a compression-decompression algorithm based on colorization to produce a possible colorized image that is synchronized with the human visual system (HVS).
- The proposed algorithm has tried to produce the best plausible colorized image wherein the image information can be correctly predicted through the human eye. Although we are not successful in predicting the exact color in the ground truth, at least the proposed algorithm has not disturbed the salient information present in the image.
- The proposed algorithm does not need to store the additional side information, which ultimately increases the bytes to be stored.

The remaining part of the paper is organized as follows. Section 2 discusses the proposed methodology details. Section 3 contains the experimental setup and hyper-parameters details. Section 4 represents a discussion on empirical results and their comparison with the state-of-the-art methods. Section 5 finally summarizes the work by showing advantages of the work dropping a further scope for the extension of the work.

2 Methodology

Given an input luminance channel $I_G^{(i)} \in \mathbb{R}^{H \times W \times 1}$, colorization entails generating a color image $I_C^{(i)} \in \mathbb{R}^{H \times W \times 2}$ from a grayscale image. Assuming a training set of N examples, where $I_G^{(i)} \in \mathbb{R}^{H \times W \times 1}$ represents the grayscale version of the ith- training image, $I_C^{(i)} \in \mathbb{R}^{H \times W \times 2}$ contains its two color channels, and H and W are the image dimensions, the task is to estimate $I_C^{(i)}$ from $I_G^{(i)}$ by learning a mapping $I_C^{(i)} = \mathcal{F}(I_G^{(i)})$, where $\mathcal{F}$ is the function to be learnt. The proposed algorithm aims to compress the luminance component of the image with the compression-decompression network, which is then further fed to the colorization network in order to generate two other chrominance channels. Then, the final RGB image is reconstructed after combining the luminance component obtained after decompression and the other two generated chrominance components using a colorization algorithm. It must be noted that the compression and colorization network are trained separately for different parameters and datasets simultaneously, which is discussed in Sect. 3. Based upon the function-ality, we named the colorization-based coding network as ColCompNeT, which is shown in Fig. 1. For the task of compression, we have considered our algorithm presented in Mishra et al. [8]. The compression-decompression network is simply a wavelet-based encoder–decoder network (WDAED) consisting of a sequence of convolution layers and the ReLU activation function, which is already reported to

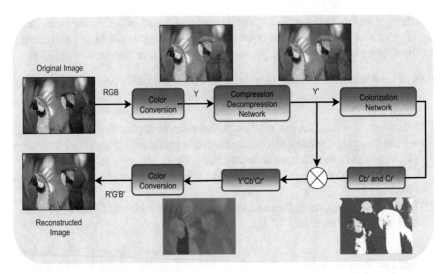

Fig. 1 Block diagram of the proposed colorization-based compression network ColCompNeT

produce better images comprised of very fine edge details. The algorithm considers only the luminance component that contains maximum information to compress and then decompress it. The colorization model utilized here is similar to Su et al. [21], where instances are initially detected using the pre-trained model. Secondly, an instance aware colorization model and full image colorization models are trained separately. Finally, a fusion module is trained to blend both types of features obtained from two models (also called fine-tuning the fusion model by freezing the weights of two models). We use the Leaky ReLU activation function instead of the simple ReLU activation function, giving better results.

3 Implementation Details

3.1 Dataset

The compression network is trained and validated with 900 and 100 images with the size of 256×256 of the ImageNet dataset, respectively. However, the colorization model is trained with 1,18,257 images of 256×256 of "Train 17" version of COCO dataset. To check the efficacy of the proposed algorithm, we have also tested some images from Live 1 [22], Set 14 [23], and Kodak dataset [24]. Live 1 dataset has twenty-nine diversified generic PNG images with sizes ranging from 438×634 to 720×480. These images contain high resolution and high-quality, sharp images. Set 14 contains fourteen low-frequency PNG images with size ranging from 250×361 to 768×512.

3.2 Hyperparameter Tuning

The wavelet-based compression network and the colorization network are trained with ℓ-1 norm loss function [25] which is widely used for the task of image classification. The compression network is trained for 50 epochs with learning rate of 0.001 as used in [8–11, 26]. Both the networks are optimized with Adam algorithm [26] using default momentum parameters as $\beta_1 = 0.9$, $\beta_2 = 0.999$, and $\epsilon = 1 \times 10^{-7}$, to find the global minima. On the other hand, colorization network is trained for 250 epochs with learning rates of 1×10^{-5}, 5×10^{-5}, and 2×10^{-5} with a batch size of 25.

3.3 Evaluation Parameters

For the assessment of the proposed algorithm and its comparison with the state-of-the-art methods, peak signal-to-noise ratio (PSNR) [27] and structural similarity (SSIM) [28] have been used as objective and subjective performance metrics, respectively. Python with Keras and TensorFlow deep learning framework has been used for compression network, while PyTorch library has been used for colorization network. This is just done to use some functionalities available in different libraries.

4 Results and Discussion

Both the compression and the colorization network are trained simultaneously with the different types of datasets. Since the complete architecture of deep autoencoder

Fig. 2 Comparison of ground truth images (first row) with the colorized images (second row) using Su's approach [21]

Table 1 Comparison of the file size (KB), PSNR, and SSIM with the different colorization-based methods

Image	Algorithm	File Size	PSNR	SSIM (Cb)	SSIM (Cr)
Pepper	Cheng et al. [14]	0.7	23.49	0.872	0.796
	Ono et al. [16]	0.744	22.04	0.781	0.757
	Lee et al. [17]	0.7	29.84	0.910	0.973
	Deshmukh et al. [20]	0.527	30.37	0.900	0.726
	Proposed	0.513	30.03	0.9105	0.7900
Cap	Cheng et al. [14]	0.7	34.37	0.970	0.971
	Ono et al. [16]	0.736	30.16	0.918	0.923
	Lee et al. [17]	0.7	38.62	0.980	0.994
	Deshmukh et al. [20]	0.527	31.81	0.801	0.969
	Proposed	0.514	30.79	0.9132	0.9722
Butterfly	Cheng et al. [14]	0.7	26.39	0.789	0.867
	Ono et al. [16]	0.904	25.17	0.751	0.722
	Lee et al. [17]	0.7	30.93	0.930	0.955
	Deshmukh et al. [20]	0.527	32.88	0.858	0.865
	Proposed	0.521	31.99	0.9124	0.8957

Table 2 Comparison of the file size (KB) and PSNR (dB) with JPEG and JPEG2000

Image	Algorithm	File Size	PSNR
Pepper	JPEG	4.22	25.72
	JPEG2000	4.22	27.87
	Lee et al. [17]	4.08	27.63
	Proposed	3.97	28.65
Cap	JPEG	3.57	28.27
	JPEG2000	3.59	30.07
	Lee et al. [17]	3.47	30.52
	Proposed	3.23	31.55
Butterfly	JPEG	4.80	23.83
	JPEG2000	4.54	25.79
	Lee et al. [17]	4.54	25.95
	Proposed	4.09	26.42

and colorization network is different, parameter sharing is not possible. Figure 2 shows the colorized images using the colorization algorithm given in Su et al. [21], which shows the efficiency of the colorization model, as the difference between the colorized and ground truth image is not so high. Using this colorized model, we tried to colorize the decompressed images and obtain good results. Table 1 shows the results for the colorized images using uncompressed images or raw gray images. More precisely, to check how much color is recovered, SSIM for red chrominance channel (SSIM (Cr)) and SSIM for blue chrominance channel (SSIM (Cb)) are calculated separately. SSIM (Cb) (or SSIM (Cr)) measures the similarity between the blue chrominance channel (or red chrominance channel) of the original and that of colorized image. Table 2 shows the results for the colorized images for the decompressed images obtained after the implementation of the compression-decompression algorithm. From Tables 1 and 2, it can infer that with the proposed algorithm, the subjective or the perceptual quality metric obtained is still higher at reduced file size. Compared to other algorithms, the reduced file size helped a lot in data saving. The reduced value of PSNR does not impact the semanticity of the image or image understanding after decompression, as the interpretation of information present in the image is still the same and not changing. Figure 3 shows some of the reconstructed images processed by colorization algorithm, which denotes approximately the same as the ground truth images. Figure 4 shows the performance comparison of the proposed algorithm with the other state-of-the-art algorithms. The image at the extreme right is the colorized reconstructed image obtained with the application of the proposed algorithm. The results obtained with the ColCompNeT network-based algorithm are found to be encouraging, which is at least not degrading the information present in the image. From all these results, the main inference is that for a learning-based colorization, the network should be trained for a specific class of images separately, e.g., facial images, medical images, cartoon images, flora, and fauna, etc., then only the network will be capable of learning a particular type of attribute of an object lying in a specific class. A single trained network cannot provide a color exactly the same as the ground truth image. Comprehensively, it can be reported that it is a good attempt of colorizing gray images which are already gone through the process of compression-decompression. As per the author's knowledge, this is the first approach using learning-based colorization for the application of image compression.

5   Conclusion and Future Scope

A learning-based network is designed, capable of coding and decoding gray images, followed by converting them into color images with the help of a colorization network. Compression using colorization helped in saving the storage space and transmission bandwidth. Based on the results obtained, it can be reported to be a better compression scheme synchronized with saliency since it focuses on salient information present in the image and does not distort the picture's quality. However, we will further extend our approach for compressing the specific type of images like facial images, medical images, cartoon images, flora, fauna, etc., to obtain the improvement in color production exactly the same as that of ground truth. In addition, in the future, the work is extended to use generative adversarial networks for parallel compression and colorization.

Fig. 3 Comparison of reconstructed images obtained after colorization using Su's approach [21] (first column) with the ground truth images (second column)

| (a) | (b) | (c) | (d) | (e) | (f) | (g) |

Fig. 4 Comparison of the reconstructed image obtained after compression-decompression process by various algorithms as **a** original, **b** gray image, **c** Cheng's [14], **d** Ono's [16], **e** Lee's [17], **f** Deshmukh's [20], and **g** proposed compression algorithm processed by colorization

References

1. Baig MH, Torresani L (2017) Multiple hypothesis colorization and its application to image compression. Comput Vis Image Understanding 164:111–123
2. Iizuka S, Simo-Serra E, Ishikawa H (2016) Let there be color!: joint end-to-end learning of global and local image priors for automatic image colorization with simultaneous classification. ACM Trans Graphics (TOG) 35(4):110
3. Zhang R, Isola P, Efros AA (2016) Colorful image colorization. In: European conference on computer vision. Springer, Heidelberg, pp 649–666
4. Singh SK, Kumar S (2011) Singular value decomposition based sub-band decomposition and multi-resolution (SVD-SBD-MRR) representation of digital colour images. Pertanika J Sci Technol (JST) 19(2):229–235
5. Singh SK, Kumar S (2009) Improved image compression based on feed-forward adaptive downsampling algorithm. Int J Image Graphics 9(04):575–589
6. Singh SK, Kumar S (2011) Novel adaptive color space transform and application to image compression. Signal Process: Image Commun 26(10):662–672
7. Singh SK, Kumar S (2010) Mathematical transforms and image compression: a review. Maejo Int J Sci Technol 4(2):235–249
8. Mishra D, Singh SK, Singh RK (2021) Wavelet-based deep auto encoder-decoder (WDAED)-based image compression. IEEE Trans Circuits Syst Video Technol 31(4):1452–1462. https://doi.org/10.1109/TCSVT.2020.3010627
9. Mishra D, Singh SK, Singh RK (2020) Lossy medical image compression using residual learning-based dual autoencoder model. In: 2020 IEEE 7th Uttar Pradesh section international conference on electrical, electronics and computer engineering (UPCON), pp 1–5. https://doi.org/10.1109/UPCON50219.2020.9376417
10. Mishra D, Singh SK, Singh RK, Kedia D (2021) Multi-scale network (MsSG-CNN) for joint image and saliency map learning-based compression. Neurocomputing 460:95–105. https://doi.org/10.1016/j.neucom.2021.07.012. https://www.sciencedirect.com/science/article/pii/S0925231221010468
11. Mishra D, Singh SK, Singh RK (2021b) Deep architectures for image compression: a critical review. Signal Process 108346. https://doi.org/10.1016/j.sigpro.2021.108346, https://www.sciencedirect.com/science/article/pii/S0165168421003832
12. Cui MY, Zhu Z, Yang Y, Lu SP (2022) Towards natural object-based image recoloring. Comput Vis Media 8(2):317–328
13. Neogi D, Das N, Deb S (2022) A deep neural approach toward staining and tinting of monochrome images. In: Advanced computing and intelligent technologies. Springer, Heidelberg, pp 25–36
14. Cheng L, Vishwanathan S (2007) Learning to compress images and videos. In: Proceedings of the 24th international conference on Machine learning. ACM, pp 161–168
15. He X, Ji M, Bao H (2009) A unified active and semi-supervised learning framework for image compression. In: 2009 IEEE conference on computer vision and pattern recognition. IEEE, pp 65–72

16. Ono S, Miyata T, Sakai Y (2010) Colorization-based coding by focusing on characteristics of colorization bases. In: 28th picture coding symposium. IEEE, pp 230–233
17. Lee S, Park SW, Oh P, Kang MG (2013) Colorization-based compression using optimization. IEEE Trans Image Process 22(7):2627–2636
18. Wallace GK (1992) The jpeg still picture compression standard. IEEE Trans Consumer Electronics 38(1):xviii–xxxiv
19. Christopoulos C, Skodras A, Ebrahimi T (2000) The jpeg2000 still image coding system: an overview. IEEE Trans Consumer Electronics 46(4):1103–1127
20. Deshmukh S, Tiwari V (2015) Colorization based compression using fuzzy c-means and smooth l0 minimization. In: 2015 international conference on computer, communication and control (IC4). IEEE, pp 1–4
21. Su JW, Chu HK, Huang JB (2020) Instance-aware image colorization. In: Proceedings of the IEEE/CVF conference on computer vision and pattern recognition, pp 7968–7977
22. Sheikh HR, Wang Z, Cormack L, Bovik AC (2005) LIVE image quality assessment database release, 2
23. Zeyde R, Elad M, Protter M (2010) On single image scale-up using sparse-representations. In: International conference on curves and surfaces. Springer, Heidelberg, pp 711–730
24. MacKnight CB (1995) Kodak Photo CD Eastman Kodak Company Kodak Information Center Department E 343 State Street Rochester, NY 14650–0811. J Comput Higher Educ 7(1):129–131. https://doi.org/10.1007/BF02946148
25. Willmott CJ, Matsuura K (2005) Advantages of the mean absolute error (mae) over the root mean square error (rmse) in assessing average model performance. Climate Res 30(1):79–82
26. Kingma DP, Ba J (2015) Adam: a method for stochastic optimization. CoRR abs/1412.6980
27. Salomon D (2004) Data compression: the complete reference. Springer Science & Business Media
28. Wang Z, Bovik AC, Sheikh HR, Simoncelli EP et al (2004) Image quality assessment: from error visibility to structural similarity. IEEE Trans Image Process 13(4):600–612

Printed Elliptical Cut Vivaldi 1 × 5 Linear Array Antenna for X and KU Band Applications

Renuka Chowdary Bezawada and Bhagya Lakshmi Munagoti

1 Introduction

The Vivaldi antenna has led to the widespread use of frequency-independent antennas, including Archimedean spirals, logarithmic spirals, for communication or electronic warfare purposes [1]. There has been a lot of attention lately given to slot antenna antennas (TSA) for its ease of printing and implementation in hardware. Using a shared aperture for both electronic support and electronic countermeasure subsystems, phased array antennas are increasingly becoming the logical choice. The reason for this is that they reduce the radar signature of various antennas. Slotting, lens loading, and steel loading (to mention a few) have all been used to increase antenna gain during the last decade [2–5]. Slot and hole designs may be altered to increase the bandwidth of Vivaldi antennas [6–9]. Radiation directors like dielectric lenses can also be included to increase bandwidth [6–9]. This frequency range has a variety of strategies to boost antenna gain while simultaneously distorting the pattern at higher frequencies. An autoexposure stimulating environment amplitude or phase distributions makes it difficult to change the return loss bandwidth of the electromagnetic wave [10]. An elliptic tapering of the slot area improves impedance matching at low frequencies by reducing the radius of the radiating aperture to a circular shape [11, 12]. This study uses a FR4 substrate with a relative permittivity of 4.4 as a reference antenna for ultra-wideband applications with a frequency band of 3.1–10 GHz. This research used HFSS to undertake a parametric examination of the model, and the findings were simulated.

R. C. Bezawada · B. L. Munagoti (✉)
Department of Electronics and Communication Engineering, V. R. Siddhartha Engineering College, Vijayawada, India
e-mail: bhagyalakshmi@vrsiddhartha.ac.in

© The Author(s), under exclusive license to Springer Nature Singapore Pte Ltd. 2023
D. S. Sisodia et al. (eds.), *Machine Intelligence Techniques for Data Analysis and Signal Processing*, Lecture Notes in Electrical Engineering 997,
https://doi.org/10.1007/978-981-99-0085-5_14

2 Single Antenna Geometry

A single antenna geometry and dimensional details are given in Fig. 1.

In terms of dimensions, its antenna is 100 mm long by 40 mm wide and 1.6 mm high. The slot's length and width are both 5 mm. The stubs is 3 mm × 3 mm in length (Table 1).

Single antenna models make use of FR4 substrates. For a single antenna, a microstrip line with a rectangular slit is utilized for feeding.

Fig. 1 Single antenna geometry

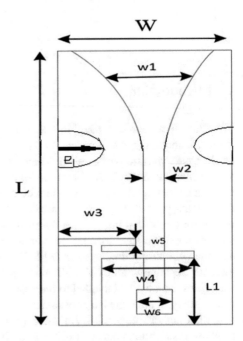

Table 1 Single antenna dimensional details

Parameter	Dimensions in mm
L	100
W	40
E1	2.0
W1	4.0
W2	5.0
W3	1.0
W4	1.0
W5	7.0
W6	10
L1	7.0

Fig. 2 Tapered slot line design

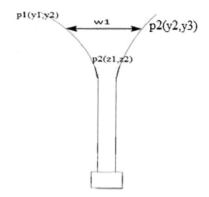

There are a number of common formulae that are used to build the tapered slot line is shown in Fig. 2. P_1 (y_1, y_2) and P_2 (y_2, y_3) are shown in the results can be achieved of the opening rate (R) (z_1, z_2).

The exponential curve was defined by

$$Y = C_1 e^{RX} + C_2 \tag{1}$$

where

$$C_1 = \frac{(y_2 - y_1)}{e_2^{RZ} - e_1^{RZ}} \tag{2}$$

$$C_2 = \frac{(y_1 e_2^{RZ} - y_2 e_1^{RZ})}{e_2^{RZ} - e_1^{RZ}} \tag{3}$$

C_1 and C_2 are constants, while R is the growth rate at which the exponential taper begins to taper.

3 Vivaldi 1 × 5 Linear Array Antenna Analysis

An array antenna is depicted below as an expansion of the single antenna. Microstrip line feed is used to provide power to the antenna. The array design is shown in Fig. 3a, b. For transmitting the high frequency and reflecting the standing waves, two elliptical cuts are employed to improve the bandwidth.

The design uses a tiny strip line wire connector for feeding. Rectangular rather than circular slots in the antenna are employed because they have a wider bandwidth. Waves reverberate in a gap just at end of the feeding line for micro-strip lines. An ellipse is used to increase the operational speed from 8 to 19 GHz inside the design requirements. With the use of the power flow theorem, a rectangular hole for reflected standing waves may enhance the design's bandwidth.

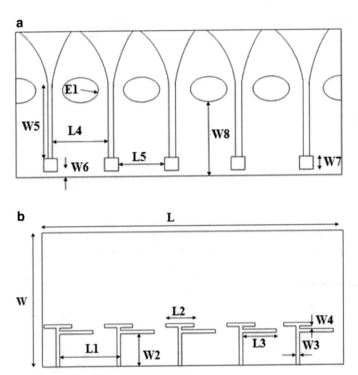

Fig. 3 **a** Tapered slot line design. **b** 1 × 5 linear array back view

Because of its large bandwidth and to reflect all standing waves, the feeding is delivered directly to an antenna in the form of a rectangular slot. The following table lists the benefits and drawbacks of circular slots over rectangular ones.

For the Vivaldi array, the design equations for a taper line are given by

$$T = S * v * \exp(u * \text{rate}) \tag{4}$$

where u equals 0.04
Its final value is 50.
'v' a starting value of -1.
'v' a final value is $+1$.

3.1 Rectangular and Circular Cavity

Antenna resistance matching and bandwidth are controlled by the circular slot of the Vivaldi antenna. When used outdoors, circular slot antennas work well because they can adjust their impedance to that of the ambient environment. Using a circular slot

Table 2 1 × 5 linear array antenna dimensional details

Parameter	Dimensions in mm
L	100
W	200
H	1.6
E1	2.0
W2	25
W3	3
W4	4.5
W5	37
W6	48
W7	10
L1	37
L2	12
L3	24
L4	38
L5	30

improves frequency response, weather resistance, and radiation resistance. Satellite dishes and military communication systems alike both benefit from a circular slot's advantages.

Because it is simple to measure all bandwidths in a rectangular slot, this shape is popular. With increased power, it is able to transmit more data at a greater frequency range. It is less effective in transmitting because it is shorter than an antenna. It can hold a wider area with much less loss due to its improved radiation pattern. An antenna's operating frequency may be increased by using an ellipse. In order to get a broader band (Table 2).

The main axis radius is 2.0 mm in this case; the feed line's dimensions are 3.08 mm * 1.6 mm, which are in the plane of the device. The feed line rectangle is 10 mm * 10 mm in size. In terms of size, the slot is 41 mm in length and 2.3 mm wide.

4 Simulation Results

Figure 4 depicts the 1 × 5 linear array simulation design model presented below (Fig. 5).

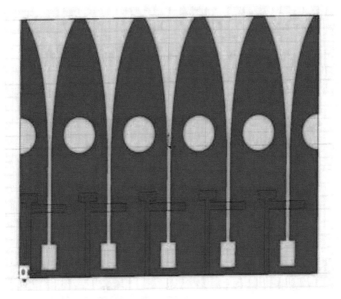

Fig. 4 1 × 5 linear array simulated design

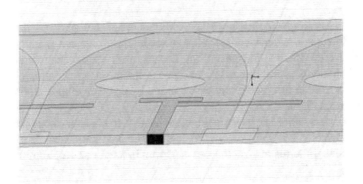

Fig. 5 Feeding model

4.1 Voltage Standing Wave Ratio (VSWR)

In an array, the central element is stimulated while the others elements terminated It has been shown in Fig. 6 that the VSWR peak across 8–19 GHz is less than 2.5. In the graph below, the VSWR is shown. Between 8 and 19 GHz, the maximum VSWR value falls below 2.5, while the lowest value is below 1.0.

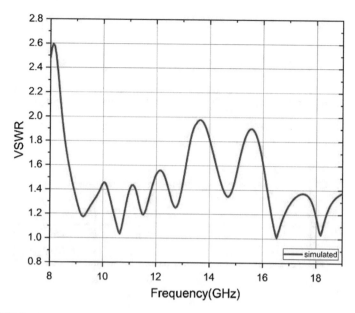

Fig. 6 VSWR plot

4.2 Gain

The center element is stimulated, and all other elements are terminated at 8–19 GHz. At lowest frequency 8 GHz, gain about 3.7 dB and at highest frequency gain about 6.2 dB (Fig. 7).

4.3 Radiation Patterns

The radiation patterns are shown in Fig. 8a–e. That have been simulated between 8 and 19 GHz for both azimuthal (E plane) and elevation plane (H plane). E plane patterns represented in red color and solid line; H plane patterns are determined in green color and dotted line.

5 Experimental Results

First, the antenna is printed on the board, and then, the center element is activated, and all other elements are terminated in the fabrication model is shown in Fig. 9 in which the frequency of operation ranges from 8 to 19 GHz.

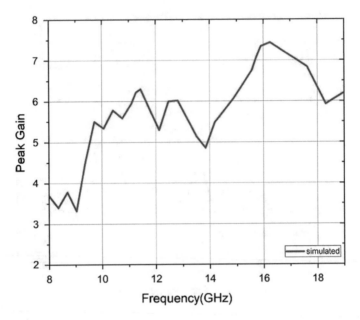

Fig. 7 Gain plot

By using vector network analyzer's designed and simulated antenna tested, observed and measured the antenna parameters, i.e., return loss, gain; Test setup is shown in Fig. 10. As an example of a typical array setting, these figures depict the bottom and front perspectives of a manufactured antenna. The core component is excited, and others are terminating in such an array.

5.1 Measured Return Loss

An antenna's return loss plot is shown in Fig. 11. When the center element is activated and the frequency ranges from 8 to 19 GHz, return loss is less than −10 dB obtained.

5.2 Gain Plot

The gain plot is shown in Fig. 12. Gain variation over 8–19 GHz is depicted, and average gain is around 4 dBi.

Antenna simulations and measurements are shown. Both simulated and experimental findings seem to be identical. It is between 8 and 19 GHz. Both the simulated and the experimental findings are acquired from the HFSS as well as the network analyzer, satisfactorily.

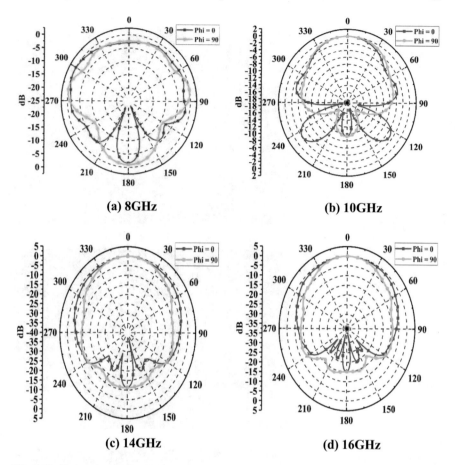

Fig. 8 Radiation patterns for E and H planes

6 Conclusion

In this design involving a rectangle slot and an elliptical cuts. An 8–19 GHz band with return loss less than −10 dB as well as high gain about 4 dB at 19 GHz as well as a gain of 2.5 dB at 11 GHz have been achieved. HFSS software was used to develop the Vivaldi 1 × 5 linear antenna array for X, Ku, and ultra-wide applications. The center element of a five-element linear array is stimulated, while the remaining elements are terminated.

Fig. 9 a 1 × 5 linear array hardware model top view. **b** 1 × 5 linear array hardware model back view

Fig. 10 Test setup

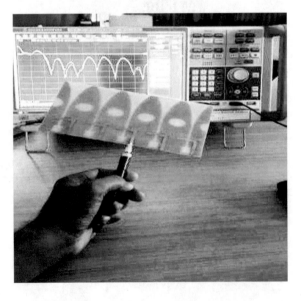

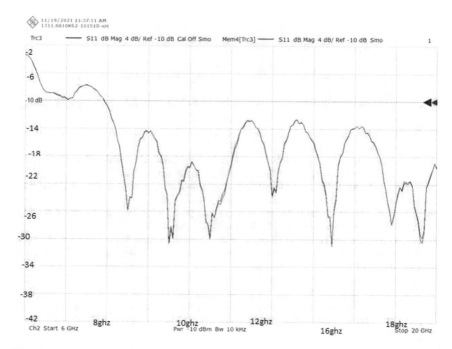

Fig. 11 Measured return loss plot

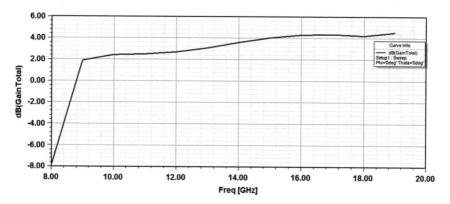

Fig. 12 Gain plot

References

1. Federal Communication Commissions first report and order revision of part 15 of the commissions rule regarding ultrawide band systems, 22 Apr 2002
2. Ma K, Zhao ZQ, Wu JN, Ellis MS, Nie ZP (2014) A printed Vivaldi antenna with improved radiation patterns by using two pairs of eye shaped slots for UWB applications. Prog Electromagn Res 148:63–71

3. Kakhki MB, Mantash M, Kesavan A, Tahseen MM, Denidni TA (2018) Millimeter-wave beam-tilting Vivaldi antenna with gain enhancement using multilayer FSS. IEEE Antennas Wirel Propag Lett 17(12):2279–2283

4. Moosazadeh M, Kharkovsky S (2016) A compact high-gain and front-to back ratio elliptically tapered antipodal Vivaldi antenna with trapezoid shaped dielectric lens. IEEE Antennas Wirel Propag Lett 15:552–555

5. Nassar IT, Weller TM (2015) A novel method for improving antipodal Vivaldi antenna performance. IEEE Trans Antennas Propag 63(7):3321–3324

6. Dastranj A (2015) Wideband antipodal Vivaldi antenna with enhanced radiation parameters. Microw Antennas Propag 9(15):1755–1760

7. Wang W, Zheng Y (2018) Improved design of the Vivaldi dielectric notch radiator with etched slots and a parasitic patch. IEEE Antennas Wirel Propag Lett 17(6):1064–1068

8. Puskely J, Lacik J, Raida Z, Arthaber H High-gain dielectric loaded Vivaldi antennas for Ka-band applications. IEEE Antennas Wirel Propag Lett 15(2)

9. Li X, Liu G, Zhang Y (2017) A compact multi-layer phase correcting lens to improve directive radiation of Vivaldi antenna. Int J RF Microw Comput-Aided Eng 27(7)

10. Zhu S, Liu H, Chen Z, Wen P (2018) A compact gain-enhanced Vivaldi antenna array with suppressed mutual coupling for 5G mm wave application. IEEE Antennas Wirel Propag Lett 17(5):776–779

11. Ahn B-C, Gambo O A 3–20 GHz Vivaldi antenna with modified edge. Applied Electronics Laboratory, Department of Radio Communication Engineering, pp 361–763

12. Sang L, Wa S, Liu G, Wang J, Huang W (2020) High gain UWB Vivaldi antenna loaded with reconfigurable 3-D phase adjusting unit lens. IEEE Antennas Wirel Propag Lett 19(2)

Texture Classification Using ResNet and EfficientNet

Vinat Goyal, Sanjeev Sharma, and Bharat Garg

1 Introduction

The texture is the fundamental unit of an image that aids in its identification. Analysis of texture forms the foundational basis of Computer Vision tasks like image recognition and image segmentation. Images of various domains like satellite [1] and medical [6] have been identifiable because of their texture. The texture is a crucial aspect of Computer Vision. Over the years, many developments have been made in proposing models/ algorithms for texture classification. Earlier methods involved using traditional machine learning methods, which make use of handmade kernels. One of the most earlier filters to be used is the Gabor filter. These kernels are then followed by a statistical classification algorithm like Support Vector Machines (SVMs) in the final layer. They have been very effective and have attained state-of-the-art results.

Deep Learning algorithms have been there for three decades. However, these traditional approaches out shadowed them until the win of AlexNet [11] in the 2012 ImageNet large-scale visual recognition challenge. After this win, Deep Learning saw rapid growth and attention. The two main reasons for this turnaround were the availability of massive data and the availability of more computation power which were missing in the early 2000s.

V. Goyal (✉) · S. Sharma
Indian Institute of Information Technology Pune, Pune, India
e-mail: vinatgoyal19@cse.iiitp.ac.in
URL: https://deeplearningforresearch.com/

S. Sharma
e-mail: sanjeevsharma@iiitp.ac.in

B. Garg
Thapar Institute of Engineering and Technology, Patiala, India
e-mail: bharat.garg@thapar.edu

One of the most famous Deep Learning architects is the Convolutional Neural Network (CNN). CNNs are widely used. They are not restricted to Computer Vision. They are also used in other fields like Natural Language Processing. CNNs have the advantage of sparse connections, weight sharing, and local connectivity over Artificial Neural Networks (ANNs). The initial layers of a CNN learn the image's simple features, and the deeper layers learn the more abstract qualities. Novel convolution neural network models have shown better performance than the classic machine learning algorithms. This paper aims to propose models that would perform better than the previously proposed models.

This paper proposes the use of transfer learning for texture classification. Unlike the other papers, which only use accuracy as the evaluation metric, this paper also uses F1 score, precision, and recall to evaluate the proposed models. Transfer learning is a technique through which a Deep Learning model trained on extensive data can be used on a similar data set. We take a pre-trained model and use it as a feature extraction layer. This feature extraction layer is then connected to a softmax layer. This set-up is then trained on the data set. The pre-trained models used in this paper are EfficientNet-B4and ResNetV2. The presented work focuses on achieving better results on the provided benchmark data set than previous work on the same data set.

Section 2 discusses the literature survey of the related work. Section 4 covers the study of material and methods. Section 4 presents the experiments and results. At last, we are concluding work in Sect. 5.

2 Literature Review

Owing to the uses in different domains that texture classification offers, there has been a lot of research dedicated to it. Over time, researchers have proposed many different methods for this task. One of the earliest papers to do texture analysis was the Picard et al. [14]. It used two powerful algorithms, principal component analysis and multiscale autoregressive models on the Brodatz data set. This approach was later followed by energy-based models, which attained an accuracy of over 90%. One of the earliest methods used for texture classification is the statistical method. They make use of powerful feature extractor descriptors. They are the simplest and yet powerful algorithms. There have been many descriptors proposed. Some of the most widely used are the Local Binary Pattern (LBP), Local Ternary Pattern (LTP) [21], Completed Local Binary Pattern (CLBP) [5] and Completed Local Ternary Pattern (CLTP) [15].

A descriptor called the Feat-WCLTP descriptor is presented in Shamaileh et al. [17]. They tested their descriptor on various data sets, including the Kylberg data set, and got competitive results. Tang et al. [22] propose a multi-threshold corner detection and region matching algorithm based on texture classification to solve the problem of unreasonably dispersed corners in single-threshold Harris detection, as well as the high computational cost of image region matching using the normalised cross correlation (NCC) approach.

Traditional Machine Learnings have got great results in the past. However, they have the disadvantage that these feature extractors need to be handmade. Since the feature extractors are hand-designed for a particular problem, they fail to generalise or adapt to another such data set. Deep Learning models overcome this problem. Zheng et al. [23] proposed an eight feature learning model alongside a Deep Learning perceptron-based architecture. This paper showed the Deep Learning model's advantage over the other model.

CNNs were proposed three decades back by LeCun et al. [13], but didn't get popular then because of lack of data and computational power. One of the major drawbacks of Deep Learning architectures is that they require significantly more data and computational power for training than traditional Machine Learning algorithms. With everything getting digitalised today, there is a lot of data available. Computers have also developed a lot through the years. A modern-day laptop is capable of training a model on thousands of images. All these things have favoured Deep Learning in today's world. The introduction of new and effective optimisation algorithms like the stochastic gradient descent with momentum(SGDM) and RMS prop has also helped the blooming of Deep Learning. Many CNN architectures such as AlexNet [11], VGG [19], ResNet [8] and MobileNet [16] have emerged and are being used widely.

A standard Convolution Neural Network for the task of classification of images of flower and KTH data sets is presented in Simon and Vijayasundaram [18]. This paper achieved excellent results as compared to its predecessors [3] proposes another approach to classification where whale optimisation algorithm (WOA) is used along with the CNN. Results of this model on the Kylberg, Brodatz and Outex data sets are compared to the results obtained by other models on the same data set. This model gained excellent results and beat other models in comparison.

As discussed earlier, the disadvantage of traditional Machine Learning algorithms was that one model trained on a data set could not be used on another data set. Deep Learning overcomes this problem. A deep model trained on a large data set can adapt or generalise to a similar data set. This technique is called transfer learning. This paper makes use of this approach. It helps to save computation time and even helps to propose models when the data available is small.

3 Materials and Methods

Figure 1 depicts the flow chart followed. The first step was to find the problem statement. The following step was to collect the related data set to the problem statement. After the data was collected, it was pre-processed to make it of desirable format and size. The pre-processing stage also included data augmentation, which was done to avoid over-fitting the model. After pre-processing, models were designed for the problem statement and then tested on the pre-processed data set. Transfer learning models are used to classify the different data sets collected. We use the MobileNetV3 and the ResNetV2 models for the classification task.

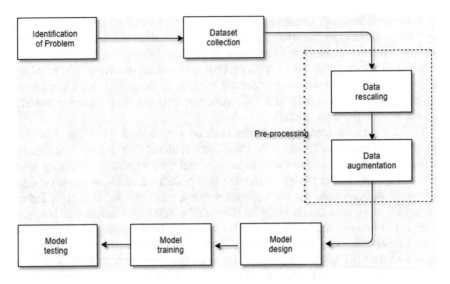

Fig. 1 Flow chart for the proposed method

3.1 Data set

This paper makes use of the standard texture classification data set, the Kylberg data set. Below is a summary of this data set.

3.1.1 Kylberg Data set

Another data set that is used widely for texture classification problems is the Kylberg data set [12]. It is used as a standard benchmark and to evaluate models. This data set has two versions. The first has images with rotation patches, whereas the second has images without rotation patches. In this paper, we have used the data set without rotation patches. Some of the classes of this data set are blanket1, ceiling1, cushion1, floor2, grass1, rice1, rice2, rug1, sand1, scarf1, scarf2,stone3, stoneslab1 and wall1. The samples of this data set are displayed in Fig. 2. The summary of this data set is given in Table 1.

3.2 Data Pre-processing and Splitting

Data pre-processing is the most vital step in any Machine Learning project. It is essential to process the data before feeding it into a model. The images were first converted to a compatible format suitable for the model. After this, the images were

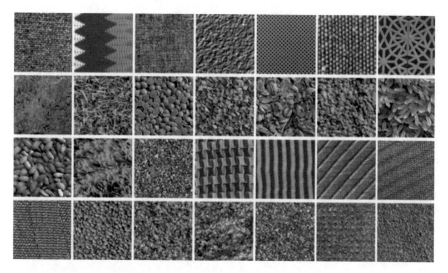

Fig. 2 Samples of the Kylberg data set

Table 1 Summary of the Kylberg data set

Features	Value
Number of classes	28
Number of samples/class	160
Total number of samples	4480
Texture patch size	576*576
Format of image	8 bit grey scale PNG
Total size of data set	1.76 GB

resized to 224*224*3, making them consistent with the pre-trained model we are using. The splitting step follows the pre-processing stage. The data sets are split in a ratio of 80:20 as training and testing data, respectively.

3.3 Data Augmentation

Data augmentation is the process of creating more data from already existing data by flipping the images, rotating the images, etc. Data augmentation is used in this paper.

3.4 Deep Learning Models

This paper makes use of transfer learning. Transfer learning is a famous approach, wherein an already trained model is used for other problems. Intuition uses the knowledge gained by a model on one problem to solve another similar situation. This approach is the perfect example of how Deep Learning models trained on large amounts of data can adapt themselves to new unseen data. This paper uses the ResNetV2 and the EfficientNet-B4 models. For each approach, the last dense layer (classification layer) of the pre-trained model is replaced with a softmax layer suitable for classifying the texture of classes of that data set. Here, we froze all the model layers and trained only the added dense layer. Here, the pre-trained model is only used as a feature extractor for the classifier.

3.4.1 Transfer Learning

Training a model from scratch comes with many constraints like computational power and unavailability of enough data. In such cases, it is advisable to use transfer learning. Transfer learning is an approach, wherein a model trained on a vast data set is used to solve a related problem. In this paper, we have used TensorFlow Hub to import such pre-trained models without their top layers. A softmax layer is then added to these layers. Only the last layer is trained.

3.4.2 ResNetV2

In Deep Learning modes, network depth plays an important role. Stacking more conventional layers to increase depth can easily result in problems of vanishing gradients. This problem is addressed by ResNet, which proposes a residual block with a shortcut connection solving the issue of vanishing gradients. This model obtained a first place at ILSVRC2015 [7]. A ResNet model with a depth of up to 152 layers, eight times deeper than a 19-layer VGG model, has lesser complexity and better accuracy than a VGG mode. He et al. [9] introduce the V2 variant of this model, which is being used in this paper. TensorFlow Hub is used to use the ResNetV2 model trained on ImageNet (ILSVRC-2012-CLS) data set. The model is used as feature extraction for the Kylberg data set. Below is the summary of the ResNetV2 model Fig. 3.

3.4.3 EfficientNet-B4

EfficientNets [20] are a family of models and scaling methods that uniformly scales all dimensions of depth/width/resolution using a compound coefficient. Unlike conventional practice that arbitrary scales these factors, the EfficientNet scaling method

Layer (type)	Output Shape	Param #	Connected to
input_3 (InputLayer)	[(None, 224, 224, 3)	0	
conv1_pad (ZeroPadding2D)	(None, 230, 230, 3)	0	input_3[0][0]
conv1_conv (Conv2D)	(None, 112, 112, 64)	9472	conv1_pad[0][0]
pool1_pad (ZeroPadding2D)	(None, 114, 114, 64)	0	conv1_conv[0][0]
pool1_pool (MaxPooling2D)	(None, 56, 56, 64)	0	pool1_pad[0][0]
...	...	...	...
...	...	...	...
conv5_block3_2_relu (Activation	(None, 7, 7, 512)	0	conv5_block3_2_bn[0][0]
conv5_block3_3_conv (Conv2D)	(None, 7, 7, 2048)	1050624	conv5_block3_2_relu[0][0]
conv5_block3_out (Add)	(None, 7, 7, 2048)	0	conv5_block2_out[0][0] conv5_block3_3_conv[0][0]
post_bn (BatchNormalization)	(None, 7, 7, 2048)	8192	conv5_block3_out[0][0]
post_relu (Activation)	(None, 7, 7, 2048)	0	post_bn[0][0]

Total params: 23,564,800
Trainable params: 23,519,360
Non-trainable params: 45,440

Fig. 3 Architecture of ResNetV2 model

uniformly scales network width, depth, and answer with a set of fixed scaling coefficients. Here, we use the EfficientNet-B4 model. TensorFlow Hub is used to operate the EfficientNet-B4 model trained on ImageNet (ILSVRC-2012-CLS) data set. The model is used as feature extraction for the Kylberg data set. Below is the summary of the EfficientNet-B4 model Fig. 4.

4 Experiments and Results

Here to simulate the results Tesla K80 GPU and 13 GB RAM used for training along with TensorFlow, Keras and Scikit-learn libraries in Google Colab, coded in Python 3.7.10. For training and testing data, the Kylberg data set is split into training data (80%) and testing data (20%). Adam optimisation and categorical cross-entropy loss functions are used in all cases. A learning rate of 0.01 has been used. In all the circumstances, the pre-trained model is used as a feature vector, and only the top added layer is trained on the training data.

Layer (type)	Output Shape	Param #	Connected to
input_2 (InputLayer)	[(None, 224, 224, 3)	0	
rescaling_1 (Rescaling)	(None, 224, 224, 3)	0	input_2[0][0]
normalization_1 (Normalization)	(None, 224, 224, 3)	7	rescaling_1[0][0]
stem_conv_pad (ZeroPadding2D)	(None, 225, 225, 3)	0	normalization_1[0][0]
stem_conv (Conv2D)	(None, 112, 112, 48)	1296	stem_conv_pad[0][0]
...	...	...	...
...	...	...	...
block7b_drop (Dropout)	(None, 7, 7, 448)	0	block7b_project_bn[0][0]
block7b_add (Add)	(None, 7, 7, 448)	802816	block7b_drop[0][0] block7a_project_bn[0][0]
top_conv (Conv2D)	(None, 7, 7, 1792)	7168	block7b_add[0][0]
top_bn (BatchNormalization)	(None, 7, 7, 1792)	0	top_conv[0][0]
top_activation (Activation)	(None, 7, 7, 1792)	0	top_bn[0][0]

Total params: 17,673,823
Trainable params: 17,548,616
Non-trainable params: 125,207

Fig. 4 Architecture of EfficientNetB4 model

4.1 Evaluation Criteria

In the prediction phase, accuracy is the quantitative performance measures computed to assess the reliability of trained models. This metric is calculated based on True Positive (TP), True Negative (TN), False Positive (FP), False Negative (FN).

$$\text{Precision} = \frac{TP}{TP + FP} \tag{1}$$

$$\text{Recall} = \frac{TP}{TP + FN} \tag{2}$$

$$\text{F1Score} = 2 * \frac{\text{Precision} * \text{Recall}}{\text{Precision} + \text{Recall}} \tag{3}$$

$$\text{Accuracy} = \frac{TP + TN}{TP + FN + TN + FP} \tag{4}$$

Weighted avg $= F1class1 * W1 + F1class2 * W2 + F1class3 * W3 + \cdots + F1classn * Wn$

$$(5)$$

F1classm : F1 score of class m

Macro avg $= F1class1 + F1class2 + F1class3 + \cdots + F1classn \quad (6)$

F1classm : F1 score of class m Cohen kappa score:

$$K = \frac{p0 - pe}{1 - pe} \tag{7}$$

p0 = relative observed agreement among raters, pe = the hypothetical probability of chance agreement.

4.2 Training Single Convolution Model

All the images in the .gif or the .ras format were converted to a compatible format. After that, all the images of the two data sets in the study were rescaled to 224*224. The images were then normalised to make the values of their pixels range from 0 to 1. Following this, data sets were subjected to data augmentation before passing them to the proposed model.

4.2.1 ResNetV2 Model

The first model is developed using the ResNetV2 model, trained on the ImageNet data set. The top layer of the pre-trained model is removed and replaced by a softmax layer with 28 classes. Only the top layer of the model was trained on the training data set, keeping the rest of the layers of the model frozen. The proposed model was trained for ten epochs on the training data set. The model achieved an accuracy of 99.78% on the testing data set. The classification report is given in Table 2. The confusion matrix of ResNetV2 on testing it on testing data is shown in Fig. 5. The accuracy vs epochs graph and the loss vs epochs graph of ResNetV2 for the data set while training is shown in Fig. 6

Table 2 Classification report for model 1 Kylberg data set

	Precision	Recall	f1-score	Support
Accuracy	–	–	1.0	896
Macro Avg	1.0	1.0	1.0	896
Weighted Avg	1.0	1.0	1.0	896

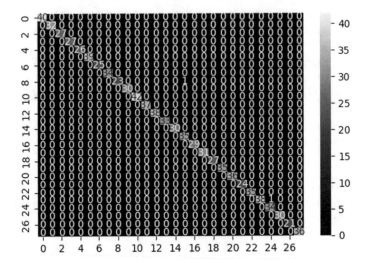

Fig. 5 ResNetV2 confusion matrix for the Kylberg data set

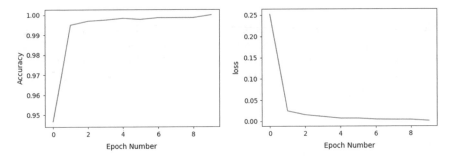

Fig. 6 ResNetV2 accuracy and losses graph for the Kylberg data set

4.2.2 EfficientNet-B4 Results

The second model is developed using the EfficientNet-B4 model trained on the ImageNet data set. The top layer of the pre-trained model is removed and replaced by a softmax layer with 28 classes. The pre-trained model was used as a feature extractor, i.e. all the layers of the pre-trained model were frozen, and only the top layer was trained on the training data set. The proposed model was trained for ten epochs on the training data set. The model achieved an accuracy of 92.97% on the testing data set. The classification report is shown in Table 3. The confusion matrix of EfficientNet-B4 on testing it on testing data is shown in Fig. 7. The accuracy vs epochs graph and the loss versus epochs graph of EfficientNet-B4 for the Kylberg data set while training is shown in Fig. 8

Table 3 Classification report for EfficientNet-B4 data set

	Precision	Recall	f1-score	Support
Accuracy	–	–	0.93	896
Macro Avg	0.94	0.93	0.93	896
Weighted Avg	0.94	0.93	0.93	896

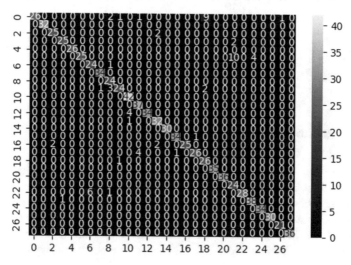

Fig. 7 EfficientNet-B4 confusion matrix for the Kylberg data set

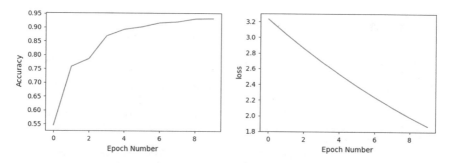

Fig. 8 EfficientNet-B4 accuracy and losses graph for the Kylberg data set

4.3 Comparative Study

The results of the 2 proposed models are compared with other recently proposed models. Table 4 shows the comparison between the two proposed models and other recently applied models on the Kylberg data set.

Table 4 Performance comparison of our models with the existing techniques for the Kylberg data set

References	Model/technique	Classification accuracy (%)
[2]	T-CNN-3	99.4 ± 0.2
[10]	KNN+nLBP(d=1)	99.64
[4]	RALBGC	99.23
[10]	LBP	97.97
[3]	Modifed CNN+WOA	99.71
Proposed model 1	ResNetV2-50	99.78
Proposed model 2	EfficientNet-B4(Feature extraction)	92.97

5 Conclusion and Future Scope

As stated, texture classification is an important area of research that has attracted many researchers to propose different models. From the comparative study, it can be seen that our models have given better results than most of the existing models for the Kylberg data set. In future, we would like to test our models on more texture data sets and even use them for other domains like medical and forestry imagery. We also wish to develop further our models to attain better accuracy for the Kylberg data set.

References

1. 1972 Index IEEE Transactions on Computers 21(12):1475. https://doi.org/10.1109/T-C.1972. 223532
2. Andrearczyk V, Whelan P (2016) Using filter banks in convolutional neural networks for texture classification. Pattern Recogn Lett 84:63–69. https://doi.org/10.1016/j.patrec.2016.08.016
3. Dixit U, Mishra A, Shukla A, Tiwari R (2019) Texture classification using convolutional neural network optimized with whale optimization algorithm. SN Appl Sci 1(6):655. https://doi.org/10.1007/s42452-019-0678-y
4. El Khadiri I, Kas M, El Merabet Y, Ruichek Y, Touahni R (2018) Repulsive-and-attractive local binary gradient contours: new and efficient feature descriptors for texture classification. Inf Sci 467:634–653 https://doi.org/10.1016/j.ins.2018.02.009, https://www.sciencedirect.com/science/article/pii/S0020025518300896
5. Guo Z, Zhang L, Zhang D (2010) A completed modeling of local binary pattern operator for texture classification. IEEE Trans Image Process 19(6):1657–1663. https://doi.org/10.1109/TIP.2010.2044957
6. Haralick RM, Shanmugam K, Dinstein I (1973) Textural features for image classification. IEEE Trans Syst Man Cybern SMC-3(6):610–621 (1973). https://doi.org/10.1109/TSMC.1973.4309314
7. He K, Zhang X, Ren S, Sun J (2015) Deep residual learning for image recognition
8. He K, Zhang X, Ren S, Sun J (2016) Deep residual learning for image recognition. In: 2016 IEEE conference on computer vision and pattern recognition (CVPR), pp 770–778. https://doi.org/10.1109/CVPR.2016.90

9. He K, Zhang X, Ren S, Sun J (2016) Identity mappings in deep residual networks
10. Kaya Y, Ertuğrul OF, Tekin R (2015) Two novel local binary pattern descriptors for texture analysis. Appl Soft Comput 34(C):728–735. https://doi.org/10.1016/j.asoc.2015.06.009, https://doi.org/10.1016/j.asoc.2015.06.009
11. Krizhevsky A, Sutskever I, Hinton GE (2012) Imagenet classification with deep convolutional neural networks. In: Proceedings of the 25th international conference on neural Iinformation processing systems, vol 1. NIPS'12, Curran Associates Inc, Red Hook, NY, USA, pp 1097–1105
12. Kylberg G (2011) The kylberg texture dataset v. 1.0. External report (Blue series) 35, Centre for image analysis, Swedish University of Agricultural Sciences and Uppsala University, Uppsala, Sweden. http://www.cb.uu.se/gustaf/texture/
13. LeCun Y, Boser B, Denker JS, Henderson D, Howard RE, Hubbard W, Jackel LD (1989) Backpropagation applied to handwritten zip code recognition. Neural Comput 1(4):541–551. https://doi.org/10.1162/neco.1989.1.4.541
14. Picard R, Kabir T, Liu F (1993) Real-time recognition with the entire brodatz texture database. In: Proceedings of IEEE conference on computer vision and pattern recognition, pp 638–639. https://doi.org/10.1109/CVPR.1993.341050
15. Rassem TH, Khoo BE (Apr2014) Completed local ternary pattern for rotation invariant texture classification. Sci World J 2014:373254. https://doi.org/10.1155/2014/373254
16. Sandler M, Howard A, Zhu M, Zhmoginov A, Chen LC (2019) Mobilenetv2: inverted residuals and linear bottlenecks
17. Shamaileh AM, Rassem TH, Chuin LS, Sayaydeh ONA (2020) A new feature-based wavelet completed local ternary pattern (feat-wcltp) for texture image classification. IEEE Access 8:28276–28288. https://doi.org/10.1109/ACCESS.2020.2972151
18. Simon P, Vijayasundaram U (2020) Deep learning based feature extraction for texture classification. Proc Comput Sci 171:1680–1687. https://doi.org/10.1016/j.procs.2020.04.180
19. Simonyan K, Zisserman A (2015) Very deep convolutional networks for large-scale image recognition. In: International conference on learning representations
20. Tan M, Le QV (2020) Efficientnet: rethinking model scaling for convolutional neural networks
21. Tan X, Triggs B (2010) Enhanced local texture feature sets for face recognition under difficult lighting conditions. IEEE Trans Image Process 19(6):1635–1650. https://doi.org/10.1109/TIP.2010.2042645
22. Tang Z, Ding Z, Zeng R, Wang Y, Wen J, Bian L, Yang C (2019) Multi-threshold corner detection and region matching algorithm based on texture classification. IEEE Access 7:128372–128383. https://doi.org/10.1109/ACCESS.2019.2940137
23. Zheng Y, Zhong G, Liu J, Cai X, Dong J (2014) Visual texture perception with feature learning models and deep architectures, vol 483. https://doi.org/10.1007/978-3-662-45646-0_41

Performance Analysis of Parametric and Non-parametric Classifier Models for Predicting the Liver Disease

Vijayalakshmi G. V. Mahesh, M. Mohan Kumar, and Alex Noel Joseph Raj

1 Introduction

Liver being the largest organ plays a significant role in various functions of the body such as blood clotting, protein production, enzyme activation, blood detoxification, bile production, and metabolism of carbohydrates, proteins, and fats. Liver also has the ability to replace the damaged cells. Taking into consideration of all the functions, it is very important to maintain a healthy liver as it indicates overall well-being of the body such as increased energy level, hormonal balance, healthier skin, better immune system, reduced allergies, and good emotional health.

The functionality of the liver can be affected due to excessive intake of drugs, supplements, alcohol, variations in metabolism, virus and genetic abnormalities which may affect the liver and leads to diseases which can be infectious and noninfectious. These include hepatitis (viral infection), cirrhosis (scarring of liver), fatty liver, cancer, Wilson's disease (abnormalities in metabolism), hemochromatosis (excess iron), and overdose of acetaminophen also causes liver failure.

The diseases are identified and diagnosed by the health experts or clinicians after examining the symptoms and the risk factors manually. With manual process, inaccuracies may occur that can be overcome by designing a machine learning-based computer-aided diagnostic system which can provide effectual results to improve

V. G. V. Mahesh
Department of ECE, BMS Institute of Technology and Management, Bangalore, India
e-mail: vijayalakshmi@bmsit.in

M. M. Kumar (✉)
Department of ECE, Yenepoya Institute of Technology, Moodbidri, India
e-mail: mohankumar@yit.edu.in

A. N. J. Raj
Department of Electronic Engineering, Shantou University, Shantou, China
e-mail: jalexnoel@stu.edu.cn

© The Author(s), under exclusive license to Springer Nature Singapore Pte Ltd. 2023
D. S. Sisodia et al. (eds.), *Machine Intelligence Techniques for Data Analysis and Signal Processing*, Lecture Notes in Electrical Engineering 997,
https://doi.org/10.1007/978-981-99-0085-5_16

the healthcare solutions. Biomarkers play a major role in identifying the key characteristics of the disease using (i) imaging techniques and (ii) blood sample analysis.
The imaging modalities such as computed tomography scans (CT), ultrasound and
magnetic resonance images (MRI) assess the liver conditions effectively. Further the
functionality tests of the liver: alanine transaminase (ALT), aspartate aminotrans-
ferase (AST), alkaline phosphatase (ALP), albumin, and bilirubin are performed
on the blood samples to decide the health of the liver by measuring the levels of
bilirubin, enzymes, and protein. These key characteristics form the patterns, and
when provided to machine learning algorithms, the algorithm learns from the data
and assists the clinicians with the diagnosis. This paper proposes a computer-aided
diagnostic method that uses the parameters measured from the liver functionality
tests as the descriptors with parametric and non-parametric machine learning algorithms to predict the presence of the liver disease. The algorithms are quantitatively
assessed using the metrics computed form the confusion matrix or contingency table.

The rest of the paper is structured as follows: Sect. 2 provides a brief review of
the state of the art methods. Section 3 reports the methodology or the process flow
of the proposed work. The simulation results and discussion are provided in Sect. 4.
Finally, Sect. 5 concludes the paper.

2   Related Work

In recent years, various experiments and analysis have been carried out considering different input modes such as imaging methods and biomarkers with machine
learning techniques to identify or predict liver diseases. Tanwar and Rahman [1]
provided a review on the progress and future trends of utilizing machine learning in
assisting physicians or clinical experts to make accurate and also timely decisions in
precisely predicting and diagnosing liver diseases. Wu et al. [2] proposed a model
to predict fatty liver disease with high risk using machine learning methods to assist
clinicians in diagnosis. The work utilized the abdominal ultrasonography images
obtained from 577 patients from New Taipei City Hospital. From these images,
nine descriptors: systolic blood pressure, HDl-C, abdominal growth, diastolic blood
pressure, glucose AC, SGOT-AST, triglyceride and SGPT-ALT were extracted and
provided to random forest (RF), artificial neural networks (ANNs), Naive Baye's
(NB), and logistic regressions (LRs) machine learning classifier models, and the
performance was validated with tenfold cross-validation with metrics AUC and accuracy. The analysis on results proved the efficiency of random forest classifier by
providing an accuracy of 87.48% and AUC of 0.925. Sartakhi et al. [3] investigated
the efficacy of using stochastic simulated annealing method for optimization with
SVM for liver disease prediction. The authors verified the proposed method on the
dataset extracted from UCI machine learning database which provided an accuracy
of 96.25% through tenfold cross-validation. Singh et al. [4] presented an intelligent
liver disease healthcare software to predict the liver disease based on the symptoms.

The development of the software included continuous interactions with the physicians to deliver accurate results. The implementation included feature selection and transformation, disease prediction using classifiers and performance evaluation. The proposed work was validated on Indian Liver Patient Dataset with predictors: age, direct bilirubin, gender, total bilirubin, SGPT, Alkphos, albumin, SGOT, and globulin ratio and LR, SMO, RF, NB, J48, and k-nearest neighbor (KNN) classifiers. The comparison of the performance of the classifiers showed the better accuracy of LR which provided 74.36% with feature selection. Liu et al. [5] explored the potential of using gut metagenomic data with machine learning to predict incident and alcoholic liver disease and the associated risk factors. The experimental result analysis of gradient boosted classifiers with performance metrics of average area under curve (AUC) provided a value of 0.834 and 0.956 for incident and alcoholic liver disease, respectively. The performance proved that the gut microbial predictors play a vital role in disease prediction. Naseem et al. [6] studied the performance of ten classifiers to find the optimal model for early detection of liver disease. The classifiers included AIDE, MLP, NB, SVM, CHIRP, KNN, CDT, Forest-PA, RF, and J48. The dataset for the proposed methodology was accessed from UCI machine learning database and GitHub repository. The performance of was assessed through the measures accuracy, F1-score, G measure, precision, specificity, recall, RRSE, and RMSE. The analysis indicated the better performance of RF with UCI and SVM with GitHub datasets with an accuracy of 72.173% and 71.355%, respectively. Singh et al. [7] worked with SVM, KNN, and LR machine learning algorithms for early detection of liver disease using the Indian Liver Patient Dataset. The dataset had 567 samples with 10 predictors representing each sample. The proposed method was tenfold cross-validated to obtain an accuracy of 73.97% for both KNN and LR classifiers. A study on the use of non-invasive biomarkers to determine fibrosis due to chronic alcoholic liver disease was conducted by [8]. The study was conducted on 221 patients whose alcohol intake is greater than 50 g per day with an examination on liver biopsy and FibroTest Fibro-Sure. The presence of fibrosis was examined using five stage histologic scale. Based on the analysis, the system suggests the screening for fibrosis at an age of 40. The survey on the research findings of predicting liver diseases emphasized on the use machine learning algorithms for prediction and classification.

3 Methodology

The proposed work aims at presenting the performance analysis of parametric and non-parametric supervised machine learning algorithms for liver disease prediction on the data accessed from UCI machine learning repository. The dataset is preprocessed to make it balanced and provided to the machine learning algorithms to learn the relation between the predictors/features and the responses/class labels to build the classifier models for disease prediction. Further the models are assessed using the metrics computed from confusion matrices. The implementation structure of the proposed work is shown in Fig. 1.

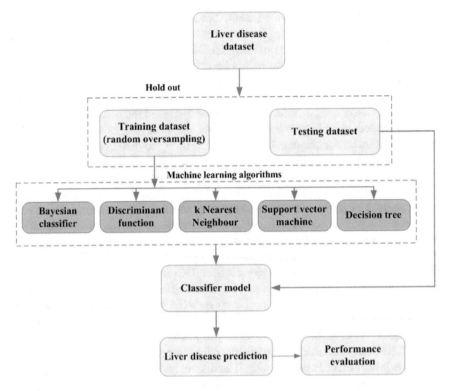

Fig. 1 Implementation structure of the work

3.1 Dataset

The work utilized the Indian Liver Patient Dataset obtained from the UCI machine learning repository [9]. The dataset is created from the observations and measurements of protein levels, liver enzymes, and bilirubin made from the blood samples recorded from 583 individuals of which 441 are male and 142 are female. These measurements are important and recommended by the clinicians to know the health of the liver. They are numerical and form the predictors or attributes that play a key or decisive role in predicting the presence of liver disease. Among 583 records, 416 were collected from patients with liver disease, and the remaining 167 are the details of non-liver disease patients. The attributes are labeled with '1' and '2.' Here, '1' indicates liver disease (LD), and '2' represents non-liver disease (NLD). The details of the attributes and their description are presented in Table 1.

Table 1 Attributes and descriptions

S. No	Attribute	Description
1	AG	Age
2	TB	Total Bilirubin: The amount of bilirubin present in the blood is indicative of the proper functionality of the liver. Smaller amount of it is normal while a high level is a warning of liver disease. The regular range of total bilirubin is 0.1–1.2 mg/dL
3	DB	Direct Bilirubin: This is also referred to as conjugate bilirubin. The level of direct bilirubin is regularly high due to various reasons like alcohol intake, autoimmune disorders, etc. Under normal conditions, it ranges from 0 to 0.3 mg/dL. A higher is a sign of liver disorder
4	ALP	Alkaline Phosphatase (ALP): It is an enzyme present in the blood, and one of the main source of it is liver. ALP level is an indicator for identifying the liver functionality. The regular range of ALP if from 20 to 140 IU/L, but it may vary depending upon gender, type of blood, and age. An unusual value indicates trouble with the liver
5	ALT	Alamine Aminotransferase (ALT): It is the enzyme created in liver, but when the liver is damaged, the enzyme is freed into blood. Thus, identifying and measuring ALT is critical to find the health of the liver. The normal value of ALT in blood ranges from 19 to 25 IU/L for females and 29–33 IU/L for males. Like ALP, ALT also varies with age and gender
6	AST	Aspartate Aminotransferase (AST): A lower level of AST in blood represents the normal functioning of the liver while a higher value of this enzyme is associated with liver damage such as cirrhosis or chronic hepatitis. The typical range of AST is 10–40 U/L for males and 9–32 U/L
7	TP	Total proteins: Total protein test is performed to observe and measure the total amount of albumin and globulin present in the body. This requires the analyses of blood sample. The total protein ranges from 6 to 8.3 g/dL but is variable with respect to age and gender. The abnormal levels (elevated or low) of total protein are a caution of liver disease
8	A	Albumin: Is one of important proteins the body requires present in blood. For proper functioning, a balanced level of albumin is required. The typical range of albumin in blood ranges from 3.4 to 5.4 g/dL. A lower level is a warning of liver disorder
9	A/G	Albumin and Globulin ratio: The ratio is significant in identifying the liver disorder such as cirrhosis. Normally, the ratio A/G is slightly greater than 1. A higher or lower ratio is considered to be abnormal and is sign liver disease

The dataset is unbalanced as the distribution of samples are skewed: where class 1 has majority of the samples and class 2 is minority with few samples. Handling this data is more challenging as the classifier models that predict the class label of a given observation may result in poor predictive performance due to the following reasons: (i) Most of the machine learning algorithms for classification are designed assuming equal distribution of samples in the classes. (ii) As the samples in minority class are fewer, it is challenging for the classifier model to learn the characteristics from the samples provided to it during training. (iii) With a large number of samples present in the majority class, the model may focus on the observations of majority class during training. The challenges are addressed by random oversampling the minority class samples during the training process for balanced classification.

3.2 Machine Learning Algorithms

In pattern recognition process, the classifiers are designed using the observations and the prior probabilities to compute the posterior probabilities. Later, the decision boundary for classification is formulated based on the posterior probabilities.

Parametric methods

These methods rely on probability distributions and approximate parameters such as mean and standard deviation from the distributions to provide firm representation of the classes [10, 11]. The algorithms under parametric methods have fixed number of parameters and are computationally faster. The algorithms include linear discriminant analysis (DF) and Bayesian classifier (BC). These methods rely on probability distributions and approximate.

Non-parametric methods

These methods are popularly known as distribution-free methods. Unlike parametric models, they do not have fixed number of parameters, but the parameters grow depending on the size of the training data which indicate the complexity of the models [10, 11]. The non-parametric models work without prior knowledge about the data, and they make few assumptions about the data. These methods include k-nearest neighbors (KNN), decision trees (DTs), and support vector machine (SVM) with RBF kernels.

3.3 Classifier Performance Evaluation

The performance of both parametric and non-parametric classifier models is estimated quantitatively using various metrics computed using the entrants of confusion matrix (CM) [12] or contingency table. The confusion matrix indicates the ability of the classifier in predicting the class labels of a sample and is a table with rows representing the actual class and columns predicted class. The elements of the CM are true positive, true negative, false positive, and false negative (Table 2).

Table 2 Confusion matrix

Actual class	Predicted class	
	True positive (TP)	False negative (FN)
	False positive (FP)	True negative (TN)

For binary classification, one of the class is considered as positive, and other one is negative. The positive results obtained from testing are tabulated in column 1, and negative results are entered in other column. With this, the entries of CM indicate the following. With this, the entries of CM indicate the following:

True Positive: for a positive sample if the prediction is positive
True Negative: actual negative sample predicted as negative
False Positive: negative sample predicted as positive
False Negative: positive sample predicted as negative.

The performance metrics computed for performance evaluation are shown in Table 3.

For disease prediction, FPs and FNs play an imperative role as they provide FPR or Type I error and FNR or Type II error. (i) FPs may trigger worry and unnecessarily direct the person toward receiving the medical treatment and (ii) FNs are more dangerous as it provides delusion of good health to the patient when actually the treatment is essential. Further chi-square test was done and statistical measures like: chi-square value, Kappa, Cramer's V, and confidence intervals (CI) of 95% were computed from the contingency table [13, 14] for evaluation of classifiers. The statistics computed for performance evaluation from chi-square test is shown in Table 4. The chi-square value, Kappa, Cramer's V, and confidence intervals (CI) of 95% are important to evaluate the performance by interpreting the confusion matrix.

Table 3 Performance metrics computed from confusion matrix

S. No	Performance metric	Equation
1	True positive rate (TPR)/sensitivity	$TP/(TP + FN)$
2	True negative rate (TNR)/specificity	$TN/(FP + TN)$
3	Precision	$TP/(TP + FP)$
4	Negative predicted value (NPV)	$TN/(TN + FN)$
5	False positive rate (FPR)	$FP/(FP + TN)$
6	False discovery rate (FDR)	$FP/(FP + TP)$
7	False negative rate (FNR)	$FN/(FN + TP)$
8	Accuracy (ACC)	$(TP + TN)/(TP + TN + FP + FN)$
9	F1-score	$(2 \times \text{Precision} \times \text{Recall})/(\text{Precision} + \text{Recall})$
10	Matthews correlation coefficient (MCC)	$(TP \times TN - FP \times FN)/\sqrt{((TP + FP) \times (TP + FN) \times (TN + FP) \times (TN + FN))}$

Table 4 Chi-square test statistics computed from confusion matrix

S. No	Performance metric	Equation
1	Chi-square (χ_c^2)-Pearson	$\sum \frac{(O_i - E_i)^2}{E_i}$, where c indicates degrees of freedom (Number of classes-1; O and E are the observed and expected values, respectively
	Chi-square(χ_c^2)-Yates	$\sum \frac{(\max(0,|O_i - E_i|-0.5)^2}{E_i}$
2	Kappa	(Po − Pe)/(1 − Pe), with Po = Accuracy (ACC) Pe = ((TP + FN) × (TP + FP) + (FP + TN) × (FN + TN))/(TP + TN + FP + FN)^2
3	Cramer's V ($\emptyset_c$)	$\sqrt{\frac{\chi_c^2}{N_s(k-1)}}$, where χ_c^2, indicates Pearson chi-square value; N_s, the number of samples and k, number of class labels/categories
4	Confidence intervals (CI) of 95%	$\exp(\text{lor} \pm Z/2 \times \sqrt{\frac{1}{TP} + \frac{1}{FP} + \frac{1}{FN} + \frac{1}{TN}})$, where lor is the log odds ratio given by $\log_e\left(\frac{TP*TN}{FP*FN}\right)$, $Z\alpha/2$ is the critical value of the normal distribution at $\alpha/2$ For a CI of 95% $\alpha = 0.05$ and $Z\alpha/2 = 1.96$

4   Experimental Results and Discussion

This section provides the details about the experimental results and analysis performed on the liver disease dataset for disease prediction. The dataset for the work was obtained from UCI machine learning repository. The dataset provides 8 biomarkers obtained during blood analysis for 583 individuals. Thus, the dataset has 583 samples of which 416 belong to patients with liver disease and 167 samples are from non-liver disease patients. These biomarkers along with age of the person are considered as the attributes or features that characterize the disease. The quality of the features decides their discriminating ability in identifying the samples of different classes.

The proposed framework is evaluated using hold out method where the dataset is randomly split into two exclusive datasets: training and testing datasets. Training dataset is used to create the classifier model while the testing dataset is used for assessing the performance of the model. As the dataset considered here is unbalanced with class 1 (LD) containing 79% and class 2 holding 29% of the samples, respectively, the classifier may provide poor performance. Thus, the methodology includes a module where the minority class is randomly oversampled to obtain the balanced dataset (1:1) during the training process. Consequently, 60% of the dataset is considered for training, and 40% is held for testing.

The training dataset is labeled and provided to both parametric: Bayesian classifier, linear, quadratic, pseudolinear, pseudoquadratic discriminant function, and non-parametric: KNN, SVM (Linear and RBF kernels), and DT-CART algorithms for model creation. The algorithms learned from the samples and found the model that minimized the cost function. The models are tested, and the performance is evaluated using the performance metrics sensitivity, specificity, NPV, FPR, FDR, FNR, ACC, F1-score, and MCC. From the results obtained from both parametric and non-parametric classifier models, confusion matrices are framed for each classifier. From the elements of the matrices, the performance metrics sensitivity, specificity, NPV, FPR, FDR, FNR, ACC, F1-score, and MCC are calculated. The same is shown in Table 5.

Further, the performance metrics of both the classifier categories are compared. The comparison in regard to accuracy, FNR and FPR are displayed in Fig. 2. From Fig. 2, it can be noticed that parametric algorithms have performed well by providing high accuracy, low FPR, and low FNR indicating less miss classifications. Further the value of MCC is observed which normally lies between $[-1 \; +1]$. With a good model MCC $= +1$ whereas a poor model gives MCC $= -1$. The results in table show that the value of MCC is greater than 0.69 for parametric classifiers also a value of 0.9189 obtained for SVM-RBF indicates its better performance in discriminating the samples with liver disease and without liver disease.

Later, the analysis is continued with the chi-square value, Kappa, Cramer's V, and confidence intervals (CI) of 95% obtained from all the classifier models. Table 6 indicates the metrics obtained from both parametric and non-parametric classifier models. The performance considered Yates and Pearson chi-squares for analyzing the correlation between two categories of data. Under a chi-square test, it is important to notice the difference between the observed and expected outcomes after testing the classifier model and the p value obtained. The results obtained from the testing set of the proposed work provided a p value less than 0.0001 with degrees of freedom $=$ 1 indicating a significance association between the target and predicted class labels.

Later, the measures of association between the target and predicted class labels were found using Cramer's V and Kappa value which actually lie in the range of $[0 \; 1]$. The results obtained indicate that the Cramer's V and Kappa value obtained from parametric classifiers is greater than 0.68 indicating a high association. Also, the particular values of Cramer's V $= 0.9189$ and Kappa $= 0.916$ were provided by the SVM-RBF classifier that indicated the excellent association.

Finally, the performance of the classifier models was analyzed using the most widely used 95% confidence interval which is clinically significant. The CI indicates the magnitude of difference between classifier models. The plot of CI of 95% displayed in Fig. 3 indicates the better performance of SVM-RBF classifier. The results and their analysis considering various performance metrics calculated from confusion matrix indicated the better performance of parametric class of classifiers-KNN, SVM, and DT-CART.

Table 5 Performance analysis of results from parametric algorithms

Performance metrics	BC	DF-linear	DF-quadratic	DF-pseudolinear	DF-pseudoquadratic	kNN	SVM-linear	SVM-Rbf	DT-CART
Sensitivity (%)	45.78	58.4	45.97	58.87	48.39	72.29	81.13	100	75.9
Specificity (%)	96.39	88.71	95.2	86.4	97.6	98.8	87.95	91.57	92.77
Precision (%)	92.68	83.91	90.48	81.11	95.24	98.36	81.13	92.22	91.3
NPV (%)	64	67.9	63.98	67.92	65.59	78.1	87.95	100	79.38
FPR (%)	3.61	11.29	4.8	13.6	2.4	1.2	12.05	8.43	7.23
FDR (%)	7.32	16.09	9.52	18.89	4.76	1.64	18.87	7.78	8.7
FNR (%)	54.22	41.6	54.03	41.13	51.61	27.71	18.87	0	24.1
ACC (%)	71.08	73.49	76.08	72.69	73.09	85.54	85.29	95.78	84.34
F1-score	0.613	0.6887	0.6096	0.6822	0.6417	0.834	0.812	0.959	0.829
MCC	0.489	0.494	0.4735	0.4712	0.5289	0.738	0.691	0.919	0.697

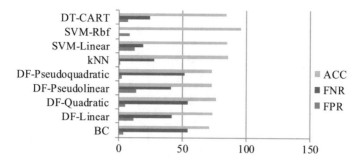

Fig. 2 Comparison of FNR and FPR of both parametric and non-parametric classifier models and accuracy provided by both parametric and non-parametric classifier models

5 Conclusion

This paper proposes a framework to predict the presence of liver disease using machine learning algorithms. The presented work utilized the dataset obtained from Indian Liver Patient Dataset which has 583 samples with liver disease and non-liver disease. Each sample of the dataset is described using the parameters measured during the series of blood sample tests. These parameters form the features or attributes that are fed to both parametric (Bayesian classifier, Discriminant function) and non-parametric classifiers (k-nearest neighbor, support vector machine, decision tree) to learn the data and derive the classifier models. The classifiers models are validated, and their performance is assessed using the metrics computed from the confusion matrix. The analysis of the results obtained from parametric and non-parametric classifiers demonstrated the better performance of non-parametric classifiers in discriminating the sample with liver disease from non-liver disease with lower error rates.

Table 6 Chi-square statistics obtained from parametric and non-parametric classifier models

Performance metrics	BC	DF-linear	DF-quadratic	DF-pseudolinear	DF-pseudoquadratic	kNN	SVM-linear	SVM-Rbf	DT-CART
Kappa	0.422	0.47	0.434	0.422	0.398	0.711	0.916	0.691	0.687
Chi-square yates	37.44	37.97	36.84	31.43	30.39	87.19	136.5	62.03	77.78
Chi-square Pearson	39.68	39.99	38.97	33.31	32.32	90.22	140.2	64.91	80.58
Cramer's V	0.489	0.491	0.485	0.448	0.441	0.737	0.919	0.691	0.697

Fig. 3 Confidence intervals of 95% of all the classifier models with 1: BC, 2: DF-linear, 3: DF-quadratic, 4: DF-pseudolinear, 5: DF-pseudoquadratic, 6: KNN, 7: SVM-RBF, 8: SVM-linear, and 9: DT-CART

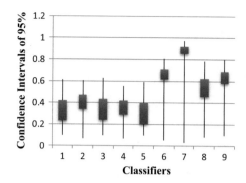

References

1. Tanwar N, Rahman KF (2021) Machine learning in liver disease diagnosis: current progress and future opportunities. IOP Conf Ser Mater Sci Eng 1022(1). IOP Publishing
2. Wu C-C et al (2019) Prediction of fatty liver disease using machine learning algorithms. Comput Methods Prog Biomed 170:23–29
3. Sartakhti JS, Zangooei MH, Mozafari K (2012) Hepatitis disease diagnosis using a novel hybrid method based on support vector machine and simulated annealing (SVM-SA). Comput Methods Prog Biomed 108(2):570–579
4. Singh J, Bagga S, Kaur R (2020) Software-based prediction of liver disease with feature selection and classification techniques. Procedia Comput Sci 167:1970–1980
5. Liu Y, Meric G, Havulinna AS, Teo SM, Ruuskanen M, Sanders J, Zhu Q et al (2020) Early prediction of liver disease using conventional risk factors and gut microbiome-augmented gradient boosting. medRxiv 2020
6. Naseem R, Khan B, Shah MA, Wakil K, Khan A, Alosaimi W, Uddin MI, Alouffi B (2020) Performance assessment of classification algorithms on early detection of liver syndrome. J Healthcare Eng
7. Singh AS, Irfan M, Chowdhury A (2018) Prediction of liver disease using classification algorithms. In: 2018 4th international conference on computing communication and automation (ICCCA). IEEE, pp 1–3
8. Naveau S, Raynard B, Ratziu V, Abella A, Imbert–bismuth F, Messous D, Beuzen F et al (2005) Biomarkers for the prediction of liver fibrosis in patients with chronic alcoholic liver disease. Clin Gastroenterol Hepatol 3(2):167–174
9. Lichman M (2013) UCI machine learning repository. University of California, School of Information and Computer Science, Irvine, CA
10. Dougherty G (2012) Pattern recognition and classification: an introduction. Springer Science & Business Media
11. Larose DT, Larose CD (2014) Discovering knowledge in data: an introduction to data mining, vol 4. Wiley
12. Han J, Pei J, Kamber M (2011) Data mining: concepts and techniques. Elsevier
13. Howell DC (2011) Chi-square test: analysis of contingency tables, pp 250–252
14. Kateri M (2014) Contingency table analysis. In: Methods and implementation using R, 1st ed. Editorial Advisory Board, Aachen, Germany

angularParameter: A Novel Optimization Technique for Deep Learning Models

Shubhankar Bhakta, Utpal Nandi, Chiranjit Changdar, and Moirangthem Marjit Singh

1 Introduction

Deep learning (DL) [1, 2] emulates the works like the human brain to conclusion making, composing patterns, and process data formulated on artificial neurons (AN). To do this work, ANN requires lots of time and complex calculations formulated on predictive models and statistics. Nowadays, machine performs like the human brain exclusively but there are some strain, curb, and fault of accuracy. To reduce these deterrents, a lot of research works are going on. The human brain uses biological neurons but DL uses ANN for carrying out several works like classification (text, image) [3], machine translation [4], character and speech recognition [5], question answering, etc. ANN consists of different layers, various parameters like weight (primacy of input data), biases, and fully connected multi-layer perceptrons (MLPs). The performance of ANN is assessed by determining the perfection of the loss function.

The optimizer takes the leading role to reduce the loss function's value. Stochastic gradient-dependent optimizers are the major approaches for deep neural networks' conquest. The gradient furnishes knowledge about direction and variation of step size. The SGD [6, 7] method modifies equal step size for every parameter, and it is also non-aligned with gradient nature like a local change of gradient value or the angular coefficient of the gradient. To conquer equal step size trouble of stochastic

S. Bhakta · U. Nandi (✉)
Department of Computer Science, Vidyasagar University, West Bengal, India
e-mail: nandi.3utpal@gmail.com

C. Changdar
Department of Computer Science, Belda College, West Bengal, India

M. Marjit Singh
Department of Computer Science and Engineering, North Eastern Regional Institute of Science and Technology, Arunachal Pradesh, India

© The Author(s), under exclusive license to Springer Nature Singapore Pte Ltd. 2023
D. S. Sisodia et al. (eds.), *Machine Intelligence Techniques for Data Analysis and Signal Processing*, Lecture Notes in Electrical Engineering 997,
https://doi.org/10.1007/978-981-99-0085-5_17

201

gradient-based methods, adaptive methods are employed. Adam [8] is a widely used adaptive method, but it also can't take the benefit of local change of gradient. This problem is tackled by diffGrad optimization method [9]. However, it obviates the angular coefficient of gradients for a new update.

angularGrad [10] is a new optimization method that takes the benefit of angular coefficient of the previous two successive gradients and reaches global minima [11, 12] smoothly. However, it does not consider the angular concept for the coefficient of the parameter. In this paper, a novel optimization method (termed as angularParameter) has been presented that adapts the step size and update-parameter depending on the angular coefficient of the previous two consecutive parameters. As a result, it offers a smoother trajectory. The method can apply either the cosine or tangent function to determine the angular information of the parameter. The rest of the paper is structured as follows:

The related works are given in Sect. 2. The proposed method has been discussed in Sect. 3. The empirical analysis is shown in Sect. 4m. The results are illustrated in Sect. 5 with a discussion and a conclusion is made in Sect. 6.

2 Related Work

In ANN, parameters (biases and weights) are adapted by the optimizer to amend loss or cost function. The gradient descent (GD) [13, 14] is a simple algorithm which compacts with three variants (i) how much data is used to calculate the gradient of an objective function, (ii) hang on the quantity of data precision of parameter modernize, and (iii) time get hold of to modify the parameters. The method calculates gradient for the entire data set to make just one update as given in Eq. (1) that causes several deficiencies like (i) it requires powerful computational memory for few iterations also [15], (ii) it is computationally methodical but not rapid due to fixed learning rate (Lr).

$$P_{\text{new}} = P_{\text{old}} - a \cdot \frac{\delta l}{\delta P(\text{old})} \tag{1}$$

Stochastic GD (SGD) [6, 7], other favoured algorithm which performs only one parameter modification for each training example $S_x^{(i)}$ and label $S_y^{(i)}$ given in Eq. (2).

$$P = P - a . \nabla_P J(P, S_x^{(i)}; S_y^{(i)}) \tag{2}$$

Although this algorithm needs less computational power but conducts frequent updates with gusty variance and shows excessive zigzag path with rowdy updates, and also time-consuming. To bring down disagreement of parameter updates, and steady convergence, another method introduced that is mini-batch SGD where 'B' $(1 < B <= n)$ no of value update is done in every back-propagation, given in equation (3).

$$P = P - a . \nabla_P J(P, S_x^{(i,i+B)}; S_y^{(i,i+B)}) \tag{3}$$

However, there are several shortcomings like (i) initially perfect Lr selection is so perplexing, (ii) using unvarying Lr for every epoch to upgrade parameter, and (iii) due to noisy update, it is also time-consuming. A method accelerates mini-batch SGD to proper direction using a momentum variable (λ) that is known as SGD with momentum (SGDM) [16]. The parameter upgrade is done using Eq. (4).

$$P_t = P_{t-1} - a_t.G_t \tag{4}$$

where P_t, P_{t-1}, and a_t are upgrade parameter afterwards t-th epoch, most recently past parameter, and step size, respectively. And G_t denotes gradient-based variable calculated as Eq. (5).

$$G_t = \lambda G_{t-1} + (1 - \lambda).\nabla_P \psi(P_t) \tag{5}$$

where $\psi(P_t)$ indicates loss function.

There is distinct deficiency such as time absorbing and racket renovation in the case of prearranged lr, to conquer those issues adaptive lr is initiated in optimization algorithms. Adam optimizer bargains with those issues and brings down them. It utilizes exponential decaying average (EAD) of previous gradient (c_t) and EAD of previous squared gradient (d_t). c_t and d_t mark out 1st and $2^{n}d$ order moments as Eqs. (6) and (7) respectively.

$$c_t = r_1 * c_{t-1} + (1 - r_1) * gr_t \tag{6}$$

$$d_t = r_2 * d_{t-1} + (1 - r_2) * gr_t^2 \tag{7}$$

To direct EDA, two hyper-parameter (r_1 and r_2, typically set 0.9 and 0.999 proportionately) are operated. Adam does two bias corrections, i.e. $\hat{c}_t$ and $\hat{d}_t$ as given in Eqs. (8) and (9), respectively. It updates parameters by Eq. (10)

$$\hat{c}_t = \frac{c_t}{1 - r_1^t} \tag{8}$$

$$\hat{d}_t = \frac{d_t}{1 - r_2^t} \tag{9}$$

$$P_t = P_{t-1} - \frac{a_t}{\sqrt{\hat{d}_t} + \epsilon}.\hat{c}_t \tag{10}$$

To keep away from division by 0, ϵ (in a general set as 10^{-8}) is commenced, and a_t represents lr. There is also the possibility of an adaptive lr-based algorithm that ceases learning at the local optimum. To repress this problem, RAdam [17] is launched instinctively restraint variance pruning term s_t as given in Eq. (11).

$$s_t = \sqrt{\frac{(F_t - 4)(F_t - 2)F_\infty}{(F_\infty - 4)(F_\infty - 2)F_t}} \tag{11}$$

where $F_\infty = \frac{2}{(1-r_2)} - 1$ and $F_t = F_\infty - \frac{2tr_2^t}{(1-r_2^t)}$. But, the algorithm applied a condition, i.e. if F_t is greater than 4, then the parameter is updated by Eq. (12); otherwise, it does precisely like SGD.

$$P_t = P_{t-1} - \frac{a_t s_t \hat{c}_t}{\hat{d}_t} \tag{12}$$

Adam and RAdam can't grip the gain of gradient change. diffGrad in another adaptive lr base optimizer that grips the profit of changing gradient. As a result, it keeps away from the noisy upgrade and it is faster than Adam to reach the global optimum. It illustrates vast step size for hurriedly gradient change and small step size for slight gradient change. Friction coefficient χ is launched here in obedience to present gradient and recently past gradient, calculated as Eq. (13), and parameter renovated by diffGrad as Eq. (14).

$$\chi_t = \frac{1}{1 + e^{-|gr_{t-1} - gr_t|}} \tag{13}$$

$$P_t = P_{t-1} - \frac{a_t \chi_t}{\sqrt{\hat{d}_t} + \epsilon} . \hat{c}_t \tag{14}$$

where t indicates t-th epoch.

But in the case of diffGrad, the convergence carves bows to yet zigzag track. To bring down this difficulty, angularGrad optimizer employs the angular coefficient(κ) which depends on the angle between two successive gradients, calculated as in equation(15).

$$\kappa_t = \tanh(|\forall(Angl_{min})|).\varrho_1 + \varrho_2 \tag{15}$$

where ϱ_1 and ϱ_2 lies between 0 and 1, and $\forall$ points to $<$cos or $<$tan and $\tanh(x)$ points to non-liner function in the facts of x between -1 and 1 , $Angl_t$ is angle between gr_t and gr_{t-1} at t-th step. Similarly, $Angl_{t-1}$ is angle between two previous successive gradients at $(t-1)$-th step, and $Angl_{min} = min(Angl_{t-1}, Angl_t)$. Angularagrad renovates parameter as in Eq. (16)

$$P_t = P_{t-1} - \frac{a_t \kappa_t}{\sqrt{\hat{d}_t} + \epsilon} . \hat{c}_t \tag{16}$$

Although the convergence curve of angularGrad is found well than Adam or diffGrad, there are a few limitations like (i) angularGrad renovates the parameter based on the difference between the angle of current gradient and recently past gradient but can't consider the changing value of the parameter, and (ii) the algorithm also decides the minimum tangent value between $Angl_t$ and $Angl_{t-1}$, that is the extra worry. To conquer those difficulties, a novel optimization technique termed angularParameter has been proposed.

3 The Proposed Optimizer

In this portion, talk about our raised optimization method, which is encouraged by angularGrad. Yet it enables to utterly take out the variance of convergence curve but finer than angularGrad. Here, it is not needed to decide the minimum tangent value of two consecutive gradients' angles also. The angularParameter method renovates parameters not only based on gradient modernization of presently gradient and recently past gradient, but also based on previous two parameters. First, we calculate a fraction value for_tan as in Eq. (17).

$$for\_tan = abs\left(\frac{diffp}{(1 + P^2)}\right) \qquad (17)$$

where 'diffp' is calculated by equation (18) ans P is previous parameter.

$$diffp = \left(\frac{abs(P_{new} - P)}{2}\right) \qquad (18)$$

Now calculate tangent of angular coefficient κ_t as in equation (19).

$$\kappa_t = tanh(for\_tan) * \varrho_1 + \varrho_2 \qquad (19)$$

It reflects that a variable for_cos defined in equation (20) may also be used just like in the angularGrad optimizer.

$$for\_cos = \frac{1}{\sqrt{1 + for\_cos^2}} \qquad (20)$$

The angularParameter optimization method also enumerates 2nd order moment c_t and d_t, respectively, and bias adjustment $\hat{c}_t$ and $\hat{d}_t$ like angularGrad. It renovates parameter as in Eq. (21).

$$P_t = P_{t-1} - \frac{a_t \kappa_t}{\sqrt{\hat{d}_t} + \epsilon} . \hat{c}_t \qquad (21)$$

The flow diagram of the proposed method has been illustrated in Fig. 1.

Fig. 1 angularParameter method

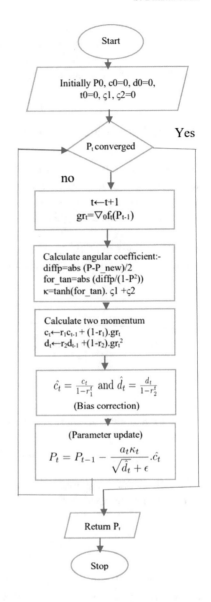

4 Empirical Analysis

In this section, three functions $f_1(22)$, $f_2(23)$, and $f_3(24)$ are employed to evaluate regression loss(RL) and parameter renovation with varying number of epochs. These are non-convex [18, 19] and one-dimensional functions where input $m \in (+\infty, -\infty)$.

$$f_{-1}(m) = \begin{cases} (m + 0.3)^2, & \text{for } m \leq 0 \\ (m - 0.2)^2 + 0.05, & \text{for } m > 0 \end{cases} \tag{22}$$

$$f_{-2}(m) = \begin{cases} -40m - 35.15, & \text{for } m \leq -0.9 \\ m * sin(8m) + m^3 + 0.85, & \text{for } m > -0.9 \end{cases} \tag{23}$$

$$f_{-3}(m) = \begin{cases} m^2, & \text{for } m \leq -0.5 \\ 0.75 + m & \text{for } -0.5 < m \leq -0.4 \\ -\frac{7m}{8}, & \text{for } -0.4 < m \leq 0 \\ \frac{7m}{8}, & \text{for } 0 < m \leq 0.4 \\ 0.75 - m, & \text{for } 0.4 < m \leq 0.5 \\ m^2, & \text{for } m < 0.5 \end{cases} \tag{24}$$

The proposed method angularParameter (cos and tan) has been juxtaposed with Adam, angularGrad(cos and tan) and diffGrad as shown in Fig. 2 based on RL and parameter renovation.

The shapes of functions f_{-1}, f_{-2}, and f_{-3} are illustrated in Fig. 2a–c, respectively, where first one retains single minima (lm) and rest two have two lm, $m \in (+1, -1)$. For these evaluations, different variables are adjust as $r_1 = 0.95$, $r_2 = 0.999$, c = 0, t = 0, a =0 .1, P = −1, and the number of iteration = 300. The plots of RL with varying number of iteration are depicted in Fig. 2d–f for functions f_{-1}, f_{-2}, and f_{-3}, respectively. The parameter renovation vs number of iteration curves for functions f_{-1}, f_{-2}, and f_{-3} are shown in Fig. 2g–i, respectively. For function $f_{-1}(m)$, Fig. 2g shows that Adam overreaches the global optima but angularGrad(cos and tan) and angularParameter(cos and tan) apts to gain global optima. From this analysis, it is observed that the proposed optimization method apts to gain global optima and put back zero loss more definitely with slight noise.

5 Experimental Results and Discussion

The different experiments are continued on the proposed method angularParameter to validate, test performance, and compare it with its counterparts as compactly recapitulate bellow:

- ○ Firstly, to appreciate behaviour and effectiveness of angularParameter method assess on Rosenbrock function [20, 21] as given in Sect. 5.1.
- ○ secondly, experiment done by image classification to quantify the perfection of angularParameter method on CIFAR10 and CIFAR100 [22, 23] as given in Sect. 5.2.

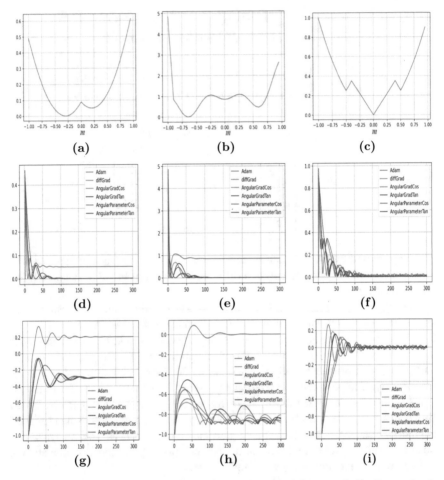

Fig. 2 Empirical Analysis's result over f_1, f_2, and f_3 using Adam, angularGrad(cos and tan), diffGrad, and angularParameter (cos and tan)

5.1 Convergence Analysis Using Rosenbrock Function

For convergence analysis of different optimizers on 2D form, the Rosenbrock (also called Bannana or Valley) function [20, 21] are used. The global minima(GM) of this model lies in a parabolic valley with a narrow path, for this causes it tries to see how smoothly reach the GM using an optimizer.

The function is defined in Eq. (25).

$$f(R) = f(R_1, R_2, R_3, \ldots, R_N) = \sum_{j=1}^{N-1} \left[h.(R_{j+1} - R_j^2)^2 + (g - R_j)^2 \right] \quad (25)$$

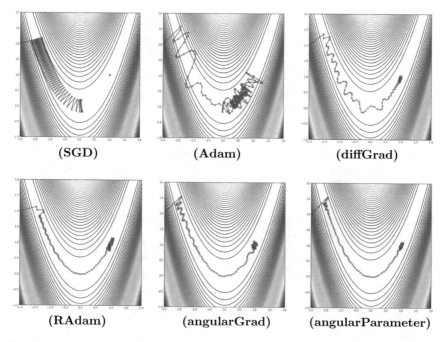

(SGD)   (Adam)   (diffGrad)

(RAdam)   (angularGrad)   (angularParameter)

Fig. 3 Convergence analysis using Rosenbrock function on SGD, Adam, diffGrad, RAdam, angularGrad, angularParameter optimization method

where $R = [R_1, R_2, R_3, \ldots, R_N] \in \Re^N$ is an input and N is positive integers. Two hyper-parameter g and h are initialized with constant values 1 and 100, respectively.

The proposed angularParameter method converges the path to reach the GM how easily or smoothly along other standard optimizers Adam, SGD, diffGrad, RAdam, and angularGrad are shown in figure (3) for comparative purpose.

From the Fig. 3, it is realized that SGD is ineffectual to reach GM. However, other methods can reach GM but follow a high zigzag path with noisy movement. However, the proposed angularParameter method can reach GM smoothly with less noisy movement comparatively.

5.2 Image Classification Results on CIFAR10 and CIFAR100 Data sets

In this section, the proposed optimizer angularParameter method carries out image classification screening using five CNN deep model (VGGNet16 [24], ResNet34, ResNet18, ResNet50 [25], and DenseNet121 [26]) on CIFAR 10 and also CIFAR 100 data set. The classification accuracy of angularParameter(cos and tan) has been presented agenst other methods like SGDM, Adam, diffGrad, and angularGrad(cos

Table 1 Classification outcome with VGGNet16, ResNet34, ResNet18, ResNet50, DenseNet121 deep network on CIFAR10 data set

Optimizers	VGG16	ResNet34	ResNet18	ResNet50	DenseNet121
SGDM	92.36	93.40	93.18	92.93	93.73
Adam	92.55	93.83	93.58	93.11	94.15
diffGrad	92.64	93.96	93.64	93.78	94.21
angularGrad(cos)	92.86	**94.13**	93.80	94.21	94.38
angularGrad(tan)	91,77	94.04	93.71	94.20	94.40
angularParameter(cos)	**92.87**	94.12	**93.84**	**94.23**	94.37
angularParameter(tan)	92.78	94.05	93.72	94.19	**94.41**

Best outcomes are highlighting in bold

Table 2 Classification outcome with VGGNet16, ResNet34, ResNet18, ResNet50, DenseNet121 deep network on CIFAR100 data set

Optimizers	VGGNet16	ResNet34	ResNet18	ResNet50	DenseNet121
SGDM	**69.75**	72.78	72.89	73.58	74.91
Adam	67.93	71.92	72.73	73.81	74.96
diffGrad	67.97	72.80	72.98	74.25	74.98
angularGrad(cos)	69.10	73.41	73.01	75.69	75.51
angularGrad(tan)	69.08	73.59	**73.24**	75.51	75.47
angularParameter(cos)	69.62	**73.63**	73.11	75.65	**75.53**
angularParameter(tan)	69.59	73.61	73.16	**75.71**	75.33

Best outcomes are highlighting in bold

and tan). Here, the SGDM is imposed with lr (a) = 0.01, but others are imposed with $a = 1e - 3, r_1 = 0.9, r_2 = 0.999$ and $\epsilon = 1e - 8$. The networks have been trained with 100 iterations.

The performance on CIFAR10 and also CIFAR 100 data set of different methods has been compared with angularParameter(cos and tan) as given in Tables 1 and 2, respectively. For CIFAR10, the proposed optimization method angularParameter(cos or tan) offers the best performance on VGGNet16, ResNet18, ResNet50, and DenseNet121. However, the angularGrad(cos) achieves the best result on ResNet34 network. For CIFAR100, the angularParameter(cos or tan) achieves the best performance on ResNet34, ResNet50, and DenseNet121. However, the SGDM yields the best performance on the VGG16 network, and angularGrad(tan) offers the best result on the ResNet18 network.

6 Conclusion

This paper proposes an optimization method angularParameter that adapts the step size and updates parameter depending on the angular coefficient of the previous two consecutive parameters and offers a smoother trajectory. The method can apply either the cosine or tangent function for determining the angular information of the parameter. The comparative empirical analysis shows that the proposed angularParameter method can reach GM smoothly with less noisy movement comparatively. The image classification performances of the proposed method on CIFAR10 and CIFAR100 data sets are significantly better compared to other methods most of the time.

However, the classification performances on some networks of the proposed method on CIFAR10 and CIFAR100 data sets are not satisfactory. Further investigations are required to deal with this problem.

References

1. LeCun Y, Bengio Y, Hinton G (2015) Deep learning. Nature 521:436–44. https://doi.org/10.1038/nature14539
2. Goodfellow I, Bengio Y, Courville A (2016) Deep learning. MIT Press. http://www.deeplearningbook.org
3. Krizhevsky A, Sutskever I, Hinton GE (2012) Imagenet classification with deep convolutional neural networks. In: Pereira F, Burges CJC, Bottou L, Weinberger KQ (eds) Advances in neural information processing systems, vol 25. Curran Associates, Inc, pp 1097–1105. http://papers.nips.cc/paper/4824-imagenet-classification.pdf
4. Stahlberg F (2020) Neural machine translation: a review. J Artif Intell Res 69:343–418. https://doi.org/10.1613/jair.1.12007
5. Hwang K, Sung W (2016) Character-level incremental speech recognition with recurrent neural networks. In: 2016 IEEE international conference on acoustics, speech and signal processing (ICASSP). http://dx.doi.org/10.1109/ICASSP.2016.7472696
6. Bottou L (2010) Large-scale machine learning with stochastic gradient descent. In: Lechevallier Y, Saporta G (eds) Proceedings of COMPSTAT'2010. Physica-Verlag HD, Heidelberg, pp 177–186
7. Robbins H, Monro S (1951) A stochastic approximation method. Ann Math Statistics 22(3):400–407. http://www.jstor.org/stable/2236626
8. Kingma DP, Ba J (2015) Adam: A method for stochastic optimization. In: Bengio Y, LeCun Y (eds) 3rd international conference on learning representations, ICLR 2015, San Diego, CA, USA, 7–9 May 2015, Conference track proceedings. http://arxiv.org/abs/1412.6980
9. Dubey SR, Chakraborty S, Roy SK, Mukherjee S, Singh SK, Chaudhuri BB (2020) diffGrad: an optimization method for convolutional neural networks. IEEE Trans Neural Netw Learn Syst 31(11):4500–4511. https://doi.org/10.1109/TNNLS.2019.2955777
10. Roy SK, Paoletti ME, Haut JM, Dubey SR, Kar P, Plaza A, Chaudhuri BB (2021) AngularGrad: a new optimization technique for angular convergence of convolutional neural networks. arXiv preprint arXiv:2105.10190, https://arxiv.org/abs/2105.10190
11. Lacotte J, Pilanci M (2020) All local minima are global for two-layer relu neural networks: the hidden convex optimization landscape. https://arxiv.org/abs/2006.05900
12. Kawaguchi K, Kaelbling LP (2020) Elimination of all bad local minima in deep learning. Proc Mach Learn Res 108:853–863
13. Ruder S (2017) An overview of gradient descent optimization algorithms. 1609.04747
14. Zhang J (2019) Gradient descent based optimization algorithms for deep learning models training. CoRR abs/1903.03614, http://arxiv.org/abs/1903.03614, 1903.03614
15. Ruder S (2016) An overview of gradient descent optimization algorithms. Comment: Added derivations of AdaMax and Nadam. http://arxiv.org/abs/1609.04747

16. Sutskever I, Martens J, Dahl G, Hinton G (2013) On the importance of initialization and momentum in deep learning. In: Proceedings of the 30th international conference on international conference on machine learning, vol 28, JMLR.org, ICML'13, pp III-1139–III-1147
17. Liu L, Jiang H, He P, Chen W, Liu X, Gao J, Han J (2020) On the variance of the adaptive learning rate and beyond. In: Proceedings of the eighth international conference on learning representations (ICLR 2020)
18. Jain P, Kar P (2017) Non-convex optimization for machine learning. Foundations Trends Mach Learn 10(3–4):142–336. https://doi.org/10.1561/2200000058
19. Danilova M, Dvurechensky PE, Gasnikov AV, Gorbunov EA, Guminov S, Kamzolov D, Shibaev I (2012) Recent theoretical advances in non-convex optimization, CoRR abs/2012.06188. arXiv:2012.06188. https://arxiv.org/abs/2012.06188
20. Rosenbrock HH (1960) An automatic method for finding the greatest or least value of a function. Comput J 3(3):175–184. https://academic.oup.com/comjnl/article-lookup/doi/10.1093/comjnl/3.3.175
21. https://github.com/jettify/pytorch-optimizer
22. Krizhevsky A (2009) Learning multiple layers of features from tiny images. Master's thesis, University of Tront
23. https://www.cs.toronto.edu/kriz/cifar.html
24. S. Liu, W. Deng (2015) Very deep convolutional neural network based image classification using small training sample size. In: 2015 3rd IAPR Asian conference on pattern recognition (ACPR), pp 730–734. https://doi.org/10.1109/ACPR.2015.7486599
25. He K, Zhang X, Ren S, Sun J (2016) Deep residual learning for image recognition. In: Proceedings of the IEEE conference on computer vision and pattern recognition (CVPR)
26. Huang G, Liu Z, van der Maaten L, Weinberger K (2017) Densely connected convolutional networks. https://doi.org/10.1109/CVPR.2017.243

Efficient Prediction of Annual Yield from Stocks Using Hybrid Deep Learning

Ashish Papanai⊙

1 Introduction

The stock market is an area of high profit and high risks, and this is considered while devising and generating a quantitative investment strategy to predict and judge the stock's future price using the historical stock data. The market governed by various financial and non-financial factors poses a new challenge to the researchers to develop a best-fit solution for predicting the stock prices or the annual yield by investing in a particular stock. The three significant factors measuring the outcome of investing in a company or business are Environment, Social, and Governance (ESG). Considering ESGs while making investing decisions leads to a favorable outcome in most cases [1].

The environment of an investment market involves assessment and investment vehicles, financial markets, market structure, market intermediaries, investment process, regulation, and economy. The Indian Financial markets include the credit market, money market, foreign exchange market, debt market, and capital market. The social factors include the interactions between neighbors, friends, colleagues, advice given by analysts, planners, bankers [2]. Social media and interactions are prominent areas to analyze and predict the market trends based on recent activities of significant investors, executives, and influencers [3]. Corporate governance drafts the rules and regulations for the companies and the shareholders and assists in the smooth day-to-day functioning of trade. The governance also handles illegal trade practices like Insider Trade, Security Frauds, etcetera [4].

This study follows the idea of conquering one step (or problem) at a time. It focuses on providing a solution to predict the fluctuations on a stock based on the environment (the Historical Data). With the availability of a massive amount of

Present Address:
A. Papanai (✉)
Maharaja Agrasen Institute of Technology, Delhi, India
e-mail: ashishpapanai00@gmail.com

© The Author(s), under exclusive license to Springer Nature Singapore Pte Ltd. 2023   213
D. S. Sisodia et al. (eds.), *Machine Intelligence Techniques for Data Analysis and Signal Processing*, Lecture Notes in Electrical Engineering 997,
https://doi.org/10.1007/978-981-99-0085-5_18

data, the challenge is to use it and draw meaningful conclusions from it. We have analyzed the data from the date of the initial issue to the date of making predictions. The two deep learning algorithms introduced in this study include a short-term long memory (LSTM), a recurrent neural network architecture for time series prediction [5] and a convolution neural network (CNN) and LSTM mix architecture for making the predictions. The baseline comparison of the deep learning strategy is with two traditional stock trading strategies, the buy and hold system and the moving averages.

The deep learning model is trained end-to-end on new data (between a training window of the past six years from the day of making the predictions). This strategy prevents any sudden old fluctuations in the data from contaminating the projections. All outliers are scaled using a min–max scaler to stop them from dominating in the results.

Our results suggest a substantial promise in integrating the traditional and computer-generated strategies for developing an improved quantitative investment strategy. The algorithms outperform the baselines set by the conventional approach, and the enhanced LSTM-CNN mix model provides better performance with reduced computational cost. Integration of this model with other deep learning strategies to handle other governing factors like governance and social can bring a new revolution in algorithmic trading using deep learning.

2 Related Work

Our work connects several relevant pieces of literature. Recent work highlights how deep learning can be used in algorithmic trading. With researchers working on individual factors affecting the stock prices, as the social factor, environment, and governance [3, 4, 6, 7], a new vision is being added to the analysis and projection of stock prices based on the contributing factors. Reinforcement learning libraries like Financial Reinforcement Learning (finRL) and Technical Analysis Library in Python (TA-Lib) have set a new benchmark in artificial intelligence for finance.

The use of machine learning algorithms like support vector machine (SVM) [8] and a hybrid feature selection method provided a detailed parameter adjustment procedure with performance under different parameter values. The performance of this algorithm is significantly less than the state-of-art LSTMs.

Various other LSTM-based models on long-term data fail to address the computational complexities and the efforts required to train it [9]. Our study focuses on the training aspect of the model and tries to reduce the training time even in a limited computational environment.

The limitation of these developments is that they fail to scale because of high computational requirements [10]. The creation of new replicas of the trading environment and dummy data generation fail to use the existing data. The current deep learning solutions using LSTMs and LSTM-CNN mix are tested only on short-term predictions ranging from 1 to 10 days [11].

The main contribution of our work to this problem is to:

- Use existing, publicly available historical data of stocks to train and test the model.
- Develop a computationally efficient, less complex, and simple deep learning solution to improve performance and scalability.
- Comparing various investment strategies (Traditional and deep learning) makes it easier for the user to create a rational decision.

Our work uses the stock market to focus on a crucial issue, to align the human-devised strategies with the computer-generated strategy for algorithmic trading and quantitative finance.

3 Data and Background

3.1 Problem Formulation

The current study aims to develop a deep learning-backed automated in and out trading strategy to maximize the annual yield from a single stock. The in and out trading strategy involves two import decisions:

1. In: Trade in the market if the stock price is predicted to rise for the current month.
2. Out: Sell out and exit the market if the prize is predicted to decrease for the current month.

This study introduces a neural network architecture that learns the historical stock data and predicts the price of the fluctuation in the stock price for the current month thus, solving the market's most important question: "Will the stock soar or crash?" The model receives eight features (highest price, lowest price, opening price, closing price, the volume traded, the first day of the current month, moving average for 12 months, moving average for 24 months) for each month in the period of six years considered. In the training phase, the model tries to find a pattern in the historical data, which is tested in the test data. Finally, the model makes an in and out prediction for the next 12 months from the prediction date to provide the user with an investment strategy. This is later used to predict an efficient annual and gross yield. The yield predicted by the deep learning strategy is compared with baseline and popular statistical prediction and trading strategies, which explains the deep learning strategy's efficiency.

3.2 Background

Our work leverages three radically different approaches. The moving average method is mathematical, and the buy and hold strategy is positional and confidence based. In contrast, stockDL focuses on two deep learning architectures, a black-box, and tries to understand the pattern in the historical data and derive meaningful insights from it.

Moving Average Strategy. It follows a set of rules to decide whether to trade in or stay out of the market for the month considered. This strategy focuses on smoothening out the price trends by filtering out the noise from short-term predictions. It acts as a support in case of an uptrend of the time frame taken. This method helps understand where moving averages will offer support and resistance. Support indicates a price level where we can expect a downtrend, whereas resistance suggests an increase in the price level, that is, an uptrend. Traditional investors have developed various tools to use the moving average to indicate upcoming trends in the prices.

$$\frac{\sum_{i=m}^{n} \text{Price}}{m - n} \tag{1}$$

In the formula (1), 'm' is the starting date (Six years before the current date), and 'n' is the final date (date of making the prediction, Current Date).

Buy and Hold Strategy. The investor using this strategy buys stocks and holds them for a long time irrespective of the market fluctuations. This approach is generally long term and relies on the confidence of the investor. It is a passive investment strategy, and the investor might not sell the possessions at the optimal time.

Deep Learning (LSTM Strategy). LSTM is the backbone of the two architectures defined in this study. LSTM learns to keep the relevant information and forget non-relevant data. RNNs can retain the information at time t about the input seen many timestamps before t; this fails in practical implementation due to the problem of vanishing gradients (the gradient gradually becomes zero because of multiplication of long series of numbers less than zero). LSTM saves the information for later and prevents the older signals from being lost during the processing. The LSTM cells allow the past information to be reinjected later, overcoming the vanishing gradient problem.

Figure 1 shows the architecture of an LSTM network; it gives a pictorial representation of the activation functions used in the network and the flow of computation of carry by incorporating the past data into the calculating formula.

3.3 Data

The data module of stockDL collects the data required for the study. The data are retrieved from the Yahoo Finance API based on a unique symbol provided to each stock (this unique symbol is called the stock ticker), the starting date, and the end date.

The data table considered in the study (shown as Table 1) includes information such as opening rate (of the day), the highest rate (of the day), the lowest rate (of the

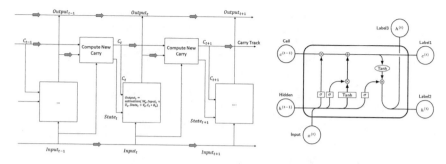

Fig. 1 Anatomy of an LSTM with three input cells [12], single LSTM cell

day), the closing rate (of the day), and the volume traded (on that day). As we consider dividends and stock split in calculating the annual yield, the columns comprising it are dropped (Table 2).

We split the monthly data (shown in Table 1, Plotted in Fig. 2 as candle plots and Fig. 3 as Line Curves) into training and testing data to train and evaluate the model. The split data were normalized separately to prevent any data leak. We normalized the data to values between −1 and 1 using min–max scaling, to ensure that no feature is falsely prioritized based on its value.

Normalization replaces the value in each column with the following formula:

$$m = \frac{x - x_{\min}}{x_{\max} - x_{\min}} \tag{2}$$

Table 1 First five rows of the historical stock data for further preprocessing of HDFC Bank Limited

Date	Open	High	Low	Close	Volume
1996-01-01	2.458312	2.458312	2.373122	2.417745	350,000
1996-01-02	2.417746	2.454255	2.393406	2.413689	412,000
1996-01-03	2.413689	2.429916	2.393406	2.421802	284,000
1996-01-04	2.421802	2.417746	2.385293	2.405576	282,000
1996-01-05	2.405576	2.417746	2.393406	2.401519	189,000

Table 2 Stock symbols (ticker) of the stocks used in the study

Stock name	Stock symbol (Ticker)	Stock exchange	Currency
Alphabet Inc. (Google)	GOOG	NASDAQ global select market	USD
Apple Inc.	AAPL	NASDAQ global select market	USD
HDFC Bank Limited	HDFCBANK.NS	National stock exchange	INR
Reliance Industries Limited	RELIANCE.NS	National stock exchange	INR

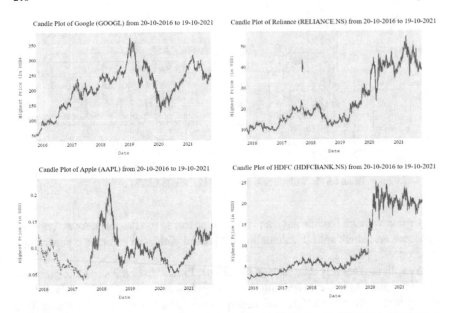

Fig. 2 Candle plots for the stocks considered Apple Inc. (AAPL), Google (GOOGL), HDFC Bank Limited (HDFCBANK.NS), and Reliance Industries Limited (RELIANCE.NS)

Monthly Opening Prices of The Chosen Stocks

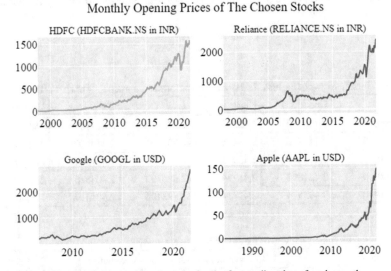

Fig. 3 Opening prices of the considered stocks for the first trading day of each month

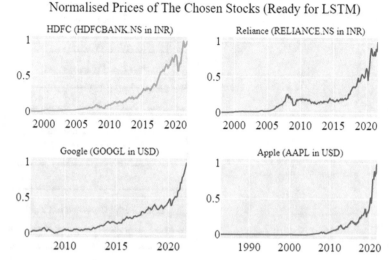

Fig. 4 Normalized monthly opening prices

In formula 2, m = new cell value, x = initial cell value, x_{max} = maximum column value, x_{min} = minimum column value. Figure 4 shows the plot of normalized monthly opening prices.

4 Method

stockDL uses a long short-term memory network. This study explains the two approaches of using the LSTM network. The accuracy is tested against activation functions like hyperbolic tangent (tanh), rectified linear unit (ReLU), leaky rectified linear unit (Leaky ReLU). The following flowchart represents the pipeline of predictions from stockDL (Fig. 5).

stockDL trains two models for comparisons with the baseline statistical models; the pure LSTM model is computationally expensive and slower than the ensembled CNN and LSTM model.

4.1 LSTM Network

The preprocessed data are scaled and fed to the model's input layer. The following image represents the architecture of the LSTM model used in stockDL (Fig. 6).

The input later of the LSTM model (as shown in Fig. 7) focuses on six data features: date, opening price, highest trading price, lowest trading price, closing price, and the volume traded in a day. After each LSTM layer, a dropout layer is

Fig. 5 Pipeline of stockDL

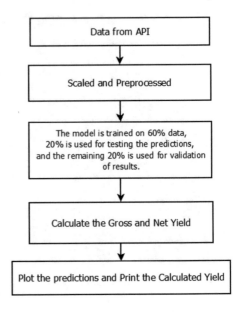

added, which drops 50% of the parameters. This step is added to ensure that the complexity of the model doesn't result in the model overfitting the data while it is trained. The model takes around 90 s to train daily data for the 6-year time frame and another 30 s for validation.

Various loss functions (Mean squared error, quadratic loss, L2 loss) are tested for the model, and mean squared error loss comes out to be the best parameter for testing the divergence of the predictions from the actual value.

The formula for mean squared error (MSE):

$$\sum_{i=1}^{D} (x_i - y_i)^2 \tag{3}$$

4.2 Hybrid CNN-LSTM Model

This is the novel approach introduced in this study which is ten times faster than the pure LSTM model. The LSTM model takes time in understanding the pattern in the time series data. To reduce the preliminary time taken by LSTM in pattern recognition, a layer of convolution neural network is employed for pattern recognition, and subsequent LSTM layers follow this layer for time series prediction; the result of the prediction is further used in the pipeline to decide if the model should trade in or out of the market for the particular month to maximize the annual yield.

The following image shows the model architecture for the hybrid CNN-LSTM model used by stockDL.

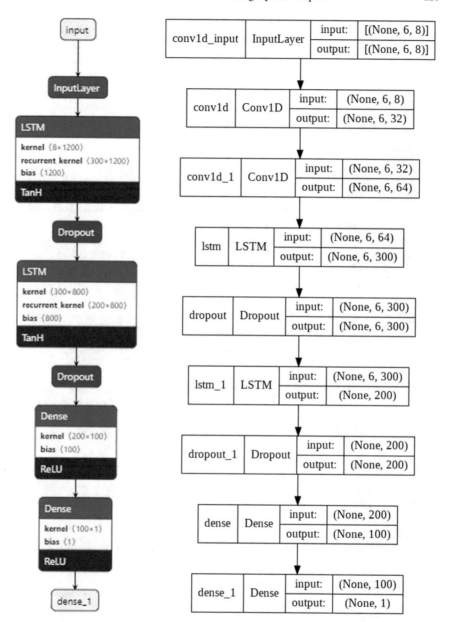

Fig. 6 stockDL pure LSTM architecture

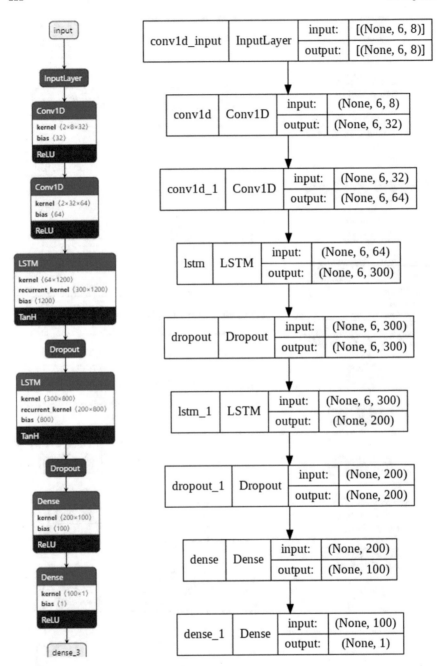

Fig. 7 Hybrid CNN-LSTM model architecture of stockDL

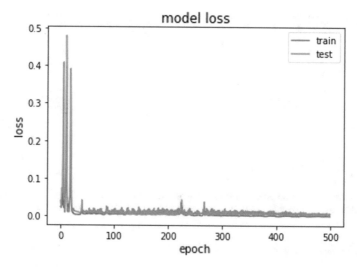

Fig. 8 Model loss for the hybrid CNN-LSTM network

The additional CNN layers are shown in Fig. 8, are added to the model to improve feature extraction and model training time. Dropout layers are added after the LSTM network to prevent overfitting. This model also uses the MSE loss function to check the actual and predicted value variation for providing better annual and gross yields.

5 Performance and Evaluations

5.1 Training Performance

The training time of training the LSTM model of stockDL is 90 s using NVIDIA K90 GPU on Google Colab, which is reduced to 30 s for the CNN-LSTM hybrid network. The following plot shows the training plot of the hybrid CNN-LSTM network.

The training loss stabilizes at 0.0001, whereas the testing loss plateaus at 0.0003. The above plot (Fig. 9) validates that the model isn't overfitting on the data, based on comparable training and testing loss. The following predictions plots of the model testify that stockDL is a novel, efficient and improved version of all existing LSTM deep learning strategies that can be employed for the stock market or any regression-based time series predictions.

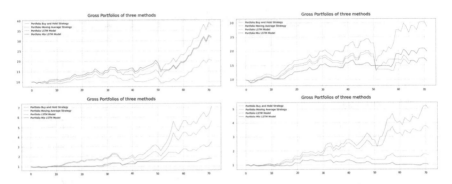

Fig. 9 Comparison of the predictions made by the strategies on the stocks of GOOGL (top left), HDFCBANK.NS (top right), AAPL (bottom left), RELIANCE.NS (bottom right)

5.2 Evaluations

The performance of the trained model is evaluated on four stock options as explained in the introduction of the study:

The methods employed for the predictions predicted a similar strategy for deciding in–out trading months for the stocks of GOOGL, HDFCBANK.NS, AAPL, and RELIANCE.NS.

It can be inferred from Table 3. The buy hold strategy, if incorporated, is predicted to provide the investor with the best yield for the stocks of Alphabet Inc. (GOOGL). For HDFC Bank, the moving average strategy predicts the annual yield to be 10.9%, whereas the buy hold, LSTM, and hybrid CNN-LSTM have a comparable prediction with 20.25%, 19.97%, and 19.99%, respectively. The stocks of Apple Inc. provided similar net yield from the four methods, with LSTM and hybrid CNN-LSTM providing 19.97% and 19.99%, respectively, and the buy and hold, moving average providing comparable net yields of 20.25% and 20.90%, respectively. For Reliance (RELIANCE.NS) stocks, the LSTM and hybrid CNN-LSTM network provide similar net yields of 19.17 and 19.31%. The buy hold strategy again predicts the maximum net yield of 28.42%.

Table 3 Predicted net yield for each method used in predicting the trading strategy

Method	Predicted net yield (GOOGLE)	Predicted net yield (HDFC)	Predicted net yield (APPLE)	Predicted net yield (RELIANCE)
Buy and hold	23.25	20.25	20.25	28.42
Moving average	10.9	10.9	20.9	21.14
LSTM	19.97	19.97	19.97	19.17
Hybrid CNN-LSTM	17.99	19.99	19.99	19.31

From Fig. 9, both deep learning strategies have similar predictions of the net yield, and this provides the investors with a trade-off between minimizing the training time or maximizing the net yield from the stock. The average moving strategy predicts 10.9% as net yield, the lowest compared to the other statistical method and the two deep learning methods.

6 Conclusion and Future Work

stockDL displays prominent and breakthrough results for efficient and accurate time series predictions on historical stock data, which is evident with the comparisons of the black-box deep learning predictions with statistical baselines. Buy hold strategy appears to predict maximum net yield in almost all cases considered in this study, but this method is prone to market fluctuations, and this method relies on faith that the price will eventually rise in the long term. This assumption/faith can be risky for investors relying blindly on this strategy. The two deep learning strategies, LSTM and the novel hybrid CNN-LSTM, predict similar results. However, the hybrid architecture introduced in this study is computationally inexpensive and takes much less training and prediction time than the existing LSTM architecture for time series prediction.

The findings of this study are interesting and surprising because of the accuracy of improved predictions with reduced/incomparable risks compared to the two manual statistical methods tested. Future work of this study can be implementing trading an alternate stock in the month we trade out from the market to improve the yield. This multi-stock trading will help investors make better AI-supported decisions, reducing financial losses, and improving the net yield. Another enhancement to stockDL or other such financial libraries can be done by developing a self-improving neural network architecture that learns from the gains or losses made by the past transaction and aims to run as humanly as possible to maximize the yield from investments on single or multiple stock portfolios.

References

1. Friede G, Busch T, Bassen A (2015) Journal of sustainable finance and investment ESG and financial performance: aggregated evidence from more than 2000 empirical studies. J Sustain Financ Invest 5(4):210–233. https://doi.org/10.1080/20430795.2015.1118917
2. Shanmugham R, Ramya K (2012) Impact of social factors on individual investors' trading behaviour. Procedia Econ Financ 2:237–246. https://doi.org/10.1016/S2212-5671(12)00084-6
3. Chen Z, Du X (2013) Study of stock prediction based on social network. Proc Soc 2013:913–916. https://doi.org/10.1109/SOCIALCOM.2013.141
4. Prabowo B, Rochmatulaili E, Rusdiyanto, Sulistyowati E (2020) Corporate governance and its impact in company's stock price: case study. Utop. y Prax. Latinoam. 25(Extra10):187–196. https://doi.org/10.5281/ZENODO.4155459

5. Hochreiter S, Schmidhuber J (1997) Long short-term memory. Neural Comput 9(8):1735–1780. https://doi.org/10.1162/NECO.1997.9.8.1735
6. Liu X-Y et al (2021) FinRL: a deep reinforcement learning library for automated stock trading in quantitative finance. Accessed: 20 Oct. 2021. [Online]. Available https://github.com/AI4Finance-LLC/FinRL-Library
7. Chen H, De P, Hu Y, Hwang BH (2011) Sentiment revealed in social media and its effect on the stock market. In: Proceedings of 2011 IEEE statistical signal processing workshop (SSP), pp 25–28. https://doi.org/10.1109/SSP.2011.5967675
8. Lee MC (2009) Using support vector machine with a hybrid feature selection method to the stock trend prediction. Exp Syst Appl 36(8):10896–10904. https://doi.org/10.1016/J.ESWA.2009.02.038
9. Fischer T, Krauss C (2018) Deep learning with long short-term memory networks for financial market predictions. Eur J Oper Res 270(2):654–669. https://doi.org/10.1016/J.EJOR.2017.11.054
10. Bao W, Liu X (2019) Multi-agent deep reinforcement learning for liquidation strategy analysis, June 2019. Accessed: 20 Oct. 2021. [Online]. Available https://arxiv.org/abs/1906.11046v1
11. Shen J, Shafiq MO (2020) Short-term stock market price trend prediction using a comprehensive deep learning system. J Big Data 7(1):1–33. https://doi.org/10.1186/S40537-020-00333-6
12. Chollet F (2018) Deep learning with Python

Multitask Deep Learning Model for Diagnosis and Prognosis of the COVID-19 Using Threshold-Based Segmentation with U-NET and SegNet Classifiers

S. NagaMallik Raj⊙, S. Neeraja⊙, N. Thirupathi Rao⊙, and Debnath Bhattacharyya⊙

1 Introduction

COVID-19 is responsible for the deaths of over 3000 [1] people every day. An early detection method for the disease had already been identified, and the method for treating infected trees had been developed. The large number of COVID-19 patients is putting a strain on the health care systems of many countries. It would be extremely beneficial to have an automated method for detecting and quantifying infected lung regions. Using CT lung images, radiologists were able to identify three abnormalities: GGO, consolidation, and pleural effusion (pleural fluid) [2]. The use of a tool that quantifies the three irregularities in medical pulmonary images of COVID-19 patients could aid in the semantic segmentation of the images. In the event of a pandemic, it would be beneficial to those dealing with overcrowded hospitals. DL is used to represent multi-layered systems and to perform at a human-level performance. Tumours were selected as a direct target for image segmentation using deep learning techniques.

S. NagaMallik Raj (✉) · N. Thirupathi Rao
Department of Computer Science and Engineering, Vignan's Institute of Information Technology (A), Visakhapatnam, AP, India
e-mail: mallikblue@gmail.com

N. Thirupathi Rao
e-mail: nakkathiru@vignaniit.edu.in

S. Neeraja
Department of Computer Science and Software Engineering, Lendi Institute of Engineering and Technology, Jonnada, Vizianagaram, AP, India
e-mail: neerajasreerama@gmail.com

D. Bhattacharyya
Department of Computer Science and Engineering, KLEF, Greenfield, Vaddeswaram, Guntur 522502, India
e-mail: debnathb@kluniversity.in

© The Author(s), under exclusive license to Springer Nature Singapore Pte Ltd. 2023
D. S. Sisodia et al. (eds.), *Machine Intelligence Techniques for Data Analysis and Signal Processing*, Lecture Notes in Electrical Engineering 997,
https://doi.org/10.1007/978-981-99-0085-5_19

To reduce the number of false positives in low-dose CT lung cancer screening, DL was used to develop a new tool. Another method for segmenting brain tumours from MRI images was to use a hybrid network of researchers from the U-NET and SegNet [3]. With the help of GANs, breast tumours were segmented with 0.90 accuracy (GANs). (CNNs). The researchers divided the reins into four categories: lungs, hepatitis, brain, and time. Numerous COVID-19 studies involving the use of digital lumineers (DL) to analyse medical imagery such as X-rays and CT scans are currently being conducted, with encouraging results. In these images, visual classification has proven to be less successful than semantic segmentation in terms of accuracy. When medical researchers were looking for COVID-19 in medical images, they used deep learning image analysis (DLIA). In a recent study, X-ray images were used to train a CNN binary classifier and a multiclass model, with classifiers as high as 0.98 and 0.87, respectively. While scanning with a target accuracy of 0.99, the researchers discovered COVID-19 on the Exception and ResNet 50V2 networks, which they then used to confirm their findings.

The reference values obtained through medical [4] imaging and DL systems ranged from 0.83 to 0.98 [5]. Recently, semantic similarity between medical images of COVID-19 patients was determined by analysing some of the patients' medical images. A tool like this would be required by a system that prioritizes [6] patients based on their severity. Segment bounding boxes are unable to provide those parameters. As a result, determining the severity of the disease is difficult. Using a deep CNN as a binary segment, one study compared it to other frameworks (FCN, U-NET, VNET, and U-NET+), with the results showing that it performed better. The DL can be used to separate binary data and received Sorensen–Dice accuracy scores of 0.73, 0.75, and 0.73 on the Sorensen–Dice accuracy test. The dials used in the study had a standard deviation of 0.78, a precision of 0.86, and a sensitivity of 0.94. Both the FCN and the U-NET have demonstrated excellent precision [7]. However, they have been ineffective in their efforts to remind. DNN structures for semantic regions included Inf-Net and Semi-Inf-Net DNN structures. They both used the same set of data. These values are obtained through the process of binary segmentation. In this instance, it was 0.64. The results of segmentation were values of 0.54, 0.56, and 0.97.

2   Related Works

This article discusses how to diagnose COVID-19 pneumonia using chest CT scan [8] and how to do this classification. Segmented lesions may aid in determining the severity of a patient's pneumonia and whether the patient need follow-up care. According to this study, a technique for detecting and segmenting COVID-19 patients may be performed using a full CT scan [9]. This is an original piece of writing. Three decoders, one common encoder, and three jobs are included in our system. The performance of the model is being assessed on a total of 1369 patients (including 449 COVID-19 patients, 425 normal people, 98 lung cancer patients, and 397 extra patients). Because these individuals are well-balanced in terms of gender, age, and health, they serve as an excellent standard against which to compare other patients. With these findings, our approach works admirably, with a segmentation coefficient of 0.88 and a receiver operating characteristic (ROC) of more than 97%.

Neural networks are being used more and more in a variety of areas, including healthcare, agriculture, education, and research. It also contributes to the development of a multidisciplinary approach to categorizing the current coronavirus outbreak (COVID-19). CNNs are a kind of neural network that has demonstrated significant effectiveness in the current COVID-19 epidemic. In this study [10], CNN models were employed to detect, categorize, and segment COVID-19 pictures, which were then analysed. The inception and extreme inception models are the most fundamental models used for detection and classification in computer science. The picture was segmented using U-Nets and voxel-based wide learning networks, which were both developed by Google. COVID-19, despite its small size, was able to integrate computational methods. To confirm our results, we used an experimental CNN to categorize the COVID-19 CT scan images. Our F_1-score was 0.93, which corresponded to 93% accuracy. It is evident that CNNs are very helpful for detecting and predicting COVID-19, especially as additional high-quality medical imaging data sets become accessible to researchers.

COVID-19, a new coronavirus that has been identified as a potential pandemic illness, has been found. At the time of writing, the disease had infected approximately 3.3 million people and claimed the lives of more than 4.4 million. A novel diagnostic technique is needed to slow the spread of COVID-19 and to assist decrease death rates. COVID-Lite is a system that integrates white balance, CLAHE [11], and a detachable CNN into a single unit (DSCNN). CLAHE is used to improve the visibility of CXR images during pre-processing, while DSCNN produces 8.4 MB of data without the need of quantization. When the COVID-Lite technique was used, custom DSCNN [12] beats conventional DSCNN in terms of performance. The accuracy of binary classification was 99.58% when using this technological approach, and the accuracy of multiclass classification was 96.43% when using this technological approach. COVID-Lite [4, 13–15] was finally operational. With an over a million confirmed cases, COVID-19's rapid spread has sparked widespread concern around the world. CT scans and other imaging tests may be used to detect the spread of the disease. COVID-19 pneumonia necessitates the use of automated detection. We explain in this paper the advanced research approach we utilized to detect COVID-19 patients and thoracic lesions in COVID-19-positive CT scans, which was developed by us. Linked operations data will be utilized to improve segmentation and classification via segmentation and classification [16, 17]. We use a multi-layer perceptron for the purposes of reconstruction and segmentation. Tens of thousands of individuals, including 449 with COVID-19, 100 healthy persons, 98 with lung cancer, and 397 with other diseases will be tested in this study [6, 18], which will use current fragmentation and classification methods as well as novel approaches. The model is presently under investigation. In our study, we found that our technique covered 93% of the region and had a segmentation coefficient of 0.78.

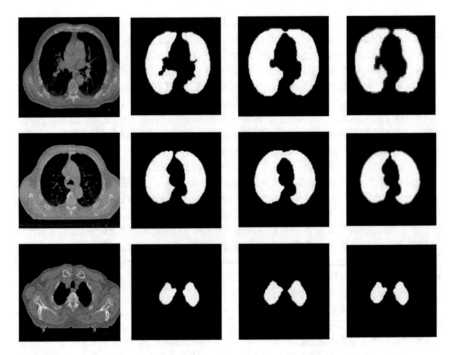

Fig. 1 Region-based segmentation with respective to the watershed segmentation

3 Methodology and Analysis

3.1 Region-Based Segmentation

It has four distinct stages: automatic selection, algorithm stoppage, area growth, and lung remediation. Automatic selection is the first stage. This algorithm is applied to all axial slices with a thickness of 2.5 mm or greater. This was shown in Fig. 1.

3.2 Threshold Segmentation

Threshold segmentation is a sophisticated parallel segmentation technique that is used in image processing. One archive contains files with grey fees that are higher than the average, while the other contains files with grey fees that are average or lower. Other methods include the minimum error method, the co-occurrence matrix method, on-site statistical methods, and hazardous entertainment methods, among others.

The threshold strategy is more frequently used because it is simpler to calculate and execute than other strategies. When intention and heritage come into conflict, segmentation results. The draw-off limit has been reduced because of the difficulty in accruing fines and penalties. While the image has a significant amount of greyscale overlap, which makes segmentation more difficult, it does not have a significant amount of greyscale or contrast range. The grey archives are part of the spatial archives of the image. This is supported by Fig. 2.

The field of image segmentation in Fig. 3 is still in its infancy. As new theories and ideas emerged across a wide range of disciplines, many convenient and useful image segmentation strategies were combined to create a more comprehensive picture. The term 'variety' refers to a collection of variables that are comparable to one another. Positive and negative requirements are grouped together to make it easier to categorize issues. When we use feature clustering, we can organize attributes into groups and assign them to specific points in the image. The k-means algorithm is one of the most widely used algorithms. The samples were clustered using non-convex clustering, which was accomplished using the k-means strategy as shown in Fig. 3.

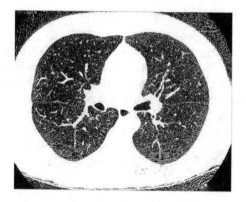

Fig. 2 Threshold segmentation after greyscale overlap

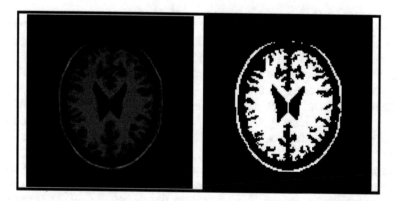

Fig. 3 Threshold segmentation after applying the convex clustering

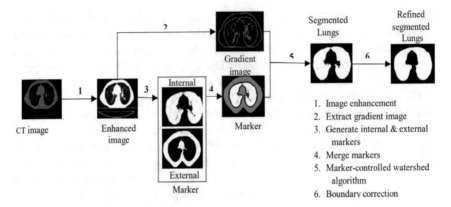

Fig. 4 Flow chart of the proposed architecture

3.3 Region Growing in Proposed System

To detect features such as greyscale, texture, colour and shape, the image is segmented using pixel-to-pixel correlation, and segmentation algorithms. When there are several similar house types in an image, a group of pixels must be used to separate the different areas. There are four steps involved in creating a region. The location of the seed pixel crew is indicated by the letter I in the image. Create a stop rule that will be activated when the stage becomes completely grey or when the coloration stops. More pixels that meet the area's inclusion criteria should be removed from the map. The proposed architecture was shown in Fig. 4.

3.4 Active Contour Models

The active contour models developed for interactive 2D image segmentation were first used in 1987. Rather than starting from the beginning, a top-down approach. Edge features are detected and associated with the shapes of the boundaries. In the preceding model, a contour is a time-varying curve, and segmentation is the optimization of an adequate energy function that is appropriate for the situation. Specifically, the COVID-19 classification method is employed to segment the market (FFBP). Before processing, identify COVID-19 by using image annotation and contrast editing techniques. After you have sectioned the image, you can remove an injection region.

An improved imaging workflow with AI functionality has been developed for better scanning and lower radiation doses administered to patients. In another instance, patient-determined X-ray exposure parameters are robot-optimized to ensure that the scan does not use any more radiation than is required by the manufacturer. Because early medical pixels produce weaker radiological signals than later

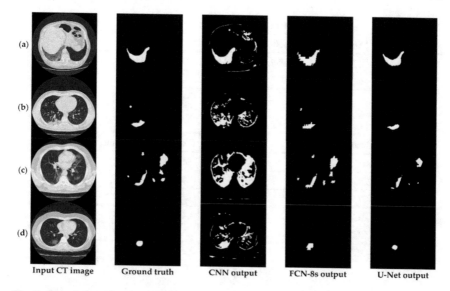

Input CT image   Ground truth   CNN output   FCN-8s output   U-Net output

Fig. 5 Output after the segmentation with U-Net

medical pixels, understanding the disease's early stages is essential for avoiding misdiagnosis. Most of the artificial intelligence segmentation and forecasting studies use small sample sizes, which increases the clinical utility of the findings as shown in Fig. 5.

4 Results and Discussions

The results of both binary classifier models are shown in Fig. 8. Our networks are accurate in all cases, with an accuracy rate greater than 0.90 and the highest accuracy rate of 0.95. The standard deviation for experiment 4 is 0.029. The accuracy of U-NET experiment 4 is 0.95, while the default accuracy is 0. 043. Both networks have similar sensitivity and G-mean values to one another. The outcomes are listed in Table 2. The outcome also determines which experiment is the most effective for each multi-classification network. SegNet experiment 7 has a default value of 0.06 and is 97% accurate, according to the results. It was the fourth U-NET experiment, with a standard deviation of 0.065, that had the best standard deviation, which was 0.908, that had the best standard deviation. The accuracy of the other experiments is all greater than 0.8.

The results of binary segmentation show in Tables 1 and 2 which experiments performed best with which architectures, and which experiments did not. Both networks had difficulty identifying the C3 classification. They are successful in achieving C1 and C2 objectives. In addition, each class has a high rate of

Table 1 Performance metrics of the SegNet and U-Net

Net	Sens	Spec	Dice	G-mean	F_2
SegNet	0.956	0.9542	0.749	0.955	0.861
U-NET	0.964	0.948	0.733	0.956	0.856

Table 2 Class of the classifiers and dice coefficient

Net	Class	Sens	Spec	Dice	G-mean	F_2
SegNet	C1C1	0.638	0.0.952	0.479	0.780	0.562
	C2C2	0.672	0.965	0.454	0.806	0.564
	C3C3	0.574	0.988	0.121	0.753	0.231
U-NET	C1C1	0.804	0.930	0.483	0.865	0.0.636
	C2C2	0.694	0.983	0.597	0.826	0.652
	C3C3	0.684	0.993	0.225	0.824	0.377

specialization among its members. However, the U-NET architecture provided more non-specifics than the other architectures.

4.1 Binary Classification Issues

As given in Table 2, SegNet has a significant advantage over U-NET. Both of the 'not infected' networks have a high percentage of true positives. The results demonstrate how DNN models distinguish between categories of non-infected and infected individuals (lung diseases). This finding is likely to be confirmed in a larger study, which is currently underway. The best network (SegNet) for the job (0.956) successfully simulates the performance of a skilled radiologist (0.945). Abnormality: The standard deviation ranges from 0.060 to 0.086 in Table 1. The values demonstrate that the test partition of the data set is extremely stable. SegNet outperforms Inf-Net and Semi-Inf-Net in terms of dice, specificity, and other metrics. In addition, U-NET outperforms its competitors in terms of sensitivity and responsiveness.

The two data sets are the same. Border information is treated as a binary segment that represents a computation in the Inf-Net model. The parallel decoder of the network makes use of critical features to draw attention to minor structural elements as well. Even though the SegNet was trained on lung-related images, the database was not included. Semi-infinite neural networks are used to generate training data, with a small margin of pseudo-labelled data included in the mix. The pseudo-labelling technique was used to create the 1600 labels. The COVID-SegNet metrics, which are based on dice, are out of bounds. To summarize, it is necessary to compare the training and testing results of both architectures on the same data set. A decrease in network performance can be observed when the mini-batch size is increased; however, further investigation is required to determine why. It has previously been investigated the role of mini-batch size in the convergence of the VGG16 network. Training results are improved by using smaller miniatures and lower learning rates. Another study discovered that smaller miniatures produce more stable ResNet networks because of the higher frequency with which gradient calculations are performed.

5 Conclusion

To diagnose and make assumptions about medical imaging images, it is necessary to conduct an analysis of the images. Image segmentation is included in this. These techniques aid in the location and identification of the target area (ROI). As a result of the COVID-19 pandemic, effective segmentation methods are required to assist in the diagnosis of pandemic diseases. In conclusion, this study proposes a method for segmenting COVID-19 CT images that is both faster and more accurate. In this study, two deep learning networks (SegNet and U-NET) were used to detect diseased areas in the lungs of COVID-19 patients using medical images taken of their lungs. In these images, SegNet was able to distinguish between healthy and infected tissue after being trained on them. In addition, the ability of the two networks to classify infected areas in lung images was compared. The findings demonstrated that the U-NET network is capable of distinguishing between these locations. Deep learning can be used to detect COVID-19 in pulmonary CT scans, according to this study.

References

1. Doppala BP, Naga Mallik Raj S, Stephen Neal Joshua E, Thirupathi Rao N (2021) Automatic determination of harassment in social network using machine learning. https://doi.org/10.1007/978-981-16-1773-7_20. Retrieved from www.scopus.com
2. Eali SNJ, Bhattacharyya D, Nakka TR, Hong S (2022) A novel approach in bio-medical image segmentation for analyzing brain cancer images with U-NET semantic segmentation and TPLD models using SVM. Traitement Sig 39(2):419–430. https://doi.org/10.18280/ts.390203
3. Eali SNJ, Rao NT, Swathi K, Satyanarayana KV, Bhattacharyya D, Kim T (2018) Simulated studies on the performance of intelligent transportation system using vehicular networks. Int J Grid Distrib Comput 11(4):27–36. https://doi.org/10.14257/ijgdc.2018.11.4.03
4. Singh VK, Rashwan HA, Romani F, Akram F, Pandey N, Sarker MMK et al (2020) Breast tumor segmentation and shape classification in mammograms using generative adversarial and convolutional neural network. Expert Syst Appl 139:112855
5. Joshua ESN, Battacharyya D, Doppala BP, Chakkravarthy M (2022) Extensive statistical analysis on novel coronavirus: towards worldwide health using apache spark. https://doi.org/10.1007/978-3-030-72752-9_8. Retrieved from www.scopus.com
6. Sun Y, Shi H, Zhang S, Wang P, Zhao W, Zhou X et al (2019) Accurate and rapid CT image segmentation of the eyes and surrounding organs for precise radiotherapy. Med Phys 46(5):2214–2222. pmid: 30815885
7. Joshua ESN, Bhattacharyya D, Chakkravarthy M (2021) Lung nodule semantic segmentation with bi-direction features using U-INET. J Med Pharm Allied Sci 10(5):3494–3499. https://doi.org/10.22270/jmpas.V10I5.1454

8. Joshua ESN, Bhattacharyya D, Chakkravarthy M, Kim H (2021) Lung cancer classification using squeeze and excitation convolutional neural networks with grad cam++ class activation function. Traitement Sig 38(4):1103–1112. https://doi.org/10.18280/ts.380421

9. Joshua ESN, Chakkravarthy M, Bhattacharyya D (2021) Lung cancer detection using impro-vised grad-cam++ with 3D CNN class activation. https://doi.org/10.1007/978-981-16-177 3-7_5. Retrieved from www.scopus.com

10. Neal Joshua ES, Rao NT, Bhattacharyya D (2022) Managing information security risk and internet of things (IoT) impact on challenges of medicinal problems with complex settings. In: Multi-chaos, fractal and multi-fractional artificial intelligence of different complex systems, pp 291–310. https://doi.org/10.1016/B978-0-323-90032-4.00007-9. Retrieved from www.sco pus.com

11. Neal Joshua ES, Thirupathi Rao N, Bhattacharyya D (2022) The use of digital technologies in the response to SARS-2 CoV2-19 in the public health sector. In: Digital innovation for healthcare in COVID-19 pandemic: strategies and solutions, pp 391–418. https://doi.org/10. 1016/B978-0-12-821318-6.00003-7. Retrieved from www.scopus.com

12. Rao NT, Neal Joshua ES, Bhattacharyya D (2022) An extensive discussion on utilization of data security and big data models for resolving healthcare problems. In: Multi-chaos, fractal and multi-fractional artificial intelligence of different complex systems, pp 311–324. https:// doi.org/10.1016/B978-0-323-90032-4.00001-8 Retrieved from www.scopus.com

13. Havaei M, Davy A, Warde-Farley D, Biard A, Courville A, Bengio Y et al (2017) Brain tumor segmentation with deep neural networks. Med Image Anal 35:18–31. pmid: 27310171

14. Pu J, Roos J, Chin AY, Napel S, Rubin GD, Paik DS (2008) Adaptive border marching algorithm: automatic lung segmentation on chest CT images. Comput Med Imaging Graph 32(6):452–462. https://doi.org/10.1016/j.compmedimag.2008.04.005

15. Zhao W, Jiang D, Queralta JP, Westerlund T (2020) MSS U-Net: 3D segmentation of kidneys and tumors from CT images with a multi-scale supervised U-Net. Inf Med Unlocked 19:100357

16. Zhu H, He H, Xu J, Fang Q, Wang W (2018) Medical image segmentation using fruit fly optimization and density peaks clustering. Comput Math Methods Med

17. Shakeel PM, Burhanuddin M, Desa MI (2019) Lung cancer detection from CT image using improved profuse clustering and deep learning instantaneously trained neural networks. Measurement 145:702–712

18. Sushma D, Thirupathi Rao N, Bhattacharyya D (2021) A comparative study on automated detection of malaria by using blood smear images. https://doi.org/10.1007/978-981-15-951 6-5_1. Retrieved from www.scopus.com

Deep Neural Transfer Network Technique for Lung Cancer Detection

B. Dinesh Reddy⬤, N. Thirupathi Rao⬤, and Debnath Bhattacharyya⬤

1 Introduction

Image processing may be a computer vision manipulation of images. There are several options for manipulating images with the occurrence of technology. In certain places, text recognition plays a significant role. But such a task on a computer is challenging to attempt. We must train the machine to understand the text. The acquisition, role extraction, classification, and identification of character require many steps. Handwriting recognition is a machine's ability to obtain and to perceive handwritten data from an outside source such as an image. The project's main objective is to design a framework that can identify successful character of a neural network format efficiently [1–4]. Neural computing may be a relatively new area, and thus style components are less well defined than the others. Neural computers use parallel data. The operation of a neural machine is distinct from the processing of a standard computer. Neural computers have been trained and do not seem to be coded. To compare the provided data to the trained system and to provide the user with the acceptable output text, it is also simpler to encode, store, edit, index, and locate in digital documents than to spend hours browsing through handwritten documents. And secondly, it is not only time-consuming to scan for anything in a large non-digital text, but more possible that we will overlook the information as we manually

B. Dinesh Reddy (✉) · N. Thirupathi Rao
Department of Computer Science and Engineering, Vignan's
Institute of Information Technology (A), Visakhapatnam, AP, India
e-mail: dinesh4net@gmail.com

N. Thirupathi Rao
e-mail: nakkathiru@vignaniit.edu.in

D. Bhattacharyya
Department of Computer Science and Engineering, Koneru Laksmaiah Education Foundation,
Greenfield, Vaddeswaram, Guntur 522502, India
e-mail: debnathb@kluniversity.in

navigate the document. Keep up the outstanding work for us. Every day machines are better at completing people's jobs, they also work better than us.

Text detection strategies are essential for the detection of image text and the formation and bounding of text around the portion of the image. Standard methods for objection identification are also trying to deploy here. By utilizing the sliding window technique, they can form the bounding box around the document. This is, though, a costly task for machines. In this method, like a co-evolutionary neural network, a glass window moves through the image to sense the text inside the window. They tried to prevent the text part of various sizes of different window sizes. The sliding window has a fully convolutional influence that can minimize computational time. There are single-shot systems developed such as YOLO (you only look once) and regionally dependent image text detection techniques.

They only pass the image once to detect text in the region. Unlike the sliding glass, YOLO is single-shot approaches. Region-based operates in two measures on the other side. First, the network suggests an area that should be checked and then classifies whether the region has the text. The next step is to identify the text until we detect bounding boxes with the text. The text is understood by many strategies. CRNN is the hybrid for image-based sequence recognition functions, such as scenes text recognition and OCR, which includes the losses of CNN, RNN, and connections temporal classification (CTC). Including the extraction, sequence modelling, and transcription into a single structure, this designs for the neural network. This model involves no segmentation of the character. The neural network derives the input representation from the text detected region. The deep bidirectional repetitive neural network predicts the mark series for a certain interaction between the characters. The transcript layer translates RNN per frame into a sequence of labels. There seem to be two transcription modes, lexicon free, and lexicon based. The highest possible mark series would be expected in the lexicon-based method.

In the recent years, the object recognition has become a testing centre and is the core of many smart transport systems. Although deep learning has quickly advanced, it speeds identification of artefacts up. Many technologies for identification of deep learning artefacts have created tremendous breakthroughs and outstanding results. It may split item identification into methods of one level and two methods of stage. Two-stage methods target detection algorithm typically comprise two stages. The first item that is to receive regional proposals is the original image. Secondly, the area proposals are observed by the classification and regression networks such as the R-CNN [5] sequence. The one-stage process object detection algorithm involves just one step. One-stage approaches may execute the tasks of classification and bounding box regression explicitly without separately finding geographic proposals.

In short, the inputs are: (1) for multiple artefacts, we suggest an object-text detection model which can simultaneously detect texts and objects, (2) we suggest a system for text recognition that integrates text detection and recognition effectively, and (3) the approach that we introduced can detect several object types and of instantiating the identification based on the text labels found. They often use the text label as a valid object identity. Paper is organized as follows, Sect. 2 is related works, Sect. 3 is method and methodologies, and Sect. 4 is results and discussions.

2 Related Works

Some of the relevant and recent papers are mentioned here with their critical findings. In [6], the authors implemented deep convolutional neural networks for open text readings in end-to-end text on historical maps. The word-bound boxes in arbitrary directions and sizes are predicted on a text sensing network. The images of the observed term with a stable recognition network are then normalized. Because correct recognition involves vast amounts of training data but very little manually labelled data, we implement dynamic map text synthesizers that include an almost overabundance of training data. Results on a labelled 30-maps data collection of over 30,000 text labels shall be analysed.

In [7], a new neutral ICPT-CNN network was created. The learning and text/non-text classification roles within the ICPT-CNN structure are combined into one full mechanism contributing to an efficient text detection model. To enhance the learning capability of the network, they employed local response standards and the Leaky ReLU feature. They test the suggested solution with standard benchmarks and produce a variety of contemporary performance. F-scores of their system increased by 1%, provided other specialized approaches maintain the same accuracy. Further character recognition tests reveal that our ICPT-CNN is gradually descriptive.

In [8], the authors suggested an algorithm for deep neural network (DNN) that can solve the spatial objective structural characteristics identification problem. For extracting features without distinguishing degrees, they have used fractal characteristics for the extraction of the fractal features of the RCS sequences, and for selecting conventional features, they used the Fisher decision rate. In addition, they started the processing of the DNN algorithm and data. Finally, the algorithm was tested with a series of simulation test results. The results of the study show that the DNN algorithm is stable and exact to resolve using the RCS series to evaluate the goal structure.

In [9], the authors showed that the very problem be overcome by using the STN-OCR proper model of deep neural and spatial transformer networks (DNN) (STNs). This study's network design comprises two components: the network translation and the network identification. In the localization network, text areas are identified and placed and sampling grid is created. Whereas it enters text regions into the recognition network, this network learns to identify text with short, curved, and multi-oriented resolution. Deep learning methods need multiple data for successful testing, so they have used two benchmark data sets in this analysis.

In order to recognize and instantly classify ticket messages, authors suggested a system of deep learning [10]. Two aspects of this process are identification and recognition of ticket texts. The text detection model CTPN is first employed to detect the ticket's horizontal messages. Then, the observed text image is cut to grey by its coordinate location and resized. After that, the improved CRNN model is used to identify text with including self-attention in role extraction layer. Experimental results show that the approach suggested will identify and understand text in the ticket automatically. These addresses the problem that the conventional approach requires a large period and consumes too much resources to extract functionality.

3   Methodology and Methods

This paper's network construction comprises two components: the detection network for object-text and the network for text recognition. We use YOLOv3 architecture to detect objects and texts on real-scene images using a complete convolutional neural network [11]. In order to extract the characteristics from the image in several scales, the convolutional network. The regression classification and bounding box networks directly output the object scoring, the object class and the offsets for the object coordinates on multiple functional maps. We'll remove redundancy bounding boxes with large overlapping of the same object with [12] not the maximum deletion (NMS). We adopt [13] convolutional recurrent neural network (CRNN) as the successful scene text recognition algorithm together with object-text detection. We cut the text regions off from the original image under the text type coordinates that are output from the object-text detection network [14–17]. To extract features from the text regions, we use a convolutional neural network. The feature maps extracted must be scaled to a standard height with a fixed aspect ratio. We encode the feature sequences of the feature maps and CTC using our recurring model to decode the encoded sequence. Figure 1 shows the network structure we propose.

Architecture of the Object-Text Detection Network

Pre-processing: Pre-treatment: Pre-processing means translating raw data to a correct network input format. We only have to resize the picture and encode the label one hot for the image classification task. But for YOLO v3, things are a little more difficult. Recognize that we have said network output is $52 \times 52 \times 3 \times$ and is three separate scales $(4 + 1 + \text{num classes})$? We must also format our ground truth into such matrix, since we must calculate the delta between ground truth and prediction, first. We must select the best scale and anchor for each box that borders on the ground truth. A tiny kite in the sky, for instance, must be small (52×52). And if the dragonfly is more like a square on the picture, we must choose also the square anchor on this scale. The author offers 9 anchors for three levels in YOLO v3. We

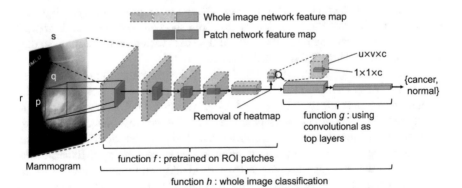

Fig. 1 Overall architecture of the proposed network architecture

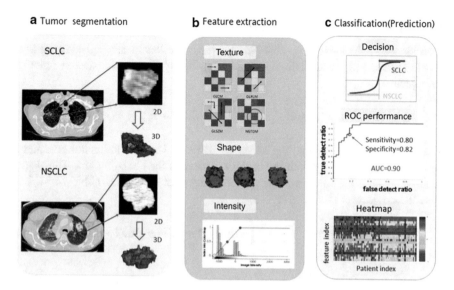

Fig. 2 Feature extraction network

only choose the one which most suits our basic truth box. We know the YOLO V3 function extractor as Darknet-53. ResNet introduced the idea of skip connections to help deeper layers propagate without decreasing the level. Darknet-53 takes this idea and expands the network successfully between 19 and 53 layers, as can be seen from Fig. 2.

Take layers as residual blocks in each rectangle. The network infrastructure is a multi-block chain with 2 Conv layers between steps in order to reduce the size. The structure of the bottle (1×1 followed by 3×3) and a skip connection exists inside the block. If the objective is to classify the multi-class layers as ImageNet does, an average pooling and 1000 ways is added. We will not include this classification head, however, with object detection. Therefore, we will add to this feature extractor a "detection" head. And as YOLO V3 is a multi-scale detector, we also need multi-scale features. Features of the last three residual blocks are therefore used for subsequent detection. The input is 416×416 in the diagram below so that 3 vectors would be 52×52, 26×26 and 13×13.

We define the method that transforms offsets to bounds as:

$$b_x = \sigma(t_x) + c_x,$$
$$b_y = \sigma\left(t_y\right) + c_y,$$
$$b_w = P_w e^{t_w},$$
$$P_h = P_h e^{t_h}, \tag{1}$$

where b_x, b_y, b_w, and b_h are the coordinates of the bounding box, c_x, c_y, p_w, and p_h are the coordinates of the anchor box, and $\sigma(\cdot)$ signifies the sigmoid function.

Loss Function of the Object-Text Detection Network

Confidence in objects is the probability that the object-text exists in the anchor box. Confidence loss abjectness adopts cross-entropy binary. The confidence loss function of objects is

$$L_{\text{conf}}(0, C) = -\sum \big(o_i \ln(\hat{c}_i)\big) + (1 - o_i)\ln\big(1 - \hat{c}_i\big), \ \hat{c}_i = \text{sigmoid}(c_i), \quad (2)$$

where $o_i \in \{0,1\}$ represents the existence of the object-text in anchor box and $\hat{c}$ represents the sigmoid probability of the existence of the bounding box. The classification value of the object-text is the probability of the class of the object-text. The loss class object-text function is

$$L_{\text{cls}}(O, C) = -\sum_{i \in \text{Pos}} \sum_{j \in \text{cls}} \Big(o_{ij} \ln(\hat{C}_{ij}) + (1 - o_{ij})\ln(1 - \hat{C}_{ij}) \Big),$$

$$\hat{C}_{ij} = \text{Sigmoid}(c_{ij}), \quad (3)$$

where $O_{ij} \in \{0,1\}$ signifies the presence of the object-texts' class j in anchor box i and $\hat{c}_{ij}$ signifies the sigmoid probability of the class j of the bounding box i.

The detection model of object-text predicts the offsets for co-ordinates among anchors and bounding boxes. For converting offsets to the bounding box coordinates, Eq. (1) is used. Object-text location loss takes the GioU model to obtain the error between bounding box and ground truth. The GioU is the generalized intersection over Union method. The algorithm of GIoU Loss is set to Algorithm 1. $L_{\text{loc}} = $ GIoU_Loss is used to form the loss of position of the object-text. The overall loss function can be,

$$L_{\text{tol}} = \alpha L_{\text{conf}} + \beta L_{\text{cls}} + \gamma L_{\text{loc}} \quad (4)$$

where α, β, and c are the weights of each loss. We empirically set $\alpha = \beta = c = 1$.

The area under the PR curve is named average precision (AP) can be written as

$$\text{AP} = \int P(R)\mathrm{d}R \quad (5)$$

where $P(R)$ is the measured precision against recall mAP. The mAP is the average of all categories of AP.

Anchor Box

The intent is to get a boundary box and its class. Object detection bounding in the xmin, ymin, xmax, and ymax formats usually is normalized. The top left corner in the box is the middle of the image, e.g. $0.5x$ min and $0.5y$ min. It confronts us with an intuitive regression problem if we want to get a numeric value like 0.5. We can also just predict values in the network and use the Mean Square Error (MSE) to compare

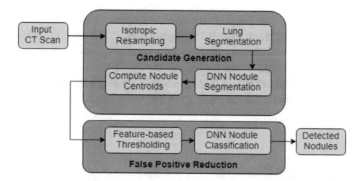

Fig. 3 Comparison of non-maximum suppression

the ground. That being said, given the large variation in the dimension and aspect ratio of boxes, researchers found it difficult to converge on the network if we only use this "brute force" to find a bounding box. We therefore suggest the concept of an anchor box in Faster-R-CNN.

Post-processing on Maximum Suppression (NMS)

A post-processor is the final component of this detection system. Post-processing is usually only trivial things, such as substituting machine-readable class ID with a readable text of a human class. However, we have one more important step to take regarding object detection to get final human-readable results. We refer this to as non-maximum deletion. We will not penalize nonbloc with false initiatives that have a huge overlap with the basic truth. In order to facilitate the training, we are empowering the network to predict close results. Even if not used in YOLO, multiple windows could expect the same object when the sliding window approach is used. Smart investigators designed an algorithm known as non-maximum deletion (NMS) to eliminate those duplicate results (Fig. 3).

NMS is straightforward. Find the first confidence detection box and add it to the result and then remove all other IOU boxes with the best box at some threshold. Next, we have chosen another case with the best faith in the rest of the boxes and do the same until nothing remains.

Text Recognition. Based on the text class, we select the bounding boxes for the text when the object positions are detected from the object detection network. First, the text extractor extracts the text regions that are the text bounding coordinates of the object-text detection module text boxes. Then, the text recognition module processes the extracted text regions by transforming them into a convolutional neural network, before it feeds them into. We scale the text regions to (32, 100, 3) where 32 is fixed, 100 is maximum, 3 is the maximum, and 3 is the number of the image channel. Eventually, the scaled area of the text an input for the coevolutionary strata. As our textual recognizes, we adopt the CRNN model.

First, feature maps from the pre-processed text space extract the co-evolutionary layers. From left to right, the function maps extract a pattern of feature vectors. The recurring layers will then be inserted into each frame of the sequences, which represents a vertical region corresponding to the original text image. The recurring layers use the deep bidirectional LSTM to encode the sequence of function vectors. After this, we implement CTC to predict the text label that matches recurrent layers sequences.

3.1 Training and Testing Data Splitting

The most accurate CNN model for recognizing opacity on chest X-rays was examined to determine which was the most accurate. Because this subject was separated into two sections, we divided our talk into two halves. From the start, all three class pictures were created and used in the training, validation, and testing sets. After then, the number of class photographs was reduced to two. As given in Table 1, the training data was used to train a large number of CNN models. The model's performance was compared using the test set. In phase 2, the model was put to the test to [18] see whether it could distinguish between samples with and without lung opacities (positive and negative). This model was not the first to use a similar number of validation rounds during the training phase. Each level was constructed in the same manner as the one before it. Deep CNN architectures like as Inception V3, Mobile Net, and ResNet152 were built from the ground up and tested using phase 1 and phase 2 test sets. Each image is reduced in size to 224 × 224 pixels before being included in the collection. The model was previously assessed using a 384 × 384 image.

The results of the two evaluation matrices were almost identical. Using photographs that are half the size of the original may allow you to save time in the classroom (224 × 224). When both the valid and test sets were available, the training set was sample wise normalized to account for the inconsistencies between them. The data was normalized [19] depending on the characteristics of the samples. When the size of the training set is increased, model generalization and overfitting are reduced to a lesser extent. Random rotations of 5° or less, as well as horizontal flips, were employed during the early stages of training. Because learning technology developments, only the AUC ratings of Xception, Densenet 201, InceptionV3, ResNet152, and Mobile Net increased. During training, the convolutional layers are not altered in any way.

Table 1 Performance of different detection frameworks

Detection frameworks	mAP
Fast R-CNN	72.2
YOLOv3 544 × 544	78.5
Proposed	81.4

This layer is provided by a global spatial average pooling layer, which is the last totally connected layer of each model. The last layer of each training model was linked together. In its place, a new, equal layer has been added. Aside from that, during phase 1, this completely connected layer produces three-dimensional output for three classes, but binary output for two classes during phase 2. An upgraded version of the loss was constructed and updated using a weighted binary cross-entropy function (Sect. 3). As a consequence, Adam was always tinkering with his network configuration. If the validation loss persisted after two epochs, the initial learning rate would be reduced by a factor of ten. Each cycle includes data for both forward and backward training. The epochs were divided into 16 groups of 16, totalling 32 epochs. The early stopping method was used to reduce overfitting and save weight. After each epoch, the check's accuracy is double checked. These approaches, as well as weight reduction, were not employed in this study.

4 Results and Discussions

We start the model on the COCO data set with pre-trained weights. We split the training process into two steps. During the first stage, we repair the network structure and only train the regression and classification network. We train the entire network in the second level (Fig. 4).

We primarily display identification findings from transportation test images as seen in Fig. 5. It may contain multiple objects in a single image in detection models. It works well with both small and large objects. Data set of text recognition includes tables, arrows, road signs, etc. Any texts occur and may be occluded in dynamic settings.

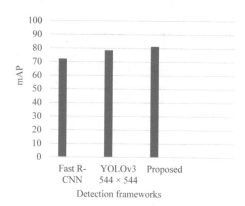

Fig. 4 Comparison of various models for measured average precision

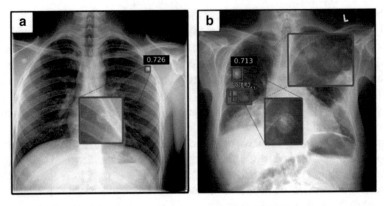

Fig. 5 Illustrations of the object detection result on PACSAL VOC2007

5 Conclusions

It was decided to utilize the RsNA pneumonia data set to train and evaluate a number of well-known deep CNN architectures for the detection of lung opacity in chest X-rays. In terms of AUC and accuracy, Xception has the greatest scores, scoring 91%. Ninety-one % (91%) the vast majority of the time. 91% of the population in addition, there is an extra 83.95%. Dense Net 201 also had the highest sensitivity (73.48%), highest net present value (87.85%), and highest F1-score of all the models tested (74.02%). The deep CNN architectures would have been unable to deliver greater sensitivity under these circumstances (above 80%). In order to enhance sensitivity, further study is necessary. With its binary data set (Lung Opacity vs. Normal Classification), exception's machine learning system attained the highest sensitivity (97.19%), AUC (99.1%), NPV (97.49%), F_1 (95.46%), and accuracy (95.46%) of any machine learning system ever developed (95.46%). Approximately 95.46% of the time the majority of the time (95.71%), the goal of future research will be to improve the sensitivity and specificity of CNN. This finding may pave the way for the development of an automated system for identifying lung opacity on chest radiographs in future.

References

1. Zhang F, Luan J, Xu Z, Chen W (2020) DetReco: object-text detection and recognition based on deep neural network. Math Prob Eng 2020:15, Article ID 2365076. https://doi.org/10.1155/2020/2365076
2. Doppala BP, Naga Mallik Raj S, Stephen Neal Joshua E, Thirupathi Rao N (2021) Automatic determination of harassment in social network using machine learning. https://doi.org/10.1007/978-981-16-1773-7_20. Retrieved from www.scopus.com
3. Eali SNJ, Bhattacharyya D, Nakka TR, Hong S (2022) A novel approach in bio-medical image segmentation for analyzing brain cancer images with U-NET semantic segmentation and TPLD

models using SVM. Traitement Sig 39(2):419–430. https://doi.org/10.18280/ts.390203

4. Eali SNJ, Rao NT, Swathi K, Satyanarayana KV, Bhattacharyya D, Kim T (2018) Simulated studies on the performance of intelligent transportation system using vehicular networks. Int J Grid Distrib Comput 11(4):27–36. https://doi.org/10.14257/ijgdc.2018.11.4.03

5. Joshua ESN, Battacharyya D, Doppala BP, Chakkravarthy M (2022) Extensive statistical analysis on novel coronavirus: towards worldwide health using apache spark. https://doi.org/10.1007/978-3-030-72752-9_8. Retrieved from www.scopus.com

6. Joshua ESN, Bhattacharyya D, Chakkravarthy M (2021) Lung nodule semantic segmentation with bi-direction features using U-INET. J Med Pharm Allied Sci 10(5):3494–3499. https://doi.org/10.22270/jmpas.V10I5.1454

7. Joshua ESN, Bhattacharyya D, Chakkravarthy M, Kim H (2021) Lung cancer classification using squeeze and excitation convolutional neural networks with grad cam++ class activation function. Traitement Sig 38(4):1103–1112. https://doi.org/10.18280/ts.380421

8. Joshua ESN, Chakkravarthy M, Bhattacharyya D (2021) Lung cancer detection using improvised grad-cam++ with 3D CNN class activation. https://doi.org/10.1007/978-981-16-1773-7_5. Retrieved from www.scopus.com

9. Neal Joshua ES, Rao NT, Bhattacharyya D (2022) Managing information security risk and internet of things (IoT) impact on challenges of medicinal problems with complex settings. In: Multi-chaos, fractal and multi-fractional artificial intelligence of different complex systems, pp 291–310. https://doi.org/10.1016/B978-0-323-90032-4.00007-9. Retrieved from www.scopus.com

10. Neal Joshua ES, Thirupathi Rao N, Bhattacharyya D (2022) The use of digital technologies in the response to SARS-2 CoV2-19 in the public health sector. In: Digital innovation for healthcare in COVID-19 pandemic: strategies and solutions, pp 391–418. https://doi.org/10.1016/B978-0-12-821318-6.00003-7. Retrieved from www.scopus.com

11. Rao NT, Neal Joshua ES, Bhattacharyya D (2022) An extensive discussion on utilization of data security and big data models for resolving healthcare problems. In: Multi-chaos, fractal and multi-fractional artificial intelligence of different complex systems, pp 311–324. https://doi.org/10.1016/B978-0-323-90032-4.00001-8. Retrieved from www.scopus.com

12. Akcay S, Breckon T (2022) Towards automatic threat detection: a survey of advances of deep learning within X-ray security imaging. Pattern Recogn 122. https://doi.org/10.1016/j.patcog.2021.108245

13. Chavaillaz A, Schwaninger A, Michel S, Sauer J (2019) Expertise, automation and trust in X-ray screening of cabin baggage. Front Psychol 10(Feb). https://doi.org/10.3389/fpsyg.2019.00256

14. Chavaillaz A, Schwaninger A, Michel S, Sauer J (2018) Automation in visual inspection tasks: X-ray luggage screening supported by a system of direct, indirect or adaptable cueing with low and high system reliability. Ergonomics 61(10):1395–1408. https://doi.org/10.1080/00140139.2018.1481231

15. Halder A, Chatterjee S, Dey D (2022) Adaptive morphology aided 2-pathway convolutional neural network for lung nodule classification. Biomed Sig Process Control 72. https://doi.org/10.1016/j.bspc.2021.103347

16. Mishra P, Swain BR, Sweta Padma A (2022) A review of cancer detection and prediction based on supervised and unsupervised learning techniques. https://doi.org/10.1007/978-981-16-5304-9_3. Retrieved from www.scopus.com

17. Thanetian C, Sekar KR (2022) Target projection feature matching based deep ANN with LSTM for lung cancer prediction. Intell Autom Soft Comput 31(1):495–506. https://doi.org/10.32604/IASC.2022.019546

18. International conference on medical imaging and computer-aided diagnosis, MICAD 2021 (2022). Retrieved from www.scopus.com

19. Lokhande NL, Jaware TH (2022) Lung CT image segmentation: a convolutional neural network approach. https://doi.org/10.1007/978-981-16-0739-4_37. Retrieved from www.scopus.com

Criminal Tendency Identification Using Deep Learning Approaches: A Novel Approach for Security Protection

Eali Stephen Neal Joshua⦿, **N. Tirupati Rao**⦿, **Debnath Bhattacharyya**⦿, and **Nakka Marline Joys**⦿

1 Introduction

Deep learning is the approach that is most often utilized [1–3]. It has the potential to be utilized to apprehend and deter criminals. Many programs are predicated on the idea that the majority of crimes are predictably committed. The capacity to filter through enormous volumes of data is required for the enforcement of laws. This data analysis would have [4] been impossible to do just a few decades ago, given the technological advances (Fig. 1).

Fortunately, recent advancements in machine learning have made this kind of investigation viable. While just 13% of shootings are reported to the police, sexual assaults are reported to the authorities in India at a rate of 97%. Only 20% [5–8] of shootings in the United States were reported to 911. An automated system that can detect crimes while also providing information about the crime scene and potential suspects would be quite beneficial in this situation. In the event of a crime,

E. Stephen Neal Joshua (✉)
Department of Computer Science and Engineering, Gandhi Institute of Technology and Management, Gandhi Nagar, Rushikonda, Visakhapatnam, AP, India
e-mail: seali@gitam.edu

N. Tirupati Rao
Department of Computer Science and Engineering, Vignan's Institute of Information Technology (A), Visakhapatnam, AP, India
e-mail: nakkathiru@vignaniit.edu.in

D. Bhattacharyya
Department of Computer Science and Engineering, Koneru Laksmaiah Education Foundation, Greenfield, Vaddeswaram, Guntur 522502, India
e-mail: debnathb@kluniversity.in

N. M. Joys
Department of Computer Science and Engineering, Anil Neerukonda Institute of Technology & Sciences, Sangivalasa, Bheemunipatnam, AP 531162, India
e-mail: marline.cse@anits.edu.in

© The Author(s), under exclusive license to Springer Nature Singapore Pte Ltd. 2023
D. S. Sisodia et al. (eds.), *Machine Intelligence Techniques for Data Analysis and Signal Processing*, Lecture Notes in Electrical Engineering 997,
https://doi.org/10.1007/978-981-99-0085-5_21

Pre-Crime Behavior   Suspicious Behavior   Crime Evidence
Segment (PCBS)   Segment (SBS)   Segment (CES)

Fig. 1 Use case of computer vision in crime scene detection

this gadget can capture photographs and analyses them in order to determine what occurred and inform the appropriate authorities. In this case, a deep neural network may be beneficial. It was capable of evaluating a broad variety of parameters from its surroundings in order to offer the best and quickest findings possible. Predpol is a good illustration of this. Predpol [9] is a system that uses big data and machine learning to predict criminal activity. Their theory is that some sorts of criminal activity have a tendency to congregate over long periods of time and over large geographic areas. They can predict future crimes based on data from the previous and present criminal [10] investigations. Police should increase their presence in certain regions on a map by using technological advancements. The crime rate in Washington was lowered by 22% as a result of this. To summarize, the use of deep learning in crime detection is advantageous, and intelligent systems may be enhanced by the addition of new skills learning inevitably. This will aid society in its efforts to resolve its problems.

2 Related Works

Attacks against a broad spectrum of software and people have lately been the target of cyberattacks [11]. There has been a great deal of investigation concerning real and possible assaults. Prior to these assaults, a variety of methods were used to get critical information about the target. The data from Twitter was utilized in this research to uncover and forecast security risks. This study analyses and assesses prior attempts to gather information on present and prospective cyber-attacks on national security, as well as the effectiveness of such programs. In order to get relevant findings in the cyber environment, a number of important factors must be taken into account throughout the survey comparison. An evaluation was made of many aspects, including the quantity of data collected from a certain place and how difficult it would be to execute the technique. According to previous research, a standard degree of success for each variable has been established. Both SYNAPSE and DataFreq need more development in order to provide more accurate forecasts in the future. Combining multiple surveillance studies, the goal of this research is to foresee cyber-security concerns in the future. Researchers will be able to make more precise cyber-attack predictions in the future as a result of this.

Criminal activity, emergency medical services, and traffic accidents must be predicted in order for emergency response and transportation firms to be successful [12]. There will be a large number of mismatched labels, and models that rely on data sufficiency will fail since these situations are, by definition, very uncommon. The presented models are insufficient for fine-grained dynamics and ineffective for real-world decision-making, as shown in this paper. Using a hidden chain-like mechanism, we are attempting to predict occurrences in a sparse environment in this work. In addition to the time dimension, other factors such as geographic location and category have a role in this chain effect. Spatial-categorical graph neural networks are a novel deep learning framework that is capable of dealing with the dynamic chain effect and fine-grained co-prediction of a number of events in real time. Three city-level event data sets were thoroughly reviewed in order to demonstrate the utility of our viewpoint and the success of the recommended strategy.

Machine learning models benefit company owners in a variety of ways [13]. Putting digital watermarks on their models during training might help model owners identify their models if they are stolen or used illegally. A malicious individual might utilize the prediction API from the original model as a training dataset to construct a new model. The extraction of models is a significant problem in this work. Due to the fact that the adversary trains the surrogate model, conventional watermarking techniques are ineffective. DAWN is our first effort at watermarking models in order to prevent model theft from occurring (Dynamic Adversarial Watermarking of Neural Networks). Prior to DAWN, watermarking systems were able to train their models in the normal way. DAWN changes the model's prediction API responses in response to a limited number of user requests (e.g., 0.5%). Customers that use the queries in this collection to train a surrogate model may see their watermarks on the model. Two fresh assaults on model extraction have been unsuccessful in their attempts to halt DAWN. However, because to DAWN's watermarking of all created surrogate models, model owners are able to validate ownership with higher certainty (with a confidence level greater than $1-2-64$) (0.03–0.5% of the population).

Digital evidence is used in a criminal investigation, and this kind of inquiry is known as "reactive forensic investigation." [14]. It makes no difference whether or not the insiders are apprehended. Unfortunately, the harm has already been done. We demonstrate how anomalies may be discovered using CAEs and a collaborative optimization network. The data from unlabeled digital evidence is retrieved with the use of CAEs and density estimation networks. It is possible that this method will assist us in anticipating insider dangers and conducting proactive forensic investigations. Insider threats may be very damaging if left unchecked. Using the data-driven bidirectional long short-term memory (BiLSTM) feature extractor, feature engineering may be completed much more quickly. BiLSTM derives a scalable feature representation from a feature set. Because insider threat detection is prone to producing false positives, a second approach is used to identify insider threats. It also has the ability to rectify hypergraphs. Our technique and methodology may be validated using publicly available benchmark data, which is readily available. In the tests, our models outperformed the best unsupervised approaches available.

This research investigates three spatiotemporal deep learning architectures that have the potential to aid in the prediction of future crime [15]. ST-ResNet, DMVST-Net, and STDN are the three network designs that are available (STD-Net). The AIs were trained using data from the Chicago police department. The RMSE and MAE measurements were used to evaluate the accuracy of the model (MAE). The STD-Net approach has the highest accuracy (0.89) and the lowest root mean square error (0.2870) of all the techniques. Both ST-ResNet and DMVST-Net have showed early signs of success. As compared to the DMVST-Net, the ST-ResNet has a smaller root mean square error as well as a lower mean absolute error. The DMVST-Net has a root mean square error of 0.4171 and a mean absolute error of 0.3455. Future versions of these algorithms will get information from a variety of sources, including weather and economic conditions. The hyperparameters of these algorithms may be improved by the use of evolutionary computing.

Intrusion Detection Systems for the Internet of Things (IoT) were developed in response to the increasing number of individuals who are employing IoT devices for a variety of purposes [16]. Intrusion Detection Systems can detect a wide range of intrusion actions, not simply the most unique. When a large number of intrusive activities can discover and fix security breaches, it is much easier to recover from security breaches. This study demonstrates how Internet of Things devices may provide uneven data that may be corrected using deep learning techniques. According to the researchers, the algorithms are unable to detect all forms of irregularities in imbalanced data unless they are oversampled first. The pre-processing step of current classification and deep learning systems use oversampling to improve performance. When it comes to predicting different invasions, the data reveals that the suggested strategy beats existing techniques in terms of accuracy, g-mean, precision, and recall.

3 Proposed Method

An ANN is used to further classify the data once it has been categorized. Artificial neural networks (ANNs) do not need to be complex or well-constructed in order to be effective. In our application, we feed the neural network 128 * 128-pixel grayscale photographs that were taken using a digital camera. To achieve the 0–1 range, divide the number of pixels in the photo by 255. We must first explain why we picked the activation function, loss function, and training technique that we did before moving on to explore the neural network's construction in detail. Gradients may be prevented from exploding by activation functions such as the sigmoid or the tanh. It is likely that the employment of non-saturated activation functions such as the rectified linear unit leads gradients to widen and become non-vanishing, as opposed to employing saturated activation functions. Both issues manifest themselves at lower synaptic weights, resulting in a reduction in network education. The presence of constraining gradients may make the detection of bursting gradients

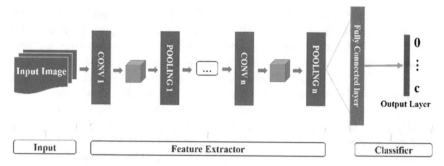

Fig. 2 Proposed architecture to classify criminals

simpler. Some of these elements may be useful in expediting the training process (Fig. 2).

As a result, ReLU was selected as the activation function. Choosing the positive interpretation, we can see that it is a linear function. $ReLU(z) = \max(0, z)$. It does this by decreasing negative inputs to zero, which is similar to dropping out of the system. The SoftMax function is sometimes referred to as $\exp(zi)/j$ in certain circles (zj). The probabilities are normalized exponential probability of one ($c\ pc = 1$). The values are normalized exponential probabilities of one (zi). Before using the cross-entropy loss function, be certain that the following conditions are met: If c were a prime number, it would be 1. You can examine how frequently a neuron in the output layer receives a 0 or 1 by using the $\log(pc)$ function (Fig. 3).

Fig. 3 Block diagram of proposed methodology

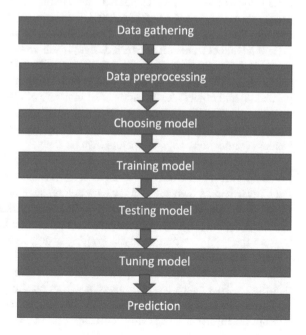

The loss function is defined as $Y \log(p) + (1\ Y) \log(p)\ (1\ p)$. Adam optimization is used to train the network in 100-batch increments. This approach extends the capabilities of stochastic gradient descent by adding a new characteristic. Adam's adaptive learning rates are governed by estimations of the first and second moments of the gradient. SGD, on the other hand, alters synaptic weights at a consistent rate. With respect to the first and second moment estimations, the decay rates were set at 0.9 and 0.999 percent of the original value, respectively. The learning rate is 0.0001, or one hundredth of a percent. Faces are classified into criminal and non-criminal categories using algorithms developed by SNN and CNN. Each of the SNN's four completely connected layers has 16,384 neurons, which is the total number of neurons in the SNN (or pixels). Each of the first three levels has a total of 512 neuron connections. Then there is the ReLU layer, which has the same number of neurons as the previous layer. This layer makes use of a SoftMax function with a magnitude that is proportional to the number of criminal and non-criminal target categories that are present.

4   Materials and Methods

Faces are classified into criminal and non-criminal categories [16–19] using algorithms developed by SNN and CNN. Each of the SNN's four completely connected layers has 16,384 neurons, which is the total number of neurons in the SNN (or pixels). Each of the first three levels has a total of 512 neuron connections. Then there is the ReLU layer, which has the same number of neurons as the previous layer. This layer makes use of a SoftMax [20–24] function with a magnitude that is proportional to the number of criminal and non-criminal target categories that are present. Recent advances in neural network design, machine learning, and image processing methods have enabled CNN to outperform alternative neural network designs, machine learning techniques, and image processing methodologies in picture categorization and object identification. The majority of CNN's [25] assumptions about how images function are correct. For example, pixel dependencies and statistical stationarity are both considered. Despite having the same number of layers, a CNN has less connections and parameters than an SNN.

Training becomes far less difficult as a result of this. In this research, two convolutional layers of networks are used, as well as two fully linked layers of networks. Each layer has its own set of filter parameters, such as size, stride, and number of filters. S_1 is one, while F_1 is three. If $f_2 = 3, 3, 8$, etc., then $n_1 = 8$. The max pooling and ReLU layers [26–28] are found in each convolution layer, and they are combined to form the final convolution layer. Pooling enables you to combine the outputs of clusters of neurons that share a kernel map in order to get a more complex result. Two-step stride and maximum pooling in the two-dimensional plane are shown by this figure. This keeps the pooling zones from clashing with one another. Smaller pooling zones have less variety, but larger pooling zones are overly generic and lack appropriate information, causing confusion (high bias). The neurons in the first layer are all connected to one another. On top of it, there is a ReLU layer. After the second entirely connected layer, a SoftMax function with the whole number of criminal and non-criminal target categories is employed to determine the final result of the algorithm. This occurs once the second layer has been completely joined.

4.1 Dataset Description

This dataset provides data from 2003 to 2017 for Vancouver, BC, based on the crime's nature (date), location (street name), and district. Many police reports are missing one or more of the four items stated below. Generally, law enforcement agencies keep two sets of reports: one for internal use and one for public use. When a publication is made public, not all material relevant to a researcher is included. The police reports used in this study ranged from 3 to 82 pages (one file had 589 pages). The brief accounts exclude vital academic information like the gunman's socioeconomic status and marital status. "I did not speak with any of the victims, witnesses, or suspects," said one officer. An initial site canvass revealed no cooperative witnesses willing to supply any information; an initial scene canvass revealed no cooperative witnesses willing to provide any information. It is also possible that different government agencies gather data in different ways, causing the research findings to be inconsistent.

Also, any police records gathered by a researcher will be kept private. It is possible that secret portions as small as a few words or as lengthy as entire paragraphs or even pages will be disclosed. When dealing with minors, it is vital to redact material to safeguard the victims' identities. In some cases, keeping private information private is possible. That' is another goal. The quantity of redactions allowed varies per state.

5 Results and Discussions

Convolutional layers are the characteristics of a CNN. The accuracy of a CNN demonstrates that the characteristics of the convolutional layers may be used to classify objects. In order to determine which facial traits are emphasized, we must examine what each convolutional layer focuses on the face. A convolutional layer is characterized by the presence of several filters. Using all of the filters together results in the formation of a feature. As a result, CNN has a better likelihood of picking high-quality photos. There are two convolution layers in our CNN. In the first, there are eight filters, and in the second, there are sixteen (Tables 1 and 2; Fig. 4).

Table 1 Performance evaluation of the proposed classifier

		Predicted_Values	
		Criminal	Non-criminal
Ground_Truth	Criminal	4881	142
	Non-criminal	192	4785

Table 2 Criminal tendency detection classifier

		Predicted	
		Criminal	Non-criminal
Truth	Criminal	4515	508
	Non-criminal	604	4373

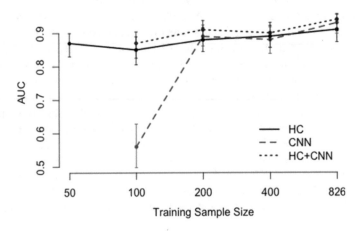

Fig. 4 ROC_AUC curve for criminal tendency with proposed architecture

6 Conclusions and Future Scope

Predicting crimes requires a great deal of effort, but the idea is straightforward. The purpose of this research was to provide assistance to academics who were forecasting crime. It provided them with assistance via the use of cutting-edge technologies. Face recognition and Sting Rays have been around for a while, but their adoption might fundamentally alter the way police officers conduct their duties. A system that is much more valuable to police officers was developed in this research via the application of machine and deep learning, as well as computer vision. It can also follow down offenders and identify them based on the voice notes they leave behind. The initial step is to design the system, and then there are considerations such as installation and operation to consider. But we have complete control over any issues that may arise in the future. We also have security systems in place that monitor the city 24 h a day, seven days a week to keep us safe. This kind of equipment may be used by police officers. Things will improve because tips and leads will be more dependable, and crimes will be investigated more quickly as a result.

References

1. Doppala BP, Naga Mallik Raj S, Stephen Neal Joshua E, Thirupathi Rao N (2021) Automatic determination of harassment in social network using machine learning. https://doi.org/10.1007/978-981-16-1773-7_20. Retrieved from www.scopus.com
2. Eali SNJ, Bhattacharyya D, Nakka TR, Hong S (2022) A novel approach in bio-medical image segmentation for analyzing brain cancer images with U-NET semantic segmentation and TPLD models using SVM. Traitement Sig 39(2):419–430. https://doi.org/10.18280/ts.390203
3. Eali SNJ, Rao NT, Swathi K, Satyanarayana KV, Bhattacharyya D, Kim T (2018) Simulated studies on the performance of intelligent transportation system using vehicular networks. Int J Grid Distrib Comput 11(4):27–36. https://doi.org/10.14257/ijgdc.2018.11.4.03
4. Joshua ESN, Battacharyya D, Doppala BP, Chakkravarthy M (2022) Extensive statistical analysis on novel coronavirus: towards worldwide health using apache spark. https://doi.org/10.1007/978-3-030-72752-9_8. Retrieved from www.scopus.com
5. Joshua ESN, Bhattacharyya D, Chakkravarthy M (2021) Lung nodule semantic segmentation with bi-direction features using U-INET. J Med Pharm Allied Sci 10(5):3494–3499. https://doi.org/10.22270/jmpas.V10I5.1454
6. Joshua ESN, Bhattacharyya D, Chakkravarthy M, Kim H (2021) Lung cancer classification using squeeze and excitation convolutional neural networks with grad cam++ class activation function. Traitement Sig 38(4):1103–1112. https://doi.org/10.18280/ts.380421
7. Joshua ESN, Chakkravarthy M, Bhattacharyya D (2021) Lung cancer detection using improvised grad-cam++ with 3D CNN class activation. https://doi.org/10.1007/978-981-16-1773-7_5. Retrieved from www.scopus.com
8. Neal Joshua ES, Rao NT, Bhattacharyya D (2022) Managing information security risk and internet of things (IoT) impact on challenges of medicinal problems with complex settings. In: Multi-chaos, fractal and multi-fractional artificial intelligence of different complex systems, pp 291–310. https://doi.org/10.1016/B978-0-323-90032-4.00007-9. Retrieved from www.scopus.com
9. Neal Joshua ES, Thirupathi Rao N, Bhattacharyya D (2022) The use of digital technologies in the response to SARS-2 CoV2-19 in the public health sector. In: Digital innovation for healthcare in COVID-19 pandemic: strategies and solutions, pp 391–418. https://doi.org/10.1016/B978-0-12-821318-6.00003-7. Retrieved from www.scopus.com
10. Rao NT, Neal Joshua ES, Bhattacharyya D (2022) An extensive discussion on utilization of data security and big data models for resolving healthcare problems. In: Multi-chaos, fractal and multi-fractional artificial intelligence of different complex systems, pp 311–324. https://doi.org/10.1016/B978-0-323-90032-4.00001-8. Retrieved from www.scopus.com
11. Bhattacharyya D, Dinesh Reddy B, Kumari NMJ, Rao NT (2021) Comprehensive analysis on comparison of machine learning and deep learning applications on cardiac arrest. J Med Pharm Allied Sci 10(4):3125–3131. https://doi.org/10.22270/jmpas.V10I4.1395
12. Matereke T, Nyirenda CN, Ghaziasgar M (2021) A comparative evaluation of spatio temporal deep learning techniques for crime prediction. Paper presented at the IEEE AFRICON conference, Sept 2021. https://doi.org/10.1109/AFRICON51333.2021.9570858. Retrieved from www.scopus.com
13. Mudgal M, Punj D, Pillai A (2021) Theoretical and empirical analysis of crime data. J Web Eng 20(1):113–128. https://doi.org/10.13052/jwe1540-9589.2016
14. Okawa M, Iwata T, Kurashima T, Tanaka Y, Toda H, Ueda N, Kashima H (2021) Deep mixture point processes spatio-temporal event prediction with external factor. Trans Japan Soc Artif Intell 36(5). https://doi.org/10.1527/tjsai.36-5_C-L37

15. Qaddoura R, Al-Zoubi AM, Almomani I, Faris H (2021) Predicting different types of imbalanced intrusion activities based on a multi-stage deep learning approach. Paper presented at the 2021 international conference on information technology, ICIT 2021—proceedings, pp 858–863. https://doi.org/10.1109/ICIT52682.2021.9491634. Retrieved from www.scopus.com

16. Safat W, Asghar S, Gillani SA (2021) Empirical analysis for crime prediction and forecasting using machine learning and deep learning techniques. IEEE Access 9:70080–70094. https://doi.org/10.1109/ACCESS.2021.3078117

17. Sushma D, Thirupathi Rao N, Bhattacharyya D (2021) A comparative study on automated detection of malaria by using blood smear images. https://doi.org/10.1007/978-981-15-9516-5_1. Retrieved from www.scopus.com

18. Sentuna A, Alsadoon A, Prasad PWC, Saadeh M, Alsadoon OH (2021) A novel enhanced Naïve Bayes posterior probability (ENBPP) using machine learning: cyber threat analysis. Neural Process Lett 53(1):177–209. https://doi.org/10.1007/s11063-020-10381-x

19. Sun J, Yue M, Lin Z, Yang X, Nocera L, Kahn G, Shahabi C (2021) CrimeForecaster: crime prediction by exploiting the geographical neighborhoods' spatiotemporal dependencies. https://doi.org/10.1007/978-3-030-67670-4_4. Retrieved from www.scopus.com

20. Szyller S, Atli BG, Marchal S, Asokan N (2021) DAWN: dynamic adversarial watermarking of neural networks. Paper presented at the MM 2021—proceedings of the 29th ACM international conference on multimedia, pp 4417–4425. https://doi.org/10.1145/3474085.3475591. Retrieved from www.scopus.com

21. Tang L, Mahmoud QH (2021) A deep learning-based framework for phishing website detection. IEEE Access. https://doi.org/10.1109/ACCESS.2021.3137636

22. Lavanaya D, Rao NT, Bhattacharyya D, Chen M (2020) Automatic identification of colloid cyst in brain through MRI/CT scan images. https://doi.org/10.1007/978-981-15-2407-3_6. Retrieved from www.scopus.com

23. Trirat P, Lee J (2021) DF-TAR: a deep fusion network for citywide traffic accident risk prediction with dangerous driving behavior. Paper presented at the the web conference 2021—proceedings of the World Wide Web conference, WWW 2021, pp 1146–1156. https://doi.org/10.1145/3442381.3450003. Retrieved from www.scopus.com

24. Varga D, Szoplák Z, Krajči S, Sokol P, Gurský P (2021) Analysis and prediction of legal judgements in the slovak criminal proceedings. Paper presented at the CEUR workshop proceedings, 2962, pp 161–170. Retrieved from www.scopus.com

25. Wang Z, Jiang R, Cai Z, Fan Z, Liu X, Kim K et al (2021) Spatiooral-categorical graph neural networks for fine-grained multi-incident co-prediction. Paper presented at the international conference on information and knowledge management, proceedings, pp 2060–2069. https://doi.org/10.1145/3459637.3482482. Retrieved from www.scopus.com

26. Wei Y, Chow K, Yiu S (2021) Insider threat prediction based on unsupervised anomaly detection scheme for proactive forensic investigation. Forensic Sci Int Digital Invest 38. https://doi.org/10.1016/j.fsidi.2021.301126

27. Wu Y (2021) The impact of criminal psychology trend prediction based on deep learning algorithm and three-dimensional convolutional neural network. J Ambient Intell Humaniz Comput. https://doi.org/10.1007/s12652-021-03455-8

28. Xie Y, Jin J, Zhang J, Yu S, Xuan Q (2021) Temporal-amount snapshot multigraph for ethereum transaction tracking. https://doi.org/10.1007/978-981-16-7993-3_10. Retrieved from www.scopus.com

Banana Ripeness Identification and Classification Using Hybrid Models with RESNET-50, VGG-16 and Machine Learning Techniques

Milan Bins Mathew, G. Surya Manjunathan, B. Gokul, K. Mohana Ganesh, S. A. Sajidha, and V. M. Nisha

1 Introduction

Bananas are one of the world's most widely consumed fruits. Having been cultivated for over 10,000 years, bananas are highly nutritious and are considered as an energy supplement during athletic events and sometimes even in medical emergencies. Well over 150 million tons of bananas are produced each year across the globe, and India leads the list with an estimated annual production of 30 million tons. But off these massive volumes of bananas produced yearly, an estimated 1.5 million bananas are thrown away everyday due to mismanagement of the product leading to an unnecessary wastage. A major problem faced by the industry is the mixing up of bananas of different ripeness together. Bananas are usually harvested when fully mature but still green. This is done so as to accommodate the factors of transportation from the plantations to the market and weather. Bananas are consumed in various forms (direct, juiced, dried, powdered, processed foods, etc.). Based on the final

M. B. Mathew (✉) · G. Surya Manjunathan · B. Gokul · K. Mohana Ganesh · S. A. Sajidha · V. M. Nisha
School of Computer Science and Engineering, Vellore Institute of Technology, Chennai 600127, India
e-mail: milanbins.mathew2019@vitstudent.ac.in

G. Surya Manjunathan
e-mail: suryamanjunatan.g2019@vitstudent.ac.in

B. Gokul
e-mail: gokul.b2019@vitstudent.ac.in

K. Mohana Ganesh
e-mail: mohanaganesh.k2019@vitstudent.ac.in

S. A. Sajidha
e-mail: sajidha.sa@vit.ac.in

V. M. Nisha
e-mail: nisha.vm@vit.ac.in

© The Author(s), under exclusive license to Springer Nature Singapore Pte Ltd. 2023
D. S. Sisodia et al. (eds.), *Machine Intelligence Techniques for Data Analysis and Signal Processing*, Lecture Notes in Electrical Engineering 997,
https://doi.org/10.1007/978-981-99-0085-5_22

utilization of the fruit, the ripeness also varies and is hence, utmost important to the food industry. Even today the commonly used technique to sort the bananas is to manually do it with humans which is done based on an experimental judgement. This is highly error prone, expensive and inefficient.

This is the problem that we will be trying to address here. Our vision is to use a computer vision enabled robot to do the sorting of the bananas as they arrive at the warehouses. This enables the entire process to be more streamlined, accurate and reliable.

While looking at some of the recent contributions similar to ours, one of the works by Sabilla et al. [1] used three machine learning algorithms, namely K-Nearest Neighbour (KNN), Support Vector Machine (SVM) and Decision Tree (DT) to determine the banana types and their ripeness levels. The banana is placed in a white background and photographed within 0.6 m with 17 different positions. The images are converted into greyscale and adjusted to 96×96 pixels in size. Principal Component Analysis (PCA) is conducted to reduce the dimensionality from 9216 pixels to 236 pixels and 128 pixels. In this research, SVM is able to provide highest accuracy compared to other methods, KNN and DT, but in contrast in our work, we used a hybrid of deep learning and machine model which is known for processing large volume of data with better accuracy. In the work by Kipli et al. [2] where the main contribution of this paper is the development of an expert system which would evaluate the ripeness stages of the banana. Here they utilize Google Cloud Platform, where the application sends the sample of banana image through Google Cloud Vision Application Programming Interface to get attribute readings from the sample image. The result of this work is then compared with application's database containing various attributes to determine the ripeness of the banana in the given sample. Here they used three types of classifiers which are Support Vector Machine (SVM), Discriminant Analysis and Rule-Based Classification. Finally, in Xiang's [3] work where they used MobileNetV2 with transfer learning to classify different fruit types. The main objective of their proposed work was to build a model which can achieve the balance of resource limitations and recognition accuracy due to the computation and storage limitations in the harvesters, which lead them used a lightweight neural network like MobileNetV2 to recognize the fruits.

Our proposed approach is to use pre-trained deep learning models VGG-16 and ResNet-50 for feature extraction. The models will be initialized with weights based on the ImageNet. In deep learning model, the Convolution Neural Networks (CNNs) are particularly efficient and superior when it comes to identifying and understanding abstract details in the data which is especially common in computer vision applications. Hence, we will be choosing the VGG-16 and the ResNet-50, two of the most popular CNN models for the purpose of feature extraction as they better understand the images and identify simple as well as complex geometric patterns. VGG-16 model has been chosen due to its effectiveness in learning patterns in images of various classes. Similarly, ResNet-50 was chosen in its ability to deal with the problem of vanishing gradients. This allows the ResNet model to better preserve or use the features extracted in early layers for the later layers so that we can make maximum usage of the data. In deep learning for multiclass classification, the commonly used classifier is a Softmax function. This function is a very simple and straight-forward approach and that is why we have proposed a techniques that combines popular machine learning classifiers such as SVM, KNN among others as they are faster and more efficient and can train on best on less but important data which will be the input in our model based on the extracted important features. The last fully connected layers of ResNet-50 and VGG-16 which act as classifiers are disabled and the output from the final block of the last convolutional layer is stored which will form the extracted feature vectors. The extracted features are then fed to the machine learning classifiers like SVM, KNN, Logistic Regression and Naive Bayes to classify the bananas based on their ripeness level taken into 4 categories (ripe, unripe, overripe and mid-ripe). The performance of each machine learning model is compared based on metrics like accuracy, R-Squared, Mean Squared Error (MSE), Mean Absolute Error (MAE), Precision, and we found that the VGG16+Logistic Regression classifier model performed consistently with better accuracy and least error compared to other models.

2 Dataset Description

The images in our dataset are collected by Fayoum University for their research on banana colour pigment. So, this dataset contains 350 images with 4 different class labels (ripe, unripe, overripe and mid-ripe). Each class has 90–120 images.

For implementing this project, we are considering 300 training images and 50 testing images. Some sample images in the dataset can be seen in Fig. 1a–d.

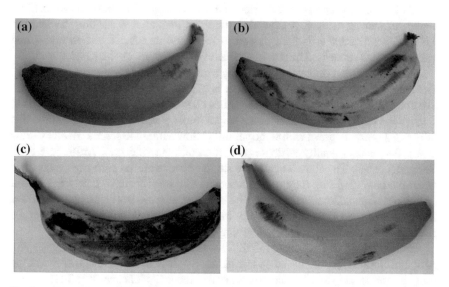

Fig. 1 **a** Green banana. **b** Mid-ripe banana. **c** Overripe banana. **d** Yellowish-green banana

3   Literature Survey

Bananas are rich in potassium Vitamins K and B6, fibre, tryptophan and amino acids [4]. Proper management of the fruit is highly essential and the problem is also seen in other fruits and vegetables as well.

Fadilah et al. [5] proposed an Artificial Neural Network (ANN) model for classifying oil palm fruit bunches according to their ripeness. The Multi-Layer Perceptron model which is a 3-layer ANN model which was used in the study. The images are segmented using K-means clustering, and the RGB segmented images are converted to HSI colour model to extract Hue values. It was reported that the proposed model had a highest accuracy of 86.67%.

El-Bendary et al. [6] suggested machine learning techniques for evaluating the ripeness of maturity. The techniques used for classification were Support Vector Machine (SVM) and Linear Discriminant Analysis (LDA) and for feature extraction Principle Component Analysis (PCA) is used. They were able to classify the fruit with an accuracy of 84%.

Mazen and Nashat [7] proposed another Artificial Neural Network model for classifying bananas based on their ripeness. The model is classified based on HSV and CIELAB characteristics, development of brown spots and Tamura texture feature. Levenberg–Marquardt backpropagation optimization algorithm is used for training the suggested artificial neural network. Sensitivity and Precision are used as performance metrics and the final results suggested that the proposed model had a correctness percentage of 97.

Sidehabi et al. [8] proposed a technique to classify passion fruit's ripeness using K-Means Clustering and Artificial Neural Network. The fruit was to be distinguished into 3 classes, namely ripe, nearly ripe and unripe stage. The proposed model first applies clustering using K-Means algorithm and then after extracting the features with RGB extraction it is then passed to the ANN that has 2 hidden layers and trained for 1000 epochs. The model attained 90% accuracy on the dataset used.

Adebayo et al. [9] suggested classifying bananas based on their optical properties. The main objective of this research paper is to predict the banana quality attributes like chlorophyll, elasticity and soluble solids content (SSC) The next objective is to classify the bananas based on ripening stages into 4 classes of different colours like fully green (unripe), green with some yellow, yellow with some green and fully yellow (ripe). For banana quality attribute prediction and classification, Artificial Neural Network is used. The above-mentioned studies have achieved such high accuracies given the condition that they use additionally extracted features through different techniques and experimental setups. The study we propose has no condition and the model itself learns the parameters required from the stock images to perform the classification to enforce true deep learning capabilities.

Saad et al. [10] in their paper titled 'Recognizing the Ripeness of Bananas using Artificial Neural Network based on Histogram Approach.' The main objective of the study was to develop a technique to classify the ripeness of bananas into 3 categories, and this was achieved through an analysis of the histogram RGB values components of the images. The dataset was composed of 60 images and achieved an accuracy of 89.2%.

Zhang et al. [11] proposed a Convolutional Neural Network (CNN) model [12], one of the first to use it for the purpose of banana ripeness classification. CNN models have been used for a variety of image classification problems in various fields such as medicine [13], road safety [14], food and agriculture [15, 16] etc. The method used a four-layer model (three convolutional and one dense layer).

In [17], Velezrivera et al. used computer vision technique to classify manila mangoes during the ripening process. The used techniques that also took into account the biochemical characteristics for its classification. The colour spaces applied in product classification are the standard RGB (sRGB) and CIELAB. sRGB can be obtained rapidly using computer vision systems. The percentage of correctness after classification was observed to be 90%.

Santi Kumari Behera [18] proposed a model to design an automation system in the fruit Industry using machine learning and deep learning techniques like ANN, CNN, SVM and random forest with adequate concepts of image processing needed to be used to provide intelligence for the automation system to classify the fruits according to its type, variety, maturity and intactness. SVM with texture features and K-means clustering outperformed other models. Srinivasan [19] defined a CNN model with 11 convolutional layers to classify the bananas based on 4 ripeness levels and compare the performance with pre-trained models like ResNet and VGG-16. They also performed data augmentation which is carried out to increase the dataset size and distribution. It duplicates the image by shifting, flipping, rotating, brightening and zooming in and out the training images. After evaluation it was clear that the proposed model performed well with better accuracy. Moreover, for Hass Avocados ripeness classification, Guerrero et al. [20], the authors proposed an automated classification system based on Fisher's Linear Discriminant Analysis and K-means algorithms. The RGB colour space has been used and the proposed system applied some filters to minimize noises. Then, an image segmentation step was applied, using Fisher's Linear Discriminant Analysis algorithm, to separate Avocado fruit from background. Finally, K-means grouping technique was employed, in order to classify Avocados into very mature avocados from mature and green avocados categories, based on pixel percentage. The proposed system achieved an accuracy of 87.85%.

The rest of the paper discusses our novel approach to solving this problem and is organized as follows. In Sect. 4 we present the details of the materials we used and our approach. Section 5 contains our experimental results and discussions. In Sect. 7, we provide our conclusion and vision for the future.

4 Techniques Employed

In our proposed work, we classify the bananas based on their ripeness stages by first performing feature extraction using pre-trained neural network models and those

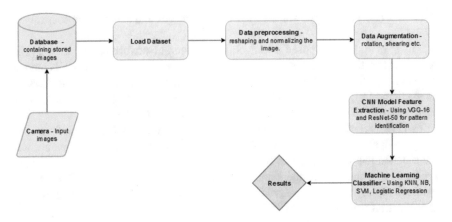

Fig. 2 BanNet—proposed model

extracted features are fed to machine learning classifiers to classify the images. Then we carry out the comparative study of performance of 2 different models for feature extraction, namely ResNet and VGG-16. The machine learning classifiers used in this project are KNN, Naive Bayes, Logistic Regression and SVM. The proposed model can be seen in Fig. 2. Deep learning architectures are model after the human brain, and the neural networks have been proved to contain more power and higher efficiency at identifying deeper patterns and information from the data than machine learning techniques.

First, we carried out data pre-processing by extracting the image and the class labels and storing them into a data frame. The images are reshaped to (200, 200) size which fits well with the ResNet-50 and VGG-16 model's input layer. We converted the images to RGB format so as to visualize them easily. In data augmentation, the images were sheared, flipped, rotated and generated images of batch size 32.

4.1 VGG-16

VGG-16 is a deep Convolutional Neural Network model first proposed by Simonyan and Zisserman [21]. It was one of the major advancements since AlexNet and was better than AlexNet by replacing the large kernel-sized filters with multiple 3 × 3 kernel-sized filters. The architecture of VGG-16 designed by Simonyan and Zisserman is given in Fig. 3.

Fig. 3 VGG-16 architecture [21]

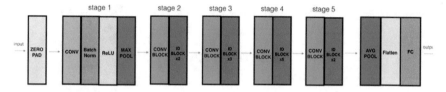

Fig. 4 ResNet-50 model [22]

4.2 ResNet-50

ResNet or Residual Network was first introduced by Kaiming et al. [22]. ResNet-50 is a residual network model with 50 layers. A major innovation of the ResNet-50 when compared to other models was the use of skip connection, i.e. the input of the previous layers is also given as input to the consecutive layer along with the output of the predecessor. Figure 4 represents the ResNet-50 architecture as first presented by Kaiming et al.

4.3 Naïve Bayes

Naive Bayes (NB) is a supervised machine learning classifier based on the Bayes' theorem with the 'naive' assumption that any two features are always conditionally independent. It is called naive because this is not a very practical assumption as most of the datasets do not have entirely conditionally independent features. The formula of Bayes Theorem is given in Eq. (1).

$$\text{Posterior} = \text{Prior} * \text{Likelihood}/\text{Evidence} \tag{1}$$

4.4 K-Nearest Neighbour

K-Nearest Neighbour or KNN is a supervised machine learning technique that classifies based on the number of data points of a particular category that satisfies the given 'k' values required. KNN is a non-parametric classifier. The distance is calculated using distance functions such as Euclidean distance, Manhattan distance, and Minkowski distance. We use Euclidean Distance for our given problem. The formula for Euclidean Distance is given in Eq. (2), where x and y represent the Cartesian coordinates of the data points and x_i and y_i represent the individual axis values.

$$d(X, Y) = \sqrt[2]{\sum_{i=1}^{k} (x_i - y_i)^2} \qquad (2)$$

4.5 Support Vector Machine

Support Vector Machine or SVM is a very popular and commonly used ML technique that has numerous uses in the field of classification, photonics, pattern recognition, etc. SVM solves problems by mapping the training examples to points in space so as to maximize the width of the gap between any two different categories. Though initially developed for binary classification, SVM can now also efficiently perform multiclass classification. Equation (3) is the function of the linear SVM model where the first term w, is the weight vector multiplied by the input sequence x, and is added to the bias β_o.

$$f(x) = \sum_{i=0}^{N} w_i^T * x + \beta_o \qquad (3)$$

4.6 Logistic Regression

Logistic Regression, despite its name, is a classification model rather than a regression model. Logistic Regression is a simple and more efficient method for binary and linear classification problems. It is a classification model, which is very easy to realize and achieves very good performance with linearly separable classes. It is an extensively employed algorithm for classification in industry. The Logistic Regression model, like the Adaline and perceptron, is a statistical method for binary classification that can be generalized to multiclass classification. The formula used in Logistic Regression is given in Eq. (4) where p is the probability of success, x is the input of the equation that is, our event, β_0 is the y-intercept and β_1 is the slope of the equation.

$$\ln(P/1 - P) = \beta_0 + \beta_1 x \tag{4}$$

5 Results

The dataset has been split in the ratio 1:4 for validation and training, respectively. This results in around 50 images for validation and 300 images for training. The results of the model have been evaluated using various performance metrics which have been given below. The evaluation is based on the validation dataset.

5.1 Accuracy

Accuracy is a commonly used performance metric that represents the fraction of the predictions that have been correctly classified by a model. Thus, accuracy can be mathematically represented as given in Eq. (5).

$$\text{Accuracy} = \frac{\text{Number of Correct Predictions}}{\text{Total Number of Predictions}} \tag{5}$$

From the above given table (Table 1), we can clearly see that the proposed model BanNet (Logistic Regression classifier + VGG-16 feature extractor) and the model with SVM classifier, and ResnNet-50 feature extractor have equivalent accuracies and outperform all other combinations considered.

Table 1 Accuracy of the different considered models

Machine learning classifiers/Pre-trained neural network feature extraction models	ResNet-50	VGG-16
KNN	85.45	76.3
Naive Bayes	87.27	80
SVM	90.9	87.2
Logistic regression	89.09	90.9 (proposed model)

Table 2 *R*-Squared score of the considered models

Machine learning classifiers/Pre-trained neural network models	ResNet-50	VGG-16
KNN	0.34	0.146
Naive Bayes	0.48	0.326
SVM	0.68	0.534
Logistic regression	0.62	0.731 (proposed model)

5.2 R-Squared

R-Squared (R^2) is a statistical measure that represents the proportion of the variance in a dependent variable that can be explained by the independent variable. *R*—It ranges between 0 and 1. The formula for computing *R*-Squared value is given in Eq. (6).

$$R^2 = 1 - \frac{\text{Unexplained Variation of the model}}{\text{Total Variation of the model}} \tag{6}$$

From Table 2, it can be observed that the proposed model BanNet (Logistic Regression classifier + VGG-16 feature extractor) has a significantly higher *R*-squared score than all other models. This suggests a better fit of the model with the dataset than others.

5.3 Mean Squared Error

Mean Squared Error or MSE of an estimator measures the average of the squares of errors. That is the average squared difference between the actual values and predicted values. It can be mathematically represented as in Eq. (7).

$$\text{MSE} = \frac{1}{n} \sum_{i=1}^{N} (y_i - \hat{y}_i)^2 \tag{7}$$

Table 3 also gives us the superiority of the proposed BanNet model over others with the BanNet model having a mean squared error of 0.309 the lowest of all considered models.

Table 3 Mean squared error of all the considered models

Machine learning classifiers/Pre-trained neural network models	ResNet-50	VGG-16
KNN	0.763	1
Naive Bayes	0.6	0.78
SVM	0.363	0.545
Logistic regression	0.43	0.309 (proposed model)

Table 4 Mean absolute errors of all considered models

Machine learning classifiers/Pre-trained neural network models	ResNet-50	VGG-16
KNN	0.254	0.454
Naive Bayes	0.272	0.381
SVM	0.181	0.254
Logistic regression	0.218	0.163 (proposed model)

5.4 Mean Absolute Error

Mean Absolute Error (MAE) is defined as the average of the difference between actual values and predicted values. This prevents equal values with opposite signs from cancelling each other out. The formula for Mean Absolute Error is given in Eq. (8) where y_i is the predicted value and x_i is the actual value and n is the total number of observations.

$$\text{MAE} = \frac{\sum_{i=1}^{N} |y_i - x_i|}{n} \tag{8}$$

The Mean Absolute Error as seen in Table 4 also proves that the proposed model has a very high accuracy and error rates than any other model that under the same training and testing conditions.

5.5 Precision

Precision gives us information regarding the proportions of the positive identifications that were actually true. It can be represented as given in Eq. (9)

$$\text{Precision} = \frac{\text{True Positive}}{\text{True Positive} + \text{False Positive}} \tag{9}$$

Table 5 Precision of all considered models

Machine learning classifiers/Pre-trained neural network models	ResNet-50	VGG-16
KNN	0.851	0.751
Naive Bayes	0.872	0.825
SVM	0.914	0.861
Logistic regression	0.894	0.899 (proposed model)

Table 6 Comparative accuracy of different models

Accuracy	BanNet (proposed model)	ANN + Histogram [10]	CNN [23]	ANN [24]
	90.9%	89.2%	76%	79.55%

Table 5 suggests that the model with SVM classifier and ResNet-50 as feature extractor has a better precision than the proposed model. But this is part of the precision-recall trade-off. Though the model has a slightly lesser precision, it accommodates the higher recall or the rate of false positives in the prediction of the model which is a more important quality in the management of foods.

5.6 Comparative Results

We will now compare our results with some other previously done work. The comparative study will be conducted with the Artificial Neural Network model based on histogram approach [10], the CNN model in [23] and the ANN model as proposed by Cho and Koeski [24]. The accuracy of the different models can be seen in Table 6.

The comparative study with other researches of the past decade shows the efficiency and effectiveness of the proposed model. From the Table 6 and Fig. 5, it can be clearly noted that the proposed model, BanNet, performs better than some of the other recent as well old models used for this purpose. The proposed model is also a much faster and simpler approach than the model in [10] which gave the second highest accuracy among the considered models.

5.7 Predictions

The testing images are fed to the feature extractor of the proposed model, i.e. the VGG-16 model as its input. The output of the VGG-16 model comprises the extracted feature vector which is generated based on the trained weights and is then passed to the Logistic Regression classifier that then classifies the image into one of the 4

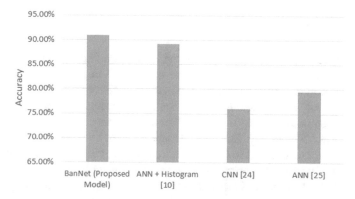

Fig. 5 Comparative accuracy of different models

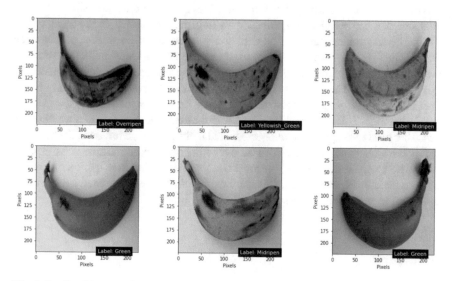

Fig. 6 Predicted images of the model with label

categories based on their ripeness. The images are tagged and presented to the user as shown in Fig. 6.

6 Compliance with Ethical Standards

Conflict of Interest The authors declare no conflict of interest.

Ethical Approval This article does not contain any studies with human participants performed by any of the authors.

7 Conclusion

With the incredibly powerful deep learning techniques and well refined machine learning techniques, hybrid models are one of the most important and dignified as one of the fastest developing research zones. A powerful computer vision system is a key goal to truly achieving Industry 4.0 and other sustainable development goals.

We have considered two of the most powerful convolutional neural network models (Resnet-50 and the VGG-16) and machine learning models (SVM, Logistic Regression, NB and KNN). The CNN models chosen are two of the most widely used models in the computer vision industry and is well known for its feature identification capabilities. It forms the base of the entire model as the extracted features play a very important if not the most important role in the performance of the model. With careful evaluation of the generated results by different performance metrics such as MAE, R-Squared, Accuracy, Precision and MSE, we can conclude with sufficient evidence that the VGG-16 combined with Logistic Regression is the ideal model for the given problem and can be explored for various other similar problems as well. With 91% accuracy, an equally high precision and a short prediction time the proposed model, BanNet, is an ideal model for the classification task discussed in this paper and several other previous paper. The model performs significantly better than almost all off its predecessors without any significant trade-off in the average prediction time. This allows it function in real-time environments such as warehouses and other food management industries and aid in the process. Moreover, the model is highly flexible and versatile as it can conduct other ripeness and similar classification tasks with a very small dataset and considerably less training time using some of the modern techniques such as transfer learning. Thus, we have designed a model that can help the entire food industry to achieving Industry 4.0 standards.

References

1. Sabilla IA, Wahyuni CS, Fatichah C, Herumurti D (2019) Determining banana types and ripeness from image using machine learning methods. In: 2019 International conference of artificial intelligence and information technology (ICAIIT), pp 407–412
2. Kipli K et al (2018) Image processing mobile application for banana ripeness evaluation. In: 2018 International conference on computational approach in smart systems design and applications (ICASSDA), pp 1–5
3. Xiang Q, Wang X, Li R, Zhang G, Lai J, Hu Q (2019) Fruit image classification based on MobileNetV2 with transfer learning technique. In: Proceedings of the 3rd international conference on computer science and application engineering (CSAE 2019). Association for Computing Machinery, New York, NY, USA, Article 121, pp 1–7
4. Sampson HA, Aceves S, Bock SA, James J, Jones S, Lang D, Nowak-Wegrzyn A, Oppenheimer J, Perry TT (2014) Food allergy: a practice parameter update—2014. J Allergy Clin Immunol 134(5):1016–1025
5. Fadilah N, Saleh JM, Ibrahim H, Halim ZA (2012) Oil palm fresh fruit bunch ripeness classification using artificial neural network. In: IEEE 2012 4th International conference on intelligent & advanced systems (ICIAS), Kuala Lumpur, Malaysia, 2012.06.12–2012.06.14

6. El-Bendary N, El Hariri E, Hassanien AE, Badr A (2015) Using machine learning techniques for evaluating tomato ripeness. Expert Syst Appl 42(4):1892–1905
7. Mazen FMA, Nashat AA (2019) Ripeness classification of bananas using an artificial neural network. Arab J Sci Eng
8. Sidehabi SW, Suyuti A, Areni IS, Nurtanio I (2018) Classification on passion fruit's ripeness using K-means clustering and artificial neural network. In: IEEE 2018 International conference on information and communications technology (ICOIACT), Yogyakarta, Indonesia, 2018.3.6–2018.3.7
9. Adebayo SE, Hashim N, Abdan K, Hanafi M, Mollazade K (2016) Prediction of quality attributes and ripeness classification of bananas using optical properties. Sci Hortic
10. Saad H, Ismail AP, Othman N, Jusoh MH, Naim NF, Ahmad NA (2009) Recognizing the ripeness of bananas using artificial neural network based on histogram approach. In: IEEE 2009 IEEE International conference on signal and image processing applications, Kuala Lumpur, Malaysia, 2009.11.18–2009.11.19, pp 536–541
11. Zhang Y, Lian J, Fan M, Zheng Y (2018) Deep indicator for fine-grained classification of banana's ripening stages. EURASIP J Image Video Process 2018(1)
12. LeCun Y, Bottou L, Bengio Y, Haffner P (1998) Gradient-based learning applied to document recognition. Proc IEEE 86(11):2278–2324
13. Vo HH, Verma A (2016) New deep neural nets for fine-grained diabetic retinopathy recognition on hybrid color space. In: IEEE International symposium on multimedia. IEEE, Los Alamitos, pp 209–215
14. Fang J, Zhou Y, Yu Y, Du S (2016) Fine-grained vehicle model recognition using a coarse-to-fine convolutional neural network architecture. IEEE Trans Intell Trans Syst PP(99):1–11
15. Sunderhauf N, Mccool C, Upcroft B, Tristan P (2014) Fine-grained plant classification using convolutional neural networks for feature extraction. Proc Congress Faons 37(2):123–130
16. Yanai K, Kawano Y (2015) Food image recognition using deep convolutional networks with pre-training and fine-tuning. In: IEEE International conference on multimedia & expo workshops. IEEE, Los Alamitos, pp 1–6
17. Velezrivera N, Blasco J, Chanonaperez J, Calderondominguez G, Pereaflores MDJ, Arzate-vazquez I, Cubero S, Farrerarebollo RR (2013) Computer vision system applied to classification of "manila" mangoes during ripening process 7(4):1183–1194
18. Kumari Behera S, Kumar Rath A, Kumar Sethy P (2020) Maturity status classification of papaya fruits based on machine learning and transfer learning approach. Inf Process Agric
19. Saranya N, Srinivasan K, Kumar SKP (2021) Banana ripeness stage identification: a deep learning approach. J Ambient Intell Humanized Comput
20. Guerrero ER, Benavides GM (2014) In: 2014 IEEE colombian conference on communications and computing (COLCOM) - automated system for classifying Hass avocados based on image processing techniques, pp 1–6
21. Simonyan K, Zisserman A (2014) Very deep convolutional networks for large-scale image recognition. ILSVRC-2014
22. He K, Zhang X, Ren S, Sun J (2016) Deep residual learning for image recognition. In: Proceedings of the IEEE conference on computer vision and pattern recognition (CVPR), pp 770–778
23. Hari Priyanka C, Shikha Rachel V, Harshith B, Moulisha R (2020) Color recognition algorithm using a neural network model in determining the ripeness of a banana. J Eng Sci 11(6)
24. Cho B-H, Koseki S (2021) Determination of banana quality indices during the ripening process at different temperatures using smartphone images and an artificial neural network. Sci Hortic 288:110382
25. Simonyan K, Zisserman A (2014) Very deep convolutional networks for large-scale image recognition. ILSVRC-2014

The Study of Effectiveness of Automated Essay Scoring in MOOCs

Younes-aziz Bachiri and **Hicham Mouncif**

1 Introduction

Anyone with an Internet connection and a desire to learn can enroll in a massive open online course (MOOC). Video presentations, computer-based examinations, and online discussion forums are just some of the web-based technologies used in MOOCs to allow thousands of students access to all the course information at the same time [1]. High enrollment MOOCs, on the other hand, mean that the amount of time an instructor spends teaching and evaluating each student is extremely low.

A critical part of the learning process is the evaluation of the learner's progress in terms of knowledge [2]. Typically, an instructor or a grader provides students with feedback on their answers to questions related to the subject matter on an exam, assignment, or quiz. But in some cases, such as in online learning environments and one-on-one or group study sessions, an instructor may not be readily available. Computer-assisted assessment must be used when students still need to be assessed on their topic knowledge.

Peer review is most frequently utilized for written work, but it can also be used for presentations, performances, posters, and videos. Additionally, it is frequently used as a method for analyzing group work and assignments [3].

The assignment and assessment criteria are explained to the students. They receive training and practice using the scoring grid and providing feedback. The final task is completed and submitted by the students. They grade and provide feedback on the assignments of more students using the marking criteria. Each student's grade is calculated by averaging the grades assigned by their peers.

Y. Bachiri (✉) · H. Mouncif
Laboratory of Innovation in Mathematics, Applications, and Information Technologies, Sultan Moulay Slimane University, Beni-Mellal, Morocco
e-mail: younes-aziz.bachiri@usms.ma

H. Mouncif
e-mail: h.mouncif@usms.ma

D. S. Sisodia et al. (eds.), *Machine Intelligence Techniques for Data Analysis and Signal Processing*, Lecture Notes in Electrical Engineering 997,
https://doi.org/10.1007/978-981-99-0085-5_23

Additionally, peer review is utilized to foster transferable skills like critical thinking, communication, and teamwork in the classroom. Peer review in MOOCs has largely been driven by resolving the evaluation and feedback problems that arise when classes have many students and by improving the student experience, with consequently no drawbacks or retention [3].

MOOC participants come from a diverse spectrum of educational backgrounds. Due to the great variety of backgrounds and abilities that students can have, they may not be true peers if they have unequal subject knowledge and language ability. Additionally, different pupils are likely to have varying educational objectives.

By employing machine learning techniques, in this paper, we investigate the feasibility of improving existing bag-of-words (BOW) approaches for creating automated peer bot reviewer grading. Additionally, to replicate the ability of human peers without impairing the pedagogical value of peer evaluation.

We are specifically looking for responses to the following questions: To begin, to what extent may machine learning be used to enhance current methods of peer review grading? Second, will the model produced perform well in terms of accuracy?

2   Related Works

Research on Automated Essay Scoring (AES) has been ongoing for nearly four decades, and several assessment systems are being employed in real-world applications as a supplement to human graders.

Ramnarain-Seetohul et al. [4] presented the results of a decade-long assessment of similarity techniques used in AES systems and discussed the effectiveness and limitations of current methods.

Tan and Tan [5] conducted an empirical investigation to determine the effect of lexical, grammatical, and semantic feature groups on the performance of Automated Essay Scoring classification models that are based on the AES system's general approach.

Yang and Zhong [6] established an auxiliary task in which they included a dynamic semantic matching block to capture the hidden properties via example-based learning.

Ma et al. [7] suggested a new hierarchical graph structure based on graph convolutional networks (GCNs) for encoding the essay's hierarchical structure.

Rahimi et al. [8] have provided an experiment into score prediction for two specific constructs within analytic text-based writing using natural language processing: (1) the students' effective use of evidence; and (2) the structure of their thoughts and evidence in support of their thesis. With the long-term goal of providing feedback to students and teachers, they created a task-dependent model for each dimension that is aligned with the scoring rubric and incorporates source material.

Vajjala [9] investigated the role of numerous linguistic variables in autonomous essay scoring using two publicly available datasets of non-native English essays written in test-taking circumstances. The linguistic qualities of learner language are modeled by encoding lexical, syntactic, discourse, and mistake kinds in the feature set.

To improve the accuracy of the vocabulary level, which is a score item in AES support, Yamamoto et al. [10] have offered a highly comprehensive Japanese vocabulary difficulty level dictionary from the Wikipedia corpus. The appearance frequency of words has been utilized thus far to determine the complexity of a word.

Bachiri and Mouncif [11] have considered developing criteria for selecting an appropriate learning management system and incorporating new automatic natural language processing techniques, such as those used in artificial intelligence, to increase the appeal of MOOCs. To accomplish this, they developed a plugin that automatically generates multilingual questions and converts the course into an engaging game complete with points and badges.

We present a review of AES systems that have seen significant commercial success. These are primarily proprietary software-based systems:

The first AES is thought to be Ellis Page's Project Essay Grader (PEG) [12]. It focuses on analyzing essays using trins and proxes based on their writing styles. PEG enables statistical regression analysis for the purpose of estimating essay scores.

To evaluate the quality of essay content, the Intelligent Essay Assessor (IEA) was introduced [13]. The IEA assigns grades to essays based on Latent Semantic Analysis (LSA), a computational distribution model used to determine the semantic similarity of texts.

IntelliMetric is a proprietary AES that has been effectively scoring essays since 1998 [14]. IntelliMetric is widely recognized as the first AES system to replicate the scoring process using Artificial Intelligence (AI) and Machine Learning (ML). It utilizes features while calculating an essay score (including semantics, syntactics, and discourses).

Since 1999, the E-rater has been created and used. It utilizes proprietary natural language processing algorithms to extract linguistic elements for the purpose of analyzing an essay's style and content. E-rater 2.0 analyzes essay aspects using syntactic, discourse, and thematic analysis modules [15].

3 Integrating Automated Essay Scoring into MOOC Platforms

Our system is made up of several interdependent parts that follow a set of rules to function as a single unit. Systems are defined by their parameters and structure as well as their purpose. They are also reflected in their work.

Evaluation can take numerous shapes and forms, which might vary depending on the learning objectives, the disciplinary setting, and the technology that is available to participants. Peer assessment is frequently distinguished using a formative or summative approach.

3.1 Proposed Evaluation System's Components

There are various components that must be included when putting together an essay assignment. A set of questions or prompts for students to respond to a set of guidelines. The assessment uses a single rubric to grade all the questions. There are various types of assessments. Review Bot Assessment can be used in conjunction with other types of assessments such as peer and staff evaluations and self-assessments as seen in Fig. 1.

Rubric

A rubric must be included in the assignment. Each response is evaluated against the same rubric, regardless of the evaluation type. The grader sees the rubric when she begins grading and compares the supplied response to the rubric. There are multiple criteria and options for each criterion in the rubric. There are a variety of criteria that can be used to determine the quality of an answer. Here are a few examples of what an answer should address or how much supporting information should be included in it [16].

Criteria have a set of options, generally a range of scores, which describe how well each response meets the requirement. "Fair," "Good," or "Excellent," e.g., can

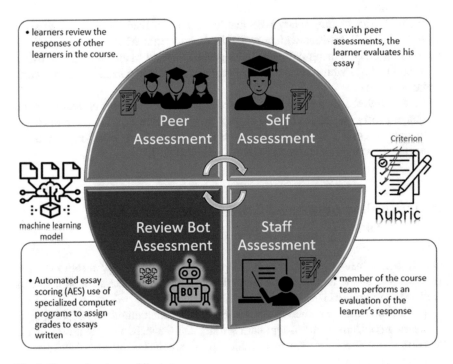

Fig. 1 Proposed system architecture

be among the choices. The number of grade points you'll receive is known as the "choice point value."

Peer Assessment

Learners examine the responses of other course participants [17]. They select one alternative for each criterion in the rubric for each response based on the response. Additionally, students can add written feedback or comments to each response.

Self-assessment

The learner is presented with his own answer and the rubric [18]. Just like with peer assessments, the learner analyzes the response by picking an option for each criterion on the rubric.

Staff Assessment

A course team member evaluates the learner's response [19]. Course team members score the response using the problem's rubric, similar to how they grade self and peer evaluations, and may provide recommendations in their evaluation.

Review Bot Assessment

The response of the student is evaluated by our Review Bot. The trained machine learning model should be used to grade the response in the same way as staff and peer assessments are done, and it can provide comments in their assessment. Students' data will be used to train the bot indefinitely and more efficiently.

Human biases such as rater tiredness, experience, scale, stereotyping, the Halo effect, rater drift, perceptual difference, and inconsistency will be eliminated, which is a major advantage of this type of assessment.

Computer-scored essays submitted by students on an inquiry topic can be used to diagnose students' progress, both to alert teachers to problematic students and to generate automated help.

4 Methods: Automated Essay Scoring

By analyzing the data of three sets of 13,000 essays from kaggle.com, we constructed a general model for automated essay grading in this study. These essays were grouped into eight distinct categories based on their setting. The dataset is transformed in a variety of ways and features are extracted. The dataset was analyzed using Linear Regression as shown in Fig. 2.

Fig. 2 High-level block diagram

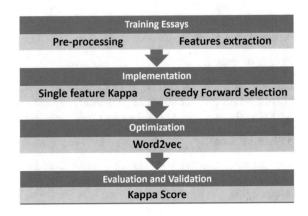

4.1 Dataset

The dataset contains 12,976 essays with manual evaluation scores. According to context, the dataset is separated into eight essay sets. It is critical to keep in mind that each essay set has a unique scoring system [20].

At first glance, the dataset appears as follows:

In 2012, the Hewlett Foundation sponsored the Automated Student Assessment Prize on Kaggle.

There are eight essay sets in this competition. Each pair of essays was generated in response to a specific prompt. Selected writings range in length from 150 to 550 words on average. Some of the essays rely on secondary sources, while others do not. All replies were written by children in grades 7 through 10. All essays were graded by hand and scored twice. Each of the eight datasets is distinct in its own way. The unpredictability is designed to push the capabilities of your scoring engine to their limits.

To train and evaluate the learning model for this research, essay sets [1–3] were utilized. The model can recognize the scale difference and generate the appropriate grade.

Table 1 summarizes the data. Essays were hand-scored on a variety of scales, depending on the prompt, with each essay receiving at least two scores.

Table 1 Essay sets [1–3] information and statistics

Essay set	Essay type	Domain	Score range	Average length	Total
1	Persuasive/Narrative/Expository	Letter writing	2–12	350	1783
2	Persuasive/Narrative/Expository	• Writing applications • Language conventions	1–6 1–4	350	1800
3	Source dependent responses	–	0–3	150	1726

Table 2 Quick overview of the used columns from datasets

essay_id	essay_set	essay	rater1	rater2	score
1	1	Dear local …	4.0	4.0	8.0
2	1	Dear@CAPS1 …	5.0	4.0	9.0
3	1	Dear, @CAPS1 …	4.0	3.0	7.0
…	…	…	…	…	…
12976	8	Many people believe	20.0	20.0	40.0

The training data is available in three formats: tab-separated values (TSV) files, Microsoft Excel 2010 spreadsheets, and Microsoft Excel 2003 spreadsheets.

Essay sets 1–6 are included in the current release of the training data. Each of these files is 28 columns in length (example, Table 2).

4.2 The Processing of the Essays

Preprocessing of the data was performed by the Stanford Natural Language Processing group. Instead, IDs consisting of the @ symbol followed by all-caps words were employed. Individuals, organizations, locations, times/dates, numbers, percentages, e-mail addresses, and money were all substituted with new names. Because the training set lacked features, several unique heuristic features were produced.

Feature Extraction

Eleven open-source libraries were used to build a total of sixteen features. The words in the essay that already had the "@Text" attribute were not taken into account when making the features.

Heuristic Characteristics

Numerous heuristic elements were produced that are likely to add to the quality of an essay. Several heuristic aspects include the following: word count, big word count (words with a length greater than six), average word length per essay, quotation mark count, character count, sentence count, and comma count.

Grammatical and Spelling Features

It is expected that a learner will make grammatical and spelling errors; grammar check created a large number of spelling and grammatical problems.

Tags for Parts of Speech (POS)

A count for most common POS tags was utilized to assist in determining a decent sentence structure, e.g., a count of nouns, verbs, adjectives, and adverbs for a specific essay using NLTK.

Additional Characteristics

Additionally, crucial metrics such as incorrect words, domain words (the quantity of words that are relevant to the essay's domain), punctuation count, and word-to-sentence ratio were generated using NLTK's WordNet library.

5   Results and Discussion

5.1   Single Feature Kappa

Following the generation of the heuristic features, each one was examined independently to determine its contribution to the learning model. We used a method to avoid overfitting. It employs Linear regression as the learning model and a fivefold cross-validation technique with quadratic weighted kappa as the evaluation metric. Kappa obtained a single feature, as illustrated in Fig. 3.

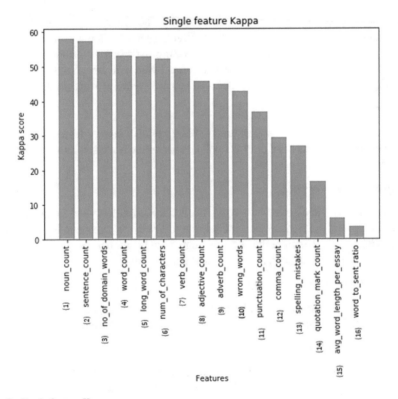

Fig. 3 Single feature Kappa scores

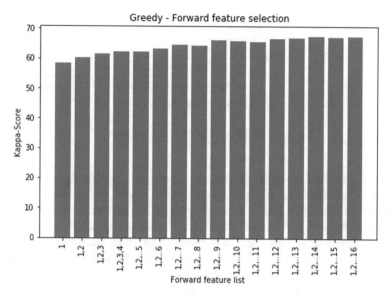

Fig. 4 Forward feature Kappa score (refer to Fig. 3 for indexing)

5.2 Greedy Forward Selection

According to the single kappa scores obtained, the features were ordered in decreasing order. We used the forward selection approach to select the feature set with the greatest Kappa score (Set [1–15]) as seen in Fig. 4. A Kappa score of 0.67 was the highest obtained. This score exceeded our predefined cutoff of 0.5. With the exception of the final feature (word_to_sent_ratio), which was not included in the feature set with the greatest Kappa score, we can conclude that practically all of the characteristics contributed to the model's increased Kappa score.

5.3 Optimization: Word2vec Model

Since its inception in 2013, the Word2vec model has demonstrated interesting applications in the Natural Language Processing (NLP) field [21]. To improve the model's accuracy, I used the Gensim library's word2vec model and combined the resulting word vectors with the previously built heuristic feature vectors. Any improvement in accuracy was welcomed, as relying exclusively on heuristic features is no longer preferable for NLP tasks. The produced sentence list was used to construct a word2vec model with 300 features (a 300-dimension word vector). Vectors were created for training and testing purposes. These vectors were then concatenated with the previously created heuristic feature vectors.

Table 3 Results

Model	Kappa score
Dummy regressor	0.00
Linear regression with heuristic features	0.67
Linear regression with word2vec model	0.90
Linear regression using Word2vec and heuristic features	0.93
Kaggle competition's winning team	0.81

- Dummy Regressor: We ran the model via Sklearn's Dummy Regressor to determine whether the learnt model outperforms a model that always guesses one answer. It received a Kappa score of 0, suggesting that there was no agreement between the projected values and the grades assigned by human graders.
- Linear Regression: Using the concatenated vectors of the word2vec model and the heuristic features, a Linear Regression model with fivefold cross-validation was then trained. After evaluating the model with the test vectors, an average Kappa score of 0.93 was obtained during cross-validation.

The model's outputs are shown in Table 3.

The final model, which was constructed using the Word2vec model's generated word vectors and heuristic features, performed remarkably well. The score increased by 0.26 points when compared to employing purely heuristic features, which was the method taken by a large number of competitors during the competition. The final score of 0.93 outperforms the benchmark by a factor of 0.43, and the model outperformed the winning team's score, showing the utility of utilizing word2vec models for NLP tasks.

A review of previous work in this area [22–24] demonstrates that Linear Regression is an effective technique for grading essays.

The Deep Learning technique is popular at the moment, and by utilizing TensorFlow and its great visualization features, the deep learning algorithms performed admirably on this task. According to Nguyen and Dery's study [25], employing a two-layer neural network that trains word vectors simultaneously, Deep Learning approaches generated an average Kappa score of 0.9447, which is higher than this model.

6 Conclusion

We studied the viability of upgrading existing methodologies for establishing automated peer bot reviewer grading using machine learning techniques. Additionally, to duplicate the ability of human peers without jeopardizing peer evaluation's pedagogical value.

For this portion of the project, only essay sets 1, 2, and 3 were used. Adding the next five essay sets to the model should have no effect on its efficiency, as the essays in these five sets contain essays comparable to those in essay set 3, which were created by a single prompt. The heuristic qualities were developed by considering the broad characteristics of an effective essay; these characteristics should be applicable to any essay in any domain. Similarly, a huge corpus was used to train the word2vec model. This should verify that the model is applicable to any essay input domain on which it was trained.

References

1. Aleven V, Sewall J, Popescu O, Ringenberg M, van Velsen M, Demi S (2016) Embedding intelligent tutoring systems in MOOCs and e-learning platforms. In: Intelligent tutoring systems. Cham, pp 409–415. https://doi.org/10.1007/978-3-319-39583-8_49
2. Deng R, Benckendorff P, Gannaway D (2019) Progress and new directions for teaching and learning in MOOCs. Comput Educ 129:48–60. https://doi.org/10.1016/j.compedu.2018.10.019
3. Garcia-Loro F, Martin S, Ruipérez-Valiente JA, Sancristobal E, Castro M (2020) Reviewing and analyzing peer review inter-rater reliability in a MOOC platform. Comput Educ 154:103894. https://doi.org/10.1016/j.compedu.2020.103894
4. Ramnarain-Seetohul V, Bassoo V, Rosunally Y (2022) Similarity measures in automated essay scoring systems: a ten-year review. Educ Inf Technol. https://doi.org/10.1007/s10639-021-108 38-z
5. Tan JS, Tan IKT (2021) Feature group importance for automated essay scoring. In: Multidisciplinary trends in artificial intelligence. Cham, pp 58–70. https://doi.org/10.1007/978-3-030-80253-0_6
6. Yang Y, Zhong J (2021) Automated essay scoring via example-based learning. In: Web engineering. Cham, pp 201–208. https://doi.org/10.1007/978-3-030-74296-6_16
7. Ma J, Li X, Chen M, Yang W (2021) Enhanced hierarchical structure features for automated essay scoring. In: Information retrieval. Cham, pp 168–179. https://doi.org/10.1007/978-3-030-88189-4_13
8. Rahimi Z, Litman D, Correnti R, Wang E, Matsumura LC (2017) Assessing students' use of evidence and organization in response-to-text writing: using natural language processing for rubric-based automated scoring. Int J Artif Intell Educ 27(4):694–728. https://doi.org/10.1007/s40593-017-0143-2
9. Vajjala S (Mar.2018) Automated assessment of non-native learner essays: investigating the role of linguistic features. Int J Artif Intell Educ 28(1):79–105. https://doi.org/10.1007/s40593-017-0142-3
10. Yamamoto M, Umemura N, Kawano H (2020) Proposal of Japanese vocabulary difficulty level dictionaries for automated essay scoring support system using rubric. J Oper Res Soc China 8(4):601–617. https://doi.org/10.1007/s40305-019-00270-z
11. Bachiri Y, Mouncif H (2020) Applicable strategy to choose and deploy a MOOC platform with multilingual AQG feature. In: 2020 21st International Arab conference on information technology (ACIT), pp 1–6, Nov 2020. https://doi.org/10.1109/ACIT50332.2020.9300051
12. Page EB (1966) The imminence of... grading essays by computer. Phi Delta Kappan 47(5):238–243
13. Foltz PW, Laham D, Landauer TK (1999) Automated essay scoring: applications to educational technology, pp 939–944. Accessed: 15 Jan 2022. [Online]. Available: https://www.learntech lib.org/primary/p/6607/

14. Edelblut P (2020) Realizing the promise of AI-powered, adaptive, automated, instant feedback on writing for students in grade 3–8 with an IEP. In: Adaptive instructional systems. Cham, pp 283–292. https://doi.org/10.1007/978-3-030-50788-6_21
15. Dikli S (2006) An overview of automated scoring of essays. J Technol Learn Assess 5(1), Art. no. 1. Accessed: 15 Jan 2022. [Online]. Available: https://ejournals.bc.edu/index.php/jtla/article/view/1640
16. Lu J, Law N (2012) Online peer assessment: effects of cognitive and affective feedback. Instr Sci 40(2):257–275. https://doi.org/10.1007/s11251-011-9177-2
17. Kane JS, Lawler EE (1978) Methods of peer assessment. Psychol Bull 85(3):555–586. https://doi.org/10.1037/0033-2909.85.3.555
18. Ward M, Gruppen L, Regehr G (2002) Measuring self-assessment: current state of the art. Adv Health Sci Educ Theory Pract 7(1):63–80. https://doi.org/10.1023/A:1014585522084
19. Luo H, Robinson A, Park J-Y (2014) Peer grading in a MOOC: reliability, validity, and perceived effects. Online Learn J 18(2). Accessed: 16 Jan 2022. [Online]. Available: https://www.learntechlib.org/p/183756/
20. Zhao S, Zhang Y, Xiong X, Botelho A, Heffernan N (2017) A memory-augmented neural model for automated grading. In: Proceedings of the fourth (2017) ACM conference on learning @ scale, New York, NY, USA, Apr 2017, pp 189–192. https://doi.org/10.1145/3051457.3053982
21. Mikolov T, Sutskever I, Chen K, Corrado GS, Dean J (2013) Distributed representations of words and phrases and their compositionality. In: Advances in neural information processing systems, vol 26. Accessed: 19 Feb 2022. [Online]. Available: https://proceedings.neurips.cc/paper/2013/hash/9aa42b31882ec039965f3c4923ce901b-Abstract.html
22. Mahana M, Johns M, Apte A (2012) Automated essay grading using machine learning. Mach Learn Session 5
23. Valenti S, Neri F, Cucchiarelli A (2003) An overview of current research on automated essay grading. J Inf Technol Educ Res 2(1):319–330
24. Attali Y, Burstein J (2006) Automated essay scoring with e-rater® V.2. J Technol Learn Assess 4(3), Art. no. 3, Feb 2006. Accessed: 19 Feb 2022. [Online]. Available: https://ejournals.bc.edu/index.php/jtla/article/view/1650
25. Nguyen H, Dery L (2016) Neural networks for automated essay grading. CS224d Stanford reports, pp 1–11

Remote Authentication of IoT Devices Based Upon Fog Computing

Manabhanjan Pradhan and Sujata Mohanty

1 Introduction

The interconnection between different devices through a network is popularly known as the Internet of Things (IoT) [11]. Kevin Ashton first proposed the IoT technology in 1999, which has become a promising research area nowadays. For instance, smart bulb, smart TV, and smart assistance, such as Alexa, have become a part of our daily life. The IoT devices work independently with dedicated sensors, known as nodes which gather information from the environment [1]. With a predefined set of rules, they execute the instructions. In general, IoT devices consist of three parts, namely tags, sensors, and RFIDs. As many IoT devices are connected and communicated with each other, security is the utmost concern. A breach in security may damage the functionalities and also resulting personal data theft [8]. To mitigate this, these devices can be associated with mutual authentication for each communication, and thereby, any adversary could not get access to the central architecture.

Blockchain was first introduced by Satoshi Nakamoto, who proposed a decentralized technique for digital cryptocurrency named Bitcoin [9]. He proposed a technique that overcomes the trust issues involving the third party. Here, two parties can communicate and share their resources, whereas the decisions are made by the nodes involved in the network. Blockchain can be used for access management, authentication, and many more real-life applications. It provides basic security services, such as authentication, data integrity, data confidentiality, and non-repudiation. A blockchain consists of a database, miners, and network of nodes. After 2009, many blockchains have been implemented. Every blockchain has a different mechanism for mining a new node, while Bitcoin needs Proof of Work (PoW), Ethereum needs Proof of Stake (PoS), and Hyperledger needs Practical Byzantine Fault Tolerance (PBFT) approach.

M. Pradhan (✉) · S. Mohanty
National Institute of Technology, Rourkela, Odisha, India
e-mail: mannpradhan@protonmail.com

© The Author(s), under exclusive license to Springer Nature Singapore Pte Ltd. 2023
D. S. Sisodia et al. (eds.), *Machine Intelligence Techniques for Data Analysis and Signal Processing*, Lecture Notes in Electrical Engineering 997,
https://doi.org/10.1007/978-981-99-0085-5_24

Authentication in blockchain technology is a promising field, and much research is going on in this area [2, 3, 5]. Hammi et al. [4] proposed an authentication mechanism based on blockchain with a focus on various security issues. Khalid et al. [10] developed a blockchain-based authentication mechanism for IoT. But later, it was found to be susceptible to malicious requests. Nguyen et al. [6] proposed a prototype implementation in a real data sharing scenario for secure Electronic Health Records (EHRs). M. Tahir et al. [12] proposed an authentication framework using a probabilistic model.

We present a fog-based system model for remote authentication. It ensures mutual authentication between the communicating devices and rejects any requests which are found to be malicious. In the proposed model, only the authentic users can access IoT devices through the nearest fog node. Every time a device associated with a user sends a request, it is duly verified by the blockchain-enabled fog node. The proposed scheme can withstand passive and active attacks, such as spoofing attack, man-in-the-middle attack, and fog node impersonation attack [7]. Also, the proposed work is compared with some competent schemes and found to achieve security features, such as data integrity, availability, and mutual authentication. The proposed work is of low computational cost along with less communication overhead as compared to existing literature.

The rest of the paper is organized as follows. Section 2 demonstrates the proposed system model for remote authentication. Section 3 discusses and analyzes the experiment with results. Finally, the paper concludes in Sect. 4.

2 Proposed System Model

According to the Open Web Application Security Project (OWASP), one of the major vulnerabilities in IoT communication is weak authentication and inefficient mobile interfaces. We proposed a blockchain-based model to overcome the problem stated earlier that provides a secure environment for communication between IoT devices in a network. We devised a fog-based system model for IoT architecture. The model consists of interconnected fog nodes which are controlled by a fog node controller. The layout is shown in Fig. 1. Each fog node is associated with users and IoT devices, as shown in Fig. 2. The users act as an interface between the fog node and IoT devices. If one IoT device wants to communicate with another device, it must be authenticated by the user who has access to that IoT device. The fog nodes and users are blockchain enabled.

The following assumptions are considered for the proposed blockchain-based IoT architecture.

- Every device has a unique identity (ID), which is a combination of the device's MAC address and private key.
- The fog node controller is trusted.

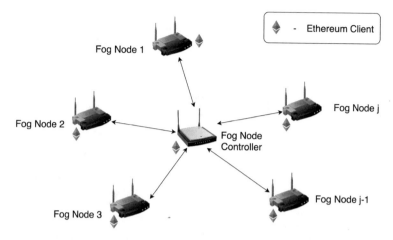

Fig. 1 Proposed system model (a)

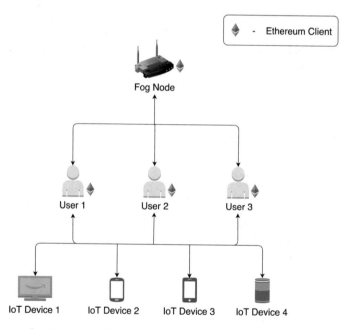

Fig. 2 Proposed system model (b)

Table 1 Notations used in transactions

S. No.	Notations	Meaning
1	C	Fog node controller
2	F	Fog node
3	U	User
4	D	IoT device
5	$X_i Id$	Unique ID of i-th instance of X Node
6	$X_i Pu$	Public Key of i-th instance of X Node
7	$X_i Pr$	Private Key of i-th instance of X Node

The model is designed in such a way that each and every device in the network except the fog node controller must register with the unique ID. After registering the device, whenever a device wants to communicate with another device, authentication is confirmed before the communication. The proposed model is divided into three phases, namely initialization phase, authentication phase, and communication phase. Table 1 shows the notations used in the proposed modelß. Details of each phase are described below.

2.1 Initialization Phase

The initialization process consists of the registration of the fog node, user, and IoT device. In fog node registration, each fog node F registers itself with the fog node controller C, which is associated with a unique identity. The fog node can be registered only when the fog node controller is in the active state. F will send its ID to C encrypted with C's public key. It will be considered as the first transaction named as ϕ_a. This process is described in Eq. 1.

$$Step1: \quad F \rightarrow C: \quad \phi_a = Enc\left[C_i Pu\left(F_j Id\right)\right] \tag{1}$$

Next, C will decrypt ϕ_a with its private key to obtain F's ID.

$$Step2: \quad F_j Id \leftarrow Dec\left[C_i Pr\left(\phi_a\right)\right] \tag{2}$$

A certificate $Cert_a$ will be generated by C and sent to F along with F's ID, encrypted with F's public key, and this will be considered as next transaction ϕ_b.

$$Step3: \quad C \rightarrow F: \quad \phi_b = Enc\left[F_i Pu\left(F_j Id, Cert_a\right)\right] \tag{3}$$

F will decrypt ϕ_b using its private key and will obtain ID and certificate. F will verify ϕ_b by comparing that ID with its own ID, if both IDs are same, then it will store the certificate $Cert_a$ and fog node will be registered. The fog node registration phase is depicted in Algorithm 1.

$$Step4: \quad F_j Id, Cert_a \leftarrow Dec[F_i Pr(\phi_b)] \qquad (4)$$

Algorithm 1: Fog Node Registration

```
begin
if (fog_node_controller.active() = true) then
    if (FNID_Exists(fog_node.id, block_chain) = true) then
    |   return error()
    end
    else
    |   FNID_Register(fog_node.id, fog_node_controller.id, block_chain)
    |   return success()
    end
else
|   return error()
end
```

In the user device registration, user U first registers itself to the fog node F. The fog node provides certificate to user, which is used to authenticate itself with other devices attached to it. U will send its ID to F encrypted with F's public key in transaction named as ϕ_c. This process is described in Eq. 5

$$Step1: \quad U \rightarrow F: \quad \phi_c = Enc[F_k Pu(U_l Id)] \qquad (5)$$

Next, F will decrypt ϕ_b with its private key to obtain U's ID.

$$Step2: \quad U_l Id \leftarrow Dec[F_k Pr(\phi_c)] \qquad (6)$$

A certificate $Cert_b$ will be generated by F and sent to U along with U's ID, encrypted with U's public key and will be considered as next transaction ϕ_d.

$$Step3: \quad F \rightarrow U: \quad \phi_d = Enc[U_l Pu(U_l Id, Cert_b)] \qquad (7)$$

U will decrypt ϕ_d using its private key and will get ID and certificate. U will verify ϕ_d by comparing that ID with its own ID, if both IDs are same, then it will store the certificate $Cert_b$ and user will be registered.

$$Step4: \quad U_l Id, Cert_b \leftarrow Dec[U_l Pr(\phi_d)] \qquad (8)$$

The details of this process are shown in Algorithm 2.

Algorithm 2: User Device Registration

---

begin
if *(FNID_Exists(fog_node.id, block_chain) = true)* **then**
  if *UserID_Exists(user_device.id, block_chain) = true* **then**
  | return error()
  else
  | UserID_Register(user_device.id, fog_node.id, block_chain)
  | return success()
  end
else
| return error()
end

In the IoT device registration, each IoT device sends request to its respective user. Then the user approves the request and sends to blockchain-based fog node for registration. After successful registration of IoT device, fog node notifies the user that the device is registered. IoT device D will send its ID to User U encrypted with U's public key in transaction named as ϕ_e. This process is described in Eq. 9

$$Step1: \quad D \rightarrow U: \quad \phi_e = Enc\left[U_x Pu\left(D_y Id\right)\right] \tag{9}$$

Next, U will decrypt ϕ_e with its private key to obtain D's ID.

$$Step2: \quad D_y Id \leftarrow Dec\left[U_x Pr\left(\phi_e\right)\right] \tag{10}$$

U will forward D's ID to fog node F by encrypting it with F's public key.

$$Step3: \quad U \rightarrow F: \quad \phi_f = Enc\left[F_z Pu\left(D_y Id\right)\right] \tag{11}$$

F will decrypt ϕ_e with its private key to obtain D's ID.

$$Step4: \quad D_y Id \leftarrow Dec\left[F_z Pr\left(\phi_f\right)\right] \tag{12}$$

A certificate $Cert_c$ will be generated by F and will be sent to D along with D's ID, encrypted with D's public key, and this will be considered as transaction ϕ_g. The same certificate $Cert_c$ is sent to user U encrypted with U's public key considered as transaction ϕ_h.

$$Step5: \quad F \rightarrow D: \quad \phi_g = Enc\left[D_y Pu\left(D_y Id, Cert_c\right)\right]$$
$$F \rightarrow U: \quad \phi_h = Enc\left[U_x Pu\left(D_y Id, Cert_c\right)\right] \tag{13}$$

D and U will decrypt respective ϕ_g and ϕ_h using their private key and will get ID and certificate. Both D and U will verify the D's ID, and then only IoT device D will be registered. The details of this process are shown in Algorithm 3.

$$Step6: \quad D_y Id, Cert_c \leftarrow Dec\left[D_y Pr\left(\phi_g\right)\right]$$
$$D_y Id, Cert_c \leftarrow Dec\left[U_x Pr\left(\phi_h\right)\right] \tag{14}$$

Algorithm 3: IoT Device Registration

---

begin
if *(FNID_Exists(fog_node.id, block_chain) = true)* **then**
   if *(UserID_Exists(user_device.id, block_chain) = true)* **then**
      if *(IoTID_Exists(iot_device.id, block_chain) = true)* **then**
      | return error()
      else
         IoTID_Register(iot_device.id, user_device.id, fog_node.id, block_chain)
         return success()
      end
   else
   | return error()
   end
else
| return error()
end

2.2 Authentication Phase

In this phase, first the fog node authentication takes place, followed by the user device authentication and by the IoT device authentication.

In the authentication phase of fog node, the certificate received from the registration process as shown in Algorithm 1 is used to authenticate the fog node. Fog node F initiates the transaction ϕ_i to fog node controller C that is encrypted using C's public key. It contains F's ID and the certificate $Cert_a$ received in Eq. 4.

$$Step1: \quad F \rightarrow C: \quad \phi_i = Enc\left[C_i Pu\left(F_j Id, Cert_a\right)\right] \tag{15}$$

Then, ϕ_i is decrypted by C using its private key to obtain certificate $Cert_a$ and F's ID as shown in Eq. 16.

$$Step2: \quad F_j Id, Cert_a \leftarrow Dec\left[C_i Pr\left(\phi_i\right)\right] \tag{16}$$

Now blockchain will verify by comparing the IDs and certificates from Eq. 4 with the IDs and certificates from Eq. 16, if both are same, then authentication is valid else not. The details of this process are shown in Algorithm 4.

Algorithm 4: Fog Node Authentication

---

begin
if *(fog_node_controller.active() = true)* **then**
    if *(FNID_Exists(fog_node.id, block_chain) = true)* **then**
        if *(FNID_Auth(fog_node.id, fog_node_controller.id, block_chain) = true)* **then**
        | return success()

else
  | return error()
end

In the authentication phase of user, the certificate received from the registration process as shown in Algorithm 2 is used to authenticate the user device. Here user U will initiate the transaction ϕ_j to fog node F that is encepted using F's public key. It contains U's ID and the certificate $Cert_b$ received in Eq. 8.

$$Step1: \quad U \rightarrow F: \quad \phi_j = Enc\left[F_k Pu\left(U_l Id, Cert_b\right)\right] \tag{17}$$

Then, ϕ_i is decrypted by F using its private key to obtain certificate $Cert_b$ and U's ID as shown in Eq. 18.

$$Step2: \quad U_l Id, Cert_b \leftarrow Dec\left[F_k Pr\left(\phi_j\right)\right] \tag{18}$$

Now blockchain will verify by comparing the IDs and certificates from Eq. 8 with the IDs and certificates from Eq. 18, if both are same, then authentication is valid else not. The details of this process are shown in Algorithm 5.

Algorithm 5: User Device Authentication

---

begin
if *(FNID_Exists(fog_node.id, block_chain) = true)* **then**
    if *(UserID_Exists(user_device.id, block_chain) = true)* **then**
        if *(UserID_Auth(user_device.id, fog_node.id, block_chain) = true)* **then**
        | return success()

else
  | return error()
end

In the authentication phase of IoT device, the certificate received from the registration process as shown in Algorithm 3, is used to authenticate the IoT. Here user D will initiate the transaction ϕ_k to user U that is encepted using U's public key. It contains D's ID and the certificate $Cert_c$ received in Eq. 15.

$$Step1: \quad D \rightarrow U: \quad \phi_k = Enc\left[U_x Pu\left(D_y Id, Cert_c\right)\right] \tag{19}$$

Then, ϕ_k is decrypted by U using its private key to obtain certificate $Cert_c$ and D's ID as shown in Eq. 19.

$$Step2: \quad D_y Id, Cert_c \leftarrow Dec\left[U_x Pr\left(\phi_k\right)\right] \tag{20}$$

Now blockchain will verify by comparing the IDs and certificates from Eq. 14 with the IDs and certificates from Eq. 20, if both are same, then authentication is valid else not. The details of this process are shown in Algorithm 6.

Algorithm 6: IoT Device Authentication

---

begin
if *(FNID_Exists(fog_node.id, block_chain)* = *true)* **then**
    if *(UserID_Exists(user_device.id, block_chain)* = *true* **then**
        if *(IoTDID_Exists(iot_device.id, block_chain)* = *true)* **then**
            if *(IoTDID_Auth(iot_device.id, fog_node.id, block_chain)* = *true)* **then**
             | return success()

else
  | return error()
end

2.3 Communication Phase

In this phase, the communication rule for each device is established. If one IoT device initiates for communication, then the corresponding fog node checks if it is a legitimate device or not. Also, it verifies if the request is malicious or not. After that, it checks for the device to which it wants to communicate. If all the conditions are satisfied, then only a secure communication is established.

3 Experiment and Result Discussion

This section presents an overview of experiments that have been performed to validate the proposed model starting with experimental setup to the results obtained and the discussion on the obtained results. Fog node controller, fog node and user are connected to blockchain-enabled IoT network. Blockchain simulation is achieved with

---

Algorithm 7: Device Communication

---

begin
  if *(FNID_Exists(fog_node.id, block_chain) = true)* **then**
    if *(IoTDID_Exists(iot_device.id, block_chain) = true)* **then**
      if *(IoT_Device_Authentication() = true)* **then**
        if *Request_Malicious() = false)* **then**
          if *(FNID_Exists(fog_node.id, block_chain) = true)* **then**
            if *(IoTDID_Exists(iot_device.id, block_chain) = true)* **then**
              | Secure Communication Established

  else
    | return error()
  end

---

the help of true Ganache framework and MetaMask connected with web3 interface. Communication between the nodes has been done using JsonRPC library.

3.1 Time and Power Consumption

In this section, we compared time and power consumption for registration of node in millisecond and to send data message in milliwatt with existing schemes. We have presented results that are based on 100 experiments. Comparative analysis of Hammi et al. [4] with our proposed model for time consumption is given in Table 2 and for power consumption is given in Table 3.

Table 2 Time comparison of different operations

Approach	Time needed			
	For registration of node (ms)		To send data message (ms)	
	Average	SD	Average	SD
Bubbles of Trust [4]	1.56	0.13	0.04	0.001
Proposed Model	1.44	0.11	0.04	0.001

3.2   Security Analysis

In this section, the security analysis of the proposed architecture is done, which is illustrated as follows.

- **Data Integrity**: In the proposed architecture, all transactions are stored in blockchains. Each block in the blockchain consists of many parameters, including the previous block hash. If someone tries to change transaction data, they have to change each and every block present in the blockchain, which is very difficult as each client has a copy of the blockchain. So, data integrity is maintained.
- **Non-Repudiation**: After registration of any node, whenever a transaction happens, the sender will use its private key to encrypt the message and sends it to the receiver. Thereby, in every transaction, the sender's private key will be used. Hence, the sender cannot refute its initialization of the message as it can be verified by its public key.
- **Mutual Authentication**: Every time a device sends a request, it will be verified by the blockchain-enabled fog node, and every time a fog node sends a response, it will be verified by the device. For example, in Algorithm 1, the fog node sends its ID as Eq. 1. The fog node controller verifies it by decrypting the transaction using its private key shown in Eq. 2. This ID is searched in the blockchain whether it already exists or not. After that, the fog node controller sends the certificate to the fog node. This transaction is verified by the fog node with its private key, as shown in Eq. 3. It is further verified by the public key of the fog node controller as shown in Eq. 4. Hence, mutual authentication is achieved.
- **Availability**: Users are ethereum clients, which means they can access their respective IoT devices from anywhere in real time. Only the authentic and legitimate users can access IoT devices through the nearest fog node, thereby they are available to authentic users.
- **Spoofing Attack**: When the attacker in the network pretends to be some other entity, it is known as spoofing attack. In the proposed model, every device has a unique ID, i.e., $C_i Id$, $F_i Id$, $U_i Id$ and $D_i Id$, which are stored in blockchains. Since every device's private key is unknown to every other device, it is very difficult to obtain the unique ID; hence, the proposed model is resistant to spoofing attack.
- **Man-in-the-Middle Attack (MIMA):** In the authentication phase, every device uses its own public and private keys. As long as they do not share these keys with

Table 3 Power comparison of different operations

Approach	Power needed			
	For registration of node (mW)		To send data message (mW)	
	Average	SD	Average	SD
Bubbles of Trust [4]	9.76	2.04	3.35	0.87
Proposed Model	8.00	2.02	3.10	0.85

Table 4 Performance evaluation with existing models

Author(s) & Year(s)	S1	S2	S3	S4	S5	S6	S7	S8	S9
Hammi et al. 2018 [4]	Yes	Yes	Yes	Yes	Yes	No	No	Yes	No
Nguyen et al. 2019 [6]	Yes	Yes	No	Yes	No	No	No	No	No
Khalid et al. 2020 [10]	Yes	Yes	Yes	No	Yes	No	No	Yes	No
Tahir et al. 2020 [12]	Yes	Yes	Yes	Yes	Yes	No	No	Yes	No
Proposed Model	Yes	Yes	Yes	Yes	Yes	Yes	Yes	Yes	Yes

Note S1—Data Integrity, S2—Non-Repudiation, S3—Mutual Authentication, S4—Availaibility, S5—Spoofing Attack, S6—Man-In-The-Middle Attack (MIMA), S7—Denial of Service, S8—Message Replay Attack, S9—Fog Node Impersonation Attack

third parties, MIMA is not possible. The keys used in proposed model are $C_i Pu$, $C_i Pr$, $F_i Pu$, $F_i Pr$, $U_i Pu$, $U_i Pr$, $D_i Pu$, and $D_i Pr$.

- **Denial of Service**: In the proposed model, the request is checked if it is malicious or not using $Request\_Malicious()$. If the same request with the same parameters is sent many times, the model will discard those requests to prevent a DOS attack. Hence, all recourses are available to authentic participants of the system.
- **Message Replay Attack**: Since all the transactions are done by the devices with the help of their unique IDs, the same transaction cannot happen back to back, else the request will be considered malicious. The system will reject the request. Hence, replay attack is prevented.
- **Fog Node Impersonation Attack**: It may happen that someone poses as a fog node to access the details of user devices and IoT devices. Every transaction related to the fog node involves $F_i Pu$ or $F_i Pr$. The $F_i Pr$ is not available publicly; thereby, no one can decrypt the transaction except the original fog node. The public key of the fog node is available in the network. One can use $F_i Pu$ for the verification of the transaction or send some message to the fog node by first encrypting with their private key. Hence, it is not possible to impersonate a user or IoT device in the proposed architecture.

3.3 Performance Evaluation

In this section, we compared the proposed architecture with some of the existing schemes. The performance evaluation is depicted in Table 4. Hammi [4] and Khalid [10] models do not check if the request is being malicious or not, which makes their model vulnerable to intruder attack as well as social engineering. Our proposed model overcomes these shortcomings.

4 Conclusion

IoT is a promising area of research, and IoT devices have become a part of day-to-day life. Combining blockchain with IoT gives a lot of promising work to make IoT more secure. In the proposed model, IoT devices take less time and consume less power for communicating as compared to the existing scheme which can be used in real-life scenarios. The proposed scheme can withstand passive and active attacks, such as spoofing attack, man-in-the-middle attack, and fog node impersonation attack. In order to make the proposed model lightweight, the future work would focus on certificate less authentication mechanism.

References

1. Atzori L, Iera A, Morabito G (2010) The Internet of Things: a survey. Comput Netw 54(15):2787–2805
2. Bahga A, Madisetti VK (2016) Blockchain platform for industrial Internet of Things. J Softw Eng Appl 9(10):533–546
3. Dorri A, Kanhere SS, Jurdak R, Gauravaram P (2017) Blockchain for IoT security and privacy: the case study of a smart home. In: 2017 IEEE international conference on pervasive computing and communications workshops (PerCom Workshops), IEEE, pp 618–623
4. Hammi MT, Hammi B, Bellot P, Serhrouchni A (2018) Bubbles of trust: a decentralized blockchain-based authentication system for IoT. Comput Security 78:126–142
5. Kamran M, Khan HU, Nisar W, Farooq M, Rehman SU (2020) Blockchain and Internet of Things: a bibliometric study. Comput Electrical Eng 81:1–12
6. Khalid U, Asim M, Baker T, Hung PC, Tariq MA, Rafferty L (2020) A decentralized lightweight blockchain-based authentication mechanism for IoT systems. Cluster Comput 23(3):1–21
7. Khan MA, Salah K (2018) IoT security: review, blockchain solutions, and open challenges. Future Gener Comput Syst 82:395–411
8. Lin J, Yu W, Zhang N, Yang X, Zhang H, Zhao W (2017) A survey on Internet of Things: architecture, enabling technologies, security and privacy, and applications. IEEE Internet Things J 4(5):1125–1142
9. Nakamoto S (2019) Bitcoin: a peer-to-peer electronic cash system. Technical report
10. Nguyen DC, Pathirana PN, Ding M, Seneviratne A (2019) Blockchain for secure ehrs sharing of mobile cloud based e-health systems. IEEE Access 7:66792–66806
11. Ray P (2018) A survey on Internet of Things architectures. J King Saud Univ—Comput Inf Sci 30(3):291–319
12. Tahir M, Sardaraz M, Muhammad S, Saud Khan M (2020) A lightweight authentication and authorization framework for blockchain-enabled IoT network in health-informatics. Sustainability 12(17):1–23

Comparison of Different Denoising Networks on Motion Artifacted MRI Scans

Vijay Tripathi, Manish Tibdewal, and Ravi Mishra

1 Introduction

Medical images such as magnetic resonance imaging (MRI), X-ray, and computer tomography (CT) scans are employed to view the internal anatomy. Each medical imaging technique captures varying levels of detail about the internal anatomy of humans and other organisms. In an MRI machine, radio waves are aimed at the patient's body, and the radio waves generated in response are recorded [5]. It was discovered that the radio waves reflected from cancer cells were distinct and unlike the radio waves emitted by healthy cells [2]. By varying the sequence of pulses provided to the MRI machine, various types of MRI scans can be generated such as T1-weighted, T2-weighted, and proton density. Any kind of movement of the patient during an MRI generates blurry scans [10]. Such blurry MRI scans are called motion artifact-induced MRIs. Motion artifacts can be introduced due to two types of patient movements, namely voluntary and involuntary. As the MRI scan takes a lot of time to complete, some patients tend to squirm and adjust their position during the scan. Such movements by the patients are called voluntary movements. The subtle movement of the torso during breathing and the micro-pulsation of the blood being pushed in the veins, these movements are categorized as involuntary motions. Physical restrains, patient training, and global anesthesia can be used to effectively suppress voluntary motions. However, involuntary motions are not in the control of

V. Tripathi (✉)
G. H. Raisoni University, Amaravati 444701, India
e-mail: vijayrtripathi@rediffmail.com

M. Tibdewal
Shri Sant Gajanan Maharaj College of Engineering, Shegaon 444203, India
e-mail: mntibdewal@ssgmce.ac.in

R. Mishra
G. H. Raisoni Institute of Engineering and Technology, Nagpur 440028, India
e-mail: ravi.mishra@raisoni.net

© The Author(s), under exclusive license to Springer Nature Singapore Pte Ltd. 2023
D. S. Sisodia et al. (eds.), *Machine Intelligence Techniques for Data Analysis and Signal Processing*, Lecture Notes in Electrical Engineering 997,
https://doi.org/10.1007/978-981-99-0085-5_25

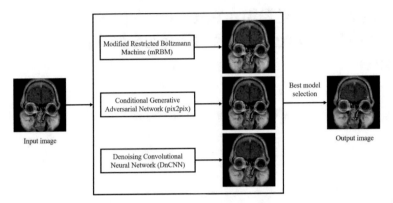

Fig. 1 Concept diagram for comaparing the performance of three methods, denoising convolutional neural network (DnCNN), modified restricted Boltzmann machine (mRBM), and conditional generative adversarial network (cGAN), for denoising blurry-input MRI scan. When an input image is presented to the proposed system, its corresponding output image generated by all three methods is compared. The output of the model that yields the most optimal error parameters is selected as the output image

the patient and hence cannot be curbed by external means or sedation. Currently, the standard procedure for dealing with motion artifact-induced MRI scans is to repeat the scan entirely [9]. However, rescanning consumes time and resources. Hence, to save resources and time for the patient and the radiologists, we are proposing two methods for rectifying a blurry MRI scan that does not require rescanning. Both the methods use deep learning to process an input blurred MRI scan and produce the denoised MRI scans as outputs. The first method uses a modified restricted Boltzmann machine (mRBM) as the deep learning model, whereas the second method used a conditional generative adversarial network (cGAN) called pix2pix. The models were trained on a dataset consisting of 218 pairs of blurry MRI scans, and their corresponding clear versions were sourced from local hospitals. The performance of these methods was compared with the standard MATLAB model for denoising MRI scans called denoising convolutional neural network (DnCNN). DnCNN and the two proposed methods were compared using parameters such as root mean square Error (RMSE), peak signal-to-noise ratio (PSNR), mutual information index (MII), structural similarity index measure (SSIM), accuracy, and error, as shown in Fig. 1. As input, the system is given the coronal view of a T1-weighted MRI scan of the skull. The corresponding RMSE values from DnCNN, mRBM, and cGAN were 0.0014, 0.0063, and 0.0075, respectively. Because DnCNN produced the lowest RMSE, its output was chosen as the system's output, as seen in Fig. 1. Figure 2 depicts the proposed system's data flow diagram. The 218 MRI scan dataset is first converted from RGB to grayscale and saved in datastore D1. The transformed dataset was used to train the two proposed models, mRBM and cGAN. Following the completion of the training process, the network parameters of the trained mRBM and cGAN were saved in another datastore D2. A new blurry MRI scan can be denoised using the

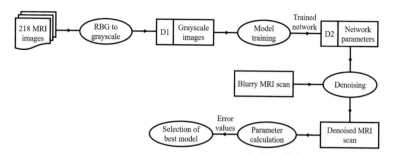

Fig. 2 Data flow diagram of the proposed system. The dataset of 218 MRI scans is first converted to grayscale and then used for training the three models, DnCNN, mRBM, and cGAN. After the training process, the network parameters of the trained models are used when a new MRI scan is presented to the models to generate the corresponding outputs for all three models. Certain parameters such as RMSE, PSNR, and MII are computed for the three outputs, and based on them, the most optimal output is selected

trained network parameters. The outputs of mRBM, cGAN, and DnCNN were used to compute parameters like RMSE, PSNR, MI, and pixel deviation between input and output images.

2 Literature Review

Kustner et al. [6] have developed a deep learning-based model for retrospective correction of motion artifact-induced MRI scan. By using a GAN, they were able to generate a high-quality denoised version of a blurred MRI scan, which was very close to the actual clear version of the MRI image obtained by rescanning. Their model was able to achieve a similarity index (SI) of 0.8 RMSE of 0.08 and normalized mutual information (NMI) of 0.9. However, their system would occasionally deform the shape of the internal organs in the MRI scan. At times, some vital features from input MRI scans were eliminated. Lyu et al. [7] employed a recurrent neural network (RNN) to reduce the distortion caused due to cardiac-related motion artifacts in MRI scans. Multiscale convolution was combined with a bi-directional convolutional long short-term memory (LSTM) model. They also attempted to enhance temporal resolution by recovering missing frames in a set of MRI scans. Kidoh et al. [3] had employed two convolutional neural networks to effectively reduce blurriness in MRI scans of the brain. A denoising convolutional neural network and a shrinkage convolutional neural network were used. They were able to achieve a SI of 0.83 on input images that had 10% noise engrained in them. Their models were able to reach the highest PSNR of 33.7 dB. Unsupervised deep learning was used by Oh et al. [8] to correct motion artifacts in MRI images. A cyclic GAN along with outlier-rejecting bootstrap aggregation was utilized. The highest PSNR of 34.13 was achieved on input images with dimensions of 320 × 320 pixels. Furthering the trend of utilizing deep learning for denoising MRI scans, Uetani et al. [12] applied the same on three-dimensional MRI scans rather than the commonly used two-dimensional MRI scans. They tried their approach on MRI scans of 35 patients and were able to attain an SI of 0.63 along with an SNR of 21.81. Cycle GAN was also used by Armanious et al. [1] to rectify motion artifact-induced. They received an MSE of 375.01 for the images with dimensions of 256×256 pixels. They utilized 1101 pictures from twelve individuals in all. Their recorded PSNR was 24.47, which was higher than the value given by pix2pix. Kim et al. [4] demonstrated how deep learning can be employed to improve arterial spin labeling. Their mean square error (MSE) was reduced by 40% after generating corrected MRI scans from blurry input images. They tested their convolutional neural network (CNN) using 114 photographs in total. Tamada et al. [11] enhanced MRI scans of the liver using a CNN. A sizeable reduction in motion artifact was discovered by them after using a multichannel convolutional neural network. They were able to eliminate the blurring caused by the respiratory motion. About 268,423 learning parameters of the CNN were fine-tuned. Their RMSE was larger than 0.01 after 80 epochs. They trained their model using the data of fourteen patients.

3   Methodology

3.1   Dataset

The dataset used for training and evaluating the two proposed models consisted of 218 MRI scans sourced from local hospitals. Each image had the dimension of 512×512 pixels. The dataset was split into training and testing sets in the ratio of 70/30. The images consisted of MRI scans of various body parts such as neck, spine, brain, ankle, and abdomen. Various views are also included such as cervical view, axial view, and coronal view. The dataset was first converted into grayscale. Then they were converted from the standard uint-8 format to the double format.

3.2   Modified Restricted Boltazmann Machine (mRBM)

By keeping in mind the difference between denoised MRI scans and the corresponding ground truth images, the parameters of the restricted Boltzmann machine (RBM) were modified. The network structure of mRBM is shown in Fig. 3a. By selecting the most optimal training parameters, the performance of a deep learning model can be enhanced. Hence, to find the most optimal set of learning parameters, the modified restricted Boltzmann machine (mRBM) was trained using several sets of training parameters. Parameters such as the number of maximum iterations, Gibbs sampling steps, number of nodes in the hidden layer, learning rate, mini-batch size,

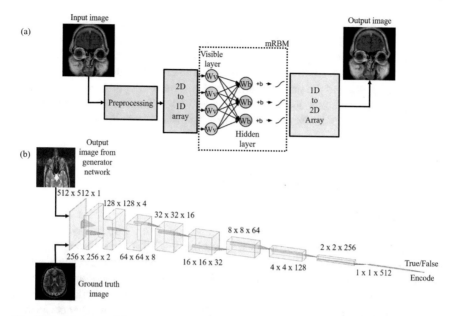

Fig. 3 Architecture of the proposed models. **a** Structure of the modified restricted Boltzmann machine (mRBM). **b** Structure of the discriminator network of the conditional generative adversarial network (cGAN)

and the size of the input image were varied. After various experimentations, 900, 80, 0.02, 13, and 256 × 256 pixels were selected as the number of maximum iterations, Gibbs sampling steps, learning rate, mini-batch size, and the size of the input image, respectively.

3.3 Conditional Generative Adversarial Network (cGAN)

Pix2pix is a conditional generative adversarial network (cGAN). While denoising a blurry MRI using pix2pix, the input image was first passed through a basic building block network. The basic building network consisted of layers such as a 2D convolutional layer, batch normalization, dropout, and a leaky rectified linear unit (ReLU) activation function. The output of the basic building block network was passed on to the generator network. The generator network had a U-Net structure, where the input image was downsized to 32 × 32-pixel dimensions, then it was upsampled back to the original dimensions of the input image. The upsampling layers are connected to the corresponding downsampling layers. The output generated by the generator network was compared with the actual ground truth image, by the discriminator network. The architecture of the discriminator network is shown in Fig. 3b The PatchNet concept was employed in the architecture of the discriminator network. The output of the generator and the ground truth image were provided as inputs. They were downscaled by a factor of 2 in each consecutive layer until they reached the dimensions of 1×1 pixels. The final layer of the discriminator gave a binary output that would signify if the two input images were the same or not. The error between the output of the generator network and the ground truth image was used to optimize the weights of each layer in the generator and discriminator networks. The weights were optimized such that the output of the generator matched the ground truth image.

3.4   Denoising Convolutional Neural Network (DnCNN)

The denoising convolutional neural network (DnCNN) was proposed in 2016. Its internal structure consisted of repeated instances of one group of layers. Layers such as a convolutional layer, a batch normalization layer, and a rectified linear unit layer (ReLU) were grouped. They were repeated one after the other nineteen times. The convolutional layers performed 64, $3 \times 3 \times 64$ convolutions. The batch normalization layer consisted of 64 layers. The final layers of the network consisted of a convolutional layer and a regression layer. The final convolutional layer performed a single $3 \times 3 \times 64$ convolution. DnCNN accepted only grayscale input images and produced their denoised versions at the output. Apart from denoising, DnCNN could also be used for removing JPEG compression artifacts and to increase image super-resolution.

4   Results and Discussions

While comparing the performance of denoising convolutional neural network (DnCNN), mRBM, and cGAN, we have considered many parameters such as root mean square error (RMSE), peak signal-to-noise ratio (PSNR), mutual information index (MII), structural similarity index measure (SSIM), and pixel deviation, as shown in Tables 1 and 2.

4.1   Root Mean Square Error (RMSE)

Concerning RMSE, mRBM was the best performing algorithm among the three, whereas cGAN was comparable to mRBM with only 4% difference between the two, and DnCNN was almost 60% in performance compared to the other two. Only in the case of ankle and foot images (Image 6), DnCNN outperformed the other two algorithms. Saggital T2-weighted MRIs of the spine (Image 7) can be best corrected with the cGAN network. Also for ankle images (Image 6), cGAN was 30% better than mRBM. In all other cases, especially for the brain (Images 1, 4, 8, 9) and abdomen

Table 1 Comparison of mRBM, cGAN, and DnCNN based on root mean square error (RMSE), peak signal-to-noise ratio (PSNR), and mutual information index (MII)

Input images	RMSE			PSNR			MII		
	DnCNN	mRBM	cGAN	DnCNN	mRBM	cGAN	DnCNN	mRBM	cGAN
Image 1	0.02	0.009	0.007	28.36	19.27	16.63	0.51	0.78	0.49
Image 2	0.02	0.007	0.01	17.87	14.75	13.89	0.67	0.7	0.88
Image 3	0.02	0.01	0.02	18.16	22.83	27.97	0.66	0.77	0.55
Image 4	0.009	0.01	0.007	20.47	9.58	20.55	0.56	0.3	0.78
Image 5	0.02	0.01	0.01	12.86	24.46	14.35	0.35	0.40	0.75
Image 6	0.001	0.01	0.007	20.14	22.81	17.18	0.54	0.64	0.83
Image 7	0.02	0.013	0.004	29.15	21.06	21.5	0.38	0.54	0.5
Image 8	0.018	0.009	0.009	18.58	19.20	14.62	0.44	0.63	0.97
Image 9	0.001	0.006	0.007	30.51	22.5	25.74	0.42	0.52	0.44
Image 10	0.02	0.006	0.007	29.32	19.67	22.45	0.58	0.64	0.52

On average, mRBM had the lowest RMSE values out of the three models

(Images 2, 3, 10), mRBM was the best algorithm to denoise the motion artifacts. It was quite evident from RMSE that mRBM is the choice of the algorithm because of its low computational cost during training. The highest RMSE was 0.0201, and the lowest was 0.0014 which was achieved for T1-weighted MRI of the skull with its coronal view (Image 9), as shown in Fig. 4.

4.2 Peak Signal-to-Noise Ratio (PSNR)

From Fig. 4, it can be observed that the average PSNR with all three algorithms could go as high as 20 dB. PSNR value represents the maximum possible power of the signal and the power of the corrupting noise that affects the fidelity that can hamper the visual representation. Ideal PSNR should lie between 20 and 50 dB, but acceptable quality can be considered about 20 dB. For PSNR, DnCNN was almost 13% better than mRBM and 15% better than cGAN. The highest SNY obtained for any image was 30.51 dB, and the lowest PSNR was 9.57 dB.

4.3 Mutual Information (MI)

Mutual information (MI) is a way of representing the matching between the two images. It is the measure of how much one can predict the other image. The mutual information could reach up to 67% in the case of cGAN, whereas mRBM and DnCNN were 58% and 51%, respectively, as seen in Fig. 5. The mutual information parameter ideally should be 1 for images without noise and can be very low value for images with heavy noise. In the current scenario, all the images were having a moderate

Table 2 Comparison of mRBM, cGAN, and DnCNN based on structural similarity index measure (SSIM), pixel deviation, and error. cGAN had the highest average pixel deviation of 0.0509

Input images	SSIM			Pixel deviation			Error		
	DnCNN	mRBM	cGAN	DnCNN	mRBM	cGAN	DnCNN	mRBM	cGAN
Image 1	0.82	0.58	0.49	0.01	0.03	0.13	3.619×10^{-4}	0.003	0.01
Image 2	0.79	0.35	0.42	0.01	0.13	0.17	4.0491×10^{-4}	0.01	0.01
Image 3	0.49	0.73	0.69	0.001	0.01	0.002	3.78×10^{-4}	0.001	3.9577×10^{-4}
Image 4	0.81	0.19	0.57	0.006	0.09	0.02	0.002	0.03	0.002
Image 5	0.68	0.77	0.34	0.007	0.01	0.05	0.0032	8.877×10^{-4}	0.009
Image 6	0.85	0.72	0.61	0.004	0.03	0.03	2.39×10^{-4}	0.001	0.004
Image 7	0.83	0.65	0.69	0.01	0.03	0.03	3.01×10^{-4}	0.002	0.002
Image 8	0.71	0.64	0.33	0.005	0.02	0.05	3.43×10^{-4}	0.003	0.008
Image 9	0.76	0.76	0.70	0.003	0.01	0.02	2.2×10^{-4}	0.001	6.616×10^{-4}
Image 10	0.5	0.8	0.81	0.0008	0.01	0.005	2.88×10^{-4}	0.002	0.001

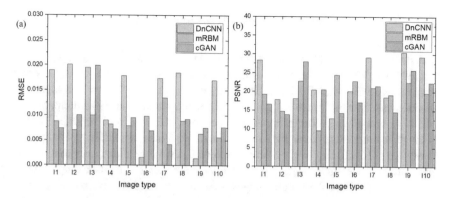

Fig. 4 Evaluating parameters like root mean square error (RMSE) and peak signal-to-noise ratio (PSNR) of the denoised images produced by mRBM, cGAN, and DnCNN for a set of ten sample input images. **a** Plot of RMSE values for the three methods. mRBM had the lowest RMSE values on average, while DnCNN had the highest RMSE values. **b** Plot of PSNR values using three methods. In terms of PSNR, DnCNN performed the best followed by mRBM and cGAN

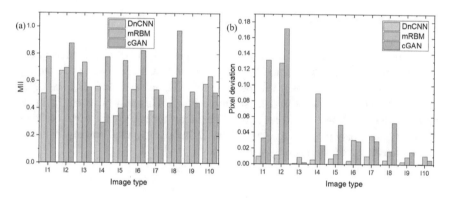

Fig. 5 Evaluating parameters like mutual information (MI) and pixel deviation (accuracy) of the denoised images produced by mRBM, cGAN, and DnCNN for a set of ten sample input images. **a** Plot of MI values for the three methods. cGAN, mRBM, and DnCNN had average MI values of 67%, 58%, and 51%, respectively. Based on MI, none of the three methods can be declared better than the other. **b** Plot of pixel deviation values for the three methods. cGAN had achieved the highest value of 0.0509 units per pixel

noise level, and hence, predicting which algorithms perform best in terms of mutual information was hard to decide. An average MI of 58% can be considered as the best case, where it will preserve the original image content as well as remove the noisy one. Giving any conclusive statement how much is the denoising based on MI is out of the scope of this manuscript as the noises were naturally occurred and not manually added, and hence, it was not possible to quantify them. Hence, MI is not particularly useful here.

4.4   Structural Similarity Index Measure (SSIM)

The structural similarity index measure (SSIM) was found to be the highest with DnCNN with 72% and the lowest with cGAN with 56.6%. With mRBM, we could achieve a 61.9% similarity index. It is very difficult to conclude which algorithm was best based on the structural similarity index measure as the original input image had noise as well as useful information. For input images with no noise, the output of the denoising process was the same as the input; hence, the SSIM and MI should ideally be 1. If SSIM or MI is calculated to be 0, then it would signify that a lot of information of the input image was lost while generating the denoised output image. An SSIM and MI of 0.5 would suggest that the denoised output image has 50% information of the input image and that the input image had 50% noise. Hence, it is very difficult to ascertain which algorithm performs better based on SSIM and MI alone.

4.5   Pixel Deviation

It was found that overall pixel deviation was maximum for cGAN with an average of 0.0509 units per pixel when normalized, whereas for DnCNN, it was 0.0078 units per pixels when normalized. The pixel deviation for mRBM was found out to be 0.0372 units per pixel when normalized, as shown in Fig. 5. This term was important as it predicts how much was the average pixel deviation for each pixel value compared to the original noisy input image. cGAN was found out to be the best algorithm that can deviate the pixel value easily within 5%.

4.6   Best Performing Model for MRI Scans of Various Parts of the Body

By observing Fig. 6, it was found that for brain images with the axial view (Image 1), it is preferred to use mRBM as it is faster and provides the best RMSE value of 0.008 even with images with arteries (Image 4) irrespective of whether it is T1- or T2-weighted. mRBM provided the PSNR of 19.26 dB which was very close to the acceptable PSNR of 20 dB and 77% of MI and 58% of SSIM; we can easily conclude that mRBM is the algorithm of choice. For images of the abdomen, preferably with an axial view again, mRBM was the algorithm of choice because with an average RMSE of 0.0073 and average PSNR of 19 dB with 65% MI and 62% SSIM. For the cervical view of the neck and spine (Image 5), mRBM outperformed the other two algorithms with 24.46 dB PSNR and 0.007 RMSE with 40% MI and 76% SSIM, as seen in Fig. 7. The performance of DnCNN was poor for the cervical view of the neck and spine with 0.017 RMSE and only 12 dB PSNR. On the other hand, DnCNN

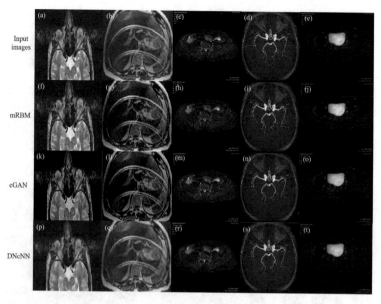

Fig. 6 Comparison of the performance of mRBM, cGAN, and DnCNN on some sample blurry MRI scans. **a–e** Input images of motion artifact-induced MRI scan, of the axial view (AV) of the brain **a**, AV of the abdomen **b, c**, (AV) of brain with arteries **d**, and (AV) of the lower abdomen **e**. **f–j** Corresponding mRBM outputs. **k–o** Corresponding cGAN outputs. **p–t** Corresponding DnCNN outputs

performed exceptionally well in the case of MRI of the ankle and foot. It achieved almost 0.002 RMSE with 20.14 dB PSNR. The performance of mRBM was really bad for the T2-weighted sagittal view of the spine (Image 7) with 0.013 of RMSE and PSNR just over 20 dB. For the same case, cGAN performed the best with 0.004 RMSE with 21.5 dB of PSNR. Hence, it is recommended that for the T2-weighted sagittal view of the spine, cGAN is the most optimal network. For the coronal view of T1-weighted MRI of a skull (Image 9), DnCNN was the most preferred algorithm with a huge PSNR of 30 dB and RMSE of 0.0014. Overall, we found that in terms of PSNR, DnCNN was the best algorithm. In terms of RMSE, mRBM is the algorithm of choice. In terms of MI, mRBM provides a decent match between preserving the information and removing the noise. In terms of SSIM, cGAN was the most powerful algorithm that can rebuild the noise-free image from scratch. In terms of pixel deviation for abdominal MRIs, cGAN outperformed the other two algorithms. Whereas for the coronal view MRI of the skull, the most preferred algorithm was DnCNN. cGAN is the most optimal method for denoising brain MRI scans. For the other types of MRI, mRBM was the most efficient algorithm to denoise a blurry MRI scan with respect to RMSE. Through our analysis, we have discovered that for certain previously mentioned types of MRI scans, cGAN or DnCNN performs better in terms of some evaluation parameters. But only in terms of RMSE, mRBM is the best performing model.

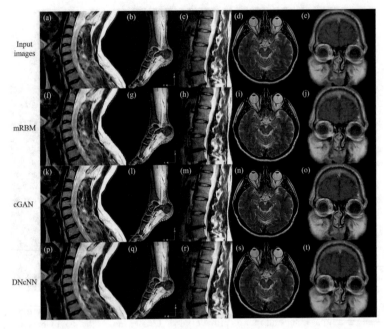

Fig. 7 Comparison of the performance of mRBM, cGAN, and DnCNN on some sample blurry MRI scans. **a–e** Input images of motion artifact-induced MRI scan, of the cervical view of neck and spine **a**, ankle and foot **b**, sagittal T2-weighted MRI of the spine **c**, axial view of T2-weighted MRI of the brain **d**, and coronal view of T1-weighted MRI of the skull **e**. **f–j** Corresponding mRBM outputs. **k–o** cGAN outputs. **p–t** DnCNN outputs

5   Conclusion

In this manuscript, we have compared the denoising performance of three methods, namely denoising convolutional neural network (DnCNN), modified restricted Boltzmann machine (mRBM), and conditional generative adversarial network (cGAN), on a set of ten diverse input blurry MRI scans. The three methods were compared based on various parameters, such as RMSE, PSNR, MI, SSIM, pixel deviation, and error. DnCNN yields the best PSNR value. SSIM and pixel deviation values generated by cGAN were the most desirable. mRBM produced the best RSME and MI values. We found that for MRI scans of the brain, abdomen, and neck, mRBM performs the most optimal denoising. DnCNN should be preferred while denoising blurry MRI scans of the foot and ankle, along with the coronal view of T1-weighted MRI of a skull, while cGAN produced the best denoising performance for the T2-weighted sagittal view of the spine. Overall except for a few select cases, mRBM is the choice of the algorithm for denoising blurry MRI scans.

Declarations

Funding Information

No funding was involved in the present work.

Conflicts of Interest

Authors V. R. Tripathi, M. N. Tibdewal, and R. Mishra declare that there has been no conflict of interest.

Consent to Participate

Informed consent was obtained from patients whose images are used in this work. All the necessary permissions were obtained from Institute Ethical committee.

References

1. Armanious K, Tanwar A, Abdulatif S, Küstner T, Gatidis S, Yang B (2020) Unsupervised adversarial correction of rigid MR motion artifacts. In: 2020 IEEE 17th international symposium on biomedical imaging (ISBI). IEEE, pp 1494–1498
2. Gordon R, Hines J, Gordon D (1979) Intracellular hyperthermia a biophysical approach to cancer treatment via intracellular temperature and biophysical alterations. Med Hypotheses 5(1):83–102
3. Kidoh M, Shinoda K, Kitajima M, Isogawa K, Nambu M, Uetani H, Morita K, Nakaura T, Tateishi M, Yamashita Y et al (2020) Deep learning based noise reduction for brain MR imaging: tests on phantoms and healthy volunteers. Magn Reson Med Sci 19(3):195
4. Kim KH, Choi SH, Park SH (2018) Improving arterial spin labeling by using deep learning. Radiology 287(2):658–666
5. Kruger RA, Kiser Jr WL, Miller KD, Reynolds HE, Reinecke DR, Kruger GA, Hofacker PJ (2000) Thermoacoustic CT: imaging principles. In: Biomedical optoacoustics, vol 3916. International Society for Optics and Photonics, pp 150–159
6. Küstner T, Armanious K, Yang J, Yang B, Schick F, Gatidis S (2019) Retrospective correction of motion-affected MR images using deep learning frameworks. Magnetic Reson Med 82(4):1527–1540
7. Lyu Q, Shan H, Xie Y, Kwan AC, Otaki Y, Kuronuma K, Li D, Wang G (2021) Cine cardiac MRI motion artifact reduction using a recurrent neural network. IEEE Trans Med Imaging
8. Oh G, Lee JE, Ye JC (2020) Unsupervised MR motion artifact deep learning using outlier-rejecting bootstrap aggregation. arXiv preprint arXiv:2011.06337
9. Satterthwaite TD, Elliott MA, Gerraty RT, Ruparel K, Loughead J, Calkins ME, Eickhoff SB, Hakonarson H, Gur RC, Gur RE et al (2013) An improved framework for confound regression and filtering for control of motion artifact in the preprocessing of resting-state functional connectivity data. Neuroimage 64:240–256

10. Smith TB (2010) MRI artifacts and correction strategies. Imaging Med 2(4):445
11. Tamada D, Kromrey ML, Ichikawa S, Onishi H, Motosugi U (2020) Motion artifact reduction using a convolutional neural network for dynamic contrast enhanced MR imaging of the liver. Magn Reson Med Sci 19(1):64
12. Uetani H, Nakaura T, Kitajima M, Yamashita Y, Hamasaki T, Tateishi M, Morita K, Sasao A, Oda S, Ikeda O et al (2021) A preliminary study of deep learning-based reconstruction specialized for denoising in high-frequency domain: usefulness in high-resolution three-dimensional magnetic resonance cisternography of the cerebellopontine angle. Neuroradiology 63(1):63–71

Plant Species Recognition Using Custom-Developed Neural Network with Optimized Hyperparameters

S. Anubha Pearline⬚ and V. Sathiesh Kumar⬚

1 Introduction

Global Biodiversity is steadily deteriorating as a consequence of direct and indirect human activities. A crucial prerequisite in preserving plant biodiversity is the identification of the presence of plant species around the globe. This warrants the development of an efficient and robust real-time plant species recognition system. Botanists manually examine a wide range of plant characteristics to identify and recognize the plant species. However, this method is incredibly hard for non-professionals such as farmers, gardeners and individuals. As an alternative approach, plant species recognition could be performed by using computer vision and machine learning techniques. Numerous plant species recognition systems employ leaves as an input to identify the plant species owing to the prevalence of the leaves throughout the year. The real-time classification of plant species based on the characteristics (geometry, shape, colour, texture, vein pattern, leaf margin and leaf form–simple or compound) of leaves is more challenging due to the large diversity in plant species [1]. Computer-based plant species recognition methods also encounter challenges due to variation in camera viewpoint, ambient light changes, scale modification and cluttered background [2, 3]. Recently, Neural Networks play a major role in several applications of computer vision. Neural Networks are capable of feature learning and classification.

Numerous research studies have been reported by different research groups across the world to tackle these aforementioned issues. Conventional image processing

S. Anubha Pearline
Department of Computer Science Engineering, SRM Institute of Science and Technology, Vadapalani Campus, Chennai, Tamilnadu 600026, India
e-mail: anubhaps@srmist.edu.in

V. Sathiesh Kumar (✉)
Department of Electronics Engineering, MIT Campus, Anna University, Chennai, Tamilnadu 600044, India
e-mail: sathieshkumar@annauniv.edu

© The Author(s), under exclusive license to Springer Nature Singapore Pte Ltd. 2023
D. S. Sisodia et al. (eds.), *Machine Intelligence Techniques for Data Analysis and Signal Processing*, Lecture Notes in Electrical Engineering 997,
https://doi.org/10.1007/978-981-99-0085-5_26

methods [4–6], Neural Networks and deep learning-based architectures are used in the context of plant species recognition.

The main contributions of this research paper are as follows:

- The pre-processed leaf images are utilized as direct input (pixel intensities) to the recognition system. Such processing of leaf images gives access to complete information of the leaf rather than extracting specific feature information.
- Investigated the performance of a custom-developed neural network (four layers– one input and output layer, two hidden layers) with different activation functions (ReLU, PReLU, ELU and L-ReLU).
- Hyperparameters such as optimizer, learning rate, activation function, alpha in PReLU and number of epochs are tuned to achieve the best performance metrics.
- Extensive experiments are performed on two standard (Flavia and Swedish leaf) and one real-time (Leaf 12) dataset.
- The proposed method (Neural Network with PReLU activation function) has lower model complexity as well as lower computational time compared to deep neural networks.

The rest of the paper is organized as follows. A short review of the existing literature is presented in the next section. Section 3 discusses the proposed methodology. Section 4 describes the experimental results and discussion. Also, Sect. 4 contains the comparison between the proposed method and other existing methods. Section 5 concludes this research article.

2   Related Works

Kumar et al. [5] proposed a method to perform a plant species recognition process using a Multi-layer Perceptron with Adaboosting technique. Flavia dataset is used in the evaluation of the architecture. The images from the dataset are pre-processed and the morphological features are extracted. The extracted features are classified using machine learning classifiers such as K-Nearest Neighbour (K-NN), Decision Tree and Multi-layer Perceptron with Adaboosting technique.

Another method for plant species recognition is described by Saleem et al. [6]. The authors utilized the handcrafted features with machine learning classifiers. The images from the dataset are pre-processed, colour converted (RGB to Lab colour space) and thresholded (Otsu method) to form a binary image. From which, the shape features (11), statistical features (7) and venation features (5) are extracted. Principal Component Analysis (PCA) is used to reduce the number of features before it is classified using machine learning classifiers such as K-NN, Decision Tree, Naive Bayesian and Multi-Support Vector Machine. Two datasets namely, Flavia and self-collected are used to evaluate the method. This method exhibited better performance when compared to AlexNet Convolutional Neural Network.

Plant species recognition based on the GIST texture feature extraction method is demonstrated by Kheirkhah et al. [7]. The dimensionality reduction of features is performed using PCA followed by classification using Patternnet Neural Network, SVM and Cosine-based K-NN classifiers. The method is evaluated using the leaf datasets, namely Flavia, ImageCLEF and LeafSnap. The authors observed that the utilization of Cosine-based K-NN classifier resulted in achieving higher performance metrics compared to other methods. The reported accuracies are 98.7, 88.8 and 74.5% on Flavia, ImageCLEF and LeafSnap datasets, respectively.

Yousefi et al. [8] proposed a neural network approach to carry out a plant species recognition. The shape, morphological and textural features are extracted from the images of Flavia dataset. The shape feature (leaf contour) is extracted by Rotation Invariant Wavelet Descriptor (RIWD). The extracted features are then classified using a Multi-layer Perceptron (MLP) with 6 layers (inclusive of 4 hidden layers). The authors reported that the utilization of the RIWD feature along with morphological and texture features resulted in achieving higher accuracy (97.5%) compared to Invariant Elliptic Fourier Descriptor (IEFD).

A method to classify the medicinal plant species is suggested by Naresh et al. [9] wherein the feature extraction process is implemented using a Modified Local Binary Pattern (M-LBP). The M-LBP method uses the mean and standard deviation values instead of threshold values to generate the binary pattern. The Nearest Neighbour classifier is used to classify the medicinal plant species. Five datasets (UoM medicinal plant dataset, Flavia, Foliage, Swedish Leaf and Outex) are used. Anubha et al. proposed several ensemble techniques such as bilateral Convolutional Neural Network (CNN) [10], dual deep learning architecture [11] and Bi-channel Convolutional Network [12] for plant species recognition.

In a study by Lu et al. [13], a method is proposed to classify five different species of Camelia Genus plant. The total number of images/classes in the dataset is about 93. The considered features are comprised of architectural and morphological characteristics. The classification is performed by different methods such as Learning Vector Quantization Artificial Neural Network (LVQ-ANN), Dynamic Architecture for Artificial Neural Network (DAN2) and SVM. It is found that the classification using the DAN2 method resulted in achieving higher accuracy compared to other methods.

Pandolfi et al. [14] used a back-propagation neural network (BPNN) to classify 17 different species of tea plants from Vietnam. The Neural Network contains three layers inclusive of a hidden layer. The hidden layer has 50 optimal hidden neurons which use the Logistic Sigmoid activation function. Fourteen morphological features are extracted. In addition to it, the BPNN outputs are investigated through a cluster analysis (Unweighted Pair Group Method Analysis–UPGMA) to form a dendrogram.

Plant species recognition system employing a Probabilistic Neural Network (PNN) model is proposed by Wu et al. [15]. PNN includes an input layer, a radial basis layer and a competitive layer. Flavia dataset is used to determine the performance analysis of the PNN. Among twelve morphological features extracted, five important features are selected using PCA and fed as an input to the PNN (classification). The authors reported a classification accuracy of 90%.

Fu et al. [16] utilized the leaf vein pattern to perform the classification of plant species. The segmentation and threshold process are carried out using the Sobel and

Laplacian operators. The extracted features (4 gradient features from Sobel operator, local pixel contrast and 5 statistical features) are fed as an input to the Artificial Neural Network (classification).

Based on the extensive literature survey, it has been identified that most of the reported literature utilize a feature extraction method (to extract the features) and classifier or Neural Network (classification). Further, the considered dataset images are heavily pre-processed even before the methodology is tested. Also, most of the Neural Networks reported in the literature consist of three layers (inclusive of one hidden layer) with a lesser number of neurons per layer. On the other hand, studies that involve feeding raw pixel intensities to the Neural Network are sparse.

Hence, in this paper, a custom-developed Neural Network with Parametric Rectified Linear Unit (PReLU) activation function is proposed and evaluated for plant species recognition. Three datasets (Flavia (D1), Swedish Leaf (D2) and custom-developed Leaf 12 (D3)) are considered in the process of model evaluation. The model hyperparameter (Optimizers, Learning Rate, Activation Function and Epochs) are tuned to achieve higher classification accuracy. Since the raw pixel intensities are used in the process of classification, the proposed architecture is better suited to perform a real-time plant species recognition.

3   Methodology

The proposed plant species recognition methodology consists of a Neural Network comprising of four layers inclusive of two hidden layers with the back-propagation algorithm. The schematic representation of the proposed methodology is shown in Fig. 1. The first hidden layer has 500 neurons while the second one has about 250 neurons. The transfer of information from one layer to the other is performed by the activation function in the hidden layer neuron. The hidden layers are tested with different activation function such as Rectified Linear Unit (ReLU) [17], Parametric Rectified Linear Unit (PReLU) [18], Exponential Linear Unit (ELU) [19] and Leaky Rectified Linear Unit (L-ReLU) [20].

Three datasets namely, Flavia (D1) [15], Swedish Leaf (D2) [21] and custom-developed Leaf 12 (D3) [22] are considered. The total number of images and number of images/class for the above-specified datasets are listed in Table 1. A few sample images of the plant species from each of the datasets are shown in Fig. 2. The images in the dataset are resized to 32×32 pixels using the Nearest Interpolation method. Resizing is performed by maintaining the aspect ratio of the images. The normalized pixel intensities of the image are provided as an input ($32 \times 32 \times 3 = 3072$–pixel intensities) to the input layer neuron of the network. The number of output layer neurons depends on the number of output classes. A softmax activation function is used in the output layer neurons.

The output prediction (y_i) is carried out in the output layer, resulting in the determination of the Categorical Cross-Entropy Loss function [23] as given in Eq. (1).

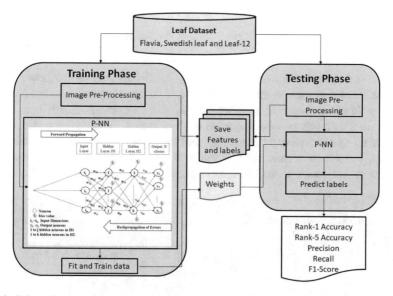

Fig. 1 Schematic methodology representation to perform real-time plant species recognition

Table 1 Dataset description

Dataset	Number of images/class	Number of classes	Total number of images	Image size
D1–Flavia	50	32	1600	1600×1200
D2–Swedish Leaf	75	15	1125	799×1554
D3–Leaf 12	320	12	3840	1920×1080

Fig. 2 Sample leaf images: (i) D1–Flavia, (ii) D2–Swedish Leaf and (iii) D3–Leaf 12

$$E_{CC} = -\frac{1}{N} \sum_{i=1}^{N} \sum_{c=1}^{C} p_{ic}(\log(y_{ic})) \tag{1}$$

where N is the number of pairs $((x_1, t_1)$ to $(x_N, t_N))$ in the training data, $c = 1$ to C is the number of classes, x_i is the input vector, t_i is the target vector, $p_{ic} \in c$ denotes the indicator function for the ith training pattern, $y_{ic} \in c$ represents the predicted probability for the ith observation. Based on the loss values, the interconnection weight and bias terms of the Neural Network are updated in the back-propagation phase [24] resulting in the minimization of loss values, thereby improving the performance of the system. The total number of trainable network parameters are 1,664,762.

Algorithm 1 Inputs: Image, I; pixel intensity, $I(p)$; Weights, w_i; bias, b; net input, y_{in}

Output: Class labels, 0 to N number of classes

1. Resize the image, I to $32 \times 32 \times 3$ pixels
2. Normalize the pixel intensity $I(p)$
3. Calculate $y_{in} = b_i + \sum I(p).w_i$
4. Apply activation function (PReLU or ReLU or ELU or L-ReLU)
5. Repeat steps 3 and 4, until no more hidden layer neurons are available for computation
6. Calculate Error function (Categorical Cross-Entropy Loss function) using Eq. (1)
7. Update w_i and b_i
8. Repeat steps 3–7, until a certain number of epochs
9. Evaluate each of the images, I belonging to 0 to N class to determine the performance metrics.

The steps involved in the process of plant species recognition using Neural Network with back-propagation method is represented in Algorithm 1. The optimum hyperparameters (Optimizers, Learning Rate, Activation Function and Epochs) are determined to improve the performance of the plant species recognition system. Optimizers are used in the updation of the weight interconnection between the layers and bias term in the neuron, during the back-propagation phase. The optimizers considered in the present work include Stochastic Gradient Descent (SGD) [25], Adam [26], Adamax [26] and Adadelta [27]. The learning rate is varied from 0.01 to 0.00001, while the number of Epoch are set from 1 to 200.

The custom-created Neural Network is developed using a Python framework in Windows 10 64-bit OS with Intel Core i7-4790 CPU and NVIDIA Titan X GPU with 3584 CUDA cores. The Python libraries utilized during the implementation process are Numpy, Keras [28] with backend as Tensorflow, Scikit-learn [29], h5py and Matplotlib.

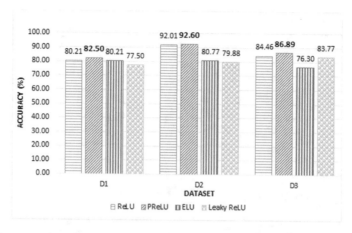

Fig. 3 Accuracy chart obtained by varying the activation function of the network applied on different datasets

4 Results and Discussion

4.1 Determination of Activation Function

Initially, the custom-developed architecture is evaluated to determine a suitable activation function (ReLU, PReLU, ELU, L-ReLU) for the network to perform a real-time plant species recognition. The selection of the activation function is carried out by fixing the optimizer (Adam), learning rate (0.001), Epoch (50) and Batch size (32). The accuracy chart for the three datasets, obtained by varying the activation functions is presented in Fig. 3. It is observed that the PReLU activation function with an α value of 0.02 exhibited the highest accuracy (Top-1: Flavia–82.50%, Swedish Leaf–92.60%, Leaf 12–86.89%).

The influence of the tuneable hyperparameter 'α' of the PReLU activation function on the accuracy of the system is examined. The α values are varied from 0.02 to 0.35 and the corresponding accuracies are listed in Table 2. Highest accuracy (Top-1: Flavia–84.38%, Swedish Leaf–94.28%, Leaf 12–87.50%) is observed by setting an α value of 0.1.

4.2 Determination of Optimizer

The PReLU activation function with an α value of 0.1 is used in the hidden layers of the Neural Network. The learning rate is set to the default values as specified by the Keras library package (SGD–0.01, Adam–0.001, Adamax–0.002, Adadelta–1.0). The number of Epochs and Batch Size is set to 50 and 32, respectively. Figure 4

Table 2 Model accuracy (%) obtained by tuning α value in PReLU activation function

Alpha, α	D1–Flavia	D2–Swedish Leaf	D3–Leaf 12
0.02	82.50	92.60	86.89
0.05	82.71	91.72	87.33
0.1	**84.38**	**94.28**	**87.50**
0.15	80.00	89.94	86.46
0.2	81.88	79.88	86.55
0.25	75.21	93.79	85.42

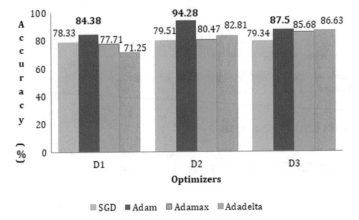

Fig. 4 Accuracy chart obtained by varying the optimizer of the network (PReLU with α = 0.1) applied on different datasets

shows the accuracy chart obtained by varying the model optimizers for the three datasets (Flavia (D1), Swedish Leaf (D2) and Leaf 12 (D3)). It is observed that the Adam optimizer performed better (Top-1 accuracy: Flavia–84.38%, Swedish Leaf–94.28%, Leaf 12–87.50%) compared to other optimizers, irrespective of the dataset. The Top-5 accuracies obtained for Flavia, Swedish Leaf and Leaf 12 datasets are 98.75%, 100% and 99.41%, respectively. Hence, in the subsequent optimization of other hyperparameters, the Adam optimizer is utilized.

4.3 Determination of Learning Rate

The identification of the optimizer (Adam) is followed by determining an appropriate learning rate. The learning rate is varied from 0.01 to 0.00001. The other parameters such as α value of the PReLU activation function, the number of Epoch and Batch size are maintained at 0.1, 50 and 32, respectively.

Figure 5 shows the accuracy chart obtained by varying the learning rate of the network applied on three datasets. It is observed that the learning rate of 0.001 resulted

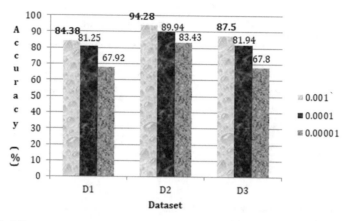

Fig. 5 Model's accuracy chart obtained by varying the learning rate of the Adam optimizer (network with PReLU ($\alpha = 0.1$)) applied on different datasets

in better performance (Top-1 accuracy: Flavia–84.38%, Swedish Leaf–94.28%, Leaf 12–87.5%) compared to other learning rates. The learning rate of 0.01 resulted in low accuracy for plant species recognition and hence is not included in Fig. 5. A similar result in which the Adam optimizer with a learning rate of 0.001 exhibited higher accuracy in comparison to other optimizers is reported by Kingma et al. [26].

4.4 Determination of Epoch

Finally, the optimization of the number of Epochs is carried out. Using Adam optimizer with a learning rate at 0.001 and α value at 0.1, the number of Epochs is varied between 50 and 200. Table 3 lists the values of accuracies obtained by varying the number of Epochs. It is observed that the best accuracy (D1–Flavia: Top-1–85.83%, Top-5–98.75%; D2–Swedish Leaf: Top-1–94.97%, Top-5–100%) is obtained at 200 Epoch. However, in the case of D3–Leaf 12 dataset, the best accuracy (Top-1–87.50%, Top-5–99.41%) is obtained at 50 Epochs.

Table 4 list the performance metrics obtained by the custom-created Neural Network with optimized parameters. The performance metrics include Top-1 accuracy, Top-5 accuracy, precision, recall and F1-score.

The results obtained by the proposed PReLU-based Neural Network with back-propagation algorithm after hyperparameter optimization is compared with those of the existing literature. From the comparison presented in Table 5, it is observed that the proposed PReLU-based Neural Network exhibits better performance in plant species recognition, compared to the existing methods.

Table 3 Model (PReLU with α = 0.1, Adam, learning rate = 0.001) accuracy (%) obtained by varying the number of Epoch

Number of Epochs	Accuracy (%)					
	D1–Flavia		D2–Swedish Leaf		D3–Leaf 12	
	Rank-1	Rank-5	Rank-1	Rank-5	Rank-1	Rank-5
50	84.38	98.75	94.38	100	87.5	99.41
100	79.58	98.12	94.08	100	84.38	99.22
150	85.83	98.54	94.28	100	85.76	98.96
200	**85.83**	**98.75**	**94.97**	**100**	86.63	99.22

Table 4 Performance metrics obtained by the optimized model (PReLU with α = 0.1, Adam, learning rate = 0.001) on three datasets

Dataset	Accuracy (%)						Computation time (s)	
	Rank-1	Rank-5	Precision	Recall	F1-Score	Epoch	Training	Test
D1–Flavia	85.83	98.75	0.86	0.86	0.85	200	31.66	0.06
D2–Swedish Leaf	94.97	100	0.95	0.95	0.95	200	23.28	0.06
D3–Leaf 12	87.50	99.41	0.88	0.88	0.87	50	18.72	0.10

5   Conclusion

In this paper, a PReLU-based Neural Network (PNN) with back-propagation algorithm is proposed and demonstrated to perform recognition of plant species. The model's hyperparameters are optimized to achieve high performance metrics. From the experimental analysis, it is observed that the PReLU-based Neural Network with optimized hyperparameters performs better compared to other existing methodologies. The optimized hyperparameters of the PNN are identified as Adam optimizer, a learning rate of 0.001 and PReLU activation function with an α value of 0.1. The model attained higher performance metrics at 200 (D1–Flavia), 200 (D2–Swedish Leaf) and 50 (D3–Leaf 12) epochs. The proposed model obtained an accuracy of 85.83%, 94.97% and 87.50% when it is subjected to Flavia, Swedish Leaf and Leaf 12 datasets.

Table 5 Comparison of accuracies obtained by the proposed PReLU-based Neural Network with existing literature

Method	Reference	D1–Flavia	D2–Swedish Leaf	D3–Leaf 12
Artificial Neural Network	Söderkvist [21]	–	82.40	–
Probabilistic Neural Network	Kadir et al. [30]	81.56	–	–
Gabor-based method	Wang et al. [31]	–	85.75	–
Fuzzy integral method	Wang et al.[31]	–	89.25	
Texture and shape features	Zhang et al. [4]	–	91.14	–
Shape and colour features	Zhang et al. [4]	–	91.66	–
Locally linear discriminant embedding	Zhang et al. [4]	–	91.86	–
Wavelet-based method	Wang et al.[31]	–	91.37	–
Shape features + K-Nearest Neighbour	Priya et al. [33]	78.00	–	–
2D-linear discriminant analysis + bagging classifier	Gaber et al. [34]	85.00	–	–
Random forest classifier	Anubha Pearline et al. [22]	85.62	87.87	82.38
Colour shape and texture features	Zhang et al. [4]	–	92.44	–
Singular value decomposition sparse representation	Zhang et al. [4]	–	92.59	–
Proposed PReLU-based Neural Network		85.83	94.97	87.50

Acknowledgements The authors would like to thank NVIDIA for providing NVIDIA Titan X GPU under the University Research Grant Programme.

References

1. Yanikoglu B, Aptoula E, Tirkaz C (2014) Automatic plant identification from photographs. Mach Vis Appl 25(6):1369–1383
2. Mata-Montero E, Carranza-Rojas J (2016) Automated plant species identification: challenges and opportunities. In: IFIP world information technology forum. Springer, Cham, pp 26–36

3. Wäldchen J, Rzanny M, Seeland M, Mäder P (2018) Automated plant species identification—trends and future directions. PLoS Comput Biol 14(4):1–19
4. Zhang S, Zhang C, Wang Z, Kong W (2018) Combining sparse representation and singular value decomposition for plant recognition. Appl Soft Comput J 67:164–171
5. Kumar M, Gupta S, Gao XZ, Singh A (2019) Plant species recognition using morphological features and adaptive boosting methodology. IEEE Access 7:163912–163918
6. Saleem G, Akhtar M, Ahmed N, Qureshi WS (2019) Automated analysis of visual leaf shape features for plant classification. Comput Electron Agric 157:270–280
7. Mostajer Kheirkhah F, Asghari H (2019) Plant leaf classification using GIST texture features. IET Comput Vis 13(4):395–403
8. Yousefi E, Baleghi Y, Sakhaei SM (2017) Rotation invariant wavelet descriptors, a new set of features to enhance plant leaves classification. Comput Electron Agric 140:70–76
9. Naresh YG, Nagendraswamy HS (2016) Classification of medicinal plants: an approach using modified LBP with symbolic representation. Neurocomputing 173:1789–1797
10. Pearline SA, Kumar VS (2022) Performance analysis of real-time plant species recognition using bilateral network combined with machine learning classifier. Eco Inform 67:101492
11. Pearline SA, Sathiesh Kumar V (2020) High performance ensembled convolutional neural network for plant species recognition. In: International conference on computer vision and image processing. Springer, Singapore, pp 526–538
12. Pearline SA, Vajravelu SK (2019) DDLA: dual deep learning architecture for classification of plant species. IET Image Proc 13(12):2176–2182
13. Lu H, Jiang W, Ghiassi M, Lee S, Nitin M (2012) Classification of Camellia (Theaceae) species using leaf architecture variations and pattern recognition techniques. PLoS ONE 7(1):e29704
14. Pandolfi C, Mugnai S, Azzarello E, Bergamasco S, Masi E, Mancuso S (2009) Artificial neural networks as a tool for plant identification: a case study on Vietnamese tea accessions. Euphytica 166(3):411–421
15. Wu SG, Bao FS, Xu EY, Wang YX, Chang YF, Xiang QL (2007) A leaf recognition algorithm for plant classification using probabilistic neural network. In: 2007 IEEE international symposium on signal processing and information technology. IEEE, Giza, Egypt, pp 11–16
16. Fu H, Chi Z (2006) Combined thresholding and neural network approach for vein pattern extraction from leaf images. IEE Proc-Vis Image Sig Process 153(6):881–892
17. Arora R, Basu A, Mianjy P, Mukherjee A (2016) Understanding deep neural networks with rectified linear units. arXiv preprint arXiv:1611.01491
18. He K, Zhang X, Ren S, Sun J (2015) Delving deep into rectifiers: surpassing human-level performance on imagenet classification. In: Proceedings of the IEEE international conference on computer vision. IEEE, Santiago, Chile, pp 1026–1034
19. Clevert DA, Unterthiner T, Hochreiter S (2016) Fast and accurate deep network learning by exponential linear units (ELUs), pp 1–14. arXiv 2015. arXiv preprint arXiv:1511.07289
20. Maas AL, Hannun AY, Ng AY (2013) Rectifier nonlinearities improve neural network acoustic models. In: ICML workshop on deep learning for audio, speech and language processing, p 28
21. Söderkvist O Computer vision classification of leaves from Swedish tree. Master's thesis, Linkoping University
22. Anubha Pearline S, Sathiesh Kumar V, Harini S (2019) A study on plant recognition using conventional image processing and deep learning approaches. J Intell Fuzzy Syst 36(3):1997–2004
23. Rusiecki A (2019) Trimmed categorical cross-entropy for deep learning with label noise. Electron Lett 55(6):319–320
24. Sivanandam SN, Deepa SN (2007) Principles of soft computing. Wiley, New Delhi, India
25. Ruder S (2016) An overview of gradient descent optimization algorithms, pp 1–14. Retrieved from http://arxiv.org/abs/1609.04747
26. Kingma DP, Ba JL (2015) Adam: a method for stochastic optimization. In: 3rd international conference on learning representations, ICLR 2015—conference track proceedings, pp 1–15
27. Zeiler MD (2012) ADADELTA: an adaptive learning rate method. Retrieved from http://arxiv.org/abs/1212.5701

28. Chollet F Keras documentation. https://keras.io. Accessed 17 Mar. 2021
29. Pedregosa F et al (2011) Scikit-learn: machine learning in Python. J Mach Learn Res 12:2825–2830
30. Kadir A, Nugroho LE, Susanto A, Santosa PI (2012) Performance improvement of leaf identification system using principal component analysis. Int J Adv Sci Technol 44:113–124
31. Wang X, Liang J, Guo F (2014) Feature extraction algorithm based on dual-scale decomposition and local binary descriptors for plant leaf recognition. Dig Sig Process Rev J 34:101–107
32. Ling H, Jacobs DW (2007) Shape classification using the inner-distance. IEEE Trans Pattern Anal Mach Intell 29(2):286–299
33. Priya CA, Balasaravanan T, Thanamani AS (2012) An efficient leaf recognition algorithm for plant classification using support vector machine. In: International conference on pattern recognition, informatics and medical engineering, PRIME 2012. IEEE, Salem, India, pp 428–432
34. Gaber T, Tharwat A, Snasel V, Hassanien AE (2015) Plant identification: two dimensional-based vs. one dimensional-based feature extraction methods. In: 10th international conference on soft computing models in industrial and environmental applications. Springer, Cham, pp 375–385

Implementation of Legal Documents Text Summarization and Classification by Applying Neural Network Techniques

Siddhartha Rusiya and Anupam Jamatia

1 Introduction

A legal document is a document where two or more parties enter into an agreement, and it is confirmed by the placement of their signatures at the end. It has been observed that it is tough to analyse legal documents. So, we work on problems of rhetorical role labelling and text summarization and automate these for saving human efforts. Rhetorical role labelling is the classification of legal judgements into seven different labels, namely Facts, Ruling by Lower Court, Arguments, Statute, Precedent, Ratio of the Decision, and Ruling by Lower Court. Apart from classification, rhetorical role labelling can be used in solving various other problems like searching similar legal documents, summarization of legal documents, etc. Since, rhetorical role labelling has various uses. So, it is necessary to do research work on this problem. We did various experiments for the rhetorical role labelling task by using deep learning methods like BERT, RoBERTa, XLNet, and differently preprocessed datasets. In automatic text summarization, we have to automate the task of generating a summary of legal documents. Automatic text summarization is divided into two parts—the first part is finding relevant sentences in legal documents for summary generation, and the second part is to generate a summary from those relevant sentences of legal documents. Text summarization can help in tasks like building searching functionality in huge storage of legal documents. Through text summarization, we can get keywords for a particular legal document, and from keywords, we can provide search results. Since, we saw searching functionality proves to very beneficial in every application. So, it is important to work on the research problem of automatic text summarization. Just like the rhetorical role labelling task, for finding relevant sentences in a legal document, we also train various models on a differently preprocessed dataset and use deep learning methods like BERT, RoBERTa, and XLNet to observe the effect of

S. Rusiya (✉) · A. Jamatia
National Institute of Technology, Agartala, 799046 Tripura, India
e-mail: siddhartharusiya84@gmail.com

© The Author(s), under exclusive license to Springer Nature Singapore Pte Ltd. 2023   329
D. S. Sisodia et al. (eds.), *Machine Intelligence Techniques for Data Analysis and Signal Processing*, Lecture Notes in Electrical Engineering 997,
https://doi.org/10.1007/978-981-99-0085-5_27

preprocessing on different methods for our corpus. For generating the summary from those relevant sentences, we use a summarizer from BERT-extractive-summarizer to generate the summary of a legal document. This paper presents the methodologies used for classification and summarization. Many prior attempts to automate the identification of rhetorical roles and text summarization of legal documents rely on hand-crafted features like linguistic cue phrases indicative of a particular rhetorical role, phrases that make the judgement to be summary-worthy. Some of these features, e.g. indicator cue phrases, summary-worthy phrases, etc. are largely dependent on legal-expert knowledge which is expensive to obtain. Previous hand-crafted features work still accumulated in the particular domains only. The deep learning methods can automatically learn features from a sufficient amount of training data and perform better classification and summarization.

2 Related Works

In this section, we discuss the various prior works related to legal judgement classification and legal document summarization.

Automatic labelling of the rhetorical role of sentences relies heavily on manual annotation. While papers that aim to automate the task of semantic labelling also perform annotation analysis [1, 2], other works focus on the process of annotation—developing a manual/set of rules for annotation, inter-annotator studies, creation of a gold standard corpus, and so on. Methodological issues during the design and the development of TEMIS, a syntactically and semantically annotated corpus of Italian legislative texts, give deep insights into the process of rhetorical role labelling [3]. An in-depth annotation study and creation of a gold standard corpus for the task of sentence labelling can be found in the paper by Wyner [4], where assessor agreement was low for labels like facts and reasoning outcomes. Towards automating the annotation task, an initial methodology is discussed using NLP tools on 47 criminal cases drawn from the California Supreme Court and state court of appeals [5].

Most of the works related to the rhetorical role labelling of sentences were implemented using hand-crafted features. In comparison to those in this thesis, we used modern deep learning models to perform the task of rhetorical role labelling of sentences. Deep learning methods are being more frequently in the fields of the legal domain, e.g. classification of factual and non-factual statements in a legal document [9], crime classification [6, 7], and many other works. Some domain-general approaches to segmenting texts into multi- paragraph passages by topic are based on statistical similarity and lexical cohesion, the repetition of similar words incoherent segments, and the tendency for vocabulary to change across segment boundaries. Segmenting legal texts into topics or, as in our project, into functional sections or parts, has required the application of more legal domain-specific knowledge [8]. For instance, one must become familiar with the type of functional sections present in the legal texts of interest, such as courts' legal decisions. One approach to segmentation has focused on automatically identifying the rhetorical roles of sentences. The work on rhetorical role labelling based on machine learning using conditional random field had explored various important features during their analysis [9].

There have been several prior attempts to automatically summarize the legal documents. An undirected graphical model has been trained to segment the documents along with different rhetorical structures. To represent the documents for this work, features like cue words, state transition, named entity, position, and other local and global features were used [10]. Automatic summarization has been developed based on the approach of Teufel and Moens. The core component of this is a statistical classifier that categorizes the sentences in an order in which they might be seen as text parts to be used in a summary [11]. Useful features mainly include word frequency and other highly informative linguistic properties of sentences. Legal text summarizer (LetSum) is a prototype system, which determines the thematic structure of a judgement in four themes introduction, context, juridical analysis, and conclusion. Then it identifies the relevant sentences for each theme [12]. A system is designed and implemented which produces summaries of newspaper articles from the Daily Telegraph Corpus (approx. 15 million words) by extracting a predefined number of sentences of which a relevance score is computed based on combining the weights of the content words in the sentences [13]. In some of the works for generating summary, a two-phase process has been done: scoring and selection. Firstly, a scoring model has been trained using deep learning networks like convolutional neural networks with our proposed objectives for optimizing the model. The document features depend on only the document sentences and are therefore independent of the gold catchphrases which are not available for new documents. The optimization task is similar to the learning to rank task where catchphrases are ranked higher than normal sentence phrases. Secondly, phrases having the highest anchor scores are selected as output catchphrases [14].

Some measures can be used to test the quality of the summary generated by certain methods. ROUGE stands for Recall-Oriented Understudy for Gisting Evaluation. It mainly includes comparing the generated summary with the actual one. These measures count the number of overlapping units such as n-gram, word sequences, and word pairs between the computer-generated summary to be evaluated and that created by humans. To assess the effectiveness of ROUGE measures, the correlation between ROUGE-assigned summary scores and human-assigned summary scores has been compared. The intuition is that a good evaluation measure should assign a good score to a good summary and a bad score to a bad summary. The ground truth is based on human-assigned scores. Acquiring human judgements are usually very expensive; fortunately, The DUC 2001, 2002, and 2003 evaluation data include human judgements for various single-document summaries, very short summaries, single-document very short summaries, and multi-document summaries [15].

CaseSummarizer [16] is a tool for automated text summarization of legal documents which uses standard summary methods based on word frequency augmented with additional domain-specific knowledge. Summaries are then provided through an informative interface with abbreviations, significance heat maps, and other flexible controls. It is evaluated using ROUGE and human scoring including summary text and feedback provided by domain experts.

Many previous attempts towards classifying the legal judgements were using conditional random fields (CRF). This paper presents the benefits of using deep learning models over conditional random fields. Some works were also used deep learning methods like LSTM and Bi-LSTM. We used more trained deep learning methods (like BERT, RoBERTa, and XLNet) that are increasingly being applied for the classification of factual and non-factual sentences in a legal document as deep learning methods evaluate the significance of words more accurately. These deep learning methods reduce the need for feature engineering as well as gave more accurate results.

3 Dataset

There were two separate datasets provided for both the tasks—the training and testing datasets. The training dataset for the text summarization task contains 400 legal documents in training data, and the test data for the text summarization task contain 100 legal documents in test data. The training dataset for the rhetorical role labelling task contains 60 files having 11,285 pieces of training data, and the test data for the rhetorical role labelling task contains 10 files having 697 pieces of test data. There are seven rhetorical labels in which all legal judgements are classified. The number of judgements for labels differs from as low as 341(3%) to as high as 4211(37%) as shown in Table 1.

In the dataset of automatic text summarization, overall relevant sentences in the training dataset are 13,557(23%) and non-relevant sentences are 46,323(77%). The class weights of different classes are relevant (2.208453) and non-relevant (0.646331).

It has been observed that both datasets are quite unbalanced through the percentage comparison of different labels. As legal judgements contain majorly legal terms, we have to counter the significance of every word of legal judgement in legal significance. It has also been observed that some of the words have very little significance in the process of classification.

Table 1 Distribution among labels in rhetorical role labelling dataset

Label	Frequency	Percentage (%)	Class weight
Ratio of the decision	4211	37	0.449
Facts	2622	23	0.724
Precedent	1787	16	1.060
Argument	0939	08	1.358
Statute	0902	08	1.429
Ruling by lower court	0483	04	1.552
Ruling by present court	0341	03	2.682

One more notable thing is that during classification, dates and numbers have no significant role to play. There is a need to filter the legal judgements before modelling the classifier. During preprocessing of datasets, we change all letters to lowercase and then remove punctuation marks, stopwords from our sentences. We used tokenization so that we can apply lemmatization or stemmation as these operations help in better training of our model. We use Python's split function for the tokenization of our sentences. We used PortStemmer for stemmation and WordNetLemmatizer for lemmatization for conversion words to their root form. The number of sentences was counted using the value count function of the data frame. For the dataset of the rhetorical role labelling task, we observed that the difference between majority judgements containing labels and minority containing label is high. So, we use an oversampling technique to reduce this difference. During oversampling, we take some predefined threshold values and based on the range in which they lie to decide the degree of oversampling. The total judgements are 26,198 in the balanced rhetorical role labelling dataset. The total judgements are 73,437 in the balanced text summarization dataset.

4 Experiments

For rhetorical role labelling and text summarization, various experiments had been done using these deep learning models, namely BERT, RoBERTa, and XLNet. Different experiments are a variation of preprocessing or sampling techniques on different models. The size of the development set is 2620 legal judgements for the rhetorical role labelling task and 40 legal documents for the text summarization task. The value of hyperparameters decides through feature analysis of the dataset and unbiased evaluation of validation set of dataset.

4.1 BERT Model

For the rhetorical role labelling task, three different experiments have been done using BERT model. The fine-tuned parameters of the BERT model are selected according to the classification accuracy of the BERT on the development set. The final BERT fine-tuned parameters are batch size = 32, learning rate = $5e^{-5}$, verbose = 1, and epoch = 4. In the first experiment, the corpus has legal judgements which are cleaned, i.e. free from stopwords, punctuation marks, dates, etc. In this experiment, it was observed that the training loss factor decreases from 1.0337 to 0.1184 in subsequent epochs. The validation loss factor in the first epoch decreases from 0.8365 to 0.3072 in subsequent epochs. In the second experiment, the corpus has legal judgement sentences which are cleaned, i.e. free from stopwords, punctuation marks, dates, etc. These sentences are also lemmatized. In this experiment, it was observed that the training loss factor decreases from 1.1514 to 0.1466 in subsequent epochs. The validation loss factor decreases from 0.8091 to 0.3015 in subsequent epochs. In the third experiment, the corpus has legal judgement sentences which are cleaned, i.e. free from stopwords, punctuation marks, dates, etc. These sentences are also stemmed. In this experiment, it was observed that the training loss factor decreases from 1.2990 to 0.1626 in subsequent epochs. The validation loss factor decreases from 1.000 to 0.3614 in subsequent epochs.

For the automatic text summarization task, two different experiments have been done using BERT model. The fine-tuned parameters of the BERT model are selected according to the classification accuracy of the BERT on the development set. The final BERT fine-tuned parameters are batch size $= 32$, learning rate $= 5e^{-5}$, verbose $= 1$, and epoch $= 3$. In the first experiment, we started with removing stopwords, punctuation marks, dates, etc. from the legal judgements of a dataset. In this experiment, it was observed that the training loss factor decreases from 0.6485 to 0.4228 in subsequent epochs. The validation loss factor decreases from 0.7694 to 0.6615 in subsequent epochs. In the second experiment, we did a cleaning of a dataset and then lemmatize the text. In this experiment, it was observed that the training loss factor decreases from 0.6648 to 0.4872 in subsequent epochs. The validation loss factor decreases from 0.6147 to 0.5783 in subsequent epochs.

4.2   RoBERTa Model

For the task of rhetorical role labelling, three different experiments have been done using RoBERTa model. The fine-tuned parameters of the RoBERTa model are selected according to the classification accuracy of the RoBERTa on the development set. Since the maximum sentence length of 128 can cover all of the sentence lengths, sequence length is directly set to 128, and no parameter adjustment is required. The final RoBERTa fine-tuned parameters are batch size $= 32$, learning rate $= 5e^{-5}$, verbose $= 1$, and epoch $= 5$. We used cleaned sentences containing dataset, lemmatized sentences containing dataset, and stemmed sentences containing dataset, respectively, in subsequent three experiments. During the training of models, the loss factor in the first experiment reduces from 1.7023 to 0.3957 in subsequent epochs, loss factor in the second experiment reduces from 1.5667 to 0.3384, and loss factor in the third experiment reduces from 1.4342 to 0.3031.

For the task of automatic text summarization, two different experiments have been done using RoBERTa model. The fine-tuned parameters of the RoBERTa model are selected according to the classification accuracy of the RoBERTa on the development set. Since the maximum sentence length of 128 can cover all of the sentence lengths, sequence length is directly set to 128, and no parameter adjustment is required. The final RoBERTa fine-tuned parameters are batch size $= 32$, learning rate $= 5e^{-5}$, verbose $= 1$, and epoch $= 2$. We used cleaned sentences containing dataset and lemmatized sentences containing dataset, respectively, in subsequent three experiments. During the training of models, the loss factor in the first experiment reduces from 0.6937 to 0.6341 in subsequent epochs, and the loss factor in the second experiment reduces from 0.6938 to 0.6141.

4.3   XLNet Model

For the task of rhetorical role labelling, three different experiments have been done using XLNet model. The fine-tuned parameters of the XLNet model are selected according to the classification accuracy of the XLNet on the development set. Since the maximum sentence length of 128 can cover all of the sentence lengths, sequence length is directly set to 128, and no parameter adjustment is required. The final XLNet fine-tuned parameters are batch size = 32, learning rate = $5e^{-5}$, verbose = 1, and epoch = 3. We used cleaned sentences containing dataset, lemmatized sentences containing dataset, and stemmed sentences containing dataset, respectively, in subsequent three experiments. During the training of the model, the loss factor in the first experiment reduces from 1.9053 to 0.5616 in subsequent epochs, loss factor in the second experiment reduces from 1.5222 to 0.4616, and loss factor in the third experiment reduces from 1.2812 to 0.3146.

For the task of automatic text summarization, two different experiments have been done using XLNet model. The fine-tuned parameters of the XLNet model are selected according to the classification accuracy of the XLNet on the development set. Since the maximum sentence length of 128 can cover all of the sentence lengths, sequence length is directly set to 128, and no parameter adjustment is required. The final XLNet fine-tuned parameters are batch size = 32, learning rate = $5e^{-5}$, verbose = 1, and epoch = 3. We used cleaned sentences containing dataset and lemmatized sentences containing dataset, respectively, in subsequent three experiments. During the training of the model, the loss factor in the first experiment reduces from 0.7049 to 0.5209 in subsequent epochs, and the loss factor in the second experiment reduces from 0.7152 to 0.5424.

5   Results Analysis

For evaluating the performance of our models for rhetorical role labelling classification and finding relevant sentences for text summarization classification, we used micro-, macro-, and weighted average precision, recall, and F1-score as our performance metrics. The micro-version representing the performance of the models on the classification of individual instances is calculated. The average is calculated using the weighted F1 method, which takes into account the contributing portion of each class in the sample. The macro-version is similar to the weighted F1; however in macro-averaging, the performance of the model is calculated separately for each class, and then the mean of the obtained scores is calculated [17]. Since in the macro-averaged, performance is calculated individually and then averaged, we used these scores as they prove to be more effective for comparison of models.

For evaluating the performance of our generated summary for the text summarization task, we used ROUGE-1, ROUGE-2, and ROUGE-1 scores as our ROUGE metrics. The ROUGE-1 scores measure the overlap of unigrams (any single word

Table 2 Rhetorical role labelling: evaluation scores of different experiments

Model		Accuracy	Precision	Recall	F1-score
BERT$_{Cleaned}$	Training	0.94	0.95	0.94	0.94
	Validation	0.91	0.92	0.91	0.91
	Testing	0.56	0.57	0.65	0.58
BERT$_{Lemmatized}$	Training	0.93	0.93	0.92	0.93
	Validation	0.91	0.92	0.90	0.91
	Testing	0.55	0.53	0.65	0.57
BERT$_{Stemmed}$	Training	0.91	0.92	0.91	0.91
	Validation	0.89	0.90	0.89	0.89
	Testing	0.46	0.50	0.59	0.50
RoBERTa$_{Cleaned}$	Training	0.86	0.87	0.85	0.86
	Validation	0.89	0.90	0.88	0.89
	Testing	0.57	0.57	0.59	0.57
RoBERTa$_{Lemmatized}$	Training	0.83	0.84	0.81	0.83
	Validation	0.83	0.84	0.82	0.83
	Testing	0.54	0.53	0.57	0.53
RoBERTa$_{Stemmed}$	Training	0.80	0.81	0.80	0.81
	Validation	0.82	0.83	0.81	0.82
	Testing	0.52	0.50	0.60	0.52
XLNet$_{Cleaned}$	Training	0.83	0.84	0.82	0.83
	Validation	0.84	0.85	0.83	0.84
	Testing	0.51	0.53	0.64	0.54
XLNet$_{Lemmatized}$	Training	0.72	0.72	0.70	0.71
	Validation	0.80	0.81	0.80	0.80
	Testing	0.54	0.54	0.60	0.55
XLNet$_{Stemmed}$	Training	0.76	0.77	0.74	0.76
	Validation	0.80	0.81	0.79	0.80
	Testing	0.45	0.53	0.59	0.50

from a sentence) between the system summary and reference summary. The ROUGE-2 scores measure the overlap of bigrams (consecutive two words from a sentence) between the system and reference summaries. The ROUGE-1 scores measure the longest matching sequence of words between system and reference summaries [18].

For the rhetorical role labelling task, the macro-averaged precision, recall, and F1-scores recorded during training, validation, and testing are shown in Table 2. The best-performing model for comparing criteria F1-score is BERT$_{Cleaned}$ with accuracy (0.56), precision (0.57), recall (0.65), F1-score (0.58), and its category-wise evaluation scores are presented in Table 3. The reason for the good performance of BERT is the masking of each sentence in ten different ways, and as a result, our model trained better using these ten variations of sentences [19]. If we compare based on accuracy, RoBERTa$_{Cleaned}$ scores the highest. The reason for the performance is that

Table 3 Rhetorical role labelling: categorical scores of best-performing model

Label	Precision	Recall	F1-score
Arguments	0.52	0.79	0.62
Facts	0.56	0.71	0.63
Precedent	0.36	0.72	0.48
Ratio of the decision	0.75	0.37	0.49
Ruling by lower court	0.28	0.40	0.33
Ruling by present court	0.87	0.82	0.85
Statute	0.64	0.72	0.68
Overall	0.57	0.65	0.58

Table 4 Text summarization: evaluation scores of different experiments

Model		Accuracy	Precision	Recall	F1-score
$BERT_{Cleaned}$	Training	0.79	0.79	0.79	0.79
	Validation	0.70	0.70	0.70	0.70
	Testing	0.57	0.55	0.51	0.53
$BERT_{Lemmatized}$	Training	0.75	0.75	0.75	0.75
	Validation	0.66	0.66	0.66	0.66
	Testing	0.57	0.58	0.61	0.57
$RoBERTa_{Cleaned}$	Training	0.24	0.87	0.85	0.24
	Validation	0.22	0.90	0.88	0.22
	Testing	0.50	0.50	0.50	0.47
$RoBERTa_{Lemmatized}$	Training	0.38	0.84	0.81	0.38
	Validation	0.22	0.84	0.82	0.22
	Testing	0.50	0.50	0.50	0.47
$XLNet_{Cleaned}$	Training	0.72	0.84	0.82	0.72
	Validation	0.62	0.85	0.83	0.62
	Testing	0.64	0.60	0.63	0.59
$XLNet_{Lemmatized}$	Training	0.71	0.72	0.70	0.71
	Validation	0.71	0.81	0.80	0.71
	Testing	0.66	0.69	0.66	0.58

it avoids the same training mask for each training instance; rather, it dynamically updates the training mask after each training instance.[1]

For finding relevant sentences in the text summarization task, the macro-averaged precision, recall, and F1-scores recorded during training, validation, and testing are shown in Table 4. The best-performing model for finding relevant sentences part of automatic text summarization is $XLNet_{Cleaned}$ with accuracy (0.64), precision (0.60),

---

[1] https://iq.opengenus.org/roberta/.

Table 5 Text summarization: ROUGE scores of different experiments

Model		Precision	Recall	F1-score
BERT$_{Cleaned}$	ROUGE-1	0.39	0.98	0.54
	ROUGE-2	0.30	0.96	0.44
	ROUGE-L	0.39	0.98	0.54
BERT$_{Lemmatized}$	ROUGE-1	0.59	0.59	0.57
	ROUGE-2	0.46	0.44	0.42
	ROUGE-L	0.57	0.57	0.54
RoBERTa$_{Cleaned}$	ROUGE-1	0.45	0.78	0.55
	ROUGE-2	0.34	0.63	0.42
	ROUGE-L	0.44	0.75	0.53
RoBERTa$_{Lemmatized}$	ROUGE-1	0.44	0.77	0.55
	ROUGE-2	0.33	0.62	0.41
	ROUGE-L	0.43	0.75	0.53
XLNet$_{Cleaned}$	ROUGE-1	0.53	0.77	0.61
	ROUGE-2	0.42	0.66	0.49
	ROUGE-L	0.52	0.75	0.60
XLNet$_{Lemmatized}$	ROUGE-1	0.54	0.71	0.59
	ROUGE-2	0.42	0.57	0.46
	ROUGE-L	0.53	0.69	0.57

recall (0.63), and F1-score (0.59). The reason for the good performance of XLNet is it enables learning bidirectional contexts by maximizing the expected likelihood over all permutations of the factorization order [20].

In generating summary part of text summarization task, the ROUGE scores recorded during experiments are presented in Table 5. The best scoring model for the generating summary part is the model that used XLNet$_{Cleaned}$ for finding relevant sentences and BERT-extractive-summarizer for generating summary with ROUGE-1 score of average precision (0.53), average recall (0.77), and average F1-score (0.61); ROUGE-2 score of average precision (0.42), average recall (0.66), and average F1-score (0.49); and ROUGE-L score of average precision (0.52), average recall (0.75), and average F1-score (0.60) as presented in Table 5.

It is necessary to observe the nature of corpus and study model architecture for getting good results.

5.1 Comparison of Different Models

The comparison results of different models for the task of rhetorical role labelling are shown in Table 2. BERT and RoBERTa models on a cleaned dataset performed better than the other models. The comparison results of different models for finding relevance in the text summarization task are shown in Table 4. In that, XLNet model on a cleaned dataset performed better than other models. We observe the negative impact of lemmatization or stemmation on our models as there exists the possibility of losing the significance of words during lemmatization or stemmation. We observe the positive impact of oversampling in rhetorical role labelling, but no such impact on finding relevance in text summarization as the dataset is quite balanced there.

5.2 Error Analysis

To get a better understanding of our prepared models, this section gives a short qualitative analysis of the misclassification and hypothesis for the potential reasons for the errors. In our prepared models, the worst-performing class in almost all the prepared models for the task of rhetorical role classification is Ruling by Lower Court. One of the reasons for the poor performance of some classes in rhetorical role labelling is they lack in terms of the number of samples due to which the model(s) are not much trained on these minority classes. Classes constituting the corpus could be very closely related. Thus, having many common features makes it complex for the model to accurately classify the correct label. Error(s) caused by humans also contribute to errors produced by the language models. These errors including mislabelling of judgements by legal advisors also lead to such errors. One of the reasons for not getting very good-performing models in finding relevant judgements for text summarizer tasks is the difference in relevance of judgements from one legal document to another as a legal document is a combination of legal judgements. There exists in the variation of opinions for assigning relevance to judgements for generating the summary.

6 Conclusion

In this work, we explore the various methods for automatic text summarization and rhetorical role labelling of legal documents. The rhetorical role labelling of sentences consisted mainly of classification into seven roles. Throughout the experiment, we see that selection of models and techniques like oversampling plays a vital role in both tasks. We proposed three models for rhetorical role labelling of sentences and summarization of legal documents. The proposed models are the BERT model, RoBERTa model, and XLNet model. In the task of rhetorical labelling, we observed that preprocessing methods such as lemmatization and stemmation can affect the results achieved by our models. Following this, we performed lemmatization and stemmation on each of our models, through which we see the variety of models using the same model architecture, one being through lemmatized and others being through stemmed text and the cleaned text. All our proposed models were trained and tested on a set of datasets provided by the Artificial Intelligence for Legal Assistance

(AILA)[2] Shared Task organizer. Among all of our proposed models, we found that the BERT model that is using a cleaned dataset is the most consistent and best-performing model, while all other models also achieved significantly good results.

Automatic text summarization of legal documents is divided into two parts—the first part is finding relevant sentences in legal documents for summary generation, and the second part is to generate a summary of those relevant sentences of legal documents. In this task, we did various experiments to observe the effect of normalization and lemmatization on the performance of the model. While experimenting, we develop various models using different model architecture, namely BERT, RoBERTa, and XLNet. All our proposed models were trained and tested on a set of datasets provided by the AILA. Among all proposed models, XLNet that is using a cleaned dataset performed best in both parts of the task. XLNet that is using cleaned text has the highest evaluation scores as well as the ROUGE scores.

Conclusively, we can say that the effect of lemmatization or stemmation depends on the nature of the dataset. A deep analysis of the dataset is necessary for preparing good-performing models. Deep learning models are a suitable choice for text classification in the dataset where the significance of words depends upon their surrounding words. In our dataset, the number of words in legal judgements varies from 40 to 120. Due to the large sentences, we choose deep learning models over machine learning models as they perform better. Considering all the above factors, we did various experiments and achieved a satisfactory result for both the tasks of rhetorical role labelling and summarization of legal documents.

Acknowledgements The authors would like to thank all the anonymous reviewers for reviewing this work and also AILA Shared Task organizer for providing these task datasets.

References

1. Savelka J, Ashley KD (2018) Segmenting us court decisions into functional and issue specific parts. In: JURIX, pp 111–120
2. Shulayeva O, Siddharthan A, Wyner A (2017) Recognizing cited facts and principles in legal judgements. Artif Intell Law 25(1):107–126
3. Venturi G (2012) Design and development of TEMIS: a syntactically and semantically annotated corpus of Italian legislative texts. In: Proceedings of the workshop on semantic processing of legal texts (SPLeT 2012), pp 1–12
4. Wyner AZ, Peters W, Katz D (2013) A case study on legal case annotation. In: JURIX, pp 165–174
5. Wyner AZ (2010) Towards annotating and extracting textual legal case elements. Informatica e Diritto Special Issue Legal Ontol Artif Intell Tech 19(1–2):9–18
6. Wang P, Yang Z, Niu S, Zhang Y, Zhang L, Niu S (2018) Modeling dynamic pairwise attention for crime classification over legal articles. In: The 41st international ACM SIGIR conference on research and development in information retrieval, pp 485–494

---

[2] https://sites.google.com/view/aila-2021.

7. Wang P, Fan Y, Niu S, Yang Z, Zhang Y, Guo J (2019) Hierarchical matching network for crime classification. In: Proceedings of the 42nd international ACM SIGIR conference on research and development in information retrieval, pp 325–334

8. Vanderbeck S, Bockhorst J, Oldfather C (2011) A machine learning approach to identifying sections in legal briefs. In: MAICS, pp 16–22

9. Walker VR, Pillaipakkamnatt K, Davidson AM, Linares M, Pesce DJ (2019) Automatic classification of rhetorical roles for sentences: comparing rule-based scripts with machine learning. In: ASAIL@ ICAIL

10. Saravanan M, Ravindran B, Raman S (2008) Automatic identification of rhetorical roles using conditional random fields for legal document summarization. In: Proceedings of the third international joint conference on natural language processing: Volume-I

11. Grover C, Hachey B, Hughson I, Korycinski C (2003) Automatic summarisation of legal documents. In: Proceedings of the 9th international conference on artificial intelligence and law, pp 243–251

12. Farzindar A (2004) Atefeh farzindar and guy lapalme,'letsum, an automatic legal text summarizing system in the seventeenth annual conference. In: Legal knowledge and information systems: JURIX 2004, the seventeenth annual conference, vol 120. IOS Press, pp 11–18

13. Zechner K (1995) Automatic text abstracting by selecting relevant passages. Master's thesis, Centre for Cognitive Science, University of Edinburgh

14. Tran V, Nguyen ML, Satoh K (2018) Automatic catchphrase extraction from legal case documents via scoring using deep neural networks. arXiv preprint arXiv:1809.05219

15. Lin C-Y (2004a) ROUGE: a package for automatic evaluation of summaries. In: Text summarization branches out. Association for Computational Linguistics, Barcelona, Spain, pp 74–81

16. Polsley S, Jhunjhunwala P, Huang R (2016) Casesummarizer: a system for automated summarization of legal texts. In: Proceedings of COLING 2016, the 26th international conference on computational linguistics: system demonstrations

17. Haneczok J, Jacquet G, Piskorski J, Stefanovitch N (2021) Fine-grained event classification in news-like text, pp 179–192

18. Lin C-Y (2004b) ROUGE: a package for automatic evaluation of summaries. In: Text summarization branches out, pp 74–81

19. Devlin J, Chang M-W, Lee K, Toutanova K (2018) Bert: pre-training of deep bidirectional transformers for language understanding. arXiv preprint arXiv:1810.04805

20. Yang Z, Dai Z, Yang Y, Carbonell J, Salakhutdinov RR, Le QV (2019) Xlnet: generalized autoregressive pretraining for language understanding. In: Advances in neural information processing systems, vol 32

Robust Image Watermarking Using Arnold Map and Adaptive Threshold Value in LWT Domain

Sushma Jaiswal and Manoj Kumar Pandey

1 Introduction

The rapid enhancement and development of the Internet and use of digital device have enormously augmented the ease of formation and sharing of digital content over the internet. Illegal copying of data, tempering, manipulation of digital data, and protecting copyright has become a serious concern with all this. Adding digital watermark has been emerged as one of the possible solutions for these problems. Watermarking is the process of adding signals to the digital data and, then proving the authorship by extracting signals from digital data. The watermarking can be performed on text, digital image, audio, and video. Watermarking can be of two categories, i.e., spatial domain and frequency domain/transform domain. The spatial watermarking is done based on the pixel value of the digital cover image [1, 2], and frequency domain-based watermarking is done based on various wavelet transform of cover images [3, 4]. The reviewed literature reveals that pixel-based watermarking is less robust as compared to wavelet-based watermarking. Transform domain, has been extensively used for image watermarking and the majority of these watermarking methods claim to be much robust against various watermarking attacks (for example Ali et al. [5]; Mishra et al. [6]). Previously Lin et al. [7] presented a watermarking process based upon wavelet tree of the cover image, and in this scheme, a tree is decomposed into two parts for embedding binary watermarking bits 0 and 1, and bits are added based on changing the two smallest value of wavelet tree. An adaptive threshold-based scheme is used for extraction of watermark. Another watermarking method based on wavelet tree, is given by Run et al. [8], in this method

S. Jaiswal (✉) · M. K. Pandey
CSIT Department, Guru Ghasidas Central University, Bilaspur, India
e-mail: jaiswal1302@gmail.com

© The Author(s), under exclusive license to Springer Nature Singapore Pte Ltd. 2023
D. S. Sisodia et al. (eds.), *Machine Intelligence Techniques for Data Analysis and Signal Processing*, Lecture Notes in Electrical Engineering 997,
https://doi.org/10.1007/978-981-99-0085-5_28

maximum coefficient of wavelet tree is used for watermark embedding and an adaptive threshold is applied for extraction of watermark. Both schemes [7, 8] demonstrate more robustness but not showing robustness against JPEG attack with a high compression ratio.

The new generation of wavelet transform which is more efficient as compared to other transform is lifting wavelet transform (LWT), proposed by Sweldens in the year 1996 [9]. Various lifting-based watermarking algorithms have been proposed by Verma et al. [10], Islam et al. [11], Makbol et al. [12], and Li et al. [13]. Islam et al. [11] have done a study on various attacks on different sub-band in the LWT domain using SVM and found that sub-band 7 provides optimum performance against various attacks. Using LWT-SVD, a watermarking idea was proposed by Makbol et al. [12]. In overall, the majority of the existing watermarking methods shows robustness against some of the attacks but fails to demonstrate robustness against most of the attacks. The study of the state of art suggests that the tasks that need to be address more precisely, is to keep up the balance between imperceptibility and robustness. Aiming at providing maximum security and robustness against common attacks, a watermarking method is proposed that not only maintains adequate image quality but also outperforms other methods.

1.1 Important Contribution

The proposed watermarking is motivated by method developed by Lin et al. [7], Li et al. [13], Verma et al. [10], and Islam et al. [11] with aiming at the providing robustness and security of watermarked images against most of the attacks. In the proposed scheme the sub-band HL3 is selected, as used in Islam's method [11]. The idea of selecting a region for embedding is taken from method of Li et al. [13]. In method of Li the region for embedding is selected using the pixel error (PE) but in proposed idea, embedding spot is selected using the least coefficient difference of HL3 sub-bands and, the dissimilarity between the proposed scheme and Lin's scheme [7] is that in method of Lin, binary bits are added by changing the lowest coefficient of small frequency sub-band of the tree, whereas in proposed method maximum coefficient of higher frequency sub-band is used for embedding using quantization of coefficients. The difference between Verma's methods [10] and this idea is that, in Verma's method LH3 sub-band is selected for watermark embedding but in this work HL3 sub-band is selected for embedding, and in addition different watermark embedding rules using threshold value has been used, and Arnold map has been applied additionally for enhancing the robustness and security of the system. In this method, threshold value (T) is selected based on a cumulative density function (CDF) of lowest coefficient difference in the blocks. The applying of CDF makes scheme stronger as, for the most number of blocks, the lowest difference of the coefficient, for the majority of the blocks comes under a narrow range and for all this reason threshold (T) is selected, such that $N_w <= N_b$, where N_w and N_b represent the number of bits used for watermarking and the number of bits in important region

(IR), respectively. This scheme provides an adequate number of bits for watermark embedding and improves security as well. The proposed method uses randomization at two separate levels, block selection level and wavelet level and the watermark is twisted using Arnold cat map which makes the proposed idea more secure.

The motivation of introducing block identification method (BIM) is to randomly embed watermark bits, which helps to enhance the security beside intentional and unintentional attacks, and in the proposed work lifting wavelet is employed instead of conventional wavelets because LWT is quicker in terms of computation and the energy compacted property of LWT makes it a good contender for watermarking. The important contributions of the watermarking idea are as follows:

1. This is a blind scheme so it doesn't require an original cover image or watermark image.
2. Different secret keys at different levels and the use of Arnold map before watermark embedding make it more secure and robust and use of LWT provides enough robustness because of its advantage over other methods.
3. The scheme provides good robustness as compare to other similar method.

This manuscript is structured as follows: Sect. 2 explained about LWT. Section 3 shows the proposed watermarking technique and Sect. 4 contains outcome and comparison. Section 5 contains conclusion and future scope.

2 Preliminaries

This section shows the basic concepts for the further discussion and it start with the brief introduction to LWT and Arnold map.

2.1 Lifting Wavelet Transform (LWT)

Integral wavelet transforms [14] or lifting wavelet transform was introduced by Sweldens [9] in the year 1996, in this the decomposition problem is treated in integer domain rather than floating-point domain as experience in other conventional wavelets. The LWT ease the reversibility by looking problem in integer domain. Generally, LWT is comprised of three basic steps, i.e., split, predict, and update.

LWT based watermarking is popular among the researchers because

(i) LWT transform is composed of the integer coefficients because of lifting step.
(ii) LWT-based watermarking needs less memory and time as compared to other transformation.
(iii) The energy compaction property of LWT helps to attain secure and robust watermarking.

2.2 *Arnold Cat Map (ACM)*

The Arnold cat map is introduced in the year 1960 and it is a 2-D chaotic map that shuffles bits of image using a key-value and also reverses the shuffling using the key when needed. ACM is applied additionally to enhance the protection of the proposed scheme as done by [15]. The applicability of ACM make it good applicant for using in the proposed method for enhancing the security.

3 Proposed Method

This part shows the method of developing robust watermarking using LWT and Arnold map. For watermark adding, first BIM concept is used to get the IR and NIR (non-important region). Then changed (using Arnold) bits of watermark are added to chosen block of size 2×2 matrix of IR by quantifying the two greatest coefficients of the block. The step of embedding is discussed in a later section. For extraction, largest coefficient difference of such selected blocks is calculated, and compared against the adaptive threshold value (Υ). The watermark extraction is discussed in a later section.

3.1 *Block Identification Method (BIM)*

First of all the cover image is decomposed using 3 level LWT transform and HL3 sub-band is further chosen for watermark adding. The size of HL3 sub-band is (64×64) which shuffled using secret seed key (key1). Then the entire separate coefficients are arranged to form blocks of size 2×2. Then blocks are arbitrarily shuffled using a different seed key (key2). In each single block calculates the CD_{min} (minimum coefficient difference) and CD_{max} (maximum coefficient difference) using formula (1) and (2).

$$CD_{min} = x(2) - x(1); \tag{1}$$

$$CD_{max} = x(n) - x(n-1); \tag{2}$$

where $x(1)$, $x(2)$, ..., $x(n-1)$ and $x(n)$ is smallest, second smallest, second largest, and largest coefficient of the block. Here CD_{max} is maximum coefficient difference value used for watermark embedding, and CD_{min} is minimum coefficient difference value used to select blocks for watermark embedding. The selection criterion for IR and NIR is as follows:

$$\text{if } CD^i \min <= T, \quad i\text{th block belongs to IR} \tag{3}$$

$$\text{if } CD^i \text{ min} > T, \quad i\text{th block belongs to NIR} \tag{4}$$

where T = threshold value.

Here, we have presented the analysis of "Lena" of size (512×512) for BIM. We have observed in the experiment that HL3 sub-bands has a 4096 (64×64) coefficient, and after considering block size of (2×2) total of 1024 blocks have been obtained. One can observe that applying $T = 38$; there is a sufficient 915 IR blocks generated. The number of blocks generated for IR is dependent on the value of T, and for embedding a watermark of size 32×16 (512 bits), the IR must contain at least 512 blocks.

3.2 Watermark Adding Procedure

This part contains the proposed digital watermark adding procedure, and the formula for watermark embedding is as follows:

For every watermark bit, 1 or 0 quantify the max coefficient of related blocks in IR by applying Eqs. (5) and (6).

If watermark bit = 1,

$$x(n)_i = x(n)_i + T, \quad \text{if } CD^i \text{ max} < \max(\sigma, T), \quad \text{Else } x(n)_i - CD^i \text{ max}, \tag{5}$$

$$\text{And, if the watermark bit} = 0 \quad x(n)_i = x(n)_i \tag{6}$$

Here T represents threshold, and CD^i max shows the variation among two largest values of the respective ith block. The averaging difference of coefficient value σ for all N_b blocks is shown as:

$$\sigma = \frac{\sum_{i=1}^{N_b} CD^i \text{ max}}{N_b} \tag{7}$$

For random shuffling of HL3 sub-band coefficient and block, the shuffle scheme is applied based on Fisher-Yates; more details on this can be found at Knuth et al. [16].

The step for watermark embedding is given by follows:

Step 1 The cover image is decomposed up-to third level using LWT, and HL3 sub-band coefficients are randomly shuffled using **shuffle** (HL3; Key1; HL3').

Step 2 Successive non overlapped coefficients of the HL3 sub-band are joined to make blocks of size (2×2), and all the blocks are arbitrarily mixed-up using **randomize** (Blocks; Key2; Blocks').

Step 3 Arbitrarily mixed-up blocks are passed and further process using BIM, to form IR and NIR.

Step 4 Then average difference of coefficient of the entire N_b blocks of IR is calculated using (7).

Step 5 Watermark image W is scrambled using Arnold map and changed to a 1-dimensional array of length N_w.

Step 6 For all N_w bits of the watermark, perform the following:

   6.1. Randomly select 512 blocks from the IR (where $N_w < N_b$) and find the two most extensive coefficient values, $x(n)$ and $x(n-1)$ of the blocks.
   6.2 If the watermarking bit is equal to 1, modify $x(n)$ using (5).
   6.3 If the watermarking bit is equal to 0, modify $x(n)$ using (6).

Step 7 All the updated and other blocks of IR and NIR are restored to its original index value. Then, the entire blocks along with updated coefficients of the HL3 band are inversely restored with the similar keys as used in Step 1 and Step 2.

Step 8 Redesign the modified HL3 sub-band to its novel dimension and do inverse lifting transform to get the watermarked image.

3.3   The Decoder Design

The proposed watermarking method is blind; therefore it does not necessitate the original cover and original watermark image for extraction. Here the adaptive threshold-based method is used for extraction of watermark. The adaptive threshold-based approach Υ is expressed as

$$\Upsilon = \frac{\sum_{i=1}^{Nw \, x \, \alpha} \phi'i}{Nw \; x \; \alpha} \qquad (8)$$

where $\phi'i$ is a collection of $\{CD'_{\max 1}, CD'_{\max 2}, \ldots, CD'_{\max i}\}$, for $i = 1, 2, \ldots, N_w$. Here all the maximum coefficient difference is re-arranged in increasing manner. In Eq. (8), scaling factor α is applied to ensure the minimum Υ value for extraction of watermark, where $0 < \alpha \leq 1$.

The watermark bits are extracted as:

$$\text{water mark bit} = 1, \text{if}(CD'_{\max i}) \geq \min(\Upsilon, T)$$
$$0, \text{otherwise} \qquad (9)$$

3.4 *Watermark Extraction Method*

The process of watermark extraction is shown in Fig. 2. The step for extraction is as discussed below.

Step 1 The HL3 sub-band values of watermarked image are fetched as done in the embedding step.

Step 2 With the confidential keys Key1 and Key2, the coefficients value and respective blocks of the HL3 sub-band are shuffled randomly as done in the embedding step.

Step 3 Save the position of blocks, in which watermark bits are added in the embedding process.

Step 4 For all N_w block, perform the following step:

    4.1 For every block of coefficient, determine the two maximum coefficients $x(n)'$, $x(n-1)'$ to determine $CD'_{max\,i}$.
    4.2 Bits for watermark formation are extracted using Eq. (9).

Step 5 The obtained values are saved in 1-D array W' having length N_w, and then, inversely operation is performed using the Arnold map as done by embedding step.

Step 6 Reform the obtained array, as the size of the original watermark (Fig. 1).

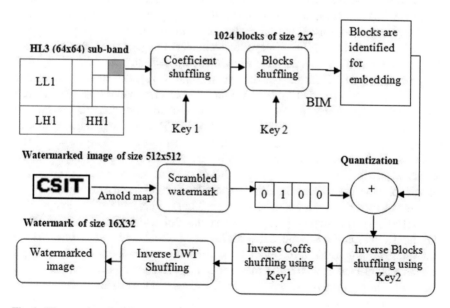

Fig. 1 Watermark embedding procedure

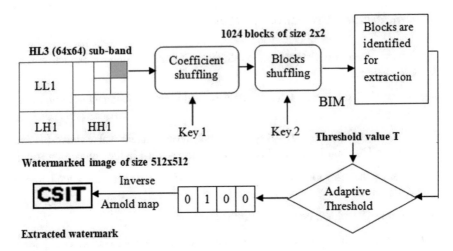

Fig. 2 Watermark extraction procedure

4   Results and Discussion

This section contains the experiment result and its performance correlate with some of the similar watermarking schemes. For the experiment purpose, ten standard grayscale images, including Lena, mandril, and peppers, of size (512 × 512), 8bit/pixels have been used, but for making it brief, results of only Lena standard image is shown in Table 1. Lena image and watermark image of size (16 × 32) is used for experiment. All images used for experiment are collected from the CVG_UGR image dataset and USC-SIPI image dataset available at [17]. The proposed algorithm is implemented using matlab 2016a and i5 Intel processor, window 7(64-bit). Figure 3 contains the original standard image of Lena, watermark, watermarked Lena image and retrieved watermark image with NC = 1 and BER = 0. The parameter applied for watermark adding and extraction on the Lena image is $T = 38$ and $\alpha = 0.4$ as it gives good balance between imperceptibility and robustness. Figure 4 represent the standard gray level mandril and peppers image used for the research purpose. The correlation of proposed scheme with other similar existing scheme is shown in Fig. 5 in terms of PSNR value for standard images. Figure 6 depicts the imperceptibility value (PSNR) on the standard image of Lena, mandril, and peppers for the different threshold values and it is noticed that for lower threshold the PSNR is high but the robustness is compromised therefore threshold and watermark embedding strength (α) value is selected in such a way that it exhibits good stability between PSNR and NC value. During the experiment it has been notices that the proposed scheme gives good balance between the imperceptibility (PSNR) and robustness (NC and BER).

The reliability of the proposed watermarking method is analyzed in terms of PSNR and NC. Peak signal to noise ratio (PSNR) is used for calculation of the dissimilarity between the original cover image and watermark image, and the PSNR is calculated using

Table 1 Comparison of robustness in terms of NC against different attacks

Attacks	MD (3 × 3)	SPLN (0.01)	SLP (0.01)	HE	Cr (50%)	Cr (25%)	Cr (10%)	Jpeg (60)	Jpeg (50)	Jpeg (40)	Jpeg (30)	Jpeg (20)	Jpeg (10)
[8]	0.93	X	X	0.86	X	0.68	X	0.99	0.96	0.93	0.85	0.68	0.6
[10]	0.95	0.746	0.753	0.92	X	0.93	X	1.0	1.0	1.0	0.99	0.94	0.87
[11]	X	0.890	0.864	0.984	X	0.757	0.897	0.99	0.98	0.97	0.98	0.94	X
[18]	X	0.881	0.807	0.588	X	X	X	0.95	X	X	0.92	X	0.69
[19]	0.985	X	0.863	X	X	X	X	X	0.99	X	0.97	0.97	0.89
PSNR	32.01	32.13	38.12	32.21	32.34	32.23	35.39	31.92	32.89	32.57	31.94	38.80	32.54
NC	0.589	0.860	0.921	0.933	0.2380	0.457	0.661	1.0	0.97	0.98	0.99	0.98	0.97
BER	0.205	0.069	0.039	0.033	0.3810	0.271	0.169	0	0.011	0.009	0.001	0.009	0.013
Watermark	CSIT	CSIT	CSIT	CSIT	CSIT	CSIT	CSIT	CSIT	CSIT	CSIT	CSIT	CSIT	CSIT

Fig. 3 **a** Standard Lena image, **b** original watermark image, **c** watermarked image with PSNR = 38.98 dB using $T = 38$ and $\alpha = 0.4$, **d** extracted watermark with BER = 0 and NC = 1

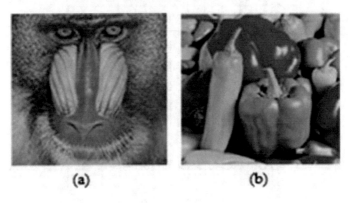

Fig. 4 Standard images **a** mandril, **b** peppers

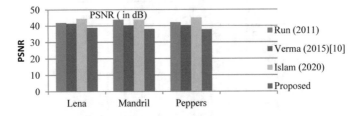

Fig. 5 Comparing PSNR value of images with other similar methods

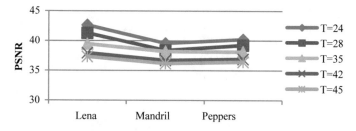

Fig. 6 PSNR value for different thresholds on standard image Lena, mandril, and peppers

$$PSNR = 10 \log_{10} \frac{255^2}{MSE} \tag{10}$$

where MSE = mean-square error.

$$MSE = \frac{1}{M * N} \sum_{ij=0}^{MN} CI(i, j) - WI(i, j) \tag{11}$$

$$NC = \frac{\sum_i W_{ij} \sum_j W'_{ij}}{h \times w} \tag{12}$$

The proposed blind watermarking method is tested against most of the standard image attacks like compression cropping, histogram equalization (HE), salt and pepper noise (SLP), speckle noise (SPLN), median filter (MD) (3×3), and found to be robust against most of the attacks. The proposed watermarking method is also compared with some other similar existing methods in terms of NC value for different attacks in Table 1. The PSNR and BER are also shown along with the NC value, and the bold value signifies that the proposed watermarking method outperforms other similar existing methods. Table 2 shows the comparison with [10, 11] in terms of BER on Lena image. One can notice that for jpeg (QF = 10) attacks, proposed scheme is extracting watermark with NC of 0.97 and PSNR of 32.54, and for cropping of 50%, the scheme is extracting watermark with NC of 0.2380 and PSNR of 32.34, which undoubtedly shows the proposed method outperforms the other similar existing watermarking method. Table 3 shows the comparison with [20] in terms of time taken (in second) for embedding and extraction process for image Lena, mandril, and peppers. The simplicity of the proposed idea and good results in case of most of the image attacks, make it significant as compare to other existing scheme.

Table 2 Performance comparison with [10] and [11] in terms of BER on Lena

Attacks	[11]	[10]	Proposed
JPEG (10)	X	0.05	0.0134
JPEG (20)	X	0.025	0.0098
JPEG (30)	0.0098	0.01	0.0019
JPEG (40)	0.0117	X	0.0098
JPEG (50)	0.0059	X	0.0117
SLP (0.01)	0.0488	X	0.0390
HE	0.0156	X	0.0332
Crop (25%)	0.04	X	0.2715

Table 3 Time taken in second for embedding and extraction

Test Images	Embedding time (s)		Extraction time (s)
	Prop.	Other method [20]	Prop.
Lena	0.781	1.93	0.763
Mandril	0.758	1.91	0.750
Peppers	0.804	1.93	0.747

5 Conclusion and Future Scope

The given blind watermarking method is implemented using LWT and Arnold map with BIM concept. BIM is applied for generating IR and NIR regions. The energy saving property of LWT and the use of secret key and Arnold map provide an extra security layer to the system. The significance of the proposed method is that it perform well and take less time (refer Table 2) as compared to other watermarking method [20], so it can be applied where the resources are less or limited. The idea can be extended to other real-life examples for ensuring security and integrity. Adequate PSNR approx to 39 dB and a good NC value ensure that the proposed method is good enough to resist common attacks and it outperformed other existing methods. One of the limitations of the proposed watermarking scheme is that it not performs well in case of cropping and median filter attacks. In the future, machine learning method can be applied for improving the results.

References

1. Singh A, Sharma N, Dave M, Mohan A (2012) A novel technique for digital image watermarking in the spatial domain. In: Proceedings of 2012, 2nd IEEE international conference on parallel, distributed and grid computing, PDGC, pp 497–501
2. Fridrich J, Goljan M, Du R (2001) Detecting LSB steganography in color and grayscale images. IEEE Multimedia Special Issue Sec 22–28
3. Solachidis V, Pitas I (2001) Circularly symmetric watermark embedding in 2-D DFT domain. IEEE Trans Image Process 10(11):1741–1753

4. Pereira S, Pun T (2000) Robust template matching for affine resistant image watermarks. IEEE Trans Image Process 9(6):1123–1129
5. Ali M, Ahn CW, Pant M, Siarry P (2015) An image watermarking scheme in wavelet domain with optimized compensation of singular value decomposition via artificial bee colony. Inf Sci 301:44–60
6. Mishra A, Agarwal C, Sharma A, Bedi P (2014) Optimized grayscale image watermarking using DWT–SVD and firefly algorithm. Exp Syst Appl 41:7858–7867
7. Lin W-H, Wang Y-R, Horng S-J (2009) A wavelet-tree-based watermarking method using distance vector of binary cluster. Exp Syst Appl 36:9869–9878
8. Run R-S, Horng S-J, Lin W-H, Kao T-W, Fan P, Khan MK (2011) An efficient wavelet-tree-based watermarking method. Exp Syst Appl 38:14357–14366
9. Sweldens W (1996) The lifting scheme: a custom-design construction of biorthogonal wavelets. Appl Comput Harmon Anal 3:186–200
10. Verma VS, Jha RK, Ojha A (2015) Significant region-based robust watermarking scheme in lifting wavelet transform domain. Exp Syst Appl 42(21):8184–8197
11. Islam M, Roy A, Laskar RH (2020) SVM-based robust image watermarking technique in LWT domain using different sub-bands. Neural Comput Appl 32:1379–1403
12. Makbol NM, Khoo BE (2014) A new robust and secure digital image watermarking scheme based on the integer wavelet transform and singular value decomposition. Digital Sig Process 33:134–147
13. Li X, Li J, Li B, Yang B (2013) High-fidelity reversible data hiding scheme based on pixel-value-ordering and prediction-error expansion. Sig Process 93:198–205
14. Zhu T, Qu W, Cao W (2022) An optimized image watermarking algorithm based on SVD and IWT. J Supercomput 78:222–237
15. Jithin KC, Sankar S (2020) Color image encryption algorithm combining Arnold map, DNA sequence operation and a Mandelbrot set. J Inform Sec Appl 50
16. Knuth DE (1997) The art of computer programming. In: Seminumerical algorithms, vol 2, 3rd edn. Addison-Wesley, MA
17. http://www.imageprocessingplace.com/root_files_V3/image_databases.htm. Accessed 10 Dec. 2021
18. Singh AK (2019) Robust and distortion control dual watermarking in LWT domain using DCT and error correction code for medical image. Multimedia Tools Appl 78(21):30523–30533
19. Verma VS, Jha RK (2015) Improved watermarking technique based on significant difference of lifting wavelet coefficients. VIP 9(6):1443–1450
20. Sinhal R, Jain DK, Ansari IA (2021) Machine learning based blind color image watermarking scheme for copyright protection. Patter Recogn Lett 145:171–177

Handcrafted and Deep Features for Micro-expressions: A Study

Ankita Jain and Dhananjoy Bhakta

1 Introduction

The expressions that are depicted on the human face and can be identified by naked human eyes are known as facial expressions (FE). One special category of expressions is also present on the face known as micro-expressions (MEs). It is difficult to detect MEs since the study suggested that it occurs as a leakage, only when a high-stake situation is involved [1, 2]. Darwin's theory of involuntarily facial action stated, "if you cannot make an action voluntarily, then you will not be able to prevent it when involuntary processes such as emotion instigate it" [3]. This theory validates the criteria of MEs revealing true emotion and happening involuntarily on the face. Earlier, the ME study was only restricted to human behavior and psychological domain, and gradually, it became an interesting topic for computer vision and machine learning. There are six universal emotions identified [3], viz., happiness, anger, fear, sadness, disgust, and surprise. All emotions are coded with the Facial Action Coding System (FACS) [4] which describes emotions with the combinations of facial muscles called action unit (AU). The 32 atomic AUs are divided in the upper and lower region of face depicted in Fig. 1. Table 1 illustrates the expressions with the combinations of AUs. For example, happy expression is coded by AU6 + AU12 + AU25.

The researchers are spotting ME in videos and classifying them by applying various computer vision tools. Feature extraction is a prime step for ME analysis as these are subtle and low-intensity expressions. There are two methods of feature interception: handcrafted feature and deep feature. Both the features contain spatial and temporal information of ME samples. In literature, the researchers pull out infor-

A. Jain (✉) · D. Bhakta
Indian Institute of Information Technology, Ranchi 834010, India
e-mail: ajain.rs@iiitranchi.ac.in

D. Bhakta
e-mail: bhaktadhananjoy@iiitranchi.ac.in

© The Author(s), under exclusive license to Springer Nature Singapore Pte Ltd. 2023
D. S. Sisodia et al. (eds.), *Machine Intelligence Techniques for Data Analysis and Signal Processing*, Lecture Notes in Electrical Engineering 997,
https://doi.org/10.1007/978-981-99-0085-5_29

Table 1 Six universal expressions with AU association

Expression	AU association
Happiness	AU6, AU12, AU25
Sadness	AU1, AU4, AU6, AU11, AU15, AU17
Anger	AU4, AU5, AU7, AU10, AU17, AU22–AU26
Fear	AU1, AU2, AU4, AU5, AU20, AU25, AU26, AU27
Disgust	AU9, AU10, AU16, AU17, AU25, AU26
Surprise	AU1, AU2, AU5, AU26, AU27

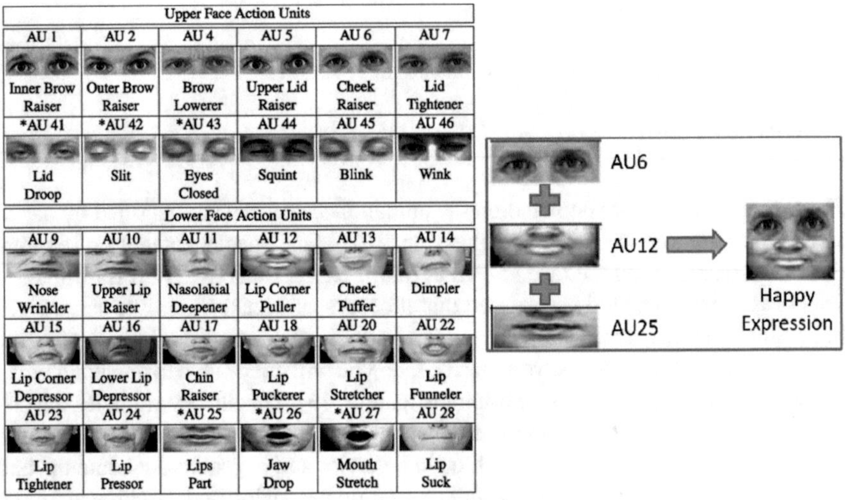

Fig. 1 Action unit (left). AU(6+12+25) associated with happy expression (right)

mation from an image either by using one of these two features. In this paper, our focus is to classify the research study based on handcrafted and deep features along with contributions and results, in reverse chronological order. The goal is to assist beginners in comprehending current research, problems, and future potential in this discipline.

2 Micro-expression: Duration and Database

Micro means extremely small. In this context, "small" means less amount of time. Duration is a key feature that distinguishes it from macro- and facial expression [1]. Different groups [6, 9, 28, 34, 42, 45–47] considered different time span, but generally 1/5 s is considered as an upper limit. In most articles, ME duration is considered ranging from 1/25 to 1/5 s.

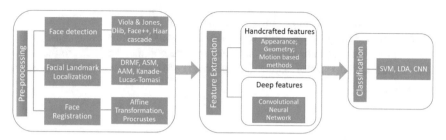

Fig. 2 Stages of ME recognition with some methods specified in each category

The first discussion of ME dataset was reported in [7, 8]. Though only eight ME datasets are published to date, and among them, only six are publicly available. The York-DDT [7], which contains only 18 samples, was the first spontaneous database that detected lies from emotional and unemotional cues. The Polikovsky's dataset [8] is not publicly available, and most of the information is missing like the total number of ME samples. The USF-HD database [9] contains 181 macro- and 100 micro-samples. The SMIC dataset [10] has three variants, namely: HS, VIS, and NIR. It has 164 samples in SMIC-HS and 71 each in VIS and NIR datasets. The CASME [11], CASMEII [12], and CAS(ME)^2 [13] datasets have 195, 247, and 357 samples, respectively. The SAMM [14] has 159 micro-movements categorized into 7 classes. The latest two databases [13] and [14] have both macro- and micro-expressions in it. All the above-mentioned datasets have following things in common: (i) they are produced in an enclosed lab atmosphere, and (ii) they have limited number of samples.

3 Stages of Micro-expression Detection and Identification

The researcher detects ME periods in a video before classifying them. The whole process of either spotting or classifying is split into three main steps: preprocessing, feature extraction, and classification. The subsequent sections cover all three processes, but the prime object is to describe the current research with respect to handcrafted and deep methods. Figure 2 depicts ME recognition process using some of the strategies stated in each of the three categories.

3.1 Preprocessing

The ME video contains irrelevant head and eye movements which act as noise in ME analysis. The preprocessing step involves face detection, registration, facial landmark identification, and alignment. The Viola and Jones [15], OpenCV, dlib library, Haar cascade, and pre-trained functions in MATLAB are used for face detection. The ME occurs in a limited area of face regions; thus, facial landmarks are detected to cover those areas known as region of interest (ROI). Facial landmarks detection identifies the distinct face locations as a set of facial points such as commissure (lip-corners) and canthus (inner eye corner). In this category, the popular methods are the active shape model (ASM), active appearance model (AAM), and discriminative response map fitting (DRMF).

3.2 Feature Extraction

Feature extraction implies to obtain the pattern and information by means of hand-crafted and deep features. It is a critical stage for MEs as these are subtle and low-intensity expressions.

3.2.1 Handcrafted Features

Handcrafted features are extracted through various algorithms. For instance, to detect edges, blobs, and colors of an image, different filters are used. The specific filter selection to intercept particular information is crucial for ME samples. The existing feature extractor like histogram of oriented gradients (HOG), local binary pattern (LBP), and local binary pattern for three orthogonal plane (LBP-TOP) may not work properly for the ME dataset. Thus researchers are creating new descriptors specific to subtle changes. Table 2 compares the studies of handcrafted features under several points like feature selection method, classifier, performance, and contribution of each paper.

1. **Appearance-based**: Appearance aspects describe the texture and color of the face region. These properties, however, are susceptible to non-frontal head posture and illumination changes. The fundamental feature descriptors are LBP [17, 28] and LBP-TOP [24, 25, 28], and some recent descriptors based on the same principle are LBP-MOP, LBP-MOP* [48], CBP-TOP [49], ADLBPTOP, and RDLBPTOP [50]. In the work [16, 27], authors used principle of motion magnification for subtle motion changes by analyzing the phase variations between frames obtained from the Riesz pyramid. Moreover, spatiotemporal local binary pattern based on a revisited integral projection (STLBP-RIP) [21] and spatiotemporal local binary pattern with integral projection) (STLBP-IP) [26] are proposed by Huang et al. Both the methods preserve shape attributes with texture and motion information by applying robust principal component analysis (PCA) and Laplacian method to extract the discriminative information of ME samples.
2. **Geometry-based**: These features capture face geometry and represent them as shapes, patches, and positions of facial landmarks. ROI is defined by geometric features, and most of the recent studies [18, 22, 31] adopted this method for ME recognition. In the work [25], ME classification is accomplished by considering only the FACS-based action units.

3. **Optical flow-based**: This feature extraction method attracts the researchers mostly because of their ability to catch minute changes. This technique captures the motion of a point in subsequent frames which helps to spot MEs. The sparse main directional mean optical flow (MDMO) [32] works on the principle of optical flow domain where they have reduced the number of features by obtaining the mean histograms in the principal direction. The work [19] proposed an optical flow features from apex frame network (OFF-ApexNet). It calculates the horizontal and vertical flow using TV-L1 technique. In this series, fuzzy histogram of optical flow orientations (FHOFO) [20], histogram of oriented optical flow (HOOF) [23], fusion of motion boundary histogram (FMBH) [29], and bi-weighted oriented optical flow (Bi-WOOF) [30] are some other feature descriptors. Figure 3 illustrates one of the feature extraction methods based on optical flow (OF) domain.

3.2.2 Deep Features

They are automatically intercepted by neural network like convolutional neural network (CNN). The CNN works as an end-to-end learning algorithm, where ME image data is fed to the architecture and prediction is made available as one of the expression category. There are multiple convolutional layers present which try to learn the local and global features. The local features are obtained by focusing upon specific area of interests. For instance, for happy expression, lips, cheeks, and eye regions are focused. Generally, for the local features, the face is segmented into different parts, whereas for global feature extraction, whole face area is examined at once by convolutional layers.

1. **Pre-trained Convolutional Neural Network**: In preliminary studies [42–45], it was found that ME analysis was performed on existing networks. These networks are already trained with different datasets but tested on ME database. However, this practice may not give good results because those networks learn different patterns than MEs. So to address this issue, researchers started to employ transfer learning where they train networks on an FE dataset and test on ME samples. This, too, is not a good practice.
2. **Novel Convolutional Neural Network**: As the deep learning approach progressed, researchers started to propose their new architectures, specifically designed for MEs, at a later stage. In [33], the author presented a CNN architecture named micro-expression recognition by analyzing spatial and temporal characteristics (MERASTC). It considers the action unit, gaze features, and appearance-based features for ME classification. Takalkar et al. [35] developed CNN architecture local and global attention network (LGAttNet) uses an attention mechanism [41] to focus on the specific portion of an image to spot MEs in videos. Verma et al. [39] created lateral accretive hybrid network (LEARNet) incorporating 4 streams of convolutional layers working simultaneously to classify MEs based on dynamic image sequences. In [40], authors exploited both appearance and

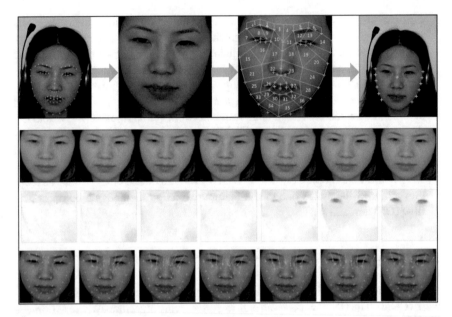

Fig. 3 Overview of optical flow-based feature extractor. First row: facial landmark localization, face cropping, ROI formation, and static point detection. Second row: frame alignment based on static points. Third row: optical flow calculation. Fourth row: optical flow vector construction [32]

geometric properties of MEs by means of spatiotemporal recurrent convolutional networks (STRCN). The recurrent neural network (RNN) variant long short term memory (LSTM) to capture time-series information is utilized by work [42, 45]. The comparison study of deep features from different articles is mentioned in Table 3.

3.2.3   Handcrafted Versus Deep Features

The handcrafted feature requires extra knowledge and human intervention to extract ME features, whereas deep networks automatically do this with the convolutional operation. As the size of the ME databases is small, thus shallow CNN works fine with them. The CNNs are capable of learning complex patterns hidden inside the data with aid of convolutional operation with specified kernel size. Thereby results suggest that they are more accurate than machine learning techniques for both feature extraction and classification tasks. However, a deep learning network requires a significant amount of data to train on, and the ME samples are currently restricted. Also selecting a large kernel size may degrade CNN's performance, while the pros of the handcrafted feature are: the better the knowledge, the more accurate be the extracted feature.

3.3 Classification

This step decides the ME class based on the extracted features. The several classification methods like support vector machine (SVM) [16–18, 20, 21, 24–26, 29, 30, 32] with different kernels, K-mean cluster [20, 28], extreme learning machines, random forest, multiple kernel learning, AdaBoost, and many more are used by machine learning techniques. The CNN [19, 33, 36–40, 44, 45] can also be used as a classifier, where the output layer has an equal number of nodes as the number of emotion categories is to be identified.

Different research teams presented their performance using various evaluation metrics. The majority of them addressed the accuracy. Some of the metrics are: area under ROC curve (AUC), F1-score, precision, recall, unweighted average recall (UAR), and weighted average recall (WAR). The authors also demonstrated their experiments by varying block size, having multiple normalization frames, and using different feature descriptors. But in this study, we only included significant results from each paper and mentioned the same in Tables 2 and 3. For further results, readers can look into respective papers.

4 Issues and Implications

In this section, some of the challenges with possible suggestions are mentioned.

1. The feature extraction process possesses high importance for the ME analysis. It is so because the extracted feature describes ME patterns. But the selection of appropriate feature descriptors is a concern. Thus, researchers are developing the new descriptors according to the ME specifications.
2. The most five recent datasets, viz., SAMM, CAS(ME)^2, CASMEII, CASME, and SMIC have a limited number of samples. Hence, they are not adequate for deep learning approach. Thus, researchers are adopting transfer learning and data augmentation (rotation, and flipping) techniques. Furthermore, in order to develop a new database, one must have a thorough knowledge of FACS coding and MEs' characteristics.
3. The handcrafted feature extraction method requires a great understanding of the ME dataset. Since MEs are subtle and low-intensity expressions, the researchers should know the prominent area of the face where the ME occurrence is frequent. So they can choose the feature extraction scheme wisely.
4. There is an open issue to recognize MEs in wild/natural scenarios. Till now, databases are created in a controlled laboratory environment. If we succeeded in ME recognition in wild, then this can be a break-through and can be used as an important practical application in criminal investigation and lie detection during the interrogation process.

Table 2 Comparison of feature selection method, classifier, performance, and contribution of handcrafted feature-based work

AY	SR	FE	Cs	Contribution	Performances (Accuracy)
[16] 2020	R	Riesz pyramid	SVM	Mean-oriented Riesz features to get the mean-oriented phase element of the monogenic signal	CASMEII: 62.20% SMIC: 65.45%
[17] 2020	R	LBP	SVM	Proposed optical flow filtering mechanism: OFF2BD and OFF3WD based on two-branch and three-way decisions	CASMEII: 61.57%SMIC: 65.41%
[18] 2020	R	FACS	SVM	Features are extracted from graphs drawn by AU groups associated with FACS	CASMEII: 75.04% CAS(ME)^2: 81.85% SAMM: 87.33% SMIC: 76.67%
[19] 2019	S + R	OF	CNN	Proposed 2D-CNN, OFF-Apexnet, is a shallow CNN and learns only 1.3 million parameters	CASMEII: 88.28% SAMM: 68.18% SMIC: 67.68%
[20] 2019	R	FHOFO	KNNSVMLDA	Proposed two new feature extractor: FHOOF and FHOFO based on optical flow domain	CASME: 67.01% CASMEII: 56.64% SMIC: 51.83%
[21] 2019	R	STLBP- RIP	SVM	Proposed STLBP-RIP, while preserving shape attribute and discriminative information between 2 ME classes	CASME: 64.33% CASMEII: 64.78% SMIC: 63.41%
[22] 2019	S	PCA	–	The local temporal pattern machine learning method is extended for long videos by using a sliding temporal window	**F1-score:** CAS(ME)^2: 0.0179 SAMM: 0.0316
[23] 2019	S	HOOF	–	ME interval detection in long videos by RNN on HOOF features	**Recall:** 0.4654 **F1-score:** 0.0821 **Precision:** 0.0450
[24] 2019	R	LBP-TOP	SVM	New weight matrix formation to eliminate confusion between noise and ME weight	CASMEII: 62.89% SMIC-HS: 63.25%
[25] 2018	R	LBP-TOPHOOFHOG-3D	SVM	ME classification is done solely based on action units	CASMEII: 86.35% SAMM: 81.93%
[26] 2018	R	STLBP- IP	SVM	Extracting the spatiotemporal features from a video with different grid densities by proposing a hierarchical spatiotemporal descriptor	CASMEII: 65.18% SMIC: 66.46%

(continued)

Table 2 (continued)

AY	SR	FE	Cs	Contribution	Performances (Accuracy)
[27] 2018	S	Riesz pyramid	–	Employed Riesz pyramid with non-causal zero-phase FIR filter to remove noise and detect subtle motion information	**AUC**: CASMEII: 95.13% SMIC-HS: 89.80%
[28] 2018	R	LBP LBP-TOP	KNN	Transfer knowledge learning method is used to recognize ME	**Recognition rate**: 0.655
[29] 2018	R	FMBH	SVM	Proposed a new feature extractor FMBH and a face mask algorithm	CASME: 61.33% CASMEII: 69.11% CAS(ME)^2: 73.67% SMIC: 71.95%
[30] 2018	S + R	Bi-WOOF	SVM	Apex frame spotted by using the divide-and-conquer strategy	CASMEII: 58.85% SMIC-HS: 62.20%
[31] 2018	S	PCA	–	Exploited local temporal pattern of ME to differentiate it from other facial movements	CASME-A: 77.90% CASME-B: 82.61% CASMEII: 65.07%
[32] 2018	R	MDMO	SVM	Proposed sparse MDMO considers both local statistic motion information and its spatial location	**F1-score**: CASME: 74.98% CASMEII: 69.11% SMIC: 70.41%

Here, AY: Article and year, S: Spotting, R: Recognition, FE: Feature extractor, Cs: Classifier

Table 3 Comparison of feature selection method, classifier, performance, and contribution of deep feature-based work

AY	SR	FE	Cs	Contribution	Performances (Accuracy)
[33] 2021	R	open face algorithm	CNN	Proposed MERASTC based on gaze, action unit, landmark, and appearance features of ME	CASMEII: 85.4% CAS(ME)^2: 91.20% SAMM: 83.8% SMIC-HS: 79.3%
[34] 2021	S	CNN	–	The clip proposal and classification regression network spots ME intervals of various scales	CAS(ME)^2: 91.67% SAMM: 97.73%
[35] 2020	S	CNN	–	Proposed CNN architecture based on local and global attention mechanisms	CASME: 93.3% CASMEII: 94.3% CAS(ME)^2: 86.5% SAMM: 86.8%
[36] 2020	R	CNN	CNN	Proposed a multi-stream CNN by adding an over-sampler after dense layers of ResNet-18	CASMEII: 56.5% SAMM: 43.04%
[37] 2020	R	CNN	CNN	Proposed teacher and student networks, and apply knowledge transfer learning	CASME: 81.8% CASMEII: 72.61% SAMM: 86.74% SMIC-2: 76.06%
[38] 2020	R	CNN	CNN	Proposed TSNN architecture comprises 2D-CNN and 3D-CNN, based on intermediate fusion and late fusion, respectively	F1-score: CASMEII: 0.6142 SMIC: 0.6921
[39] 2019	R	CNN	CNN	Proposed four-stream parallel hybrid CNN architecture trained on a dynamic image sequence	CASME: 80.62% CASMEII: 76.57% CAS(ME)^2: 76.33% SMIC: 81.60%
[40] 2019	R	CNN	CNN	Proposed network has 2 variants: STRCN-A extracts appearance-based information, whereas STRCN-G exploits geometric information	CASMEII: 80.3% SAMM: 78.6% SMIC: 72.3%
[41] 2019	R	NN	NN	Introduced capsule network architecture.	UF1-score: CASMEII: 0.7068S AMM: 0.6209 SMIC: 0.5820
[42] 2018	R	RNN	NN	Proposed two networks: (i) spatial dimension enrichment by channel-wise stacking of input data and (ii) temporal dimension enrichment by feature-wise stacking of features	CDE: UAR: 0.39 WAR: 0.57 F1-score: 0.4107
[43] 2018	R	NN	NN	Residual network is integrated with attention mechanism without increasing the parameters	CASMEII: 65.9% SAMM: 48.5% SMIC: 49.4%
[44] 2018	R	CNN	CNN	ResNet10 is pre-trained using ImageNet and fine-tuned on macro- and micro-expression data	CASMEII: 75.68% SAMM: 70.59%
[45] 2018	R	CNN	CNN	Proposed model comprises CNN and LSTM which works on transfer learning	76.17%

Here, AY: Article and year, S: Spotting, R: Recognition, FE: Feature extractor, Cs: Classifier, NN: Neural network, CDE: Cross-database evaluation

5 Conclusion and Future Scope

Micro-expressions are involuntary, and low-intensity and short duration occur as a leakage when a person suppresses the true emotion. The survey paper focuses on feature extraction through handcrafted and deep approaches. We have discussed the state of the art at different stages, along with contributions and results. According to the findings, both the techniques do well in their respective fields. However, the deep learning approach does not require human expertise but requires large data. The literature in reverse chronology indicates that researchers are growing toward a deep learning approach.

This paper discusses a limited number of articles and also organizes studies only in a handcrafted and deep feature viewpoint. Therefore, we intend to further extend this work by incorporating more articles and dividing them according to spotting and recognition tasks. We would also conduct a comparative analysis of handcrafted and deep feature methods. Moreover, there are two major areas where researchers can focus on, (i) creating new datasets or extending the existing ones and (ii) developing new feature extractors. The perceived knowledge of MEs can be used in applied research, e.g., children's engagement during activity.

References

1. Ekman P, Friesen W (1969) Nonverbal leakage and clues to deceptions. Psychiatry 32(1):88–106
2. Ekman P (2009) Lie catching and microexpressions. In: The philosophy of deception. Oxford, New York, NY, USA, pp 118–133
3. Ekman P (2003) Darwin, deception, and facial expression. Ann New York Acad Sci 1000:205–221
4. Ekman P, Friesen WV (1978) Facial action coding system, vol 1. Consulting Psychologists Press, Washington
5. FACS—measuring emotions on the face. https://clue-lab.com.br/2018/01/04/facs-medindo-as-emocoes-na-face/
6. Yan WJ, Wu Q, Liang J, Chen YH, Fu X (2013) How fast are the leaked facial expressions: the duration of microexpressions. J Nonverbal Behav 37(4):217–230
7. Warren G, Schertler E, Bull P (2009) Detecting deception from emotional and unemotional cues. J Nonverbal Behav 33(1):59–69
8. Polikovsky S, Kameda Y, Ohta Y (2009) Facial microexpressions recognition using high speed camera and 3D gradient descriptor. In: Proceedings of 3rd international conference on crime detection prevention, pp 1–6
9. Shreve M, Godavarthy S, Goldgof D, Sarkar S (2011) Macro-and micro-expression spotting in long videos using spatio-temporal strain. In: Face and gesture, pp 51–56
10. Li X, Pfister T, Huang X, Zhao G, Pietikäinen M (2013) A spontaneous micro-expression database: inducement, collection and baseline. In: 2013 10th IEEE international conference and workshops on automatic face and gesture recognition (FG). IEEE, pp 1–6
11. Yan WJ, Wu Q, Liu YJ, Wang SJ, Fu X (2013) CASME database: a dataset of spontaneous micro-expressions collected from neutralized faces. In: 10th Proceedings of international conference on automatic face and gesture recognition (FG2013). Shanghai, China. IEEE

12. Yan WJ, Li X, Wang SJ, Zhao G, Liu YJ, Chen YH, Fu X (2014) CASMEII: an improved spontaneous microexpression database and the baseline evaluation. PloS one 9(1)
13. Qu F, Wang SJ, Yan WJ, Li H, Wu S, Fu X (2017) CAS(ME)^2: a database for spontaneous macroexpression and micro-expression spotting and recognition. IEEE Trans Affect Comput 9(4):424–436
14. Davison AK, Lansley C, Costen N, Tan K, Yap MH (2016) Samm: a spontaneous micro-facial movement dataset. IEEE Trans Affect Comput 9(1):116–129
15. Viola P, Jones MJ (2004) Robust real-time face detection. Int J Comput Vis 57(2):137–154
16. Duque CA, Alata O, Emonet R, Konik H, Legrand AC (2020) Mean oriented Riesz features for micro expression classification. Pattern Recogn Lett 135:382–389
17. Wu J, Xu J, Lin D, Tu M (2020) Optical flow filtering based micro-expression recognition method. Electronics 9(12):2056
18. Buhari AM, Ooi CP, Baskaran VM, Phan RC, Wong K, Tan WH (2020) FACS-based graph features for real-time micro-expression recognition. J Imaging 6(12):130
19. Gan YS, Liong ST, Yau WC, Huang YC, Tan LK (2019) Off-apexnet on micro-expression recognition system. Sig Process Image Commun 74:129–139
20. Happy SL, Routray A (2017) Fuzzy histogram of optical flow orientations for micro-expression recognition. IEEE Trans Affect Comput 10(3):394–406
21. Huang X, Wang SJ, Liu X, Zhao G, Feng X, Pietikäinen M (2017) Discriminative spatiotemporal local binary pattern with revisited integral projection for spontaneous facial micro-expression recognition. IEEE Trans Affect Comput 10(1):32–47
22. Li J, Soladie C, Seguier R, Wang SJ, Yap MH (2019) Spotting micro-expressions on long videos sequences. In: 2019 14th IEEE international conference on automatic face and gesture recognition (FG 2019). IEEE, pp 1–5
23. Verburg M, Menkovski V (2019) Micro-expression detection in long videos using optical flow and recurrent neural networks. In: 2019 14th IEEE international conference on automatic face and gesture recognition (FG 2019). IEEE, pp 1–6
24. Wang L, Xiao H, Luo S, Zhang J, Liu X (2019) A weighted feature extraction method based on temporal accumulation of optical flow for micro-expression recognition. Sig Process Image Commun 78:246–253
25. Davison AK, Merghani W, Yap MH (2018) Objective classes for micro-facial expression recognition. J Imaging 4(10):119
26. Zong Y, Huang X, Zheng W, Cui Z, Zhao G (2018) Learning from hierarchical spatio-temporal descriptors for micro-expression recognition. IEEE Trans Multimedia 20(11):3160–3172
27. Duque CA, Alata O, Emonet R, Legrand AC, Konik H (2018) Micro-expression spot-ting using the Riesz pyramid. In: 2018 IEEE winter conference on applications of computer vision (WACV). IEEE, pp 66–74
28. Jia X, Ben X, Yuan H, Kpalma K, Meng W (2018) Macro-to-micro transformation model for micro-expression recognition. J Comput Sci 25:289–297
29. Lu H, Kpalma K, Ronsin J (2018) Motion descriptors for micro-expression recognition. Sig Process Image Commun 67:108–117
30. Liong ST, See J, Wong K, Phan RCW (2018) Less is more: micro-expression recognition from video using apex frame. Sig Process Image Commun 62:82–92
31. Li J, Soladie C, Seguier R (2018) LTP-ML: microexpression detection by recognition of local temporal pattern of facial movements. In: 2018 13th IEEE international conference on automatic face and gesture recognition (FG 2018). IEEE, pp 634–641
32. Liu YJ, Li BJ, Lai YK (2018) Sparse MDMO: learning a discriminative feature for micro-expression recognition. IEEE Trans Affect Comput 12(1):254–261
33. Gupta P (2021) MERASTC: micro-expression recognition using effective feature encodings and 2D convolutional neural network. IEEE Trans Affect Comput
34. Wang SJ, He Y, Li J, Fu X (2021) MESNet: a convolutional neural network for spotting multiscale micro-expression intervals in long videos. IEEE Trans Image Process 30:3956–3969

35. Takalkar MA, Thuseethan S, Rajasegarar S, Chaczko Z, Xu M, Yearwood J (2021) LGAttNet: automatic micro-expression detection using dual-stream local and global attentions. Knowl-Based Syst 212:106566
36. Liu J, Li K, Song B, Zhao (2020) A multi-stream convolutional neural network for micro-expression recognition using optical flow and EVM. arXiv preprint arXiv:2011.03756
37. Sun B, Cao S, Li D, He J, Yu L (2020) Dynamic micro-expression recognition using knowledge distillation. IEEE Trans Affect Comput
38. Wu C, Guo F (2021) TSNN: three-stream combining 2D and 3D convolutional neural network for micro-expression recognition. IEEJ Trans Electr Electron Eng 16(1):98–107
39. Verma M, Vipparthi SK, Singh G, Murala S (2019) LEARNet: dynamic imaging network for micro expression recognition. IEEE Trans Image Process 29:1618–1627
40. Xia Z, Hong X, Gao X, Feng X, Zhao G (2019) Spatiotemporal recurrent convolutional networks for recognizing spontaneous micro-expressions. IEEE Trans Multi-media 22(3):626–640
41. Quang NV, Chun J, Tokuyama T (2019) CapsuleNet for micro-expression recognition. In: 2019 14th IEEE international conference on automatic face and gesture recognition (FG 2019). IEEE, pp 1–7
42. Khor HQ, See J, Phan RCW, Lin W (2018) Enriched long-term recurrent convolutional network for facial micro-expression recognition. In: 2018 13th IEEE international conference on automatic face and gesture recognition (FG 2018). IEEE, pp 667–674
43. Wang C, Peng M, Bi T, Chen T (2018) MicroAttention for micro-expression recognition. arXiv preprint arXiv:1811.02360
44. Peng M, Wu Z, Zhang Z, Chen T (2018) From macro to micro expression recognition: deep learning on small datasets using transfer learning. In: 2018 13th IEEE international conference on automatic face and gesture recognition (FG 2018). IEEE, pp 657–661
45. Wang SJ, Li BJ, Liu YJ, Yan WJ, Ou X, Huang X, Xu F, Fu X (2018) Micro-expression recognition with small sample size by transferring long-term convolutional neural network. Neurocomputing 312:251–262
46. Shen XB, Wu Q, Fu XL (2012) Effects of the duration of expressions on the recognition of microexpressions. J Zhejiang Univ Sci B 13(3):221–230
47. Yan WJ, Wang SJ, Liu YJ, Wu Q, Fu X (2014) For micro-expression recognition: database and suggestions. Neurocomputing 136:82–87
48. He J, Hu JF, Lu X, Zheng WS (2017) Multitask mid-level feature learning for micro-expression recognition. Pattern Recogn 66:44–52
49. Guo Y, Xue C, Wang Y, Yu M (2015) Microexpression recognition based on CBP-TOP feature with ELM. Optik 126(23):4446–4451
50. Guo C, Liang J, Zhan G, Liu Z, Pietikäinen M, Liu L (2019) Extended local binary patterns for efficient and robust spontaneous facial micro-expression recognition. IEEE Access 7:174517–174530

Multiclass Text Emotion Recognition in Social Media Data

Nirmal Varghese Babu and E. Grace Mary Kanaga

1 Introduction

Sentiment analysis [1] is a subfield of Natural Language Processing that focuses on the examination of subjective views or sentiments acquired from numerous sources regarding a certain issue. It is a collection of tools for recognising, extracting, and utilising opinions for the benefit of company operations. Emotion detection apperception (EDR) is a technique for detecting and apperceiving human emotions [2] that uses technological capabilities such as facial apperception, verbalisation, and voice apperception, biosensing, machine learning, and pattern apperception to extract opinions for business operations. The practise of acquiring and analysing data from social networks like as Facebook, Instagram, LinkedIn, and Twitter is referred to as social media analytics. Marketers regularly employ it to monitor Internet discussions about products and companies. Businesses [3] can utilise social media analytics to listen in on, monitor, and examine these conversations to make smarter decisions. The categorisation of emotions inside data using text analysis techniques is known as sentiment analysis. Using deep learning, sentiment analysis models may be trained to grasp text beyond simple definitions, read for context, sarcasm, and so on, and identify the writer's true mood and feelings. Deep learning is a branch of machine learning that use "artificial neural networks" to process information in a manner like that of the human brain. Deep learning and machine learning [4] are terms that are used interchangeably at times. Deep learning is like machine learning, but it is more sophisticated. When basic machine learning makes a mistake, it requires human intervention to fix it—to adjust the output and "push" the model to learn.

N. V. Babu · E. Grace Mary Kanaga (✉)
Department of Computer Science and Engineering, Karunya Institute of Technology and Sciences, Coimbatore 641114, Tamil Nadu, India
e-mail: grace@karunya.edu

N. V. Babu
e-mail: nirmalvarghese@karunya.edu.in

© The Author(s), under exclusive license to Springer Nature Singapore Pte Ltd. 2023
D. S. Sisodia et al. (eds.), *Machine Intelligence Techniques for Data Analysis and Signal Processing*, Lecture Notes in Electrical Engineering 997,
https://doi.org/10.1007/978-981-99-0085-5_30

However, with deep learning, the neural network may learn to repair itself using its sophisticated algorithm chain.

The initial training of a deep learning model takes a long time and frequently millions of data points before it can learn on its own. To continue the connection to the human brain, consider how long it takes a kid to master proper sentence construction or fundamental math. Machine learning models may do incredible things after they have been thoroughly taught to educate themselves successfully. Text analysis, for example, employs natural language processing (NLP) to break down and comprehend words in the same way that a person would: subject, verb, object, and so on. The computer cannot proceed to additional analytical procedures until it has mathematically broken down a statement. Of course, it's far more complicated than just breaking down a phrase into subject, verb, and object and moving on. It takes years to develop successful NLP models. Deep learning models, on the other hand, may employ NLP to divide phrases, paragraphs, and entire publications into individual opinion units.

Emotion Detection [5] will play an important role in the field of Artificial Intelligence, particularly in the creation of Human–Machine Interfaces. A variety of parameters should be considered when using artificial intelligence to detect emotions. To identify emotions in a human, several approaches such as facial expressions, bodily movements, blood pressure, heart rate, and textual information are employed. This article focuses on detecting emotions in textual data. It's intriguing to extract emotion from different aims, such as those of business. For example, in the case of luxury goods [6], emotional factors such as brand, personality, and prestige are far more important than technical, practical, or financial considerations. In such circumstances, the customer is willing to pay a premium price for a product. Emotion selling seeks to imitate emotions in clients to tie them to a brand and so enhance service/product sales.

1.1 Objectives

1. To improve the classification accuracy of the sentiment analysis in emotion recognition for healthcare applications using Social Media Data.
2. To analyse various sentiment analysis techniques and enhance the performance on the analysis.
3. To design an algorithm which provides precise classification using multiclass classifier using deep learning algorithms.

1.2 Emotion Recognition

The technique of detecting human emotions is known as emotion recognition. The capacity of people to recognise the emotions [7, 8] of others differs substantially. Most

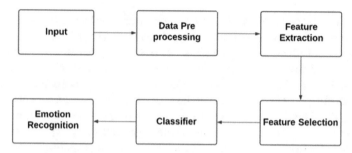

Fig. 1 Emotion recognition steps

of the research has been on automating face expression detection from video, spoken expression recognition from audio, written expression recognition from text, and wearable physiology. The accuracy of emotion identification is frequently increased when the study of human expressions [9] from multimodal sources such as texts, physiology, audio, or video is combined. Data from facial expressions, bodily movement and gestures, and voice are used to determine various emotion kinds. Existing ways to classifying distinct emotion types in emotion recognition may be split into three categories: knowledge-based techniques, statistical methods, and hybrid approaches. Figure 1 depicts the emotion recognition steps.

1.3 Natural Language Processing

NLP is a discipline of artificial intelligence concerned with analysing, interpreting, and creating the natural languages that humans use to interact with computers in both written and vocal situations [10], utilising genuine human languages rather than computer languages. Natural language processing (NLP) enables computers to interpret natural language in the same manner that people do. Natural language processing, whether spoken or written, uses artificial perspicacity to take real-world data, process it, and make sense of it in a way that a computer can understand. Computers have programmes to read and microphones to gather sounds, much as humans have diverse sensors, such as auditory perceivers to detect auricular impulses and ocular perceivers to detect optical signals. In addition, just as humans have an encephalon to process input, computers have a software to do the same. The input is transformed to computer-readable code at some point throughout the processing.

1.4 Social Media Analytics

The process of unearthing hidden insights from structured and unstructured social data in order to make educated judgements is known as social media analytics [11]. It may also investigate online media outlets such as news websites, blogs, and forums. The three primary processes in social media analysis are data identification, data analysis, and information interpretation. Analysts can establish a question that has to be addressed to maximise the value derived at each stage of the process. Data identification refers to the process of identifying subsets of accessible data to focus on for analysis [12, 13]. Raw data can be useful once it has been interpreted. After analysing the data, it may be utilised to communicate a message. Information is defined as any data that sends a meaningful message. Data analysis [14] is a collection of actions that help in the translation of raw data into insight, resulting in an increasing basis of knowledge and economic value. In other words, data analysis is the process of taking filtered data and converting it into information that analysts can utilise.

Section 1 Concerns the paper's introduction.

Section 2 Describes the related works.

Section 3 This section describes the proposed system.

Section 4 Describes the Experimental Results.

Section 5 The summary, conclusion, and future work are all depicted.

Section 6 Concerns itself with the Compliance of Ethical Standards.

2   Related Work

According to Sethi et al. [2], people have begun to acquire conflicting sentiments about the situation because to the quick increase in infection and death rate caused by COVID-19. As a result, the study's sole focus is to analyse the emotions expressed by people using social media platforms such as Twitter and others. The purpose of this research is to present a domain-specific approach to understanding the sentiments expressed by people all over the world about this situation. Corona-specific tweets are obtained from the Twitter platform to accomplish this. After collecting the tweets, they are labelled, and a model is developed that can detect the true sentiment behind a COVID-19-related tweet. In [2], social media comprises of numerous types of emotions and sentiments expressed by its users in the form of electronic media. Analysing the emotions or thoughts of people on a certain post is likewise a difficult undertaking. Our project seeks to automate the work of evaluating reactions and postings and producing a report based on the results. A unique social media network is proposed, which would allow users to undertake actions such as posting, like, commenting, and sharing.

To extract sentiment from social media, Cheng et al. [5] offers a unique senti-ment analysis framework based on deep learning models. Collected information and assemble it into a dataset. Following the analysis of these specific words, want to create a semantic dataset for future research. The gathered data will be valuable for a variety of future applications. Crawling multiple social media networks yielded the experimental data. In [6], the DWRs are concatenated using a weighted method using Recurrent Neural Network (RNN) variations combined with Convolutional Neural Network (CNN) variants, with weighted attentive pooling (WAP). CNNs with standard pooling procedures, on the other hand, have numerous layers that are only capable of capturing adequate characteristics. Experiments show that DWRs, in conjunction with the suggested concatenation, can fix the concerns with modest hyper-parameter setups. Our design, which did not stack multiple layers, produced a moderate accuracy of 89.67% via DECR-Bi-GRU-CNN (WAP) on IMDB, compared to 81.11% by random initialisation on SST. Jabreel and Moreno [14] discusses the creation of a revolutionary deep learning-based system that handles the multiple emotion categorisation challenge in Twitter. Proposed a unique way for transforming it to a binary classification issue and used a deep learning methodology to solve the modified problem. This system beats state-of-the-art algorithms, attaining an accuracy score of 0.59 on the difficult SemEval2018. In [15], the suggested multi-view deep network performs classification using intermediate characteristics collected from convolutional and recursive neural networks. According to the findings of the trials, the suggested multi-view deep network not only beats single-view deep neural networks, but it also outperforms single-view deep neural networks in terms of efficiency and generalisation performance.

Dragoni et al. [16] offers a method for building sentiment models that allow polarity inference for documents from any domain by leveraging linguistic overlap. Word embeddings and a deep learning architecture have been included into the NeuroSent tool to enable the development of multi-domain sentiment models. The suggested approach is validated by adhering to the Dranziera procedure to facilitate experiment reproducibility and result comparability. ABCDM will extract both past and future contexts by considering temporal information flow in both directions, according to Basiri et al. [17]. ABCDM will extract both past and future contexts by employing two independent bidirectional LSTM and GRU layers. Furthermore, the attention mechanism is used on the outputs of the bidirectional layers of ABCDM to place focus on particular words. ABCDM employs convolution and pooling algorithms to decrease feature dimensionality and extract position-invariant local features. ABCDM's performance is measured using sentiment polarity detection, which is the most common and important activity in sentiment analysis. Five review datasets and three Twitter datasets were used in the experiments. Harb et al. [18] present an analytical methodology for investigating the emotional reactions to large traumatic occurrences on Twitter and utilise it to draw conclusions regarding eight mass shooting events. The framework includes crawling of pre/post-event tweets to compare emotional reactions, classification of sentiment in terms of Ekman's basic emotions, and the use of data extracted from Twitter users' profiles to understand these reactions in relation to users' demographics (age and gender), proximity to the event, and number of victims. We ran trials with three deep learning algorithms to categorise emotions: CNN, biLSTM, and BERT, with the former producing the best results.

For improved categorisation, the suggested technique in [19] employs many syntactic characteristics such as n-grams, POS-related features, negation-related features, level-related features, and so on. The classifier divides Malayalam texts into several emotion classes such as happy, sad, angry, fear, normal, and so on, along with level information such as high, low, and so on. It also specifies if the sentence is a conversation or an inquiry in order to improve the reading experience using a voice synthesiser. Huang et al. [20] suggested a recurrent neural network based on emojis for sentiment analysis in Chinese microblogs. To begin, use pre-trained word embedding and a new sentiment lexicon to distinguish between ambiguous and explicit emojis. Then, prove that users' information can, to some extent, erase the ambiguity of ambiguous emojis and confirm the emotion polarity of ambiguous emojis. On that basis, obtain emoji representations by leveraging the location vector, semantic vector, and sentiment vector of emojis, and then feed the emoji representations into the Bi-directional gated recurrent unit (BiGRU) neural network model to perform sentiment analysis. The experimental findings on a Chinese microblog dataset show that, when compared to other baselines, the proposed model may greatly enhance sentiment analysis accuracy.

3 Proposed System

The proposed system addresses Multiclass Emotion Recognition in Social Media Data Using CNN, LSTM, and GRU [15, 21–23], which assists in identifying the various emotions and sentiments of people in various situations of their lives using data posted in social media such as Twitter in the form of tweets or comments. Sentiment scores are utilised in addition to textual data and emoticons for emotion identification. Figure 2 depicts the proposed system architecture.

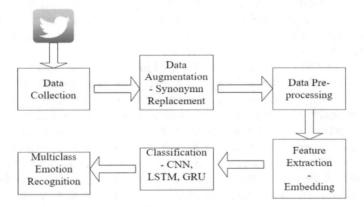

Fig. 2 Proposed system

3.1 Architecture

The text and emoticon-based data acquired from online social media platforms such as Twitter will be forwarded to the Data Pre-Processing Stage. During the Data Pre-Processing step, Numbers, Hashtags, White Spaces, URLs, and Special Symbols were extracted, and the dataset was stemmed and tokenised. The embedding layer [16] is utilised in the Feature Extraction to extract the features. The feelings associated with the tweets were also translated into binary numbers. CNN, LSTM and GRU Classification algorithms were employed in the classification step. In addition, a data categorisation procedure has been implemented to solve the dataset's over-fitting problem. Following classification, the precision of each method was determined, as well as the loss of each algorithm. The Validation Dataset will be used to validate the model, and the precision and score values will be recorded. Here, the data will be separated into X and Y categories. The tweets are represented by X, and the feelings connected with each tweet are represented by Y. Among the 13 categories were Anger, Boredom, Surprise, Relief, Neutral, Happiness, Love, Worry, Enthusiasm, Hate, Fun, Empty, and Sadness.

Data Collection The dataset for this study was gathered using the Twitter API [17, 24, 25]. This dataset also includes tweets from many people. The overall size of the dataset was 4.2 MB. Approximately 40,000 tweets were downloaded from Twitter. The dataset is in CSV format, and it includes columns for the Tweet id, Sentiment, Author Name, and Content. Because emoticons contain sentiment value, the collection includes not just textual data but also emoticons, which will help with emotion or sentiment recognition. Tables 1 and 2 shows the data collected with total count of their sentiments.

3.2 Module Description

Data Augmentation, Data Pre-Processing, Feature Extraction, Classification, Multiclass Emotion Recognition, and Results Extraction and Evaluation are among the modules included in the suggested system.

Table 1 Data collected with their sentiments count

Sentiments	Happiness	Sadness	Worry	Boredom	Anger	Surprise	Relief
Count	2699	1395	42	17	36	33	31

Table 2 Data collected with their sentiments count (contd.)

Sentiments	Neutral	Love	Hate	Fun	Empty	Enthusiasm
Count	31	34	31	31	30	41

Text Data Augmentation

Data augmentation techniques are used in data analysis to expand the quantity of data by combining scarcely changed copies of previously existing data or incipiently developed synthetic data from previously existing data. It functions as a regulariser and helps to avoid over-fitting [26] while training a machine learning model. In data analysis, it is equivalent to oversampling. To enhance the size of a genuine dataset, data augmentation techniques build false variations of it. Computer vision and natural language processing (NLP) models use data augmentation methodologies to deal with data scarcity and inadequate data variety.

Over-fitting occurs when a model learns the information and noise in the training data to the point that it severely impairs the model's performance on fresh data. This implies that the model picks up on noise or random oscillations in the training data and learns them as ideas. Another method for reducing overfitting on models is data augmentation, which involves increasing the quantity of training data by using only information from the training data. The major strategies are classified as data warping, which is a strategy that aims to directly supplement the model's input data in data space.

Synonym Replacement The process of replacing a term with one of its synonyms is known as synonym replacement. To find appropriate synonyms, we use WordNet, a vast linguistic database. This first function finds and pre-processes synonyms for a given term. The synonyms are then substituted at random in the original statement. Simply locate synonyms of a term and associate a probability with each synonym of a word for data augmentation. Then, using a probability distribution, replace the original term with a synonym chosen at random. If the replaced word is a keyword, the keyword has also been changed. Sentiment was preserved. The total number of data collected initially was around 40,000 and with the help of data augmentation, the data raised to 78,000.

Data Pre-processing

Data pre-processing [27] is the process of preparing raw data for use with a machine learning model. It is the first and most important stage in developing a machine learning model. Real-world data typically contains noise, missing values, and may be in an unsuitable format that cannot be utilised directly for machine learning models. Data pre-processing is a necessary job for cleaning the data and preparing it for a machine learning model, which improves the accuracy and efficiency of the machine learning model. Data pre-processing betokens the procedure of processing of the input data afore the analysis process. Ascertain that the emoticons are not abstracted during this step. The various steps involved in this procedure are:

1. Removal of Numbers—Numbers present in the data set which is of no avail to the analysis will be removed.
2. Removal of Hashtags—Hashtags are acclimated to extract or download the data from the Twitter. Utilising the Keyword Matching, the data will be extracted.

After the extraction procedure, these hashtags are of not utilise so it will be abstracted during the pre-processing step.

3. Removal of White spaces—The White spaces between the words in the tweets will additionally be removal as it increments the processing time of the entire procedure.

4. Removal of URL's—URLs doesn't implicatively insinuates any construal or doesn't play any part in the sentiment analysis procedure, so it will additionally be removed afore the analysis.

5. Removal of Special Symbols—The processing of special symbols takes immensely colossal duration, and it doesn't have any inner meaning unless it's an emoticon. Emoticons [?]-[?] have an inner meaning and these emoticons only must be kept during the pre-processing step.

6. Stemming—Stemming denotes the process of finding the root forms of the words in the tweets by abstracting the -es, ed, ing forms ie by removal the suffixes and prefixes.

7. Tokenisation [18]—Tokenisation refers to the process of tokenising or separating the words in tweets into different words in order to facilitate future word processing or analysis. The tokenised words are represented by the symbol ' '. Tokenisation was carried out in this case using both the NLTK Word Tokenizer and the Word Punctuation Tokenizer. The disadvantage of the Word Tokenizer is that it considers each word separately and the emoticons are split during the tokenisation, whereas in Word Punctuation Tokenizer, the emoticons are also considered as a single word and will not be split during the procedure.

Feature Extraction

The purpose of feature extraction [19] is to minimise the number of features in a dataset by extracting new ones from old ones (and then discarding the original features). The bulk of the information in the original set of characteristics should then be able to be summarised by the new reduced set of features. A summarised version of the original features may be generated by merging the original set. Feature Extraction is simply a dimensionality reduction procedure that divides raw data into understandable groupings. The fact that these enormous datasets include many variables and demand a great number of computational resources to handle differentiates them. As a result, in this scenario, Feature Extraction might be advantageous in terms of picking certain variables as well as merging some of the related variables, hence lowering the quantity of data.

In this scenario, the embedding layer is employed to extract features. The embedding layer is one of the possible layers in Keras. This is mostly utilised in Natural Language Processing for applications such as language modelling, but it may also be used for other neural network-based activities. Neural network embeddings, also known as eternal vectors, are learnt low-dimensional representations of discontinuous data. These embeddings circumvent the restrictions of typical encoding methods and may be utilised for tasks such as locating the closest neighbours, supplying input to another model, and making visualisations. The first step in putting an embedding layer together is to encode this statement using indices. In this situation, an index is allocated to each unique word. The embedding matrix [?] is then generated. Decide how many "latent factors" to apply to each index. This essentially tells how long we want the vector to be. Lengths of 32–50 feet are popular use cases.

GloVe is an acronym that stands for global vectors for word representation [28]. It is a Stanford-developed unsupervised learning approach for producing word embeddings by aggregating a corpus's global word-word co-occurrence matrix. The obtained embeddings reveal fascinating vector space linear substructures of the word. The embedding layer here use the GloVe to convert the words in the datasets to the matching word vectors. This aids in the categorisation step for data training.

Classification

The Classification [29, 30] method is a Supervised Learning approach that uses training data to identify the category of new observations. In Classification, the programme learns from a given dataset or observations and then classifies additional observations into one of many classes or groupings. Classes can also be referred to as targets/labels or categories. A discrete output function (y) is transferred to an input variable in a classification method (x).

Varied Artificial Neural Network [20] methods will be utilised to categorise tweets and compute Precision and Score under various scenarios. To calculate the Precision and Score, an algorithm combining text and emoticon analysis will be constructed, and the values will be compared to the existing machine learning techniques. The model will be evaluated using the Test Dataset. The X depicts the actual tweet, whilst the Y represents the numerous Sentiments. Both the content and the Sentiments sections will be used to teach the system. In this scenario, CNN, LSTM, and GRU were employed.

1. **CNN** Convolutional Neural Networks are a form of deep network utilised for visual image processing. Regularised versions of these are multilayer perceptrons. When compared to other classification methods, a ConvNet requires substantially less pre-processing. These filters/characteristics can be learned by ConvNets. A ConvNet may successfully capture the Spatial and Temporal relationships in a image by using proper filters. It performs better fitting since the number of parameters involved is minimised.

2. **LSTM** Long Short-Term Memory networks (LSTMs) are a form of RNN that can learn long-term dependencies. Hochreiter and Schmidhuber (1997) pioneered them, and several authors developed and popularised them in following work. They are currently frequently utilised and function phenomenally well on a wide range of difficulties. LSTMs are expressly intended to prevent the issue of long-term reliance. It is basically their default behaviour to remember information for lengthy periods of time. All recurrent neural networks are composed of a series of repeating neural network modules. In conventional RNNs, this repeating module will have a relatively basic structure, such as a single tanh layer.

3. **GRU** Many solutions have been devised to handle the Vanishing-Exploding gradients problem, which is typically experienced while running a simple Recurrent Neural Network. One of the most well-known versions is the Long Short-Term Memory Network (LSTM). A lesser known but as powerful variation is the Gated Recurrent Unit Network (GRU). In contrast to LSTM, it just has three gates and does not maintain an Internal Cell State. The information recorded in an LSTM recurrent unit's Internal Cell State is included into the Gated Recurrent Unit's concealed state.

Tuning deep neural network hyperparameters is tough since deep neural networks are slow to train and have several parameters to customise.

1. **Learning rate**: The learning rate determines how frequently the weight in the optimisation method is updated. Depending on the optimizer we employ (SGD, Adam, Adagrad, AdaDelta, or RMSProp), we can utilise a constant learning rate, a progressively declining learning rate, momentum-based approaches, or adaptive learning rates.
2. **Number of epochs**: The number of epochs is the number of times the full training set is processed by the neural network. We should increase the number of epochs until there is only a modest difference between the test and training errors. Batch size: Mini-batch is frequently preferred in the convnet learning process. A test range of 16–128 is a suitable choice. It should be noted that convnet is batch size sensitive.
3. **Activation function**: The activation function adds non-linearity to the model. Typically, rectifier and convnet function nicely together. Other activation functions to consider depending on the job include sigmoid, tanh, and others.
4. **Number of hidden layers and units**: It's typically a good idea to keep adding layers until the test error no longer improves. The trade-off is that training the network is computationally costly. Having a minimal number of units may result in underfitting, although having more units is typically not hazardous with proper regularisation.
5. **Weight initialisation**: To avoid dead neurons, we should start the weights with tiny random integers that are not too small to result in a zero gradient. Uniform distribution is typically effective. Dropout for regularisation: To avoid overfitting in deep neural networks, dropout is a recommended regularisation strategy. The approach simply removes units from the neural network based on the desired probability. A default value of 0.5 is a reasonable starting point for testing.

Multiclass Emotion Recognition

The complete dataset, which comprises of content and their accompanying sentiments, will be separated into Training, Testing, and Validation data, with 70% of the data designated as Training and the remainder designated as Testing and Validation data. The classification algorithms will be used to train the training data using the content and sentiments. Each content will be trained using their sentiments, and a model will be built. Regardless, the trained model was tested on the test dataset and verified on the validation dataset. Anger, Boredom, Surprise, Relief, Neutral, Happiness, Love, Worry, Enthusiasm, Hate, Fun, Empty, and Sadness were among the 13 sentiments [31]. Following the classification, the sentiments were predicted using the trained model.

Results Extraction and Evaluation

An evaluation is a judgement or assessment made by comparing what transpired to what was anticipated. The evaluation of decisions is based on the study of data. The text or tweets are accompanied by the accompanying sentiments. Various deep learning methods are used to classify them. Using the deep learning algorithms, the emotions or sentiments of each tweet or piece of information were recognised.

Emotions are psychological states caused by neurophysiological changes that are related with varied thoughts, sensations, behavioural reactions, and a degree of pleasure or dissatisfaction. A human being can experience a variety of emotions. Research can help scientists who study how the brain creates emotion comprehend how different emotions may be represented in different brain areas. The amygdala, a tiny structure embedded deep within either side of the brain, between the ears, is one brain area known to be involved in emotion, particularly in conditions of fear. Various human emotions are Anger, Boredom, Surprise, Relief, Neutral, Happiness, Love, Worry, Enthusiasm, Hate, Fun, Empty, and Sadness.

Depression is frequently associated with a variety of emotions. Emotions will have an impact on every element of the life. It may be difficult, if not impossible, to find happiness in anything, including activities and people you formerly enjoyed. These emotions may be utilised to identify and recognise the many phases of depression in a person's life. Depression may be classified into several stages. The strength of these human emotions may be utilised to determine the various stages of depression in a certain person. Scientists studying mood disorders such as anxiety and depression can utilise research to better understand the variety of emotions these individuals experience in their daily lives. It is possible that two people with the same condition, such as depression, will experience distinct emotional patterns and react to various types of treatment.

4 Experimental Results

Convolutional Neural Network, Long Short-Term Memory, and Gated Recurrent Unit methods were used to create the proposed system. Following the analysis and classification technique, the data was displayed based on the appropriate sentiments, and the correctness of the findings acquired while utilising different algorithms or methodologies was validated. Based on the sentiments, the total number of tweets relating to the associated sentiments was categorised. The accuracy values of several algorithms were determined. For the multiclass sentiment analysis and text classification, the accuracy was measured using multiple techniques. When compared to

Table 3 Accuracy scored with different deep learning algorithms

Algorithm	Accuracy (%)	
	Without data augmentation	With data augmentation
CNN	45	89
LSTM	33	85
GRU	36	70

other techniques, the Convolutional Neural Network had the best accuracy. Table 3 shows the accuracy values obtained in the process.

The outcome was seen to be captured in two stages, without and with text data augmentation. The total quantity of data rows was initially approximately 40,000. Because of the over-fitting problem, the accuracy values during classification for multiclass emotion detection utilising various deep learning algorithms were comparably extremely poor. Over-fitting occurs when a model learns the information and noise in the training data to the point that it severely impairs the model's performance on fresh data. This implies that the model picks up on noise or random oscillations in the training data and learns them as ideas. Without data augmentation, the CNN model had an accuracy of roughly 45%, LSTM had a score of 33%, and GRU had a score of 36% (Fig. 3).

A technique known as data augmentation has been devised to alleviate the problem of over-fitting. A process known as synonym replacement was utilised, in which the synonyms of the terms in the content of the original data were discovered and new data was produced. The number of rows in the data has been expanded to 78,000 because of the data augmentation procedure. The entire procedure is then repeated using the new enriched data. The accuracy of the deep learning algorithms utilised appears to have improved following data augmentation. Data augmentation aided in improving the accuracy of many models. CNN won by an 89% margin, LSTM by an 85% margin, and GRU by a 70% margin.

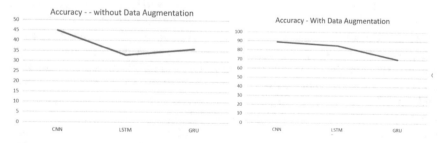

Fig. 3 Accuracy scores—without and with data augmentation

5 Summary and Conclusions

Analysis of Public Opinion/Sentiment Analysis are used to assess human behaviour, mining include the feelings, replications, and judgements acquired or retrieved from texts or other data used in data analysis or mining, web mining, and social media analytics. They can be categorised as good, negative, or neutral. It detects, transmits, and categorises division-sapient views. The data obtained throughout the procedure, evaluating their feelings, eliminating characteristics, relegating sentiments, and computing sentiment polarity definitively. It is particularly valuable for corporate product reviews, stock market fluctuations, people's reading habits, and political argument perspectives. Sentiments can be marginalised in several ways. In the Machine Learning technique, there are two forms of learning: supervised learning and unsupervised learning. Unlabelled Learning is the process of learning from unlabelled data to identify the provided data, whereas Supervised Learning is the process of developing a model using previously learnt data and predicting the target class for the given data. Unsupervised learning is the process of learning from unlabelled data to identify the supplied input data, whereas supervised learning is the process of developing a model using previously learnt data and predicting the target class for that data. Deep understanding is a vital field in Machine Learning for assimilation of awareness about various feature portrayal levels via approaches and activities.

In Multi class Classification, the data was categorised into several subclasses based on sentiment polarity, and we may anticipate an accurate or exact classification. Social media data also include emoticons and emojis, which have sentiment score values and may be employed in the sentiment analysis or classification process. The first and most critical step is to guarantee that the emoticons, which hold the emotion value, are not lost during the pre-processing data stage. It may also be used to analyse sentiment. To extract features from pre-processed data, feature extraction techniques are utilised. Using a classification technique done by multiple machine learning and deep learning algorithms, the data was split into sentiment classes based on sentiment values. Various deep learning techniques are used in the proposed system to recognise multiclass emotions in social media. The data source will be data obtained from the well-known social media platform, Twitter. To alleviate the problem of overfitting, a data augmentation phase has been implemented. Following data augmentation, a data pre-processing step was performed to remove unnecessary data from the data. The feature extraction phase is used to extract features. For the word representation, an embedding layer was employed in conjunction with GloVe. Deep learning algorithms such as CNN, LSTM, and GRU were then utilised in the classification stage. The classification methods were employed to identify the emotions in the text. Based on the retrieved characteristics and classification techniques, the models were trained. It was discovered that CNN outperformed LSTM and GRU with and without data augmentation.

Other data sources might be utilised to detect emotions in the future. Biometric data, for example, user facial expressions, user voice signals, and EEG signals. These data, together with the social media data, served as an auxiliary for the emotion analysis. A mixture of algorithms may also be developed to assess the precision value under different settings and with varied data. Moreover, these emotions can be identified to detect the depression in the users.

6   Compliances with Ethical Standards

- Funding: No Funding
- Conflict of Interest: No Conflict of Interest
- Ethics approval and consent to participate: This article does not contain any studies with human or animal participants performed by any of the authors.
- Consent of Participation: Not Applicable
- Availability of data and materials: Research involving human participants and/or animals: This article does not contain any studies with human or animal participants performed by any of the authors.
- Authors Contributions: The entire work and the paper was done by the main author, Nirmal Varghese Babu and all the review and guiding was done by Dr. Grace Mary Kanaga E.
- Acknowledgements: This paper and the research behind it would not have been possible without the exceptional support of my supervisor, Dr. Grace Mary Kanaga E. Her enthusiasm, knowledge and exacting attention to detail have been an inspiration and kept my work on track. The generosity and expertise of one and all have improved this study in innumerable ways and saved me from many errors; those that inevitably remain are entirely my own responsibility.

References

1. Chen B, Cheng L, Chen R, Huang Q, Chen Y-PP (2018) Deep neural networks for multi-class sentiment classification. In: IEEE 20th international conference on high performance computing and communications, IEEE 16th international conference on smart city, IEEE 4th international conference on data science and systems, pp 854–859
2. Sethi M, Pande S, Trar P, Soni P (2020) Sentiment identification in COVID-19 specific tweets. In: International conference on electronics and sustainable communication systems (ICESC 2020), pp 509–516. https://doi.org/10.1109/ICESC48915.2020.9155674
3. Ruz GA, Henriquez PA, Mascareno A (2020) Sentiment analysis of Twitter data during critical events through Bayesian networks classifiers. Future Gener Comput Syst 106:92–104
4. Tanna D, Dudhane M, Sardar A, Deshpande K, Deshmukh N (2020) Sentiment analysis on social media for emotion classification. In: International conference on intelligent computing and control systems (ICICCS 2020), pp 911–915. https://doi.org/10.1109/ICICCS48265.2020. 9121057
5. Cheng L-C, Tsai S-L (2019) Deep learning for automated sentiment analysis of social media. In: IEEE/ACM international conference on advances in social networks analysis and mining, pp 1001–1004. https://doi.org/10.1145/3341161.3344821

6. Abid F, Li C, Alam M (2020) Multi-source social media data sentiment analysis using bidirectional recurrent convolutional neural networks. Comput Commun 157:102–115
7. Tadessi MM, Lin H, Xu B, Yang L (2019) Detection of depression-related posts in Reddit social media forum. IEEE Access 7:44883–44893. https://doi.org/10.1109/ACCESS.2019.2909180
8. Syarif I, Ningtias N, Badriyah T (2019) Study on mental disorder detection via social media mining. IEEE, pp 1–6. https://doi.org/10.1109/CCCS.2019.8888096
9. Katchapakirin K, Wongpatikaseree K, Yomaboot P, Kaewpitakkun Y (2018) Facebook social media for depression detection in the Thai community. In: 15th international joint conference on computer science and software engineering (JCSSE), pp 1–6. https://doi.org/10.1109/JCSSE.2018.8457362
10. Kumar A, Sharma A, Arora A (2019) Anxious depression prediction in real-time social data. In: International conference on advanced engineering, science, management and technology—2019 (ICAESMT19)
11. Nalinde PB, Shinde A (2019) Machine learning framework for detection of psychological disorders at OSN. Int J Innov Technol Explor Eng (IJITEE) 8(11). ISSN 2278-3075
12. Tajuddin M, Kabeer M, Misbahuddin M (2020) Analysis of social media for psychological stress detection using ontologies. In: Fourth international conference on inventive systems and control (ICISC 2020). IEEE Xplore Part Number: CFP20J06-ART; ISBN 978-1-7281-2813-9
13. Baheti RR, Kinariwala S (2019) Detection and analysis of stress using machine learning techniques. Int J Eng Adv Technol (IJEAT) 9(1). ISSN 2249-8958
14. Jabreel M, Moreno A (2019) A deep learning-based approach for multi-label emotion classification in tweets. Appl Sci 9(6):1123. MDPI
15. Sadr H, Pedram MM, Teshnehlab M (2020) Multi-view deep network: a deep model based on learning features from heterogeneous neural networks for sentiment analysis. IEEE Access 8:86984–86997. https://doi.org/10.1109/ACCESS.2020.2992063
16. Dragoni M, Kessler FB (2017) A neural word embeddings approach for multi-domain sentiment analysis. IEEE Trans Affect Comput 8(4):457–470. https://doi.org/10.1109/TAFFC.2017.2717879
17. Basiri ME, Nemati S, Abdar M, Cambria E, Acharrya UR (2021) ABCDM: an attention-based bidirectional CNN-RNN deep model for sentiment analysis. Future Gener Comput Syst 115:279–294
18. Harb JGD, Ebeling R, Becker K (2020) A framework to analyze the emotional reactions to mass violent events on Twitter and influential factors. Inform Process Manage 57(6):102372
19. Jayakrishnan R, Gopal GN, Santhikrishna MS (2018) Multi-class emotion detection and annotation in Malayalam novels. In: 2018 international conference on computer communication and informatics (ICCCI-2018), 04–06 Jan. 2018, Coimbatore, pp 1–5. https://doi.org/10.1109/ICCCI.2018.8441492
20. Huang S, Zhao Q, Xu X-Z, Zhang B, Wang D (2019) Emojis-based recurrent neural network for Chinese microblogs sentiment analysis. In: IEEE international conference on service operations and logistics, and informatics (SOLI) 2019, pp 59–64. https://doi.org/10.1109/SOLI48380.2019.8955016
21. Bouzazi M, Ohtsuki T (2017) A pattern-based approach for multi-class sentiment analysis in Twitter. IEEE Access 5:20617–20639. https://doi.org/10.1109/ACCESS.2017.2740982
22. Yang L, Li Y, Wang J, Sherrarat RS (2020) Sentiment analysis for E-commerce product reviews in Chinese based on sentiment Lexicon and deep learning. IEEE Access 8:23522–23530. https://doi.org/10.1109/ACCESS.2020.2969854
23. Salur MU, Aydin I (2020) A novel hybrid deep learning model for sentiment classification. IEEE Access 8:58080–58093. https://doi.org/10.1109/ACCESS.2020.2982538
24. Liang H, Ganeshbabu U, Thorne T (2020) A dynamic Bayesian network approach for analysing topic-sentiment evolution. IEEE Access 8:54164–54174. https://doi.org/10.1109/ACCESS.2020.2979012
25. Khan R, Shrivastava P, Kapoor A, Tiwari A, Mittal A (2020) Social media analysis with AI: sentiment analysis techniques for the analysis of Twitter COVID-19 data. J Crit Rev 7(09). ISSN 2394-5125

26. Burdisso SG, Errecalde M, Montes-y-Gómez M (2019) Text classification framework for simple and effective early depression detection over social media streams. Exp Syst Appl 133:182–197
27. Birjalia M, Beni-Hssane A, Erritali M (2017) Machine learning and semantic sentiment analysis based algorithms for suicide sentiment prediction in social networks. Procedia Comput Sci 113:65–72
28. Wang T, Lu K, Chow KP, Zhu Q (2020) COVID-19 sensing: negative sentiment analysis on social media in China via BERT model. IEEE Access 8:138162–138169. https://doi.org/10.1109/ACCESS.2020.3012595
29. Shahare FF (2017) Sentiment analysis for the news data based on the social media. In: International conference on intelligent computing and control systems (ICICCS 2017), pp 1365–1370. https://doi.org/10.1109/ICCONS.2017.8250692
30. Solakidis GS, Vavliakis KN, Mitkas PA (2014) Multilingual sentiment analysis using emoticons and keywords. In: IEEE/WIC/ACM international joint conferences on web intelligence (WI) and intelligent agent technologies (IAT) 2014, pp 102–109. https://doi.org/10.1109/WIIAT.2014.86
31. Babu NV, Kanaga EGM (2022) Sentiment analysis in social media data for depression detection using artificial intelligence: a review. SN Comput Sci 3:74

Machine Learning-Based Technique for Phishing URLs Detection from TLS 1.2 and TLS 1.3 Traffic Without Decryption

Munish Kumar, Alwyn Roshan Pais, and Routhu Srinivasa Rao

1 Introduction

As Internet services are increasing day by day, phishing attacks are also increasing. Phishing attacks have now become a major issue. Because of phishing attacks, victims face financial, and sensitive information losses. Attackers use various techniques to trick the victim to get sensitive information such as credit card details, bank account details. Attackers create websites that are similar to legitimate sites and spread them. Victims are not able to identify whether the website is legitimate or phishing and lose sensitive information. So detecting the phishing attack and alerting the target is very important. There exists various anti-phishing techniques which work at the application layer. These techniques are content-based [1–3], blacklist-based [4], white-list-based [5, 6], and URL-based [7–9]. Content-based approaches use the content of the website such as images, source code, and links for the classification. In blacklist-based approaches, the existence of the suspicious link is checked in the list of phishing sites. In URL-based approaches, features are extracted from Hostname, subdomain, and full URL to classify if the given URL is phishing or not.

All these anti-phishing techniques are fast but they have some limitations. Some techniques are not able to detect zero-day phishing attacks. Furthermore, due to the absence of the domain age, some approaches wrongly identifies the new legitimate sites as phishing. Also, to improve the detection rate, various existing strategies

M. Kumar (✉) · A. R. Pais
National Institute of Technology Karnataka, Surathkal, India
e-mail: 1995munish17@gmail.com

A. R. Pais
e-mail: alwyn@nitk.edu.in

R. S. Rao
GMR Institute of Technology, Rajam, Andhra Pradesh 532127, India
e-mail: srinivas.r@gmrit.edu.in

D. S. Sisodia et al. (eds.), *Machine Intelligence Techniques for Data Analysis and Signal Processing*, Lecture Notes in Electrical Engineering 997,
https://doi.org/10.1007/978-981-99-0085-5_31

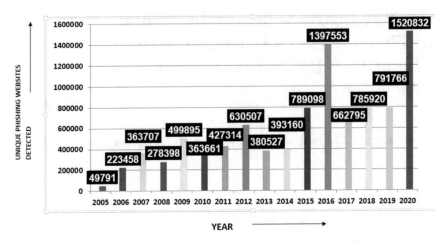

Fig. 1 Phishing sites that are unique and detected during 2005–2020 as per APWG report

rely on third-party services, such as WHOIS records, search engine indexing, and domain rankings. However, using these services slows down the detection process and prevents them from detecting websites hosted on compromised servers.

Figure 1 shows phishing websites that are unique and detected during 2005–2020 as per Anti-Phishing Working Group (APWG) report.[1] Figure 1 shows that phishing attacks are increasing very fast and have become a major threat nowadays.

From the past statistics, we can see the importance of efficient phishing detection techniques that can prevent the end-users from financial and sensitive information losses.

There are no approaches to detect phishing URLs from the encrypted traffic without decryption. So this paper proposed an anti-phishing technique based on features extracted from TLS 1.2 and TLS 1.3 encrypted traffic without decryption. In our work, a novel ML-based approach is proposed to classify URLs based on the features extracted from ClientHello message of TLS 1.2 and TLS 1.3 encrypted traffic without decryption.

The remainder of the paper is organized in the following manner: Sect. 2 covers a number of current anti-phishing solutions. The proposed work is described in Sect. 3. Section 4 contains experimentation and results. Finally, we conclude our paper in Sect. 5.

2 Literature Survey

Many anti-phishing techniques exist which reduce the risk of financial loss and sensitive information loss. These techniques use different methods for detecting phishing

---

[1] https://apwg.org/trendsreports/.

URLs. Some techniques use content-based approaches which use the content of the website such as images, source code, and links. Some techniques use a list-based approach such as a list of URLs which are blacklisted and white-listed and some techniques use features extracted from URLs such as features from hostname and full URL.

2.1 Content-Based Methods for Phishing Detection

Content-based phishing techniques extract the features from the content of websites such as images, source code, and links. Techniques in [1–3, 10–15] uses content-based methods for the phishing detection. Nakayama et al. [1] content-based technique is used in which keywords from a target web page and the pages linked with the target page are extracted. Using these keywords, the corresponding legitimate sites are retrieved. The target site is classified as a phishing site if the domain of the target page does not match the retrieved site. Dunlop et al. [2] image of the web page is captured, with the help of optical character recognition to convert the image to text, then it uses the third-party service which is the Google Page Rank algorithm to make the decision on the validity of the site. Wenyin et al. [3] phishing and legitimate websites are compared based on visual similarity, and metrics that can be utilized for classification are developed. The method decomposes web pages based on visual clues before calculating the similarity between two web sites. If the visual similarity between the web page and legitimate web page is higher than a threshold, then the web page is considered a phishing page.

Rao and Pais [11] Content-based techniques extract features from the web page's source code such as anchor links, title, domain, and with the help of third-party services get the information about these extracted features. Information includes page ranking, WHO-IS, etc.

Adebowale et al. [16] features of text, images, and frames of legitimate as well as non-legitimate sites are used to detect the phishing sites.

2.2 List-Based Methods for Phishing Detection

For phishing detection, the majority of anti-phishing tools use list-based methods. Techniques for phishing detection in [4–6, 17–19] are list-based. The blacklist-based detection technique is one of the list-based approaches. PhishTank and Google have prepared a blacklist of phishing URLs. Anti-phishing techniques use this list to identify whether or not a URL is phishing. The white-list-based technique is another list-based technique which is a collection of legitimate URLs. Phishing detection techniques use this list to identify whether the URL is legitimate or not.

Rao and Pais [17] two levels of filtering has been used for phishing detection. This approach first detects phishing sites based on the visual similarity of the site with the blacklisted sites. The suspicious URLs that are not detected at the first level are detected at the second level. At the second level, URL and source code-based features are used for the detection of phishing sites.

Some popular browsers have integrated the blacklist-based anti-phishing solution to help the users not get tricked by phishing attacks. Microsoft's Internet Explorer browser now includes a blacklist-based anti-phishing solution. The Google Safe Browsing API also uses phishing URLs blacklists to identify phishing sites. However, the approach has a drawback in the detection of zero-day phishing attacks.

Rao and Pais [19] an application named as Jail-phish is proposed, which enhance the search engine-based techniques for phishing detection. This technique not only detects the newly registered sites but also detects the phishing sites hosted on compromised servers based on the similarity score. This technique calculates similarity score by comparing matched domain search results with the suspicious site.

2.3 URL Feature-Based Methods for Phishing Detection

To detect phishing URLs, URL-based classification algorithms extract information from the URL. To determine if suspicious URLs are legitimate or phishing, these techniques extract count-based and binary information from URLs. Attackers control these features to make the victim believe that the URL is legitimate. In [7–9, 20–26] URL feature-based methods are used for phishing detection.

Nguyen et al. [7] URL feature-based technique is used to detect phishing. This approach checks the similarity of phishing sites URLs and legitimate sites URLs and also considers ranking of sites for phishing detection.

Rao et al. [8] URL feature-based approach is used in which features are extracted from the URLs are divided into two lists: a list of features based on hostname and another one based on full URL. Table 1 lists features based on hostname, while Table 2 lists features based on full URL. Also, the features are separated into two categories: count-based and binary. The legitimate and phishing sites dataset is used, and features for each URL are extracted. Various machine learning algorithms are trained using the dataset to classify the URL. To identify a suspicious input URL this approach first extracts the features from URL then does the feature vectorization. The machine learning algorithm gets the feature vector as input. The output of the machine learning algorithm classifies the given URL as phishing or legitimate.

3 Proposed Work

The proposed model classify URLs based on features extracted from TLS 1.2 and TLS 1.3 encrypted traffic without decryption. A dataset of features for legitimate as well as phishing sites is prepared from TLS 1.2 and TLS 1.3 traffic. Machine learning algorithms are trained on this dataset. As shown in Fig. 2, for the suspicious input

Table 1 Hostname-based features

HF1: @ in the Hostname	HF10: Count of dots in Hostname
HF2: Count of Digits in the Hostname	HF11: Count of hyphens in Hostname
HF3: Average word length in the Hostname	HF12: Count of underscores in the Hostname
HF4: Word with the longest length in Hostname	HF13: Presence of IP in the Hostname
HF5: Length of the Hostname	HF14: Validate Top-level domain
HF6: Phish-Hinted word in the Hostname	HF15: Multiple Top-level domains are present in the Hostname
HF7: Brand name in the subdomain	HF16: Underscores presence in the Hostname
HF8: Count of digits in the domain	
HF9: Domain length	

Table 2 Full URL-based features

UF1: @ is present in the URL	UF10: Count of HTTPS in the path
UF2: Count of digits in the path	UF11: $ is present in the path
UF3: Average word length in the path	UF12: * is present in the URL
UF4: Length of the longest word in the path	UF13: OR symbol present in the URL
UF5: Length of the base URL	UF14: The semicolon is present in the base URL
UF6: Phish-Hinted words are present in the path	UF15: White space is present in the URL
UF7: The brand name is present in the path	UF16: Hyphens ratio
UF8: Count of? in the path	UF17: HTTPS is present in the base URL
UF9: Count of slash in the path	

URL first proposed model capture TLS 1.2 and TLS 1.3 traffic and then convert the captured network traffic into a JSON file to extract features. The extracted features vector is given as input to the trained machine learning algorithm. The output of the trained model classifies whether the suspicious input URL is legitimate or phishing.

The different steps of the proposed architecture are explained below.

- **Input URL:** Input URL is the suspicious URL, which is classified by the trained proposed model whether it is phishing or not.
- **Capture TLS 1.2 and TLS 1.3 traffic:** A python program has been implemented, which capture TLS 1.2 and TLS 1.3 traffic for suspicious input URL.
- **Converting Captured TLS 1.2 and TLS 1.3 traffic file to JSON file:** The captured TLS 1.2 and TLS 1.3 traffic file of suspicious input URL is converted into JSON file for features extraction.
- **Extract features from JSON File:** Transport Layer Security (TLS) protocol is a technology for online privacy. TLS is a cryptographic protocol that enables secure communication between client and server applications over the Internet.

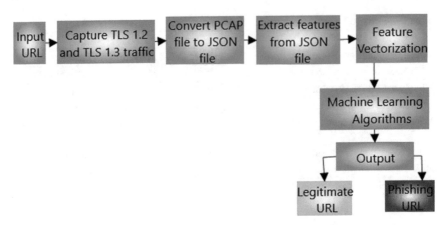

Fig. 2 Architecture of the proposed work

TLS is designed to prevent eavesdropping, tampering, and message forgery. TLS is implemented on top of the common HTTP protocol for web browsing. The extension of HTTP protocol known as HyperText Transfer Protocol Secure (HTTPS) is the usage of TLS over HTTP. In TLS handshake clientHello message contains ordered list of offered cipher suites and list of supported extensions such as server_name, max_fragment_length, status_request, supported_groups, signature_algorithms, signed_certificate_timestamp, supported_versions, etc. In the proposed work, unencrypted server_name extension is used. From the server_name extension, features are extracted to classify suspicious URLs. Server_name extension contains Server Name Indication sub extension. Server Name Indication extension contains Server Name List length, Server Name Type, Server Name length, Server Name. In proposed work, Server Name is used. From the Server Name value, a list of count-based features is prepared.

Count-based features from Server Name Indication: From the value of the Server Name Indication extension, Table 1 count-based features are derived. These features are HF2, HF3, HF4, HF5, HF10, HF11, and along with these features one more feature is added which is a Number of Servers (NoS). All these features help in the classification of the suspicious URL whether it is phishing or not.

- **Feature Vectorization:** Count-based features which are extracted from Server Name Indication are used to prepare features vector. Dataset of the features vectors used for training of the proposed model.
- **Machine Learning algorithms:** Various machine learning algorithms such as Random Forest (RF), Logistic Regression (LR) and Support Vector Machine (SVM), XGBoost (XGB), and Light GBM (LGBM) are used to train our proposed model. To compare the performance and determine the best classifier for our dataset, we have used different classifiers. For implementing different classifiers, the Scikit-learn package is used.

Table 3 Dataset used in our experimentation

Type	Source	Sites
Legitimate	Alexa database	1516
Phishing	PhishTank	1510

4 Experimentation and Results

In this section, the details of customized dataset generation and classification using the proposed model is provided.

4.1 Dataset

Dataset for the experiment has been prepared by capturing TLS 1.2 and TLS 1.3 traffic for the legitimate as well as phishing sites. The Alexa database is used to collect the legitimate websites and valid phishing sites are collected from the PhishTank. For preparing the dataset a python program has been implemented which takes the URLs of legitimate and phishing sites from the text file and captures the TLS 1.2 and TLS 1.3 traffic in the PCAP files. The captured PCAP files are converted into JSON files. These JSON files are used to extract the features to prepare the dataset. The proposed model is trained using the prepared dataset presented in Table 3.

Two experiments are conducted on our dataset to study the performance of the proposed model. In experiment 1, LR, SVM, RF, XGBoost, and LGBM classifiers are used to classify URLs. In experiment 2, individual feature contributions to URL classification have been studied.

For evaluating the performance of the proposed model, True Positive Rate (TPR), True Negative Rate (TNR), False Positive Rate (FPR), False Negative Rate (FNR), Accuracy (Acc), Precision (Pre), and F-measure (F) are used. All these traditional metrics are calculated as given below

$$TPR = \frac{\text{\# of phish URLs classified as phish}}{\text{Total\# of phish URLs}} \tag{1}$$

$$TNR = \frac{\text{\# of legit URLs classified as legit}}{\text{Total\# of legit URLs}} \tag{2}$$

$$FPR = \frac{\text{\# of legit URLs classified as phish}}{\text{Total\# of legit URLs}} \tag{3}$$

$$FNR = \frac{\text{\# of phish URLs classified as legit}}{\text{Total\# of phish URLs}} \tag{4}$$

$$Acc = \frac{\text{\# of correctly classified URLS as phish and legit}}{\text{Total \# of URLs}} * 100 \qquad (5)$$

$$Pre = \frac{\text{\# of phish URLs classified as phish}}{\text{Total \# of URLs classified as phish}} * 100 \qquad (6)$$

$$F = 2 * \frac{Pre * TPR}{Pre + TPR} \qquad (7)$$

We have used 80% of dataset for training the different machine learning classifiers and 20% of dataset for testing in both experiment 1 and experiment 2.

Experiment 1: In this experiment, different machine learning classifiers have been applied to the dataset. Table 4 shows the results of the different machine learning classifiers on the dataset. In Table 4, LR gives 82.5% accuracy, SVM gives 87.2% accuracy, RF gives 88.2% accuracy, XGBoost and LGBM give 89.6% accuracy. From the results, it has been observed that XGBoost outperforms Light GBM in terms of TPR, TNR, FPR, and FNR.

Experiment 2: In this experiment, the accuracy for each feature of the dataset has been calculated and presented in Table 3 using Eq. 5. This experiment has been performed to study the importance of each feature for the classification. Table 5 shows the results of this experiment on the dataset. In Table 5, all the classifiers give 63.9% accuracy for feature HF2. For feature HF3, the highest accuracy of 69.9% has been obtained for the RF classifier, and other classifiers have accuracy greater than 58%. RF, XGBoost, and LGBM classifiers give the highest accuracy of 60.7% for

Table 4 Experiment results for different classifiers on our dataset

Classifier	TPR (%)	FPR (%)	TNR (%)	FNR (%)	Accuracy (%)	Precision (%)	F-measure(%)
LR	74.9	9.4	90.5	25.0	82.5	89.2	81.4
SVM	84.6	10.1	89.8	15.3	87.2	89.7	87.1
RF	87.1	10.5	89.4	12.8	88.2	89.6	88.3
XGB	88.2	9.0	90.9	11.8	89.6	91.3	89.7
LGBM	86.0	10.1	89.8	13.9	89.6	91.1	89.6

Table 5 Individual feature accuracy on our Dataset

Classifier	HF2(%)	HF3(%)	HF4(%)	HF5(%)	HF10(%)	HF11(%)	NoS(%)
LR	63.9	58.6	55.9	65.8	49.7	64.6	65.8
SVM	63.9	60.2	54.3	66.4	69.2	64.6	66.4
RF	63.9	69.9	60.7	66.9	69.2	64.6	66.9
XGBoost	63.9	69.8	60.7	65.7	69.2	64.6	67.1
LGBM	63.9	69.2	60.7	67.1	69.2	64.6	67.1

feature HF4. For feature HF5, the highest accuracy of 67.1% has been obtained for the LGBM classifier and other classifiers have accuracy greater than 65%. From the results in Table 5, it has been observed that features HF2, HF3, HF4, and HF5 have a significant role in the classification for all the classifiers. Feature HF10 has 69.2% accuracy for SVM, RF, XGBoost, and LGBM classifiers whereas 49.7% accuracy for LR classifier. Feature HF10 plays an essential role in the classification for SVM, RF, XGBoost, and LGBM but has less importance for LR classifier for the classification. For feature HF11, all classifiers give 64.6% accuracy. The accuracy results of HF11 for different classifiers shows that it plays a significant role in the classification. For newly added feature NoS, the highest accuracy of 67.1% has been obtained for XGBoost and LGBM classifiers and other classifiers have accuracy greater than 65%. This shows that newly added feature has an important role in the classification.

5 Conclusion

In this work, an efficient machine learning-based model has been proposed for detecting phishing URLs from the TLS 1.2 and TLS 1.3 encrypted traffic. The proposed technique achieves an accuracy of 89.6% for the XGBoost and LGBM classifiers. We have also proposed a new feature for classification. The proposed technique is independent of any third-party services; hence it can be used for phishing URL detection in any environment.

In future work, we will investigate traffic and find new features to improve the accuracy of our model. Also, new models based on deep learning will be proposed to achieve better accuracy.

References

1. Nakayama S, Echizen I, Yoshiura H (2009) Preventing false positives in content-based phishing detection. In: 2009 Fifth international conference on intelligent information hiding and multimedia signal processing. IEEE, pp 48–51
2. Dunlop M, Groat S, Shelly D (2010) Goldphish: using images for content-based phishing analysis. In: (2010) Fifth international conference on internet monitoring and protection. IEEE pp 123–128
3. Wenyin L, Huang G, Xiaoyue L, Min Z, Deng X (2005) Detection of phishing webpages based on visual similarity. In: Special interest tracks and posters of the 14th international conference on World Wide Web, pp 1060–1061
4. Hong J, Kim T, Liu J, Park N, Kim S-W (2020) Phishing url detection with lexical features and blacklisted domains. In: Adaptive autonomous secure cyber systems. Springer, pp 253–267
5. Jain AK, Gupta BB (2016) A novel approach to protect against phishing attacks at client side using auto-updated white-list. EURASIP J Inf Secur 1:1–11
6. Han W, Cao Y, Bertino E, Yong J (2012) Using automated individual white-list to protect web digital identities. Expert Syst Appl 39(15):11861–11869

7. Nguyen LAT, To BL, Nguyen HK, Nguyen MH (2014) A novel approach for phishing detection using url-based heuristic. In: 2014 International conference on computing, management and telecommunications (ComManTel). IEEE, pp 298-303
8. Rao RS, Vaishnavi T, Pais AR (2020) Catchphish: detection of phishing websites by inspecting urls. J Ambient Int Humanized Comput 11(2):813–825
9. Butnaru A, Mylonas A, Pitropakis N (2021) Towards lightweight url-based phishing detection. Future Internet 13(6):154
10. Hannousse A, Yahiouche S (2021) Towards benchmark datasets for machine learning based website phishing detection: an experimental study. Eng Appl Artif Intell 104:104347
11. Rao RS, Pais AR (2019) Detection of phishing websites using an efficient feature-based machine learning framework. Neural Comput Appl 31(8):3851–3873
12. Tan CL, Chiew KL, Wong K et al (2016) Phishwho: phishing webpage detection via identity keywords extraction and target domain name finder. Decision Support Syst 88:18–27
13. Zhang Y, Hong JI, Cranor LF (2007) Cantina: a content-based approach to detecting phishing web sites. In: Proceedings of the 16th international conference on World Wide Web, pp 639–648
14. Jain AK, Gupta BB (2019) A machine learning based approach for phishing detection using hyperlinks information. J Ambient Int Humanized Comput 10(5):2015–2028
15. Castano F, Fidalgo E, Alegre E, Chaves D, Sanchez-Paniagua M (2021) State of the art: content-based and hybrid phishing detection. ArXiv preprint arXiv:2101.12723
16. Adebowale MA, Lwin KT, Sanchez E, Hossain MA (2019) Intelligent web-phishing detection and protection scheme using integrated features of images, frames and text. Expert Syst Appl 115:300–313
17. Rao RS, Pais AR (2019) Two level filtering mechanism to detect phishing sites using lightweight visual similarity approach. J Ambient Int Humanized Comput 1–20
18. Teraguchi NCRLY, Mitchell JC (2004) Client-side defense against web-based identity theft. Comput Sci Dept Stanford University. Available: http://crypto.stanford.edu/SpoofGuard/webspoof.pdf
19. Rao RS, Pais AR (2019) Jail-phish: an improved search engine based phishing detection system. Comput Secur 83:246–267
20. Gowtham R, Krishnamurthi I (2014) A comprehensive and efficacious architecture for detecting phishing webpages. Comput Secur 40:23–37
21. Wang W, Shirley K (2015) Breaking bad: detecting malicious domains using word segmentation. ArXiv preprint arXiv:1506.04111
22. Feng F, Zhou Q, Shen Z, Yang X, Han L, Wang J (2018) The application of a novel neural network in the detection of phishing websites. J Amb Intell Humanized Comput 1–15
23. Suleman MT, Awan SM (2019) Optimization of url-based phishing websites detection through genetic algorithms. Autom Cont Comput Sci 53(4):333–341
24. Sahingoz OK, Buber E, Demir O, Diri B (2019) Machine learning based phishing detection from urls. Expert Syst Appl 117:345–357
25. Rao RS, Pais AR, Anand P (2021) A heuristic technique to detect phishing websites using twsvm classifier. Neural Comput Appl 33(11):5733–5752
26. Gupta BB, Yadav K, Razzak I, Psannis K, Castiglione A, Chang X (2021) A novel approach for phishing urls detection using lexical based machine learning in a real-time environment. Comput Commun 175:47–57

Robust Pipeline for Detection of Adversarial Images

Ankit Vohra, Natesh Reddy, Kritika Dhawale, and Pooja Jain

1 Introduction

Deep neural networks like Recurrent Neural Networks (RNN's), Convolutional Neural Networks (CNN's), Long Short-Term Memory networks (LSTM's), etc. have shown great promise for various image-related tasks such as classification of objects [1, 2], tracking of motion [3], recognition of action [4], estimating human pose [5], and semantic segmentation [6].

Despite this huge success, it has been shown that these networks are vulnerable to adversarial examples [7]. These samples (which are visually imperceptible to human beings) are deliberately generated to fool models and cause them to change the output. Figure 1 shows an image of a 'Castle' being misclassified as 'Gloves' with high confidence by the MobileNetV2 CNN. A thing to note is that the attacked image seems similar to the original image, i.e., they are quasi imperceptible to a human observer. In the cyber-security domain, terrorists could tamper with autonomous cars or pedestrian detection systems. Deepfakes could be employed to generate fake news or commit financial fraud [8, 9].

Many methods have been proposed to create such adversarial examples. Some techniques seek to change each pixel by a small amount and can be found using many optimization strategies such as DeepFool [10], Fast Gradient Sign Method (FGSM) [11], L-BFGS [7], Projected Gradient Descent (PGD) [12], and the recent Logit-space Projected Gradient Ascent (LS-PGA) [13] for discretized inputs.

A. Vohra · N. Reddy (✉) · P. Jain
Department of Computer Science and Engineering, Indian Institute of Information Technology, Nagpur, India
e-mail: natesh1199@gmail.com

K. Dhawale
Department of Electronics and Communication Engineering, Indian Institute of Information Technology, Nagpur, India

© The Author(s), under exclusive license to Springer Nature Singapore Pte Ltd. 2023
D. S. Sisodia et al. (eds.), *Machine Intelligence Techniques for Data Analysis and Signal Processing*, Lecture Notes in Electrical Engineering 997,
https://doi.org/10.1007/978-981-99-0085-5_32

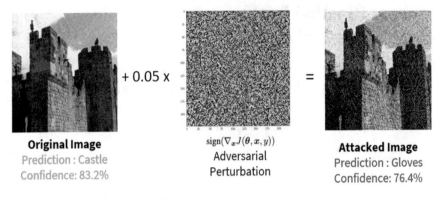

Original Image
Prediction : Castle
Confidence: 83.2%

$\mathrm{sign}(\nabla_x J(\theta, x, y))$
Adversarial
Perturbation

Attacked Image
Prediction : Gloves
Confidence: 76.4%

Fig. 1 Example of adversarial attack

While some attacks insert a small 'patch' at a fixed position in the image [14], others like the Jacobian-based saliency map [15] modify a few pixels in the image to create distortions. This has serious consequences as shown by the authors in [16]. Image detection systems in self-driving cars can be deceived by creating 'toxic signs' that are adversarially attacked images of signboards. The system may detect a 'stop sign' as a 'speed limit 100 sign' causing accidents.

Many detection strategies have been proposed involving statistical methods [17], augmentation of model [18, 19], novel training techniques [20], augmentation of training data [11], and feature squeezing [21].

We propose a simple end-to-end pipeline that can detect such adversarially-crafted images. We train our pipeline on the ILSVRC 2012 dataset. To improve the robustness of our proposed method we incorporate adversarial examples generated using the Fast Gradient Signed Method (FGSM) attack, Projected Gradient Descent (PGD) attack, and Patch attack into the training process.

2 Related Work

Attacking DNN's by adding imperceptible carefully chosen noise (adversarial perturbations) to images was first investigated by [22]. The researchers found out that the trained model incorrectly classified instances that too with high confidence. While the authors [23] have outlined that adversarial examples can be constructed for an unknown network by training an auxiliary network on related examples and exploiting the transferability of adversarial samples. The researchers [16, 24] have investigated how adversarial examples can also be transferred to the real world. They show this by creating adversarial printed road signs with graffiti-like art on top. Most of the work on addressing adversarial attacks focuses on increasing the robustness of the model itself for the detection of adversarial examples. The authors [11] have proposed a technique that augments the training set with adversarially attacked examples. During training, they try to minimize the loss for the attacked and non-attacked examples, while spoofed samples are chosen to fool the DNN's.

Techniques like adversarial training [20] and knowledge distillation [25] have been shown to be effective methods to tackle adversarially attacked images. Their drawback is that they suffer from the problem of generalization.

The researchers [18] have trained a binary classifier that separates the adversarial part from the clean data. They receive good accuracy [26] scores of 0.99 but their results are sensitive to epsilon values, dropping to 0.003 for epsilon $= 0.001$.

The authors [17] have demonstrated a unique way to identify adversarial examples using statistical tests. They state that adversarial examples are not drawn from the same distribution as the original data, and can be identified using statistical tests. They augment their machine learning model with additional input to classify fake examples. Their model performs well but it fails to give good accuracies for large sample sizes on the MNIST dataset and for some attacks like the decision tree attack.

The researchers [19] model the problem as a binary classification task of distinguishing between real and adversarially crafted data. They extend their neural network by adding a small 'detector' subnetwork that is trained on the said task. Their network generalizes to other adversaries as well. They propose a new attack method to generate adversarial perturbations that deceives both the detector subnetwork and the classifier. In addition to this, they come up with a novel training procedure for the detector that counters this attack.

The authors [21] propose a novel strategy, feature squeezing to improve the detection capability of DNN's. They compare the DNN's prediction on the original versus the squeezed inputs and show that feature squeezing detects adversarial examples with high accuracy. They also explore two feature squeezing methods, first by reducing the color bit depth of each pixel and second by using spatial smoothing. We aim to address the above-mentioned limitations by proposing an attack-independent deep learning pipeline that is capable of detecting adversarial examples from a set of samples. During training, we experiment with different epsilon values and attack methods (FGSM, PGD, Patch attack) to evaluate the robustness of our pipeline.

3   Background Work

3.1   Fast Gradient Signed Method (FGSM) Attack

According to the authors [11], owing to the deep neural network's linear nature, they are vulnerable to adversarial perturbations. They propose the Fast Gradient Sign Method (FGSM) to efficiently generate adversarially attacked samples. FGSM keeps track of the cost of an attack by assuming the strength of attack at every feature dimension is equal. It essentially measures the perturbations $(\partial)(x, x')$ using the L∞-norm. The FGSM attack comes under the hood of untargeted attacks and

| Epsilon = 0.01 | Epsilon = 0.1 | Epsilon = 0.3 |

Fig. 2 Generated FGSM examples with different epsilon values

the adversarial perturbations are calculated by using the gradients of the loss,

$$(\partial)(x, x') = (\epsilon)(.)\text{sign}((\partial)x J(g(x), y)) \tag{1}$$

Here, $J(\cdot, \cdot)$ is the loss function that has been used to train the deep neural model and y is the actual value for x. An example showing the attacked images for different values of epsilon is shown in Fig. 2.

3.2 Patch Attack

The authors have [14] introduced a patch-based attack that comes under the hood of targeted attacks. In this method, a patch 'p' is obtained by training using the Expectation over Transformation (EOT) framework proposed by [27] and optimizing the objective function.

$$\hat{p} = \arg\max_{p} E_{x \sim X, t \sim T, l \sim L}[\log \Pr(\hat{y}|A(p, x, l, t)] \tag{2}$$

where T denotes the distribution over transformations of the patch, L is the distribution over locations in the image and X comprises the set of training images. The patches can be added to any image which causes the model to ignore the other parts of the image and concentrate on the patch itself, thus predicting the chosen target class. An example generated using a patch attack is shown in Fig. 3. It causes the MobileNet-V2 model to misclassify a 'hartebeest' as a 'goldfish'.

3.3 Projected Gradient Descent (PGD) Attack

The PGD or the Projected Gradient Descent attack comes under the umbrella of a white-box attack. A white-box attack is essentially when the gradients of the model

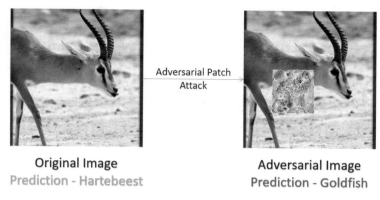

Original Image
Prediction - Hartebeest

Adversarial Image
Prediction - Goldfish

Fig. 3 Image generated with adversarial Patch

are accessible to the attacker. This is in contrast to transfer attacks which result in human-visible adversarial perturbations. Hence, the white-box attack is more powerful as compared to black-box attacks as the attacker can specifically craft the attack in order to deceive the DNN. An example generated using a PGD attack is shown in Fig. 4. This causes MobileNet-V2 to misclassify the image as a 'soccer ball'.

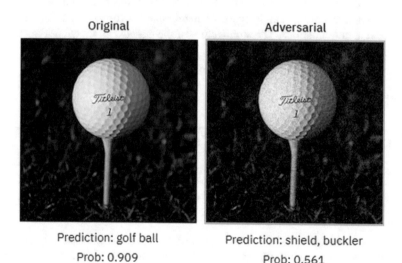

Prediction: golf ball
Prob: 0.909

Prediction: shield, buckler
Prob: 0.561

Fig. 4 Misclassified PGD attacked image

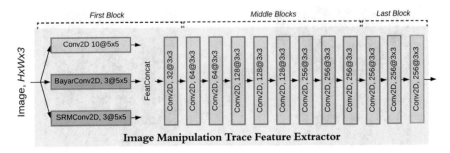

Fig. 5 Mantra-Net model

3.4 Mantra-Net

It is a deep neural architecture proposed by [28] which is an end-to-end framework that performs both detection and localization without extra preprocessing and postprocessing. Mantra-Net's robustness lies in its ability to handle arbitrarily sized images and other forgery types like splicing, copy-move, removal, enhancement, and even unknown types. It is able to detect manipulated pixels by identifying local anomalous features, hence making it independent of the forgery or manipulation type used. Experiments have shown generalizability, robustness, and superiority of Mantra-Net, not only in single types of manipulations/forgeries but also in their complicated combinations. The architecture of the network is shown in Fig. 5.

4 Methodology

We have modeled our problem as a binary classification problem, i.e., to determine whether an image is adversarially attacked or not.

4.1 ILSVRC 2012 Dataset

The ImageNet Large Scale Visual Recognition Challenge 2012 dataset [29] is a subset of the ImageNet dataset (10,000,000 labeled images, 10,000+ object categories) with 10,000 classes and 50,000 samples. We have used the ILSVRC 2012 dataset for benchmarking the pre-trained models like VGG19, ResNet50, and MobileNetV2 on original images. This 1000 class data set is the validation set of the Imagenet challenge. All the images in this dataset are of varying dimensions. We used the built-in processing pipeline of Tensorflow for different architectures of CNN. Furthermore, all the attacks such as FGSM, PGD, Patch-based attacks were performed on the same

50,000 images. The current SOTA method only tested their work on Dataset MNIST, CIFAR10, and SVHN which are very basic.

4.2 Workflow

The SOTA has only trained a simple binary classifier without any preprocessing resulting in a high accuracy difference when the epsilon value is reduced from 0.03. To overcome this we have built a 2 stage model which is depicted in Fig. 6. The training pipeline consists of 3 steps. First is generating adversarially attacked images using different attacking methods (FGSM, PGD, and Patch-based attack) from the ILSVRC 2012 dataset. Samples are generated for different values of epsilon for the FGSM attack and for the PGD attack we experiment with different placements of the adversarial patch on the image.

A subset of the generated adversarial samples (100K samples) is considered as an input to the next step. Next, we pass the generated samples through the Mantra-Net that identifies perturbations and adds manipulation masks to the affected areas of the image. A new dataset consisting of the outputs from the previous step are the images with manipulation masks and the corresponding original images are created. Finally, a binary classifier is trained on the new dataset to differentiate between real and fake images.

As shown in Fig. 7, the testing pipeline consists of 2 stages. An image is passed through Mantra-Net to generate images with manipulation masks and then the trained binary classifier is used to predict whether the image is adversarially attacked.

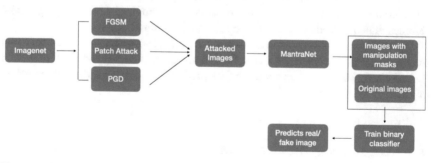

Fig. 6 Training pipeline

Fig. 7 Testing pipeline

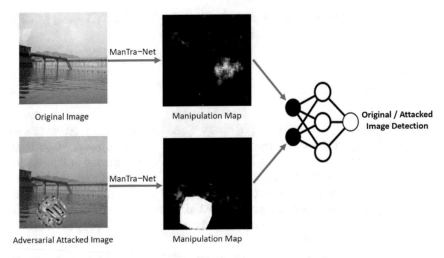

Fig. 8 Proposed pipeline

4.3 Creation of Adversarial Image Dataset

The first task was to build the training pipeline which includes collecting adversarial Samples. Since there was no benchmark dataset available incorporating multiple types of attacks like FGSM, PGD, Patch Attack, Augmentation using goggles/mask, etc., we collected the ILSVRC2012. Validation data and performed multiple types of attacks on the dataset. 50,000 samples consisting of adversarial perturbations were generated using FGSM (25,000) and PGD (25,000) attacks. We also incorporated around 10,000 samples of patch-based attacks into the dataset. The total 100K samples thus obtained comprise the input dataset to Mantra-Net.

Figure 8 shows our proposed pipeline. To differentiate between fake and real images, we generate the manipulation masks using Mantra-Net and input them to a trained binary classifier. The classifier then predicts whether the image was attacked or not.

4.4 Extraction of Manipulation Masks

Once we have the input dataset, the next step is to identify the manipulations in the image. We pass the dataset through Mantra-Net which detects the manipulation masks on the image, as shown in Fig. 9. Using the manipulation mask a binary classifier will be trained to identify whether the given image has artificial perturbations or not. We encountered problems with images of dimensions less than 150*150 px. For such images with small sizes, the pixels are very hazy and the change in pixel values is

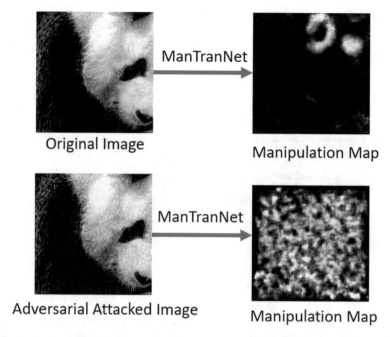

Fig. 9 Generation of manipulation mask

high, hence Mantra-Net wasn't able to detect the manipulation masks accurately in this case.

4.5 Training a Binary Classifier

Once we have collected a good amount of manipulation masks from the Mantra-Net, we segregated the masks of attacked and original images and used Tensorflow to train a binary classifier that takes input from the manipulation masks. The detailed architecture for our binary classification is described in Fig. 10.

The network takes in a 3 channel image of 224*224 pixels and uses MobileNetV2 as a feature extractor. MobileNetV2 gives a flatten output of (1280, 1) dimension which is then passed to a feed-forward layer for classifying the image into original and attacked.

In the input layer, the model uses the preprocessing functions of MobileNetV2 provided by Tensorflow. For the next sequential layer, MobileNetV2 CNN is used as a feature extractor. The final sequential layer for classification comprises a fully connected neural network using the sigmoid activation function as output.

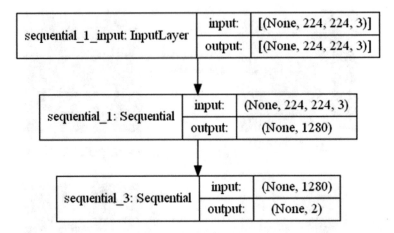

Fig. 10 Binary detector network

5 Results and Discussion

The ILSVRC dataset has 1000 different classes. The ILSVRC dataset is a subset of the ImageNet challenge on which all SOTA CNN architectures are benchmarked. We selected some SOTA architectures like VGG16, VGG19, MobileNetV1, MobileNetV2, ResNet50, and benchmark them on ILSVRC validation data.

We generated a dataset of all the attacked images which comprises 15,000 FGSM attacked images having Random Epsilon values from the list: [0.01, 0.1, 0.15, 0.3], 10,000 PGD attacked samples were generated from the original dataset, and 10,000 patches were generated which were of varying classes of the ImageNet challenge. The state-of-the-art models were then tested on the attacked data. Our findings are mentioned in Table 1, where we tried to highlight the fact that the models were unable to classify the adversarial images but performed well on the original validation data.

Figure 11 shows the different kinds of images generated using FGSM and PGD attacks from the test set and also their prediction on MobileNetV2. We can observe that the models were unable to correctly classify the images with minor perturbations.

Table 1 Top-1 accuracy for original and adversarial samples

Model	Original samples (%)	Adversarial samples (%)
VGG16	71.3	10.4
VGG19	73.2	11.6
MobileNetV1	70.2	9.8
MobileNetV2	70.6	9.7
ResNet50	74.6	13.4
InceptionNetV3	78.8	14.6

Fig. 11 Perturbed versus original images

The images appear to be very close to the original samples but they were able to fool the state-of-the-art architectures.

The entire attacked data was passed through Mantra-Net and the manipulation masks were stored irrespective of the type of attack. The binary classifier gave 98.6% accuracy on the test set which was trained on the manipulation mask. It was interesting to observe that the pipeline was able to identify images having low epsilon values like 0.01.

Figure 12 shows the variation of accuracy per epoch on the training as well as validation data. The model came to a point of equilibrium after 15 epochs. The model performs exceptionally well in the training as well as a testing phase. The model is pretty robust on different types of attacks and is also able to handle attacks that are specifically designed to fool face match algorithms. The confusion matrix as depicted in Fig. 13 elaborates the performance of the model.

We compared our results with the previous state-of-the-art technique to detect adversarial samples and concluded that the approach devised in this paper is able to capture attacked images with lower epsilon values with high accuracy. The comparison is mentioned in Table 2.

Fig. 12 Model accuracy
versus epoch graph

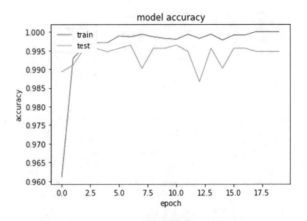

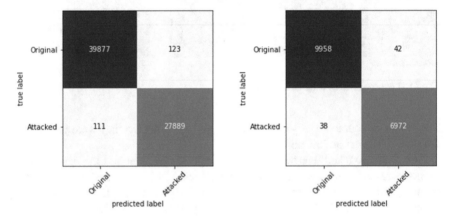

Fig. 13 i Training confusion matrix. **ii** Testing confusion matrix

Table 2 Comparison with SOTA

	Previous SOTA	Basic binary classifier	Binary classifier with Mantra-Net
Accuracy of detecting adversarial sample	99.9% for $\epsilon \geq 0.03$ 0.05% for $\epsilon < 0.03$	68%	**98.6% Irrespective of** ϵ

6 Conclusion and Future Work

In this new era of technological advancement, it becomes a necessity to look after the safety measures also. People are worrying that Artificial Intelligence (AI) may be destructive in the future as it has enormous potential to overcome human performance. One wrong prediction can lead to the loss of thousands of dollars. Imagine attackers using adversarial attacks to fool face recognition systems by impersonating someone's identity. In this paper, we introduce a simple method to detect spoofed images, irrespective of the attack method used. We obtained an accuracy [26] score of 0.986. To further improve results a dataset can be annotated to train Mantra-Net for the specific task of detecting adversarial samples. Morphological operations can be tried to avoid misclassifying raw images into an attacked one. Furthermore, our pipeline fails to identify a single-pixel attack. Though the pipeline has been evaluated on augmented images (Ex: addition of glasses), experiments need to be carried out on a benchmark dataset to further evaluate the pipeline. This model can be used as a front-line defense to any production model to bypass all the attacked images.

References

1. Ouyang W, Zeng X, Wang X, Qiu S, Luo P, Tian Y, Li H, Yang S, Wang Z, Li H et al (2016) Deepid-net: object detection with deformable part-based convolutional neural networks. IEEE Trans Pattern Anal Mach Intell 39(7):1320–1334
2. Diba A, Sharma V, Pazandeh A, Pirsiavash H, Van Gool L (2017) Weakly supervised cascaded convolutional networks. In: Proceedings of the IEEE conference on computer vision and pattern recognition, pp 914–922
3. Mrabti W, Baibai K, Bellach B, Haj Thami RO, Tairi H (2019) Human motion tracking: a comparative study. Procedia Comput Sci 148:145–153. The second international conference on intelligent computing in data sciences (ICDS2018)
4. Kalfaoglu ME, Kalkan S, Alatan AA (2020) Late temporal modeling in 3D CNN architectures with BERT for action recognition. CoRR, vol. abs/2008.01232
5. Bulat A, Kossaifi J, Tzimiropoulos G, Pantic M (2020) Toward fast and accurate human pose estimation via soft-gated skip connections. CoRR, vol. abs/2002.11098
6. Tao A, Sapra K, Catanzaro B (2020) Hierarchical multi-scale attention for semantic segmentation. arXiv preprint arXiv:2005.10821
7. Szegedy C, Zaremba W, Sutskever I, Bruna J, Erhan D, Goodfellow I, Fergus R (2014) Intriguing properties of neural networks. In: International conference on learning representations
8. Evtimov I, Eykholt K, Fernandes E, Kohno T, Li B, Prakash A, Rahmati A, Song D (2017) Robust physical-world attacks on machine learning models. 2(3):4. arXiv preprint arXiv:1707.08945
9. Liu Z, Qi Z, Torr PH (2020) Global texture enhancement for fake face detection in the wild. In: Proceedings of the IEEE/CVF conference on computer vision and pattern recognition (CVPR), June
10. Moosavi-Dezfooli S-M, Fawzi A, Frossard P (2016) Deepfool: a simple and accurate method to fool deep neural networks. In: Proceedings of the IEEE conference on computer vision and pattern recognition (CVPR), June
11. Goodfellow IJ, Shlens J, Szegedy C (2014) Explaining and harnessing adversarial examples. arXiv preprint arXiv:1412.6572
12. Madry A, Makelov A, Schmidt L, Tsipras D, Vladu A (2017) Towards deep learning models resistant to adversarial attacks. arXiv preprint arXiv:1706.06083
13. Buckman J, Roy A, Raffel C, Goodfellow I (2018) Thermometer encoding: one hot way to resist adversarial examples. 1(1):2–2
14. Brown TB, Mané D, Roy A, Abadi M, Gilmer J (2017) Adversarial patch. arXiv preprint arXiv:1712.09665
15. Papernot N, McDaniel P, Jha S, Fredrikson M, Celik ZB, Swami A (2016) The limitations of deep learning in adversarial settings. In: 2016 IEEE European symposium on security and privacy (EuroS P), pp 372–387

16. Sitawarin C, Bhagoji AN, Mosenia A, Chiang M, Mittal P (2018) Darts: deceiving autonomous cars with toxic signs. arXiv preprint arXiv:1802.06430
17. Grosse K, Manoharan P, Papernot N, Backes M, McDaniel P (2017) On the (statistical) detection of adversarial examples. arXiv preprint arXiv:1702.06280
18. Gong Z, Wang W, Ku W-S (2017) Adversarial and clean data are not twins. arXiv preprint arXiv:1704.04960
19. Metzen JH, Genewein T, Fischer V, Bischoff B (2017) On detecting adversarial perturbations. arXiv preprint arXiv:1702.04267
20. Kurakin A, Goodfellow IJ, Bengio S (2016) Adversarial machine learning at scale. CoRR, vol. abs/1611.01236
21. Xu W, Evans D, Qi Y (2017) Feature squeezing: detecting adversarial examples in deep neural networks. arXiv preprint arXiv:1704.01155
22. Szegedy C, Zaremba W, Sutskever I, Bruna J, Erhan D, Goodfellow I, Fergus R (2013) Intriguing properties of neural networks. arXiv preprint arXiv:1312.6199
23. Papernot N, McDaniel P, Goodfellow I, Jha S, Celik ZB, Swami A (2017) Practical black-box attacks against machine learning. In: Proceedings of the 2017 ACM on Asia conference on computer and communications security, pp 506–519
24. Kurakin A, Goodfellow I, Bengio S et al (2016) Adversarial examples in the physical world
25. Papernot N, McDaniel P, Wu X, Jha S, Swami A (2016) Distillation as a defense to adversarial perturbations against deep neural networks. In: 2016 IEEE symposium on security and privacy (SP). IEEE, pp 582–597
26. Hossin M, Sulaiman MdN (2015) A review on evaluation metrics for data classification evaluations. Int J Data Mining Knowl Manage Process 5(2):1
27. Athalye A, Engstrom L, Ilyas A, Kwok K (2018) Synthesizing robust adversarial examples. In: Dy J, Krause A (eds) Proceedings of the 35th international conference on machine learning, vol. 80 of proceedings of machine learning research, PMLR, 10–15 July 2018, pp 284–293
28. Wu Y, AbdAlmageed W, Natarajan P (2019) ManTra-Net: manipulation tracing network for detection and localization of image forgeries with anomalous features. In: Proceedings of the IEEE conference on computer vision and pattern recognition (CVPR), June 2019
29. Russakovsky O, Deng J, Su H, Krause J, Satheesh S, Ma S, Huang Z, Karpathy A, Khosla A, Bernstein M, Berg AC, Fei-Fei L (2015) ImageNet large scale visual recognition challenge. Int J Comput Vis (IJCV) 115(3):211–252

A Hybrid Approach Using Wavelet and 2D Convolutional Neural Network for Hyperspectral Image Classification

Apoorv Joshi, Roopa Golchha, Ram Nivas Giri, Rekh Ram Janghel, Himanshu Govil, and Saroj Kumar Pandey

1 Introduction

Hyperspectral imaging analyses a broad light spectrum and does not assign primary colors, i.e., red, green and blue to each pixel. In HSI one continuous spectrum is measured for each pixel ranging from near-infrared to visible wavelength range [1]. Each pixel in HSI can be seen as a high-dimensional vector whose elements correspond to the spectral reflectance at a particular wavelength [2]. In remote sensing, object information is obtained without making any contact with that object mainly from aircraft or satellites. HSI classification categorizes each hyperspectral pixel into a particular class (or label) based on its spectral characteristics. Urban development, land change detection, military applications, land cover analysis, crop detection and others are some applications of HSI [3]. The fact that HSIs carry both spectral and spatial information is a distinguishing property of HSI from other models.

Classification methods which are used earlier rely on spectral data, which is usually divided into two parts: feature classifiers and feature engineering. Feature engineering aim is to obtain different features or bands while reducing the HSI dimensions. In feature engineering, two typical strategies are extraction of features

A. Joshi (✉) · R. Golchha · R. N. Giri · R. R. Janghel
Department of Information Technology, NIT Raipur, Raipur, India
e-mail: apoorvjoshi23@gmail.com

R. R. Janghel
e-mail: rrjanghel.it@nitrr.ac.in

H. Govil
Department of Applied Geology, NIT Raipur, Raipur, India
e-mail: hgovil.geo@nitrr.ac.in

S. K. Pandey
Department of Computer Engineering and Applications, GLA University, Mathura, India
e-mail: saroj.pandey@gla.ac.in

© The Author(s), under exclusive license to Springer Nature Singapore Pte Ltd. 2023
D. S. Sisodia et al. (eds.), *Machine Intelligence Techniques for Data Analysis and Signal Processing*, Lecture Notes in Electrical Engineering 997,
https://doi.org/10.1007/978-981-99-0085-5_33

413

and selecting important features [4]. Feature extraction is used to lower the high-dimensional space data to low-dimensional and using this the categories may be differentiated more readily. Factor Analysis (FA), Principal Component Analysis (PCA) are examples of feature extraction approaches [5, 6]. Feature selection, on the other hand, maintains the spectral information of bands that gives the most information from the raw HSI while rejecting those that aren't useful for categorization [7]. Wavelet is a type of image compression. Wavelet transformations are divided into several groups. In this Wavelet CNN model, we apply the Wavelet Haar transformation. This type of transformation has the advantage of capturing both frequency and location information for data. In signal processing and analysis, wavelet analysis is widely employed. The wavelet analysis approach is referred to as a mathematical microscope, and it is a useful tool for zooming in on the sound, image and other elements [8–10].

2 Literature Survey

Chakraborty et al. [1], proposed a 2D CNN model for classification along with wavelet CNN. Datasets used are the benchmark dataset. Classification results obtaining on 30% samples for three datasets are for IP, OA is 99.86%, AA is 99.98% and Kappa is 99.84%, for UP, OA is 99.99%, AA is 99.98% and Kappa is 99.98% and for SA, OA is 100%, AA is 100% and Kappa is 100%.

Ghaderizadeh et al. [11], suggested a hybrid approach and the result of the classification using 100 samples from each class is OA is 99.07%, AA is 99.54% and Kappa is 98.96%.

Paoletti et al. [5], suggested a 3D CNN technique. The Classification Results from each class for the dataset is given as for IP, OA is 90.11%, AA is 96.12% and Kappa is 88.81%, for PU, OA is 96.83%, AA is 97.50% and Kappa is 95.83%.

Melgani et al. [2], evaluated the capabilities of SVM classifiers in hyperdimensional feature spaces. The IP dataset is used for evaluating the performance with Training samples used is 4754 and Testing Samples used is 4588 and the classification gives the result as OA is 87.10%.

Chen et al. [3], proposed a 3D CNN-based feature extraction model combined with regularization. The dataset used are IP, PU, Kennedy Space Center with accuracies as for IP, OA is 97.56%, AA is 99.23% and Kappa is 97.02%, for PU, OA is 99.54%, AA is 99.66% and Kappa is 99.41% and for KSC, OA is 96.31%, AA is 94.68% and Kappa is 95.90%.

Vaddi et al. [4], provides Feature Extraction using Probabilistic Principal Component Analysis and Gabor filtering and Classification is done using the CNN framework. The accuracy results obtained are as for IP, OA is 99.02%, AA is 99.17% and Kappa is 98.90%, for PU, OA is 99.94%, AA is 99.92%, Kappa is 99.01% and for SA, OA is 99.94%, AA is 99.08% and Kappa is 98.03%.

Yang et al. [12], proposed the model using 2D CNN, recurrent 2D CNN, 3D CNN and recurrent 3D CNN for HSI classification. The dataset used is Indian Pines and the classification results for all the approaches is given as, for 2D CNN, OA is 97.08%, AA is 96.37%, for 3D CNN, OA is 98.92%, AA is 97.31%, for Recurrent 2D CNN, OA is 99.19%, AA is 97.03% and for Recurrent 3D CNN, OA is 99.50% and AA is 99.42%.

Mohan et al. [6], proposed a hybrid 2D and 3D CNN classification technique. The classification results are obtained as follows for IP, OA is 99.80%, AA is 99.71% and Kappa as 99.75%, for PU, OA is 99.99%, AA is 99.98% and Kappa is 99.99% and for SA, OA is 100%, AA is 100% and Kappa is 100%.

Roy et al. [13], suggested a hybrid model as HybridSN which is 3D CNN followed by 2D CNN. The dataset and their classification results are given as for IP, OA is 99.75%, AA is 99.63% and Kappa is 99.71%, for PU, OA is 99.98%, AA is 99.97% and Kappa is 99.98%, for SA, OA is 100%, AA is 100% and Kappa is 100%.

Li et al. [14], proposes a hybrid model. Here, the IP dataset is used with the classification result as OA is 94.59% and AA is 93.00%.

Plaza et al. [15], uses SVM and Transductive Support Vector Machine (TSVM) for HSI classification. The dataset used is IP and the classification results obtained as OA is 88.55% and Kappa is 87.00%.

Ji et al. [16], suggested a 3D CNN model. It proposed a HSI method in which the relationship between pixels is expressed as a hypergraph structure. The classification results obtained are for IP, OA is 75.00%, for PU, OA is 82.00% and for SA, OA is 99.00%.

Fujieda et al. [17], proposed a Wavelet CNN model. The results obtained are for Kth-tips2-b, accuracy is 74.0% and for DTD, accuracy is 59.8%.

Okwuashi et al. [7], uses a DSVM for HSI classification. DSVM is implemented with four kernel functions and the accuracy is given as the mean of accuracies from all kernel functions. The dataset used is PU and IP and the accuracies obtained are 98.86% and 98.17%, respectively.

Hang et al. [18], proposed a cascaded RNN model to investigate the redundant information of HSI. The classification results contained are for IP, OA is 91.79%, AA is 95.94% and Kappa is 90.62% and for PU, OA is 90.30%, AA is 87.97% and Kappa is 86.26%.

Bandos et al. [19], suggested a LDA with 2D CNN. The dataset used here are IP, KSC and Botswana and the accuracies obtained for them is for IP, OA is 55.26% and for KSC, OA is 90.00% and for Botswana, OA is 95.00%.

Mou et al. [20], suggested a new RNN model for assessing HSI as sequential input and then used network reasoning to determine information categories. The dataset used are IP and PU and they provide the accuracies as OA is 88.63% and AA is 85.26% and Kappa is 73.66% and OA is 88.85%, AA is 86.33% and Kappa is 80.48% for PU.

3   Proposed Methodology

In Fig. 1, the architecture of our proposed model using a hybrid approach of 2D and wavelet CNN is described. The features extracted at each layer are concatenated channel by channel and fed into the hybrid CNN's dense classification layers for further classification and the detailed description of the model is explained below.

3.1   Dataset Description

The proposed model is implemented using benchmark dataset Salinas Scene dataset. The dataset has 512×217 spatial dimensions and 224 spectral bands in the 360–2500 nm wavelength range and has a total of 16 classes [1]. The samples are divided for the testing and training purpose in the ratio of 10:90 which is shown in Table 1.

3.2   Description of the Model

The input HSI patch is decomposed into four levels using the model architecture. The size of the input kernel is 3×3 with 1×1 padding used to reduce feature map size. The wavelet transformed features are channel-wise added. With 1×1 convolutions, projection shortcuts are used to keep the gradient from vanishing. The output is transferred to the fully connected layers with dropout neurons after an average pooling layer is implemented globally. The wavelet CNN model proposed here works by applying the wavelet transform on the Salinas scene Dataset. In this model, the Haar wavelet is used as the mother wavelet. The information provided in an image is divided using wavelet analysis into various detailed sub-signals. The approximation and sub-signal depicts the overall trend in pixel values, as well as three detailed sub-signals for vertical, horizontal and diagonal details. The signal is

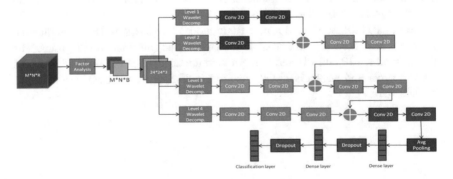

Fig. 1   Architecture of the 2D and wavelet CNN for HSI classification

Table 1 Training and testing samples for Salinas dataset

Class No	Class name	Training sample	Testing sample
1	Brocoli_green_weeds_1	201	1808
2	Brocoli_green_weeds_2	372	3354
3	Fallow	197	1779
4	Fallow_rough_plow	139	1255
5	Fallow_smooth	268	2410
6	Stubble	396	3563
7	Celery	358	3221
8	Grapes_untrained	1127	10,144
9	Soil_vinyard_develop	620	5583
10	Corn_senesced_green_weeds	328	2950
11	Lettuce_romaine_4wk	107	961
12	Lettuce_romaine_5wk	193	1734
13	Lettuce_romaine_6wk	91	825
14	Lettuce_romaine_7wk	107	963
15	Vinyard_untrained	727	6541
16	Vinyard_vertical_trellis	181	1626
Total samples		5412	48,717

processed by two filters: high pass and low pass filter. The signal is then divided into two parts, i.e., approximation part which comes when the signal passes through low frequency and detailed part which comes when the signal passes through high frequency. The low filter will produce a sub-signal with a high frequency equal to half of the original.

The HSI cube given as input is of dimension $512 \times 217 \times 204$ is first given to the layer of Factor Analysis (FA) to reduce the dimension into $512 \times 217 \times 3$. By applying FA, i.e., by reducing the dimension it is seen that training time decreases generally by 60%. FA preserves the spectral dimensions, i.e., 512×217, only the bands are reduced from 204 to 3. Using FA as a pre-processing step in HSI is very beneficial because using FA the variability among the different correlated and overlapping spectrum bands can be described, thereby making the model capable of classifying similar examples in a better way [1]. The window size for patch extraction for the Salinas Scene dataset is 24×24. The class category of the central pixel determines the truth values for these patches. A stride of 2 was used to replace pooling layers in between convolutions. Before transmitting into the dense layer, a global mean pooling was used at the conclusion of all the convolution layers to prevent the model from overfitting. For more efficient use of the wavelet modified data, dense connections and projection shortcuts were used. Dense connections and channel-wise concatenation of deconstructed data ensure that all features flow to the model's conclusion. As the number of samples in HSI is very small, there is a substantial risk of overfitting in order to avoid this, the model looked at two dropout layers as well as batch normalization.

4 Experimental Result and Discussion

4.1 Performance Measurement Criteria

Three performance metrics named OA, AA and Cohen Kappa are considered to measure the model performance [21–23].

Overall Accuracy (OA): It tells us what percentage of all the reference sites were accurately mapped. The overall accuracy is expressed as

$$OA = \frac{\text{Area correctly labelled}}{\text{Total number of labels}} \tag{1}$$

Average Accuracy (AA): It is calculated as the mean of the Overall Accuracy measured over each category and given as

$$AA = \frac{\text{Overall accuracy of each class}}{\text{Total number of classes}} \tag{2}$$

Cohen Kappa: It examines the level of agreement between two raters who categorize N objects into C mutually exclusive classes.

$$K = \frac{Po - Pe}{1 - Pe} \tag{3}$$

where Po is relative observation and Pe is the hypothetical chance agreement probability. Values of K define the model as follows, i.e., when K equals to 1 it signifies full agreement and $K = 0$ shows random agreement whereas $K < 0$ shows there is no effective agreement.

4.2 Experimental Results

Classwise accuracies are also calculated for Salinas dataset using our proposed hybrid model with both Sigmoid and Softmax as the activation function and the results are shown in Table 2. In Fig. 2, (a) the Groundtruth image and (b) Classified image of the Salinas valley using Wavelet 2D CNN is shown.

In Table 3, the results are compared to the results obtained using state-of-the-art techniques [24, 25]. The findings are examined for a set of 10:90 random train test

Table 2 Detailed classification report on Salinas scene dataset

Class No	Performance using Wavelet 2D CNN	
	Wavelet 2D CNN + Softmax	Wavelet 2D CNN + Sigmoid
1	100	100
2	100	100
3	100	100
4	99.92	100
5	99.91	99.62
6	100	99.97
7	99.93	99.59
8	99.83	98.66
9	100	100
10	99.96	99.96
11	99.79	100
12	100	100
13	100	100
14	99.89	99.89
15	100	100
16	100	100
Average accuracy (%)	99.88	99.85
Overall accuracy (%)	99.87	99.67
Cohen kappa score (%)	99.85	99.63

split in SA dataset. The suggested model outperformed all state-of-the-art models as evidenced by the findings.

5 Conclusion

In this paper, we came up with a Wavelet 2D CNN model for the HSI classification of images using the Salinas Scene dataset. This hybrid approach takes both spectral and spatial information present in a high-dimensional HSI cube. The results from our model when compared with state-of-the-art approaches illustrate that our proposed hybrid model performs better. While implementing our work we found that this work can also be extended by experimenting and implementing it by using various other activation methods and also using other datasets with more training samples.

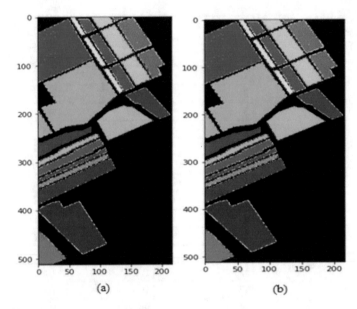

Fig. 2 **a** Groundtruth image and **b** classified image of Salinas scene dataset using Wavelet 2D CNN

Table 3 Comparison with the state-of-the-art approaches

	Hybrid deep ResNet and inception model [24]	Semi-supervised 3D deep neural network [25]	Wavelet CNN + Softmax	Wavelet CNN + Sigmoid
AA (%)	97.69	98.28	99.88	99.85
OA (%)	94.11	98.29	99.87	99.67
Kappa (%)	–	98.16	99.85	99.63

References

1. Chakraborty T, Trehan U (2021) SpectralNET: exploring spatial-spectral WaveletCNN for hyperspectral image classification. IEEE Trans Geosci Rem Sens 1(7)
2. Melgani F, Bruzzone L (2004) Classification of hyperspectral remote sensing images with support vector machines. IEEE Trans Geosci Rem Sens 42(8):1778–1790. https://doi.org/10.1109/TGRS.2004.831865
3. Chen Y, Jiang H, Li C, Jia X, Ghamisi P (2016) Deep feature extraction and classification of hyperspectral images based on convolutional neural networks. IEEE Trans Geosci Rem Sens 54(10):6232–6251. https://doi.org/10.1109/TGRS.2016.2584107
4. Vaddi R, Manoharan P (2020) Hyperspectral image classification using CNN with spectral and spatial features integration. Infrared Phys Technol 107. https://doi.org/10.1016/j.infrared.2020.103296
5. Paoletti ME, Haut JM, Plaza J, Plaza A (2018) A new deep convolutional neural network for fast hyperspectral image classification. ISPRS J Photogramm Rem Sens 145:120–147. https://doi.org/10.1016/j.isprsjprs.2017.11.021

6. Mohan A, Venkatesan M (2020) HybridCNN based hyperspectral image classification using multiscale spatiospectral features. Infrared Phys Technol 108. https://doi.org/10.1016/j.inf rared.2020.103326
7. Okwuashi O, Ndehedehe CE (2020) Deep support vector machine for hyperspectral image classification. Pattern Recogn 103. https://doi.org/10.1016/j.patcog.2020.107298
8. Aria SEH, Menenti M, Gorte BGH (2017) Spectral region identification versus individual channel selection in supervised dimensionality reduction of hyperspectral image data. J Appl Rem Sens 11(04):1–1. https://doi.org/10.1117/1.jrs.11.046010
9. Jia S, Shen L, Zhu J, Li Q (2018) A 3-D Gabor phase-based coding and matching framework for hyperspectral imagery classification. IEEE Trans Cybern 48(4):1176–1188
10. Hong D, Wu X, Ghamisi P, Chanussot J, Yokoya N, Zhu XX (2020) Invariant attribute profiles: a spatial-frequency joint feature extractor for hyperspectral image classification. IEEE Trans Geosci Rem Sens 58(6):3791–3808
11. Ghaderizadeh S, Abbasi-Moghadam D, Sharifi A, Zhao N, Tariq A (2021) Hyperspectral image classification using a hybrid 3D–2D convolutional neural networks. IEEE J Sel Topics Appl Earth Observ Rem Sens 14(19):7570–7588. https://doi.org/10.1109/JSTARS.2021.3099118
12. Yang X, Ye Y, Li X, Lau RYK, Zhang X, Huang X (2018) Hyperspectral image classification with deep learning models. IEEE Trans Geosci Rem Sens 56(9):5408–5423. https://doi.org/10.1109/TGRS.2018.2815613
13. Roy SK, Krishna G, Dubey SR, Chaudhuri BB (2020) HybridSN: exploring 3-D-2-D CNN feature hierarchy for hyperspectral image classification. IEEE Geosci Rem Sens Lett 17(2):277–281. https://doi.org/10.1109/LGRS.2019.2918719
14. Li J et al (2015) Multiple feature learning for hyperspectral image classification. IEEE Trans Geosci Rem Sens 53(3):1592–1606. https://doi.org/10.1109/TGRS.2014.2345739
15. Plaza A et al (2009) Recent advances in techniques for hyperspectral image processing. Rem Sens Environ 113(Suppl. 1). https://doi.org/10.1016/j.rse.2007.07.028
16. Ji R, Gao Y, Hong R, Liu Q, Tao D, Li X (2014) Spectral-spatial constraint hyperspectral image classification. IEEE Trans Geosci Rem Sens 52(3):1811–1824. https://doi.org/10.1109/TGRS.2013.2255297
17. Fujieda S, Takayama K, Hachisuka T (2018) Wavelet convolutional neural networks. Proc IEEE Trans Geosci Rem Sens
18. Hang R, Liu Q, Hong D, Ghamisi P (2019) Cascaded recurrent neural networks for hyperspectral image classification. IEEE Trans Geosci Rem Sens 8:5384–5394. https://doi.org/10.1109/TGRS.2019.2899129
19. Bandos TV, Bruzzone L, Camps-Valls G (2009) Classification of hyperspectral images with regularized linear discriminant analysis. IEEE Trans Geosci Rem Sens 47(3):862–873. https://doi.org/10.1109/TGRS.2008.2005729
20. Mou L, Ghamisi P, Zhu XX (2017) Deep recurrent neural networks for hyperspectral image classification. IEEE Trans Geosci Rem Sens 55(7):3639–3655. https://doi.org/10.1109/TGRS.2016.2636241
21. Roy SK, Krishna G, Ram Dubey S, Chaudhuri BB Supplementary Material Paper Title: 'HybridSN: exploring 3D-2D CNN feature hierarchy for hyperspectral image classification'
22. Xia J, Bombrun L, Berthoumieu Y, Germain C, Du P (2016) Spectral spatial rotation forest for hyperspectral image classification. In: Proceedings of IEEE international symposium on geoscience and remote sensing, pp 5126–5129
23. Imani M, Ghassemian H (2020) An overview on spectral and spatial information fusion for hyperspectral image classification: current trends and challenges. Inform Fusion 59:59–83. https://doi.org/10.1016/j.inffus.2020.01.007
24. Alotaibi B, Alotaibi M (2020) A hybrid deep ResNet and inception model for hyperspectral image classification. PFG 88:463–476. https://doi.org/10.1007/s41064-020-00124-x
25. Sellami A, Farah M, Farah IR, Solaiman B (2019) Hyperspectral imagery classification based on semi-supervised 3-D deep neural network and adaptive band selection. Exp Syst Appl.https://doi.org/10.1016/j.eswa.2019.04.006

26. Gogineni R, Chaturvedi A (2020) Hyperspectral image classification in processing and analysis of hyperspectral data. J Appl Rem Sens. IntechOpen
27. Liu J-W, Zuo L, Guo Y-X, Li T-Y, Chen J-M (2015) Research on improved wavelet convolutional wavelet neural networks. https://doi.org/10.1007/s10489-020-02015-5/Published
28. Zhang L, Zhang L, Du B (2016) Deep learning for remote sensing data: a technical tutorial on the state of the art. IEEE Geosci Rem Sens Mag 4(2):22–40
29. Cao X, Zhou F, Xu L, Meng D, Xu Z, Paisley J (2018) Hyperspectral image classification with Markov random fields and a convolutional neural network. IEEE Trans Image Process 27(5):2354–2367
30. Hao S, Wang W, Ye Y, Li E, Bruzzone L (2018) A deep network architecture for super-resolution-aided hyperspectral image classification with classwise loss. IEEE Trans Geosci Rem Sens 56(8):4650–4663
31. Li X et al (2020) A wavelet transform-assisted convolutional neural network multi-model framework for monitoring large-scale fluorochemical engineering processes. Processes 8(11):1–17. https://doi.org/10.3390/pr8111480
32. Roy SK, Krishna G, Dubey SR, Chaudhuri BB (2020) HybridSN: exploring 3D–2D CNN feature hierarchy for hyperspectral image classification. IEEE Trans Geosci Rem Sens 7(2):277–281
33. Bera S, Shrivastava VK (2020) Analysis of various optimizers on deep convolutional neural network model in the application of hyperspectral remote sensing image classification. Int J Rem Sens 41(7):2664–2683

Assessment of Visual Stress Using EEG Signals

Durlov Jyoti Buragohain, Jupitara Hazarika, and K. Shankar

1 Introduction

Emotional and physical tension gives rise to stress. It can emerge from anger, frustration, or nervousness. In fact, the human body is designed to feel and respond to stress. A close relationship exists between stress and mental health [1]. Psychological stress may result in anxiety, anger, and depression. Moderate stress can be good, as it helps to avoid danger or meet deadlines. Positive stress can keep us alert and motivated. But when it persists for long, it results in chronic stress and can be detrimental to a person's health. Some physical symptoms of stress include increased blood pressure, heightened muscle preparedness, increased sweating, reduced alertness, dizziness, digestion problems, weak immune system, etc. [2].

Assessment and monitoring of stress is challenging since its effect varies from person to person. Psychometric and medical tests are helpful in assessing stress. Since biomedical signals like electroencephalogram (EEG) are widely used in clinical diagnosis of mental health and in bioengineering research, it can provide real-time information in stress assessment [2]. Electroencephalogram (EEG) is the recording of potential from the scalp of the brain. Compared to other brain imaging techniques like MRI, etc., EEG is fast, cheap, non-stationary, has high precision and high temporal resolution [3].

EEG-based assessment has been used in many applications including psychophysiology, psychology, video games, etc. [2]. In such applications, the frequency pattern of EEG plays a vital role. Mostly 1–100 Hz frequency content of EEG is considered to be of interest and carries the brain activation-related information. Sub-division of this frequency range results in five significant frequency sub-bands, namely, delta (1–4 Hz), theta (4–8 Hz), alpha (8–13 Hz), beta (13–30 Hz), and gamma (30–100 Hz)

D. J. Buragohain · J. Hazarika (✉) · K. Shankar
Department of Electronics and Instrumentation Engineering, National Institute of Technology Silchar, Silchar, Assam 788010, India
e-mail: jupitara@ei.nits.ac.in

[4]. Low-frequency delta and theta waves are observed in deep sleep and represent drowsiness [5], whereas high-frequency beta is considered to be a marker of attention [6]. The alpha activity is found to be prominent with eyes open, relaxed wakefulness [5], and the gamma frequency band correlates cognitive task execution [7].

Considering the literature, the present study has considered alpha and beta frequency bands of EEG to examine. This paper aims to develop an algorithm to correlate the stress-induced activation pattern with these two frequency bands.

This paper is structured as follows: Sect. 2 includes the methodology describing the considered and proposed algorithms. Section 5 presents the results of processing and analyses. Finally, conclusions are given in Sect. 6.

2 Methodology

As a low-intensity signal, EEG is prone to different noises and artefacts like ECG, EMG, etc., and extraction of clean EEG is a challenge. Also, for getting information from EEG, we need to work on perfect feature extraction and selection. Here statistical tools can be handy for this purpose. As shown in Fig. 1, the raw EEG data, first being pre-processed by using a high pass filter of 0.15 Hz so that low-frequency noise gets eliminated. The EEG signal that we have has maximum frequency content of 28 Hz, as it is first resampled at 64 Hz to reduce the size. After that, on the application of two sets of band-pass filters (8–13) Hz and (13–28) Hz, the alpha and beta frequency bands (brain waves) of EEG are extracted.

2.1 Data Set

The data set is collected from an open-source database available in physinet.org [12]. It contains EEG data of 11 healthy participants. Stress is induced in the participants by rapidly presenting images at 5, 6, and 10 Hz presentation rates with three difficulty levels of Rapid Serial Visual Presentation (RSVP) task. A total of 8-channel EEG data was collected mainly from the parietal and occipital lobe of the brain namely PO8, PO7, PO3, PO4, P7, P8, O1, and O2. These signals are being sampled at 2048 Hz at

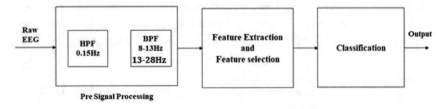

Fig. 1 Block diagram for signal analysis

the time of acquisition, hence re-sampled at 64 Hz to reduce the size. Hence, each EEG signal has 0.15–28 Hz. To our interest, we have used data taken at 5 and 10 Hz presentation rates from RSVP task.

2.2 Feature Extraction

Different statistical features are being calculated like mean, mean frequency, energy, band power, relative band power, skewness, interquartile range, mean absolute deviation, the median absolute deviation for both alpha and beta frequency bands. The mathematical formulae are as follows:

Mean. Mean of the signal, x gives the average of all the samples as

$$\text{Mean}(\mu) = \frac{\sum_{n=1}^{M} x_n}{M} \tag{1}$$

where M is the length of the signal.

Energy. The energy of any signal, x is defined mathematically as,

$$\text{Energy} = \sum_{n=1}^{M} |x_n|^2 \tag{2}$$

Band power. The power of signal, x is the sum of the absolute squares of its time-domain samples divided by the signal length and is defined mathematically as,

$$\text{Power} = \frac{\sum_{n=1}^{M} |x_n|^2}{M} \tag{3}$$

Mean frequency. This is calculated by the sum of the product of the spectrogram intensity (in dB) and the frequency divided by the total sum of the spectrogram intensity.

Mathematically,

$$f_{\text{mean}} = \frac{\sum_{n=1}^{m} I_n f_n}{\sum_{n=1}^{m} I_n} \tag{4}$$

where

f_{mean} mean frequency.
m total numbers of frequency terms present in the spectrum.
f_n frequency of nth term.
I_n nth spectrogram intensity (dB).

Relative band power. Relative band power (RBP) is calculated as

$$RBP = \frac{\sum x_\alpha[n]}{\sum x[n]} \tag{5}$$

where

$\sum x_\alpha[n]$ sum of individual band power
$\sum x[n]$ sum of total band power.

Skewness. Skewness is the measure of asymmetry in a probability distribution. Skewness is negative if the distribution is towards left and positive while towards right.

$$S = \frac{E(x - \mu)^3}{\sigma^3} \tag{6}$$

where

μ mean of the signal.
σ standard deviation of the signal.

Interquartile range. The interquartile range is one of the methods to measure dispersion in a given data set. It is the difference between the upper quartile Q_3 and lower quartile Q_1 from a given data set.

$$QR = Q_3 - Q_1 \tag{7}$$

Mean absolute deviation. Mean absolute deviation calculates the variability of the data set, i.e. on average how is the data point from the mean.

$$\text{Mean absolute deviation} = \frac{\sum_{i=1}^{N} |x_i - \mu|}{N} \tag{8}$$

where

N length of the signal
μ mean of the signal.

Median absolute deviation. Median absolute deviation calculates how far the sample from the central tendency (median).

$$\text{Median absolute deviation} = \text{median}|x_i - \text{median}| \tag{9}$$

3 Hypothesis Testing

After feature extraction, hypothesis testing is used to determine the features which can present significant statistical difference between brain activation patterns related to 5 and 10 Hz visual representation. The statistical tool Wilcoxon signed-rank test is used for the purpose of feature reduction. It is a nonparametric statistical test used to check for a difference in the mean (or median) of two set of observations [8].

Considering a 5% significance level, all the extracted features are tested. Features of EEG alpha and beta frequency bands which satisfies the criteria of 'significant difference' ($p < 0.05$) are then further used in machine learning classification algorithms.

4 Classification

In this block, we tried to build a model that predicts different stress levels associated with RSVP task using machine learning algorithms. In this project, we are using two classifiers, namely Logistic Regression and Linear SVM classifier. In literature, multiple EEG studies have successfully used these two algorithms for EEG analysis [9, 10].

Logistic Regression is a supervised learning classification algorithm that uses probability to predict the target variable. A linear support vector machine (SVM) classifier is also a supervised learning classification algorithm that uses optimum hyperplane for classification.

Firstly, the data set is being prepared with both the classes along with the class label before feeding into the classification model. Tenfold cross-validation is used to check model accuracy, where approximately 30% of data used as testing data set and 70% as training data set. For a better understanding of the performance of the model, F1-scores are also calculated.

5 Results and Discussions

The extracted features from the alpha frequency band during the 5 and 10 Hz RSVP task are tested first using the statistical tool and obtained results are displayed here in Table 1. Similar experimentation is performed with the beta band features derived from the 5 and 10 Hz RSVP task. In Table 1, results are displayed.

It is seen from the results of hypothesis testing (shown in Table 1) that most of the EEG feature obtained from the alpha frequency band, except mean frequency, shows significant statistical difference, i.e. p-value is less than 0.05. Whereas a few features like mean, energy, band power, and skewness of beta frequency band reveals significant statistical difference.

Table 1 Result of Wilcoxon signed-rank test (5 Hz vs 10 Hz)

Features	*p*-value	
	Alpha (8–13) Hz	Beta (13–28) Hz
Mean	**0.0488**	**9.2453 $\times$ 10^{-4}**
Energy	**1.1040 $\times$ 10^{-21}**	**1.3805 $\times$ 10^{-15}**
Band power	**5.6285 $\times$ 10^{-11}**	**0.0196**
Relative band power	**5.0435 $\times$ 10^{-6}**	0.0947
Mean frequency	0.9079	0.1539
Skewness	**0.0058**	**0.0053**
IQR	**4.5429 $\times$ 10^{-11}**	0.4216
Mean absolute deviation	**1.3435 $\times$ 10^{-12}**	0.0878
Median absolute deviation	**4.6043 $\times$ 10^{-11}**	0.4297

In the field of machine learning algorithms, confusion matrix or error matrix is a specific table structure that demonstrates performance of an algorithm. This matrix (Fig. 2) gives a proper understanding of the model behaviour.

Performance of a machine learning algorithm is often characterized by classification accuracy. This parameter narrates the performance of a dividing model as the number of correct predictions divided by the total number of predictions. We can calculate accuracy as,

$$\text{Accuracy} = \frac{\text{TP} + \text{TN}}{\text{TP} + \text{TN} + \text{FP} + \text{FN}} \times 100\% \qquad (10)$$

Another parameter F-score which is the harmonic mean can be calculated as,

$$\text{F1-Score} = 2 \times \frac{\text{Precision} \times \text{Recall}}{\text{Precision} + \text{Recall}} \qquad (11)$$

Fig. 2 Confusion matrix

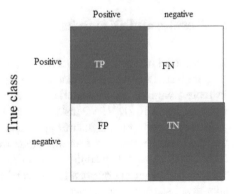

where

$$Recall = \frac{TP}{TP + FN}$$

$$Precision = \frac{TP}{TP + FP}$$

It is clear from the Wilcoxon signed-rank test which features are to be selected for the machine learning algorithms. Statistical features including mean, energy, band power and skewness are considered for further processing since they demonstrate significant difference in both alpha and beta frequency ranges. Accuracies along with F1-scores for different combination of features are given in Tables 2 and 3 for logistic regression algorithm and for SVM, respectively.

Results shown in Tables 2 and 3 demonstrates that the statistical features mean, energy, band power and skewness are competent in producing better classification accuracies with both the machine learning algorithms, i.e. Lite logistic regression and linear SVM. Although classification accuracies seem to get improved without skewness parameter. The best result is given by the logistic regression model with the features energy and band power of beta EEG frequency band.

Table 2 Classification results using logistic regression

Features	Accuracy		F1-score	
	Alpha (8–13) Hz (%)	Beta (13–28) Hz (%)	Alpha (8–13) Hz	Beta (13–28) Hz
Mean, energy, band power, skewness	59.4	95.1	0.4935	0.9492
Mean, energy, band power	95.1	99.7	0.9488	0.9964
Energy, band power	84.4	100	0.8525	1

Table 3 Classification results using Linear SVM

Features	Accuracy		F1-score	
	Alpha (8–13) Hz (%)	Beta (13–28) Hz (%)	Alpha (8–13) Hz	Beta (13–28) Hz
Mean, energy, band power, skewness	77.1	88.5	0.8035	0.8911
Mean, energy, band power	94.8	94.4	0.9469	0.9411
Energyandband power	95.8	97.6	0.9565	0.9751

Again, on comparing the results of alpha band and beta band, it is seen that a significant difference exists when visual images are presented to the subjects at 5 and 10 Hz frequency. This may be because visual attention particularly effects the activation pattern of EEG in alpha band [11] and beta band [6]. By increasing the speed of visual representation, a stress has been imposed on the brain for faster information processing which in-turn induced EEG activity. Compared to alpha, better accuracy has been observed in the beta band.

6 Conclusion

The combination of energy and band power features have enabled the classifiers to detect and classify the visual attention related task induced stress. The stress features in the subject group is confirmed by the Wilcoxon signed-rank test. Feature selection improves the performance of the classifier. If the features are reliable, classifier will produce high classification accuracy. Hence, it is important to ensure that only the appropriate features are selected for the classifier. It is clear from the results that both the classifier (Logistic Regression and support vector machine) is almost equally effective in classifying different stress levels using selected signal features. Farther, high classification accuracy confirms rise in stress with the increased task levels.

This study examined a simple and robust way for feature selection and stress detection. The process incorporates superior computational efficiency. Future work may include development of a real-time monitoring system to help medical practitioners to detect the cause of various modern-day problems like anxiety and stress.

References

1. Seo SH, Lee JT (2010) Stress and EEG. In: Convergence and hybrid information technologies, vol 27
2. Hou X, Liu Y, Sourina O, Tan YRE, Wang L, Mueller-Wittig W (2015) EEG based stress monitoring. In: 2015 IEEE international conference on systems, man, and cybernetics. IEEE, pp 3110–3115
3. Barlow JS (1993) The electroencephalogram: its patterns and origins. MIT Press
4. Crone NE, Korzeniewska A, Franaszczuk PJ (2011) Cortical gamma responses: searching high and low. Int J Psychophysiol 79(1):9–15
5. Louis EKS, Frey LC, Britton JW, Hopp JL, Korb P, Koubeissi MZ, Lievens WE, Pestana-Knight EM (2016) The normal EEG. In: Electroencephalography (EEG): an introductory text and atlas of normal and abnormal findings in adults, children, and infants
6. Wróbel A (2000) Beta activity: a carrier for visual attention. Acta Neurobiol Exp 60(2):247–260
7. Fitzgibbon SP, Pope KJ, Mackenzie L, Clark CR, Willoughby JO (2004) Cognitive tasks augment gamma EEG power. Clin Neurophysiol 115(8):1802–1809
8. Woolson RF (2007) Wilcoxon signed-rank test. In: Wiley encyclopedia of clinical trials, pp 1–3
9. Arora A, Lin JJ, Gasperian A, Maldjian J, Stein J, Kahana M, Lega B (2018) Comparison of logistic regression, support vector machines, and deep learning classifiers for

predicting memory encoding success using human intracranial EEG recordings. J Neural Eng 15(6):066028

10. Isa NM, Amir A, Ilyas MZ, Razalli MS (2019) Motor imagery classification in Brain computer interface (BCI) based on EEG signal by using machine learning technique. Bull Electr Eng Inform 8(1):269–275

11. Sauseng P, Klimesch W, Stadler W, Schabus M, Doppelmayr M, Hanslmayr S, Gruber WR, Birbaumer N (2005) A shift of visual spatial attention is selectively associated with human EEG alpha activity. Eur J Neurosci 22(11):2917–2926

12. https://physionet.org/content/ltrsvp/1.0.0/. Last accessed 2022/01/12

A Novel Recommendation System for Vaccines Using Hybrid Machine Learning Model

Nishant Singh Hada, Sreenu Maloth, Chandrashekar Jatoth, Ugo Fiore, Sangeeta Sharma, Subrahmanyam Chatharasupalli, and Rajkumar Buyya

1 Introduction

Advancement in medical science over the centuries has gifted humanity with a crucial boon. Vaccines are one such offering that has brought a much-needed change in the medical field. Both non-infectious and infectious diseases are now within the realm of vaccinology [1]. The journey from the discovery of the first vaccine by Edward Jenner in 1796 [2] to the continuous work being done by medical researchers over the continents on COVID-19 vaccines is remarkable. The virus disturbed the world economies and impacted millions of people. Luckily, with constant development, trials and testing by the R&D departments, several vaccines are in use currently to

N. S. Hada · S. Maloth · S. Sharma
National Institute of Technology Hamirpur, Hamirpur, HP 177005, India
e-mail: hadanis.singh@gmail.com

S. Maloth
e-mail: sreenu@nith.ac.in

S. Sharma
e-mail: sangeetas@nith.ac.in

C. Jatoth (✉)
National Institute of Technology Raipur, Raipur, CG 492010, India
e-mail: chandrashekar.jatoth@gmail.com

U. Fiore
Federico II University of Naples, Naples, Italy
e-mail: ugo.fiore@unina.it

S. Chatharasupalli
Union Public Service Commission, New Delhi, India
e-mail: subrahmanyamch.1981@gov.in

R. Buyya
The University of Melbourne, Parkville VIC 3010, Australia
e-mail: rbuyya@unimelb.edu.au

© The Author(s), under exclusive license to Springer Nature Singapore Pte Ltd. 2023
D. S. Sisodia et al. (eds.), *Machine Intelligence Techniques for Data Analysis and Signal Processing*, Lecture Notes in Electrical Engineering 997,
https://doi.org/10.1007/978-981-99-0085-5_35

protect us from COVID-19. The first mass vaccination programme started in early December 2020 [3]. WHO issued an Emergency Use Listing for the Pfizer COVID-19 vaccine (BNT162b2), two versions of the AstraZeneca/Oxford COVID-19 vaccine and the vaccine Ad26.COV2.S, developed by Janssen (Johnson & Johnson) [3].

Not just COVID-19, but for other diseases and outbreaks there have been multiple vaccines from different manufacturers [16]. The efficacy rate of vaccines varies from region to region and from person to person because of substantial variation in how an individuals' immune response to the vaccine. The study conducted by Zimmerman et al. [4] investigates and talks about various factors that influence humoral and cellular vaccine responses in humans. Some of these factors are age, sex, gestational age, extrinsic factors like medical history, allergies, pre-existing immunity, infections and antibiotics. There are the vaccine factors as well such as vaccine type, product, adjuvant and dose that governs how effective a vaccine will be on the host. Studying all these factors can help medical officers in suggesting the right manufacturer of vaccine that shows the highest efficacy based on the previous records of patients having similar factors to the current host.

Over the last decade, we saw various recommendation systems for drugs and medicines based on different algorithms and techniques [5–11]. The common goal behind the development of these systems is to build a platform that helps the medical officers in decision-making. Considering all the factors quantitatively can be a tedious process which these systems can perform swiftly and help the doctors in making more concise decisions. This field of intelligent recommendation systems is unexplored for vaccines. In our paper, we propose an unprecedented recommendation system for vaccines that produce decisions after considering host-based and vaccine-based factors. This is a new step towards helping doctors in suggesting the right vaccine for the patient.

In our system, we consider data points like age, sex, medical history, allergies of the host and recovery rate, death rate, after vaccination symptoms of the vaccine to make a recommendation. The system stands upon a score-based algorithm that utilizes machine learning to recommend the right vaccine.

The rest of the paper is organized as follows. Section 2 contains the background literature of previously developed recommendation systems for drugs. We brief about the preliminaries in Sect. 3. Section 4 gives an overview of the recommendation system is provided. Section 5 displays the varying experimental results based on changes made in the input. In Sect. 6, the paper is concluded by mentioning the future scope of improvement in the work.

2   Related Work

In the literature, we have various drug recommendation systems that with the help of leading algorithms and techniques like machine learning consider all the factors quantitatively and suggest the right drug. Stark et al. [5] presented an approach for a drug recommendation system using Neo4J, a graph database with high scalability.

The system considered the individual features of the patient and assigned scores to the drugs. The drugs with the highest relevance scores were recommended. This would help physicians to know which drug fits the patient best based on related factors. In 2019, Stark et al. described and compared various recommendation systems based on various features [7]. Hossain et al. in 2020 implemented a drug recommender system that applied sentiment analysis on drug reviews to generate ratings on drugs. They carried out the rating generation using decision tree, K-nearest neighbours and linear support vector classifier algorithm. Finally, linear support vector classifier was selected for rating generation to obtain a good trade-off among model accuracy, efficiency and scalability and the hybrid model for recommendation [6]. Garg also proposed recommendation system based on sentiment analysis [10].

Bao and Jiang in 2016 implemented a recommendation system for medicines using data mining models like ID3 decision tree algorithm, BP neural network and SVM. Finally, the SVM model was selected to obtain a good trade-off. They also proposed a mechanism to ensure the diagnosis accuracy and service quality and showed that their results had excellent accuracy [8]. Abbas et al. in 2020 proposed a drug supply chain management and recommendation system based on blockchain and machine learning. The N-gram and LightGBM models were used to recommend the medicines. These models were trained on the publicly available drug reviews dataset provided by the UCI [9]. Yong et al. in 2020 developed a system based on blockchain and machine learning to support vaccine traceability and smart contract functions [15]. Chen et al. in 2018 proposed a diagnosis system for diseases and also a system to recommend treatment. They introduced density-peaked clustering analysis algorithm to cluster the disease symptoms and perform association analyses based on D-D and D-T rules. They achieved high performance and low latency using the parallel solution provided by Apache Spark [11].

3 Proposed System

In Sect. 2, we discussed various recommendation system for drugs. In this paper, we introduce a novel recommendation system for vaccines that considers age, sex, allergies and medical history to recommend a vaccine to a subject.

3.1 Dataset Collection

In our system, we use the VAERS dataset that is a national early warning system to detect possible safety problems in U.S. licensed vaccines. Healthcare professionals and vaccine manufacturers are required to report adverse events that come to their attention. We use those adverse effects to recommend the best possible vaccine for a patient depending on his age, sex, allergies and medical history. We get a tremendous amount of information from the dataset and have to preprocess it to extract the

Table 1 Selected attributes for the system

Attribute name	Attribute meaning
vaers_id	Unique ID
Recvdate	Received date
age_yrs	Age of patient
Sex	Sex of patient
History	Medical history of the patient
Allergies	Allergies of the patient
symptom_text	Post-vaccination symptoms
Died	Post-vaccination death
Recovd	Post-vaccination recovery
vax_type	Disease type
vax_manu	Vaccine variant

important information for our system. The next section explains the preprocessing of data.

3.2 Data Preprocessing and Analytics

To create an efficient system, we extract the important attributes from the dataset shown in Table 1. In our system, we have three preprocessing stages. In the first stage, we remove data rows where the vaccination information is not available for any user. In the second stage, we remove entries where vaccine manufacturer or type is unknown. In the third stage, we restructure the dataset to improve the performance. The dataset contains various vaccines for different diseases. There are various vaccine types for FLU3, flu4, hepa, flun(h1n1), flux(h1n1), chol, COVID-19, etc.

We further in the process to recommend the vaccine, normalize the data to remove any disadvantage or advantage of having more data as compared to another vaccine. After the preprocessing and analysis, we apply our scoring-based algorithm to recommend the vaccine based on the patient's information. This process is explained in the next section.

3.3 Recommendation Methodology

The user is asked for age, sex and type (disease/outbreak type). Medical history is an optional input. Using the input and processed dataset, the algorithm recommends the vaccine by following the steps given in sections below. Various elements of the system are shown in Fig. 1.

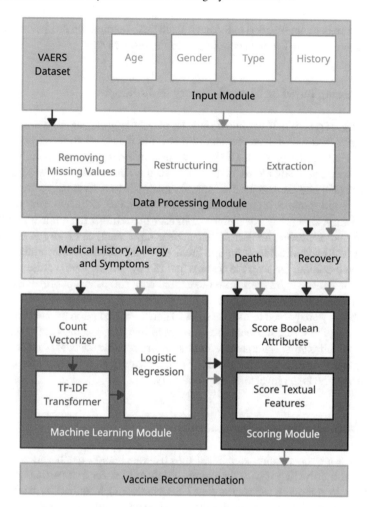

Fig. 1 Vaccine recommendation system design

3.3.1 Data Extraction

We remove all the vaccination data other than the type (disease/outbreak type) that user inputs. The user input age is converted to a range from age −5 to age +5. Next, we remove all the data rows that are of the opposite sex and lie outside the age range.

3.3.2 Scoring Based on Type, Age and Sex

After we get the required data for the recommendation system, we score different vaccines for the given type based on defined criteria. Finally, the vaccine with the highest score is recommended to the user. Each post-vaccination death and recovery contributes -2 and $+1$ to the score of a vaccine, respectively.

3.3.3   Scoring Based on Medical History and Allergies

Based on general idea, if we know that the patient is going through a problem or has allergies, we will not recommend the vaccine that leads to those problems or allergies. Next in the process, all the post-vaccination symptoms and allergies for different vaccines available as textual data in the dataset are fed into the Count Vectorizer [12] and then to TF-IDF transformer [13] to extract the features. These features are unigrams or bigrams based on the vocabulary extracted from user input patient history. Extracting vocabulary from user input limits the number of features extracted and improves the performance. User input medical history is also fed to the same vectorizer and transformer to break down into tokens. As an additional step before vectorization, we strip the accents and remove all the stop words.

After the feature extraction from the textual data and user input medical history, the data is fitted to a logistic regression classifier [14]. The features extracted from the user history are passed to the classifier to generate prediction probabilities. Most probable vaccine will contribute the least score towards vaccine recommendation. The reason behind this statement is that the most probable vaccine is most likely to give negative side effects to the patient. So, the scores generated from the classifier for each vaccine is the inverse of its probability.

3.3.4   Normalization and Recommendation

The score for each vaccine calculated by post-vaccination death and recovery is normalized by dividing it with the count of data entries for that particular vaccine. This helps in removing the merits that a vaccine could have had by having more data than others. Find and store the maximum normalized score over all vaccines. The scores for each vaccine calculated after logistic regression classification are also normalized by bringing them in range [0, 1] and then multiply by the maximum stored earlier. Finally, the two scores are added for each vaccine and are brought again to the range [0.1, 1] by rescaling. Finally, the vaccine with 1.0 score is recommended by the system for the particular patient.

4   Performance Evaluation

The results were calculated on a computer with a 1.8 GHz Dual-Core Intel Core i5 processor and 8 GB 1600 MHz DDR3 RAM. The results were produced by creating simulated patient information. For each experiment, a single input was varied to

Table 2 Normalized scores of COVID-19 vaccines in experiments 1–8

	Exp. 1	Exp. 2	Exp. 3	Exp. 4	Exp. 5	Exp. 6	Exp. 7	Exp. 8
Pfizer\BioNTech	0.1	0.1	0.1	0.1	0.1	0.1	0.1	0.1
Moderna	0.1128	1.0	0.7196	0.2815	0.3427	0.2537	0.1889	0.3230
Janssen	1.0	0.2445	1.0	1.0	1.0	1.0	1.0	1.0

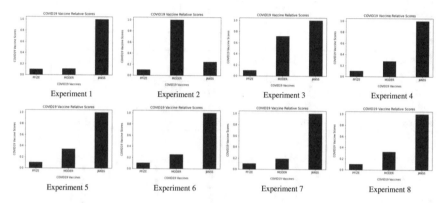

Fig. 2 Visualization of normalized scores of COVID-19 vaccines in experiments 1–8

observe the change in vaccine scores. A total of 16 experiments were conducted on COVID-19 and FLU3 vaccines for eight different kinds of simulated patient's data.

Experiments 1–8 were conducted for the COVID-19 vaccine, and the normalized scores for the different vaccines are shown in Table 2. The graphical representation of the results is shown in Fig. 2.

- Experiment 1: age = 32, sex = M. Result = Janssen.
- Experiment 2: age = 32, sex = F. Result = Moderna.
- Experiment 3: age = 52, sex = M. Result = Janssen.
- Experiment 4: age = 52, sex = F. Result = Janssen.
- Experiment 5: age = 32, sex = M, history = 'I have problem with penicillin also I usually have high headache and mild fever'. Result = Janssen.
- Experiment 6: age = 32, sex = F, history = 'I have problem with penicillin also I usually have high headache and mild fever'. Result = Janssen.
- Experiment 7: age = 52, sex = F, history = 'I have problem with penicillin also I usually have high headache and mild fever'. Result = Janssen.
- Experiment 8: age = 52, sex = M, history = 'I have problem with penicillin also I usually have high headache and mild fever'. Result = Janssen.

Experiments 9–16 were conducted for the FLU3 vaccine, and the normalized scores for the different vaccines are shown in Table 3. The graphical representation of the results is shown in Fig. 3.

- Experiment 9: age = 32, sex = M. Result = Novartis Vaccines and Diagnostics.

Table 3 Normalized scores of FLU3 vaccines in experiments 9–16

	Exp. 9	Exp. 10	Exp. 11	Exp. 12	Exp. 13	Exp. 14	Exp. 15	Exp. 16
Sanofi Pasteur	0.4541	0.4382	0.1895	0.4519	0.2873	0.4389	0.1448	0.4519
CSL Limited	0.9844	1.0	0.2328	1.0	0.6421	1.0	0.1671	1.0
GlaxoSmithKline Biologicals	0.9015	0.7287	0.2928	0.6477	1.0	0.7342	0.2393	0.6518
Novartis Vaccines and Diagnostics	1.0	0.6537	0.232	0.5007	0.6392	0.6567	0.1764	0.5042
Pfizer\Wyeth	0.1	0.1	1.0	0.1	0.1	0.1	1.0	0.1
Connaught Laboratories	0.1	0.1	0.1	0.1	0.1	0.1	0.1	0.1
Medeva Pharma, Ltd	0.1	0.1	0.1	0.1	0.1	0.1	0.1	0.9997
Aventis Pasteur	0.1	0.1	0.1	0.1	0.1	0.1	0.1	0.1

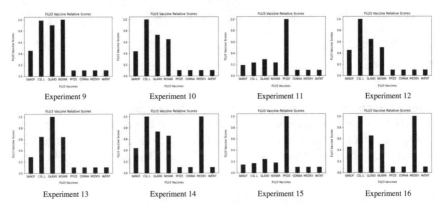

Fig. 3 Visualization of normalized scores of FLU3 vaccines in experiments 9–16

- Experiment 10: age = 32, sex = F. Result = CSL Limited.
- Experiment 11: age = 52, sex = M. Result = Pfizer\Wyeth.
- Experiment 12: age = 52, sex = F. Result = CSL Limited.
- Experiment 13: age = 32, sex = M, history = 'I have problem with penicillin also I usually have high headache and mild fever'. Result = GlaxoSmithKline Biologicals.
- Experiment 14: age = 32, sex = F, history = 'I have problem with penicillin also I usually have high headache and mild fever'. Result = CSL Limited.
- Experiment 15: age = 52, sex = M, history = 'I have problem with penicillin also I usually have high headache and mild fever'. Result = Pfizer\Wyeth.
- Experiment 16: age = 52, sex = F, history = 'I have problem with penicillin also I usually have high headache and mild fever'. Result = CSL Limited.

5 Conclusions and Future Work

Recommendation systems are becoming a boon for the medical officers and are helping in considering factors that were difficult to consider by humans while suggesting a vaccine. The system is based on scoring that is supported by machine learning modules containing Count Vectorizer, TF-IDF transformer and N-grams. Our system is the first of its kind that recommends vaccines based on the patient's age, sex, medical history and allergies. These are one of the most important factors that decide how a vaccine will perform after entering the patient's body. The score-based algorithm displays the score of not just the recommended vaccine but also provides the results for other vaccines to allow medical officers and other researchers to compare the vaccines for different patients. In the results, we demonstrated the working of our system for COVID-19 and FLU3 vaccines. We can clearly observe the difference in scores generated by the system for different patient data. As a future work, we will increase the dataset involving vaccination data from various other countries. We will collaborate with pharmaceutical companies and laboratories to test the performance and validity of our system and further improve our algorithm in terms of accuracy and recommendation results.

References

1. Plotkin S (2005) Vaccines past, present and future. Nat Med 11(4):S5–S11
2. Stewart A, Devlin P (2016) The history of the smallpox vaccine. J Infect 52(5):329–334
3. WHO Coronavirus disease (COVID-19): vaccines. https://www.who.int/news-room/q-a-detail/coronavirus-disease-(covid-19)-vaccines. Last accessed 2021/04/17
4. Zimmermann P, Curtis N (2019) Factors that influence the immune response to vaccination. Clin Microbiol Rev 32(2):e00084-e118
5. Stark B, Knahl C, Aydin M, Samarah M, Elish K (2017) BetterChoice: a migraine drug recommendation system based on Neo4J. In: Proceedings of the 2nd IEEE international conference on computational intelligence and applications, Beijing, China, pp 382–386
6. Hossain MD, Azam MS, Ali MJ, Sabit H (2020) Drugs rating generation and recommendation from sentiment analysis of drug reviews using machine learning. In: Proceedings of the 2020 emerging technology in computing, communication and electronics, Bangladesh, pp 1–6
7. Stark B, Knahl C, Aydin M, Elish K (2019) A literature review on medicine recommender systems. Int J Adv Comput Sci Appl 10(8)
8. Ni J, Huang Z, Cheng J, Gao S (2021) An effective recommendation model based on deep representation learning. Inform Sci 542:324–342. ISSN 0020-0255, https://doi.org/10.1016/j.ins.2020.07.038
9. Abbas K, Afaq M, Ahmed Khan T, Song WC (2020) A blockchain and machine learning-based drug supply chain management and recommendation system for smart pharmaceutical industry. Electronics 9(5):852
10. Garg S (2021) Drug recommendation system based on sentiment analysis of drug reviews using machine learning. In: 2021 11th international conference on cloud computing, data science and engineering (confluence), pp 175–181. https://doi.org/10.1109/Confluence51648.2021.9377188
11. Chen J, Li K, Rong H, Bilal K, Yang N, Li K (2018) A disease diagnosis and treatment recommendation system based on big data mining and cloud computing. Inf Sci 435:124–149

12. Count Vectorizer. https://www.studytonight.com/post/scikitlearn-countvectorizer-in-nlp. Last accessed 2021/04/17
13. TF-IDF Transformer. https://scikit-learn.org/stable/modules/generated/sklearn.feature_extraction.text.TfidfTransformer.html. Last accessed 2021/04/17
14. Wikipedia n-Grams. https://machinelearningmastery.com/logistic-regression-for-machine-learning/. Last accessed 2021/05/04
15. Yong BB, Shen J, Liu X, Li F, Chen H, Zhou Q An intelligent blockchain-based system for safe vaccine supply and supervision. Int J Inform Manage 52:102024
16. Top 10 Vaccine Manufacturers in the World 2020. https://blog.technavio.org/blog/top-10-vaccine-manufacturers. Last accessed 2022/02/15

SiamLBP: Exploiting Texture Discrepancies for Deepfake Detection

Staffy Kingra, Naveen Aggarwal, and Nirmal Kaur

1 Introduction

Ever since the development of digital media, there has always been a need of its manipulation. These manipulation operations were primarily performed for enhancement purpose. Soon, malicious users started using it for concealing the truth. AI advancement, especially generative adversarial network (GAN), increased the possibility of altering digital media or creating media from scratch with high realism. Deepfakes are one of the products of GAN models. Unlike traditional image manipulation techniques such as Photoshop, deepfakes are created by deep neural networks and are rarely distinguishable from the real counterparts. In general, deepfakes refer to synthetic media generated either by swapping the face of source person with target in a video or synchronizing the lip movement of target person in video with a particular audio clip. Besides manipulating existing videos, deepfake technology can also put a new life into existing images by mapping motion from source person's face onto the target image. These advancements are equally impressive and alarming. However, this technology, in wrong hands, can spread misinformation and also undermine public trust.

Since their inception in 2017, deepfakes have been in great demand. After some fake pornography videos of Hollywood celebrities [1], former US President Barack Obama [2] and Facebook CEO Mark Zuckerberg [3] were the targets of deepfakes. In these viral deepfakes, victims were made to say and/or do anything in a video that were not actually said/performed. Apart from image and video deepfaking, one of the malicious user deepfaked audio of CEO of some European company and made a fraudulent transfer of more than 2 lakh dollars. Such generation mechanisms are getting advanced at such a high rate that human eye won't be able to pick certain artifacts from generated media. Moreover, easy availability of user-friendly and open-source

S. Kingra (✉) · N. Aggarwal · N. Kaur
UIET, Panjab University, Chandigarh, India
e-mail: staffysk@gmail.com

© The Author(s), under exclusive license to Springer Nature Singapore Pte Ltd. 2023
D. S. Sisodia et al. (eds.), *Machine Intelligence Techniques for Data Analysis and Signal Processing*, Lecture Notes in Electrical Engineering 997,
https://doi.org/10.1007/978-981-99-0085-5_36

deepfaking applications worsens the situation more. Such faking mechanisms can lead to serious security concerns and social repercussions. Seeing the advancement in deepfake generation mechanisms, there is a need to focus on advancing detection approaches also.

In spite of such advancements, there are still some discrepancies between original image and deepfaked image. Many approaches for deepfake detection have analyzed various artifacts and provided efficient results. For instance, one of the researchers [4] proposed a CNN model to analyze frequency domain artifacts of face region and generated an accuracy of 85.24 and 66.50% on deepfake videos of FF++ [5] and Celeb-DF [6] dataset. However, the technique was found inefficient for neural-textured videos. Recently, texture-based approach [7] for deepfake detection is developed that utilized LBP-coded histograms and reported an AUC of 77.1% on Celeb-DF dataset (after fine-tuning). In combination, researchers utilized features extracted from HRNet [8].

The technique proposed in this paper analyzes facial discrepancies by comparing suspected image with original image of same person. Most of the existing GAN models are not able to produce high-quality deepfakes either due to lack of training or simplicity of the technique. This inefficiency results in irregular contours around face region, blurred image or ghost artifacts. These artifacts distort the normal texture of deepfaked face. The proposed technique compares the texture features of the suspected face with original face of same person. To compare the texture artifacts, a Siamese-based network is proposed by considering ResNet18 as base model for feature extraction. The proposed model is tested on FF++ and Celeb-DF dataset and provided efficient performance in comparison with state-of-the-art techniques.

Major contributions of proposed technique are as follows:

1. A novel Siamese-based model, SiamLBP, is presented to compare texture features for deepfake detection.
2. The proposed model is trained and tested on different deepfake specified datasets such as FF++ and Celeb-DF which generated an accuracy of 99% and 91%, respectively. SiamLBP also provides efficient results irrespective of highly compressed videos.
3. A novel dataset, focused on Indian subjects, is developed for testing the proposed technique. Experimental results revealed the accuracy of 98% and 99% on dataset developed using FOM and FaceApp, respectively.
4. Generalizability of proposed model is tested for different manipulation types at different compression levels. The proposed model also outperformed state-of-the-art techniques such as fCNN [4] and iCaps-DFake [7].

This section explains the concept of deepfaking and social problems concerned with the same. The section also introduced the technique proposed in this paper along with other contributions. The existing deepfake detection approaches are discussed in Sect. 2. In Sect. 3, the proposed technique for deepfake detection is explained in detail including feature extraction and model architecture. The implementation details and performance results of proposed technique are provided in Sect. 4 followed by conclusion in Sect. 5.

2 Related Work

Since it is important to detect deepfakes that can harm the dignity of any person, various researchers have proposed different approaches for the same. On the one hand, there are traditional approaches utilizing a combination of handcrafted features and neural network to perform classification based on extracted features. On the other hand, some approaches proposed a new variant of deep neural network to perform deepfake detection in black-box. To normalize evaluation of deepfake detection algorithm, numerous deepfake datasets have also been released such as FF++ [5], Celeb-DF [6] and DFDC [9].

Considering classification based on handcrafted features, one of the researchers analyzed the frequency of eyeblinking and train a CNN-LSTM based on the same. The technique was tested on self-created dataset and provided an AUC of 99%. Afterward, some researchers exploited traces left by GAN-based deepfake generators through EM clustering [10] and frequency domain analysis [11] which generated an accuracy of 90.22 and 90%. One of the early approaches also explored frequency domain features of facial region [12] which provided an accuracy of 85.24% on FF++ [5] and 66.50% on Celeb-DF [6] dataset. Another technique [13], that analyzed eye color difference, eye and teeth inconsistency and irregular contouring, generated an AUC of 86.6%. Meanwhile, another researcher analyzed optical flow [14] of suspected videos, on which an accuracy of 81.61% was obtained. Recently, a technique based on texture artifacts for deepfake detection provided an AUC of 77.1% on Celeb-DF dataset.

Researchers also proposed different deep architectures for deepfake detection [15–21]. For instance, FakeSpotter [21] detect deepfaked-based activation behavior of neurons in a model and reported an average accuracy of 90.6%. Adaptive residual extraction network (ARENnet) [16], focused on image residuals, generated an average accuracy of 93.99%. Another deepfake detection model was trained to detect inconsistency in human behavior and appearance features [15] and generated an accuracy of 94.14%. Similar research was followed in [20] where inconsistency between audio and visual features was analyzed using Siamese architecture. However, the technique provided an AUC of 84.4%. Meanwhile, another researcher proposed to train ResNet18 models on spatial face regions and obtained an accuracy of 96.75%. One of the researchers also explored unsupervised way of deepfake detection [17] through variational autoencoder (VAE)-based architecture. VAE is trained only on real videos and detects fake videos as anomaly with an accuracy of 88.28%.

Recently, some approaches, focused on temporal nature of the video, are proposed for deepfake detection. For instance, a 3D-CNN model [22] utilized 16 frames per sequence and generated an accuracy of 99.33% on deepfake videos of FF++ [5] dataset. Another model, ResNet50 [23], was trained using optical flows of video sequences and generated an accuracy of 97.35% on deepfake videos of FF++ dataset.

One such approach [24] utilized a combination of EfficientNet-b5 [25] followed by a weighting layer and bidirectional GRU.[1] The model provided an accuracy of 92.61% on DFDC datasets.

Although numerous researchers provided efficient results for deepfake detection, most of those utilized a blind approach for the same. Usually, a complex deep network is developed and trained on large amount of data for automatic extraction of deepfake artifacts. In this paper, a simple yet efficient model is designed by focusing on certain GAN-generated artifact only. It has been observed that popular deepfake generation models cause texture irregularity around deepfaked face due to mix-and-match of source and target images. Due to inefficient training of GAN either due to less number of iterations or small amount of training data, intra-frame and inter-frame consistency of deepfaked faces gets disturbed. Thereby, a novel architecture called SiamLBP is proposed in this paper to compare the texture features of deepfaked face and original face of target person.

3   Proposed Methodology

Deepfake generation mechanisms, though map face onto source person's image in a realistic manner, may introduce discrepancies at some level. This paper exploits texture differences between original and deepfaked face of a person for deepfake detection. Due to its relevance in most of the state-of-the-art techniques, local binary pattern (LBP) [26] is utilized to compute texture features which is explained in Sect. 3.1. Section 3.2 demonstrates the proposed model to compare texture features of original and deepfaked face extracted using LBP.

3.1   LBP-Based Feature Extraction

Local binary pattern is utilized to analyze texture pattern of 2D image through local pixels. It became popular due to its relevance in various computer vision [26, 27] and face recognition [28–30] applications. It represents the comparison of each pixel in an image with its surrounding ones. The comparison process between a pixel and its neighbor outputs a binary value 1 if former pixel is higher than the later or 0 otherwise. As each pixel is surrounded by eight neighbors in a 2D image, comparison with all results in a 8-bit sequence. Decimal equivalent of resulted sequence replaces the center pixel. This procedure is demonstrated in Fig. 1 where center pixel is being compared with its neighbors. Pixel being compared is termed as anchor pixel here. It is to be noted that LBP-coded image is obtained by considering each pixel as anchor pixel one by one.

---

[1] Gated Recurrent Unit.

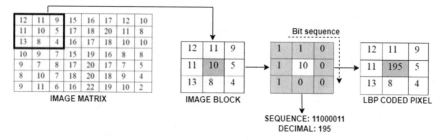

Fig. 1 Demonstrating local binary pattern (LBP) computation from one image block

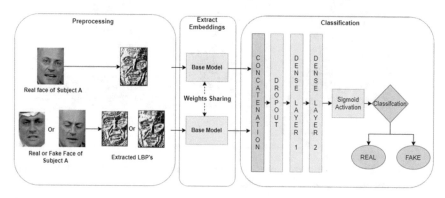

Fig. 2 Demonstrating framework of proposed technique. (For base model, ResNet18 is utilized)

3.2 Siamese Training for Texture Feature Comparison

The proposed technique is based on the idea that texture features of original face are not fully mapped to its deepfaked version and thereby can be analyzed for deepfake detection. The procedure to compute texture features is explained in Sect. 3.1. Since the technique is supposed to detect deepfaked face of person by comparing its suspected face with the original one, Siamese training is adopted for the process. Figure 2 demonstrates the framework of the proposed technique which performs deepfake detection in three phases as explained here.

Preprocessing: Since deepfaking is usually performed on facial area of target person's images or videos, the model is trained on facial areas only. For facial extraction, various face recognition methods were tried. However, best results are reported by MTCNN [31] classifier and are utilized here. This face recognition classifier outputs three values, i.e., key landmarks positions, probability of face detected and coordinates of face bounding box. Bounding box coordinates are utilized to crop the facial region from whole image/frame, which is resized to (224, 224, 3). Next step is to compute LBP of extracted faces as explained in Sect. 3.1. Computed LBPs of real–real and real–fake pairs are used for training SiamLBP.

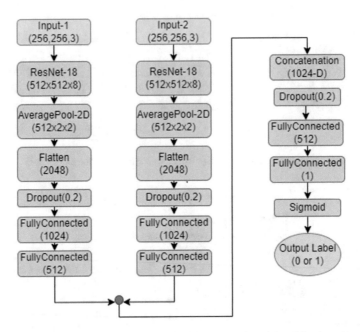

Fig. 3 SiamLBP: proposed Siamese-based model for texture-based deepfake detection. (Output shapes of each layer are provided along with)

Embedding Extraction: The next step is to extract embedding vector from texture features for comparison. Comparison is performed through proposed Siamese-based model, SiamLBP. The basic nature of Siamese model is to train a single network on pair of images through dual-stream and measure similarities or dissimilarities in respective pair. Both streams must have same architecture and also share weights with each other but trained on different images simultaneously. Due to weight sharing, these streams are supposed to analyze same set of features from both images. For deepfake detection in this paper, texture features are adopted for Siamese training. SiamLBP utilizes ResNet18 as base model. To extract features from base model, the last two layers of existing ResNet18 are popped off which provides 4096-D feature vector. In place of these, pooling layer and dropout layer followed by two dense layers are inserted with an output size of 1024 and 512, respectively.

Classification Second phase provides 2 512-D embeddings for two images of an input pair. For binary classification, feature vectors computed from both images are concatenated and 1024-D vector is obtained as demonstrated in Fig. 2. However, detailed architecture of SiamLBP with output shapes of each layer is described in Fig. 3. Afterward, two fully connected layers are employed as the final classification layers with an output size of 512 and 1 neurons, respectively. In addition, a dropout layer is employed before these to reduce overfitting on training data. However, sigmoid activation is employed at the end for binary classification (Fig. 4).

Real faces   Generated using FaceApp   Generated using FOM

Fig. 4 Dataset of Indian celebrities generated using FaceApp and FOM

3.3 Datasets

Due to the network structure, a pair of images is needed at training time. One of the images in the pair is original face of particular person, while other one can be real or deepfaked face of same person. For training, an image pair is labeled as real if both images are original, while it is labeled fake if a pair exhibits original and deepfaked face of respective person. There are only few datasets, such as FF++ and Celeb-DF that provide information of person targeted in deepfaked image. Also, a new dataset is created which also provides information of person being deepfaked. Detailed information of datasets is provided as follows:

1. **FaceForensics++ (FF++)** [5]: FF++ is one of the early generation deepfaked datasets, which contains an organized collection of original videos and four types of manipulated videos. Manipulated set contains Deepfakes (DF), FaceSwap (FS), Face2Face (F2F) and NeuralTextured (NT) videos. Each set exhibits 1000 videos. DF and FS set contain face swapped videos, while F2F and NT contain face reenacted videos. However, DF and NT manipulations are performed using deep networks, while FS and F2F manipulations are performed using computer graphic techniques. Proposed technique is evaluated on each set separately by considering first 700 videos for training and 300 for testing. For real–real pair, 15 random frames are paired with 15 other frames of same video. On the other hand, 15 frames from fake video of same target id are paired with their real counterparts for real–fake pair.

2. **Celeb-DF** [6]: Considered as the most realistic deepfake dataset, deepfake videos in this exhibit very few visible faking artifacts. The dataset is composed by the collection of real YouTube videos of 59 different celebrities. From 590 original videos, 5639 deepfake videos were generated through face swapping. For this technique, real–real image pairs are generated by selecting ten frames from two original videos of same person. Frames from one of the video are paired with fake video of same person for real–fake pair. To prevent subject overfitting, different identities are used for training and validation.

3. **FaceApp dataset**[2]: Through a simple mobile application named "FaceApp", 1000 real and 1000 fake images are generated. Since the dataset is focused on Indian subjects, a Bollywood celebrity dataset is collected from Kaggle. Keeping in mind

---

[2] https://www.faceapp.com/.

the use of open-source and user-friendly applications by common users, only such applications are utilized to fake data.

4. **FOM dataset** [32]: First-order motion model has been made popular recently by performing animations on image through any driving video. This application is utilized, in this paper, to deepfake face region by mapping the face movements of another person from another video. However, driving videos are recorded first and then applied onto celebrity images which generated deepfaked videos of respective celebrities. For training, one frame from each generated video is included in fake set, while Bollywood celebrity dataset is considered in real part.

4 Implementation and Performance Evaluation

The proposed model is implemented in PyTorch with existing PyTorch model, ResNet18 [33], as the base model. Training is performed on NVIDIA P5000 GPU with 16 GB memory. To prove the efficacy, the model is tested on different deepfake specified datasets. Hyperparameters of proposed model are fixed after set of experiments. Moreover, various in-dataset and cross-dataset experimentations are performed.

4.1 Hyperparameters

Proposed model is end-to-end trained with a batch size of 16. It was observed from different experiments that validation performance is comparatively less stable if training is performed on batches of more than 16 images. Contrary to this, validation loss stuck at local optimum value if batch size is less than 16. Next hyperparameter to optimize is learning rate. Random values of learning rate are tried in the range of $1e-1$ to $1e-5$. However, the best performance is noted at a learning rate of $1e-4$ with Adam optimizer. For more optimization, learning rate is decayed with a factor of 0.005 after every ten epochs.

4.2 Evaluation Results

After training the proposed model on real–real and real–fake pairs of different datasets, trained models are tested on pairs of same and different datasets. To classify the suspected image as original or deepfaked, it must be accompanied by an original image of same subject for texture comparison through proposed model. FF++ and Celeb-DF datasets exhibit such information and are, thereby, utilized for evaluating SiamLBP.

Table 1 Performance of proposed model on FF++ dataset at different resolutions (in %)

Dataset	Compression	Accuracy	Precision	Recall	FPR	FNR
FF++ (DF)	Raw	99.3	1.0	98.7	0.006	0.019
	C23	95.4	94.05	96.1	0.05	0.03
	C40	86.39	85.6	86.02	0.10	0.18
FF++ (FS)	Raw	97.3	96.0	98.4	0.01	0.01
	C23	92.3	91.8	92.4	0.06	0.07
	C40	79.2	78.0	75.7	0.17	0.24
FF++ (F2F)	Raw	96.9	95.9	97.5	0.03	0.04
	C23	91.9	91.6	91.4	0.07	0.08
	C40	79.4	78.8	74.9	0.16	0.25
FF++ (NT)	Raw	94.3	93.2	96.17	0.07	0.03
	C23	85.6	80.9	89.4	0.174	0.105
	C40	67.53	65.2	61.9	0.27	0.38

Models are trained and tested separately for different manipulation methods and compression levels

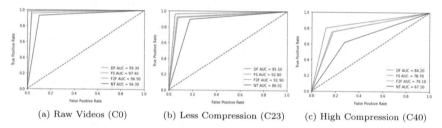

(a) Raw Videos (C0) (b) Less Compression (C23) (c) High Compression (C40)

Fig. 5 AUC-ROC comparison at varied compression levels for different tampering types (Reported results demonstrates the performance of FF++ dataset)

In-dataset experiments: First set of experiments are performed on four different subsets of FF++ dataset. Since DeepFake and FaceSwap manipulations are performed on the entire face, there are high chances of texture variations in such faces. Thereby, the highest performance is reported on DeepFake manipulations followed by FaceSwap as given in Table 1. On the other hand, Face2Face and NeuralTexture manipulations target only lip portion and contain less texture discrepancies. Since FF++ dataset contains videos of different manipulation types at varied compression levels, model is trained and tested on each set separately. Experiments are also performed on Celeb-DF dataset, results of which are reported in Table 2. ROC curves provided in Fig. 5 also prove the efficiency of proposed model on different datasets. However, the performance slightly deteriorates for highly compressed videos.

Cross-dataset experiments: Generalizability ensures that model trained on one dataset can also prove efficient for another data. For the said purpose, experiment is performed by testing the model on dataset that is different from the one used in training as reported in Table 3. Different subsets of FF++ dataset are considered separately to test if the model is generalizable from one manipulation type to another. It

Table 2 Performance on Celeb-DF and self-generated datasets

Dataset	Accuracy	Precision	Recall	FPR	FNR
FaceApp	99.5	99.1	1.0	0.009	0.0
FOM dataset	98.1	99.1	97	0.009	0.08
Celeb-DF	91.58	90.2	95.02	0.12	0.049

Table 3 Generalization ability of FF++ dataset across different manipulation types (rows represent training dataset, while columns represent testing data)

C0	DF	FS	F2F	NT
DF	99	56.3	61.9	62.6
FS	57.2	97.6	57.1	57.7
F2F	82.1	55.6	96.8	64.3
NT	89.2	54.7	67.5	94.3
C23	**DF**	**FS**	**F2F**	**NT**
DF	95.4	52.6	57.4	66.8
FS	52.7	92.3	53.3	52.1
F2F	67.5	56	91.5	59.1
NT	81.2	52.1	66.6	85.6
C40	**DF**	**FS**	**F2F**	**NT**
DF	86.39	57.3	55.3	55.8
FS	55.7	79.2	53.4	51.2
F2F	58.5	56.0	79.4	56.1
NT	61.1	52.3	60.9	67.5

DF DeepFake, *FS* FaceSwap, *F2F* Face2Face, *NT* NeuralTexture

Table 4 Performance comparison of proposed model on Celeb-DF dataset

Approach	Accuracy
fCNN [4]	66.50
FakeSpotter [21]	66.8
Two-branch [18]	73.41
iCaps-Dfake [7]	86.0
Proposed	91.5

is observed that model trained on lip-sync videos is adaptable toward face manipulation also, but the reverse is not same. The reason being that model trained on face manipulated images learn more broad artifacts which may not be present in videos where only lip movement is being manipulated.

Comparison with state of the art: To prove the efficiency of proposed technique, it is compared with state-of-the-art approaches. Results demonstrated in Table 4 and Fig. 6, reveal that SiamLBP achieved significant performance for deepfake detection

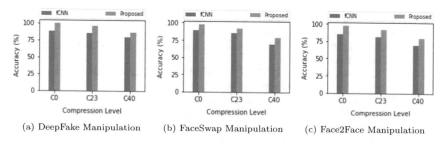

(a) DeepFake Manipulation   (b) FaceSwap Manipulation   (c) Face2Face Manipulation

Fig. 6 Performance of proposed model on FF++ dataset as compared to state of the art (in terms of accuracy)

and outperform various existing approaches. Table 4 provided the comparative analysis of Celeb-DF dataset. Similarly, Fig. 6 reported its comparison on FF++ videos at different compression levels and for different tampering types.

5 Conclusion

Rapid advancement in deepfake generation technologies is increasing the threat of deepfaking by defaming common man, sharing misinformation and promoting fake campaigning in the society. As GANs are getting advanced, the need to develop more deepfake detection mechanisms increases. Although researchers have proposed various deepfake detection mechanisms, this cat-and-mouse race is a never ending problem. Therefore, this paper proposed a novel technique, SiamLBP, for deepfake detection that performs texture-based comparison of deepfaked face with original face using Siamese model. The proposed model provided an accuracy of 99 and 91.9% on deepfake videos of FF++ and Celeb-DF dataset which outperforms the state-of-the-art techniques. In addition, SiamLBP has obtained an accuracy of more than 98% on dataset generated using deepfaking applications. In future, different base models can be considered for feature extraction in SiamLBP.

References

1. Cole S (2017) Ai-assisted fake porn is here and we're all fucked. https://www.vice.com/en_us/article/gydydm/gal-gadot-fake-ai-porn
2. Vincent J (2018) Jordan peele use ai to make barack obama deliver a psa about fake news. https://www.theverge.com/tldr/2018/4/17/17247334/ai-fake-news-video-barack-obama-jordan-peele-buzzfeed
3. Posters B (2018) Bill posters on Instagram. Artificially generated video of mark zuckerberg https://twitter.com/PressSec/status/1060374680991883265
4. Kohli A, Gupta A (2021) Detecting deepfake, faceswap and face2face facial forgeries using frequency cnn. Multimedia Tools Appl 1–18

5. Rössler A, Cozzolino D, Verdoliva L, Riess C, Thies J, Nießner M (2019) Faceforensics++: learning to detect manipulated facial images. ArXiv preprint arXiv:1901.08971
6. Li Y, Yang X, Sun P, Qi H, Lyu S (2019) Celeb-df: a new dataset for deepfake forensics. ArXiv preprint arXiv:1909.12962
7. Khalil SS, Youssef SM, Saleh SN (2021) icaps-dfake: an integrated capsule-based model for deepfake image and video detection. Fut Internet 13(4):93
8. Sun K, Zhao Y, Jiang B, Cheng T, Xiao B, Liu D, Mu Y, Wang X, Liu W, Wang J (2019) High-resolution representations for labeling pixels and regions. ArXiv preprint arXiv:1904.04514
9. Dolhansky B, Bitton J, Pflaum B, Lu J, Howes R, Wang M, Ferrer CC (2020) The deepfake detection challenge dataset. ArXiv preprint arXiv:2006.07397
10. Guarnera L, Giudice O, Battiato S (2020) Deepfake detection by analyzing convolutional traces. In: Proceedings of the IEEE/CVF conference on computer vision and pattern recognition workshops, pp 666–667
11. Durall R, Keuper M, Keuper J (2020) Watch your up-convolution: Cnn based generative deep neural networks are failing to reproduce spectral distributions. In: Proceedings of the IEEE/CVF conference on computer vision and pattern recognition, pp 7890–7899
12. Kohli A, Gupta A (2021) Detecting deepfake, faceswap and face2face facial forgeries using frequency cnn. Multimedia Tools Appl 1–18
13. Matern F, Riess C, Stamminger M (2019) Exploiting visual artifacts to expose deepfakes and face manipulations. In: 2019 IEEE winter applications of computer vision workshops (WACVW). IEEE, pp 83–92
14. Amerini I, Galteri L, Caldelli R, Del Bimbo A (2019) Deepfake video detection through optical flow based cnn. In: Proceedings of the IEEE international conference on computer vision workshops, pp 0–0
15. Agarwal S, El-Gaaly T, Farid H, Lim SN (2020) Detecting deep-fake videos from appearance and behavior. ArXiv preprint arXiv:2004.14491
16. Guo Z, Yang G, Chen J, Sun X (2020) Fake face detection via adaptive residuals extraction network. ArXiv preprint arXiv:2005.04945
17. Khalid H, Woo SS (2020) Oc-fakedect: Classifying deepfakes using one-class variational autoencoder. In: Proceedings of the IEEE/CVF conference on computer vision and pattern recognition workshops, pp 656–657
18. Masi I, Killekar A, Mascarenhas RM, Gurudatt SP, AbdAlmageed W (2020) Two-branch recurrent network for isolating deepfakes in videos. ArXiv preprint arXiv:2008.03412
19. Mehra A (2020) Deepfake detection using capsule networks with long short-term memory networks. Master's thesis, University of Twente
20. Mittal T, Bhattacharya U, Chandra R, Bera A, Manocha D (2020) Emotions don't lie: a deepfake detection method using audio-visual affective cues. ArXiv preprint arXiv:2003.06711
21. Wang R, Juefei-Xu F, Ma L, Xie X, Huang Y, Wang J, Liu Y (2020) Fakespotter: a simple yet robust baseline for spotting ai-synthesized fake faces. In: International joint conference on artificial intelligence (IJCAI)
22. Nguyen XH, Tran TS, Nguyen KD, Truong DT et al (2021) Learning spatio-temporal features to detect manipulated facial videos created by the deepfake techniques. Forensic Sci Int: Digital Invest 36(301):108
23. Caldelli R, Galteri L, Amerini I, Del Bimbo A (2021) Optical flow based cnn for detection of unlearnt deepfake manipulations. Pattern Recogn Lett (2021)
24. Montserrat DM, Hao H, Yarlagadda SK, Baireddy S, Shao R, Horváth J, Bartusiak E, Yang J, Güera D, Zhu F et al (2020) Deepfakes detection with automatic face weighting. ArXiv preprint arXiv:2004.12027
25. Tan M, Le QV (2019) Efficientnet: rethinking model scaling for convolutional neural networks. ArXiv preprint arXiv:1905.11946
26. Ojala T, Pietikäinen M, Harwood D (1996) A comparative study of texture measures with classification based on featured distributions. Pattern Recogn 29(1):51–59
27. Zhang W, Shan S, Gao W, Chen X, Zhang H (2005) Local gabor binary pattern histogram sequence (lgbphs): A novel non-statistical model for face representation and recognition. In:

Tenth IEEE international conference on computer vision (ICCV'05) Volume 1, pp. 786–791. IEEE (2005)

28. Huang D, Shan C, Ardabilian M, Wang Y, Chen L (2011) Local binary patterns and its application to facial image analysis: a survey. IEEE Trans Syst Man Cybern Part C (Appl Rev) 41(6):765–781

29. Ojala T, Pietikainen M, Maenpaa T (2002) Multiresolution gray-scale and rotation invariant texture classification with local binary patterns. IEEE Trans Pattern Anal Mach Intell 24(7):971–987

30. Xiao B, Wang K, Bi X, Li W, Han J (2018) 2d-lbp: an enhanced local binary feature for texture image classification. IEEE Trans n Circ Syst Video Tech 29(9):2796–2808

31. Xiang J, Zhu G (2017) Joint face detection and facial expression recognition with mtcnn. In: 2017 4th International conference on information science and control engineering (ICISCE). IEEE, pp 424–427

32. Siarohin A, Lathuilière S, Tulyakov S, Ricci E, Sebe N (2019) First order motion model for image animation. Adv Neural Inf Proc Syst 32:7137–7147

33. He K, Zhang X, Ren S, Sun J (2016) Deep residual learning for image recognition. In: Proceedings of the IEEE conference on computer vision and pattern recognition, pp 770–778

Rainfall Prediction Using Machine Learning

K. Prathibha, G. Rithvik Reddy, Harsh Kosre, K. Lohith Kumar, Anjali Rajak, and Rakesh Tripathi

1 Introduction

Rainfall is a necessary aspect of a climatic system, and its unpredictable nature has a direct impact on water resource management, agriculture, forestry, and biological systems. Climate change is an important issue with worldwide ramifications. Climate change is having major effects on people all around the world. Climate change refers to a shift in the global or regional climate, particularly as it relates to the mid-to late-twentieth century. All of this is occurring as a result of high levels of pollution in the atmosphere, which is partly due to the increased levels of atmospheric carbon dioxide created by the burning of fossil fuels. Weather change, often known as global warming, is a phenomenon that occurs as a result of human activity. The term "global warming" implies an increase in the average surface temperature of the world, as a result of which the entire planet is altering. Since the previous decade, scientists, engineers, and researchers have successfully developed different models for making correct predictions in various fields. Machine learning (ML) technique is used to make predictions and classify objects. To deal with this unpredictability, we used several machine learning models to make accurate predictions. We implement models such as Random Forest classifier, K-Nearest Neighbor, Extra Tree classifier, Gradient Boosting classifier, Adaboost, Decision Tree classifier, Gaussian NB, Multilayer Perceptron. For calculation purposes, we used RMS Error, Accuracy, Precision, F1 Score, AUC as performance metrics. Data is collected from day-to-day weather records from different Australian weather stations. It contains weather and rainfall

K. Prathibha · G. Rithvik Reddy · H. Kosre · K. Lohith Kumar · A. Rajak (✉) · R. Tripathi
Department of Information Technology, National Institute of Technology Raipur, Raipur 492010, India
e-mail: arajak.phd2021.it@nitrr.ac.in

R. Tripathi
e-mail: rtripathi.it@nitrr.ac.in

© The Author(s), under exclusive license to Springer Nature Singapore Pte Ltd. 2023 457
D. S. Sisodia et al. (eds.), *Machine Intelligence Techniques for Data Analysis and Signal Processing*, Lecture Notes in Electrical Engineering 997,
https://doi.org/10.1007/978-981-99-0085-5_37

observations of over 10 years in Australia. The dataset has around 145,460 records with 23 features.

The rest of this paper is organized as follows. Section 2 describes the literature review of related work; Sect. 3 describes the dataset used and ML algorithms applied on it after processing the data; Sect. 4 describes the brief discussion of the experimental results followed by concluding the paper in Sect. 5.

2 Literature Review

Rainfall plays a vital role. Approaches that are commonly used to predict rainfall. One option is to examine the large volumes of data accumulated over time to obtain information on future rains. Creating mathematical equations by defining various parameters and substituting values to achieve the result is the other. One of the most recent studies included in which researchers stated that rainfall expectation refers to the assurance of precipitation patterns for a certain place [1]. It is vital for the horticultural sector as well as other companies. This is the first attempt, as far as anyone is concerned, in applying someplace down in anticipation of monthly precipitation. The suggested approach is a gauging design offered by the meteorology service, as opposed to the Australia Public weather report and Earth System Prediction model. The recommended method was compared against the Australian Weather and Earth System Prediction Model, a prediction model provided by the weather service, and a prediction model provided by the weather department. The proposed Nash–Sutcliffe efficiency coefficient, mean absolute error, Pearson's correlation coefficient, MAE and RMSE have all been improved. According to another study, months with greater annual rainfall averages outperform months with lower annual rainfall averages. The efficiency is great, and the application has a lot of potentials. There are three stages to the deep learning process. The correct algorithm must be determined in the first phase. The algorithm is selected based on the dataset supplied, such as which algorithm best suits the given data. The second phase is all about figuring out which one performs best with the algorithm. In the final phase, various Metrics products for predictive precipitation modeling, false report percentages, and risk percentages are being deployed. The first two phases' computations are based on the expected time sequence, which includes negative values [2]. Thirumalai et al. [3] analyzed crop seasons to analyze rainfall quantities in prior years and anticipate rainfall for upcoming years. The season of crops is Zaid, kharif, and rabi. A linear regression technique is employed for early prediction. If one variable is accessible, linear regression can be used to predict the other. As variables, Rabi and Kharif were employed. The standard deviation and mean were calculated for future crop season predictions. With the help of this implementation, farmers will be able to choose which crop to yield based on the season of crops. Dash et al. [4] showed that AI approaches such as the ANN, the Extreme Learning Machine, and the Knowledge-Based Neural Network have been used to predict post-monsoon and summer rainfall. The Indian Institute of Tropical Meteorology (IITM) provided time-series data for Kerala since 1871.

The data is pre-processed and then standardized before being divided into training and testing groups. The training set included data before 2010, while the test set included data from 2011 to 2016. Geetha et al. [5] developed a model that predicted meteorological events such as thunderstorms, rain, cyclones, and fog allowing people to take preventative from these events. Mining techniques and an application called Rapid miner were used to model the decision trees. Day, temperature, dew point, pressure, and other variables are included in the Trivandrum data set. Using the decision tree technique dataset is divided into two sets that is training and testing sets. The precision of the measurement is calculated, and the actual and predicted value is compared. Different soft computing algorithms include neural networks, genetic algorithms, and fuzzy logic might be applied to improve accuracy. MAE and RMSE were used to calculate the performance of the previously described techniques.

3 Rainfall Prediction Using Machine Learning Technique

3.1 Dataset Description

The "weatherAus" dataset is used in this study [9]. The dataset includes the weather and rainfall observations of over 10 years in Australia. The dataset has around 145,460 records with 22 independent features and 1 dependent variable or target feature (Table 1).

3.2 Rainfall Prediction Using Machine Learning Technique

We determined various classifiers each belonging to a different model class. The following classifier has been used to build models for our experiment.

Decision Tree: Decision Trees are a supervised non-parametric learning method that might be employed for both classification as well as regression applications. It is similar to a flowchart in which the internal nodes specify the test on the feature, i.e., if the rain occurs or not. While the leaf nodes represent the decision made after all the features are computed. Branches specify the combinations of different features which lead to the class labels/the outcome in the leaf nodes. Classification rules are represented by the pathways from the root to the leaf node. It gives a high variance.

Random Forest: Random forest is a classification algorithm that uses numerous decision trees to classify data. When creating individual trees, it employs feature randomization to generate an uncorrelated forest of trees whose committee prediction is more accurate than that of any one tree.

Extra Trees: It is an ensemble machine learning technique that is imported using Scikit-learn ensemble. The base foundation of this classifier is the decision trees.

Table 1 Features of the dataset

Feature	Description
Date	The date on which weather was observed
Location	Weather station location
MinTemp	Min temperature in °C
MaxTemp	Max temperature in °C
Rainfall	Rainfall recorded on the day. It is measured in mm
Evaporation	Twenty-four hours leading up to 9 a.m., Class A pan evaporation. It is measured in mm
Sunshine	Hours in the day the sun is shining brightly
WindGustDir	Strongest wind gust direction in 24 leading up to midnight
WindGustSpeed	The highest wind gust speed in 24 h leading up to midnight was measured in km/h
WindDir9am	It gives the wind direction at 9 a.m.
WindDir3pm	It gives the wind direction at 3 p.m.
WindSpeed9am	Wind speed (km/h). This speed is averaged over 10 min before 9 a.m.
WindSpeed3pm	The wind speed is km per hour
Humidity9am	At 9 a.m. shows humidity (percent)
Humidity3pm	At 3 p.m. shows humidity (percent)
Pressure3pm	The measure of (hPa) atmospheric pressure was reduced to the mean sea level at 3 p.m.
Cloud9am	At 9 a.m., clouds covered a portion of the sky. This is expressed in "oktas," an eight-unit. It keeps track of how many 8 of the sky are hidden by clouds and 0 denotes a totally clear sky, whereas an 8 suggests a completely clouded sky
Cloud3pm	At 3 p.m., clouds covered a portion of the sky. This is expressed in "oktas," an eight-unit
Temp9am	The measure of temperature at 9 a.m. in °C
Temp3pm	A measure of temperature at 3 p.m. in °C
RainToday	It is a Boolean value. The amount of precipitation is measured in 24 h leading up to 9 am. This value is in mm 1—if precipitation exceeds 1 mm 0—if precipitation is below 1 mm
RainTomorrow	The quantity of rain expected the next day in millimeters. Creates a response feature RainTomorrow. A type of "risk" assessment

Similar to Random Forest Classifier it uses random decision trees and reduces over-fitting or over-learning of the data. Compared to Decision Tree and Random Forest it gives low variance.

K-Nearest Neighbors (K-NN): This is a supervised learning technique. The method retains all existing information and classifies a new data point based on its similarity to the available. This means that new information can be effectively sorted into a well-defined group using this method. This approach can be used for both regression and classification, but it is more commonly used for the classification process.

AdaBoost: It is a meta-estimator that starts by trying to fit a classifier on the actual data, then fits more replicas of the classifier on the same dataset, while adjusting the weights of poorly classified samples so that future classifiers can focus more on complex problems.

Gradient Boosting: Each predictor in gradient boosting corrects the error of its previous predictor. Unlike Adaboost, the training sample loads are not adjusted; instead, each predictor is learned using the predecessor's residual errors as labels. Classification and Regression Trees (CART) is the base learner in a technique called Gradient Boosted Trees.

Gaussian Naive Bayes: Gaussian Naive Bayes is a Naive Bayes variation that uses a Gaussian normal distribution and can handle sequential data. Unlike Naive Bayes which assumes the data features are independent, GaussianNB assumes each feature follows Gaussian distribution. The Bayes theorem provides the basis for a collection of supervised machine learning classification algorithms known as Naive Bayes.

Multilayer Perceptron (MLP): This belongs to a class of feed-forward artificial neural networks. It has three-layer input, output, and a greater number of hidden layers with neurons are connected. Neurons in MLP can select any arbitrary activation function such as sigmoid and ReLU. Recognition, prediction, approximation, and pattern categorization, are some common applications of MLPs.

3.3 Performance Evaluation

We predicted the rainfall using the open-source dataset "weatherAUS" [8]. In this section, we discussed the experimental configuration, and then the performance of various algorithms using the same dataset is comprehensively compared (Fig. 1).

Data Preprocessing: During the data preprocessing stage, it is found that the features evaporation, cloud9am, and some other features have the same value repeated or NA values. We fill the null values of our object columns with the mode and numeric columns with a median. The location and date features are not influencing the target variable RainTomorrow, these features are dropped from the dataset. From the dataset, we observed that some features like WindGustDir, WindDir9am,

Fig. 1 Flow chart of the proposed working model

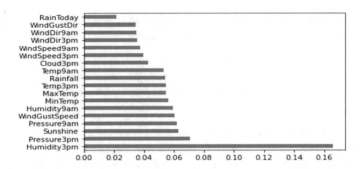

Fig. 2 Topmost feature

WindDir3pm, RainToday, and RainTomorrow are in string format. As the string input cannot be processed by machine learning techniques, we utilize the label encoder from Scikit-learn preprocessing to transform them to numeric format.

Feature Importance: A horizontal bar plot is plotted to identify the most contributing features for RainTomorrow. In Fig. 2 we observed that Humidity3pm, Wind-GustSpeed, Pressure3pm, Humidity9am, Pressure9am, and RainToday are highly contributing features.

Data Analysis: The dataset is spitted into two parts X and Y. X has all the features except for the target variable RainTomorrow is the independent variable that comprises the target variable. A scatter plot is plotted between MaxTemp and MinTemp on and x and y axes, respectively, with RainTomorrow as hue. From Fig. 3a we observed that MinTemp and MaxTemp have a linear relationship. The occurrence of rain the next day is high when the temperatures increase. A scatter plot is plotted between Humidity9am and Temp9am on and x and y axes, respectively, with Rain-Tomorrow as hue. Figure 3b shows that the occurrence of rainfall increases with the increase in humidity. We noticed that temperatures are at a moderate range when the humidity is high for the rain to occur the next day. We plot a heatmap to understand the correlation among the features in our dataset. In Fig. 3c Heatmap shows that the values close to 1 have the highest correlation. That is to notice that the lighter the block the higher the features are connected. We noticed MinTemp and MaxTemp are highly correlated. RainTomorrow is highly connected with the parameters Humidity3pm, Humidity9a.m, rain today, Rainfall, and WindGustSpeed. Dataset is divided in ratio 80:20 for training and testing and the model selection and cross-val score are imported from Scikit-learn for cross-validating the data. Different baseline models are built and threefold cross-validation is performed. Table 2 shows the result obtained from cross-validation. In k-fold cross-validation, the data is split into k parts for each iteration. In these k parts, k-1 parts are used as training data while 1 part is used as testing data. The number of splits is given by the attribute n splits.

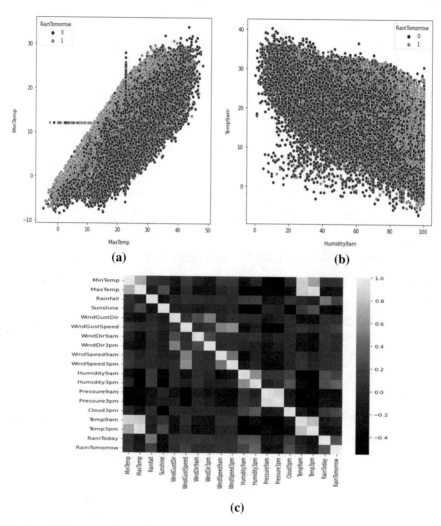

Fig. 3 **a, b** Shows the relation between the features and **c** Heat Map

3.4 Description of Evaluation Metrics

In this section to evaluate the performance of the proposed model, a detailed result analysis is performed using various evaluation metrics. The proposed model is also compared with other works found in the literature.

The performance metrics are as follows:

True Positive (TP): Defines true predictions of the rain on the next day.

True Negative (TN): Define true predictions of no rain on the next day.

Table 2 Result obtained from cross-validation

Model	Accuracy
KNN7	0.8397
KNN9	0.8419
KNN11	0.8430
NB	0.8082
AB	0.8423
GBM	0.8464
RF_Ent100	0.8541
RF_Gini100	0.8539
ET100	0.8526
ET500	0.8541

False Positive (FP): Defines false predictions of the rain on the next day.

False Negative (FN): Defines false prediction that the rain does not occur on the next day.

Accuracy (ACC): This is used to determine how near the measured value is to the standard known value. Accuracy is calculated as follow:

$$ACC = \frac{TP + TN}{TP + TN + FP + FN} \tag{1}$$

Precision (PRE): This measure is used to define, the closeness of two or more measurements with each other. It is represented as

$$PRE = \frac{TP}{TP + FP} \tag{2}$$

Recall (REC): This calculates the value of correct hits that were found.

$$REC = \frac{TP}{FN + TP} \tag{3}$$

F1 Score: This defined the mean of recall and precision for the rainfall prediction and can be defined as

$$F1\ Score = \frac{(2 * Prec * Rec)}{Prec + Rec} \tag{4}$$

Roc (AUC) score: It is the area under the ROC curve. A higher value of this score indicates that the algorithm is more normalized. The model with a high ROC value is effective at predicting rainfall.

Root Mean Square Error (RMSE): Defined by the square root of the mean of the square of all errors.

$$RMSE = \sqrt{\frac{1}{n} \sum_{i=1} (s_i - o_i)^2} \qquad (5)$$

4 Experimental Results

In this section, we showed the result related to all scenarios in our models. The performance of the classifier is measured by the evaluation metrics, which are trained and tested on the open-source "weatherAUS dataset" from Kaggle. Python is used for implementation with the required libraries, such as Pandas, Matplotlib, Numpy, and Scikit-learn. Experiments are performed on Intel Core i7, 16GB RAM, CPU — 2.06 GHz processor, for effective analysis, the model is tested in different cases. We divided the dataset in the ratio of 80:20 for training and testing purposes. The dataset contains 145,460 instances. The total number of features is 23. It has 22 independent features and 1 dependent or target feature. Table 3 shows the performance analysis of the different classifiers in terms of performance matrix Accuracy, Precision, Sensitivity, Specificity, Fl-score, ROC, RMS error. Figure 4 depict the ROC AUC Curve is plotted for Random Forest, Gradient Boosting, and Extra Trees Classifier, Fig. 5 depict the Precision-Recall curve is plotted for Random Forest, Extra Trees Classifier, and Gradient Boosting Classifier and Fig. 6a, b depict the Confusion Matrix for Random Forest and Extra Trees Classifier.

Our work was compared against the existing works by Sarvani et al. [20] who used the Support Vector Machine and obtained an accuracy of 80% and by Jalgaonkar et al. [21] who worked with Linear Regression and obtained an accuracy of 85%.

5 Conclusion

In this study, we have used several machine learning models to forecast rainfall depending on different weather parameters. The highest accuracy is obtained by selecting the Random Forest and Extra Tree classifier as compared to another model. In our future work, the performance of the predicted model could be improved by deep learning models and feature selection methods to eliminate unnecessary features. This model will be incorporated into a mobile app which will be helpful for the

Table 3 The result shows the performance measure of the various classifier

Classifier	Accuracy	Precision	Sensitivity	Specificity	F1 score	ROC	RMS error
Random Forest	85.5	75.7	50.4	95.4	60.5	72.94	0.14
K-NN	84.3	72.04	46.9	94.8	56.8	70.8	0.15
Extra Tree	85.6	76.9	49.4	95.8	60.2	72.6	0.14
Gradient Boosting	84.9	73.9	48.2	95.2	58.4	71.7	0.15
Adaboost	84.4	72.2	47.4	94.8	57.2	71.1	0.15
Decision Tree	78.5	50.9	52.1	85.8	51.5	69.0	0.21
Gaussian NB	81.1	56.9	57.6	87.7	57.3	72.2	0.18
MLP	84.5	69.9	51.3	93.8	59.2	72.5	0.15

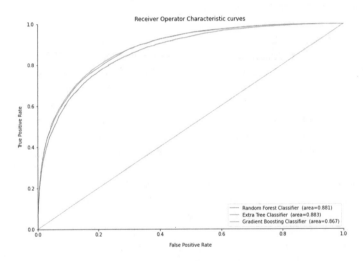

X axis-True positive Rate
Y axis-False positive Rate

Fig. 4 ROCAUC curve

farmers to predict rain and plan the work accordingly. This helps in reducing the loss of crop production if any unnecessary flooding occurs due to sudden rainfall.

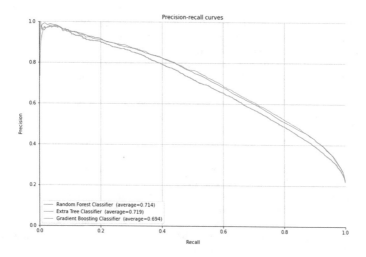

X axis-Precision
Y axis-Recall

Fig. 5 Precision recall curve

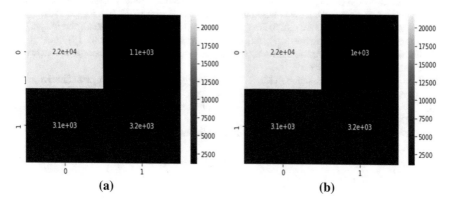

Fig. 6 **a** Confusion Matrix for Random Forest model, **b** Extra Trees model

References

1. Maspo N-A et al (2020) Evaluation of Machine Learning approach in flood prediction scenarios and its input parameters: a systematic review. IOP Conf Ser Earth Environ Sci 479(1). IOP Publishing
2. Aswin S, Geetha P, Vinayakumar R (2018) Deep learning models for the prediction of rainfall. In: 2018 international conference on communication and signal processing (ICCSP). IEEE
3. Thirumalai C et al (2017) Heuristic prediction of rainfall using machine learning techniques. In: 2017 international conference on trends in electronics and informatics (ICEI). IEEE
4. Dash Y, Mishra SK, Panigrahi BK (2018) Rainfall prediction for the Kerala state of India using artificial intelligence approaches. Comput Electr Eng 7066–73

5. Geetha A, Nasira GM (2014) Data mining for meteorological applications: decision trees for modeling rainfall prediction. In: 2014 IEEE international conference on computational intelligence and computing research. IEEE

6. Pedregosa F et al (2011) Scikit-learn: machine learning in Python. J Mach Learn Res

7. Schaul T, Bayer J, Wierstra D, Sun Y, Felder M, Sehnke F, Rückstieß T, Schmidhuber J (2010) PyBrain. J Mach Learn Res

8. Cramer S, Kampouridis M, Freitas AA (2018) Decomposition genetic programming: an extensive evaluation on rainfall prediction in the context of weather derivatives. Appl Soft Comput

9. Oswal N (2019) Predicting rainfall using machine learning techniques. arXiv preprint arXiv: 1910.13827

10. Ridwan WM et al (2021) Rainfall forecasting model using machine learning methods: case study Terengganu, Malaysia. Ain Shams Eng J

11. Sumi SM, Zaman MF, Hirose H (2012) A rainfall forecasting method using machine learning models and its application to the Fukuoka city case. Int J Appl Math Comput Sci

12. Basha CZ et al (2020) Rainfall prediction using machine learning and deep learning techniques. In: 2020 international conference on electronics and sustainable communication systems (ICESC). IEEE

13. Naidu D, Majhi B, Chandniha SK (2021) Development of rainfall prediction models using machine learning approaches for different agro-climatic zones. In: Handbook of research on automated feature engineering and advanced applications in data science. IGI Global

14. Gowtham Sethupathi M, Sai Ganesh Y, Ali MM (2021) Efficient rainfall prediction and analysis using machine learning techniques. Turk J Comput Math Educ (TURCOMAT)

15. Adnan RM et al (2021) Short term rainfall-runoff modelling using several machine learning methods and a conceptual event-based model. Stochast Environ Res Risk Assess 35(3)

16. Diez-Sierra J, del Jesus M (2020) Long-term rainfall prediction using atmospheric synoptic patterns in semi-arid climates with statistical and machine learning methods. J Hydrol

17. Refonaa J et al (2019) Machine learning techniques for rainfall prediction using neural network. J Comput Theor Nanosci

18. Van Klompenburg T, Kassahun A, Catal C (2020) Crop yield prediction using machine learning: a systematic literature review. Comput Electron Agric

19. Choubin B et al (2019) Snow avalanche hazard prediction using machine learning methods. J Hydrol

20. Sarvani K et al (2021) Rainfall analysis and prediction using machine learning techniques. J Eng Sci

21. Jalgaonkar M, Kulkarni D (2022) Rainfall prediction using regression and multiple algorithms

Explainable AI Model to Minimize AI Risk and Maximize Trust in Malignancy Detection of the Pulmonary Nodules

Mahua Pal and Sujoy Mistry

1 Introduction

Modern intelligent systems such as image-based medical diagnostic AI systems are becoming the most accurate techniques for predicting cancer. But these AI models due to its unexplainable black box problem lack transparency and create a barrier for clinical implementation due to trust issues. Resolving this issue, recent development in the area of XAI shows the way to overcome the problem of deep learning architectures by providing after the fact explanations. XAI output is not only used for validating the prediction, but also recommends some correction in the prediction model and explores some new potential biomarkers. These explanations can be local, global post-hoc, ante-hoc, visual and textual. Many research works have been conducted using CNN, DNN models in the medical fields such as dermatology, ophthalmology and radiology. Few AI models [1] were also developed using different ML algorithms in diagnosing lung cancer where pulmonary nodules are the abnormal growths in one or both lungs that show up on CT scans. Lung cancer is the foremost contributor to cancer-related mortality, resulting in 1.3 million cancer deaths per year worldwide [2]. Due to the lack of the adequate number of qualified medical doctors and other healthcare professionals in India, there is a major concern and associated burden towards the cancer morbidity and mortality. Here, AI-based CAD models could be acting like an augmented doctor [3], an assistant system to the clinical experts and medical practitioners where medical practitioners control the process. Henceforth, this CAD model is a human-in-the-loop (HITL) model [4] where interpretations could be made afterwards (post-hoc) or interpretations could be derived

M. Pal (✉)
Department of Sciences and Commerce, J. D. Birla Institute, Kolkata, India
e-mail: mahuag@jdbikolkata.in

S. Mistry
Department of Computer Science and Engineering, Maulana Abul Kalam Azad University of Technology, Kolkata, India

© The Author(s), under exclusive license to Springer Nature Singapore Pte Ltd. 2023
D. S. Sisodia et al. (eds.), *Machine Intelligence Techniques for Data Analysis and Signal Processing*, Lecture Notes in Electrical Engineering 997,
https://doi.org/10.1007/978-981-99-0085-5_38

in their architecture (ante-hoc). Usually an ante-hoc method-based AI model suffers from lower accuracy than that of post-hoc methods, as its modelling capacity may be limited due to the architectural restriction.

An XAI model could be utilized for the preliminary investigation of the states of pulmonary nodules by identifying the reasons as explanations to doctors and patients. Developing an XAI model using SHAP and LIME tools in detection of lung cancer from the CT scan reports of patients is a novel attempt that could assist the medical practitioners in their first step of the lung cancer medical treatment process. With this perspective, this XAI model was developed using a binary AI classification algorithm—XGBoost and two XAI tools were applied on the top of this classifier for global and local post-hoc interpretation. The ante-hoc explanation feature of the tree explainer 'XGBoost' was also taken into account for the explanations of the malignancy of the pulmonary nodules. This intelligent automation could reduce the stress of the preliminary examination and detection process taken by the healthcare system to identify the malignant cases, and thus, medical experts' teams can concentrate on the emergency cases. Many a time the experimental result of AI model looks uncertain for unrelated data features of an image to detect pulmonary nodules besides providing the correct prediction, and sometimes there may be serious contrariety to give proper explanation in how different deep learning models give the same prediction. Our proposed model focuses on the fact that XAI can provide trust in AI systems through it is an elevated predictability and understandability. We have arranged this research paper in the next few sections in the following order with the background motivation of this work along with some literature reviews in Sect. 2, proposed model with methodology in Sect. 3, implementations and result discussions in Sect. 4 and finally the conclusion with future work in Sect. 5.

2 Background

The importance and different methods of explainability for this kind of AI intelligent model for detection of the disease such as lung cancer based on medical images are thoroughly discussed in this section. Visual heat maps are the most popular methods for explaining and interpreting image-based classification. Local explanations are useful for explaining a single data sample at a time as an assistive diagnosis system, whereas global-level explanations are useful for understanding the model performance as a whole to identify decision biases. Class activation mapping (CAM) [5], GRAD-CAM [6], gradient-based methods like saliency, integrated gradient, DeepLift [7] or methods based on mathematical decomposition like layerwise-relevance propagation (LRP), agglomerative contextual decomposition (ACD) and Shapley additive explanations (SHAP) [8, 9] methods access the model parameters to understand the attribution of the classification results to the individual pixels and the model architecture. Model-agnostic perturbation-based methods can be used for model independent explanation without knowledge of their internals. LIME [10], occlusion, RISE and extremal perturbation are the examples which differ in the occlusion strategy (procedure and perturbation) [11].

Siddhartha et al. [12] proposed an XAI model to predict the postoperative life expectancy in the lung cancer patients after surgery by using SHAP and LIME techniques on the top of ML model built using random forest algorithm. They had examined the data of those patients who were already qualified for surgery after the detection of lung cancer. Bartczak et al. [13] proposed another XAI model to predict the postoperative life expectancy in the lung cancer patients after surgery by using SHAP and Ceteris Paribus methods on the top of ML model built using hyperparameters cross validation and logistic regression. Venugopal et al. [14] had developed a model with 20-layer deep residual CNN and applied occlusion technique on a nodule and thus clinical attribution heat maps on the nodules provided aid to the radiologists to identify features aiding classification. Ahmed et al. [15] applied XAI techniques in stack ensemble ML framework which was built using generalized linear model (GLM), random forest (RF), gradient boosting machine (GBM), extreme gradient boosting machine (XGBoost), deep neural network (DNN) and visualized the risk factors of lung and bronchus cancer (LBC) mortality from the stack ensemble model's output in global and local scales after considering the input features such as air-pollution and socio-economic status. So far, a very insignificant amount of research works have been conducted to enforce trust in the healthcare system to accept the AI CAD model required for preliminary automated investigations of lung cancer malignancy and the predictions with minimal AI risk. These models will not replace doctors, but facilitate them to provide faster and better service. Henceforth, there needs to be a lot of research work, discovery of new explainable algorithms and fine-tunings which motivated us to work in this field. The XAI model in this paper explained the reasons while predicting the state of cancerous cells in lungs through heat maps by checking the contributions of the input biomarkers obtained from CT scan report of a patient, and this XAI model was fine-tuned by rebuilding it using the important input biomarkers and prediction was verified with the help of the outputs of SHAP and LIME methods.

3　Proposed Model with Methodology

In this research work, we propose an XAI-based CAD model that uses SHAP and LIME techniques on the top of an XGBoost classifier to classify benign and malignant states of lung cancer. The main goal of this work was to employ AI-based techniques to detect infectious pulmonary nodules and to use XAI to understand and compare the strategy of each model to increase trust. Accordingly, XAI techniques were applied to detect lung cancer using the knowledgebase of radiological biomarkers from LIDC-IDRI's diagnostic data and lung cancer screening thoracic computed tomography (CT) scans with marked-up annotated lesions [16]. The flow diagram of the proposed model is shown in Fig. 1.

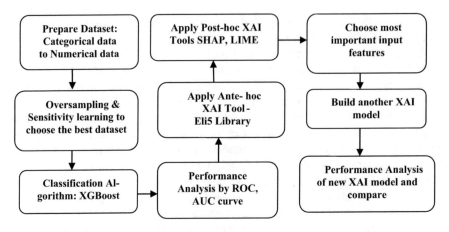

Fig. 1 Flow diagram of proposed model

The XAI techniques were applied to interpret the predictions made by the XGBoost classification model. Finally, based on the highlighted features visualized in the heat maps, the important biomarkers were selected that had higher influence to predict the malignancy of pulmonary nodules and unnecessary biomarkers were excluded to build another model that could speed up the execution time of the model.

The Lung Image Database Consortium image collection (LIDC-IDRI) provides a web-accessible international resource for development, training and evaluation of intelligent CAD models for lung cancer detection and diagnosis. In this study, 1307 data which were released in October 2011 were investigated. This dataset contained images from a clinical thoracic CT scan and an associated XML file that recorded the results of a two-phase image annotation process performed by four experienced thoracic radiologists. In each CT scan, the marked lesions belong to one of three categories.

Nodule > or = 3 mm	Nodule < 3 mm	Non-nodule > or = 3 mm
May be malignant	Could not be annotated; not malignant	Not malignant

Features listed in the dataset for conducting this work with categorical values were **Calcification, Internal _Structure, Lobulation, Margin, Sphericity, Spiculation, Subtlety, Texture, Number_of _Nodules, Nodule_Size_More_than_3 mm, Malignancy.** We classified malignancy into two states—malignant (high-risk) and benign (low-risk). The annotated grading 1 and 2 under malignancy, featured in the pylidc documentation [17] were considered as **benign** class and the grading 3, 4 and 5 were considered as **malignant** class. After referring to the pylidc documentation, we arranged the dataset with categorical values under each input feature attribute. The categorical data were converted to numerical. Since the dataset had 493 low-risk/benign data and 814 high-risk/malignant data, the problem related to the imbalance classes was handled by comparing the original dataset performance with the oversampled dataset performance. The oversampling of the dataset was done using the ADASYN, ROS and SMOTE algorithm separately, and finally, the ADASYN method was chosen to balance the dataset. The 80% data were used for training the model, and 20% data were used for testing. XGBoost classifier supports binary classification. After oversampling and sensitivity analysis, these ten different features or biomarkers were fed into the supervised binary classification AI model built using XGBoost tree-explainer classifier. The SHAP and LIME XAI tools were applied on the top of this classification model to highlight the important features with their corresponding risk factors from their visual heat maps. Thus, an XAI model was produced. The important input features which contributed more to the prediction of malignancy have been taken into account and that initial AI model has been reconstructed (the classifier algorithm remained the same, i.e. XGBoost classifier). Thus, the first model got fine-tuned as the performance of this reconstructed model was found better than that of the previous one. SHAP and LIME tools were again applied to validate the model prediction by observing the output of SHAP and LIME methods, and these interpretations were found consistent with biological intuition which enforce trust to the system.

4   Implementation and Results Discussions

The performance of the XGBoost AI model for detecting lung cancer at its initial stage was illustrated in this section. XGBoost [18] is a popular scalable machine learning system for tree boosting. Deep neural networks (DNNs) are a state-of-the-art model which is popularly used in medical diagnosis and needs a humongous amount of data to show their relevance, so we preferred the XGBoost model for this problem with 1307 data. The coding of the flow of tasks (Fig. 1) was done using Python language in Google Colaboratory platform which provides high-end support. This dataset had an imbalance-dataset problem which was resolved by using an oversampling algorithm. Synthetic minority oversampling technique (SMOTE) and adaptive synthetic (ADASYN) are the methods that generate synthetic data. The XGBoost classification model's performances with those oversampled data were separately evaluated based on the different metrics such as AUC value, F1 score, precision and specificity. We used a fivefold cross validation technique here. Area under the ROC curve (AUC) which was 0.952 signified aggregate measure of performance across all possible classification thresholds was outstanding. Therefore, ADASYN [19] was finally chosen as the oversampling technique for this model based on the AUC value and other metrics (Table 1), and finally, the total number of data was 1466 in the dataset. Based on the performance value of important metrics (Table 1) after dividing the dataset into 80–20, 70–30, 60–40 for training and testing, 80–20 train-test division of the dataset was implemented in this model.

Table 1 Comparing SMOTE versus ADASYN methods

Method	Train-test (%)	Metrics	AUC	F1 score
ADASYN	80–20	Accuracy = 0.929 Precision = 0.908 Recall (TPR) = 0.977 Specificity (TNR) = 0.861	0.952	0.941237
	70–30	Accuracy = 0.928 Precision = 0.913 Recall (TPR) = 0.940 Specificity (TNR) = 0.915	0.934	0.926303
	60–40	Accuracy = 0.939 Precision = 0.931 Recall (TPR) = 0.946 Specificity (TNR) = 0.933	0.940	0.93844
SMOTE	80–20	Accuracy = 0.930 Precision = 0.883 Recall (TPR) = 0.994 Specificity (TNR) = 0.841	0.950	0.935218
	70–30	Accuracy = 0.942 Precision = 0.896 Recall (TPR) = 0.994 Specificity (TNR) = 0.849	0.944	0.942459
	60–40	Accuracy = 0.915 Precision = 0.904 Recall (TPR) = 0.890 Specificity (TNR) = 0.903	0.947	0.934048

4.1 Machine Learning Model

The performance of the XGBoost classification model (Table 1) showed that the AUC of the model under the ROC curve was 0.952. The accuracy of the model is 0.929, and specificity is 0.861 (Fig. 2).

Accuracy = 0.929
Precision = 0.908
Recall (TPR) = 0.977
Specificity (TNR) = 0.861
Fallout (FPR) = $1.393e{-}01$
gmean = 0.9168644872558049.

4.2 Explainable AI

Ante-hoc Methods: The critical risk factors for the detection of the state of the pulmonary nodules using the XGBoost model were explained by its inherent feature

Fig. 2 Confusion matrix of XGBoost classification model

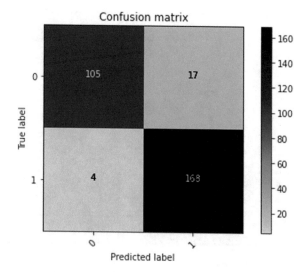

explanation methods. XGBoost has its default library, eli5, which was used to find out the impact of the individual feature over the prediction by the model (Table 2). It was observed that Calcification was the input feature contributing maximum in the prediction of the state of the nodules.

Post-hoc Methods

SHAP: The SHAP tool was used to highlight the risk factors for explanations and interpretations of the predictions. Once the result is explained and interpreted, the trusts of the end-users will automatically build up. Figure 3 illustrated the SHAP summary plot which validated the result obtained by the inherent feature explanation methods of XGBoost classifier (Table 2).

Table 2 Feature versus weight contribution list

Feature	Weight
Calcification	0.512
Internal structure	0.2331
Lobulation	0.0848
Margin	0.0584
Sphericity	0.0307
Spiculation	0.0291
Subtlety	0.0277
Texture	0.0242
Number of nodules	0
Nodule_Size More_than_3 mm	0

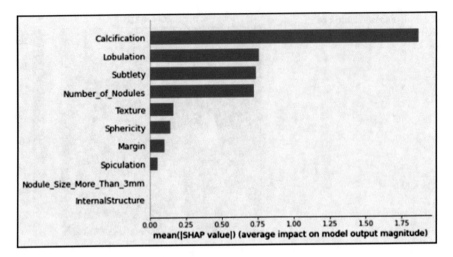

Fig. 3 SHAP summary plot

The feature contribution summary plot (Fig. 4) illustrated top feature contributions and also illustrated data point distribution to provide visual indicators of how feature values affect predictions. Calcification, Number_of_Nodules, Texture of the nodules and Subtlety had a high positive impact while detecting the state of the pulmonary nodules as malignant. Here, the red colour signified the positive influence on declaring the nodule was malignant, whereas the blue colour implied the negative influence.

The dependency plots (Fig. 5) illustrated the relationship of the calcification of the nodules with the texture of the nodules. Similarly, few other dependency plots

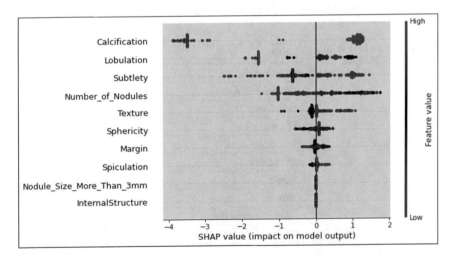

Fig. 4 SHAP feature contribution summary plot for global explanation

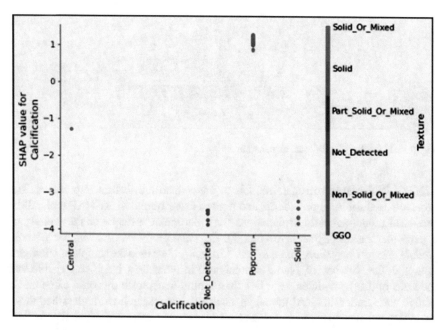

Fig. 5 SHAP dependency plot

were created. The higher the SHAP value of an input feature or biomarker, the higher the influence of that biomarker in the detection of cancerous cells, thus the nodules with 'popcorn' calcification pattern and irregular shape with solidified texture were more prone to be of malignant in nature (cancerous cells). In this way, dependency plots could explain any dependency in between the features which can lead to some unseen properties that could be revealed by these plots.

The partial dependency plot visualized the relationship of a biomarker with its SHAP value to understand its overall impact in the final prediction of the system. The higher the SHAP values implicated, the higher the risk factors.

The force plot was used for local prediction, and it was formed based on any single record by calculating the SHAP values of the biomarkers of that individual patient (one specific single record). Here, the force plot (Fig. 6) visualized -2.43 was the model output for the sample with index 1 using the following coding in Python:

```
shap.force_plot(shap_explainer.expected_value,          test_shap_values[1,:],
X_test_disp.iloc[1,:])
```

All of the features' (Calcification, Sphericity, Texture, etc.) values led to the prediction value of -2.43, which was then converted to a value of 0 class (Benign). SHAP plotted the top most influential features for the sample under study. Features in red colour influenced positively, i.e. dragged the prediction value closer to class 1 (malignant), features in blue colour influenced the opposite. The force plot is very helpful to detect individual patients' status from their CT scan report.

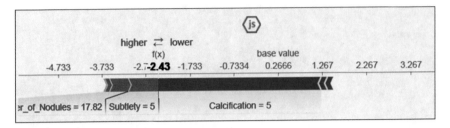

Fig. 6 SHAP force plot for local explanation

LIME: LIME makes surrogate models to understand the feature importance. The plots are used for local prediction, and it gives faster output than SHAP tool. LIME produced model-agnostic explanations for a particular instance or the vicinity of a particular instance, and here (Fig. 7), the values of *Calcification, Lobulation, Nodule_Size_More_Than_3mm, Internal Structure* classified the sample as 'Benign'. Although the Number_of_Nodules influenced in other way, but it was visible that the sizes of all the nodules were less than 3 mm, henceforth this case came under benign class under this XAI model. Henceforth, our research work classified those physiological attributes as a high-risk factors for lung cancer, for which the medical practitioners required a post-hoc clarification and provided explanations consistent with biological intuition.

Similarly, another refined AI model using XGBoost classifier was created with few biomarkers and the performance of the new model was evaluated based on the result of important metrics. Here, the second refined AI model was built by selecting seven important features only, such as **'Calcification', 'Lobulation', 'Margin',**

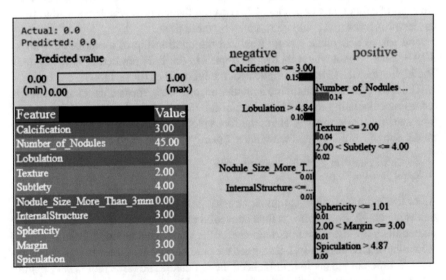

Fig. 7 LIME plot for model—agnostic local explanation

Fig. 8 Confusion matrix of refined classification model

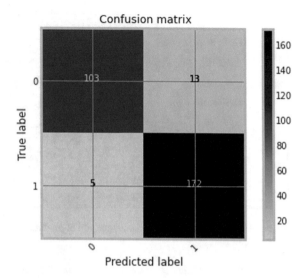

'Sphericity', 'Subtlety', 'Texture', 'Number_of_Nodules', 'Malignancy'. The values of the important metrics of the second AI model are shown in Fig. 8.

Accuracy = 0.939
Precision = 0.930
Recall (TPR) = 0.972
Specificity (TNR) = 0.888
Fallout (FPR) = 1.121e−01.

It was understood that after selecting seven features out of eleven features, the performance of the refined model was better than that of the initial model. The results of SHAP for the new model were displayed in Fig. 9. The influence of biomarkers remained the same to predict the malignancy of pulmonary nodules. Henceforth, any new test data can also be fed into this XAI model and the model can predict along with highlighting those biomarkers as reasons for prediction which endow with the confidence to the medical practitioners, doctors and patients to trust the model.

5 Conclusion with Future Work

In this research work, the prediction of XGBoost AI binary classification model was successfully interpreted by XAI tools and the interpretations achieved by using the XAI tools were applied to rebuild and reform the classification model by selecting only the most important input features and that made the execution speed of the new model faster than the previous one. Through experimental results we observed that, many times AI approaches looks indecisive for extraneous oversampling data features of an image to detect pulmonary nodes (although it may correctly predicted),

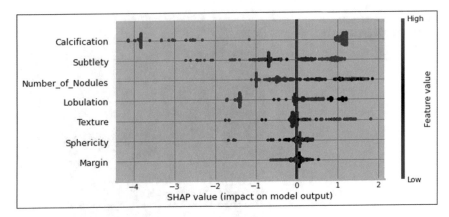

Fig. 9 SHAP feature contribution summary plot of final refined classification model

and also sometime there may be serious contrariety to give proper explanation in how different deep learning models give the same prediction. Our proposed model focuses on the fact that XAI can provide trust in AI systems especially that are used in high-stake decision process and can minimize AI implementation risk factors. Similar XAI models could be built to diagnose other diseases as well. Here, the XAI model was formed using XGBoost's inherent architecture and two post-hoc XAI tools SHAP and LIME. We could explore other new XAI methods and different statistical approaches on hybrid AI models for assured predictions, and those trustable prediction models could be extended to multi-class detection models from a binary classification model to grade cancer stages.

References

1. Zhu P, Ogino M (2019) Guideline-based additive explanation for computer-aided diagnosis of lung nodules. In: Interpretability of machine intelligence in medical image computing and multimodal learning for clinical decision support. Springer, Cham, pp 39–47
2. The global burden of disease 2004 update. WHO. https://www.who.int/healthinfo/global_bur den_disease/GBD_report_2004update_full.pdf. Last accessed 2008
3. AMA passes first policy recommendations on augmented intelligence. https://www.ama-assn. org/press-center/press-releases/ama-passes-first-policy-recommendations-augmented-intell igence. Last accessed 2018/06/14
4. Singh A, Sengupta S, Lakshminarayanan V (2020) Explainable deep learning models in medical image analysis. J Imaging 6(6):52
5. Zhou B, Khosla A, Lapedriza A, Oliva A, Torralba A (2016) Learning deep features for discrim-inative localization. In: Proceedings of the IEEE conference on computer vision and pattern recognition, pp 2921–2929
6. Selvaraju RR, Cogswell M, Das A, Vedantam R, Parikh D, Batra D (2017) Grad-cam: visual explanations from deep networks via gradient-based localization. In: Proceedings of the IEEE international conference on computer vision, pp 618–626

7. Shrikumar A, Greenside P, Kundaje A (2017, July) Learning important features through propagating activation differences. In: International conference on machine learning. PMLR, pp 3145–3153

8. An introduction to explainable AI with Shapley values. https://shap.readthedocs.io/en/latest/overviews.html. Last accessed 2020/12/19

9. Lundberg SM, Lee SI (2017, December) A unified approach to interpreting model predictions. In: Proceedings of the 31st international conference on neural information processing systems, pp 4768–4777

10. Tulio Ribeiro M, Singh S, Guestrin C (2016) Why should I trust you? Explaining the predictions of any classifier. arXiv e-prints, pp.arXiv-1602. https://doi.org/10.1145/2939672.2939778

11. Lucieri A, Bajwa MN, Dengel A, Ahmed S (2020) Achievements and challenges in explaining deep learning based computer-aided diagnosis systems. arXiv preprint arXiv:2011.13169

12. Siddhartha M, Maity P, Nath R (2020) Explanatory artificial intelligence (XAI) in the prediction of post-operative life expectancy in lung cancer patients. Int J Sci Res 8

13. Bartczak M, Partyka M Chapter 8 Story lungs: eXplainable predictions for post operational risks. Available at https://pbiecek.github.io/xai_stories/story-lungs.html

14. Venugopal VK, Vaidhya K, Murugavel M, Chunduru A, Mahajan V, Vaidya S, Mahra D, Rangasai A, Mahajan H (2020) Unboxing AI-radiological insights into a deep neural network for lung nodule characterization. Acad Radiol 27(1):88–95

15. Ahmed ZU, Sun K, Shelly M, Mu L (2021) Explainable artificial intelligence (XAI) for exploring spatial variability of lung and bronchus cancer (LBC) mortality rates in the contiguous USA. Sci Rep 11(1):1–15

16. The cancer imaging archive (TCIA) public access. LIDC-IDRI. Available at wiki.cancerimagingarchive.net/display/Public/LIDC-IDRI

17. Hancock M Pylidc. MIT. https://pylidc.github.io/tuts/annotation.html

18. Rathe A Random forest vs XGBoost vs deep neural network. Kaggle. https://www.kaggle.com/arathee2/random-forest-vs-xgboost-vs-deep-neural-network. Last accessed 2017/05/18

19. He H, Bai Y, Garcia EA, Li S (2008, June) ADASYN: adaptive synthetic sampling approach for imbalanced learning. In: 2008 IEEE international joint conference on neural networks (IEEE world congress on computational intelligence). IEEE, pp 1322–1328

Empirical Analysis of Hybrid Classical Variational Quantum Neural Networks for Target Classification from SAR Data

Pranshav Gajjar, Aayush Saxena, Divyesh Ranpariya, Pooja Shah, and Anup Das

1 Introduction

Image classification has always been one of the most traffic-generating topics in the field of machine learning. It has reconstituted countless advancements in numerous fields like medical imaging, pattern recognition, and satellite imaging to name a few [1, 2]. It is mainly used to extract some useful information from a heap of data and act as an assistant for a variety of tasks, especially the ones concerned with images. Now manually doing the whole process of checking every image, identifying the pattern and classification would be a laborious task. Hence, automation of computer vision tasks is an important endeavor.

There exist plenty of methods and approaches for deep learning tasks in the modern literature [3–5]. Many methods that were developed matched the human accuracy in identifying the traffic signs and handwritten text using neural networks [6]. Presently, there exists several artificial intelligence approaches inspired by human intelligence itself; one of these approaches is transfer learning [7]. It is based on the concept

P. Gajjar (✉) · D. Ranpariya · P. Shah
Institute of Technology, Nirma University, Ahmedabad, India
e-mail: 19bce060@nirmauni.ac.in

D. Ranpariya
e-mail: 18bec083@nirmauni.ac.in

P. Shah
e-mail: pooja.shah@nirmauni.ac.in

A. Saxena
SRM Institute Of Science and Technology, Chennai, India
e-mail: as3368@srmist.edu.in

A. Das
Space Applications Centre, ISRO, Ahmedabad, India
e-mail: anup@sac.isro.gov.in

© The Author(s), under exclusive license to Springer Nature Singapore Pte Ltd. 2023
D. S. Sisodia et al. (eds.), *Machine Intelligence Techniques for Data Analysis and Signal Processing*, Lecture Notes in Electrical Engineering 997,
https://doi.org/10.1007/978-981-99-0085-5_39

that when we learn something from one situation it can be directly or indirectly applied into a completely different situation altogether. This same principle of transfer learning was used to perform reinforcement learning tasks, simply put, the data used to learn one task were taken and applied to learn a different, but related, task, which made learning the latter much easier [8]. Transfer learning approaches were also applied to natural language processing [9]; furthermore, a unified framework [10] was created to simplify problems requiring transfer learning. Everything considered transfer learning can be a viable approach for designing neural networks that fetch cutting-edge accuracy [11].

There also have been developments in leveraging quantum computing for machine learning tasks, as research on quantum machine learning models [12] to solve complicated tasks increases day by day, the combinational architectures of conventional deep neural networks/classical networks might be an effective solution to various prediction oriented tasks. There also exist numerous paradigms, which are notoriously difficult for deep learning implementations and one such domain is SAR data [13]. Synthetic Aperture Radar (SAR), is a radar that adopts the procedure of generating high-resolution images by using the passage of the antenna over a great span of distance, thus essentially creating a broad "synthetic" antenna aperture [14]. The problem statements associated with SAR images have also seen advancements by the use of machine learning algorithms, refer to Table 1. This paper aims to incorporate quantum neural networks as an extension to the existing technologies for an improved classification experience.

The relevance and usability of quantum networks can be understood by using metrics like effective dimensions [22] which represents the size that a model occupies in model space (space consisting of all possible functions associated with a particular model class). The paper [22] showed that quantum-based networks showed a higher effective dimension than its classical counterparts, hence justifying their use for SAR classification.

The methods proposed in the paper [23] leveraged transfer learning for quantum networks, as a procedure to obtain a hybrid or a dressed quantum circuit, which can be a useful prediction model. The summary of this paper is to apply hybrid quantum

Table 1 Recent studies related to machine learning and SAR data

Approach	Dataset	Result
Sparse representation-based classification [15]	MSTAR dataset	The average recognition accuracy of 95%
A-ConvNets architecture [16]	MSTAR dataset	An average accuracy of 96%
Boundary based Edge description [17]	Data from sentinel 1	Mean accuracy of 90%
OSCNet and VGG-16 [18]	Data obtained from envisat	Accuracy of 95.09 %
MultiEdge Detection [19]	ERS-1 SAR archive	73% Accuracy was recorded
Oil-Fully ConvNet [20]	Data from sentinel 1	An accuracy of 98.8%
CNN with ResNet-101 [21]	Data from sentinel 1	Highest accuracy of 89.3%

networks on SAR imaging using the moving and stationary target acquisition and recognition (MSTAR) dataset [24]. The existing state-of-the-art convolutional neural networks like ResNet-18 [25] and DenseNet-161 [26] are assessed in the form of a dressed quantum circuit, and a baseline SqueezeNet [27]-inspired architecture is also used to obtain a harsher comparison setting for hybrid networks and their classical counterparts. The paper contributes to novel architectures and offers experiments and inferences on quantum-enhanced neural networks.

2 Hybrid Classical-Quantum Circuits

The quantum network used here can be explained as a variational circuit or a parametrized quantum circuit [23]. A variational quantum circuit of depth q can be explained as a concatenation of many quantum layers, where each quantum layer can be a sequence of single-qubit rotations followed by a fixed sequence of entangling gates [28, 29] or, in the case of optical modes, some active and passive Gaussian operations followed by single-mode non-Gaussian gates [30].

For the task of transfer learning, classical to quantum networks are used, so analogous to Fig. 1, the generic networks/encoders are trained on the ImageNet dataset [31], and the specific dataset used here is a preprocessed MSTAR subset. The portion B is a dressed variational circuit [23] with a depth of 2 and 4 qubits [23] per layer, which is meant to replace the fully connected portion of the encoder.

For a comparative analysis, transfer learning is also tested with a classical network. Non-pretrained variations are also used for an unbiased comparison of the pipelines.

For a stricter comparison of hybrid networks and classical networks, a vanilla SqueezeNet with reduced layers is used and a hybrid quantum-squeezeNet abbreviated as HQuantSN is created, and its classical variant replaces the quantum circuit with a linear layer counterpart, where the quantum layer structure is replicated and the model is abbreviated as SCN. The lightweight nature of the SqueezeNet enables a better comparison of the quantum circuit and the linear layers; all the experimental results are thoroughly explained in Sect. 3.

Fig. 1 Sample transfer learning pipeline [23]

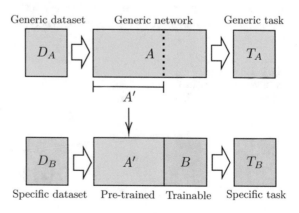

2.1 Encoders

Two encoder architectures mainly ResNet-18 and DenseNet-161 are tested; for each network, four experiments are considered:

1. Pretrained-Encoder + Quantum Circuit.
2. Pretrained-Encoder + Classical Net
3. Encoder + Quantum Circuit
4. Encoder

The equivalent classical layer is a linear layer that outputs the total number of classes. The third category resembles a hybrid network where the encoder layers are not pretrained. The 4th category represents the vanilla architecture with no pretrained weights and a modified final layer for the total number of classes.

ResNet-18 This subsection provides a summarized understanding of ResNets as defined in the original paper [25]. The residual networks more commonly known as ResNets are one of the proposed encoders that we are going to use. The uniqueness that ResNets offer, over plain networks, is that they address the degradation problem that was exposed when the overly deep networks started to converge. The ResNets introduced identity mapping which meant that the input from a previous layer was taken and passed to another layer as a shortcut.

FIgure 2 shows the identity mapping feature and the notion of skip connection.

Mostly, the 34-layer and 18-layer ResNets are used as they fetch less error and better accuracy as compared to their plain competitors. The 34-layer ResNet displays fairly decreased training error and handles the degradation problem that is observed in its plain compeer, thus high accuracy is gained from increased depths. Not to overlook the fact that the 18-layer ResNet also fetched a better accuracy than its plain compeer; the 18-layer ResNet was able to achieve convergence faster and obtain good solutions on smaller datasets [25]. For the experiments in this paper, the ResNet-18 variant of residual networks is used.

DenseNet-161 This subsection provides a summarized understanding of the DenseNet architecture as defined in the original paper [26]. The dense convolution network, also known as DenseNet, is another variety of encoders used in the proposed architecture. The advantage of using this particular type of network is that each of its

Fig. 2 Constituent architecture for Residual network functionality [25]

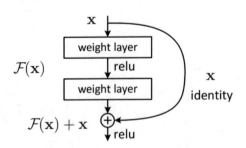

layers gathers supplementary inputs from all of the layers before it. The data are concatenated, so essentially every single layer obtains the cumulative intelligence of all the antecedent layers. Therefore, when each layer obtains feature maps from the previous layers, it makes the complete network more compressed, which means that the total channels will be fewer.

The contrasting detail that separates the DenseNets from the ResNets is that they use the parameters more dexterously. Outwardly, both networks are quite similar; the only major difference is that the DenseNets concatenate the inputs while the process of summation is what happens in ResNets. Although this seems like a small adjustment, it brings out a rather considerable change in behavior between them both. Adding to that fact DenseNets crave extensively less number of parameters and computational power to obtain highly accurate and cutting-edge performances, and results with better accuracy can be achieved when the hyperparameters are tuned with attention to detail. In this regard, we will be using the DenseNet-161 model; it is one of the high accuracy models of the DenseNet group; the size of this model is considerably larger than its other variants at 100 MB [26].

Figure 3 displays the principle working of DenseNet in which all the layers take the previous feature maps as the input [26].

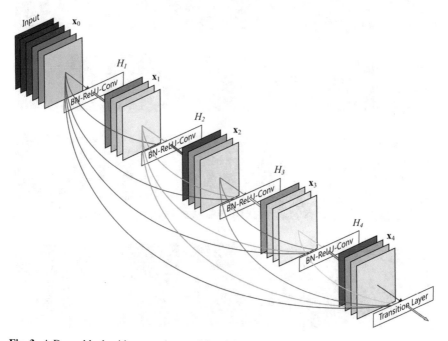

Fig. 3 A Dense block with a growth rate of 4 and 5 constituent layers

3 Dataset

The MSTAR dataset used here is a collection of SAR images collected by the San-
dia National Laboratory (SNL) X-band SAR sensor and was jointly sponsored by
Defense Advanced Research Projects Agency and Air Force Research Laboratory
[24, 32]. Due to the computationally expensive nature of quantum networks and
apparatus-centric bottlenecks, data from two classes, namely BMP2 and T72, are
used, and the images are resized and normalized according to the ImageNet standards. For a better understanding of a methodologies performance on a smaller
dataset, a 60–40 train-test split is used. The same train-test split is used across all the
methods to ensure an unbiased comparison.

Figure 4 depicts an MSTAR preview, where (a) and (b) represent the visible light
images and (c) and (d) are the respective SAR images for the targets measured at an
azimuth angle of 45 °C. The classes related to (a) and (b) are BMP2, BTR70, T72,
BTR60, 2S1, BRDM2, D7, T62, ZIL131, and ZSU23/4, in the same chronology [33].

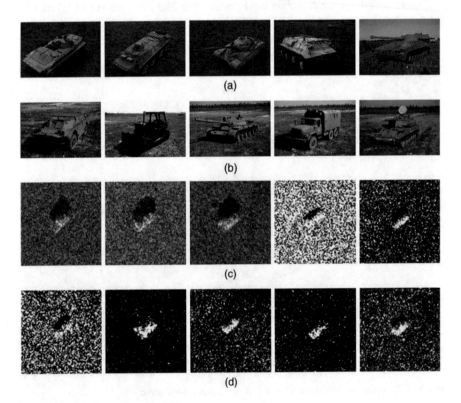

Fig. 4 MSTAR dataset samples [33]

4 Results and Discussion

The experiments as mentioned in Sect. 2 are performed using Pennylane [34] and PyTorch [35]; the Table 2 contains the classification results of the ResNet-18 and DenseNet-161 experiments, metrics like percentage accuracy, F1-score, precision, and recall [36] are used.

It can be conclusively said that for tasks centric to transfer learning, using a quantum network can show an increase in accuracy as when we compare rows 1, 3 there is an accuracy boost (0.49%). The same result is also observed in the DenseNet experiment (0.2%).

When the ResNet setup is considered, the quantum-enhanced model was outperformed by the ResNet, but still, a comparable accuracy was obtained; however, the DenseNet experiment showed an inverse result. It can also be inferred that networks involving transfer learning had an inferior performance than their non-pretrained counterparts.

To further understand the behavior of quantum-enhanced networks on a smaller backbone network, HQuantSN and SCN were trained; the experimental results obtained are mentioned in Table 3. It can be concluded that when a small classical encoder is used in conjunction with a quantum model, it pales in comparison when its classical equivalents are considered. The experimented networks can also be validated against the existing literature by referring the Table 1.

Table 2 Results for transfer learning and hybrid models

Model	Accuracy (%)	Precision	F1-Score	Recall
Pretrained ResNet-18 QuantumNet	82.60	0.8249	0.8265	0.8281
ResNet-18 + QuantumNet	92.67	0.9105	0.9258	0.9416
Pretrained ResNet-18 + ClassicalNet	82.11	0.7860	0.8157	0.8470
ResNet-18	**99.61**	**0.9942**	**0.9961**	**0.9980**
Pretrained DenseNet-161 + QuantumNet	82.31	0.7996	0.8195	0.8405
DenseNet-161 + QuantumNet	**99.22**	**0.9883**	**0.9922**	**0.9961**
Pretrained DenseNet-161 + ClassicalNet	82.11	0.7860	0.8153	0.8470
DenseNet-161	98.83	0.9903	0.9883	0.9864

All the models are optimized using Adam with a learning rate of 0.0004, which is reduced periodically for 15 epochs

Table 3 All the models are optimized using Adam with a learning rate of 0.0001, which is reduced periodically for 10 epochs

Model	Accuracy (%)	Precision	F1-score	Recall
HQuantSN	78.69	0.793	0.7859	0.7937
SCN	**95.99**	**0.9553**	**0.9599**	**0.9646**

5   Conclusion and Future Work

The paper aimed to leverage parametrized quantum circuits or quantum neural networks as an enhancement for the existing deep neural architectures, and find probable use cases and application-based utilities for SAR images. We used a subset of a publicly available dataset (MSTAR) and provided a novel empirical analysis. From the conducted experiments, it can be deduced that quantum-enhanced networks have their significant uses, and for a favorable scenario with properly pretrained classical or quantum networks, the use of highly accurate transfer learning encoders and quantum networks can be recommended to obtain the desired outcomes.

Furthermore, we aim to work on more complex quantum circuits and fully quantum convolutional networks, another way of development can be segmentation and other deep learning applications involving already existing forefront technologies that improve the accuracy and training time for quantum models. We also aim to assess the limitations related to the computationally expensive nature of quantum architectures and quantum enhancements.

References

1. Viswanathan J, Saranya N, Inbamani A (2021) Deep learning applications in medical imaging. In: Deep learning applications in medical imaging. IGI Global, pp 156–177
2. Mehta N, Shah P, Gajjar P (2021) Oil spill detection over ocean surface using deep learning: a comparative study 16(3–4):213–220
3. Gallego FJ (2004) Remote sensing and land cover area estimation 25(15):3019–3047
4. Lu D, Weng Q (2007) A survey of image classification methods and techniques for improving classification performance 28(5):823–870
5. Pal M, Mather PM (2003) An assessment of the effectiveness of decision tree methods for land cover classification 86(4):554–565
6. Ciresan D, Meier U, Schmidhuber J (20212) Multi-column deep neural networks for image classification. In: 2012 IEEE conference on computer vision and pattern recognition. IEEE
7. Weiss K, Khoshgoftaar TM, Wang DD (2016) A survey of transfer learning. J Big Data 3(1):1–40
8. Taylor ME, Stone P (2007) Cross-domain transfer for reinforcement learning. In: Proceedings of the 24th international conference on Machine learning—ICML '07. ACM Press
9. Ruder S, Peters ME, Swayamdipta S, Wolf T (2019) Transfer learning in natural language processing. In Proceedings of the 2019 conference of the North. Association for Computational Linguistics
10. Dai W, Jin O, Xue G-R, Yang Q, Yu Y (2009) Eigentransfer. In: Proceedings of the 26th annual international conference on machine learning—ICML '09. ACM Press
11. Pan SJ, Yang Q (2010) A survey on transfer learning 22(10):1345–1359
12. Verdon G, McCourt T, Luzhnica E, Singh V, Leichenauer S, Hidary J (2019) Quantum graph neural networks
13. Balnarsaiah B, Rajitha G (2021) Denoising and optical and SAR image classifications based on feature extraction and sparse representation
14. Gao X, Roy S, Xing G (2021) MIMO-SAR: a hierarchical high-resolution imaging algorithm for mmwave FMCW radar in autonomous driving
15. Song H, Ji K, Zhang Y, Xing X, Zou H (2016) Sparse representation-based SAR image target classification on the 10-class MSTAR data set 6(1):26

16. Haipeng W, Sizhe C, Xu F, Ya-Qiu J (2015) Application of deep-learning algorithms to MSTAR data. In: IEEE international geoscience and remote sensing symposium (IGARSS). IEEE
17. Ronghua S, Junkai L, Licheng J, Xiaohui Y, Yangyang L (2020) Superpixel boundary-based edge description algorithm for SAR image segmentation 13:1972–1985
18. Zeng K, Wang Y (2020) A deep convolutional neural network for oil spill detection from spaceborne SAR images 12(6):1015
19. Fjortoft R, Lopes A, Marthon P, Cubero-Castan E (1998) An optimal multiedge detector for SAR image segmentation 36(3):793–802
20. Bianchi FM, Espeseth MM, Borch N (2020) Large-scale detection and categorization of oil spills from SAR images with deep learning 12(14):2260
21. Orfanidis G, Ioannidis K, Avgerinakis K, Stefanos Vrochidis, and Ioannis Kompatsiaris. A deep neural network for oil spill semantic segmentation in sar images. In: 2018 25th IEEE international conference on image processing (ICIP). IEEE
22. Abbas A, Sutter D, Zoufal C, Lucchi , Figalli A, Woerner S (2021) The power of quantum neural networks 1(6):403–409
23. Mari A, Bromley TR, Izaac J, Schuld M, Killoran N (2020) Transfer learning in hybrid classical-quantum neural networks 4:340
24. Cristian C, Renre Thaens A (2018) Deep learning SAR target classification experiment on MSTAR dataset. In: 2018 19th international radar symposium (IRS). IEEE
25. He K, Zhang X, Ren S, Sun J (2016) Deep residual learning for image recognition. In: 2016 IEEE conference on computer vision and pattern recognition (CVPR). IEEE
26. Que Y, Lee HJ (2018) Densely connected convolutional networks for multi-exposure fusion. In: 2018 international conference on computational science and computational intelligence (CSCI). IEEE
27. Iandola FN, Han S, Moskewicz MW, Ashraf K, Dally WJ, Keutzer K (2016) Squeezenet: alexnet-level accuracy with 50x fewer parameters and <0.5MB model size
28. Maria S, Alex B, Krysta SM, Nathan W (2020) Circuit-centric quantum classifiers 101(3):3
29. Sim S, Johnson PD, Aspuru-Guzik A (2019) Expressibility and entangling capability of parameterized quantum circuits for hybrid quantum-classical algorithms 2(12):1900070
30. Killoran N, Bromley TR, Arrazola JM, Schuld M, Quesada N, Lloyd S (2019) Continuous-variable quantum neural networks 1(3):10
31. Krizhevsky A, Sutskever I, Hinton GE (2017) Imagenet classification with deep convolutional neural networks 60(6):84–90
32. Gu Y, Jiahui T, Lipeng F, Hui W (2021) Using VGG16 to military target classification on MSTAR dataset. In: 2nd China international SAR symposium (CISS). IEEE
33. Vasuki P, Roomi SMM (2013) Automatic target classification of man-made objects in synthetic aperture radar images using gabor wavelet and neural network 7(1):073592
34. Bergholm V, Izaac J, Schuld M, Gogolin C, Alam MS, Ahmed S, Arrazola JM, Blank C, Delgado A, Jahangiri S, McKiernan K, Meyer JJ, Niu Z, Száva A, Killoran N (2018) Pennylane: automatic differentiation of hybrid quantum-classical computations
35. Mishra P (2019) Introduction to neural networks using pytorch. In: PyTorch Recipes. Apress, pp 111–126
36. Sebastian B, Marika K, Christoph L, Florian K, Thomas V, Dirk L (2019) Application of an interpretable classification model on early folding residues during protein folding 12(1):1

COVID-19 Social Distancing Detection and Email Violation Mechanisms

Divyanshi Bhojak, Tarushi Jat, and Dinesh Naik

1 Introduction

The very first case of COVID-19 in India in Kerala, India on January 27, 2020. Within a few months, the rampant virus had spread over the world, impacting millions of individuals. The World Health Organization on March 11, 2020 labeled this condition as a pandemic. On October 8, 2020, the first-ever horrific amount of deaths in over 200 countries was 1,056,000.

The Ministry of Health and Family Welfare of India has requested the people of India to maintain social distancing of a minimum of 6 ft from each other while in a crowded area or any public place. Figure 1 [1] depicts how social distancing is implemented in public places in India to fight the global pandemic. By maintaining social distancing, the number of physical contacts in public areas between humans can be reduced, and thus it reduces the spread of infectious disease.

In March 2021, India had its unfathomable second wave of COVID-19, which claimed a catastrophic toll when the case count surpassed 1 lakh, triggering the pandemic's most explosive phase, during which the daily count of new infections peaked at 4.14 lakhs on May 6, 2021.

According to current research, persons who have been infected with the new coronavirus infection but have little or no symptoms may also be carriers of the virus. Social detachment is critical, especially for people at increased risk of serious

D. Bhojak (✉) · T. Jat · D. Naik
Department of Information Technology, National Institute of Technology Surathkal, Surathkal, Karnataka, India
e-mail: divyanshib.202it007@nitk.edu.in

T. Jat
e-mail: tarushijat.202it029@nitk.edu.in

D. Naik
e-mail: din_nk@nitk.edu.in

D. S. Sisodia et al. (eds.), *Machine Intelligence Techniques for Data Analysis and Signal Processing*, Lecture Notes in Electrical Engineering 997,
https://doi.org/10.1007/978-981-99-0085-5_40

Fig. 1 Social distancing scenario in India [1]

illness, such as adults over the age of 80 and newborns. Spread and disease severity can be considerably lowered by lowering the likelihood of transmission of the virus from an infected person to a healthy one.

2   Literature Survey

When the first case of COVID-19 was reported in Wuhan, China, many studies have been done in various aspects to take precautions against the virus. Authors in [2] suggested a deep learning-based framework for leveraging surveillance footage to automate the task of monitoring social separation. To eliminate individuals from the backdrop and track the recognized pedestrians, their approach employs the YOLOv3 object detection paradigm. Authors in [3] proposed a DNN model in conjunction with an updated inverse perspective mapping (IPM) technique and the SORT tracking algorithm, which resulted in reliable person detection and social distancing monitoring. Their model was trained on the Microsoft Common Objects in Context (MS COCO) and Google Open Image datasets, which are the research's two most comprehensive datasets. Authors in [4] proposed the YOLOv3 object recognition framework, but this model does not perform well with input videos having challenging outdoor environments. Their results show that object detection and tracking in distant and exterior environments are quite a difficult task, so inspired by this, we proposed work using YOLOv3 and YOLOv4 models. Authors in [5] proposed using IoT and multi-access edge computing is being used to provide a service that monitors and notifies users in real time if they are not breaking social distancing guidelines. Authors in [6] recommended social distancing in public spaces by using a video feed to continuously check the distance between people and alerting the responsible person so that the necessary steps can be done.

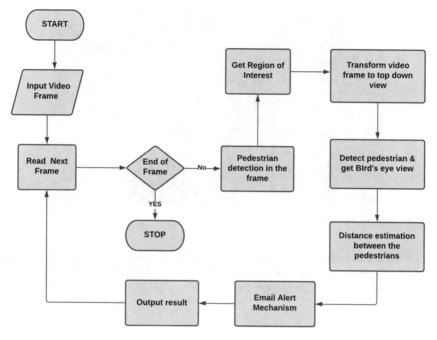

Fig. 2 Workflow architecture of social distancing detection

3 Methodology

The proposed methodology is a comparative analysis of CNN-based pre-trained [7] YOLOv3 and [8] YOLOv4. It is a three-stage model involving object detection that is detecting pedestrians in an input video frame, then monitoring distance and estimation that is monitoring the social distance between the pedestrians then distance estimation, and lastly alert mechanisms through email when distance is violated. The social distance detection and estimation are computed using pre-trained models, and the proposed methodology is discussed in the subsequent sections where the inputted videos each frame after detecting the pedestrians their region of interest and bird's eye view is demonstrated then distancing between each pedestrian is estimated and finally their distancing violations through email alert as demonstrated in Fig. 2.

3.1 Object Detection in Video

Object detection is one of the challenging tasks involving a method that is used to recognize and detect different objects present in an image or video and label them to classify these. In the early phase, object detection was one of the main technical challenges. But, after 2014, with the increase in technical advancements using deep

Fig. 3 Detection of only humans in video

learning, tasks were accomplished and solved the problem. In the proposed work, from the input pre-recorded video frame will detect pedestrians with their individual localization boundary boxes in the frame.

3.2 People Class Detection

The identification of people classes and the assignment of a unique ID to each individual is the next step. Our goal was to recognize only pedestrians in the frame using the supplied dataset and pre-recorded video; therefore, other item classes were disregarded in this application using the ID assigned to people class in the [9] COCO dataset. Consequently, after unique pedestrian detection in the frame, bounding boxes with blue color are created around each pedestrian as depicted in Fig. 3. As a result, the optimal bounding box for each identified pedestrian is generated in the picture, and this data will eventually be utilized for estimation of distance.

3.3 Camera Perspective Transformation

To setup the camera, the video frame was collected at a fixed angle. Since for each video frame, top view transformation is required, the four coordinates are chosen as reference points as depicted in Fig. 4. Thereafter for a more precise estimation of distance, each frame is transformed into a two-dimensional top-down perspective. In the implementation, we assumed that all the pedestrians are all walking on the same level of the ground. The region of interest (ROI) of a picture centered on a pedestrian going along a street was converted into a top-down 2D perspective as shown in Fig. 5 that contains 480 × 480 pixels. Then we selected the four coordinate points from the frame to transform into the top-down view by plotting the points to the end

Fig. 4 Camera perspective transformation of sample videos

Fig. 5 Top-down transformation of sample videos

corners of the rectangle in the two-dimensional view of the image. This perspective transformation is important to calculate the real-world distances in the frame. Then, by localizing pedestrian video from a top-down perspective in the location of interest, we acquire a bird's eye view. Based on this bird's eye view in Fig. 6 where green circles represent that pedestrians maintain a healthy distance among them and red depicted that they haven't maintained a prescribed social distance among them.

3.4 Distance Estimation Between Pedestrians

After the perspective transformation, each pedestrian's location may be approximated to correspond to the number of pixels in earlier generated bird's eye view. In this

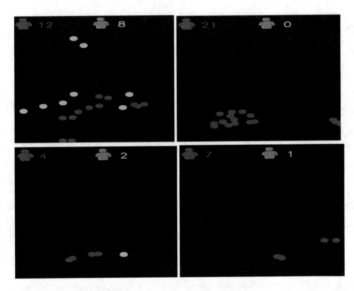

Fig. 6 Bird's eye view transformation of sample videos

phase, the bounding box position (x, y, w, h) where x and y are two dimension coordinates, w and h are the width and height of the frame view. The bottom-center point of the bounding box is used to estimate each pedestrian's position in the top-down view. Then distance is estimated between two non-compliant pairs of pedestrians using Euclidean distance as described in the below equation where x_1, x_2, and y_1, y_2 are the x and y coordinates of the center of pedestrians. The distance between two pedestrians may be calculated using the given positions (x_1, y_1) and (x_2, y_2) in an image.

$$\sqrt{(x_2 - x_1)^2 + (y_2 - y_1)^2}.$$

3.5 Bounding Box Generation

After the distance estimation between the pedestrians. Then, red and green color bounding boxes will be generated based on the distance estimation. Depending on the minimum cautious distance that has been chosen, if the distance between any two non-complaint pairs of pedestrians is less than acceptable distance will be indicated with a red bounding box for those pedestrians that serve as precautionary warnings. And when maintaining the safe and acceptable distance, then bounding box will be depicted by green color.

Fig. 7 Email alert
mechanism

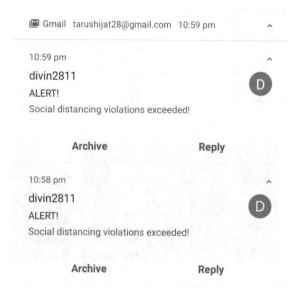

3.6 Email Alert Mechanism

After all the above phases, to create more awareness and to immediately prevent the violations made between the pedestrians. So, if the distance between the pedestrians is below the acceptable distance and the bounding box color is red, then we added a new mechanism to send an alert message through Email. After detecting violations of social distancing in video frame, alert message through Email will be sent stating the text message "Social Distancing Violations Exceeded!" that is, the number of violations exceeds the maximum violations limit. Figure 7 demonstrated the few samples resulting from emails of the email alert mechanism.

4   Experimental Analysis

For video object detection, used the two versions of YOLO pre-trained models, YOLOv3 and YOLOv4. The proposed work used models which are trained on the COCO Image dataset. We have experimented on four input videos that show the pedestrians walking on streets, malls, outside a university, and in a town. For the most precise computation of the distance between pedestrians in each of the videos, the video frame's viewpoint view is converted to a top-down view. Each of the detected pedestrians is then represented by a dot. Thereafter calculate the distance between each of the detected pedestrians which are now represented by dots. To identify which of the detected pedestrians are violating the distance criteria, we have defined a threshold value. If any of the two pedestrians are having a distance less

Table 1 Confidence scores

Video name	YOLOv3	YOLOv4
MOT20	0.844	0.750
TownCentre	0.882	0.788
PedestrianWalking	0.971	0.873
StudentVideo	0.870	0.808

Fig. 8 MOT20 output snapshot

Fig. 9 TownCentre output snapshot

than the defined threshold then they will be identified with the help of red boxes, otherwise, pedestrians will be identified with green boxes. To test the performance of the social distancing detection model, implemented with YOLOv3 and YOLOv4, on the input video with unknown ground truth values, the confidence score of each of the detected objects in the video frames is being calculated. The possibility that a deep learning model's output is correct which is represented by the confidence score and its value ranges from 0 to 1. To get the overall confidence score of each of the four input videos, we took the average of all confidence scores. Table 1 gives the confidence score of each of the four input videos.

In each of the input videos, along with the detection of pedestrians violating the social distancing criteria with the help of red and green boxes, the total number of safe people and the total number of unsafe people is also calculated. Figure 8 shows the output video snapshot of the "MOT20" input video implemented with YOLOv4 and YOLOv3 models, respectively. Figure 9 shows the output video snapshot of the "town

Fig. 10 PedestrianWalking output snapshot

Fig. 11 StudentVideo output snapshot

center" input video implemented with YOLOv4 and YOLOv3 models, respectively. Figure 10 shows the output video snapshot of the "PedestrainWalking" input video implemented with YOLOv4 and YOLOv3 models, respectively. Figure 11 shows the output video snapshot of the "StudentVideo" input video implemented with YOLOv4 and YOLOv3 models, respectively.

5 Conclusion

In this proposed work, a technique for detecting social distancing for COVID-19 using YOLOv3 and YOLOv4 models is proposed. With the help of computer vision technology, the distance between persons can be calculated, and any pair of pedestrians who fail to meet the distance requirements will be covered by red boxes. The proposed methodology is validated using four different input videos showing waking pedestrians. The visualization results revealed that the suggested technique is capable of calculating the social distance between individuals and therefore this work can be implemented in real-life environments such as universities, public transport stations and many other places as future work. Furthermore, this work can be improved by implementing many other important criteria, such as human body temperature detection and mask detection, to fight against the COVID-19 pandemic.

References

1. Social Distancing Scenario in India. India's social distancing Jugaad Amid coronavirus pandemic
2. Punn NS, Sonbhadra SK, Agarwal S (2021) Monitoring COVID-19 social distancing with person detection and tracking via fine-tuned YOLOv3 and Deepsort techniques. arXiv preprint arXiv:2005.01385
3. Rezaei M, Azarmi M (2020) DeepSOCIAL: social distancing monitoring and infection risk assessment in COVID-19 pandemic. arXiv preprint arXiv:2008.11672
4. Ahmed I, Ahmad M, Rodrigues JJPC, Jeon G, Din S (2021) A deep learning-based social distance monitoring framework for COVID-19. https://doi.org/10.1016/j.scs.2020.102571
5. Ksentini A, Brik B (2020) An edge-based social distancing detection service to mitigate COVID-19 propagation. IEEE Internet Things Mag
6. Shah J, Chandaliya M (2021) Social distancing detection using computer vision. In: 2021 5th International conference on computing methodologies and communication (ICCMC)
7. Redmon J, Farhadi A (2018) YOLOv3: an incremental improvement. arXiv preprint arXiv: 1804.02767
8. Bochkovskiy A, Wang C-Y et al (2020) YOLOv4: optimal speed and accuracy of object detection. arXiv preprint arXiv: 2004.10934
9. Lin T-Y, Maire M et al (2015) Microsoft COCO: common objects in context. arXiv preprint arXiv: 1405.0312
10. Ahamad AH, Zaini N, Latip MFA (2020) Person detection for social distancing and safety violation alert based on segmented ROI. In: 2020 10th IEEE International conference on control system, computing and engineering (ICCSCE), pp 113–118. https://doi.org/10.1109/ICCSCE50387.2020.9204934
11. Surya L, Yarlagadda RT (2021) AI Economical smart device to identify COVID-19 pandemic, and alert on social distancing who measures. Int J Creative Res Thoughts
12. Bian S et al (2021) A wearable magnetic field based proximity sensing system for monitoring COVID-19 social distancing. In: Proceedings of the 2020 international symposium on wearable computers, Sept 2021
13. Yang D, Yurtsever E et al (2020) A vision-based social distancing and critical density detection system for COVID-19
14. Brodeur A, Cook N et al (2021) On the effects of COVID-19 safer-at-home policies on social distancing, car crashes and pollution. J Environ Econ Manag

Evaluating Generative Adversarial Networks for Gurumukhi Handwritten Character Recognition (CR)

Sukhandeep Kaur, Seema Bawa, and Ravinder Kumar

1 Introduction

The modern era of digitization has greatly impacted the research trends in OCR with more focus toward the regional languages. CR can further lead to many real life vision applications such as augmented reality as well as text recognition in scenes, natural images, document image processing, script identification, digital libraries and content-based image retrieval. Deep learning with its powerful ability of automatic feature extraction has been widely adopted for CR [1, 2]. For regional languages, lack of labeled, diverse and high-quality training data makes it difficult to optimize the hyperparameters of deep networks. To create a diverse data covering, all the possible cases for each character, data augmentation techniques using distortions and additive noises have been proposed. Further to have a control over the generated data, a deep network-based GANs have been proposed to generate synthetic data looking similar to realistic data using some random data [1]. GANs have achieved an impressive results in generating character images for non Indic scripts like Latin, Chinese, etc. [1, 3–5]. Gurumukhi is a Indic script which has 35 characters (Fig. 1) and is widely spoken in the Northern part of India and in some other parts of the world like Canada, Australia, USA, etc. The research in OCR for Gurumukhi script has received high popularity in the past two decades resulting into development many training datasets for handwritten and printed data [2]. In this research, to in large the available training dataset, some efforts have been made for the efficient generation of handwritten Gurumukhi characters using variants of GANs.

S. Kaur (✉) · S. Bawa · R. Kumar
Thapar Institute of Engineering and Technology, Patiala, Punjab, India
e-mail: shergillsukhandeep@gmail.com

S. Bawa
e-mail: seema@thapar.edu

R. Kumar
e-mail: ravinder@thapar.edu

In GAN, two networks, a generator and a discriminator (binary classifier), compete with each other like two players of minmax game. The training of GAN models is considered a complex task. Both the networks are trained at the same time and one (generator) tries to fool the another (discriminator). Generator uses the random values to generate images, while discriminator tries to distinguish between the real and fake(generated) samples. Goodfellow has proposed the first GAN in 2014 gaining the immense popularity of GANs for image synthesis [6]. Apart from image synthesis in OCR, GANs have shown remarkable progress in many other applications also like object detection, dialog generation, semantic segmentation, image translation, etc. [7]. Various GANs have been reported recently to have more control over the generated data. Such as DCGAN, WGAN, LSGAN, Cycle GAN, Info GAN and CGAN.

To the best of our knowledge, we have not found any work of using GAN models for generating the synthetic dataset of Gurumukhi language. The major contributions of this paper are: (1) to generate the synthetic images of Gurumukhi handwritten characters, (2) analyzing the performance of various GANs for Gurumukhi handwritten characters and (3) to optimize the hyper parameters of GANs for efficient generation of data.

2   Related Work

Data augmentation has been recently become popular to meet the need of large training data, for deep neural network. Two types of data augmentation approaches are used: geometric transformation-based and task-specific-based data augmentation. Many approaches come under these like, flipping, rotation, scaling, elastic deformation, patch extraction, etc. However in deep networks, during character recognition, these simple image transformations are not capable to extract the deep features of an image.

For Gurumukhi script, it has been found that traditional machine learning approaches are more popular as compared to deep learning. A complete survey of traditional and deep learning methods used for Gurumukhi script has been conducted by Kaur et al. [2]. Some efforts have been made to design the benchmark datasets for Gurumukhi script. Munish et al. have created 7 datasets collected from various writers and some experimental work has been performed to evaluate the datasets [8]. Similarly Kumar et al. have conducted a performance analysis study of various classifiers used in Gurumukhi character recognition [9]. However, research in Gurumukhi has got new trends in past two decades such as text recognition in scenes, historical manuscripts, for online mode of text, etc.

To in large the scope of research, some of the researchers have used generative networks to generate character set of many Indic scripts. Simarpreet and Verma have used DCGANs for handwritten character generation of Devanagari script from Indic scripts and found promising results [10].

Fig. 1 Samples of
Gurumukhi handwritten
character

Apart from image generation, GANs have also been used for many other tasks like object detection, segmenting, etc. Rusticus et al. have used GANs as image translators for document analysis tasks to shift document domains. For domain shift, GANs have performed source to target domain translation (image to image translation). Conditional GAN and cycle GAN have been compared for domain shift with ResNet as generator and PatchGan as discriminator architecture [11]. Wu et al. have designed PixTextGAN to create synthesis dataset for license plate recognition. The structure aware-based loss function has been presented to maintain the characteristics of each character region [12]. Further, Hu et al. have proposed DCGAN to generate the large synthetic images for visual recognition. The deep visual features have been extracted from the synthetic images [13]. Dong et al. have used GANs for text to image synthesis without using human labeled text to image data. The discriminator loss function has been designed to train the generator and discriminator [14]. To recognize the camera captured document images, character-level text detection model has been proposed by Zhao et al. They have used conditional GAN which considers the text detection problem as image to image generation [15]. Shao et al. have used attention GANs for intelligent making system to recognize printed and handwritten text [16]. Kong et al. have proposed Generative Adversarial Recognition Network, a GAN-based Chinese scene character recognition network. It has two phases, i.e., to generate the synthetic dataset and classification of scene characters by recognition part. In this GAN structure, the two parts, generator and classifiers, compete in the game mechanism. The major change has been made in the classification part. In the discriminator, number of character classes are set to double the number of original character classes due to the fact that each output further can be real or fake. The proposed approach has reported better results for ICDAR, EMNIST, SVT datasets by overcoming the problem of over fitting [1]. Alonso et al. have used GAN for generating the synthetic images of handwritten text. An adversarial conditioned GAN structure with LSTM has been applied to generate the synthetic text images conditioned on sequence of characters. RNN encoder sends the input to generator as secondary input for generating the text, while discriminator and recognition modules use the adversarial and CTC loss to train the generated input [17]. Cai et al. have used GAN along with transfer learning in training the Historical Chinese character recognition. Here, discriminator uses the WGAN loss function to find the loss between generated and real target images, while generator uses the modified U-Net structure. Further in classification, transfer learning using ResNet, GoogleNet and DenseNet has been used in source domain dataset [3].

3 Types of GAN Architecture for Gurumukhi Handwritten CR

3.1 Deep Convolution GAN (DCGAN)

DCGANs proposed by Radford are considered the most popular and simple GANs, where convolution strides have been used by replacing the maxpooling in GAN networks [18]. Many applications of DCGAN can be found like, Hussain et al. have tried variants of GANs for drug discovery applications. From vanilla GAN, progressive GAN and DCGAN, they found DCGAN has generated the most efficient high-quality cellular images [19]. Similarly, Wu et al. have applied DCGAN in generating the synthesis images for tomato leaf disease identification. The increased dataset with diversity and quality from DCGAN has resulted into the leaf disease identification accuracy of 94.33% [20].

In DCGAN, generator G uses transposed convolutions and discriminator D uses convolutional operations during training as shown in Fig. 2. LeakyReLu activation has been applied in all the layers of generator and discriminator except the last layer

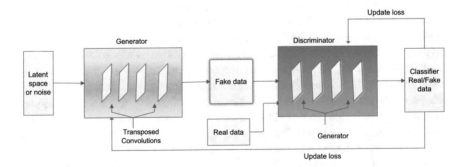

Fig. 2 DCGAN architecture

where Tanh has been used. Similarly for the optimization, stochastic gradient decent with Adam has been used. A random noise z has been passed to G as $G : z \rightarrow x$ and produces a realistic looking image x with the help of discriminator $D \rightarrow$ [real, fake]. Discriminator D maximizes the $\log D(x)$ to generate the real images of data. G is trained to make the discriminator fool for fake data by minimizing the $\log(1 - D(G(z)))$.

At discriminator the loss function is represented as:

$$D_{\text{LOSSreal}} = \log(D(x)); \quad D_{\text{LOSSfake}} = \log(1 - D(G(z))) \tag{1}$$

while the generator loss is:

$$G_{\text{LOSS}} = -\log(D(G(z))) \tag{2}$$

The complete loss function of DCGAN has a two entropy functions as shown in equation

$$\min_{G} \max_{D} V(D, G) = \Xi_x \sim P_{\text{data}(x)}[\log(D(x)] + \Xi_z \sim P_{z(z)}[\log(1 - D(G(z)))] \tag{3}$$

The first entropy D takes the real data $P_{\text{data}(x)}$ and maximizes it, while the second entropy G takes random noise and produces a fake image. Further, this generated image is passed to $D(x)$ which tries to maximize it to 0.

3.2 Wasserstein GAN (WGAN)

To solve the problem of mode collapse and for stable training of GAN, Arjovsky et al. [21] have proposed WGAN based on Earth Mover distance theory. In simple GAN, discriminator works like linear classifier as for fake images it results in 0 and for real it results into 1. Martin has changed the task of discriminator to regression problem and renamed it as critic, where EM distance measures the distance between the real and fake. Critic is updated multiple times as compared to generator in WGAN. Another notable difference in WGAN is the use of weight clipping for discriminator to clip the weights to a given range. It uses RMSprop instead of Adam optimizer. Here, Wasserstein distance called Earth Mover distance gives smooth gradient for training data. Figure 3 depicts the architecture of WGAN.

3.3 Least Square GAN (LSGAN)

Mao et al. have proposed LSGAN [22] by changing the loss function in regular GANs from sigmoid cross entropy to least square loss function (Fig. 4). The use of least

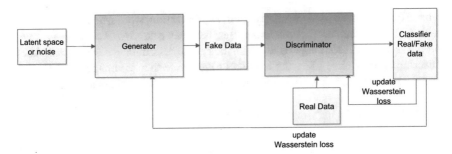

Fig. 3 WGAN architecture

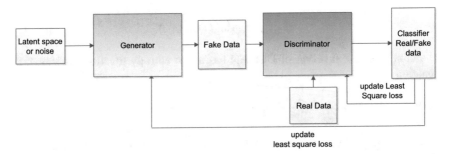

Fig. 4 LSGAN architecture

square loss function results into the more stable training of GAN as it overcomes the vanishing gradient problem in simple GANs. In regular GANs, during generator loss updation, it considers that fake data as real which is on the correct side of boundary line but far away from line resulting into no loss. Hence, objective of LSGAN is to move the fake samples toward the decision boundary. The discriminator and generator loss function are given by:

$$\min_D V_{\text{lsgan}(D)} = \frac{1}{2} \Xi_{x\ P_{\text{data}(x)}}[D(x) - b^2] + \frac{1}{2} \Xi_{z\ P_{z(z)}}[D(G(z)) - a^2] \tag{4}$$

$$\min_G V_{\text{lsgan}(G)} = \frac{1}{2} \Xi_{z\ P_{z(z)}}[D(G(z)) - c^2] \tag{5}$$

3.4 Conditional GAN

Mirza and Osindero [23] have proposed conditional GANs to generate images with some kind of control over the images. It adds some condition with the external

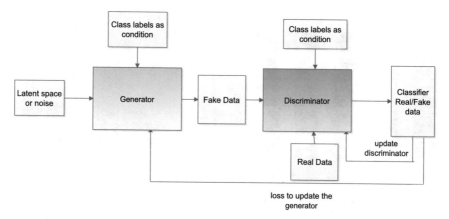

Fig. 5 CGAN architecture

data which controls the image to be generated and discriminated. Every dataset has some additional information like class labels or image (for image to image translation) which could improve the performance of GAN and targeted images will be generated. To map the relation between generated images and latent space of generator, discriminator and generator are trained to generate images of given type. Structure of CGAN is similar to DCGAN except an embedding layer. It has been inserted into discriminator and generator structure for conditional data Fig. 5.

$$\min_{G} \max_{D} V(D, G) = \Xi_x$$
$$\sim P_{\text{data}(x)}[\log(D(x \, (\text{mod} \, y)))] + \Xi_z$$
$$\sim P_{z(z)}[\log(1 - D(G(z \, (\text{mod} \, y))))] \qquad (6)$$

Here, y is the condition and x is the real input which is given to discriminator D as $D(x \, (\text{mod} \, y))$, similarly at generator G along with input z and condition y is inserted as 1-hot embedding layer as $G(z \, (\text{mod} \, y))$ to generate images based on label y.

4 Experiments and Results

We have performed experiments on four different kinds of GANs to generate Gurumukhi handwritten characters. Dataset consists of 7000 samples of handwritten Gurumukhi characters belonging to 35 classes from 100 different writers. Every writer has written a single character resulting into 7000 characters as 200 per class. Digital images collected from various writers on paper are stored in TIFF format and some preprocessing has been applied. Preprocessing of data converts the binary images into gray scale images and changes the size from 128 × 128 to 28 × 28. To evaluate the performance of DCGAN for Gurumukhi handwritten character generation,

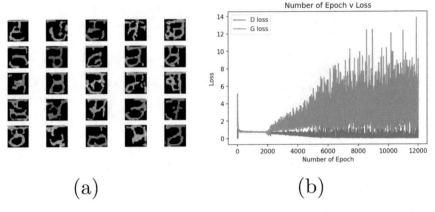

(a)   (b)

Fig. 6 Results for DCGAN: **a** images generated; **b** generator and discriminator loss for DCGAN

4 layer deep convolutional network has been implemented at discriminator and 3 layers of convolution have been applied at generator. At each layer, batch normalization has been applied except the last layer with LekyReLu activation function. Hyperparameters of both the generator and discriminator have been optimized using some hit and trial experiments and found reasonable results after 12,000 iterations with Adam optimizer having 0.0002 learning rate. Figure 6a represents the sample images generated after 12,000 iterations and Fig. 6b shows the discriminator and generator loss during the training of network. The loss graph depicts the generator loss is the highest around 1200 iteration. Hence, the quality of images generated at this point is the best.

The second variant of GAN implemented is LSGAN having 2 convolutional layers in the discriminator and transposed convolutions in generator for upsampling and downsampling. The major difference between LSGAN and normal GAN is the loss function used, i.e., minimum square error in LSGAN. The optimal value of learning rate found for Adam optimizer is 0.0001 for our dataset. Figure 7a shows the images generated using LSGAN after 20,000 iteration, while Fig. 7b shows the discriminator and generator loss corresponding to iterations. The generator loss was high around 7000 iteration. It means, at this point the discriminator was unable to distinguish between the real and fake images. The loss graph depicts that the quality of generated images goes high with increase in number of iterations however after some time it becomes stable and low.

For WGAN, we have used simple GAN structure for generator and discriminator along with weight clipping. Generated images found are more efficient with clip constraint value of 0.001, as compared to 0.01 and 0.0001. In WGAN, the critic (discriminator) has been trained more times as compared to generator. Hence, we have used 10 iterations for each critic in every iteration of complete GAN training. Instead of Adam optimizer, RMSprop optimizer having learning rate of 0.00005 with Wasserstein loss has been used. Figure 8a shows the generated images after 1000

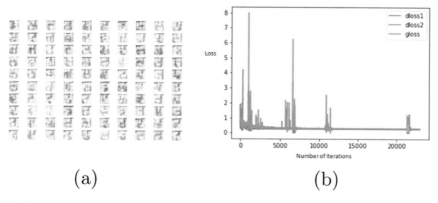

(a)   (b)

Fig. 7 Results for LSGAN: **a** images generated; **b** generator and discriminator loss for LCGAN

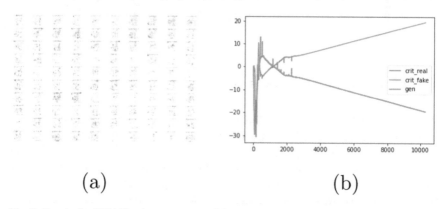

(a)   (b)

Fig. 8 Results for WGAN: **a** images generated; **b** generator and discriminator loss for WGAN

iteration of WGAN, while Fig. 8b represents the losses corresponding to iterations. The loss graph for WGAN is different from other graphs as in this the generator and discriminator loss goes in opposite direction. The highest generator loss is observed at 1000 iteration. After this, the increase in number of iteration decreases the loss.

Further, to implement the conditional GANs, the class labels of dataset have been mapped with the input of generator and discriminator network as conditions. Gurumukhi dataset has 35 classes, an embedding layer with dimension 50 has been used to map the class label with generator and discriminator. The other structure of network has been kept similar to simple GAN network. A dropout layer with value of 0.4 has been used after each convolution layer. The binary cross entropy loss function has been used in training the network with Adam optimizer with learning rate of 0.0001. Figure 9 shows the generated images using CGAN are more efficient as compared to other GANS. Moreover, images of particular classes are generated in sequence.

Fig. 9 images generated in CGAN

5 Conclusion and Future Scope

We have implemented variants of GANs for the regional script of India, i.e., Guru-mukhi script. These networks can be used to enhance the available datasets of hand-written characters for Gurumukhi script to be further used in deep networks. Results obtained from different networks are presented in the form of graph and images which depicts the efficiency of each network. The hyperparameters of each network corresponding to network structure and the dataset have been adjusted. The quality of generated images can be improved using some simple preprocessing techniques of image datasets. Out of these four different kinds of GANs, DCGAN and conditional GANs have resulted into more efficient synthetic data generation. These GANs can be further explored to improve the quality of data generated and to generate the data for some other Indic scripts.

Acknowledgements This research was supported by Council of Scientific and Industrial Research (CSIR) funded by the Ministry of Science and Technology (09/677(0031)/2018/EMR-I) as well as the Government of India.

References

1. Kong H, Tang D, Meng X, Lu T (2019) GARN: a novel generative adversarial recognition network for end-to-end scene character recognition. In: 2019 International conference on document analysis and recognition (ICDAR). IEEE, pp 689–694
2. Kaur S, Bawa S, Kumar R (2019) A survey of mono-and multi-lingual character recognition using deep and shallow architectures: indic and non-indic scripts. Artif Intell Rev 1–60

3. Cai J, Peng L, Tang Y, Liu C, Li P (2019) TH-GAN: generative adversarial network based transfer learning for historical Chinese character recognition. In: 2019 International conference on document analysis and recognition (ICDAR). IEEE, pp 178–183
4. Sinha A, Jenckel M, Bukhari SS, Dengel A (2019) Unsupervised OCR model evaluation using GAN. In: 2019 International conference on document analysis and recognition (ICDAR). IEEE, pp 1256–1261
5. Qian Z, Huang K, Wang Q-F, Xiao J, Zhang R (2020) Generative adversarial classifier for handwriting characters super-resolution. Pattern Recogn 107:107453
6. Goodfellow I, Pouget-Abadie J, Mirza M, Xu B, Warde-Farley D, Ozair S, Courville A, Bengio Y (2014) Generative adversarial nets. In: Advances in neural information processing systems, vol 27
7. Cheng K, Tahir R, Eric LK, Li M (2020) An analysis of generative adversarial networks and variants for image synthesis on MNIST dataset. Multimedia Tools Appl 79(19):13725–13752
8. Kumar M, Sharma RK, Jindal MK, Jindal SR, Singh H (2018) Benchmark datasets for offline handwritten Gurmukhi script recognition. In: Workshop on document analysis and recognition. Springer, Singapore, pp 143–151
9. Kumar M, Jindal MK, Sharma RK, Jindal SR (2020) Performance evaluation of classifiers for the recognition of offline handwritten Gurmukhi characters and numerals: a study. Artif Intell Rev 53(3):2075–2097
10. Kaur S, Verma K (2020) Handwritten Devanagari character generation using deep convolutional generative adversarial network. In: Soft computing: theories and applications. Springer, Singapore, pp 1243–1253
11. Rusticus D, Goldmann L, Reisser M, Villegas M (2019) Document domain adaptation with generative adversarial networks. In: 2019 International conference on document analysis and recognition (ICDAR). IEEE, pp 1432–1437
12. Wu S, Zhai W, Cao Y (2019) PixTextGAN: structure aware text image synthesis for license plate recognition. IET Image Process 13(14):2744–2752
13. Hu T, Long C, Xiao C (2021) A novel visual representation on text using diverse conditional GAN for visual recognition. IEEE Trans Image Process 30:3499–3512. https://doi.org/10.1109/TIP.2021.3061927
14. Dong Y, Zhang Y, Ma L, Wang Z, Luo J (2021) Unsupervised text-to-image synthesis. Pattern Recogn 110:107573
15. Zhao J, Wang Y, Xiao B, Shi C, Jia F, Wang C (2020) DetectGAN: GAN-based text detector for camera-captured document images. Int J Doc Anal Recogn (IJDAR) 23(4):267–277
16. Shao L, Liang C, Wang K, Cao W, Zhang W, Gui G, Sari H (2019) Attention GAN-based method for designing intelligent making system. IEEE Access 7:163097–163104
17. Alonso E, Moysset B, Messina R (2019) Adversarial generation of handwritten text images conditioned on sequences. In: 2019 International conference on document analysis and recognition (ICDAR). IEEE
18. Radford A, Metz L, Chintala S (2015) Unsupervised representation learning with deep convolutional generative adversarial networks. arXiv preprint arXiv:1511.06434
19. Hussain S, Anees A, Das A, Nguyen BP, Marzuki M, Lin S, Wright G, Singhal A (2020) High-content image generation for drug discovery using generative adversarial networks. Neural Networks 132:353–363
20. Wu Q, Chen Y, Meng J (2020) DCGAN-based data augmentation for tomato leaf disease identification. IEEE Access 8:98716–98728
21. Arjovsky M, Chintala S, Bottou L (2017) Wasserstein GAN. arXiv:1701.07875
22. Mao X, Li Q, Xie H, Lau RYK, Wang Z, Smolley SP (2017) Least squares generative adversarial networks. In: Proceedings of the IEEE international conference on computer vision, pp 2794–2802
23. Mirza M, Osindero S (2014) Conditional generative adversarial nets. arXiv preprint arXiv:1411.1784

Evaluating Feature Importance to Investigate Publishers Conduct for Detecting Click Fraud

Deepti Sisodia, Dilip Singh Sisodia, and Deepak Singh

1 Introduction

Due to the fast development of the Internet, online advertising campaigns play a vital role in the digital advertisement field. There are growing appeals for Pay-Per-Click (PPC) model among various pricing models in online advertising [1]. Pay-Per-Click (PPC) is a model where advertisers provide commission as per the generated clicks whether the clicks are legitimate or illegitimate. Clicks may be generated from potential publishers or through automated computer programs. The chief administrator in this process is the Advertising Commissioner which serves as a mediator/agent/dealer among advertisers and publishers. The PPC advertiser provides the advertisements to the advertising commissioner according to the planned budget and pays the commission for every generated click. Publisher communicates with the advertising commissioner for displaying ads on the web pages and incentivize accordingly. Clicks are generated either by hiring users for deliberate clicking on ads or by employing automated software that imitate human conduct [2]. This fraudulent activity is termed as click fraud and is one of the serious and challenging tasks to be investigated in the field of Internet marketing.

Click fraud has now evolved as a serious threat for the online advertising business, wherein generation of millions of illegitimate clicks and publisher's dynamic behavior complicates the task of publisher's identification. Though ensemble learning approaches [3] have more potential than the individual machine learning models, several individual machine learning methods being evaluated to analyze

D. Sisodia (✉) · D. S. Sisodia · D. Singh
National Institute of Technology, Raipur, India
e-mail: dsisodia.phd2017.cse@nitrr.ac.in

D. S. Sisodia
e-mail: dssisodia.cs@nitrr.ac.in

D. Singh
e-mail: dsingh.cs@nitrr.ac.in

© The Author(s), under exclusive license to Springer Nature Singapore Pte Ltd. 2023   515
D. S. Sisodia et al. (eds.), *Machine Intelligence Techniques for Data Analysis and Signal Processing*, Lecture Notes in Electrical Engineering 997,
https://doi.org/10.1007/978-981-99-0085-5_42

the publisher's behavior as legitimate, under-observation or illegitimate. Numerous features concerning to publishers has been extracted by Berar et al. [2], Perera et al. [4] and Sisodia et al. [5, 6] to examine the behavior of publishers by analyzing their click patterns. Vasumati et al. [7] have addressed the click-spam issue in mobile advertising, and Berar et al. [8] have designed a novel method for the generation of click patterns concerning each publisher. The use of variable elimination might have been misaddressed in this work which may enhance the performance which was slightly overcome by Taneja et al. [9]. In this context, the suspicious behaviors of publishers are being analyzed in recent past by Sisodia et al. by providing more valuable insights toward extracting 103 features using GTB [10] and designing a prototype selection based under-sampling strategy QDPSkNN [11]. They also conducted an empirical review of several data-level sampling methods resulting in a balanced dataset [12] and designed a hybrid feature selection strategy [13] which has improved the learner's ability in the identification and classification of publishers as fraud-non-fraud.

Systematically analyzing the factors at distinct granularity levels, the work reported in the literature lacks selection of robust and predictive features for effective fraud detection. Our work overcomes this limitation by evaluating feature importance scores which provides valuable insights to the dataset and model. The contributions of the current work are as follows:

- The suspicious behavior of publishers is analyzed by extracting the features per publisher using composite attributes by merging two or more attributes from the provided data.
- The dynamic behavior of publishers is investigated based on the feature importance scores provided by the standard machine learning classifiers.
- Most significant features obtained by GTB has investigated the changing conduct of publishers with an average precision score of 64.86%.
- The effectiveness of the proposed work is being compared with the existing state-of-the-art methods.

The organization of the rest of the work is as follows: methods and materials are discussed in Sect. 2 elaborating the dataset, feature extraction and selection, classification methods, validation, and evaluation measures. Section 3 presents the experimental results and discussions with a comparison of previous work with proposed work. The work is concluded in Sect. 4.

2 Methods and Materials

2.1 Block Diagram

Graphical representation of proposed methodology is outlined in Fig. 1 as block diagram. This functional view of the system depicts the flow of experiments conducted in model designing.

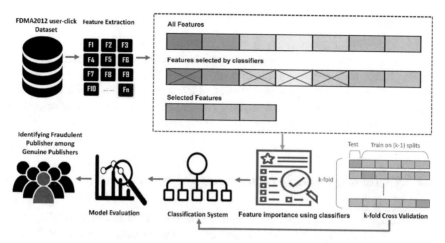

Fig. 1 Block diagram of proposed methodology

2.2 Data Acquisition

For this experiment, the data is analyzed and obtained from Fraud Detection in Mobile Advertising (FDMA) 2012 [14] competition organized by the BuzzCity Pte. Ltd. The dataset comprises of publisher dataset and click dataset which are supplied in comma-separated values (CSV) format. The publisher dataset holds the publisher details, while the click dataset holds the click details associated with each publisher. The detailed description of attributes of click and publisher dataset is discussed in detail in [10]. The publisher and click datasets are further categorized into three different sets collected at different time intervals shown in Table 1. Click dataset falls into three categories: training set (for model designing), validation set (for model selection) and testing set (for optimization and accuracy evaluation). Table 1 depicts the complete statistics of the database. All the attributes of both databases are analyzed and examined concerning its effect on the behavior of publisher.

Table 1 Statistics of dataset

Dataset	Time period	Number of clicks	Number of publishers			
			Fraud	Observation	OK	Total
Train set	9–11 FEB 2012	1,048,575	72 (2.34%)	80 (2.60%)	2929 (95.07%)	3,081
Validation set	23–25 FEB 2012	1,048,575	85 (2.77%)	84 (2.74%)	2895 (94.48%)	3,064
Test set	8–10 MAR 2012	1,048,575	82 (2.73%)	71 (2.37%)	2847 (94.90%)	3,000

2.3   Feature Extraction

Since the dataset is heterogeneous, incomplete, inconsistent and consist of some missing fields, feature extraction is thus required for data preprocessing. A total of 103 predictive features are extracted after conducting the experimentation. Since the extracted features might be irrelevant, insignificant, and noisy, selection of important features is thus required to obtain optimal set of features.

2.4   Evaluating Feature Importance Scores

Feature importance score has a significant contribution in predictive modeling, and it gives valuable insight into the data and the model, provides basis to the dimensionality reduction and variable selection which improves the efficiency of a predictive model. Feature importance assigns scores to the features toward a predictive model which defines the relative importance of every feature during prediction. In this work, GTB [10], DT [15], RF [16], LR [17] and XGB [18] are used for automated estimation of feature importance scores from the trained predictive model. Every classifier has a built-in function to identify the feature importance, wherein GTB measures the feature importance based on 'Gini importance', while DT uses the CART technique. Impurity-based feature importance is adopted by RF, and LR finds the coefficient toward predicting the output. XGB finds the important features based on permutation-based feature importance. The feature importance provided by the learners enables the selection of feature's subset by alienating those features which are less/not predictive, insignificant and redundant. Table 2 presents the feature importance scores of list of 10 most significant features provided by the classifiers, while details of selected features can be found elaborately in [10]. Figure 2 shows the graphical illustration of feature importance by different classifiers showing highly significant features among 103 features, where the x-axis represents the features and the y-axis represents the feature importance.

2.5   Classification Algorithms

This section demonstrates the proposed methods employed in experimenting for investigating click fraud. Five state-of-the-art classification [15] approaches employed in experiment are as follows: GTB [10], DT [15], RF [16], LR [17] and XGB [18]. Among several classifiers, GTB has shown superior state-of-the-art results over the dataset used. Due to its constructive approach in ensemble generation, the outcome of GTB has surpassed the other classification methods. Since GTB performed superior and due to space complexity, we have elaborately discussed only GTB than other learners.

Table 2 List of significant features provided by classifiers

GTB		XGB		RF		LR		DT	
Feature	Score	Feature	Score	Feature	Score	Feature	Score	Feature	Score
F-12	0.158	F-12	0.048	F-9	0.265	F-2	0.022	F-12	0.096
F-31	0.208	F-31	0.080	F-10	0.301	F-4	0.026	F-31	0.202
F-37	0.041	F-37	0.003	F-60	0.076	F-5	0.011	F-37	0.062
F-95	0.024	F-43	0.030	F-87	0.229	F-6	0.029	F-42	0.062
F-54	0.148	F-44	0.015	F-101	0.022	F-7	0.032	F-43	0.007
F-43	0.033	F-48	0.011	F-31	0.304	F-9	0.021	F-49	0.086
F-49	0.099	F-49	0.073	F-95	0.019	F-21	0.030	F-54	0.179
F-64	0.118	F-54	0.088	F-79	0.127	F-26	0.065	F-64	0.059
F-79	0.025	F-59	0.011	F-43	0.231	F-29	0.039	F-87	0.080
F-87	0.096	F-64	0.032	F-44	0.113	F-30	0.031	F-95	0.047

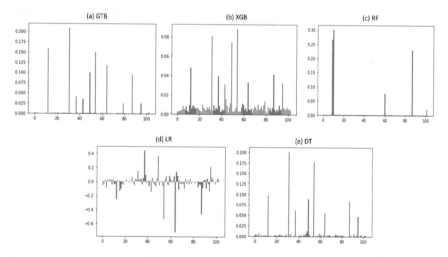

Fig. 2 Graphical illustration of feature importance scores obtained by GTB, XGB, RF, LR and DT

2.5.1 Gradient Tree Boosting

In supervised learning, let the input variable be x and an output variable be y defined through joint probability distribution $P(x, y)$. Let the training dataset is $\{(x_1, y_1), \ldots (x_n, y_n)\}$ of corresponding values of x and y. Employing this training dataset the aim is to search an estimation $\hat{F}(x)$ to a function $F(x)$ which reduces the value of loss function $L(y, F(x))$: $\hat{F} = \arg\min_y E_{x,y}[L(y, F(x))]$. The GBM [14] approach supposes a real valued variable y and searches an estimate $\hat{F}(x)$ in terms of weighted addition of function $h_i(x)$ from class $\mathcal{H}$, known as weak learners defined as: $\hat{F}(x) = \sum_{i=1}^{M} \gamma h_i(x) + \text{const}$. As per the principle of empirical risk minimization, gradient boosting reduces the empirical risk by reducing the mean value of loss function over the training dataset with an attempt by searching an estimation function $\hat{F}(x)$. It performs the function by initializing the model comprises of a constant function $F_0(x)$ and proceed it greedily with an increment:

$$F_0(x) = \arg\min_{\gamma} \sum_{i=1}^{n} L(y_i, \gamma)$$

$$F_m(x) = F_{m-1}(x) \arg\min_{\gamma h_m \in \mathcal{H}} \left[\sum_{i=1}^{n} L(y_i, F_{m-1}(x_i) + h_m(x_i)) \right]$$

where $h_m \in \mathcal{H}$ indicates a base/weak learner function. Now, selecting the appropriate function h at every step concerning loss function L leads to improbable optimization problem. Simplicity is thus required to slove the problem. Regarding the solution, the steepest descent method is applied concerning the minimization issue. If the case is continuous where $\mathcal{H}$ belongs to a random function over $\mathbb{R}$, the model is updated as follows:

$$F_m(x) = F_{m-1}(x) - \gamma_m \sum_{i=1}^{n} \nabla F_{m-1} L(F_{m-1}(x_i)),$$

$$\gamma_m = \arg\min_{\gamma} \sum_{i=1}^{n} L(y_i, F_{m-1}(x_i) - \gamma \nabla F_{m-1} L(F_{m-1}(x_i))),$$

where the derivatives are considered concerning functions F_i, for $i \in \{1, \ldots, m\}$.

2.6   Cross-validation and Evaluation Measures

In this experiment, tenfold cross-validation is employed where dataset is randomly divided into ten same sized subsamples. One subsample is used as the validation set to test the model and the rest of the nine subsamples are employed as training set for training the classifier. This process continues 10 times [19]. Conventional machine learning methodologies do not generalize well regarding accuracy when deals with imbalanced dataset due to unusual distribution of classes. Therefore, classifiers employed in this experiment are evaluated using average precision, sensitivity, specificity and geometric-mean instead of accuracy. For individual class i, the evaluation has been described as t_{p_i}-true positive for class i, f_{p_i}-false positive for class i, f_{n_i}-false negative for class i, t_{n_i}-true negative for class i and l represents total number of classes. Table 3 illustrates evaluation measures [20] for classification.

Table 3 Performance evaluation measures for classification

Evaluation measures	Formula	Description
Average precision (AP)	$\sum_{i=1}^{l} \frac{tp_i}{tp_i + fp_i}$ / l	Mean value of per class precision measure
Sensitivity (TPR)	$\frac{tp_i}{tp_i + fn_i}$	Proportion of correctly classified positive instances
Specificity (TNR)	$\frac{tn_i}{tn_i + fp_i}$	Proportion of correctly classified negative instances
Geometric-mean (GM)	$\sqrt{tp_i * tn_i}$	Maximizes true positive and true negative rate, respectively

3 Experimental Results

This section summarizes the observations and results obtained during the experiment.

To analyze the PPC fraudulent behavior of publishers, the day is categorized into four intervals, i.e., duration of four six-hour-periods—morning (6:00 a.m.–11:59 a.m.), afternoon (12:00 p.m.–5:59 p.m.), evening (6:00 p.m.–11:59 p.m.) and night (12:00 a.m.–5:59 a.m.). For example, night_click_percent can be calculated as 'percent clicks belonging to this publisher during nighttime from total clicks belonging to the publisher'. Likewise, an hour is also split/segmented into four intervals of fifteen minutes, respectively: first interval (0–14) minutes, second interval (15–29) minutes, third interval (30–44) minutes and fourth interval (45–59). For example, the feature 'first_15_minute_percent' can be defined as—'number of clicks between first to 14th minute divided by total number of clicks'. The predictive performance of several individual (LR, DT) and ensemble (GTB, XGB, RF) classification algorithms is demonstrated in Table 4 which is evaluated over the validation dataset using several adopted measures like average precision (AP), sensitivity/TPR, specificity/TNR and G-mean using all and selected features.

Table 4 Performance of individual and ensemble classifiers on FDMA2012 dataset with all and selected features

Methods	FDMA validation dataset							
	All features				Selected features			
	AP	TPR	TNR	GM	AP	TPR	TNR	GM
Gradient tree boosting	57.0	52.5	52.3	66.2	64.8	60.4	59.6	73.1
XGBoost	56.8	51.5	52.2	64.9	61.1	55.4	52.5	70.0
Random forest	56.7	51.5	52.1	64.8	58.7	53.4	52.3	66.2
Logistic regression	52.6	45.4	42.0	61.8	55.5	50.4	48.1	63.0
Decision tree	51.5	46.4	45.4	60.2	57.6	50.4	49.8	66.9

where *AP* average precision, *TPR* true positive rate, *TNR* true negative rate, *GM* geometric mean

Table 5 Comparative analysis of previous work with proposed work

Author	Classification models	Feature selection	Sampling	Result (AP) (%)
Perara et al. [4]	Bagging + Random forest	Yes	Smote	51.40
Vasumati et al. [7]	Random forest	No	No	52.30
Taneja et al. [9]	RFE + HDDT	Yes	Smote	64.07
Berrar [2]	Random forest	No	Up-down	49.99
Berrar [8]	Random forest	No	Up-down	36.20
Oentaryo et al. [14]	Random forest	Yes	No	58.84%
Sisodia et al. [10]	Gradient tree boosting	No	No	60.51
Proposed work	**Gradient tree boosting**	**Yes**	**No**	**64.86**

Bold indicates the proposed work which is being compared with the existing prior works

Since not all features are useful so, removing unwanted, duplicate, and inconsequential features might improve the accuracy of the system. Based on this concept, among individual classifiers, the performance of LR classifier yielded best than DT with an average precision of 52.6% using all features while 55.5% using selected features. As shown in the results, ensemble methods give higher average precision scores than individual classifiers. The reason behind is that ensemble methods weigh various individual classifiers and integrate them to form a classifier which surpasses each of them to enhance the accuracy of the classifiers. Results from the individual and ensemble approaches demonstrate that other than GTB, none of the algorithms were able to achieve average precision higher than 57.0% on the validation dataset with all features. Table 4 compares the consolidated results of classifiers on validation dataset with all features and with selected features. Results depict significant improvement on the dataset after the feature selection/variable elimination. Among all the classification algorithms, gradient tree boosting outperforms with selected features by achieving an average precision of 64.86% on the validation dataset. Several previous methods reported in the literature to address this issue are compared with the proposed approach. Table 5 demonstrates the results that proposed work (GTB) surpasses other existing works with the highest AP score than the state-of-the-art methods. The graphical illustration of results shown in Table 4 is illustrated in Fig. 3.

4 Conclusion

In this work, Pay-Per-Click fraud detection system has been implemented which effectively identifies fraud using several data mining techniques. This research work has focused on the detection of illegitimate publishers among legitimate publishers

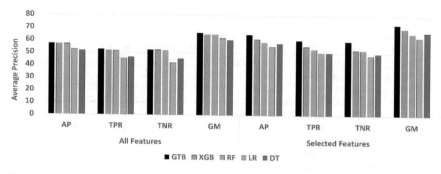

Fig. 3 Comparing the performance of all features with the selected features

based on their click profile they generated. The behavior of each publisher is addressed by evaluating feature importance scores provided by the standard machine learning learners. Noisy, irrelevant and redundant features (lowest feature importance scores) were discarded and features with highest feature importance scores were kept. An efficient predictive model is thus designed and validated by evaluating several machine learning classification approaches together with all features and selected features on the validation dataset. Performance of algorithms is analyzed on measures like average precision (AP), specificity (SP), sensitivity (SE) and G-mean (GM). The literature pertaining to final results strongly suggests that proposed model (gradient tree boosting) outperforms the validation dataset as it reduces the overall prediction error of the model by combining the succeeding model with the prior model. It has achieved an average precision score of 64.86% with selected features on the dataset. Feature importance using GTB as thus proves to be an apparent and efficacious strategy in model designing which has improved the classification performance in identifying the fraudulent conduct of publishers.

The limitation of the work lies in uneven class distribution of the dataset. So, despite significant results and investigation performed to identify fraudsters, there exists a substantial need for further work on balancing the dataset. Since the dataset is highly imbalanced, future analysis is thus pivotal to validate the methods like up-sampling, down-sampling, etc., for better results which is left uncovered in this work.

References

1. Zhang L, Guan Y (2008) Detecting click fraud in pay-per-click streams of online advertising networks. In: Distributed Computing Systems, 2008. ICDCS'08. The 28th international conference on. pp 77–84
2. Berrar D (2012) Random forests for the detection of click fraud in online mobile advertising. In: Proceedings of 2012 international workshop on fraud detection in mobile advertising (FDMA), Singapore, pp 1–10
3. Sisodia D, Sisodia DS (2022) Feature space transformation of user-clicks and deep transfer

learning framework for fraudulent publisher detection in online advertising. Appl Soft Comput 125:109142. https://doi.org/10.1016/j.asoc.2022.109142

4. Perera KS, Neupane B, Faisal MA et al (2013) A novel ensemble learning-based approach for click fraud detection in mobile advertising. In: Proceedings mining intelligence and knowledge exploration (MIKE). Springer, Berlin Heidelberg, pp 370–382

5. Sisodia D, Sisodia DS (2023) Data sampling methods for analyzing publishers conduct from highly imbalanced dataset in web advertising. In: International conference on information systems and management Science. pp 428–441

6. Sisodia D, Sisodia DS (2022) Feature distillation and accumulated selection for automated fraudulent publisher classification from user click data of online advertising. Data technol appl 56:1–24. https://doi.org/10.1108/dta-09-2021-0233

7. Vasumati D, Vani MS, Bhramaramba R, Babu OY (2015) Data mining approach to filter click-spam in mobile ad networks. International conference on computer science. Data Mining Mech Eng ICCDMME Bangkok, Thailand, pp 90–94

8. Berrar D (2016) Learning from automatically labeled data: case study on click fraud prediction. Knowl Inf Syst 46:477–490. https://doi.org/10.1007/s10115-015-0827-6

9. Taneja M, Garg K, Purwar A, Sharma S (2015) Prediction of click frauds in mobile advertising. In: International conference on contemporary computing, IC3. IEEE Computer Society. Noida, India. pp 162–166

10. Sisodia D, Sisodia DS (2020) Gradient boosting learning for fraudulent publisher detection in online advertising. Data Technol Appl 55:216–232. https://doi.org/10.1108/DTA-04-2020-0093

11. Sisodia D, Sisodia DS (2022) Quad division prototype selection-based k-nearest neighbor classifier for click fraud detection from highly skewed user click dataset. Eng Sci Technol Inter J 28:1–12. https://doi.org/10.1016/J.JESTCH.2021.05.015

12. Sisodia D, Sisodia DS (2021) Data sampling strategies for click fraud detection using imbalanced user click data of online advertising: an empirical review. IETE Tech Rev 1–10. https://doi.org/10.1080/02564602.2021.1915892

13. Sisodia D, Sisodia DS (2022) A hybrid data-level sampling approach in learning from skewed user-click data for click fraud detection in online advertising. Expert Systems 40:1–17. https://doi.org/10.1111/exsy.13147

14. Richard Oentaryo, Ee-Peng Lim, Michael Finegold, David Lo, Feida Zhu, Clifton Phua, Eng-Yeow Cheu, Ghim-Eng Yap, Kelvin Sim, Minh Nhut Nguyen, Kasun Perera, Bijay Neupane, Mustafa Faisal, Zeyar Aung WLW (2014) Detecting Click Fraud in Online Advertising : A Data Mining Approach. J Machine Learning Res 15:99–140. https://doi.org/10.1145/2623330.2623718

15. Sisodia D, Sisodia DS (2018) Prediction of diabetes using classification algorithms. Proc Comput Sci 132:1578–1585. https://doi.org/10.1016/j.procs.2018.05.122

16. Breiman L (2001) Random forests. Mach Learn 45:5–32. https://doi.org/10.1023/A:101093 3404324

17. Pregibon D (1981) Logistic regression diagnostics. Ann Stat 9:705–724

18. Thejas GS, Surya Dheeshjith SS, Iyengar NRS and PB (2021) A hybrid and effective learning approach for click fraud detection. Mach Learn Appl 3:1–10. https://doi.org/10.1016/j.mlwa.2020.100016

19. Wong TT (2015) Performance evaluation of classification algorithms by k-fold and leave-one-out cross validation. Pattern Recogn 48:2839–2846. https://doi.org/10.1016/j.patcog.2015.03.009

20. Tharwat A (2020) Classification assessment methods. Appl Comput Informatics 17:168–192. https://doi.org/10.1016/j.aci.2018.08.003

A Deep Learning Model Based on CNN Using Keras and TensorFlow to Determine Real-Time Melting Point of Chemical Substances

Anurag Shrivastava and Rama Sushil

1 Introduction

A melting-point apparatus (mpa) is scientific instrument use for determination of melting point of chemical substances for identifying purity of a chemical substances. A melting point of a solid is the temperature at which it converts into liquid after given adequate heat, the interval between starting of liquefaction and completion of the liquefaction process is called the melting-point range. Detection of melting point is tedious process in which user has to keep watching on capillary tube until unless chemical in powdery form converted into liquid form. Currently, various types of automatic melting-points detection apparatuses are being sold, types of melting devices are Thiele tube, Fisher-John's apparatus, Gallen Kamp (electronic) melting-point apparatus, automatic melting-point apparatus and digitally melting-point apparatus, digital melting-point apparatus use computer vision techniques. It does not require user's attention and determines melting point automatically. In this paper, deep learning model for melting-point detection is proposed for digital melting-point apparatus.

Computer vision techniques such as image classification, segmentation and object detection can be considered as solution of fundamental problem and forms the basis for realistic problems. Earlier process of image classification was totally based on the machine learning, but its accuracy and classification process have some limitation [1].

In this paper, a deep learning model (DL model) based on convolutional neural network (CNN) is implemented in Python using supporting library and tools such as TensorFlow and Keras have used for image classification of chemical substances'

A. Shrivastava (✉) · R. Sushil
School of Computing, DITU, Dehradun, India
e-mail: nrg.shrivastava@gmail.com

R. Sushil
e-mail: rama.sushil@dituniversity.edu.in

state to determine melting point of chemical and finally this model may be deploying on various platform such as desktop or single board computer like Raspberry Pi or Asus Tinker board.

Section II of this paper describes fundamental theory and Section III for related research. Section IV presents proposed framework and proposed DL model, in section V given implementation of proposed DL model. Section VI is shown the results, and Section VII presents conclusion and future work.

2 Fundamental Theory

2.1 CNN

CNN has been established perfectly for image classifications. Various research works dependent on CNN fundamentally improved the performance for image classification. It has good learning for the local and global structure from image dataset. It gets an input and change into hidden layers. Hidden layer comprises of a bunch of neurons, where every neuron is completely associated with all neurons in the past layer, and where neurons in a solitary layer work totally freely and do not share any associations (Fig. 1).

The last completely connected layer is known as the 'output layer' and in order of classification settings it addresses the class scores. The advantage of convolutional neural networks is its input consists of images and passes through architecture elegantly. In this proposed work, different hidden layers classify image into two class one class of solid state and other class for liquid state.

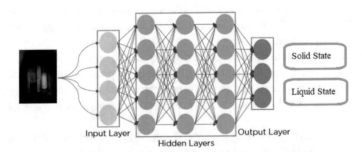

Fig. 1 CNN architecture consists of I/O layers and hidden layers. It classified images into two categories either chemical is in solid state or in liquid state

2.2 TensorFlow and Keras

TensorFlow is an adaptable and versatile programming library for mathematical computation utilizing dataflow graphs, and it is developed by google for research purpose. It can perform complex computation on different platforms like GPU, CPU and TPU in a flexible manner which trained to machine learning model and neural network designing. TensorFlow is used by company like Nvidia and Intel also. TensorFlow works on multidimensional arrays, and these arrays called tensors. In this paper, TensorFlow 1.9 version has used. It can be download easily from official google TensorFlow website. Keras is a deep learning API for TensorFlow, it is written on Python, and it was developed for fast experimentation.

2.3 ReLu and Sigmoid

Basically, activation function is used for getting output from node and it could be called as transfer function, this gives output for convolution neural network such as yes or no and maps the value between 0 and 1. This function can be divided as linear activation function and nonlinear activation function. The rectified linear unit (ReLu) is a linear activation function used in deep learning, when it gets negative input then ReLu function return zero and this returns last value for positive input a, mathematically this could be expressed as $f(a) = \max(0, a)$. The sigmoid function returns the value between 0 and 1, and this is special because it is used with different prediction model for determining the probability of output. Proposed work is using these two functions with CNN model (Fig. 2).

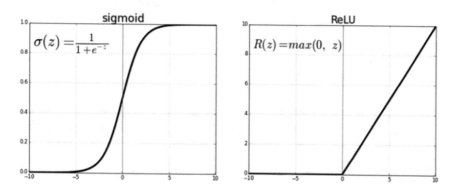

Fig. 2 Graph and equations of sigmoid and ReLu. Sigmoid very popular activation model used in neural network, its functional input transformed into values between 0.0 and 1.0, and output of ReLu will directly to input if it is positive, otherwise output will be zero

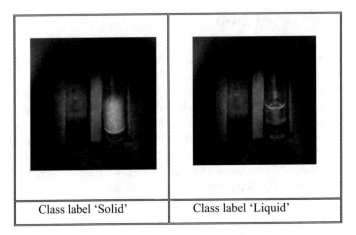

| Class label 'Solid' | Class label 'Liquid' |

Fig. 3 Sample image of DCSS dataset. Entire melting process of chemical recoded by camera, extracted images into jpg format and assigned into two class labels

2.4 Dataset

This proposed work has used own image dataset, image dataset consists images of capillary filled with chemical substances extracted from real-time video processing during melting point determination of chemical substances, and these images arranged into two class label one types of images belongs to liquid class label, while second type of image belongs to solid class. It is named as dataset of chemical substance's state (DCSS) (Fig. 3).

DCSS contains total 2500 images 1322 belongs to solid state, while 1178 images are from liquid state. Proposed deep learning model getting trained and classify images into these two class labels.

2.5 Problem Formation

The intention of this proposed work is to determine melting point of chemical substances by proposed deep learning (DL) model, it classifies images of DCSS image dataset into two categories either solid or liquid state.

3 Related Research

Azlan et al. [1] in this paper, authors have used deep learning with TensorFlow framework for classification of flowers image, thousands of flower image taken for input data and classify flower into five categories, they found that deep neural network is best option for training purpose and it produces high percentage of accuracy for rose flower category produce 90.58% accuracy and average accuracy of other type of flower category is up to 90% and above.

Sultana et al. [2] in this paper, authors have explained different convolution network (CNN) architecture for image classification, explored the benefits from LeNet-5 to latest SENet model in CNN, and they evaluated and compared the performance of these CNN model and concluded that GoogLeNet and ResNet achieved high accuracy.

Ertam et al. [3] in this paper, authors have used open-source TensorFlow library developed by Google for classifying MNIST Dataset, they have applied various activation functions such as hyperbolic tangent, exponential linear unit, rectified linear unit, sigmoid, softsign and softplus. For classification, they have used convolution neural network with softmax classifier and finally concluded that high classification accurate result can be achieved by using rectified linear unit (ReLu) activation function.

Blot et al [4] in this paper, authors have made some changes in convolutional neural network block (CNN) for transferring information from layer by layer without any variance, this becomes possible due to made certain changes in existing activation function, they used double mapping strategy between activation function, it is also known as MaxMin strategy and performed experiment on MNIST and CIFAR-10 dataset and found MaxMin strategy performed outstanding results.

Zhao et al. [5] in this paper, research on characteristics of convolution neural network layers of convolution used the rectifier functions, where mostly used activation is rectified linear unit (ReLU), they randomly taken two layers and said that it could not be best selection for best results they investigate widely used ReLu method and proposed a proportional module which keeps ratio $X:Y$ ($X > Y$) between convolution and ReLu. It may be used with all CNN network without any extra computational cost and for better results, finally their experiment result is shown that proposed model performed better results on certain benchmarks.

Li et al. [6] In this paper, authors designed customized convolution network which is used for classify lungs images their customized convolution neural network automatically taken interesting feature from lungs image which are suitable for classification their proposed architecture performed best result over medical images and classification of texture.

Fonnegra et al. [7] In this paper, authors have presented comparative study on various framework because selection of deep learning framework also plays major role when computational resource is limited. Their proposed work presented study on Theano, TensorFlow and Torch, implemented these classifiers for classifying image of MNIST and CIFAR-10 dataset and listed all performance parameter values like gradient computation time, forward time, computational costs and memory consumption are reported for both GPU and CPU settings.

Hu et al. [8] in this paper, authors have made some changes in convolution neural network (CNN) and improves CNN using pseudo derivative rectified linear unit (ReLu), ReLu is mostly used function in CNN, it is considered as an efficient active function, however, it gives zero output for negative input ReLu this becomes ReLu inactive in training, there are others ReLu versions available for resolving this issue, they proposed pseudo derivatives ReLu (PD-ReLU) in the place of original ReLu and their experiments shown that PD-ReLU performed improved results over CIFAR-10 and CIFAR-100 tests.

Gunjan et al. [9] their proposed work related to chosen of neural network technique for image classification in convolution neural network (CNN), for determining of efficient architecture they proposed a transfer learning technique it is also called fine tuning of deep learning technology.

Shiddieqy et al. [10] in this paper, authors have implemented convolution neural network (CNN) using Python for classification of images, they used thousands of animal's images which contains two class label, one class is for cat, while other is for dog, five layers are used in proposed CNN architecture, found that for getting efficient result need to increase number of layers in CNN, they developed an application and deployed it into single board computer.

Jiang et al. [11] their proposed work related to scene change detection, it is used in video application, authors presented a novel framework of deep learning for change detection. They used ResNet for training, it classified video frames into different class label, and these classified images further can be used for scene change detection. For extracting feature from image, authors have used SIFT algorithm, performed experiment on various video dataset and found result with low error rate.

Pang et al. [12] in this paper, authors reviewed on fundamental technique of neural network like convolutional neural network (CNN), multilayer perceptron and stochastic gradient. Computational cost of these methods is very high, time consuming and error prone, all of these problems overcome by the TensorFlow library which is developed by google, it accelerates research in the field of deep learning, authors reviewed on fundamental concept of TensorFlow like graph execution tool (GET), graph construction functions (GCF), and TensorFlow's visualization tool is also known as tensor board.

Motooka et al. [13] in this paper, authors presented study on functioning of filters of deep neural networks (DNNs), DNN widely used for determine the change in video frames, the background subtraction which are based DNN automatically extracts background feature from dataset. Mostly, researchers evaluate its performance but no one describe how does it works, they explored about working for background subtraction from the experiment they found that the first layer of DNN is used for background subtraction with filters and last layer classify background changes into different categories.

Mohandas et al. [14] in this paper, authors present TensorFlow architecture for real-time person detection on single board computer (Raspberry PI). Quantization model used for it, analysed the performance of quantization model, finally their study shown that interface timing of quantized model is lower than interface timing of unquantized model.

Bhat et al. [15] in this paper, authors have used extreme learning machine; they apply extreme learning machine on large dataset of melting point of chemical molecules. Their experimental results are shown that extreme learning machine gave better results than k-nearest neighbour.

4 Proposed Deep Learning Model

The proposed DL model implemented in Python, it classified images into two categories either chemical in liquid form or in solid form parallelly capillary which containing chemical substances kept over heater block getting heat continuously, when proposed DL model gets capillary images in liquid form at this moment current temperature represented as a melting point of chemical. Figure 4 is shown the proposed DL model for image classification. It consists four phases, each of these phases includes TensorFlow as the open-source library and Python as programming language. Then, the process is continued to take input of the images from DCSS image dataset, then DL model got trained and finally all images classified into two categories. Once defined image dataset for training and validation purpose then proposed DL model used convolution neural network (CNN) with maxpool and ReLu activation function, its layers sequentially fitted in listing manner, Conv2D, flatten and dense layers also used in proposed DL model, it is need to specify number of filters in each layer also size of filter with ReLu activation filter, so 16 filters of each having size of (3, 3) in Conv2D layer, 32 filters of each having size of (3, 3).

In second layer, 64 filters of each are having size of (3, 3) with ReLu activation taken. This model gets trained after several iteration of learning process, then it classified images into two categories. Its implementation and classification accuracy on different epochs with and without activation functions are discussed in the next section.

5 Proposed DL Model Implementation

In this section, proposed deep learning model implemented in Python and result analysis done. Implementation steps: Firstly, created two folder named as solid and liquid which contains images of chemical substances in powdery solid form and liquid form for training purpose, now generating image from dataset used imagedatagenerator library, labelled to all the images which are in solid and liquid folder, it is required to load images in proposed model using img = image.load_img(path) function, for knowing three dimensional shape of matrix use cv2.imread(path) function which shows size of image in the format (height, width, dimensions), e.g. (408, 402, 3), where 408 px is height and 402 px is width and three represents three-dimensional matrix.

Training and validation process consists of initialization of two class for train and validation in programming to generate images for training purpose used imagegenerator function. Although RGB pixel value ranges from 0 to 255, it is required to convert this value between 0 to 1 so rescale it and divide each value with 255 using (rescale = 1/255) function, same thing happened with validation class. Its Python code is shown in Fig. 5.

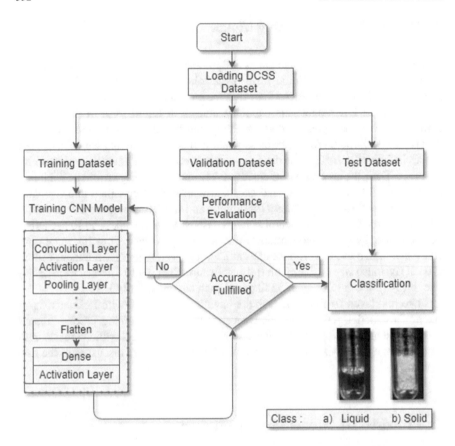

Fig. 4 Process flow diagram for proposed DL model, it started with loading of DCSS dataset, now it gets trained until maximum accuracy achieved, then classified images into two class labels

```
train= ImageDataGenerator(rescale=1/255)
validation= ImageDataGenerator(rescale=1/255)
```

Fig. 5 Python code for image generation from dataset

DL model taken training image and feed them into neural network using train.flow_from_directory, images in training class are different in size since this can be used multi-size image in neural network but it is required to resize, it using target_size function and resized images in (200, 200) also need to specify class mode, in our proposed model binary class mode used, followed same procedure for validation class. Python code for this is shown in Fig. 6.

Neural network does not label any class in string, it represents only in numeric form, so two class of images is represented by either 0 or 1, {'liquid': 0, 'solid': 1}.

```
train_dataset=train.flow_from_directory('C:/Users/nrgsh/OneDrive/Desktop/basedata/train',
            target_size=(200,200),batch_size=3, class_mode='binary')
validation_dataset=train.flow_from_directory('C:/Users/nrgsh/OneDrive/Desktop/basedata/validation',
            target_size=(200,200),batch_size=3, class_mode='binary')
```

Fig. 6 Python code for feeding images into neural network

```
model=tf.keras.models.Sequential([ tf.keras.layers.Conv2D(16,(3,3),activation= 'relu',input_shape=(200,100,3)),
            tf.keras.layers.MaxPool2D(2,2),
            #
            tf.keras.layers.Conv2D(32,(3,3),activation='relu'),
            tf.keras.layers.MaxPool2D(2,2),
            #
            tf.keras.layers.Conv2D(64,(3,3),activation='relu'),
            tf.keras.layers.MaxPool2D(2,2),
            ##
            tf.keras.layers.Flatten(),
            ##
            tf.keras.layers.Dense(512),
            ##
            tf.keras.layers.Dense(1,activation='sigmoid')
            ])
```

Fig. 7 Python code for proposed deep learning model

Once defined training and validation dataset proposed model used convolution neural network (CNN) with maxpool activation function, layers sequentially fitted by tf.keras.models.Sequential in programme. Sequential function arranged layers in listing manner, three Conv2D layers used in proposed DL model, it is need to specify number of filters in each layer also size of filter with ReLu activation filter, 16 filters of each having size of (3, 3) in first Conv2D layer, 32 filters of each having size of (3, 3) in second layer and 64 filters of each having size of (3, 3) with ReLu activation taken. Its Python code is shown in Fig. 7.

Conv2D Along with Maxpool2D that takes maximum value of pixel from pixels and flatten layer that converting multidimensional data into single dimension array used, finally dense layer that feeds all outputs from all previous layers into neurons and each neurons transfers output to other layers.

6 Results

The popularity of deep learning models is increasing day by day, it gives better results and makes effective solution for analysing big data. In my proposed work, deep learning model is used for classifying images of DCSS image dataset, and TensorFlow and Keras with various activation function are used for this purpose. For testing purpose, we have evaluated accuracy with and without using activation functions like ReLu and sigmoid on different epochs 15, 20 and 25. Total six cases considered three cases with activation function on 15, 20 and 25 iterations, respectively, likewise without activation functions on 15, 20 and 25 iterations, respectively. Finally, it is shown increasing number of epochs with activation function obtained max accuracy 99.72% and model validation max accuracy is 99.37% on DCSS image dataset.

Performance analysis of proposed DL model with activation function on 15, 20 and 25 epochs is shown in Figs. 8, 9, 10, 11, 12, 13, 14, 15 and 16.

After performance analysis of deep learning model with activation function also analysed performance without activation function on 15, 20 and 25 epochs, results are shown in Figs. 17, 18, 19, 20, 21, 22, 23, 24 and 25.

```
Epoch 1/15
3/3 [==============================] - 3s 1s/step - loss: 0.5851 - accuracy: 0.7778 - val_loss: 0.7463 - val_accuracy: 0.6040
Epoch 2/15
3/3 [==============================] - 3s 1s/step - loss: 0.7096 - accuracy: 0.5556 - val_loss: 0.6270 - val_accuracy: 0.6040
Epoch 3/15
3/3 [==============================] - 3s 1s/step - loss: 0.4205 - accuracy: 0.7778 - val_loss: 1.3311 - val_accuracy: 0.6040
Epoch 4/15
3/3 [==============================] - 3s 1s/step - loss: 0.5110 - accuracy: 0.8889 - val_loss: 1.0050 - val_accuracy: 0.6040
Epoch 5/15
3/3 [==============================] - 3s 1s/step - loss: 0.4068 - accuracy: 0.8889 - val_loss: 0.5622 - val_accuracy: 0.6040
Epoch 6/15
3/3 [==============================] - 3s 1s/step - loss: 0.5055 - accuracy: 0.7778 - val_loss: 0.5689 - val_accuracy: 0.6040
Epoch 7/15
3/3 [==============================] - 3s 1s/step - loss: 0.8123 - accuracy: 0.8571 - val_loss: 1.1731 - val_accuracy: 0.6040
Epoch 8/15
3/3 [==============================] - 3s 1s/step - loss: 0.8539 - accuracy: 0.7778 - val_loss: 0.5651 - val_accuracy: 0.6040
Epoch 9/15
3/3 [==============================] - 3s 1s/step - loss: 0.5028 - accuracy: 0.6667 - val_loss: 0.4939 - val_accuracy: 0.6040
Epoch 10/15
3/3 [==============================] - 3s 1s/step - loss: 0.3838 - accuracy: 0.6667 - val_loss: 0.9090 - val_accuracy: 0.6040
Epoch 11/15
3/3 [==============================] - 3s 1s/step - loss: 1.1451 - accuracy: 0.8889 - val_loss: 0.6726 - val_accuracy: 0.6040
Epoch 12/15
3/3 [==============================] - 3s 1s/step - loss: 0.2948 - accuracy: 0.8889 - val_loss: 0.6892 - val_accuracy: 0.6040
Epoch 13/15
3/3 [==============================] - 3s 1s/step - loss: 0.2618 - accuracy: 0.8889 - val_loss: 0.4909 - val_accuracy: 0.6040
Epoch 14/15
3/3 [==============================] - 3s 1s/step - loss: 0.2394 - accuracy: 0.7778 - val_loss: 0.4059 - val_accuracy: 0.8658
Epoch 15/15
3/3 [==============================] - 3s 1s/step - loss: 0.2335 - accuracy: 0.8889 - val_loss: 0.3505 - val_accuracy: 0.8389
dict_keys(['loss', 'accuracy', 'val_loss', 'val_accuracy'])
```

Fig. 8 Accuracy of proposed DL model on each iteration of 15 epochs with activation function. It is generated on Jupyter IDE, which is shown model loss, model accuracy, validation loss and validation accuracy on each step of 15 epochs

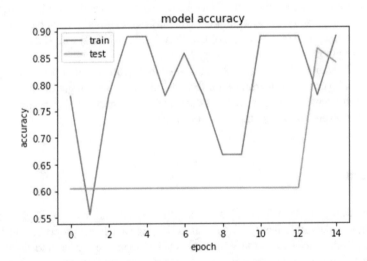

Fig. 9 Graph plotting for accuracy of proposed DL model using activation function on 15 epochs. Comparable performance of proposed model's accuracy on train and validation dataset labelled as test on each 15 epochs

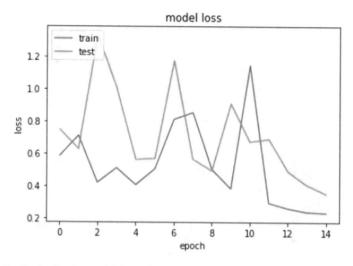

Fig. 10 Graph plotting for model loss of proposed DL model using activation function on 15 epochs. Comparable performance of proposed model's loss on train and validation dataset labelled as test on each 15 epochs

```
Epoch 1/20
3/3 [==============================] - 7s 3s/step - loss: 9.5221 - accuracy: 0.1111 - val_loss: 0.6052 - val_accuracy: 0.7531
Epoch 2/20
3/3 [==============================] - 6s 3s/step - loss: 0.2608 - accuracy: 0.9672 - val_loss: 1.2063 - val_accuracy: 0.7531
Epoch 3/20
3/3 [==============================] - 7s 3s/step - loss: 0.5347 - accuracy: 0.8889 - val_loss: 0.8348 - val_accuracy: 0.7531
Epoch 4/20
3/3 [==============================] - 7s 3s/step - loss: 0.9290 - accuracy: 0.6667 - val_loss: 0.5250 - val_accuracy: 0.7531
Epoch 5/20
3/3 [==============================] - 7s 3s/step - loss: 0.4732 - accuracy: 0.8889 - val_loss: 0.5082 - val_accuracy: 0.7531
Epoch 6/20
3/3 [==============================] - 7s 3s/step - loss: 0.7622 - accuracy: 0.5556 - val_loss: 0.5599 - val_accuracy: 0.8281
Epoch 7/20
3/3 [==============================] - 7s 3s/step - loss: 0.6557 - accuracy: 0.8889 - val_loss: 0.4890 - val_accuracy: 0.8109
Epoch 8/20
3/3 [==============================] - 7s 3s/step - loss: 0.5268 - accuracy: 0.7778 - val_loss: 0.4745 - val_accuracy: 0.7531
Epoch 9/20
3/3 [==============================] - 6s 3s/step - loss: 0.5794 - accuracy: 0.6667 - val_loss: 0.4449 - val_accuracy: 0.8062
Epoch 10/20
3/3 [==============================] - 6s 3s/step - loss: 0.1664 - accuracy: 0.9888 - val_loss: 0.6196 - val_accuracy: 0.7531
Epoch 11/20
3/3 [==============================] - 7s 3s/step - loss: 0.4296 - accuracy: 0.8889 - val_loss: 0.4182 - val_accuracy: 0.8047
Epoch 12/20
3/3 [==============================] - 6s 3s/step - loss: 0.3402 - accuracy: 0.8889 - val_loss: 0.4304 - val_accuracy: 0.7578
Epoch 13/20
3/3 [==============================] - 6s 3s/step - loss: 0.5012 - accuracy: 0.7778 - val_loss: 0.3491 - val_accuracy: 0.8250
Epoch 14/20
3/3 [==============================] - 7s 3s/step - loss: 0.3314 - accuracy: 0.8889 - val_loss: 0.4079 - val_accuracy: 0.8234
Epoch 15/20
3/3 [==============================] - 7s 3s/step - loss: 0.1875 - accuracy: 0.8889 - val_loss: 0.4976 - val_accuracy: 0.6594
Epoch 16/20
3/3 [==============================] - 6s 3s/step - loss: 0.3536 - accuracy: 0.6667 - val_loss: 0.5150 - val_accuracy: 0.8219
Epoch 17/20
3/3 [==============================] - 7s 3s/step - loss: 0.6763 - accuracy: 0.6667 - val_loss: 0.3231 - val_accuracy: 0.8109
Epoch 18/20
3/3 [==============================] - 7s 3s/step - loss: 0.3157 - accuracy: 0.8889 - val_loss: 0.3202 - val_accuracy: 0.8031
Epoch 19/20
3/3 [==============================] - 7s 3s/step - loss: 0.3298 - accuracy: 0.8889 - val_loss: 0.2040 - val_accuracy: 0.9672
Epoch 20/20
3/3 [==============================] - 7s 3s/step - loss: 0.3325 - accuracy: 0.8889 - val_loss: 0.8856 - val_accuracy: 0.7531
dict_keys(['loss', 'accuracy', 'val_loss', 'val_accuracy'])
```

Fig. 11 Accuracy of proposed DL model on each iteration of 20 epochs with activation function. It is generated on Jupyter IDE, which is shown model loss, model accuracy, validation loss and validation accuracy on each step of 20 epochs

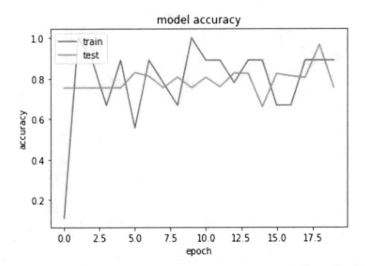

Fig. 12 Graph plotting for accuracy of proposed DL model using activation function on 20 epochs. Comparable performance of proposed model's accuracy on train and validation dataset labelled as test on each 20 epochs

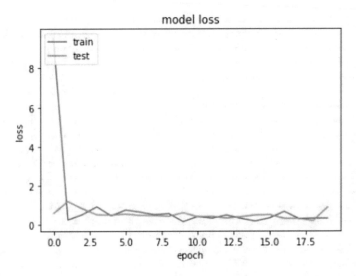

Fig. 13 Graph plotting for model loss of proposed DL model using active function on 20 epochs. Comparable performance of proposed model's loss on train and validation dataset labelled as test on each 20 epochs

```
Epoch 1/25
3/3 [==============================] - 9s 4s/step - loss: 9.9213 - accuracy: 0.4444 - val_loss: 3.2711 - val_accuracy: 0.7531
Epoch 2/25
3/3 [==============================] - 7s 4s/step - loss: 1.8440 - accuracy: 0.8889 - val_loss: 1.6128 - val_accuracy: 0.7531
Epoch 3/25
3/3 [==============================] - 7s 3s/step - loss: 0.5661 - accuracy: 0.8889 - val_loss: 1.1476 - val_accuracy: 0.7531
Epoch 4/25
3/3 [==============================] - 8s 4s/step - loss: 1.6259 - accuracy: 0.4444 - val_loss: 1.1270 - val_accuracy: 0.7531
Epoch 5/25
3/3 [==============================] - 8s 4s/step - loss: 1.5155 - accuracy: 0.5556 - val_loss: 0.5217 - val_accuracy: 0.9812
Epoch 6/25
3/3 [==============================] - 8s 4s/step - loss: 0.4884 - accuracy: 0.8889 - val_loss: 0.3582 - val_accuracy: 0.8188
Epoch 7/25
3/3 [==============================] - 8s 4s/step - loss: 0.3880 - accuracy: 0.8889 - val_loss: 0.3386 - val_accuracy: 0.8109
Epoch 8/25
3/3 [==============================] - 8s 4s/step - loss: 0.4162 - accuracy: 0.7778 - val_loss: 0.3697 - val_accuracy: 0.9516
Epoch 9/25
3/3 [==============================] - 8s 4s/step - loss: 0.3479 - accuracy: 0.8889 - val_loss: 0.2463 - val_accuracy: 0.8313
Epoch 10/25
3/3 [==============================] - 8s 4s/step - loss: 0.3194 - accuracy: 0.7778 - val_loss: 1.3281 - val_accuracy: 0.7531
Epoch 11/25
3/3 [==============================] - 8s 4s/step - loss: 1.6475 - accuracy: 0.6667 - val_loss: 0.2387 - val_accuracy: 0.9563
Epoch 12/25
3/3 [==============================] - 8s 4s/step - loss: 0.2184 - accuracy: 0.9516 - val_loss: 0.3076 - val_accuracy: 0.8047
Epoch 13/25
3/3 [==============================] - 8s 4s/step - loss: 0.4711 - accuracy: 0.6667 - val_loss: 0.2100 - val_accuracy: 0.9922
Epoch 14/25
3/3 [==============================] - 8s 4s/step - loss: 0.3247 - accuracy: 0.8889 - val_loss: 0.1979 - val_accuracy: 0.9641
Epoch 15/25
3/3 [==============================] - 8s 4s/step - loss: 0.2385 - accuracy: 0.8889 - val_loss: 0.1447 - val_accuracy: 0.9812
Epoch 16/25
3/3 [==============================] - 8s 4s/step - loss: 0.0794 - accuracy: 0.9891 - val_loss: 0.0838 - val_accuracy: 0.9906
Epoch 17/25
3/3 [==============================] - 8s 4s/step - loss: 0.0698 - accuracy: 0.9891 - val_loss: 0.1311 - val_accuracy: 0.9672
Epoch 18/25
3/3 [==============================] - 8s 4s/step - loss: 0.0348 - accuracy: 0.9891 - val_loss: 0.0784 - val_accuracy: 0.9891
Epoch 19/25
3/3 [==============================] - 8s 4s/step - loss: 0.0244 - accuracy: 0.9891 - val_loss: 0.0753 - val_accuracy: 0.9828
Epoch 20/25
3/3 [==============================] - 8s 4s/step - loss: 0.0145 - accuracy: 0.9972 - val_loss: 0.0907 - val_accuracy: 0.9937
Epoch 21/25
3/3 [==============================] - 8s 4s/step - loss: 0.0470 - accuracy: 0.9972 - val_loss: 0.0708 - val_accuracy: 0.9828
Epoch 22/25
3/3 [==============================] - 7s 4s/step - loss: 0.0091 - accuracy: 0.9972 - val_loss: 0.0591 - val_accuracy: 0.9891
Epoch 23/25
3/3 [==============================] - 8s 4s/step - loss: 0.0055 - accuracy: 0.9972 - val_loss: 0.0592 - val_accuracy: 0.9891
Epoch 24/25
3/3 [==============================] - 8s 4s/step - loss: 0.1022 - accuracy: 0.9972 - val_loss: 0.7302 - val_accuracy: 0.8266
Epoch 25/25
3/3 [==============================] - 8s 4s/step - loss: 0.5737 - accuracy: 0.8889 - val_loss: 6.7864 - val_accuracy: 0.7531
dict_keys(['loss', 'accuracy', 'val_loss', 'val_accuracy'])
```

Fig. 14 Accuracy of proposed DL model on each iteration of 25 epoch with activation function. It is generated on Jupyter IDE, which is shown model loss, model accuracy, validation loss and validation accuracy on each step of 25 epochs

In table shown that max accuracy and max validation accuracy of DL model with and without activation function, it shows increasing number of epochs with activation function obtained max accuracy 99.72% and model validation max accuracy is 99.37% on DCSS image dataset (Table 1).

7 Conclusion

TensorFlow most famous and widely used open-source library used with Keras in deep learning model for image classification purpose. In proposed work, it classified images of capillary tube filled with chemical substances in two class either chemical is in solid state or in liquid state for determining melting point of chemical substances for this purpose proposed model has used with and without using activation function, during experiment, computed results with and without using activation function on different iteration and found that activation function played a major role for getting high accurate results, it is shown 99.72% accuracy at 25 epochs with activation

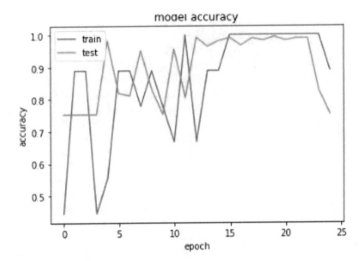

Fig. 15 Graph plotting for proposed DL model accuracy using activation function on 25 epochs. Comparable performance of proposed model's accuracy on train and validation dataset labelled as test on each 25 epochs

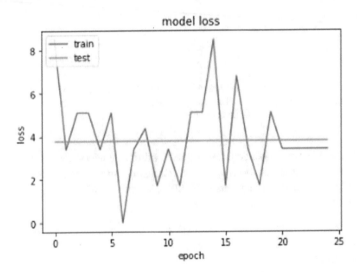

Fig. 16 Graph plotting for proposed DL model loss using activation function on 25 epochs. Comparable performance of proposed model's loss on train and validation dataset labelled as test on each 25 epochs

```
Epoch 1/15
3/3 [==============================] - 6s 3s/step - loss: 11.9972 - accuracy: 0.2222 - val_loss: 11.6169 - val_accuracy: 0.2469
Epoch 2/15
3/3 [==============================] - 7s 3s/step - loss: 11.9972 - accuracy: 0.2222 - val_loss: 11.6169 - val_accuracy: 0.2469
Epoch 3/15
3/3 [==============================] - 7s 3s/step - loss: 10.2833 - accuracy: 0.3333 - val_loss: 11.6169 - val_accuracy: 0.2469
Epoch 4/15
3/3 [==============================] - 7s 3s/step - loss: 13.7111 - accuracy: 0.1111 - val_loss: 11.6169 - val_accuracy: 0.2469
Epoch 5/15
3/3 [==============================] - 7s 3s/step - loss: 13.7111 - accuracy: 0.1111 - val_loss: 11.6169 - val_accuracy: 0.2469
Epoch 6/15
3/3 [==============================] - 7s 3s/step - loss: 13.7111 - accuracy: 0.1111 - val_loss: 11.6169 - val_accuracy: 0.2469
Epoch 7/15
3/3 [==============================] - 7s 3s/step - loss: 6.8555 - accuracy: 0.5556 - val_loss: 11.6169 - val_accuracy: 0.2469
Epoch 8/15
3/3 [==============================] - 7s 3s/step - loss: 10.2833 - accuracy: 0.3333 - val_loss: 11.6169 - val_accuracy: 0.2469
Epoch 9/15
3/3 [==============================] - 8s 4s/step - loss: 10.2833 - accuracy: 0.3333 - val_loss: 11.6169 - val_accuracy: 0.2469
Epoch 10/15
3/3 [==============================] - 7s 3s/step - loss: 4.4071 - accuracy: 0.7143 - val_loss: 11.6169 - val_accuracy: 0.2469
Epoch 11/15
3/3 [==============================] - 7s 3s/step - loss: 11.9972 - accuracy: 0.2222 - val_loss: 11.6169 - val_accuracy: 0.2469
Epoch 12/15
3/3 [==============================] - 6s 3s/step - loss: 6.8555 - accuracy: 0.5556 - val_loss: 11.6169 - val_accuracy: 0.2469
Epoch 13/15
3/3 [==============================] - 7s 3s/step - loss: 13.7111 - accuracy: 0.1111 - val_loss: 11.6169 - val_accuracy: 0.2469
Epoch 14/15
3/3 [==============================] - 7s 3s/step - loss: 10.2833 - accuracy: 0.3333 - val_loss: 11.6169 - val_accuracy: 0.2469
Epoch 15/15
3/3 [==============================] - 6s 3s/step - loss: 11.9972 - accuracy: 0.2222 - val_loss: 11.6169 - val_accuracy: 0.2469
dict_keys(['loss', 'accuracy', 'val_loss', 'val_accuracy'])
```

Fig. 17 Accuracy of proposed DL model on each iteration of 15 epoch without using activation function, it is generated on Jupyter IDE, which is shown model loss, model accuracy, validation loss and validation accuracy on each step of 15 epochs

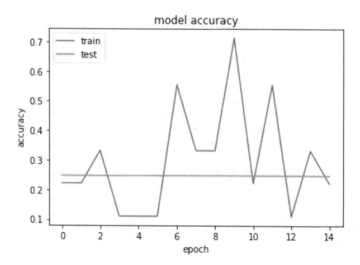

Fig. 18 Graph plotting for proposed DL model accuracy without using activation function on 15 epochs. Comparable performance of proposed model's accuracy on train and validation dataset labelled as test on each 15 epochs

function and without activation function is 77.78%. Model max validation accuracy on 25 epochs with activation function is 99.37% and without activation function is 60.40% so it is concluded that use of activation function on optimal number of epoch's high accuracy can be achieved.

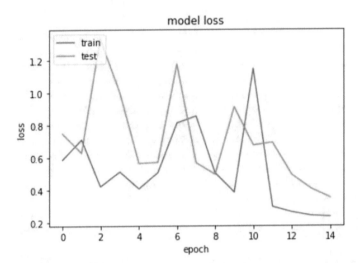

Fig. 19 Graph plotting for proposed DL model loss without using activation function on 15 epochs. Comparable performance of proposed model's loss on train and validation dataset labelled as test on each 15 epochs

```
Epoch 1/20
3/3 [==============================] - 12s 4s/step - loss: 13.7111 - accuracy: 0.1111 - val_loss: 9.3171 - val_accuracy: 0.3960
Epoch 2/20
3/3 [==============================] - 3s 1s/step - loss: 13.7111 - accuracy: 0.1111 - val_loss: 9.3171 - val_accuracy: 0.3960
Epoch 3/20
3/3 [==============================] - 3s 1s/step - loss: 10.2833 - accuracy: 0.3333 - val_loss: 9.3171 - val_accuracy: 0.3960
Epoch 4/20
3/3 [==============================] - 3s 1s/step - loss: 11.9972 - accuracy: 0.2222 - val_loss: 9.3171 - val_accuracy: 0.3960
Epoch 5/20
3/3 [==============================] - 3s 1s/step - loss: 8.5694 - accuracy: 0.4444 - val_loss: 9.3171 - val_accuracy: 0.3960
Epoch 6/20
3/3 [==============================] - 3s 1s/step - loss: 11.9972 - accuracy: 0.2222 - val_loss: 9.3171 - val_accuracy: 0.3960
Epoch 7/20
3/3 [==============================] - 3s 1s/step - loss: 8.5694 - accuracy: 0.4444 - val_loss: 9.3171 - val_accuracy: 0.3960
Epoch 8/20
3/3 [==============================] - 3s 1s/step - loss: 10.2833 - accuracy: 0.3333 - val_loss: 9.3171 - val_accuracy: 0.3960
Epoch 9/20
3/3 [==============================] - 3s 1s/step - loss: 10.2833 - accuracy: 0.3333 - val_loss: 9.3171 - val_accuracy: 0.3960
Epoch 10/20
3/3 [==============================] - 3s 1s/step - loss: 13.7111 - accuracy: 0.1111 - val_loss: 9.3171 - val_accuracy: 0.3960
Epoch 11/20
3/3 [==============================] - 3s 1s/step - loss: 8.5694 - accuracy: 0.4444 - val_loss: 9.3171 - val_accuracy: 0.3960
Epoch 12/20
3/3 [==============================] - 3s 1s/step - loss: 10.2833 - accuracy: 0.3333 - val_loss: 9.3171 - val_accuracy: 0.3960
Epoch 13/20
3/3 [==============================] - 3s 1s/step - loss: 10.2833 - accuracy: 0.3333 - val_loss: 9.3171 - val_accuracy: 0.3960
Epoch 14/20
3/3 [==============================] - 3s 1s/step - loss: 10.2833 - accuracy: 0.3333 - val_loss: 9.3171 - val_accuracy: 0.3960
Epoch 15/20
3/3 [==============================] - 3s 1s/step - loss: 10.2833 - accuracy: 0.3333 - val_loss: 9.3171 - val_accuracy: 0.3960
Epoch 16/20
3/3 [==============================] - 3s 1s/step - loss: 11.9972 - accuracy: 0.2222 - val_loss: 9.3171 - val_accuracy: 0.3960
Epoch 17/20
3/3 [==============================] - 3s 1s/step - loss: 10.2833 - accuracy: 0.3333 - val_loss: 9.3171 - val_accuracy: 0.3960
Epoch 18/20
3/3 [==============================] - 3s 1s/step - loss: 8.5694 - accuracy: 0.4444 - val_loss: 9.3171 - val_accuracy: 0.3960
Epoch 19/20
3/3 [==============================] - 3s 1s/step - loss: 8.5694 - accuracy: 0.4444 - val_loss: 9.3171 - val_accuracy: 0.3960
Epoch 20/20
3/3 [==============================] - 3s 1s/step - loss: 9.3171 - accuracy: 0.6033 - val_loss: 9.3171 - val_accuracy: 0.3960
```

Fig. 20 Accuracy of DL model on each iteration of 20 epoch without using activation function. It is generated on Jupyter IDE, which is shown model loss, model accuracy, validation loss and validation accuracy on each step of 20 epochs

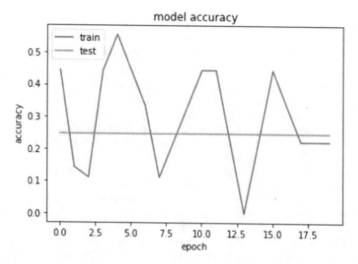

Fig. 21 Graph plotting for proposed DL model accuracy without using activation function on 20 epochs. Comparable performance of proposed model's accuracy on train and validation dataset labelled as test on each 20 epochs

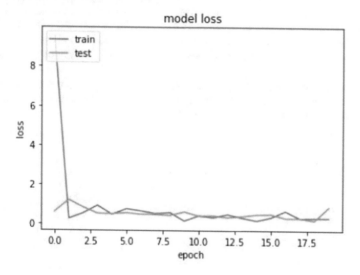

Fig. 22 Graph plotting for proposed DL model loss without using activation function on 20 epochs. Comparable performance of proposed model's loss on train and validation dataset labelled as test on each 20 epochs

```
Epoch 1/25
3/3 [==============================] - 3s 1s/step - loss: 5.0831 - accuracy: 0.6667 - val_loss: 6.0383 - val_accuracy: 0.6040
Epoch 2/25
3/3 [==============================] - 3s 1s/step - loss: 5.0831 - accuracy: 0.6667 - val_loss: 6.0383 - val_accuracy: 0.6040
Epoch 3/25
3/3 [==============================] - 3s 1s/step - loss: 0.0000e+00 - accuracy: 0.6667 - val_loss: 6.0383 - val_accuracy: 0.60
40
Epoch 4/25
3/3 [==============================] - 3s 1s/step - loss: 5.0831 - accuracy: 0.6667 - val_loss: 6.0383 - val_accuracy: 0.6040
Epoch 5/25
3/3 [==============================] - 3s 1s/step - loss: 8.4718 - accuracy: 0.4444 - val_loss: 6.0383 - val_accuracy: 0.6040
Epoch 6/25
3/3 [==============================] - 3s 1s/step - loss: 5.0831 - accuracy: 0.6667 - val_loss: 6.0383 - val_accuracy: 0.6040
Epoch 7/25
3/3 [==============================] - 3s 1s/step - loss: 3.3887 - accuracy: 0.7778 - val_loss: 6.0383 - val_accuracy: 0.6040
Epoch 8/25
3/3 [==============================] - 3s 1s/step - loss: 3.3887 - accuracy: 0.7778 - val_loss: 6.0383 - val_accuracy: 0.6040
Epoch 9/25
3/3 [==============================] - 3s 1s/step - loss: 3.3887 - accuracy: 0.7778 - val_loss: 6.0383 - val_accuracy: 0.6040
Epoch 10/25
3/3 [==============================] - 3s 1s/step - loss: 5.0831 - accuracy: 0.6667 - val_loss: 6.0383 - val_accuracy: 0.6040
Epoch 11/25
3/3 [==============================] - 3s 1s/step - loss: 5.0831 - accuracy: 0.6667 - val_loss: 6.0383 - val_accuracy: 0.6040
Epoch 12/25
3/3 [==============================] - 3s 1s/step - loss: 5.0831 - accuracy: 0.6667 - val_loss: 6.0383 - val_accuracy: 0.6040
Epoch 13/25
3/3 [==============================] - 3s 1s/step - loss: 3.3887 - accuracy: 0.7778 - val_loss: 6.0383 - val_accuracy: 0.6040
Epoch 14/25
3/3 [==============================] - 3s 1s/step - loss: 5.0831 - accuracy: 0.6667 - val_loss: 6.0383 - val_accuracy: 0.6040
Epoch 15/25
3/3 [==============================] - 3s 1s/step - loss: 3.3887 - accuracy: 0.7778 - val_loss: 6.0383 - val_accuracy: 0.6040
Epoch 16/25
3/3 [==============================] - 3s 1s/step - loss: 6.7774 - accuracy: 0.5556 - val_loss: 6.0383 - val_accuracy: 0.6040
Epoch 17/25
3/3 [==============================] - 3s 1s/step - loss: 5.0831 - accuracy: 0.6667 - val_loss: 6.0383 - val_accuracy: 0.6040
Epoch 18/25
3/3 [==============================] - 3s 1s/step - loss: 5.0831 - accuracy: 0.6667 - val_loss: 6.0383 - val_accuracy: 0.6040
Epoch 19/25
3/3 [==============================] - 3s 1s/step - loss: 3.3887 - accuracy: 0.7778 - val_loss: 6.0383 - val_accuracy: 0.6040
Epoch 20/25
3/3 [==============================] - 3s 1s/step - loss: 5.0831 - accuracy: 0.6667 - val_loss: 6.0383 - val_accuracy: 0.6040
Epoch 21/25
3/3 [==============================] - 3s 1s/step - loss: 5.0831 - accuracy: 0.6667 - val_loss: 6.0383 - val_accuracy: 0.6040
Epoch 22/25
3/3 [==============================] - 3s 1s/step - loss: 5.0831 - accuracy: 0.6667 - val_loss: 6.0383 - val_accuracy: 0.6040
Epoch 23/25
3/3 [==============================] - 3s 1s/step - loss: 5.0831 - accuracy: 0.6667 - val_loss: 6.0383 - val_accuracy: 0.6040
Epoch 24/25
3/3 [==============================] - 3s 1s/step - loss: 3.3887 - accuracy: 0.7778 - val_loss: 6.0383 - val_accuracy: 0.6040
Epoch 25/25
3/3 [==============================] - 3s 1s/step - loss: 0.0000e+00 - accuracy: 0.6667 - val_loss: 6.0383 - val_accuracy: 0.60
40
```

Fig. 23 Accuracy of proposed DL model on each iteration of 25 epoch without using activation function It is generated on Jupyter IDE, which is shown model loss, model accuracy validation loss and validation accuracy on each step of 25 epochs

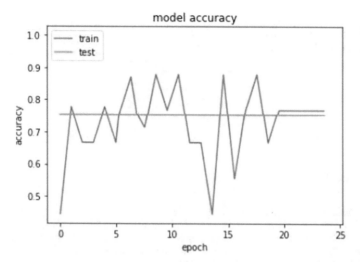

Fig. 24 Graph plotting for proposed DL model accuracy without using activation function on 25 epochs. Comparable performance of proposed model's accuracy on train and validation dataset labelled as test on each 25 epochs

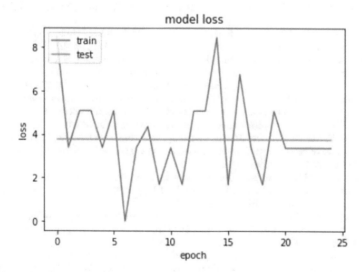

Fig. 25 Graph plotting for proposed DL model loss without using activation function on 25 epochs. Comparable performance of proposed model's loss on train and validation dataset labelled as test on each 20 epochs

Table 1 Max accuracy and max validation accuracy with and without using activation functions

Epoch	Activation	Max accuracy	Max validation accuracy
15	Yes	88.89	86.58
20	Yes	98.88	96.72
25	Yes	99.72	99.37
15	No	71.43	24.69
20	No	68.33	39.60
25	No	77.78	60.40

References

1. Abu MA, Indra NH, Abd Rahman AH, Sapiee A, Ahmad I (2019) A study on image classification based on deep learning and tensorflow. Int J Eng Res Techno 12(4) (2019):563–569. ISSN 0974–3154 © International Research Publication House. http://www.irphouse.com
2. Sultana F (2018) Advancements in image classification using convolutional neural network. 978-1-5386-7638-7/18/$31.00 © 2018 IEEE
3. Ertam F, Aydın G (2017) Data classification with deep learning using Tensorflow. Int Conf Comput Sci Eng (UBMK) 2017:755–758. https://doi.org/10.1109/UBMK.2017.8093521
4. Blot M, Cord M, Thome N (2016) Max-min convolutional neural networks for image classification. IEEE Int Conf Image Proc (ICIP) 2016:3678–3682. https://doi.org/10.1109/ICIP.2016.7533046
5. Zhao G, Zhang Z, Guan H, Tang P, Wang J (2018) Rethinking ReLU to train better CNNs. In: 2018 24th International conference on pattern recognition (ICPR), pp 603–608
6. Li Q, Cai W, Wang X, Zhou Y, Feng DD, Chen M (2014)Medical image classification with convolutional neural network. In: 2014 13th international conference on control automation robotics and vision (ICARCV), pp. 844–848. https://doi.org/10.1109/ICARCV.2014.7064414
7. Fonnegra RD, Blair B, Díaz GM (2017) Performance comparison of deep learning frameworks in image classification problems using convolutional and recurrent networks. In 2017 IEEE Colombian conference on communications and computing (COLCOM) *(2017)*, pp 1–6
8. Hu Z, Li Y, Yang Z (2018)Improving convolutional neural network using pseudo derivative ReLU. In: 2018 5th international conference on systems and informatics (ICSAI), pp 283–287. https://doi.org/10.1109/ICSAI.2018.8599372
9. Gunjan VK, Pathak R, Singh O (2019) Understanding image classification using tensorflow deep learning - convolution neural network. Int J Hyperconnect Internet of Things (IJHIoT) 3(2):19–37. https://doi.org/10.4018/IJHIoT.2019070103
10. Shiddieqy HA, Hariadi FI, Adiono, T. (2017) Implementation of deep-learning based image classification on single board computer.In: 2017 International Symposium on Electronics and Smart Devices (ISESD), pp 133–137
11. Jiang D, Kim J-W (2019) A scene change detection framework based on deep learning and image matching: MUE/FutureTech 2018. https://doi.org/10.1007/978-981-13-1328-8_80
12. Pang B, Nijkamp E, Wu YN (2020) Deep learning with TensorFlow: a review. J Edu Behav Stat 45(2):227–248. https://doi.org/10.3102/1076998619872761
13. Minematsu T, Shimada A, Taniguchi R (2017)Analytics of deep neural network in change detection. In: 2017 14th IEEE international conference on advanced video and signal based surveillance (AVSS), pp 1–6. https://doi.org/10.1109/AVSS.2017.8078550
14. Mohandas R, Bhattacharya M, Penica M, Camp, Hayes M (2020) TensorFlow enabled deep learning model optimization for enhanced realtime person detection using raspberry Pi operating at the Edge.*AICS*
15. Bhat AU, Merchant SS, Bhagwat SS (2008) Prediction of melting points of organic compounds using extreme learning machines. Ind Eng Chem Res 47(3):920–925. https://doi.org/10.1021/ie0704647

EEG Emotion Recognition Using Convolution Neural Network

Nandini Bhandari and Manish Jain

1 Introduction

Emotional state of a human being is strongly influence by his success, failures, interactions with other human beings, working atmosphere and so on. Unlike other human beings machines cannot predict complicated human emotions. Picard et al. had evolved a branch of science called affective computing, to study and design systems which can acknowledge and simulate human emotions [1, 2]. Emotion recognition study plays a crucial role in human computer interactions, diagnosis and cure of diseases generated due to emotional changes. The emotional impact can be observed through speech variation [3], facial expressions [4] and physiological signals like electroencephalography (EEG). Human can hide visible impact by developing certain habits, but he cannot deliberately change accurately generated EEG signal in the brain [5]. So, in recent time, researchers are more inclined to study emotions using EEG.

The placement of scalp electrodes is according to 10–20 system, to measure voltage fluctuations while experiencing any emotion. The EEG signal collected by each electrode is considered as one channel. A proper understanding of brain response for various emotions can help to build a more suitable model of emotion recognition. The relationship between EEG and emotions has been clearly demonstrated in a number of affective computing studies. However, with the development of less expensive wearable devices and dry electrodes [6–9], we can use EEG-based systems in real-world applications like chronic disease healing, driver fatigue detection, etc. [10].

The EEG signal has low amplitude and high noise. The information is contaminated with noise due to low SNR. These signals are non-stationary and also have temporal asymmetry [11]. So analysis of these signal is quite tedious. Feature

N. Bhandari (✉) · M. Jain
Department of Electrical and Electronics, Mandsaur University Mandsaur, M.P., Mandsaur, India
e-mail: nandiniboob@gmail.com

© The Author(s), under exclusive license to Springer Nature Singapore Pte Ltd. 2023
D. S. Sisodia et al. (eds.), *Machine Intelligence Techniques for Data Analysis and Signal Processing*, Lecture Notes in Electrical Engineering 997,
https://doi.org/10.1007/978-981-99-0085-5_44

extraction from pre-processed signal and classification is the two important steps in emotion recognition system. Generally, a pre-processed EEG signal over specific time window is used for feature extraction and classification is done using supervised machine learning. These traditional feature extraction, selection and classification systems require specific domain knowledge and system complexity depends on feature selection. Principal component analysis and Fisher projection are generally used for feature selection, whose cost multiplies quadratically with features [12]. Also, these method cannot preserve information of channels and frequency bands which are needed to find brain response.

Deep learning, emerged in machine learning [13] had a significant impact on signal processing. It helps in automatic feature selection while training a classification model, thus avoiding computation cost of feature selection part. Several deep learning models like auto-encoder [14], graph neural network, convolution neural network [15] and deep belief network [16] have been proposed. In many challenging situations, deep architecture models outperform shallow architectures like support vector machine, kNN, etc., particularly in speech and image processing [16–17]. So deep learning has recently been used in bio-signal processing and got comparable results with respect to conventional methods.

2 Related Work

To get satisfactory results of classification, salient features must be extracted from EEG signal. Generally, feature is extracted in time domain, frequency domain and time–frequency domain. Zheng et al. [18] had explored crucial frequency bands and channels with the help of power spectral density (PSD), differential entropy (DE) and brain asymmetry. They have achieved an accuracy of 61% with PSD and 86% with DE using deep belief network for three emotions. Zhong et al. [19] had used an adjacency matrix in graph neural network to describe inter channel relationship in the EEG signals. The node-wise domain adversarial training and emotion aware distribution learning were used to build a robust model. A unique dynamical graph convolutional neural network (DGCNN)-based emotion classification system was proposed by Song et al. [20]. They had used graph to characterize multichannel EEG features. DGCNN algorithm was used to dynamically learn the intrinsic relations between EEG channels by training neural network so as to get more distinguished features. Li et al. [21] calculated differential entropy at certain time interval and arranged those features in two dimensional feature maps to train hierarchical convolution neural network (HCNN) and classified three emotions from SEED dataset. Bi-hemisphere domain adversarial neural network (BiDANN) [22] was developed based on the discovery that emotional responses generated by two hemisphere are asymmetric. The model had a global domain discriminator along with two local domain discriminators. These two worked adversarial with the classifier such that distinguished emotional features were obtained from each classifier. Thus, it had made an attempt to make a model more general.

The above methods had combined prior knowledge for developing a particular framework for emotion recognition. Though deep learning techniques have made a significant progress in emotion recognition, numerous hurdles are still to overcome. One of the challenge is that a very little importance is given to electrode correlation.

In this paper, we consider differential entropy features of EEG signal as they are most suitable for emotion recognition. We organize data from 62 channels in 2D map to obtain positional relationship between electrodes. This data we give as an input to CNN model to find electrodes correlation in emotion recognition.

Our method has following contributions:

1. We extract DE features of EEG signal obtained from 62 channels at an interval of one second. Then, we use these features to produce 2D maps to preserve the information hidden in electrode placement.
2. We propose a novel CNN framework which consider spatial correlations of channels. The separable 1D convolution layer and dense layer are integrated to improve the accuracy.

The arrangement of paper is as follows. Section 3 describes feature extraction and our proposed CNN model. Section 4 describes experiment in detail. Section 5 compares and discuss about results. Finally, Sect. 6 is conclusion.

3 Method

In this section, we explain how to extract DE features. Then, we discuss our CNN model.

3.1 Feature Extraction

In EEG-based emotion recognition system, extraction of appropriate features is critical. To estimate the complexity of continuous random variable [23], differential entropy is used [24, 25]. DE has better ability to balance the discriminating EEG rhythm between energy of low and high frequency. It is calculated as

$$h(A) = - \int_a g(a) \log(g(a)) da \tag{1}$$

where A denotes random variable and $g(a)$ is probability density of A. For Gaussian distribution $N(\mu, \sigma^2)$, we can calculate DE as

$$h(A) = \frac{1}{2} \log 2\pi e \sigma^2 \tag{2}$$

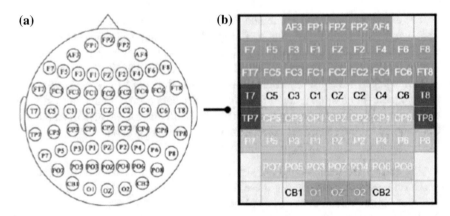

Fig. 1 Organization of differential entropy map. **a** 62 electrode placement for experiment. **b** 2D map of electrodes for considering spatial relationship of electrodes (F: frontal, C: central, P: parietal, AF: anterior frontal, T: temporal, O: occipital, Z: midline)

where e is the Euler's constant and σ is the standard deviation indicating complexity of time series. As a result, DE is estimated in five frequency bands for 62 channels namely delta (1–3 Hz), theta (4–7 Hz), alpha (8–13 Hz), beta (14–30 Hz) and gamma (31–50 Hz). To remove irrelevant components from features and to consider temporal dynamics of emotional state, we have applied linear dynamic system (LDS) [26] on extracted DE features. The DE feature map is obtained from 2D map as given in Fig. 1.

3.2 Convolutional Neural Network

Convolutional neural network (CNN) [27] has made a significant progress in computer vision [28] in recent decades. These networks are relatively expensive to train. They require more training time even with GPU when a large dataset is used. To get better accuracy, many convolutional layers are required. Number of multiplications in a convolution layer depends on input, size of kernel. For a 15 * 15 RGB image with 10 kernel size, one convolutional layer performs 45,000 multiplications.

Recently to overcome the problem of large training time, variety of depthwise separable convolution networks such as MobileNet [29, 30], ShuffleNet [31, 32] and Xception [33] are used. Figure 2 shows a depthwise separable convolution layer. In this convolution, operation is split into two viz. depthwise convolution and pointwise convolution. In depthwise convolution, a kernel is applied to each individual channel separately. It finds spatial features from each channel. The pointwise convolution is then applied to all the channels at the same time with 1×1 kernel. It helps in merging feature maps. As shown in Fig. 2, size of input is $H \times W \times M$, where H indicates number of electrodes, W indicates frequency bands and M represents

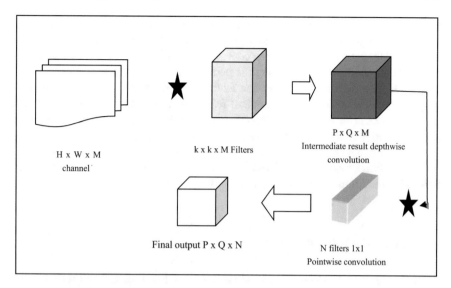

Fig. 2 Explanation of depthwise separable convolution

number of samples per film. With $k \times k$ kernel applied in depthwise convolution, output obtained is $P \times Q \times M$, which is consistent with input. 1×1 kernel with N channels is applied to this output gives feature map $P \times Q \times N$.

1. For input B of size $H \times W$, convolution kernel K of size $k \times k$, M input channels and N output channels trainable parameters in standard convolution are $Z1$ and floating point calculations are $F1$. These parameters are given by

$$z_1 = k \times k \times M \times N, \tag{3}$$

$$F_1 = k \times k \times M \times N \times P \times Q. \tag{4}$$

With separable convolution, parameters Z_2 and floating point calculations F_2 obtained are

$$z_2 = k \times k \times M + M \times N, \tag{5}$$

$$F_2 = k \times k \times P \times Q + P \times Q \times M \times N. \tag{6}$$

The ration of Eqs. (3) and (5) and Eqs. (4) and (6) shows that parameters and calculations with separable convolutions are less than standard convolution by $\frac{1}{N^2} + \frac{1}{k^2}$.

The accuracy obtained with this separable convolution is similar to traditional convolution with just 20% multiplications. Rectified linear unit (ReLU) [34] is used in CNN to prevent exponential growth of computation which leads to vanishing gradient problem. It allows back propagation of error. Pooling layer is used to down sample the features gleaned from the convolution layer, thus reducing the amount of computation and thereby improving the performance of CNN model. We use max-pooling layer which splits convolution features in several disjoint patches of N*N and selects largest value in each patch of feature map. It reduces dimension of data and avoids the problem of over fitting. When stride is used two, the data is reduced to half. A fully connected layer uses softmax function to convert all the values in output layer between zero and one, so that they can be represented as probabilities. Sofmax calculates derivative of loss function, which helps in adjusting weights to reduce loss and improve accuracy of prediction.

3.3 Implementation Details

Sparse categorical cross entropy is used to calculate loss function and for n classes it is given by

$$L_{CE} = -\sum_{m=1}^{n} t_m \log(P_m), \qquad (7)$$

where t_m is true label and pm is softmax probability of mth class. The process of optimization is done in training process till output reaches closer to true value. Adam optimizer [35] is used to update weights. Instead of adapting learning rate based on mean, Adam adds uncentered variance to find bias corrected estimate. Zero padding is added in each convolution layer to avoid information from missing at the edges. Learning rate of Adam algorithm is 0.0001. The model implementation is done with Tensor flow and Keras library.

4 Experiment

In this section, we describe SEED IV dataset. This dataset is used to estimate the performance of our model. To justify our model, results are compared with other methods used earlier.

4.1 SEED IV Dataset and Preprocessing

SEED IV dataset was developed by Zheng et al. [36]. It contains recordings from 15 subjects, and there are three sessions per subject. There are four emotions happy, sad,

Table 1 Detailed configuration of proposed CNN

Layer	Output shape	Parameters
Separable convolution 1D	(none, 64, 128)	49,088
Max-pooling 1D	(none, 32, 128)	0
Separable convolution 1D	(none, 32, 64)	8640
Max-pooling 1D	(none, 16, 64)	0
Separable convolution 1D	(none, 16, 32)	2272
Max-pooling 1D	(none, 8, 32)	0
Separable convolution 1D	(none, 8, 16)	624
Max-pooling 1D	(none, 4, 16)	0
Flatten	(none, 64)	0
Dense	(none, 128)	8320
Dense	(none, 4)	516

neutral and fear and six film clips per emotion. So, there are 24 trial in all per session. EEG signals were recorded with the help of 62 channel ESI Neuro scan system. These signal are down-sampled by 200 Hz sampling frequency. A band pass filter (1–50) Hz is used to remove artifacts. We have extracted five frequency bands from each EEG signal using a 256-point short time Fourier transform with a non-overlapping Hanning window of 1 s. Features generated per trial per subject are 62*w*5, where w is time window of each film.

4.2 Train CNN

In this paper, CNN with four separable 1D convolution layers is used to extract features from 2D maps. A kernel size of 8×8 along with zero padding is used in first convolution layer. A kernel size of 3×3 is used in other three convolution layer. After a convolution layer, we add a max-pooling layer of pool size 2 and strides 2. Flatten layer is used between convolution and dense layer to get one dimensional matrix. Dense layer is used so that inputs from all previous neurons is fed to each neuron of dense layer.

To assess the achievement of our proposed CNN model, a subject dependent data is used in our experiment. We set the batch size of 32 and learning rate of 0.001, beta1 = 0.9, beta2 = 0.999 for Adam optimizer (Table 1).

5 Result and Discussion

In recent years, various research have been done in exploring an exciting work of emotion recognition. Though there are some differences, the present techniques give valuable ideas and results. To get better accuracy, we have split data in stratified way. We have used stratified k fold cross-validation, which maintain same class ratio all over the k folds, as per ratio in dataset. This cross-validation helps us to use our data in better way and fine-tune the activation functions in the neural network. We choose the value of k as 10. For SEED IV dataset, first sixteen trials of each subject per session are used for training the model. The last eight trials having all emotions are used for testing. Average classification accuracy and standard deviation (STD.) are found for all subjects.

To authenticate the betterment of our CNN, we have compared our model with other twelve methods including SVM [37], RF [38], CCA [39], GSCCA [40], DBN [18], GRSLR [41], GCNN [42], DGCNN [20], DANN [43], BiDANN [44], Emotion-Meter [36] and BiHDAM [45]. We have taken the results for comparison of all above methods from earlier studies, to have an incontestable contrast with our method. The summarization of results is given in Table 2.

From the data in Table 2, we can conclude that our proposed CNN model performs better than all other methods on SEED IV dataset. The results of proposed method are improved by 2.2% over Emotion-Meter. Our accuracy is less than BiHDM, because they have focused on developing a model which helps to learn discrepancy relation between two hemispheres, where as we consider spatial correlation between channels. The DANN and BiDANN methods have used unlabeled testing data to improve the performance. Some baseline methods from our comparison have used only labeled training data to learn their model. In our experiment, we consider only labeled training data, to have a fair comparison with other methods. We get an accuracy of 72.68% for testing (Fig. 3).

Table 2 EEG emotion recognition subject dependent classification performance for SEED IV dataset

Method	Description	Accuracy/STD. (%)
SVM [37]	Support vector machine	56.61/20.05
RF [38]	Random forest	50.97/16.22
CCA [39]	Canonical correlation analysis	54.47/18.48
GSCCA [40]	Group sparse canonical correlation analysis	69.08/16.66
DBN [18]	Deep belief network	66.77/7.38
GRSLR [41]	Graph regularized sparse linear regression	69.32/15.42
GCNN [42]	Graph convolution neural network	68.34/15.42
DGCNN [20]	Dynamic graph convolution neural network	69.88/16.29
DANN [43]	Domain adversarial neural network	63.07/12.66
BiDANN [44]	Bi-hemisphere domain adversarial neural network	70.29/12.63
Emotion-Meter [36]	Emotion-meter	70.59/17.01
BiHDM [45]	Bi-hemispheric discrepancy model	74.35/14.09
Proposed CNN	Separable depthwise CNN	72.80/5.78

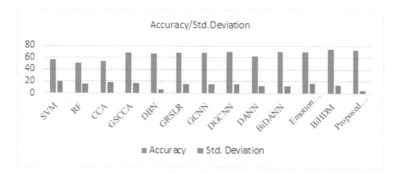

Fig. 3 Performance comparison of different models

Fig. 4 Confusion matrices of four emotions of subject dependent data for proposed CNN method on SEED IV database

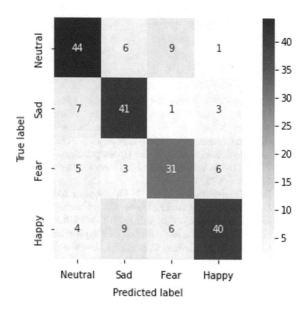

From Fig. 4, we conclude that the average emotion recognition accuracies of proposed CNN in recognizing the four emotions are 73% (neutral), (79%) sad, 69% (fear) and 68% (happy). This shows that sad and neutral emotions are much easier to recognize than both fear and happy.

6 Conclusion

Our work presents a classification model for subject dependent emotion detection using proposed convolution neural network. It is observed that differential entropy found on different frequency bands is more useful as it is a spectral feature. The immense experiments on SEED IV dataset have shown that propped CNN method improves accuracy than earlier methods. The performance is improved because the depthwise convolution works on individual channels and pointwise convolution merges the feature maps. The dense layer performs matrix vector multiplication such that parameters can be trained and upgraded using back propagation. The low value of standard deviation shows that model gives stable EEG emotion recognition.

Despite the positive experimental results obtained in this study, more research is needed on how to extract more discriminative EEG features for cross-subject emotion classification, how to select, construct and optimize deep learning models for EEG-based emotion recognition with higher accuracy, robustness and generalization. The more study should be done to incorporate emotions-based brain neurogenic analysis into the analysis of experimental results. All of these topics will be covered in-depth in our future study project.

References

1. Picard RW (2000) Affective computing. MIT Press
2. Picard R, Vyzas E, Healey J (2001) Towards machine emotional intelligence: analysis of affective physiological state. IEEE Trans Pattern Anal Mach Intell 10:1175–1191
3. Reshma C, Rajshree R (2019) A survey on speech emotion recognition. In: IEEE conference on Innovations in communication, computing and instrumentation.
4. Ko BC (2018) A brief review of facial emotion recognition based on visual information. Sensors 18(2):401–420
5. Schmidt LA, Trainor LJ (2001) Frontal brain electrical activity (EEG) distinguishes valance and intensity of musical emotions. Cogn Emotions 15(4):487–500
6. Grozea C, Voinescu CD, Fzli S (2011) Bristle-sensors-low-cost flexible passive dry EEG electrodes for neuro feedback and BCI applications. J Neural Eng 8(2):025008
7. Chi YM, Wang YT, Maier C, Jung TP, Cauwenberghs G (2012) Dry and noncontact EEG sensors for mobile brain computer interfaces. IEEE Trans Neural Syst Rehabil Eng 20(2):228–235
8. Wang LF, Liu J, Yang B, Yang CS (2012) PDMS-based low cost flexible dry electrodes for long-term EEG measurement. IEEE Sens J 12(9):2898–2904
9. Huang YJ, Wu CY, Wong AK, Lin BS (2015) Novel active combo-shaped dry electrodes for EEG measurement in hairy site. IEEE Trans Biomed Eng 62(1):256–263
10. Liu NH, Chiang CY, Hsu HM (2013) Improving driver alertness through music selection using a mobile EEG to detect brainwaves. Sensors 13(7):8199–8221
11. Paul M (1996) Nonlinearity in normal human EEG: cycles, temporal asymmetry, non-stationary and randomness, not chaos. Biol Cybern 75(5):389–396
12. Dash M, Liu H (1997) Feature selection for classification. Intell Data Anal 1(3):131–156
13. Krizhevsky A, Sutskever I, Hinton GE (2012) Imagenet classification with deep convolutional neural networks Adv Neu Inf Proc Syst 1097–1105
14. Rifai S, Vincent P, Muller X, Glorot X, Bengio Y (2011) Contractive auto-encoders: explicit invariance during feature extraction. In: Proceedings of the 28th international conference on machine learning, pp 833–840
15. LeCun Y, Bengio Y (1995) Convolutional network for images, speech and time series. Handb Brain Theory Neural Network 3361
16. Mohamed R, Yu D, Deng L (2010) Investigation of full sequence training of deep belief networks for speech recognition. Inter Speech 2846–2849

17. Jaitly N, Hinton G (2011) Learning a better representation of speech sound waves using restricted Boltzmann machines. IEEE International conference on acoustics, Speech and signal processing (ICASSP), pp 5884–5887
18. Zheng WL, Lu BL (2015) Investigating critical frequency bands and channels for EEG-based emotion recognition with deep neural networks. IEEE Trans Auton Mental Dev 7(3):162–175
19. Zhong P, Wang D, Miao C (2020) EEG-based emotion recognition using regularized graph neural network. IEEE Trans Affect Comput 11(2):99–111
20. Song T, Zheng W, Song P, Cui Z (2020) EEG emotion recognition using dynamical graph convolutional neural networks, IEEE Trans Affect Comput 11(3):532–541
21. Li J, Zhang Z, He H (2016) Implementation of EEG emotion recognition system based on hierarchical convolutional neural network. In: International conference on Brain inspired Cognitive Systems
22. Li Y, Zheng W, Zong Y, Zhang T, Zhou X (2021) A bi-hemisphere domain adversarial neural network model for EEG emotion recognition. IEEE Trans Affect Comput 12(2):494–504
23. Gibbs JW (2010) Elementary principles in statistical mechanics: developed with especial reference to the rational foundation of thermodynamics. Cambridge University Press
24. Shi LC, Jiao YY, Lu BL (2013) Differential entropy feature for EEG-based vigilance estimation. In: International conference of IEEE engineering in medicine and biology society (EMBC), IEEE, pp 6627–6630
25. Duan RN, Zhu JY, Lu BL (2013) Differential entropy feature for EEG-based emotion classification. In: 6th International IEEE/EMBS conference on neural engineering, pp 81–84
26. Lin Y, Yang Y, Jung TP (2014) Fusion of electroencephalogram dynamics and musical contents for estimating emotional responses in music listening. Front Neurosc 8:94
27. Wu J (2016) Introduction to convolutional neural networks, 1–28
28. Hongtao L, Zhang Q (2016) Applications of deep convolutional neural network in computer vision. J Data Acquisition Proc 31(1):1–17
29. Howard G, Zhu M, Chen B, Kalenichenko D, Wang W, Weyand T, Andreetto M, Adam H (2017) Mobilenets: efficient convolutional neural networks for mobile vision applications arXiv:1704.04861
30. Sandler M, Howard A, Zhu M, Zhmoginov A, Chen LC (2018) Mobilenetv2: inverted residuals and linear bottlenecks. IEEE conference on Computer vision and pattern recognition (CVPR), pp 4510–4520
31. Ma N, Zhang X, Zheng H, Sun J (2018) Shufflenetv2: practical guidelines for efficient CNN architecture design. ECCV pp 122–138
32. Zhou X, Lin M, Sun J (2018) Shufflenet: an extremely efficient convolutional neural network for mobile devices. IEEE conference on computer vision and pattern recognition (CVPR)
33. Chollet F (2017) Xception: deep learning with depthwiseseparable convolutions. IEEE conference on computer vision and pattern recognition (CVPR), pp 1800–1807
34. Glorot X, Bordes A, Bengio y (2011) Deep sparse rectifier neural networks. In: 14th international conference on artificial intelligence and statistics (AISTATS), pp 315–324
35. Kingma D, Ba J (2015) Adam: a method for stochastic optimization. In: International conference on learning representations, 1–15
36. Zheng WL, Liu W, Lu Y, Lu BL, Cichocki A (2019) EmotionMerer: a multimodal framework for recognizing human emotions. IEEE Trans Cybern 49(3):1110–1122
37. Suykens JA, Vandewalle J (1999) Least squares support vector machine classifiers. Neural Process Lett 9(3):293–300
38. Breiman L (2001) Random forests. Mach Learn 45(1):5–32
39. Thompson B (2005) Canonical correlation analysis. Encycl Stat Behav Sci
40. Zheng W (2017) Multichannel EEG-based emotion recognition via group sparse canonical correlation analysis. IEEE Trans Cogn Dev Syst 9(3):281–290
41. Li Y, Zheng W, Cui Z, Zong Y, Ge S (2018) EEG emotion recognition based on graph regularized sparse linear regression. Neural Proc Lett 1–17
42. Defferrard M, Bresson X, Vandergheynst P (2016) Convolutional neural networks on graphs with fast localized spectral filtering. Conference on neural information processing systems (NIPS), pp 3844–3852

43. Ganin Y, Ustinova E, Ajakan H, Germain P, Larochelle H, Laviolette F, Marchand M, Lampitsky V (2016) Domain-adversarial training of neural networks. J Mach Learn Res 17(59):1–35
44. Li Y, Zheng W, Cui Z, Zhang T, Zong Y (2018) A novel neural network model based on cerebral hemispheric asymmetry for EEG emotion recognition. Int Joint Conf Artif Intell, pp 1561–1567
45. Li Y, Wang L, Song T, Zheng W, Zong Y (2021) A novel Bi-hemispheric discrepancy model for EEG emotion recognition. IEEE Trans Cogn Dev Syst 13(2):354–367

CCL-Net: Complete Comprehensive Learning and Modality Preserving-Based RGBD Complex Salient Object Detection

Surya Kant Singh and Rajeev Srivastava

1 Introduction

Visual salient object detection is defined as generating conspicuous objects, similar to the prominent objects in the complex image identified by the human visual system. Recently, salient object detection has attracted lots of interest due to the emergence and the increasing popularity of depth-sensing technologies. It is used as a preprocessing or integral part in many vision-related applications, such as object classification [4], video image captioning [15], and content-based image editing [22]. Most of the existing models [8, 13, 19, 21] are only based on cross-modalities fusion and ignore the modality-dependent features. These limitations of the existing models are our motivation to propose a complete comprehensive learning model (CCL-Net) by exploring all possible features to predict the salient objects correctly. The RGBD-based early salient object detection (SOD) models [2, 3, 6, 19, 24] are based on low-level, handcrafted features. These methods have limited abilities to generalize these concepts. Therefore, CNN-based deep learning model [18] was put forward in 2017 to integrate color (RGB) saliency with depth saliency. Since then, several CNN-based models have been proposed with improved performance, while leaving a wide scope for further improvements (Fig. 1).

These models are classified according to CNN architecture into (1) single-stream models and (2) multi-stream models. RGBD-based SOD models started with single-stream network [18, 23], which is used to fuse the color and depth features to predict the saliency. In this network [18], some features are computed traditionally, which are used to guide the single-stream networks to compute the saliency. The recently

S. K. Singh (✉) · R. Srivastava
Computer Science and Engineering, Indian Institute of Technology (Banaras Hindu University) Varanasi, Varanasi, UP 221005, India
e-mail: suryakantsingh20@gmail.com

R. Srivastava
e-mail: rajeev.cse@iitbhu.ac.in

© The Author(s), under exclusive license to Springer Nature Singapore Pte Ltd. 2023
D. S. Sisodia et al. (eds.), *Machine Intelligence Techniques for Data Analysis and Signal Processing*, Lecture Notes in Electrical Engineering 997,
https://doi.org/10.1007/978-981-99-0085-5_45

(a) Input Image   (b) Depth Mask   (c) Ground Truth   (d) CCL

Fig. 1 Multi-scaled, multi-resolution, and multistage features and modality preservation for salient objects detection in complex and clutter backgrounds

proposed models are based on two streams network [8, 13, 21] or multi-stream network [5, 14]. Most of the models are based on two streams network. In this model, two separate streams are designed, dedicated to color and depth modalities. These models produce high-level semantic features independently from two streams and fuse them using the middle or late-fusion strategy. These contemporary architectures improved the performance while leaving a large scope for further improvements. The possibilities of improvements in contemporary architectures target the following limitations in the recent RGBD deep CNN model are summarized as follows:

- How to develop the strategy to explore complementary features along preserving the non-complementary features so that salient objects pop out in complex and clutter backgrounds?
- How to preserve the complementary features between deep-level and shallow-level features? Deep-level features have global contextual characteristics, and shallow-level features describe spatial and structural information.

To address the limitations mentioned above, our proposed model, complete comprehensive learning CCL-Net based on the RGBD model, provides the following distinct contributions to improve the performance of SOD.

- In this paper, novel multistage three independent streams dedicated to color, depth, and cross-modality are proposed to extract multistage raw saliencies. These raw saliencies are used to produce complementary and non-complementary features.

- The color and depth streams are used to explore non-complementary features. The fused stream is used to explore cross-complementary, and deep to shallow intra-complementary features, to generate saliency in complex scenarios. In comparison with various existing similar recent RGBD models, CCL-Net shows performance improvement.
- The concept of complete learning of most of the possible features provides robust characteristics in complex and clutter backgrounds, demonstrated by using recent evaluation metrics with state-of-the-art methods.

The rest of the paper is organized in the following sections. Section 2 provides a detailed survey of closely related deep learning-based RGBD models. Section 3 describes and defines the proposed method CCL-Net in detail. Section 4 discusses the experimental setup and demonstrates the performance of complete comprehensive learning with other state-of-the-art saliency detection methods. Section 5 describes the conclusion and the future scope of improvements in the complete comprehensive learning model.

2 Related Works

Over the last two decades, an enormously rich set of saliency detection methods (SODs) have been developed. In the past, most of the saliency methods were based on handcrafted features [2, 6, 19, 24] without learning and testing, while the recent deep learning-based methods have evolved as the first choice to enhance the performance of SODs further. The recently proposed RGBD and SODs methods are broadly classified into three categories: (a) early-fusion strategy, (b) late-fusion strategy, and (c) middle-fusion strategy.

Qu et al. [18] initiated the early-fusion strategy used handcrafted features as the inputs in CNN. In this deep fusion DF [18] model, CNN is used to integrate different low-level features with depth saliency to produce the salient object. Nevertheless, most saliency is lost in the handcrafted feature in these models, which is irrecoverable in CNN. Therefore, CNN is not fully utilized in the early-fusion model. The next representative model is based on the late-fusion strategy in which full CNN is used to generate high-level features. An adaptive fusion AFNet [21] is a representative model to use late-fusion strategy. The most recent efficient algorithm D3NET [5] is a three-stream network for RgbNet, RgbdNet, and DepthNet. These models address the issues of low depth saliency but fail in the complex and cluttered background. Because, in this model, the intra-complementary features have not been utilized. Our proposed method addresses these issues by proposing comprehensive learning from deep to shallow features aggregation.

A recent model, S2MA [13], is based on two-stream RGBD CNN to use a deep non-local network-based attention mechanism. The attention mechanism is used to improve the deep localized feature. In JL-DCF [7], the author proposed joint learning (JL) and a densely cooperative fusion (DCF) based on a shared Siamese

backbone network. The stream$_2$ of the proposed method has a similar network architecture to explore complementary features. Therefore, these models sometimes fail
in complex and clutter backgrounds. To address the limitations as mentioned above
of current SODs methods, our proposed method, CCL-Net, focuses on exploiting
modalities-dependent and -independent features. The modalities-dependent features
are extracted using color and depth streams. The proposed network CCL-Net learns
all possible saliency features to generate robust saliency in the proposed model.

3   The Proposed Method

The three-stream network is based on cross-complementary, intra-complementary,
and non-complementary features. These are extracted from backbone network VGG-
16 [20]. These features are aggregated in the backward direction by complete comprehensive learning guided by the down-sampled global loss function. Two different
parallel streams produce the modality-dependent non-complementary features to
generate raw saliency. The fused stream uses these raw saliency features to aggregate cross-complementary and intra-complementary features. These raw saliencies
are aggregated into final saliency by guided deep loss functions to predict accurate
salient objects. The input image in the second stream is composed of three channels from the RGB stream and three channels from the depth stream with resolution
$224 * 224 * 3 * 2$ and shown in Fig. 2.

3.1   Multi-scale Non-complementary Features Aggregation

The non-complementary features generated by the VGG-16 in the color and depth
streams are produced separately and simultaneously. The outputs produced by the
convolution blocks are denoted as C_i, and their corresponding raw saliency is s_i, by
applying an additional convolution block and the up-sampling layer. These features
are fused the feature s_i with the previous fused feature s_{i+1} at i and $i + 1$ scale,
respectively. Finally, these two streams produce their saliency map s^{rgb} and s^{depth}, with
modalities $m = (\text{RGB}, \text{Depth})$. The formulation of features generation and saliency
prediction using non-complementary features are defined in Eq. 1 as follows:

$$s_i = \wp \left(\varrho \left(\varrho \left(C_i \right) \right) \right) \quad 1 \leq i \leq 5$$
$$\tilde{s}_i = \begin{cases} \varrho \left(|\tilde{s}_{i+1}, \tilde{s}_i| \right) & 1 \leq i < 5 \\ s_i & i = 5 \end{cases} \tag{1}$$
$$s^m = g \left(\text{kernal}_s * \tilde{s}_i + \text{bias} \right)$$

$\wp (\ldots)$ denotes the up-sampling function to make raw saliency with the same resolution that uses bilinear interpolation. $\varrho(\ldots)$ describes convolution operation with 64

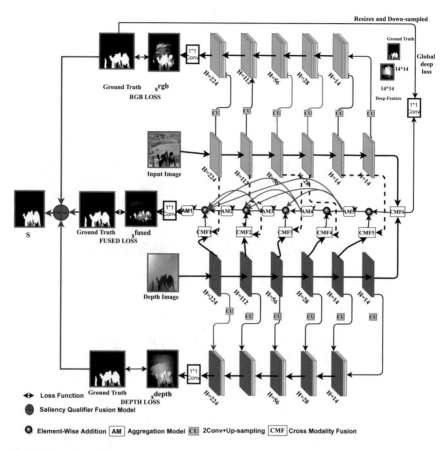

Fig. 2 Network architecture of the proposed method CCL-Net is shown here

channels. A nonlinear activation function follows it. In this convolution operation, the kennel size is 3×3 and with the stride size is 1. kernel$_s$ is 1×1 kernel and bias is bias parameter. $g(...)$ is the sigmoid function, while $*$ represents the convolution operation. $|...|$ represents a channel-wise concatenation.

3.2 Complete Comprehensive Learning Model

Complete comprehensive learning needs non-complementary, cross-complementary, and intra-complementary features. Non-complementary features are produced by stream$_1$ (color) and stream$_3$ (depth), and cross-complementary and intra-complementary features are produced in stream$_2$ (fused). The s^{rgb}, s^{depth}, and s^{fused}

are used in proposed saliency qualifier fusion model. These models are described below as follows.

3.2.1   Cross-Complementary Modalities Fusion (CMF)

The CNN backbone network in color stream$_1$ and depth stream$_3$ produces raw saliency like [9] (side output). The raw saliencies from the depth and color streams from each stage (total six stages from CMF1 to CMF6) are fused into the CMF model. In this model, the varied resolution features are compressed into smaller (fixed size equal to ch) and exact sizes. The processed features in RGB and depth modality are denoted as f_{rgb}, f_{depth}, each with equal ch channels. The output of the CMF module is defined in Eq. 2 as

$$\text{CMF}^{ch}\left(f_{rgb}, f_{depth}\right) = (f_{rgb}^{ch} \otimes f_{depth}^{ch}) \oplus (f_{rgb}^{ch} \oplus f_{depth}^{ch}) \tag{2}$$

In this cross-view fusion, "$\otimes$" and "$\oplus$" are defined as element-wise multiplication and addition, respectively.

Intra-complementary Aggregation Model In this aggregation model, coarse localized deep features and the fused features from CMF$_i$ are fed into a decoder that has a dense connection to generate the final saliency. This saliency stands out from the complex background at the shallow level from the output layer AM1. This model is mathematically formulated in Eq. 3 with width (w), height (h), and feature map I with channel ch. It is defined as follows:

$$f_{w,h}(I, \text{ch}) = f_{Con(1,1)}(I, \text{ch}/4) \circledast f_{Con(3,3)}((f_{Con(1,1)}(I, \text{ch}/2), \text{ch}/4)$$
$$\circledast f_{Con(5,5)}((f_{Con(1,1)}(I, \text{ch}/4), \text{ch}/4) \circledast f_{Max\text{-}pool(3,3)}((f_{Con(1,1)}(I, \text{ch}/4) \tag{3}$$

where $\text{Con}(i, j)$ and $\text{Max-pool}(i, j)$ are convolution operation and max-pooling operation, respectively, having stride 1 to maintain the spatial feature resolution. "$\circledast$" defines the simple concatenation. In this module, the multi-level convolutions filter with size values (i, j) $1 \times 1, 3 \times 3, 5 \times 5$ and max-pooling layer are used. This formulation is similar to the original Inception module [20], with one difference.

3.3   Saliency Qualifier Fusion Model-SQF

Finally, the output of the last purified module in each stream having finest features is fed to a $(1 \times 1, 1)$ convolution layer to generate the final stream-wise s^{rgb}, s^{depth}, and s^{fused} saliency maps in Eq. 4. During the training process, these saliency maps are further supervised with the same resolution ground truth map. Final saliency map S is produced with updated fusion.

$$S = \eta((\omega_1 \otimes (s^{fused} \oplus s^{depth})) \oplus (\omega_2 \otimes (s^{rgb} \oplus s^{fused}))) \tag{4}$$

where $\omega_1 = \eta(s^{\text{fused}} \otimes s^{\text{depth}})$ and $\omega_2 = \eta(s^{\text{fused}} \otimes s^{\text{rgb}})$ are defined as normalization cross-saliency fusion coefficient. $\otimes$ and $\oplus$ are element-wise multiplication and element-wise addition, respectively. η is used to normalized the values of saliency in range $(0, 1)$.

3.4 Loss Function

The total loss function is composed of stream-wise saliency loss and global deep loss in Eq. 5. The global deep loss function is supervised with same resolution ground truth Gt, and coarse saliency at deepest level from CMF_6 is f_{rgb} and f_{depth}. Similarly, color, depth, and fused saliency loss function are computed with their respective saliency map s^{rgb}, s^{depth}, s^{fused}, and ground truth map Gt. The total loss function is defined as

$$\mathcal{L}_{\text{total}}(s, \text{Gt}) = \sum_{k \in (\text{rgb}, \text{depth}, \text{fused})} \mathcal{L}(s^k, \text{Gt}) + \alpha \sum_{k \in (\text{rgb}, \text{depth})} \mathcal{L}(f_k, \text{Gt}) \qquad (5)$$

where α is the balancing factor between color and depth saliency during global deep loss formulation. The loss function is defined in Eq. 6 as standard cross-entropy loss, and it is defined as follows:

$$\mathcal{L}(s, \text{Gt}) = -\sum_{k} (\text{Gt}_k \log(s_k) + (1 - \text{Gt}_k) \log(1 - s_k)) \qquad (6)$$

where k is defined as pixel index in ground truth image.

4 Experiment Set-Up and Result Analysis

4.1 Dataset and Evaluation Metrics

The extensive experiments are performed on four publicly available RGBD benchmark datasets for complex salient object detection. These datasets are NLPR-1000 [17], RGBD-135 [12], STEREO [16], and NJUD-2000 [10]. For a fair comparison to the state-of-the-art method, the same data pattern [21] for training and testing is used here. In the comprehensive evaluation of the proposed method CCL-Net with other state-of-the-art methods, we used recent evaluation metrics widely used in recent comparisons. These metrics are (1) S-measure, (2) F-measure, (3) mean absolute error (MAE), and (4) E-measure (E_ψ). All these parameters are recent and adequately defined in [7].

4.2 Implementation Details

The proposed method CCL-Net is based on the Caffe.The backbone network is initialized by pre-trained parameters, which is similar to DSS [9]. The training process of the proposed network is performed in an end-to-end manner, using a widely used Adam optimizer [11]. An NVIDIA 1080Ti GPU accelerates this training process. This optimizer contains batch size 8 and a learning rate of 0.0001. The approximate training time is around 18 h/16 h, which contains 40 epochs in the VGG-16 configuration. The two extra convolution layers are added in each stream at the deepest layer to produce coarse features with the spatial size of 14×14, which is shown in Fig. 2. The pool5 has a stride of 1 which is used to augment the resolution of the coarsest feature maps. The size of convolution layers in all CMF modules is (3×3) and filter size is ch = 64. The raw saliency is up-sampled with multiple factors to maintain the exact resolution of each feature and finally up-sampling the output from FM5 to FM1 by a factor of 2, 4, 8, and 16.

4.3 Comparison and Result Analysis

To demonstrate the effectiveness of the proposed method CCL-Net, we compare the results with thirteen state-of-the-art methods with four evaluation metrics. We compare the results with the following state-of-the-art methods JL-DCF [7], S2NET [13], D3NET [5], CPFP [23] AFNet [21], CTMF [8], PCANet [1], and DF [18], which are closely related to deep learning-based RGBD methods. CDS [24], MDSF [19], DCMC [3], DES [2], and LBE [6] are traditional methods based on low-level handcrafted features. The result analysis is demonstrated through visual comparison as well as quantitative comparison.

Quantitative Comparison: The results from Table 1 illustrate that the proposed method, CCL-Net, accomplished a significant improvement on all datasets using S-measure, E-measure, and F-measure while declining in MAE significantly.

Visual Comparison: The visual assessment shown in Fig. 3 demonstrates better saliency in complex and clutter backgrounds. Most of the images shown in Fig. 3 are complex and clutter backgrounds, which are not easily recognized by the naked eye. The AFNet, CTMF, and PCAnet produce incomplete salient objects in complex and cluttered backgrounds.

4.4 Validation of Three-Stream Networks

The validation of the effectiveness of each step of the proposed method CCL-Net is essential to demonstrate the contributions in saliency. A single-stream deep network model cannot distinguish the salient regions in complex and cluttered background

Table 1 Quantitative performance of CCL-Net on four benchmark datasets compared with the deep learning-based RGBD models

Metric	NLPR [17]				NJU2K [10]				RGBD-135 [12]				STERE [16]			
	F_β^m ↑	S_α ↑	MAE ↓	E_ψ^m ↑	F_β^m ↑	S_α ↑	MAE ↓	E_ψ^m ↑	F_β^m ↑	S_α ↑	MAE ↓	E_ψ^m ↑	F_β^m ↑	S_α ↑	MAE ↓	E_ψ^m ↑
OUR	0.919	0.929	0.021	0.968	0.910	0.910	0.040	0.950	0.922	0.926	0.020	0.971	0.905	0.905	0.040	0.950
JL-DCF [7]	0.916	0.925	0.022	0.962	0.903	0.903	0.043	0.944	0.919	0.929	0.022	0.968	0.901	0.905	0.042	0.946
S2NET [13]	0.902	0.915	0.030	0.953	0.849	0.899	0.053	0.941	0.935	0.973	0.021	0.961	0.882	0.890	0.051	0.932
D3NET [5]	0.897	0.912	0.030	0.953	0.900	0.900	0.041	0.950	0.885	0.898	0.031	0.946	0.891	0.899	0.046	0.938
CPFP [23]	0.867	0.888	0.036	0.932	0.877	0.879	0.053	0.926	0.846	0.872	0.038	0.923	0.874	0.879	0.051	0.925
PCFNet [1]	0.841	0.874	0.044	0.925	0.872	0.877	0.059	0.924	0.804	0.842	0.049	0.893	0.860	0.875	0.064	0.925
CTMF [8]	0.825	0.860	0.056	0.929	0.845	0.849	0.085	0.913	0.844	0.863	0.055	0.932	0.831	0.848	0.086	0.912
AFNet [21]	0.771	0.799	0.058	0.879	0.775	0.772	0.100	0.853	0.728	0.770	0.068	0.881	0.823	0.825	0.075	0.887
DF [18]	0.778	0.802	0.085	0.880	0.804	0.763	0.141	0.864	0.766	0.752	0.093	0.870	0.757	0.757	0.141	0.847
MDSF [19]	0.793	0.805	0.095	0.885	0.775	0.748	0.157	0.838	0.746	0.741	0.122	0.851	0.728	0.719	0.176	0.809
CDS [24]	0.768	0.782	0.098	0.824	0.779	0.744	0.160	0.803	0.786	0.791	0.129	0.832	0.746	0.741	0.122	0.851
DCMC [3]	0.648	0.724	0.117	0.793	0.715	0.686	0.172	0.799	0.666	0.707	0.111	0.773	0.740	0.731	0.148	0.819
LBE [6]	0.745	0.762	0.081	0.855	0.748	0.695	0.153	0.803	0.788	0.703	0.208	0.890	0.633	0.660	0.250	0.787
DES [2]	0.681	0.702	0.125	0.700	0.704	0.713	0.189	0.754	0.666	0.682	0.143	0.770	0.566	0.582	0.193	0.670

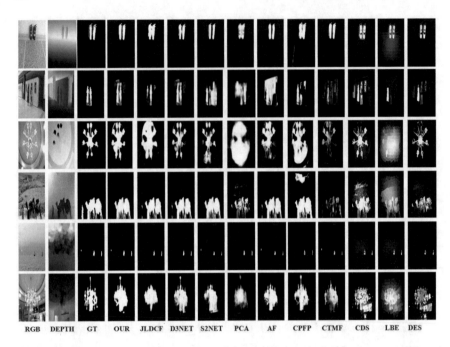

RGB   DEPTH   GT   OUR   JLDCF D3NET S2NET   PCA   AF   CPFP   CTMF   CDS   LBE DES

Fig. 3 Visual demonstration and comparison of the proposed method with other state-of-the-art methods

Table 2 Effectiveness of three-stream networks is validated using mean absolute error-MAE in the proposed method-CCL-Net

Dataset	s^{Depth}	s^{rgb}	s^{Fused}	s^{DC}	s^{DF}	s^{CF}	S
NLPR	0.0388	0.0329	0.0268	0.0368	0.0239	0.0224	0.0214
NJUD2K	0.0550	0.0485	0.0422	0.0502	0.0410	0.0405	0.0400
RGBD-135	0.0409	0.0343	0.0267	0.0360	0.0237	0.0221	0.0204
STERE	0.0591	0.0518	0.0461	0.0575	0.0431	0.0423	0.0400

images. The visual contributions of each step are shown in Fig. 4. The validation of effectiveness is measured through mean absolute error (MA), shown in Table 2. The successive contributions in saliency computations are given in Table 2, which validates the effectiveness of each stream of CCL-Net on complex RGBD datasets. In this validation, we have shown that each step achieves a reduction in MAE. There are three main combinations, s^{DF}, s^{DC}, and s^{CF} in final saliency contributions in saliency qualifier fusion model . Here, DC means fusion with depth and color saliency, DF means fusion with depth and fused saliency, and CF means fusion with color and fused saliency. In which, s^{DC} and s^{CF} are qualified for final fusion in Eq. 4 with two qualifiers ω_1 and ω_2.

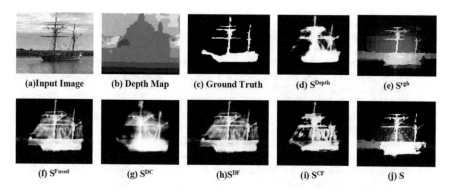

| (a)Input Image | (b) Depth Map | (c) Ground Truth | (d) S^{Depth} | (e) S^{rgb} |

| (f) S^{Fused} | (g) S^{DC} | (h)S^{DF} | (i) S^{CF} | (j) S |

Fig. 4 Visual demonstration of the contribution of three-stream network in complex image having inferior and low depth image. We define $s^{\text{DC}} = (\omega_1 \otimes (s^{\text{Depth}} \oplus s^{s^{\text{RGB}}}))$, $s^{\text{DF}} = (\omega_1 \otimes (s^{\text{fused}} \oplus s^{\text{depth}}))$, and $s^{\text{CF}} = (\omega_2 \otimes (s^{\text{rgb}} \oplus s^{\text{fused}}))$

5 Conclusion

A three-stream robust CCL-Net is proposed for complex salient object detection. The color stream, depth stream, and fused network generate modalities-dependent and modalities-independent features. The color and depth-based cross- and intra-complementary features are used to predict the saliency in complex images correctly. The modality-dependent non-complementary features-based saliency is also utilized in saliency enhancement. The innovative, comprehensive learning is designed to generate these features into three-stream networks. A three-stream CCL-Net is designed to progressively learn all possible features during feature generation to correct wrong predictions and incorrect localization by improving the low depth issues and spatial coherence. There is future scope for identifying low depth saliency and optimizing the fusion process.

References

1. Chen H, Li Y (2018) Progressively complementarity-aware fusion network for RGB-D salient object detection. In: Proceedings of the IEEE conference on computer vision and pattern recognition, pp 3051–3060
2. Cheng Y, Fu H, Wei X, Xiao J, Cao X (2014) Depth enhanced saliency detection method. In: Proceedings of international conference on internet multimedia computing and service. ACM, p 23
3. Cong R, Lei J, Zhang C, Huang Q, Cao X, Hou C (2016) Saliency detection for stereoscopic images based on depth confidence analysis and multiple cues fusion. IEEE Signal Process Lett 23(6):819–823
4. Durand T, Mordan T, Thome N, Cord M (2017) Wildcat: weakly supervised learning of deep convnets for image classification, pointwise localization and segmentation. In: IEEE Conference on computer vision and pattern recognition (CVPR 2017)

5. Fan DP, Lin Z, Zhang Z, Zhu M, Cheng MM (2020) Rethinking RGB-D salient object detection: models, data sets, and large-scale benchmarks. IEEE Trans Neural Networks Learn Syst

6. Feng D, Barnes N, You S, McCarthy C (2016) Local background enclosure for RGB-D salient object detection. In: Proceedings of the IEEE conference on computer vision and pattern recognition, pp 2343–2350

7. Fu K, Fan DP, Ji GP, Zhao Q (2020) JL-DCF: joint learning and densely-cooperative fusion framework for RGB-D salient object detection. In: Proceedings of the IEEE/CVF conference on computer vision and pattern recognition, pp 3052–3062

8. Han J, Chen H, Liu N, Yan C, Li X (2017) CNNs-based RGB-D saliency detection via cross-view transfer and multiview fusion. IEEE Trans Cybern 48(11):3171–3183

9. Hou Q, Cheng MM, Hu X, Borji A, Tu Z, Torr PH (2017) Deeply supervised salient object detection with short connections. In: Proceedings of the IEEE conference on computer vision and pattern recognition, pp 3203–3212

10. Ju R, Ge L, Geng W, Ren T, Wu G (2014) Depth saliency based on anisotropic center-surround difference. In: 2014 IEEE International conference on image processing (ICIP). IEEE, pp 1115–1119

11. Kingma DP, Ba J (2014) Adam: a method for stochastic optimization. arXiv preprint arXiv:1412.6980

12. Li N, Ye J, Ji Y, Ling H, Yu J (2014) Saliency detection on light field. In: Proceedings of the IEEE conference on computer vision and pattern recognition, pp 2806–2813

13. Liu N, Zhang N, Han J (2020) Learning selective self-mutual attention for RGB-D saliency detection. In: Proceedings of the IEEE-CVF conference on computer vision and pattern recognition, pp 13756–13765

14. Liu Z, Li Q, Li W (2020) Deep layer guided network for salient object detection. Neurocomputing 372:55–63

15. Mahasseni B, Lam M, Todorovic S (2017) Unsupervised video summarization with adversarial LSTM networks. In: Proceedings of the IEEE conference on computer vision and pattern recognition (CVPR)

16. Niu Y, Geng Y, Li X, Liu F (2012) Leveraging stereopsis for saliency analysis. In: 2012 IEEE Conference on computer vision and pattern recognition. IEEE, pp 454–461

17. Peng H, Li B, Xiong W, Hu W, Ji R (2014) RGBD salient object detection: a benchmark and algorithms. In: European conference on computer vision. Springer, Berlin, pp 92–109

18. Qu L, He S, Zhang J, Tian J, Tang Y, Yang Q (2017) RGBD salient object detection via deep fusion. IEEE Trans Image Process 26(5):2274–2285

19. Song H, Liu Z, Du H, Sun G, Le Meur O, Ren T (2017) Depth-aware salient object detection and segmentation via multiscale discriminative saliency fusion and bootstrap learning. IEEE Trans Image Process 26(9):4204–4216

20. Szegedy C, Liu W, Jia Y, Sermanet P, Reed S, Anguelov D, Erhan D, Vanhoucke V, Rabinovich A (2015) Going deeper with convolutions. In: Proceedings of the IEEE conference on computer vision and pattern recognition, pp 1–9

21. Wang N, Gong X (2019) Adaptive fusion for RGB-D salient object detection. IEEE Access 7:55277–55284

22. Wang W, Shen J (2017) Deep cropping via attention box prediction and aesthetics assessment. In: Proceedings of the IEEE international conference on computer vision, pp 2186–2194

23. Zhao JX, Cao Y, Fan DP, Cheng MM, Li XY, Zhang L (2019) Contrast prior and fluid pyramid integration for RGBD salient object detection. In: Proceedings of the IEEE conference on computer vision and pattern recognition, pp 3927–3936

24. Zhu C, Li G, Wang W, Wang R (2017) An innovative salient object detection using center-dark channel prior. In: IEEE International conference on computer vision workshop (ICCVW)

LexRank and PEGASUS Transformer for Summarization of Legal Documents

Sarthak Dalal⊙, **Amit Singhal**⊙, **and Brejesh Lall**⊙

1 Introduction

India is a country that follows the *common law* system [1]. The fundamental feature of common law is that it uses *precedents* [2] in circumstances where the parties disagree on what the law is. In such situations, the court looks to previous precedential decisions of competent courts and reconstructs the principles of those previous cases as applicable to the current facts. Thus, precedents aid a lawyer in preparation for a case by allowing them to learn how the court has handled similar matters in the past. As a result, they must analyze a large number of previous cases. To study previous cases, the lawyer must go through law reports/case judgements, which are generally verbose and contain a lot of dense legal text. Even for a legal professional, reading and comprehending the complete text of a case is challenging. Summaries of case judgements are particularly useful in situations like these. Text summarizing condenses the material of a document without sacrificing its meaning to save consumers' time and cognitive strain. From the perspective of a common citizen, these documents are crammed with legal jargon and even comprehending a summarized text might be arduous. Hence, it is necessary to paraphrase documents along with summarization for the common public.

Presently, the process of summarization of case documents is slow, laborious and expensive as documents are summarized manually by legal experts. Using ML can accelerate this process and summarize large amounts of documents rapidly in

S. Dalal
Dwarkadas J Sanghvi College of Engineering, Mumbai 400056, India

A. Singhal (✉)
Netaji Subhas University of Technology, Delhi 110078, India
e-mail: amit@nsut.ac.in

B. Lall
Indian Institute of Technology, Delhi 110016, India
e-mail: brejesh@ee.iitd.ac.in

© The Author(s), under exclusive license to Springer Nature Singapore Pte Ltd. 2023
D. S. Sisodia et al. (eds.), *Machine Intelligence Techniques for Data Analysis and Signal Processing*, Lecture Notes in Electrical Engineering 997,
https://doi.org/10.1007/978-981-99-0085-5_46

a cost-effective manner. Ordinary citizens generally do not have any access to legal experts to summarize the documents for them, and hence, using ML is the only way they can comprehend a legal document. If we have approaches that automate the process of writing case summaries as well as paraphrasing them into simpler language using ML, the user's freedom to examine legal material is considerably increased. The user may choose cases based on her preferences without the intervention of attorneys, who might hide many cases from the user. In this paper, we present a tool for unsupervised summarization and paraphrasing for legal documents. The documents are summarized using the *LexRank* method which is an unsupervised graph-based approach for extractive text summarization. Then, the summary is paraphrased using the *PEGASUS* transformer. The final output is an abstractive summary of the original document.

2 Related Work

Lots of work have been done in general summarization techniques but summarizing legal documents is a relatively unexplored avenue. In order to study previous works under this domain, we summarized six documents using the methods under this section. Apart from considering ROUGE scores, we manually surveyed other factors of the summary, such as the length, comprehensibility, grammar and punctuations.

2.1 Summarization for Legal Documents

Works for developing models specifically to summarize legal documents using ML have employed both supervised and unsupervised machine learning approaches. In [3], the authors used probabilistic graphical models to summarize documents of the legal domain. A sentence's rhetorical roles were identified using *conditional random fields*. The automatic summarization process starts by preprocessing the documents into segments, sentences and tokens.

Feature identification techniques in this paper included the understanding of abbreviated texts, section numbers and arguments specific to the structure of legal documents. K-mixture model is used as the term distribution model to assign probabilistic weights. Sentences are then re-ranked twice, first based on their weights and again based on their evolving roles throughout CRF implementation, to provide the final summary. While reproducing this summarization model, all the six summaries were around the same size, irrespective of their source document. Because of the sentence rankings, sometimes the summary does not seem to make proper sense.

A supervised model has its downsides as the model can only give accurate results for the case documents of the court it was trained on. For instance, the graphical model described above was trained on the documents of Kerala High Court, while we tested our implementations on documents of the Indian Supreme Court (which is an Indian court but still gives less accurate results). If the disparity is higher, for instance, if a model is trained on the Australian court and made to summarize documents of the Indian Supreme Court, the results would be unintelligible. In such cases, an unsupervised model designed to work on legal cases can be employed.

This can be shown by employing the *CaseSummarizer* model [4]. The authors designed the unsupervised model to work on Australian case judgements. The Natural Language Toolkit (NLTK) library was used for standard preprocessing which included stemming, lemmatization and clearing of stop words. The TF*IDF matrix was used for sentence scoring built on thousand of legal case reports. These scores are summed over each sentence and normalized by the sentence length. The model used occurrences of known entities, dates and proximity to section headings to score sentences.

Both these techniques of summarization were "extractive summarization". Extractive systems generate summaries by recognizing (and then concatenating) the most essential sentences in a document [5]. One of the main problems with this type of summarization is that the model considers all uses of the period symbol ('.') as a full stop or an end of the sentence. This results in half-cut sentences when the symbol is used to denote dates, currency units (Rs.) or designations (Dr.).

2.2 Domain Independent Summarization Techniques

We used the latent semantic analysis (LSA) summarizer available at [6]. This is also an extractive summarization technique that extracts and prepares the summary in the order in which they appear in the source document.

The first abstractive model we tried out was the T5 text-to-text transformer model from Google available at [7]. This model flushed out very small summaries, and there were problems with grammar such as punctuations and articles.

3 Paraphrasing Using Natural Language Processing (NLP)

Previous works on paraphrasing methods have considered different levels of paraphrasing granularity and different languages. While word or phrase-level paraphrasing is considered simpler, the authors in [8] used multiple sentence alignment to address sentence-level paraphrasing. Research has not been limited to the English language, as [9] presented an approach to paraphrase Hindi sentences using NLP.

4 Proposed Method

4.1 LexRank Summarization Algorithm

LexRank is an unsupervised graph-based technique for automatic text summarization. The graph method is used for sentence scoring. LexRank is used to compute sentence relevance using the notion of eigenvector centrality in a sentence graph representation [10]. This extractive algorithm works by extracting a similar set of sentences with the same intent. This technique comprehends the similarity between two sentences and generates a closeness score for each pair of sentences according to which the pair is given an equivalent weight. The total score of a sentence is finalized by accounting the weight of the edges connected with it [11]. This can be visualized in Fig. 1.

The LexRank algorithm is similar to the TextRank algorithm which uses a typical PageRank approach. LexRank considers the position and length of the sentences while TextRank does not, making LexRank very effective for legal documents.

4.2 PEGASUS Transformer

In NLP, transfer learning and pre-trained language models have pushed the boundaries of language understanding and generation. The PEGASUS model was first

Fig. 1 Graphical representation of LexRank

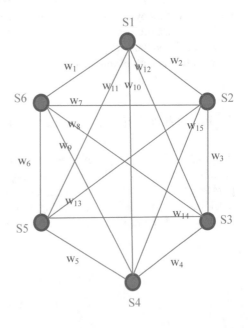

proposed in [12]. The authors built upon the success of seq2seq learning techniques and unsupervised language models like ELMo and BERT to build PEGASUS. The model uses an encoder-decoder model for sequence to sequence learning. PEGASUS also adopts the state-of-the-art transformer architecture. The way it differs from previous state-of-the-art models is the pre-training. The authors pre-trained the model on a large corpus of 350 million web pages and 1.5 billion news articles to automate the selection of important sentences using the ROUGE1-F1 metric.

4.3 Implementation

The document was first summarized using the LexRank model. The summary length was chosen as 35% of the original length of the document. We used the PEGASUS transformer model on a sentence level. The extractive summary generated by the LexRank model was split into sentences and fed to PEGASUS. The transformer then derived paraphrases for each sentence, making it simple to comprehend. The sentences were then merged to give a final abstractive level summary. The implementation flowchart can be visualized in Fig. 2.

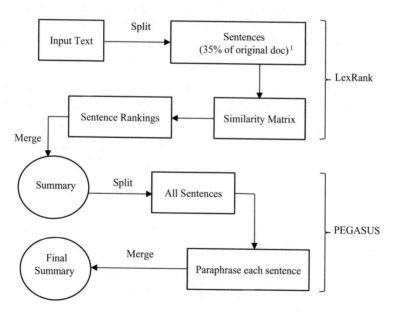

Fig. 2 Flowchart of implementation (The summary length can be n% of original document length, which is a user input)

Table 1 Recall, precision and F1-scores for all models implemented

Model	ROUGE1			ROUGEL		
	Recall	Precision	F1	Recall	Precision	F1
LSA	0.882	0.536	0.667	0.882	0.536	0.667
LexRank	0.804	**0.631**	**0.707**	0.784	**0.615**	**0.689**
CaseSummarizer	0.451	0.397	0.422	0.431	0.379	0.404
Graphical model	**0.902**	0.422	0.575	**0.902**	0.422	0.575
T5	0.176	0.237	0.202	0.157	0.211	0.179

5 Results

5.1 ROUGE Scores

We manually summarized a document for extractive summarization evaluation and used it as a reference to calculate the ROUGE scores of all the models implemented. ROUGE stands for *Recall-Oriented Understudy for Gisting Evaluation* which is a metric used for determining how good the summarization model is. From Table 1, it can be observed that while the graphical model had the highest recall, the LexRank model had far greater precision compared to others and subsequently a higher F1-score for both ROUGE1 and ROUGEL.

Recall

The recall counts the number of overlapping n-grams[1] in both the model output and the reference and divides that number by the total number of n-grams in the reference [13].

$$\text{Recall} = \frac{\text{number of } n\text{-grams found in model and reference}}{\text{number of } n\text{-grams in reference}}$$

Precision

Precision evaluates the summary based on how much relevant information is present in the summary compared to the reference text.

$$\text{Precision} = \frac{\text{number of } n\text{-grams found in model and reference}}{\text{number of } n\text{-grams in model}}$$

---

[1] An n-gram is simply a collection of tokens or words. A unigram (1-g) is made up of a single word. A bigram (2-g) is made up of two words:

F1-Score

$$\text{F1-Score} = 2 * \frac{\text{Precision} * \text{Recall}}{\text{Precision} + \text{Recall}}$$

As LexRank has the best precision value in all the models, it does the best work of extracting information that is relevant and needed than the other models.

5.2 Sample Output

We scrapped some documents from the "High Court of Bombay" from [14]. Below is a sample input and output of a document dated January 18, 2021. We picked a small document for this paper, but the results stay consistent over any large document.

Input:

1. *Heard the learned Counsel for the appellants and the respondents at a considerable length.*
2. *Perused the record.*
3. *The Appeals from Order can be disposed of in following terms in view of the peculiar circumstances of the case. 5-92265-2020-AOst-order = .doc*
4. *The operative part of the impugned order dated 20th February, 2020 passed by the learned Joint Civil Judge, Senior Division, Pune is defective in the sense that at the interlocutory stage, the relief as has been granted is as good as a final relief in the suit. As such, the said order is set aside and substituted as follows:-*

ORDER

(a) *The appellants—original defendants shall in view of its standing committee's Resolution No.62 dated 22nd June, 2020 which came to be passed after the impugned order, accept the supply of uniforms of the students for the Academic Year 2020–21.*
(b) *The learned Civil Judge, Senior Division, Pune shall expedite the suit.*
(c) *Parties shall co-operative with the expeditious disposal of the suit without seeking unnecessary adjournments.*
5. *The Appeals stand disposed of in the aforesaid terms. No order as to costs.*
6. *In view of disposal of the appeals, pending Interim Applications are also disposed of. (PRITHVIRAJ K. CHAVAN, J.)*

Output after LexRank summarization:

Heard the learned Counsel for the appellants and the respondents at a considerable length.
Perused the record.
The operative part of the impugned order dated 20th February, 2020 passed by the learned Joint Civil Judge, Senior Division, Pune is defective in the sense that

at the interlocutory stage, the relief as has been granted is as good as a final relief in the suit.

As such, the said order is set aside and substituted as follows:-

In view of disposal of the appeals, pending Interim Applications are also disposed of

Output after PEGASUS paraphrasing:

Heard the Counsel for the respondents. Used the record. The relief granted by the Joint Civil Judge, Senior Division, Pune in the impugned order dated 20th February, 2020 is not as good as the final relief in the suit. The said order is set aside and replaced. In view of the disposal of the appeals, pending interim applications are also done away with.

It can be observed that the final summary is about a quarter in length of the original document. The LexRank algorithm extracted information that is relevant and essential, while the PEGASUS transformer paraphrased each sentence in a language which is easier to comprehend for the common public. The final output acts as an abstractive summarization which has all the relevant information from the source document and is easy to understand.

6   Conclusion and Future Scope

Inculcating artificial intelligence (AI) and ML into the judiciary system has been an area of active research in the past decade. The primary virtue of the judicial system is to pass fair and just judgements. Thus, AI cannot entirely replace a human judge as human instincts are vital in determining an outcome of the case. AI can however be used to supplement the courts and help lawyers and judges deliver fair and speedy judgements. Automatic summarization of case documents can help the lawyers prepare for a case quickly and also help the common public truly understand case details and outcomes. In this project, we looked at various existing methods of summarization and proposed a novel method of abstractive summarization by using paraphrasing of summaries generated by an extractive summarization model. The summaries generated held all the relevant information from the source documents and were also in a simpler language which makes it easy for the common public to comprehend the document. The model, however, does not deal with specific legal jargons which can be a future area of study. Apart from jargons, if the document references an existing law or rule, the summary could provide a brief passage about what the rule is. There is also a need to have automatic categorization of similar court cases and verdicts which would help the lawyers. Research of infusing AI and ML into the judiciary system is still fledgling, and a lot of further research is possible to enhance the working of the judiciary.

References

1. Common Law. https://en.wikipedia.org/wiki/Common_law. Last accessed 03 Jan 2022
2. Precedent. https://en.wikipedia.org/wiki/Precedent. Last accessed 03 Jan 2022
3. Saravanan M, Ravindran B, Raman S (2006) Improving legal document summarization using graphical models. Front Artif Intell Appl 152:51
4. Polsley S, Jhunjhunwala P, Huang R (2016) Casesummarizer: a system for automated summarization of legal texts. In: Proceedings of COLING 2016, the 26th international conference on computational linguistics: system demonstrations, pp 258–262
5. Narayan S, Cohen SB, Lapata M (2018) Ranking sentences for extractive summarization with reinforcement learning. arXiv preprint arXiv:1802.08636
6. Sumy 0.9.0. https://pypi.org/project/sumy/. Last accessed 26 Dec 2021
7. Simple abstractive text summarization with pretrained T5—text-to-text transfer transformer. https://towardsdatascience.com/simple-abstractive-text-summarization-with-pretrained-t5-text-to-text-transfer-transformer-10f6d602c426. Last accessed 26 Dec 2021
8. Barzilay R, Lee L (2003) Learning to paraphrase: an unsupervised approach using multiple-sequence alignment. arXiv preprint cs/0304006
9. Sethi N, Agrawal P, Madaan V, Singh SK (2016) A novel approach to paraphrase Hindi sentences using natural language processing. Indian J Sci Technol 9(28):1–6
10. LexRank method for text summarization. https://iq.opengenus.org/lexrank-text-summarization/. Last accessed 27 Dec 2021
11. Samuel A, Sharma DK (2016) Modified lexrank for tweet summarization. Int J Rough Sets Data Anal (IJRSDA) 3(4):79–90
12. Zhang J, Zhao Y, Saleh M, Liu P (2020) Pegasus: pre-training with extracted gap-sentences for abstractive summarization. In: International conference on machine learning. PMLR, pp 11328–11339
13. The ultimate performance metric in NLP. https://towardsdatascience.com/the-ultimate-performance-metric-in-nlp-111df6c64460. Last accessed 28 Feb 2022
14. High Court of Bombay. https://bombayhighcourt.nic.in/index.php. Last accessed 02 Jan 2022

DoS Defense Using Modified Naive Bayes

Rajesh Kumar Shrivastava, Simar Preet Singh, and Abinash Pujahari

1 Introduction

Internet of Things (IoT) devices depend on Internet for reliable communication. Due to public deployment these devices are vulnerable against cyber-attacks [3]. A DoS attack is an attempt to obstruct real users from accessing services [8, 11]. In this attack, an adversary targets a server or an entire network to cause temporary or permanent unavailability by sending fake requests, consume bandwidth, or interrupt access to a specific service or a system.

Here, We have a research question: how do we capture an adversary's digital footprint. The solution to this problem is a Honeypot [7]. Honeypot is a deception system that allows an adversary to hijack and perform malicious attacks. Once an adversary does the same, the honeypot records all the activities performed by an adversary. Honeypot collects all the digital footprint and stores it in JSON format. We evaluated these files through machine learning (ML) methods. The result of ML algorithms updated the rules in intrusion detection systems (IDS).

An IDS is mainly used to prevent the network from attacks. IDS can block a malicious IP to get in. It is the best way to deal with Distributed Denial of Services (DDoS) attacks [12]. In a DDoS attack, an adversary sends many requests to the server within a short period (maybe in 1 s). The server is busy dealing with these fake requests, and regular users don't respond. This situation becomes critical if you

R. K. Shrivastava · S. P. Singh (✉)
Bennett University, Greater Noida, UP, India
e-mail: simarpreet.Singh@bennett.edu.in; dr.simarpreetsingh@gmail.com

R. K. Shrivastava
e-mail: rajesh.shrivastava@bennett.edu.in

A. Pujahari
Assistant Professor, Computer Science and Engineering, SRM University AP, Amaravati 522502, Andhra Pradesh, India
e-mail: abinash.p@srmap.edu.in

are trying to do a banking operation and your bank is not responding. So, we need a fruitful algorithm that helps us to keep away from this type of attacker.

In our solution, we deployed Cowrie honeypot into Raspberry Pi 3 board to capture attack patterns. Captured data transfers to server machine where we apply our proposed method to detect DoS attack and malicious IP.

A modified Naive Bayes algorithm is used to investigate the interaction between the honeypot and the attacker.

Our contribution is as follows:

1. Design an interaction between an attacker and a honeypot in IoT networks.
2. Design and develop of an algorithm by using Naive Bayes approach to apply in unlabeled data.
3. Design and develop interactive IDS.

The remaining paper is organize as follows: Sect. 2 discuss related work done in this research area. Section 3 disuses proposed solution, i.e., data gathering using honeypot, modified Naive Bayes algorithm, etc., Sect. 4 addresses attack analysis and also explains result of our experiments. Section 5 discuss the conclusion and the results.

2 Related Work

Irwan et al. [6] showed a honeypot based model to detect DoS attack. They used "honeyd-viz" software to visualize log collected by honeypot, but they didn't use any processing method to classify these attacks. Selvaraj et al. [5] simulated ant-based DoS detection technique in honeypots environment. Authors used virtual honeypot and collect information various attacks at various levels in the network. Author used Ant colony optimization technique to detect intruders. They also used a multi-level IP log table to detect the intruders and send this information to the honeypot. Zhan et al. [10] proposed the statistical model to analyze cyber-attacks captured by the honeypot. To demonstrate the use of mathematical model authors used a low-interaction honeypot. They exhibited the long-range dependence model over honeypot-captured attacks, and predict cyber-attacks and attack rate. The prediction of attacks alerts the defenders to prepare with their defense configurations or resource allocations. Vollmer et al. [9] proposed a dynamic virtual honeypot based passive monitoring to secure industrial networks. The proposed dynamic virtual honeypots were used to observing and attracting network intruder activities and passively examine control system network traffic. Kolias et al. [4] discussed, how Mirai and it's variant were used to execute DoS/DDoS attack in any network. As per the author gave a wake-up call, we need a secure solution to prevent IoT devices from DoS attacks. Our solution effectively prevents IoT devices from such attacks.

3 Proposed Solution

In this paper, we identify a DoS attack. For this purpose, we deployed a Cowrie honeypot and collected the attacks. Honeypot data is stored in JSON format, and it is unlabeled. To make a practical solution, we used a modified Naive Bayes Algorithm. We keep all the ping and connection requests from the database. The proposed algorithm also takes help from "virustotal.com" to verify whether captured IP is malicious or not. We applied K-Means algorithms to divide the program into different clusters. If the IP request frequency is higher than the threshold value, we consider a DoS attack. All the malicious IP's updated in the IDS.

3.1 Fingerprint Gathering

Cowrie honeypot is responsible for capturing logs in the form of JSON format, which will be available in public IP machine as specified in Fig. 1. Captured attacker data for further processing using Elastic Search, Logstash, and Kibana (ELK) [1]. Logstash works on config files. Each config file has three parameters: input, filter, and output (Fig. 2).

1. Input: It contains a parameter file which has options path, start-position, and ignore-older. The path is where log files are present and start-position indicates from which point Logstash should start reading the logs from.
2. Filter: Filter is where entire processing happens in logstash. It has many options out of which JSON option is used as the logs are in JSON format. Some more information is added to the already present logs like country, continent, latitude and longitude information which are specified in geoip option. Finally, latitude and longitude are converted into float variables for plotting on the map.

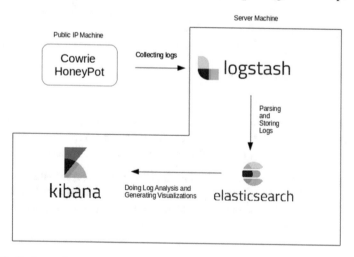

Fig. 1 Monitoring system

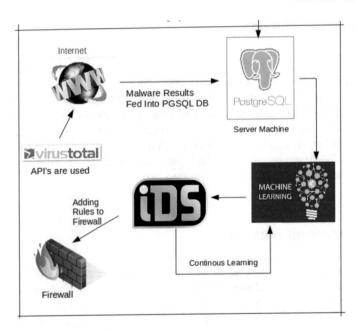

Fig. 2 Intrusion detection system

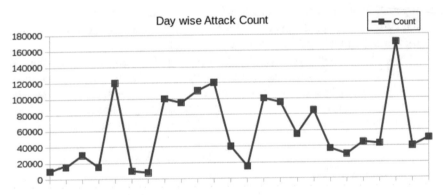

Fig. 3 Daily attacks count area graph

3. Output: It specifies options as to where the data should go after the processing is done. There are many options out of which we are using elastic search where data is stored in JSON format which will be in key-value format.

We deployed cowrie honeypot for 40 days and collected approximately 2.2 million records in JASON format. Collected data processed with kibana and visualized as shown in Fig. 3. The X-axis represents the day's, and Y-axis represents the number of attacks recorded per day.

Table 1 List of tables in PostgreSQL database

S. No.	Table name	Schema
1	Dos-Attacks	Public
2	Malicious-URLs	Public
3	Stats	Public

3.2 Intrusion Detection System

Data captured by honeypot is stored in PostgreSQL Fig. 2. We applied ML methods to classify data as benign and malicious. The output of the ML algorithm was further verified by "virustotal.com." If an IP is found malicious, we add IP to the block list. The outcome strengthens the IDS. The following section describes the detailed process.

1. Virus Total is a website where details about different malware and IP's can be fetched by using their web interface or REST API's. For our convenience, API's are used to integrate with our python programs.
2. In PostgreSQL, database named "cowrie" has been created and Table 1 created to store the required data from Elastic Search and VirusTotal.

3.2.1 K-Means Classification with Modified Naive Bayes

Honeypot collected data in JSON format, and data is unlabeled. The Naive Bayes classifier handles Probability-based classification. But Naive-based classifier needs labeled data for training and validation purposes. To bridge this gap, we modified our Algorithm 1 and applied it to unlabeled data. The Algorithm 1 explain the working model. The proposed approach is looking for the hypothesis to search-out malicious IP. The Algorithm 1 takes four inputs, respectively, Name of the database, total words, unique words, probabilities, and frequency of terms, then we look for the commands executed by individual IP. Now we segregate unique words present in the command executed by IP. these particular words are divided by the total uncommon words presented in the database. The database contains malicious commands and words used by adversaries. Then Algorithm 1 calculates the probability of each word. The following equation is used to calculate the probability of malicious IP.

Let $X =$ IP from adversary, $Y =$ Execution command by IP, Then $P(X)$ is the beginning probability that IP may be involved in the DoS attack. $P(Y)$, the probability that an adversary may executed a malicious command.

$$P(X/Y) = \frac{P(X) * P(Y/X)}{P(Y)} \tag{1}$$

Now we have the probability value of each IP. Without loss of any generality, let us assume that the initial probability of each IP is $P(X) = P(Y) = 1$.

Algorithm 1: Modified Naive Bayes Algorithms

Input: dbName, totalWords, uniqueWords, probabilities, frequency.
Output: Probability of IP, performed malicious attacks.
{ //Fetch data from honeypot
$rows \leftarrow$ fetch_from_honeypot(dbName);
while *row in rows:* **do**
  | commands $\leftarrow$ row[0].replace(";"," ").split(" ");
  | **while** *(command in commands)* **do**
    | **if** *(command != " ")* **then**
      | uniqueWords.add(command.strip());
      | totalWords.append(command.strip());
    | **end**
  | **end**
end
counts $\leftarrow$ Counter(totalWords);
while *(word in uniqueWords)* **do**
  | probability $= \frac{(counts.get(word)+1)}{(len(totalWords)+len(uniqueWords))}$;
  | probabilities.setdefault(word, math.exp(probability));
end
cur $\leftarrow$ $fetch\_from\_honeypot(IP, COMMANDS)$;
rows = cur.fetchall();
while *(row in rows)* **do**
  | TotalProb = 1.0;
  | commands = row[1].replace(";"," ").split(" ");
  | **while** *(command in commands)* **do**
    | **if** *(command != " ")* **then**
      | **if** *(probabilities.get(command) != None)* **then**
      | | TotalProb *= probabilities.get(command);
      | **end**
    | **end**
  | **end**
  | **if** *(TotalProb != 1.0)* **then**
    | frequency.append((row[0],TotalProb));
  | **end**
end
print "frequency";
while *(frequency in frequency)* **do**
  | print frequency[1];
end
}

Since each IP has the same possibility, only worried about $P(Y/X)$, the likelihood function to detect malicious words present in commands or not in a particular IP. If

$P(Y/X)$ values are closer to one, the IP is more vulnerable and may perform a DoS attack. The following equation is used to calculate the value of $P(Y/X)$. C = Count of Words, d=Unique count of Words.

$$P(Y/X) = \frac{\text{Term count of Word} + 1}{C + d} \qquad (2)$$

Our proposed method finds malicious IP by calculating the term frequency of the uncommon word used in a command executed by individual IPs.

$$P(Y/X) = \prod_{j=1}^{\text{freq}} \frac{P(\text{Word})_j \text{ in Command Executed}}{(\text{IP is harmful})} \qquad (3)$$

$$p(Y/X) = \prod_{j=1}^{\text{frequency}} \frac{c_i + 1}{C + d} \qquad (4)$$

Once we have a probability score, we go for the clustering algorithm. For this purpose, we used K-Means clustering. Algorithm 2 applied the same and performed segregation of records in two categories, benign, and malicious. Now, clustering divides whole IPs into two groups by using Euclidean Squared distance metric [2].

Algorithm 2: K-Means Clustering Algorithms to identify DoS Attack

---

Input: IP's Final Probability from Algorithm 1 as a set of probability score : P, and Cluster size: K.
Output: Cluster Center that minimizes the squared error distortion.
Select K score points randomly from P to form cluster center.
while *(No convergence in cluster)* **do**
    1. calculate minimum Euclidean Squared distance metric to assign each point P with its nearest cluster.
    2. After all probability points are assigned to their clusters, move each cluster center to mean of its assigned probability points.
    3. Recompute each cluster center.
end

---

4 Attack Analysis

4.1 Attack Statistics

Table 2 store attack statistics. Column "IP" contains the IP of the attacker or intruder and "commands" contains all the commands executed by the attacker or the intruder

Table 2 Schema of statistics table

S. No.	Column name	Column datatype
1	IP	Character varying(20)
2	Commands	Text
3	Count of commands	Integer
4	Login attempts	Integer
5	Doscluster	Text
6	Frequency	Integer

Table 3 DoS attack classification

S. No.	Cluster name	No. of IP's
1	Malicious (High)	136
2	Benign (Low)	1842
	Total	1978

till now in cowrie honeypot. "Count of commands" contains the number of commands attacker executed. "Login-attempts" contains the number of times attacker tried to login into the cowrie. "Dos-cluster" contains the cluster number to which IP is clustered which will be updated dynamically. "Frequency" contains the value of IP which is updated after running machine learning algorithms.

A Python script fetches all the commands executed per IP, appends them together separated by a semicolon and puts them in the column named "commands" in "stats" table. It also updates the "count of commands" column. The number of login attempts done by each IP from the elastic search stores in a column named "loginattempts" in "Stats" table.

4.2 DOS Attacks

The outcome of the clustering algorithm is shown in Table 3. K-means algorithm divides the whole dataset into two parts, high and low. The high cluster is directly related to the DoS attack. IDS blocks all the IP that belongs to the high cluster.

Attackers who were able to login successfully have used their access to full effect by deploying the binaries and executable files on the machine and running them remotely whenever needed. Fortunately, Cowrie happened to be a medium level interaction honeypot which would not allow running any binaries or executable files. Some of the captured binaries are sent to VirusTotal, to verify our result, and all of them are identified as serious DoS attacking binaries. Some of them are presented in Table 4. Some of the malicious URL's accessed by the attackers is captured by the cowrie. These are given in Table 5.

Table 4 List of Dos Binaries

./lPg5Am8r	./AObM55mP	./ScDrDSSt
./Tr5l603l	./yu5LvV97	./VdPacLUl
./BRB3bpfb	./PWnQ7Tcn	./GgzxWgHv
./udp4858	./kEbIZq9x	./MvlFggnh
./o3e1ROxG	./b1fNqt0C	./iUk3up10

Table 5 List of Sample Malicious URLs

http://185.165.29.196/lmao.sh
http://mdb7.cn:8081/exp
http://104.XXX.151.157/i3306m
http://23.228.113.240/do3309
http://23.228.113.240/ys808e

Table 6 Compare with other research work

Author	Unlabeled data	Method	Accuracy (%)
Irwan et al. [6]	No	Naive Bayes	90
Kolias et al. [4]	No	Naive Bayes	80
Shrivastava et al. [7]	No	Naive Bayes	96
Proposed work	Yes	Naive Bayes, K-Means	96

Based on all data listed above and tables created in PostgreSQL database and API's from VirusTotal, an IDS has been developed which continuously monitors the network where commands executed by any IP is noted and sent to the database built. If it is found in the database, it will be reported as malicious. If no information is found about the data in the database, it keeps it in a separate column and keeps monitoring. If any anomaly is found, then the database will be updated with the new one. So it is a continuous learning process for the system.

Table 6 compare our work with other researchers. Others were used Naive Algorithm by labeling dataset but we use same algorithm with unlabeled data. We achieve 96% of accuracy in detection of malicious IP. If any of the IP found malicious then IDS blocks this IP and updates the firewall if any IP is found vulnerable. The following command is used to block or unblock an IP.

Blocking: *iptables -A INPUT -s 192.XXX.218.XXX -j DROP*
Unblocking: *iptables -D INPUT -s 192.XXX.218.XXX -j DROP*

5 Conclusion

Recent developments in IoT have forced us to look at the security features microscopic and improvements to be done. Deployed IoT-based honeypot collects unlabelled data. The research challenge we face is how to assign probability scores to collected data. For this purpose, we twist a Naive Bayes algorithm. The exciting difference between traditional method and our modified algorithm is that we classify unlabelled data with a Naive Bayes algorithm. We used a frequency count of various commands and IPs to build our hypothesis. Our approach also strengthens an IDS.

References

1. Bhatnagar D, Jaya SubaLakshmi R, Vanmathi C (2020) Twitter sentiment analysis using elasticsearch, logstash and kibana. In: 2020 international conference on emerging trends in information technology and engineering (ic-ETITE), pp 1–5
2. Capó M, Pérez A, Lozano JA (2018) An efficient k-means clustering algorithm for massive data. arXiv preprint arXiv:1801.02949
3. Durumeric Z, Wustrow E, Halderman JA (2013) Fast internet-wide scanning and its security applications. Zmap. In USENIX security symposium, vol 8, pp 47–53
4. Kolias C, Kambourakis G, Stavrou A, Voas J (2017) Ddos in the IoT: Mirai and other botnets. Computer 50(7):80–84
5. Selvaraj R, Kuthadi VM, Marwala T (2016) Ant-based distributed denial of service detection technique using roaming virtual honeypots. IET Commun 10(8):929–935
6. Sembiring I (2016) Implementation of honeypot to detect and prevent distributed denial of service attack. In 2016 3rd international conference on information technology, computer, and electrical engineering (ICITACEE). IEEE, pp 345–350
7. Shrivastava RK, Bashir B, Hota C (2019) Attack detection and forensics using honeypot in IoT environment. In: International conference on distributed computing and internet technology. Springer, Heidelberg, pp 402–409
8. Sinha S, Sindhu B (2021) Impact of dos attack in IoT system and identifying the attacker location for interference attacks. In: 2021 6th international conference on communication and electronics systems (ICCES), pp 657–662
9. Vollmer T, Manic M (2014) Cyber-physical system security with deceptive virtual hosts for industrial control networks. IEEE Trans Ind Inf 10(2):1337–1347
10. Zhan Z, Xu M, Xu S (2013) Characterizing honeypot-captured cyber attacks: statistical framework and case study. IEEE Trans Inf Forensics Security 8(11):1775–1789
11. Zhang H, Cheng P, Shi L, Chen J (2016) Optimal dos attack scheduling in wireless networked control system. IEEE Trans Control Syst Technol 24(3):843–852
12. Zhu G, Yuan H, Zhuang Y, Guo Y, Zhang X, Qiu S (2021) Research on network intrusion detection method of power system based on random forest algorithm. In: 2021 13th international conference on measuring technology and mechatronics automation (ICMTMA), pp 374–379

Adaptive Threshold Peak Detection for Launch Vehicle Simulation Time Series Data Analysis

T. K. Shinoj, N. Geethu, M. Selvaraaj, P. K. Abraham, and S. Athula Devi

1 Introduction

A time series is a sequence of data points at uniform time intervals [1]. Any time series may be split into the following components:

$$\text{Time series} = \text{Base Level} + \text{Trend} + \text{Seasonality} + \text{Noise} + \text{Event}$$

Here, trend is the behavior of the time series where it has a positive or negative slope in long term. Seasonality is the distinct repeated and periodic patterns. Noise is the random behavior of the time series over the base level, and events are sudden and significant behavior variations. Peaks in a time series can be considered as typical events. Automatic detection of events in a time series data is a major area of interest with diverse applications from bioinformatics [2] to astrophysics [3].

One potential area where peak detection is significant in Launch vehicle domain is simulation results analysis. In this work, we considered Simulated Input Profile (SIP) test, which is an open loop test carried out to validate the performance of Launch vehicle onboard software by comparing the results with reference simulation

T. K. Shinoj (✉) · N. Geethu · M. Selvaraaj · P. K. Abraham · S. Athula Devi
Vikram Sarabhai Space Centre, Indian Space Research Organization, Thiruvananthapuram, Kerala, India
e-mail: shinojthrissur@gmail.com

N. Geethu
e-mail: geethu@vssc.gov.in

M. Selvaraaj
e-mail: m_selvaraaj@vssc.gov.in

P. K. Abraham
e-mail: pk_abraham@vssc.gov.in

S. Athula Devi
e-mail: s_athuladevi@vssc.gov.in

© The Author(s), under exclusive license to Springer Nature Singapore Pte Ltd. 2023
D. S. Sisodia et al. (eds.), *Machine Intelligence Techniques for Data Analysis and Signal Processing*, Lecture Notes in Electrical Engineering 997,
https://doi.org/10.1007/978-981-99-0085-5_48

software. Mainly, we have used onboard software generated control commands as SIP output. Here, one of the parameter is the time history of various commands processed by the onboard software. As part of introducing automation in the analysis, the difference between the two time series outputs are found out, and flags are raised when difference exceeds acceptable limits. The flagged instances will be then further studied.

The results from onboard software and simulation software should be identical in an ideal scenario. However, there will be minor magnitude differences due to platform dependency. Further, outputs can have definite time shifts due to execution time differences between the two platforms. The time shifts manifest as peaks in the difference plots while minor magnitude differences introduce noise in it. Hence, any automated assessment of the SIP test results should judiciously compensate for these two factors.

The ideal peak detection algorithm is desired to detect the peaks from a time series with minimum interference from the user. It should have only few parameters to tune. It would be an added benefit if the parameters are intuitive to guess. The algorithm should be able to distinguish peaks in data in the presence of trend, seasonality, noise and shift in base levels. This is important as our data set contains data with diverse characteristics.

With this aim in mind, various algorithms were explored for automatic detection of peaks in time series. Peak is defined as a very narrow region (only a few points) of high values with sharp falls on either side [4]. Standard peak detection approaches include smoothing and fitting a known function, matching a known peak shape to the time series, slope sign change between a point and its neighbors [4]. In [2, 3], wavelet-based methods for peak detection are discussed. Various approaches are available for detecting peaks in a time series data. The most basic approach will be to slide a window over the series with appropriate threshold [5, 6]. Various transformations and filtering techniques like wavelet transforms, Hilbert transforms, Hough transform, etc., can be used for peak detection [7, 8]. Studies have been done with parameters like kernel density estimation, momentum [9, 10], etc., can be used as operating variables.

In space domain, there have been many attempts to apply automated anomaly detection to spacecraft telemetry data. The limitations and improved version of Out-of-Limit (OOL) method and an assessment with ESA missions are discussed in [11]. The Fault detection, Isolation and Recovery (FDIR) project by NASA provided an automated data-driven approach for anomaly detection [12]. And they used a data-driven Inductive Monitoring System (IMS) approach for anomaly detection [13, 14]. The details of the software tools used for the data analysis are given in [15]. The Automated Telemetry Health Monitoring System (ATHMoS) uses a supervised learning algorithm for statistical detection of outliers [16]. In [17], the authors discuss about two unsupervised anomaly detection algorithms, ORCA and GritBot in context of rocket propulsion health monitoring. Concept of Shannon entropy is used in [18] to detect anomalies in sensor data of space shuttle main engines.

In our work, we adopted a Peak Transformation Function (PTF)-based strategy due to its inherent simplicity. A Peak Transformation Function transforms the input data, stripping it of all its components except peaks. From this transformed data, peaks above requisite threshold can be extracted. The strategy is further strengthened by introducing adaptive threshold setting. The algorithm parameters are tuned to be robust enough to cater for all data. The concepts of Divided Average Method (DAM) and Moving Average Method (MAM) are applied on the processed data for peak detection. Further, the performance of the algorithms was compared using standard measures derived from the Contingency matrix. A Contingency matrix provides the classification performance summary of a classifier with respect to some test data.

2 Data Characteristics

SIP test output data is a set of two univariate time series. The input data, over which the processing is carried out, is generated by taking the difference between the two time series. Ideally, the data should have a base level zero with no trend, seasonality or noise. However, the minor magnitude differences between two time series in the SIP test output data set manifest as noise in the input data. The time shift between the two series in the data sets may cause peaks in the input data. The peaks and trends can also occur due to various other factors. A sample data is shown in Fig. 1.

Fig. 1 Sample data

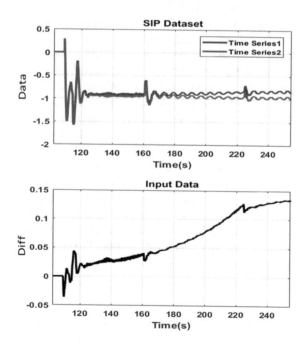

3   Implementation of Peak Detection Strategy

Visual detection of peaks in a time series can be easily carried out with graphical representations. However, this is not a feasible approach for processing a large quantity of data such as an SIP run having approximately 200 time series each of having around 5000 data points. Hence, an automatic detection of peaks is desired. For this the peak should be formally defined. The nominal notion of a peak can be defined as points outside the 3 sigma bound from the series mean. This approach has two pitfalls:

- Time series with large number of points and one or two peaks will have low standard deviation and this may lead to false detection of peaks.
- In time series with significant trends, this approach can lead to non-detection of local peaks.

Hence, more rigorous methodologies, as discussed in the introduction, are required for proper detection of peaks. In our work, initially, we selected a PTF-based algorithm. The details are discussed in Sect. 4. A PTF strips the time series of its trends, seasonality and base shifts and amplifies the peaks. The extent of amplification depends upon the local characteristics of the data as well the particular transformation function. A typical PTF transformed data is shown in Fig. 2.

After applying the PTF, we made an attempt to detect the peak based on a constant threshold. It was found that the threshold value highly depends on input data characteristics. Hence, an option was explored to filter the peaks based on the statistics of the transformed data. If we use the complete series to generate the statistical measures, it leads to the same pitfalls as discussed above. So, two methodologies based on a piecewise segmented approach: DAM and MAM were employed to generate adaptive thresholds. The details of these methodologies are discussed in Sect. 5. The threshold generation methodology is strengthened with an additional layer based on statistics of input data. For our data analysis, we had decided to neglect the peak below a magnitude of 0.05. This was added as an absolute threshold value in the detection strategy to remove the minor peaks.

To investigate the performance of different methodologies, a database of 68 input data sets is generated from the SIP data. It was ensured that the database includes series with nominal as well as erratic behavior. The outcome of the methods is visually inspected and divided in to three categories based on Contingency/Confusion matrix, and performance measures are derived. The details are provided in Sect. 6, and results are discussed in Sects. 7 and 8.

4   Algorithms for Data Transformation

Let $T(x_i, t_i)$, $i = 1, \ldots, N$ be a time series data with N elements. The PTFs S_j, $j = 1, 2, 3$ defined in this section, applied to the time series T, assigns to each element

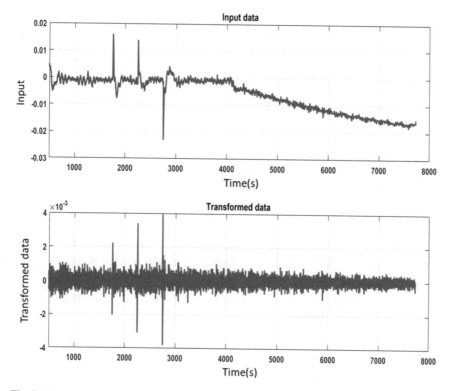

Fig. 2 Transformation using PTF

x_i, a score, based on its local position. Basic framework for peak detection using the PTF is as follows:

$$\forall x_i \in T \quad S_i = S(x_i)$$
$$\text{If } S_i > \text{threshold}(x_i) \implies x_i \to \text{Peak}$$

In [4], the authors have used three PTFs to detect peak in the annual sunspot data which are described below. Here, k is the width parameter which has selected appropriately.

1. PTF-1

$$S_1(x_i) = \frac{\max\{x_i - x_{i-1}, \ldots, x_i - x_{i-k}\} + \max\{x_i - x_{i+1}, \ldots, x_i - x_{i+k}\}}{2}$$

(1)

2. PTF-2

$$S_2(x_i) = \frac{\frac{(x_i - x_{i-1} + \cdots + x_i - x_{i-k})}{k} + \frac{(x_i - x_{i+1} + \cdots + x_i - x_{i+k})}{k}}{2} \tag{2}$$

3. PTF-3

$$S_3(x_i) = \frac{\left(x_i - \frac{(x_{i-1} + \cdots + x_{i-k})}{k}\right) + \left(x_i - \frac{(x_{i+1} + \cdots + x_{i+k})}{k}\right)}{2} \tag{3}$$

In all the three PTFs, a neighborhood of width 2k is considered around each point.

It is obvious that S3 is equivalent to S2. Hence, the PTFs S1 and S2 are selected to generate the transformed data. This transformed data is further processed to generate the threshold using the methods described in Sect. 5.

5 Piecewise Approach for Adaptive Threshold Generation

In this section, two methodologies are discussed which are used to generate adaptive threshold using data statistics.

5.1 Divided Average Method (DAM)

Here, the data is divided in to segments with pre-defined number of points. Statistics of each segments mainly mean, RMS and standard deviation are computed. Threshold for these segments are computed based on these parameters. Depending on the number of points in each segment, the threshold value will change, and hence, peak detection characteristics will vary. This is depicted in Fig. 3. So, the number of points used in DAM should be appropriately selected.

5.2 Moving Average Method (MAM)

Here, a segment with pre-defined number of points is created around each point. Statistics of the segment mainly mean, RMS and standard deviation are computed, and threshold is calculated based on these parameters. The statistics generated using this method are shown in Fig. 4. The mean and standard deviation in the figure refer to the mean and standard deviation of the segment corresponding to each point.

To analyze the performance of the methodologies discussed above, we created a Contingency matrix using the test database. The details are given in the next section.

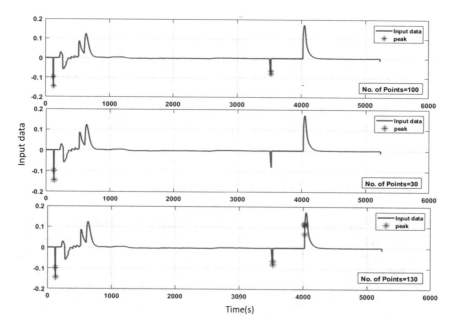

Fig. 3 Sensitivity of no. of points in DAM

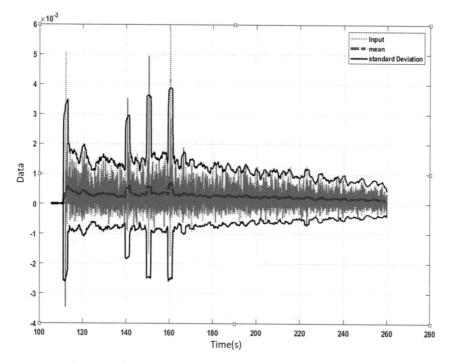

Fig. 4 Data statistics using MAM

Table 1 Contingency matrix

Serial no	Parameter	Description
1	True Positives (TP)	Points which are correctly detected as peaks
2	False Positives (FP)	Type-I error—points which are incorrectly detected as peaks
3	True Negatives (TN)	Points which are correctly detected as non-peaks
4	False Negatives (FN)	Type-II error—points which are incorrectly detected as non-peaks

6 Parameters for Performance Evaluation

The terminologies and notations used in this section are as defined in [19, 20]. A Contingency/Confusion matrix summarizes the classification performance of a classifier with respect to some test data. The Contingency matrix is having four cells (Table 1).

In context of our analysis, TN is insignificant due to large amount of data points in the input data. The following measures of classification performance are defined using the Contingency matrix:

$$\text{Sensitivity/True Positive Rate (TPR)} = \frac{TP}{TP + FN}$$

$$\text{Precision/Positive Predictive Rate (PPV)} = \frac{TP}{TP + FP}$$

$$\text{False Discovery Rate (FDR)} = \frac{FP}{TP + FP} = 1 - PPV$$

$$\text{Miss rate/False Negative Rate (FNR)} = \frac{FN}{TP + FN} = 1 - TPR$$

$$\text{F1score (Harmonic mean of Precision and Sensitivity)} = \frac{2TP}{2TP + FP + FN}$$

7 Results and Discussion

In this work, we implemented four methods combining two PTFs and two threshold generation approaches as given in the Table 2.

A typical result with M3 is shown in Fig. 5.

It is seen that the devised methods can handle time series data with trends easily. A typical example is shown in Fig. 6. The trends in the input data are removed by

Table 2 Method description

Method	PTF	Threshold generation approach
M1	S1	DAM
M2	S1	MAM
M3	S2	DAM
M4	S2	MAM

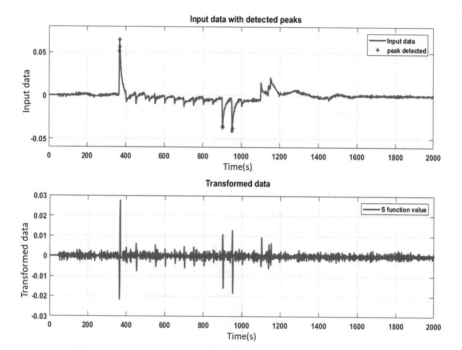

Fig. 5 Input data with detected peak

the PTF, and the peak is correctly detected even though the data points with higher magnitude than the peak are present in the series.

8   Performance Analysis Using Contingency Matrix

To compare the performance of each method, we derived the performance measures using the Contingency matrix. The results are given below (Table 3).

Among the four methods, M4 is having highest sensitivity and M2 is having highest Precision. However, F1 score is the Harmonic mean of Precision and Sensitivity, and this will give equal weightage to both. In terms of F1 score, M2 performance is found to be better than other methods.

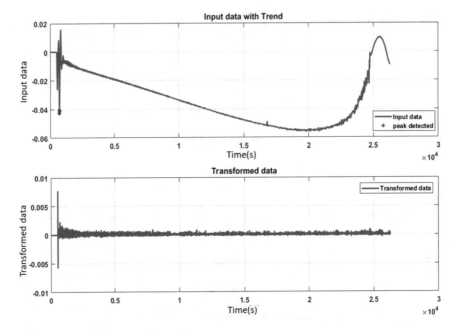

Fig. 6 Typical performance with trends

Table 3 Performance measures

	TPR	PPV	FDR	FNR	F1 score
M1	0.904	0.930	0.070	0.096	0.917
M2	0.918	0.944	0.056	0.082	0.931
M3	0.932	0.861	0.139	0.068	0.895
M4	0.945	0.885	0.115	0.055	0.914

9   Conclusion

In this work, an attempt is made to detect the peaks in a Launch vehicle simulation
time series data. A tool is developed for automated detection of peaks in a continuous
time series data. For this, various algorithms for data transformation discussed in the
literature have been explored. A piecewise approach for threshold generation using
the concepts of Divided Average Method and Moving Average Method are applied
on the transformed data for peak detection. Adaptive threshold setting is introduced
for filtering the peaks, and the performance of the algorithms was compared using
standard measures derived from the Contingency matrix. Based on F1 score which
combines the effect of Precision and Sensitivity, it is concluded that PTF S1 with
Moving Average Method shows better performance than the other. As a future work

different peak detection algorithms using kernel density functions, outlier detection methods can be explored.

Acknowledgements The authors would like to acknowledge all the members of QAMD, QRSG/SR, VSSC for their motivation, help and constructive comments. Authors are thankful to Smt. Jayalekshmy L for reviewing the paper and providing valuable suggestions.

References

1. Pal A, Prakash PKS (2017) Practical time series analysis. ISBN 978-1-78829-022-7, Packt Publishing Ltd
2. Du P, Kibbe WA, Lin SM (2006) Improved peak detection in mass spectrum by incorporating continuous wavelet transform-based pattern matching. Bioinformatics 22(17):2059–2065 (Oxford Academic)
3. Zhu Y, Shasha D (2003) Efficient elastic burst detection in data streams. In: Proceedings of the SIGKDD 2003 conference. ACM Press, pp 336–345
4. Palshikar GK (2009) Simple algorithms for peak detection in time-series. In: 1st international conference on advanced data analysis, business analytics and intelligence (ICADABAI2009), Ahmedabad, India, 6–7 June 2009
5. Jacobsen AL (2001) Auto-threshold peak detection in physiological signals. In: Proceedings of the 23rd annual international conference of the IEEE, vol 3, pp 2194–2195
6. Pan J, Tompkins WJ (1985) A real-time QRS detection algorithm. IEEE Trans Biomed Eng BME-32:230–236
7. Rabbani H, Parsa Mahjoob M, Farahabadi E, Farahabadi A (2011) R peak detection in electro-cardiogram signal based on an optimal combination of wavelet transform, Hilbert transform, and adaptive thresholding. J Med Signals Sens 1(2):91–98
8. Rao SNV, Dunn SM (1991) Hough transform based peak detection: computer-assisted reso-nance assignments in NMR spectroscopy (protein study). In: Proceedings of the 1991 IEEE 17th annual northeast bioengineering conference, pp 105–106
9. Wahid A, Rao ACS, Deb K (2018) A Relative Kernel-density based outlier detection algo-rithm. In: 12th international conference on software, knowledge, information management & applications
10. Harmer K, Howells G, Sheng W, Fairhurst M, Deravi F (2008) A peak-trough detection algorithm based on momentum. In: Congress on image and signal processing, pp 454–458
11. Martinez J, Donati A, Kirsch M, Schmidt F (2012) New telemetry monitoring paradigm with novelty detection, AIAA
12. Spirkovska L, Iverson DL, Hall DR, Taylor WM, Ann PH, Brown BL, Ferrell BA, Waterman RD (2010) Anomaly detection for next generation space launch ground operations, AIAA 2010-2182
13. Iverson DL, Martin R, Schwabacher M, Spirkovska L (2012) General purpose data-driven system monitoring for space operations. J Aerosp Comput Inf Commun
14. Iverson DL (2004) Inductive system health monitoring. In: Proceedings of the 2004 interna-tional conference on artificial intelligence (IC-AI04), CSREA, Las Vegas, Nevada
15. Iverson DL (2008) Data mining applications for space mission operations system health monitoring. In: Proceedings of the SpaceOps 2008 conference, ESA, EUMETSAT, AIAA, Heidelberg, Germany
16. O'Meara C, Schlag L, Faltenbacher L, Wickler M (2016) ATHMoS: automated telemetry health monitoring system at GSOC using outlier detection and supervised machine learning, AIAA 2016-2347

17. Schwabacher M (2005) Machine learning for rocket propulsion health monitoring. In: Proceedings of the SAE world aerospace congress, vol 114, issue 1, pp 1192–1197, Dallas, Texas, 2005-01-3370

18. Agogino A, Tumer K (2006) Entropy based anomaly detection applied to space shuttle main engines. In: Proceedings of the IEEE aerospace conference, Big Sky, MT

19. Sammut C, Webb GI (eds) (2011) Encyclopedia of machine learning. Springer, Heidelberg. https://doi.org/10.1007/978-0-387-30164-8

20. Lee W-M (2019) Python machine learning. Wiley Publications. ISBN: 978-1-119-54563-7

Clustering for Global and Local Outliers

Gouranga Duari and Rajeev Kumar

1 Introduction

Outlier detection is one of the fundamental steps in data science. Outlier identification is crucial in many areas such as health care [6], intrusion detection system [13], social network [22], and so on. A subtle concept of outlier detection is to estimate the likelihood density of unexpected input data objects, which are residing in low probability density areas. Although significant progress has been made in the domain of robust outlier detection in last few decades, it remains a challenge to identify hierarchical pattern spaces of outlier in high-dimensional data. Especially, when the dimensionality of the dataset becomes higher, it is harder to estimate the density in the original feature space, as any input data object can be a rare event with low probability to observe (Fig. 1).

Motivation to this work is from the work of Kumar and Rockett, who investigated a Learning-follows-decomposition (LFD) strategy [19] for hierarchical learning of pattern spaces using a multiobjective genetic algorithm followed by (near-) optimal learning of pattern subspaces. Their technique is a generic solution to complex high-dimensional problems where clusters are generated based on the fitness of purpose. In this paper, we use the decomposition strategy to find out the hierarchy of outliers so that we can have a better understanding of hierarchical outlier patterns at multiple levels. The clustering technique, which does not forcefully include all the data objects into the cluster, leads to the inference that a few excluded data objects are strong candidates to be outliers. So, hierarchical clusters are generated into multiple levels, which can be referred to as global-level and local-level of the outliers. This approach can be easily lead to sub-pattern spaces of outliers hierarchically.

G. Duari (✉) · R. Kumar
Data to Knowledge (D2K) Lab, School of Computer and Systems Sciences, Jawaharlal Nehru University, New Delhi 110 067, India
e-mail: gourangaduari5@gmail.com

© The Author(s), under exclusive license to Springer Nature Singapore Pte Ltd. 2023   601
D. S. Sisodia et al. (eds.), *Machine Intelligence Techniques for Data Analysis and Signal Processing*, Lecture Notes in Electrical Engineering 997,
https://doi.org/10.1007/978-981-99-0085-5_49

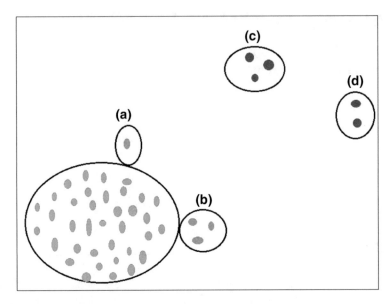

Fig. 1 olated data objects are at the global-level and these are the potential candidates to be outliers

The Learning-follows-decomposition (LFD) has the great advantage for decomposing high-dimensional pattern space into low-dimensions, this strategy can be used to decompose the hierarchical pattern space into more subspaces to find out isolated data objects from clusters. The isolated data objects are the likely candidates to be an outlier. So, we can decompose clusters in the hierarchy of outliers in a robust fashion so that we can have the generalized approach of a hierarchy of outliers. This approach of preprocessing the data helps to shrink the learning complexity and improve the classification of outliers drastically.

The rest of the paper is organized as follows. Section 2 provides the related work. Section 3 provides the problem formulation and Sect. 4 describes the experimental process. Section 5 includes results and analysis. Finally, we conclude the paper in Sect. 6.

2 Related Work

Tremendous efforts have been devoted to unsupervised outlier detection [24]. Here, the related work can be grouped as the following categories.

Aggarwal and Reddy [1] investigated many clustering-based outlier detection algorithms in the past. Guha et al. [10] proposed a hierarchical clustering-based model called CURE. In this method, a fixed number of data objects are generated by choosing well-scattered data objects from clusters, then a specified fraction of them are moved toward the center of a cluster. Here, more than one representative data object per cluster helps CURE to adjust the geometry of non-spherical shapes prudently; it also helps to reduce effects of outliers in a cluster. Karypis et al. [16] proposed a novel hierarchical clustering technique called CHAMELEON. In their work, the similarity of two clusters are measured based on a dynamic model. If the closeness and inter-connectivity between two clusters are high relative to the internal inter-connectivity of clusters and closeness of items within clusters, then two clusters are merged using a dynamic model. Kumar and Rockett [19] proposed the Learning-follows-decomposition strategy using a multiobjective genetic algorithm to partition pattern spaces into hyperspheres for subsequent mapping using a hierarchical neural network for subspace learning. This hierarchical learning is a generic solution for high-dimensional problems. This strategy decomposes a problem into a series of subproblems, then a set of function approximators are assigned to each subproblem such that each module learns to specialize in a subdomain.

Gan and Ng [9] designed an outlier detection method by introducing an iterative procedure using k-means clustering technique. They introduced an additional cluster for outliers through data clustering. Given the desired number of clusters k, this algorithm creates $k + 1$ clusters. As outliers are not compatible with these k clusters, a separate cluster is created for outliers. Chawla and Gionis [5] proposed a generalization of the K-means problem to simultaneously cluster data and discover outliers in an iterative approach to converge in a local optimum for outliers Zengyou et al. [12] proposed the Cluster-Based Local Outlier Factor (CBLOF). Here each object is assigned a degree of outlier factor by both sizes of the cluster the object belonging to and the distance between the object and its closest cluster. Jambudi and Gandhi [14] proposed k-means clustering-based algorithm to improve the performance of k-means clustering algorithm. In a two-part process, the initial clusters are broken down into sub-clusters based on criterion at the local-level, then clusters are merged according to the nearness criterion. In this work, clusters are formed by detecting and removing both global and local outliers simultaneously. The following approach converges to the optimal number of clusters automatically. Jiang and An [15] proposed the Clustering-Based Outlier Detection (CBOD), which is a two-stage process to detect outliers. In the first stage, dataset is divided into hyperspheres with almost the same radius using an one-pass clustering method. In the second stage, outlier factors are measured for all clusters and the clusters with high outlier factor are considered outliers. Dang et al. [7] proposed a well-known distance-based method, which is called k-nearest neighbor (kNN) by computing the distances between data objects. A data object that has a significantly higher distance value from its nearest neighbor based on some threshold, is regarded as an outlier. Knorr and Ng [17] proposed a distance-based method, where outliers are data objects that are far away from their nearest neighbors. An improved version of distance-based method proposed by

Ramaswamy et al. [21] following Knorr and Ng's work [17].

Gupta et al. [11] proposed a K-means clustering method for local search of out-liers. The goal is to cluster a set of data objects to minimize the variance of data objects, which are assigned to the same cluster with the freedom of ignoring a small set of data objects. These data objects can be labeled as outliers. Breunig et al. [2] designed an unsupervised outlier detection method for detecting the local outliers by introducing local outlier factors. Here, the local density deviation of a given data object is calculated for its neighbors, and a degree of being an outlier is assigned to each data object accordingly. This degree is referred to as the local outlier fac-tor (LOF) of a data object. So, a sample of data is considered as outliers, when it has a substantially lower density than their neighbors. An improved version of LOF [2] developed by Tang, which is known as Connective-based Outlier Factor (COF) [23]. Here, this approach differs from the LOF approach in the way of estimating the density of the data objects. The COF uses a chaining distance to estimate the local densities of the neighbors as the shortest path, while the LOF method uses the Euclidean distance by choosing the K-nearest neighbors for density estimation. Kriegel et al. [18] developed a more competent local density estimate approach for an outlier detection method called the Local Outlier Probabilities (LoOP) using a probabilistic oriented approach. LoOP tries to address the issue of LOF by introduc-ing an outlier score instead of an outlier probability.

3   Problem Formulation

In this section, we introduce our problem mathematically to tackle our problem in detail. We consider a dataset of n objects $X = \{x_1, x_2, \ldots, x_n\}$ having d numerical attributes. A global outlier is a data object which has a significantly large distance to its k-th nearest neighbor (usually greater than a global threshold) whereas a local outlier has a distance to its k-th neighbor that is large relative to the average distance of its neighbors to their own k-th nearest neighbors [8].

3.1   Hierarchical Clustering

Here, we have used hierarchical clustering; we consider two number of clusters. We use Euclidean distance function for hierarchical clustering between pairs of objects in X defined as: $d(x_i, x_j) = [\sum_{t=1}^{d} (x_{it} - x_{jt})^2]^{\frac{1}{2}}$. We choose to use Ward's linkage method as it measures the variance of clusters instead of measuring the distance directly.

The clustering technique [1] is used to group a homogeneous set of data objects so that similar data objects are in the same cluster. In this paper, we decompose outliers using hierarchical learning to understand the hierarchy of outliers. As the hierarchical clustering does not forcefully include data objects in clusters, so a separate cluster can be created for significantly isolated data objects (or, highly inconsistent data objects). These clusters create a hierarchy of outliers in two levels, which are referred to as global-level and local-level hierarchy for the outliers.

3.2 Isolation Forest

Isolated Forest (IForest) [20] isolates every single instance by creating a tree structure. Here, inconsistent data values are isolated nearer to the root of the tree and normal data objects are isolated at deeper end of the tree based on their susceptibility. As Isolation Forest is one of the well-known robust outlier detection techniques in the machine learning domain, we choose to use the isolation forest outlier detection technique to identify outliers in our experiment so that we can infer the global and local presence of outliers in the cluster hierarchy. It helps us to validate the hierarchy of local and global levels of outliers.

3.3 Algorithm

In this work, we adopt hierarchical clustering for detecting a hierarchy of outliers at two levels: global and local. We define an algorithm for our clustering approach for global outliers and local outliers. We describe below the algorithm for our approach.

Algorithm 1 Clustering For Global and Local Outliers Algorithm

---

Step-1: Perform hierarchical clustering,
Step-2: Clasify global and local region of outlier,
Step-3: Validate outliers using Isolation Forest, and
Step-4: Visualize the cluster with outlier hierarchy.

In our approach, hierarchical clustering decomposes data objects into two levels. Few data objects from the prime cluster are far away from the normal data distribution and they create a separate cluster of outliers. So, the prime cluster does not forcefully include the isolated data objects and these isolated data objects are more likely to be outliers. We consider these outliers as a potential global outlier. As the rest of the outliers are sharing the same space with the normal data distribution in the prime cluster, we consider those outliers as local outliers by definition [8]. This hierarchical decomposition approach successfully separates the potential global outliers in one cluster, and thus, this approach creates a hierarchy of outliers in two levels. Here, we validate outliers by isolation forest detection technique for better generalization of our approach.

4 Experiments

We have validated our approach on a real dataset. Here, we describe about the dataset and technical configuration for our experiment.

4.1 Dataset

We use publicly available benchmark datasets to demonstrate the effectiveness and competency of our approach. In this paper, we include the results of a single dataset due to paucity of space. The dataset is the PageBlocks dataset[1] [3]. This dataset, taken from UCI ML repository, has been widely used by other researchers. It contains 10 numerical attributes with 5473 objects. This dataset has 4912 (89.76%) normal objects and 560 (10.23%) outlier objects.

4.2 Experimental Setup

We run the experiments[2] on windows 10 operating system. All the algorithms are implemented on Jupiter notebook in Python programming language. We use *sci-kit learn* library for standardized the data using normalization method to make our model computation easier. We use *sci-kit learn* library for implementing hierarchical clustering and *pyod* [25] library for isolation forest outlier detection.

5 Results and Analysis

We decode the results to infer conclusions. As every outlier detection technique can have less than the expected performance due to noise [4], we will take this norm as an assumption that this dataset might have noise. The isolation forest outlier detection technique identifies 548 data objects as outliers in the dataset. As we compared Isolation Forest (IForest) with other benchmark outlier detection techniques in terms

---

[1] https://archive.ics.uci.edu/ml/datasets/Page+Blocks+Classification.

[2] https://github.com/gourangaduari1995/Clustering-for-global-and-local-outlier.

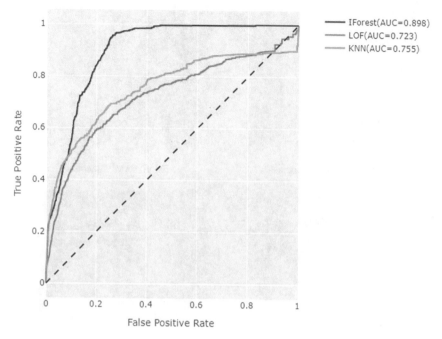

Fig. 2 ROC curve for outlier detection

of ROC curve we can see that isolation Forest is more efficient in this dataset to identify outliers than other methods. The above comparison gives us a strong reason for choosing the Isolation Forest technique in our case. Figure 2 supports the above reasons.

Here, we decipher our core concept through the number of outliers detected in each cluster and 2-D visualization. Here, we take 5 attributes out of 10 numerical attributes to visualize outliers, which are flagged by hierarchical clusters in a 2-D space. We consider the following attributes for visualization: Height, Area, Blackpix, Blackand, Wb-trans. We take four different graphs separately using five attributes which are mentioned above so that we can have a holistic view of outliers in hierarchy in different dimensions. Figure 3 shows all the data objects in the dataset. Here, dark orange diamonds are identified outliers in the global cluster. Blue diamonds are the identified outliers in the local cluster and blue circles are the identified inliers in the local cluster. Here, we can categorically visualize the potential global outliers and the local outliers and they are forming a hierarchy of outliers in two levels.

We have implemented hierarchical clustering to create two clusters to check the levels of an outlier. We can see that cluster-B has 20 isolated data objects and these are validated as outliers by isolation forest without any single inlier. These outliers are far away from the inliers in the dataset, which makes sense that these data objects

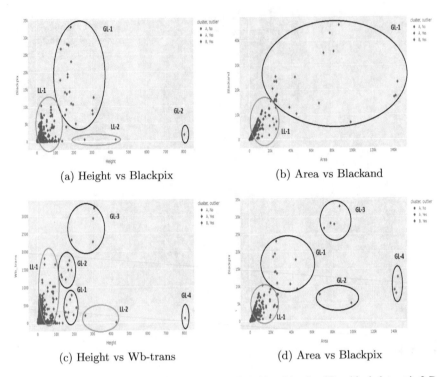

(a) Height vs Blackpix   (b) Area vs Blackand

(c) Height vs Wb-trans   (d) Area vs Blackpix

Fig. 3 Visualization of hierarchy of outliers at global- and local-levels of Pageblock dataset in 2-D space

Table 1 Hierarchy of outliers validated by Isolation Forest

Cluster	Inliers	Outliers	Global outlier (%)	Local outlier (%)
Cluster-A	4924	528	0	96.5
Cluster-B	0	**20**	**3.5**	0

are the global outliers in nature. Cluster-A has 5452 data objects. Cluster-A has 528 outliers and 4924 inliers. The outliers in cluster-A are very close to the inliers in the dataset, which gives us a sense that these data objects are local outliers.

Here, Table 1 describes the number of inliers and outliers in two clusters. It also tells us about the percentage of global and local outliers in these two clusters. Our approach enables to find a significant number of strong candidates for outliers in hierarchy. So, our approach gives us a promising alternative to find out more hierarchy of outliers.

5.1 Discussion

In 2-D visualization of outliers, we manually draw the global-level and local-level of outliers with acronym GL and LL respectively using the concept of the Learning-follows-decomposition (LFD) strategy [19]. From these results, we observe that two main categories of hierarchical clusters emerge: (a) local cluster of outliers, and (b) global cluster of outliers. Second category of cluster does not require any further classification or decomposition as it is categorically come out as potentially strong candidates to be outliers. In this approach, we have achieved to detect 3.6% of total potential outliers and this is a very significant number in terms of effectiveness. First category of cluster, which has inherent complex patterns due to noise in the dataset, is required to decompose further for micro-level analysis to understand a hierarchy of outlier patterns so that we can leverage on detection of outliers from inliers by avoiding noise-related factors.

Though global and local outliers are categorically divided with two separate sub-space as per norm but a few data objects of local outliers are in the region of global outliers. Possible reason could be noise. We can extend our research further to better understanding of such unexpected behavior.

6 Conclusion

In this paper, we have proposed clustering for global and local outlier detection to identify the hierarchy of outliers. This approach consists of two major parts: first is the detection of likely candidates at multiple levels using hierarchical learning followed by validation of the presence of likely outliers in the hierarchy using isolation forest technique. We have seen that there is a hierarchy in outliers and the excluded data objects from the prime cluster are the potential candidates to be outliers. So, we will extend our approach further to understand the hierarchy of outliers at a micro-levels how we can choose potential outlier detection techniques for each level so that we can achieve effective with reduced error rate.

References

1. Aggarwal CC, Reddy CK (2014) Data clustering. Algorithms and applications. Data mining and knowledge discovery series
2. Breunig MM, Kriegel H-P, Ng RT, Sander J (2000) LOF: Identifying density-based local outliers. In: Proceedings of the ACM international conference on management of data (SIGMOD), pp 93–104

3. Campos GO, Zimek A, Sander J, Campello RJGB, Micenková B, Schubert E, Assent I, Houle ME (2016) On the evaluation of unsupervised outlier detection: measures, datasets, and an empirical study. Data Mining Knowl Discovery 30(4):891–927
4. Cao J, Kwong S, Wang R (2012) A noise-detection based AdaBoost algorithm for mislabeled data. Pattern Recogn 45(12):4451–4465
5. Chawla S, Gionis A (2013) k-means: a unified approach to clustering and outlier detection. In: Proceedings of SIAM international conference on data mining. SIAM, pp 189–197
6. Christy A, Meera Gandhi G, Vaithyasubramanian S (2015) Cluster based outlier detection algorithm for healthcare data. Proc Comput Sci 50:209–215
7. Dang TT, Ngan HYT, Liu W (2015) Distance-based k-nearest neighbors outlier detection method in large-scale traffic data. In Proceedings of IEEE international conference on digital signal processing. IEEE, pp 507–510
8. De Vries T, Chawla S, Houle ME (2020) Finding local anomalies in very high dimensional space. In: Proceedings of IEEE international conference on data mining. IEEE, pp 128–137
9. Gan G, Kwok-Po Ng M (2017) k-means clustering with outlier removal. Pattern Recogn Lett 90:8–14
10. Guha S, Rastogi R, Shim K (1998) CURE: an efficient clustering algorithm for large databases. ACM Sigmod Rec 27(2):73–84
11. Gupta S, Kumar R, Kefu L, Moseley B, Vassilvitskii S (2017) Local search methods for k-means with outliers. Proc VLDB Endowment 10(7):757–768
12. He Z, Xu X, Deng S (2003) Discovering cluster-based local outliers. Pattern Recogn Lett 24(9–10):1641–1650
13. Jabez J, Muthukumar B (2015) Anomaly detection using outlier detection approach Intrusion detection system (IDS). Proc Comput Sci 48:338–346
14. Jambudi T, Gandhi S (2019) A new k-means based algorithm for automatic clustering and outlier discovery. In: Information and communication technology for intelligent systems. Springer, Heidelberg, pp 457–467
15. Jiang S, An Q (2008) Clustering-based outlier detection method. In: 2008 fifth international conference on fuzzy systems and knowledge discovery, vol 2. IEEE, pp 429–433
16. Karypis G, Han E, Kumar V (1999) A hierarchical clustering algorithm using dynamic modeling
17. Knorr EM, Ng RT (1997) Properties and computation. A unified notion of outliers. In: Proceedings of KDD, vol 97, pp 219–222
18. Kriegel H-P, Kröger P, Schubert E, Zimek A (2009) LoOP: local outlier probabilities. In: Proceedings of the 18th ACM conference on information and knowledge management, pp 1649–1652
19. Kumar R, Rockett P (1998) Multiobjective genetic algorithm partitioning for hierarchical learning of high-dimensional pattern spaces: a learning-follows-decomposition strategy. IEEE Trans Neural Netw 9(5):822–830
20. Liu FT, Ting KM, Zhou Z-H (2008) Isolation forest. In: Proceedings of 8th IEEE international conference on data mining. IEEE, pp 413–422
21. Ramaswamy S, Rastogi R, Shim K (2000) Efficient algorithms for mining outliers from large data sets. In: Proceedings of ACM international conference on management of data (SIGMOD), pp 427–438
22. Savage D, Zhang X, Yu X, Chou P, Wang Q (2014) Anomaly detection in online social networks. Soc Netw 39:62–70
23. Tang J, Chen Z, Fu AWC, Cheung DW (2002) Enhancing effectiveness of outlier detections for low density patterns. In: Proceedings of Pacific-Asia conference on knowledge discovery and data mining. Springer, Heidelberg, pp 535–548
24. Wang H, Bah MJ, Hammad M (2019) Progress in outlier detection techniques: a survey. IEEE Access 7:107964–108000
25. Zhao Y, Nasrullah Z, Li Z (2019) Pyod: a python toolbox for scalable outlier detection. arXiv preprint arXiv:1901.01588

Smart Timer Module for Handheld Demolition Tools

Mohmad Umair Bagali, S. Rahul, and N. Thangadurai

1 Introduction

An electromechanical tool for chipping and breaking concrete, bricks, and other hard materials is a demolition hammer. These machines are used in the construction industry to do demolition tasks [1]. When hiring demolition services, this hammer will be used, and the hammer operators will be paid on an hourly basis; additionally, the present demolition service costs roughly $300 per hour (in the year 2021). There is no gadget installed to monitor the activities of the tool currently in use. The above scenario is fantastic until the hammer users, i.e., the workers take advantage of the situation and cause the task to be delayed, demanding payment for the time they were idle. The construction process will be further slowed because of this. As a result, by rewarding demolition hammer operators for the time they worked, i.e., the time they operated the demolition hammer, a correct time recording device will aid in boosting the efficiency of the task done.

When the demolition hammer is in use, it vibrates [2–4], and we used this mechanism to determine if the hammer is in use or not. The suggested device records the length of the demolition hammer's operation while it is in use. When the demolition hammer is not in use, the suggested device automatically stops the timer for the duration of the break, and the timer resumes from that point when the hammer is used again after the break. The proposed gadget will include a simple user interface with an LCD screen, as well as the ability to save a limited number of previous time records. One will be able to calculate the real time spent on a task and demand

M. U. Bagali (✉) · S. Rahul
Department of Electronics and Communication Engineering, Faculty of Engineering and Technology, JAIN (Deemed-to-be University), Bengaluru, Karnataka, India
e-mail: umair037@gmail.com

N. Thangadurai
Sankalchand Patel University, Ambaji-Gandhinagar State Highway, Visnagar, Mehsana District, Gujarat, India

© The Author(s), under exclusive license to Springer Nature Singapore Pte Ltd. 2023　　611
D. S. Sisodia et al. (eds.), *Machine Intelligence Techniques for Data Analysis and Signal Processing*, Lecture Notes in Electrical Engineering 997,
https://doi.org/10.1007/978-981-99-0085-5_50

Fig. 1 Construction worker using demolition hammer. *Source* https://upload.wik imedia.org/wikipedia/com mons/2/25/Colombia_Jackha mmer_01.jpg

payment using the proposed device. We also believe that using such technology into demolition tasks will boost worker productivity and morale. Client satisfaction with those services will improve as a result, and the construction process will be speed up. Figure 1 shows a concrete structure being demolished using a demolition hammer.

Apart from the above-mentioned use case, the suggested device can be used to record time in any similar situation where vibration can be used as a parameter to record time for any other reason.

1.1 Working of Demolition Hammers

Demolition hammers are divided into two categories. One used an electric motor as its central component, while the other relied on pneumatics. Figure 2 depicts the demolition hammer's basic structure. The strike is delivered by the drill bit, which breaks the buildings that must be demolished. The primary drill bit is actuated by a suitable mechanism. The drill bit's movement is depicted in Fig. 2 by the red arrow. A chisel is a term used to describe a drill bit. An electric motor is at the heart of most current lightweight, extremely portable instruments. A similar system controls the main drill bit, as seen in Fig. 3. The motor as the core is powered by electricity. The shaft of this motor is coupled to a wheel-like device that has a physical contact with the drill bit that will impact the surface to be demolished on one end. The drill bit receives and converts the rotation produced by the motor. The direct contact with the drill bit causes the required movement as the wheel turns. A pulley system is utilized to drive the drill bit in a similar arrangement [3].

Fig. 2 Body of demolition hammer

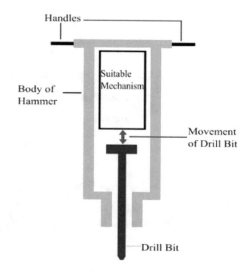

Fig. 3 Driving mechanism of electric motor

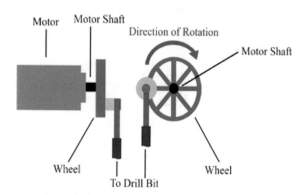

When more powerful impacts are necessary, pneumatically operated tools are available. These rely on air pressure to function. This necessitates the use of an air chamber, as shown in Fig. 4. An inlet and an outlet are available. The drill bit is first pushed downward by compressed air pumped into the air chamber. The inlet is then closed, and the outlet is opened, allowing the compressed air in the chamber to escape and mix with the surrounding air. The opening and closing of the inlet and outlet are controlled by an appropriate mechanism, as shown in Fig. 4.

While the demolition hammer is an excellent tool, there is a problem with workers delaying work for a variety of reasons while we are using demolition services. As a result of our analysis of the issue, we concluded that some type of monitoring activity is required to improve the efficiency and work ethics of construction employees. Also, the tool vibrates at about 22 Hz while in operation [2]. Some high-frequency tools vibrate up to 80 Hz [3].

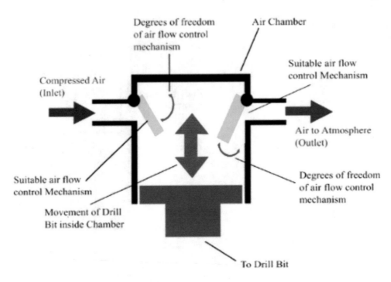

Fig. 4 Air chamber and its operation

1.2 Objective

The proposed add-on gadget's main goal is to detect vibrations and record the length of time the device vibrates.

1.3 Methodology

We used the MPU6050 and the ATMEGA328P microcontroller [5] from Atmel to achieve the objective defined in Sect. 1.2. Instead of using a general vibration sensor, the MPU6050 was chosen to make the device more robust and adaptable in terms of operation, as well as to achieve precision in time recoding.

A 16*2 character LCD was used to display information and provide a simple user interface. We utilized four pushbuttons to make the user interface basic and accessible to everyone, regardless of literacy levels: "START," "STOP," "NEXT," and "PREV." When the time device is turned on, you only need to press three buttons: START, STOP, and NEXT to start a fresh recording, stop time recording, and return to the main menu. The device's operation is simplified using an LED and a buzzer, which eliminates the requirement for the user to glance at the LCD screen to determine the device's state. As previously said, making the gadget battery operated can make it universal and compatible with all types of demolition hammers on the market, regardless of the manufacturer or model. When a user wants to refer to past values for paying, logging data elsewhere, and so on, the ability to save and retrieve a few values comes in handy. In the event of a battery failure, the recorded values will be saved in the EEPROM of the ATMEGA328P microcontroller and can be retrieved, ensuring that workers' efforts are not wasted due to battery failure or battery depletion.

As a result, Fig. 5 summarizes the device's whole operation sequence. P_n refers to the present archived value (the time duration values of earlier recordings) being displayed on the LCD in Fig. 5. When the user presses the NEXT button, $P_{(n-1)}$ refers to the next archived value that will be presented. The Arduino IDE and Arduino programming language, which is an API for Arduino, is used to program the microcontroller [6].

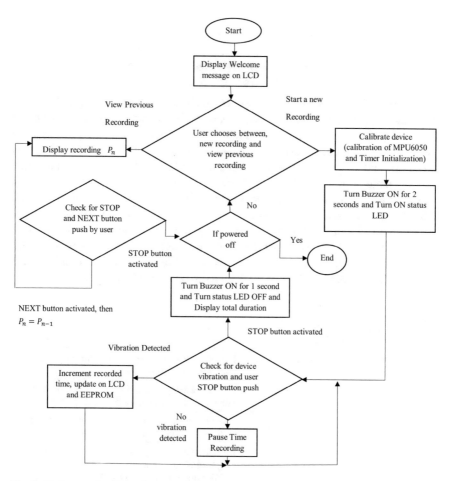

Fig. 5 Basic operation flow of proposed device

2 Literature Survey

The demolition hammer is a handheld power instrument that is mostly used in construction operations to demolish hard constructions [1]. Even though there are small changes between "demolition hammer" and "jackhammer," they both work on the same idea at their heart. The cost of this equipment varies depending on the manufacturer and specs; it ranges from Rs 6000 to Rs 150,000. There are two types of pneumatic hammers: one that uses compressed air (pneumatics) and one that uses electricity.

2.1 Early Developments in Demolition Hammer

Power-driven jackhammers have been around since at least 1849 [7]. However, there has always been a demand for a demolition instrument. A long metal rod was employed by ancient civilizations to remove undesired rock; today, a chisel and hammer are used to carve out unwanted rock on a smaller scale. Then, as technology advanced, massive drilling machines became available. These, on the other hand, were too large to be transported and operated manually. New inventions in this field were spurred by the requirement for compactness and portability. In the year 1952, O.S Coates patented one such instrument [8]. This idea entailed a device for translating rotational motion into the physical strike required to destroy hard objects. A similar technique is presently being employed in new demolition tools [9]. Pneumatics were also employed in early jackhammers; therefore, demolition hammers based on this concept exist [7]. Table 1 summarizes this information.

Table 1 Evolution of demolition hammer

Technology	Period of use	Description
Human strength	Before eighteenth century	People relied on chisels and hammers for the demolition of structures
Pneumatics	Nineteenth century	Compressed gas played an important role in the operation of such demolition tools [7]
Power driven	Twentieth century	Advancements in electrical devices, like motors, helped in driving the demolition hammer [8, 9]
Power driven	Present	While being mostly power driven, using more sophisticated techniques helped in improving the quality, safety of the device [3, 10, 11]

2.2 Electronic Design Reference

Arduino is powered by ATMEGA 328P microcontroller which can be interfaced with sensors to obtain data and process it [6]. Yusuf Abdullahi in his publication [6] provides example on sensor alarm, explaining programming digital input/output. He also explains analog input for measuring capacitance using ADC channels.

Goyal and Pabla [4] present their signal processing methods for vibration signals in their publication. Their research is like the application we intend to achieve in our gadget. Mechanical tools and machines create vibrations, and this acts to determine whether the demolition tools are being used or not.

Vasilyev et al. [4] in their publication discuss importance of adapting certain design techniques to reduce the probability of manufacturing defects in PCB by analyzing the way PCBs are fabricated. They propose these design techniques by analyzing the PCB fabrication process using mathematical models. By varying parameters suitably in those models, valuable insights can be obtained, and PCB can be designed to reduce defects at fabrication.

Al-Dahan et al. [5] in their publication discuss use of MPU6050 sensor along with Arduino to detect elderly people falling on ground. This is achieved by detecting fall by fixing a threshold acceleration value. Similar mechanism can be used in our gadget to determine vibrations.

3 Implementation

3.1 Timer Configuration and Operation

To set the 16-bit timer module of the controller, we manipulate the values in the timer's register to make the counter raise an interrupt after each second, i.e., when the counter overflows exactly at each second by picking a suitable pre-scaler value. Table 2 lists the setups that are necessary. Table 2 lists the configurations that must be completed in the same order. Table 2 value column is in hexadecimal.

Table 2 Required timer register configurations [12]

Register	Value	Description
TCCR1B	0×00	Clearing bits to clear garbage values in registers
TCCR1A	0×00	Clearing bits to clear garbage values in registers
TCCR1B	0×0D	Starting timer with pre-scalar as 1024 and WGM12
TCNT1	0×00	Initialize counter to 0
OCR1A	0×3D09	To generate an exact 1-s interrupt
TIMSK1	0×02	Enable timer interrupt

When vibrations are detected, increment the "secondRegister" for each interrupt. As a result, we get an interrupt every second, and the secondRegister is increased, and the entire time duration grows only when vibration is detected. At the needed instance, you can get that time value by accessing secondRegister. While the demolition hammer is in operation, the time value is retrieved every second, updating the count displayed on the LCD display and the time value in EEPROM at the same time. The 16-bit register (integer size) [13] allows you to store values up to 2^16 − 1 = 65,535, which is 65536 s and equals 18 h. And most demolition services will be completed within this time frame or can be restarted the next day when a new recording can begin.

3.2 MPU6050 Configuration and Operation

Configuring MPU6050 was done by referring to the register map provided by the manufacturer [14]. First, initialize I2C communication as a master device [15]. We used "Wire.h" library of Arduino to achieve I2C communication. Initially, the MPU6050's slave I2C address is passed as an argument to Wire.beginTransmission() method to initialize I2C communication as master [13, 15]. Then, configure MPU6050 sensors for normal power mode and ±4 g sensitivity range [15]. After the configuration is over, end the transmission using Wire.endTransmission() method.

We can get the acceleration, or vibration, on each axis after the MPU6050 sensor is initialized. The MPU6050 has an accelerometer and a gyroscope, but we don't use the gyroscope in this application. The acceleration data, according to the datasheet, will be saved in a set of six registers, numbered $0 \times 3B$ to 0×40 inside the sensor, as given in Table 3 [14]. Each axis contains two 8-bit registers, giving each axis a total of 16 bits for its acceleration value. The data stored in these registers can be read using the I2C bus.

However, these registers give us the raw data obtained directly from the micro-electrical mechanical system (MEMS) sensor after signal conditioning. We need to process that data to obtain the "g" value on the sensor. According to the datasheet for the range of ±4 g, the value obtained from the sensor is divided by 8192.0 to get the "g" value. Now we have our acceleration or "g" values in our controller, to detect the

Register	Bits[7:0]
$0 \times 3B$	ACCEL_XOUT[15:08]
$0 \times 3C$	ACCEL_XOUT[07:00]
$0 \times 3D$	ACCEL_YOUT[15:08]
$0 \times 3E$	ACCEL_YOUT[07:00]
$0 \times 3F$	ACCEL_ZOUT[15:08]
0×40	ACCEL_ZOUT[07:00]

Table 3 Register values from MPU6050 register map [14]

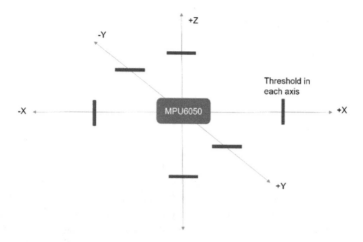

Fig. 6 Acceleration thresholds on each axis

vibrations we define certain thresholds in the X-, Y-, Z-axis which is experimentally derived, this is shown in Fig. 6. Whenever the acceleration values go beyond that threshold on each axis, the microcontroller decides whether to increment time or to pause the operation. After each iteration, the values of threshold are dynamically changed, so that even when the device is placed in improper orientation and the device is not in operation, time recording remains paused. Hence, the decision to increment time in the register is obtained by evaluating the following code,

```
if (((AccX − x) > MAX_X) | ((AccY − y) > MAX_Y) | ((AccZ − z) > MAX_Z)) {
x = AccX;
y = AccY;
z = AccZ;
*doesVibratePtr = true;
} else {
x = AccX;
y = AccY;
z = AccZ;
*doesVibratePtr = false;
}
```

Where MAX_X, MAX_Y, MAX_Z are constant values which were the experimentally determined threshold in each axis. AccX, AccY, AccZ correspond to acceleration values in the X-, Y-, and Z-axis, respectively. And x, y, z are the variables holding the previous iteration's acceleration values. Also, all the acceleration values are taken as absolute value of the acceleration values so that the orientation of the device does not affect device operation. After a lot of trial and error, the threshold was set at ±0.4 g. This range of threshold values allows for the elimination of any erroneous movements, making the gadget's observations more accurate. In addition, the MPU6050 sensor includes built-in support for 4 g acceleration limits, which aids in collecting vibrations with sufficient accuracy and resolution.

3.3 User Interface

For a simple and accessible user interface, we used four pushbuttons each with a specific function, a 16*2 LCD character display, a LED to indicate whether the device is recording time, and a buzzer to indicate user the device has started time recording. The basic operation flow of the device is mentioned in Fig. 5. The LCD was initialized and operated in 4-bit mode. The status LED is ON when the time of operation is being recorded. The buzzer is made to play a continuous tone for 2 s when the time recording started and will play a continuous tone for 1 s while the user stops the time recording by activating the STOP button. By using this combination of buzzer and status LED, the user can be aware of the device state, without having to look at the LCD. This will help the user to operate the device independent of the user's literacy level. Users with low literacy levels can be trained to use the device's combination of key activations for each operation. But at the end of the recording, when the total time duration of time recording is needed the user must use the LCD to determine the total duration. Everywhere time is displayed, HH:MM:SS format is used.

3.4 EEPROM Operations

EEPROM is incorporated into the ATMEGA328P. This non-volatile memory is important for keeping the values of the most recent recordings, as shown in the features list. The data is stored in 8-bit chucks in the 1 KB EEPROM. However, because the number we use is 16 bits long, the data is split into two portions, each of which is 1 byte (8 bits). The first half is alternately stored at even and odd EEPROM addresses. In the same method, the saved information is retrieved. The data is multi-plied with 0×FF00 and 0×00FF, respectively, to split it into higher and lower 8 bits, and then the necessary shifting operations are performed. Then for retrieving the data both multiplied with 0xFFFF and OR-ed with each other [12, 13].

3.5 Circuits and PCB

The circuit's design included a power supply, LCD mount, and MPU6050 mount, as well as a reset button, buzzer and LED unit, oscillator unit, and pushbuttons. The

Fig. 7 Power supply unit

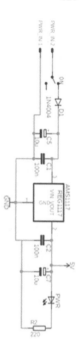

microcontroller is connected to these components. Autodesk EAGLE was used to design the schematic and PCB. From Figs. 7, 8, 9, 10, 11, 12, 13 and 14, the schematic of the proposed device is shown below. These schematics reflect the schematics of various units of the proposed device. Further details about each unit are given in Table 4.

4 Hardware and Software Tools Used

The following hardware and software tools were involved in the development of the proposed device are ATMEGA328P microcontroller, MPU6050 accelerometer cum gyroscope, 16*2 LCD, active buzzer, LEDs, push buttons, and other passive components, Arduino IDE, Autodesk EAGLE.

5 Result

The proposed gadget was constructed and tested, with the following observations. The finished gadget is depicted in Fig. 15. As shown in Fig. 5, various stages of device operation were reached. The gadget has a resolution of ± 0.4 g. The accuracy is determined by the vibration intensity, and because demolition equipment constantly

Fig. 8 User interface push buttons

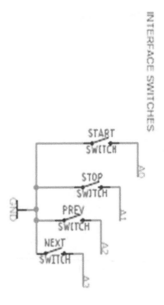

Fig. 9 Microcontroller unit

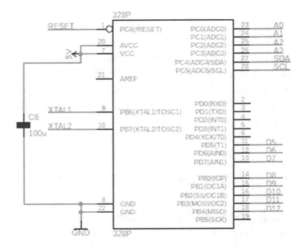

vibrate at a high intensity, low frequency, the overall accuracy will always be greater than 98%.

5.1 Boot Up

The device is powered by a 9 V battery that is linked to the device's Power IN connection. A sliding switch is used to turn on the device. The gadget also has a

Fig. 10 Oscillator unit

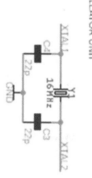

Fig. 11 LCD mount

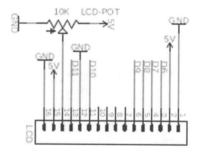

Fig. 12 MPU6050 mount

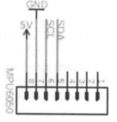

Fig. 13 Status LED and buzzer unit

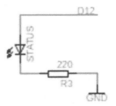

Fig. 14 Reset switch

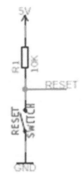

Table 4 Schematic description

Figure number	Description
Figure 7	A power supply unit with 9 V input and 5 V output using the AMS1117 regulator [16]
Figure 8	User interface pushbuttons and their connection to the microcontroller
Figure 9	Microcontroller unit with its interfaces and 100uF filtering capacitor
Figure 10	16 MHz crystal oscillator unit with 22pF capacitors at each terminal of the oscillator
Figure 11	LCD mount with a potentiometer for contrast adjustment (LCD is used in 4-bit mode) [13]
Figure 12	MPU6050 accelerometer mount with I2C interface and power pins
Figure 13	Status LED and buzzer unit for the simplified user interface
Figure 14	Reset unit in pull-up configuration

polarity safe diode, which protects the device from damage if the polarity is mistakenly reversed at Power IN. After that, an AMS1117 voltage regulator is used to lower the voltage from 9 to 5 V. However, the voltage regulator's real output voltage was discovered to be 5.07 V. Due to the existence of a tantalum capacitor of 10 F and a 100 nF ceramic capacitor connected in parallel at the input and output of the voltage regulator, the voltage is also filtered (Refer Fig. 7). This also aids in the filtering of high-frequency noise on power rails caused by the microcontroller's high-frequency

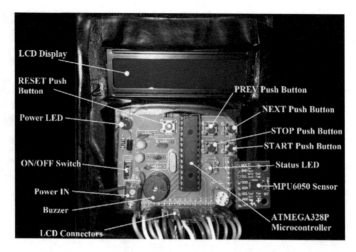

Fig. 15 Resultant gadget

Fig. 16 Device boot up and welcome message

operation. After the microcontroller is turned on, it goes through a power-up reset, and our flash program begins to run. The LCD then displays the welcome message. When the device is turned on, the power LED illuminates and remains illuminated until the device is turned off. Figure 16 depicts the above procedure.

5.2 Selection

Following the welcome message, the device advances to the following step, which is the main menu. Figure 17 depicts this. The user can select between starting a fresh recording and seeing past data at this point. The user can choose among the options

Fig. 17 Main selection
screen

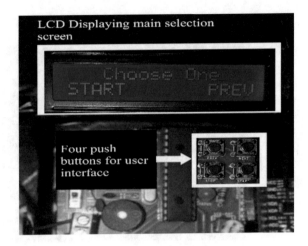

by pressing the START and PREV buttons. When the user presses START, the device
begins to calibrate itself. When the user presses the PREV button, the user is taken
to the previous values screen. To develop a user-friendly interface, only one button
activation is necessary to begin recording from the time the device is turned on. The
user returns to this screen at the end of each session. At this point, it is safe to turn
the gadget off using the sliding switch.

5.3 Time Recording

When the user wants to start a fresh recording, the gadget enters this state. When the
user presses the START button to begin a new recording, the device calibrates first.
This calibration comprises the timer initialization described in Table 2 as well as the
MPU6050 calibration described in Sect. 3.2 The buzzer is turned on for 2 s while
the gadget calibrates, informing the user that the device has begun to record time.
In addition, upon the conclusion of calibration, the green status LED illuminates.
Figure 18 shows the LCD message "Calibrating... Do not move" when calibration
is in progress.

The time elapsed is continuously updated in a periodic manner during the time
recording. When the interrupt is raised by the timer, which is a different peripheral,
the main infinite loop is designed to check for vibration using a function named
checkVibration(), which then updates a variable in the global scope. This variable is
only accessed through a pointer established for it to avoid any accidental bugs in the
code. The time register carrying the complete time value is incremented only when
the MPU6050 sensor records vibration. The EEPROM and LCD are updated when
the time is recorded. The time on the LCD is displayed in the format HH:MM:SS.
The full period is recorded in seconds. The second's value is translated to correspond

Fig. 18 Device is being calibrated

hour, minute, and second values only when it is to be updated on LCD. Figure 19 depicts the device's time recording.

The user can stop the recording after the demolition operation is completed by pressing the STOP button. The buzzer will play a tone for 1 s and the Status LED will turn off to signify to the user that the recording has been stopped. The user is then led to the overall recording summary, which displays the total recording duration, as illustrated in Fig. 20. After noting the total length, the user can proceed by pressing the NEXT button. Following that, the user is returned to the selection screen, as seen in Fig. 17.

Fig. 19 Device showing time recording

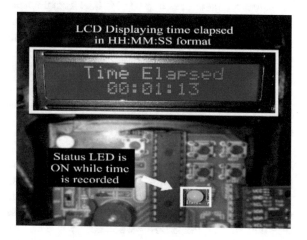

Fig. 20 Time recording completed

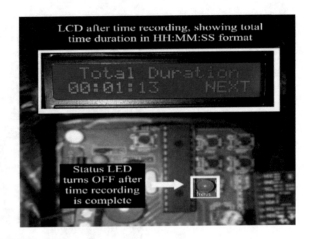

5.4 *Previous Values*

Activating PREV in the main selection menu, as shown in Fig. 17, allows the user to see past settings. EEPROM is used to retrieve previous values. Because the values are updated in EEPROM when operating time is recorded at the time of recording, the values are preserved on EEPROM even if the battery fails or the battery voltage dips. The latest recording duration is displayed when the user accesses the previous value screen. Because the device only keeps 5 records, this is designated as "P5," where P stands for previous and 5 stands for newest. When the user selects NEXT, the most recent value "P4" is displayed. The user can press STOP to exit the application and return to the main menu. Because the time values are recorded and kept in seconds, the values are transformed to HH:MM:SS format when retrieving earlier recordings. Figure 21 depicts the gadget that displays the previous value.

Based on the foregoing findings, we concluded that the device delivered the intended outcomes and could satisfy the minimal needed features as indicated in Sect. 1.2.

5.5 *Error Analysis*

There were few flaws discovered throughout extensive testing of the device. The device's maximum recording duration restriction was discovered to be 18 h, with the source of this being the size of an integer in Arduino's compiler. The maximum operational time of the device can be improved by selecting an appropriate data type and altering the EEPROM procedures to adapt to the new data type. Because the device is battery-operated, there was another limitation. The voltage across the potentiometer that controls the LCD's contrast decreases over time. As a result, the user must frequently adjust the contrast. In this situation, we may alter the contrast

Fig. 21 Previous value screen

of the LCD display using PWM rather than the potentiometer. The microcontroller itself can create these PWM waves.

The only possible false positives are when workers move the demolition tool around during breaks, as we're looking for accelerations and deaccelerations. Because demolition services are priced by the hour, these false positives amount for only 10–20 s of inaccuracy and can be ignored. Due to the write time involved in EEPROM write operations, there is also an obvious delay in the timer count being updated.

6   Conclusion

We presented our solution to the challenge indicated in the abstract in this publication. The following can be deduced based on the findings. The "smart timer module for handheld demolition tools" is a proposed gadget that can record the operating time of any handheld demolition instrument it is attached on. The suggested gadget may correctly record the working time of construction employees while they are demolishing structures, ensuring that they are making efficient use of their time and that subsequent construction work is not delayed. The proposed gadget can also be employed in circumstances where the vibrating time of any object needs to be measured. Also, because we have data on the tool's working hours, we can simply schedule demolition tool maintenance, and this data will come in handy when warranty claims, reports, and other documentation are required.

References

1. Portioli F, Unureanu V (2008) Demolition and deconstruction of building structures. In: Sustainability of constructions-integrated approach to life-time structural engineering proceedings of seminar, pp 4.124–4.131
2. Maciej T, Małgorzata W (2016) Analysis of vibration transmission in an air-operated demolition hammer. Vibrations Phys Syst 27:385–390
3. Stephen M (2019) Design and testing of a high frequency hydraulic mechanical Jackhammer. MSc thesis, McGill University, Quebec, Canada
4. Goyal D, Pabla BS (2016) The vibration monitoring methods and signal processing techniques for structural health monitoring: a review. Arch Comput Methods Eng 23:585–594
5. Al-Dahan Z, Bachache N, Bachache L (2016) Design and implementation of fall detection system using MPU6050 Arduino. In: International conference on smart homes and health telematics, pp 180–187
6. Abdullahi Y (2014) The working principle of an Arduino. In: 2014 11th international conference on electronics, computer and computation (ICECCO), pp 1–4
7. The History of Jackhammer. https://www.onestoprent.com/the-history-of-the-jackhammer/. Last accessed 21 Aug 2021
8. Coates OS (1952) Power-driven Jackhammer. US Patent no: 2597292
9. Osama A, Anil S, Hossam E, David B (1998) Concrete bridge demolition methods and equipment. J Bridg Eng 3:117–125
10. Hashimoto K, Shiratani M, Yokoyama M, Okada Y (2006) Hammer drill. US Patent no: 6988563
11. Frauhammer K, Schnerring H, Braun W, Kuhnle A (2007) Hand power tool in particular drill hammer and/or Jackhammer. US Patent no: 7303026
12. Atmega.48A/PA/88A/PA/168A/PA/328/P megaAVR® Data Sheet. https://ww1.microchip.com/downloads/en/DeviceDoc/ATmega48APA88APA168APA328P-SiliConErrataClarif-DS80000855A.pdf. Last accessed 12 May 2021
13. Arduino Reference. https://www.arduino.cc/reference/en/. Last accessed 20 May 2021
14. MPU-6000 and MPU-6050 Register Map and Descriptions Revision 4.2. https://invensense.tdk.com/wp-content/uploads/2015/02/MPU-6000-Register-Map1.pdf. Last accessed 23 May 2021
15. MPU-6000 and MPU-6050 Product Specification Revision 3.4. https://invensense.tdk.com/wp-content/uploads/2015/02/MPU-6000-Datasheet1.pdf. Last accessed 23 May 2021
16. AMS1117 Datasheet. https://robu.in/wp-content/uploads/2018/12/AMS1117-DATSHEET.pdf. Last accessed 01 June 2021
17. Vasilyev F, Isaev V, Korokov M (2021) The influence of the PCB design and the process of their manufacturing on the possibility of a defect-free production. Przegląd Elektrotechniczny 1(3):93–98

Dynamic Hand Gesture Recognition Using mYOLO-CSRT and HGCNN for Human–Machine Interaction

Manoj Kumar Sain, Joyeeta Singha, and Vishwanath Bijalwan

1 Introduction

Computer vision and image processing are related to artificial intelligence, making the full meaning interaction between machine and human. In this fieldwork, train a machine to recognize or classify the human input correctly. A large amount of data is required to train a machine for great accuracy. So, the first step in this is to collect the data. In the dynamic gesture, the main task is to detect the region of interest. There are lots of techniques are available such as background subtraction [1], color gloves detection [2], frame differencing [3], mask R-CNN [4], fast R-CNN [5], faster RCCNN [6]. The background subtraction technique is not suitable for real-time application. Color gloves have some limitations in that they are not suitable for tracking under occlusion. Frame differencing requires two or three frames to detect the object in motion. Nowadays, deep learning algorithms are commonly used for object detection. R-CNN used region proposal regions to detect the region of interest. It works on a greedy algorithm that recursively combines similar candidate regions into larger ones and finally locates the target region of interest. But, there is some problem with CNN that it takes a considerable training time(approx 1 min time for each image), and it is not much suitable for real-time application. Fast R-CNN is faster than R-CNN because the convolutional neural network does not require to feed 2000 area suggestions every time. Instead, the convolution technique, performed just once per image, is used to create a feature map. Both the R-CNN and fast R-CNN

M. K. Sain (✉) · J. Singha
The LNM Institute of Information Technology, Jaipur, India
e-mail: manoj.sain@lnmiit.ac.in

J. Singha
e-mail: joyeeta.singha@lnmiit.ac.in

V. Bijalwan
Institute of Technology, Gopeshwar, Uttarakhand, India
e-mail: vishwanath.bijalwan@itgopeshwar.ac.in

© The Author(s), under exclusive license to Springer Nature Singapore Pte Ltd. 2023
D. S. Sisodia et al. (eds.), *Machine Intelligence Techniques for Data Analysis and Signal Processing*, Lecture Notes in Electrical Engineering 997,
https://doi.org/10.1007/978-981-99-0085-5_51

algorithms (above) use selective search to find region ideas. Selective search lowers network performance because it is a slow and time-consuming procedure. YOLO [7] is orders of magnitude faster than conventional object detection algorithms (45 frames per second). Ameur Zaibi et al. [8] presented a traffic sign classification model based on pretrained model LeNet-5. The model accuracy is around 99.84%. Kaur et al. [9]proposed a classification model of abnormal brain based on VGG16. The recognition rate is around 100. Tanvir Ahmad et al. [10] modified the loss function and added a spatial pyramid pooling layer and an inception module with convolution kernels to the YOLOv1 neural network-based object detection. Srivastava et al. [11] proposed a review article related to comparing different CNN-based object detection algorithms. Paper consists of feature extraction analysis and complexity analysis of the YOLO model. Girshick et al. [12] proposed FR-CNN using R-CNN and SPPnet. Yanan et al. [13] proposed a GA to encapsulate various CNN depths, integrate skip connections to facilitate the production of deeper CNNs during evolution, and develop a parallel and a caching component to speed up fitness evaluation given a constrained CPU resource considerably. The presented approach was compared to 18 state-of-the-art peer rivals, including eight manually built CNNs, six automatic + manually tweaked CNNs, and four automatic CNN architecture discovery algorithms on two tough benchmark datasets. jmour et al. [14] presented a model for traffic sign classification. After finely tuning some training parameters, a pretrained model AlexNet was used for classification purposes. In this paper, the modified YOLO algorithm (mYOLO) is presented. The major contributions of paper are as follows:

- A new dataset LNMYOLO dataset has been created. The dataset collected in different scenarios such as

    - Hand Images having hand shape variations
    - Images with multiple hands
    - Hand images having skin color variation
    - Hand images for different age persons
    - Hand images in different Illumination conditions

- A modified residual network has been used before the final classification layer.
- Automatic hand detection instead of manual followed tracking using CSRT [15] (tracking. Channel and Spatial Reliability Tracking) tracking.
- A novel and robust CNN model named hand gesture convolution neural network (HGCNN) has been proposed for hand gesture recognition.
- The result has been evaluated and compared with LeNet [8], Alexnet [16], VGG-16 [9] and inception V3 [17] using the same dataset, and observed that the proposed model outperforms the existing literature's.

The paper is divided into different sections past by various researchers. Section 2 contains The proposed methodology, dataset description, detection and tracking algorithm, and gesture classification using HGCNN. Section 3 includes results of object tracking under various challenging conditions. Section 4 is related to the hardware application of the proposed model. Finally, the paper is concluded in the last section.

2 Proposed Methodology

The proposed methodology consists of dataset collection and labeling, hand detection, hand tracking, trajectory smoothing, classification, and hardware interaction. The proposed architecture is described in Fig. 1.

2.1 Dataset Collection and Labeling

Dataset labeling is the fundamental step for YOLO. In this step, the dataset has been primarily recorded using a Logitech Webcam in different conditions and with other variations, as mentioned in the contribution part. Some images were also gathered from the internet source and added to the LNMYOLO dataset to make the algorithm more accurate. After collection of dataset labeling of the images has been done using labelimg [18] tool. Label of the image has been done for two classes: face and hand as shown in Fig. 2. The description of the dataset is mentioned in Table 1.

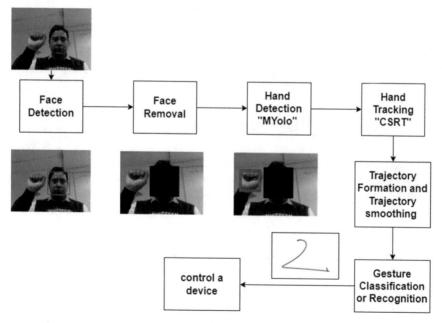

Fig. 1 Proposed model

Fig. 2 Sample of dataset with labels

Table 1 Dataset description

S. No.	Image description	Total number of datasets
1	Images with low illumination	200
2	Images with good illumination	200
3	Images with age variations	200
4	Images with skin color variations	200

2.2   Preparation of Configuration File

The next step is to prepare a configuration file for YOLO for two classes. The configuration file has classes set to 2, the number of filters set to 21, maximum batches to 6000, and steps set to 80 and 90% of the maximum batches.

2.3   Labeled Dataset Training and Hand Detection

The next step is to train the images using updated mYOLO as described in Fig. 3. mYOLO is a modification of yolov4 by using residual units to prevent vanishing gradient problems and enhance image expression of the image features. In the modification, the output feature is add to the next features that generates three different sets of estimation probabilistic feature map of distinct scale. The location and category are then forecasted using these sets of prediction feature maps. The setup file and final training weights aid in detecting the hand in the test image.

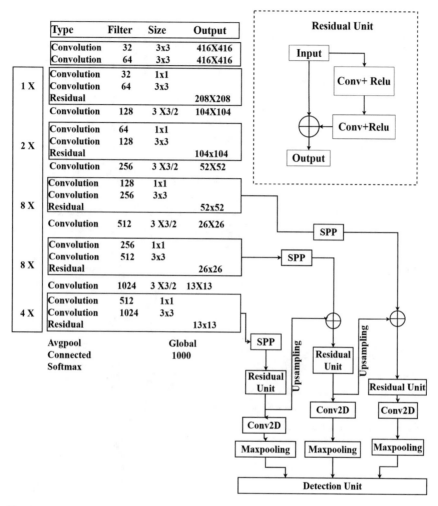

Type	Filter	Size	Output
Convolution	32	3x3	416X416
Convolution	64	3x3	416X416
Convolution	32	1x1	
Convolution	64	3x3	
Residual			208X208
Convolution	128	3 X3/2	104X104
Convolution	64	1x1	
Convolution	128	3x3	
Residual			104x104
Convolution	256	3 X3/2	52X52
Convolution	128	1x1	
Convolution	256	3x3	
Residual			52x52
Convolution	512	3 X3/2	26X26
Convolution	256	1x1	
Convolution	512	3x3	
Residual			26x26
Convolution	1024	3 X3/2	13X13
Convolution	512	1x1	
Convolution	1024	3x3	
Residual			13x13
Avgpool			Global
Connected			1000
Softmax			

Left groupings: 1 X, 2 X, 8 X, 8 X, 4 X

Residual Unit:
Input → Conv+ Relu → Conv+Relu → Output

Detection path: SPP, Residual Unit, Conv2D, Maxpooling, Upsampling, Detection Unit

Fig. 3 Proposed mYOLO architecture

2.4 Hand Tracking and Gesture Formation

Once the hand is detected in the image, apply the CSRT algorithm to track the hand in consecutive frames. It is also very much suitable for real-time application with a detection time of approximately 0.11s. The centroid of detected bounding boxes is used for the gesture trajectory formation. Next step is to smoothen the curve. For that, the Sobel filter is used. Sobel filter removes redundant points and smoothens the trajectory.

Table 2 Proposed CNN model layer description

Layers	Output shape	Param
Conv2D	(None, 222, 222, 32)	320 m
MaxPooling2D	(None, 111, 111, 32)	0
Conv2D	(None, 109, 109, 64)	18496
MaxPooling2	(None, 54, 54, 64)	0
Conv2D	(None, 52, 52, 128)	(None, 52, 52, 128)
MaxPooling2	(None, 26, 26, 128)	0
Flatten	(None, 86528)	0
Dense	(None, 128)	11075712
Dense	(None, 64)	8256
Dense	(None, 32)	2080
Dense	(None, 8)	264

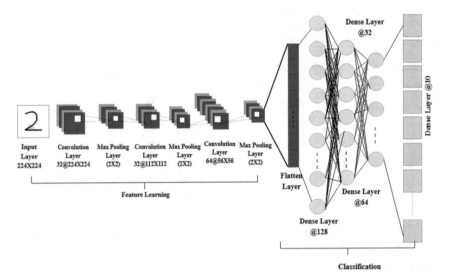

Fig. 4 Proposed CNN model

2.5 Gesture Classification

HGCNN has been proposed. The proposed CNN model consists of 11 layers. Input image size is 224 × 224, followed by convolution layers, max pooling layers, flatten layer, classification layers. The proposed model has 11 layers with 11,178,984 trainable parameters, as mentioned in Table 2. The structure of the proposed model is shown in Fig. 4. Training and validation accuracy of the proposed HGCNN model is approx 99.83 and 99.34%, which is very good compared to the other pretrained

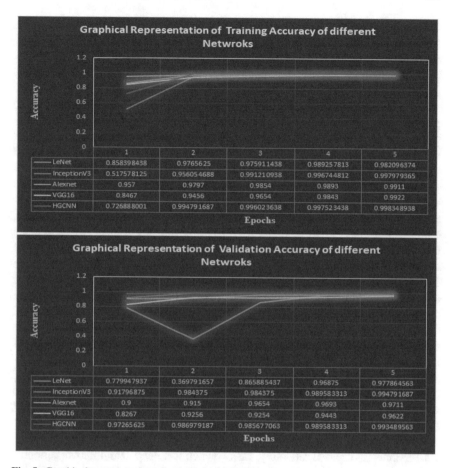

Fig. 5 Graphical representation of training and validation accuracy

networks. The resultant comparison between different networks is mentioned in Table 4. The graphical representation of training and validation accuracy for different networks is shown in Fig. 5.

3 Result Analysis

Object tracking is still challenging under various challenges such as the camera's motion and occlusion. Currently, CNN object tracking is prevalent, but in some cases, they produce miss-classifications. These miss-classifications can be overcome with the help of the channel and spatial reliability tracking (CSRT) tracker. CSRT can detect object features, classes, and locations. The CSR-DCF is implemented in

C using (Channel and spatial reference tracker). /SPP (Spatial pyramid pooling) is a pooling layer that allows a CNN to operate without a fixed-size input picture by removing the network's fixed-size limitation. The current generation of deep convolutional neural networks (CNNs) requires a fixed-size input picture. This "artificial" requirement may degrade identification accuracy for pictures or sub-images of any size or scale. To get around the constraint above, we provide the networks a more principled pooling approach called "spatial pyramid pooling." SPP is a novel network structure that can provide a fixed-length representation regardless of the size or scale of the picture. By eliminating the fixed-size limitation, we can enhance all CNN-based image classification algorithms in general.

The input of the residual network, which is the combination of the convolution layers, is the output of SPP. Residual units were incorporated into the detection system to enhance the expression of pictorial attributes. During the training process, the input size may be tweaked at any instant, allowing the trained model to adapt to varied sizes of hand pictures. From the modified YOLO, it is possible to detect the targeted object of varied size in the mYOLO previous layer feature fused in the next detected feature that improve the detection accuracy high-level features have a lot of semantic information, but low-level features have a wealth of information and location information. As you ascend from low to high level, the degree of detail decreases, but the level of semantic information grows. Location prediction necessitates more low-level feature information, whereas category prediction necessitates more high-level feature information.

Different detection methods have been employed on the self-prepared dataset, and it is found that the detection accuracy of mYOLO is much better than the YOLO and other detection methods. The comparison between all detection methods in respect to detection accuracy and detection time is shown in Table 3.

CSRT tracking algorithm is based on discriminative correlation filter tracking with channel and spatial reliability. CSRT which is a deep neural network module in opencv that has an ability to track the object in consecutive frames by calculating histogram of the target in the pevious frame into the current frame and after detection the object in the current frame it update the target. This process will continue till the last frame. Tracking rate can be calculated as (Figs. 6, 7 and 8).

$$\text{Tracking Rate} = (\text{Var1}/\text{Var2}) * 100 \tag{1}$$

where

Var1 Total tracked video Frame
Var2 Total Video Frames.

Table 3 Comparison table for different detection techniques

Detection algorithms	Detection accuracy (%)	Detection time (ms)
Frame differencing [3]	92	2000
FR-CNN [5]	93.45	1000
YOLO [7]	96.67	54
mYOLO [proposed]	98.89	34

Table 4 Comparison table for different CNN architecture

S. No.	Network	No of layers	Training accuracy	Validation accuracy
1	LeNet [8]	7	99.56	99.47
2	Inception V3 [17]	48	98.20	97.78
3	VGG16 [9]	16	98.89	98.26
4	Alexnet [16]	8	99.54	99.47
5	HGCNN [proposed]	6	99.83	99.34

Fig. 6 Tracking result under low illumination

Fig. 7 Tracking result under good illumination

Fig. 8 Tracking result under shape variation

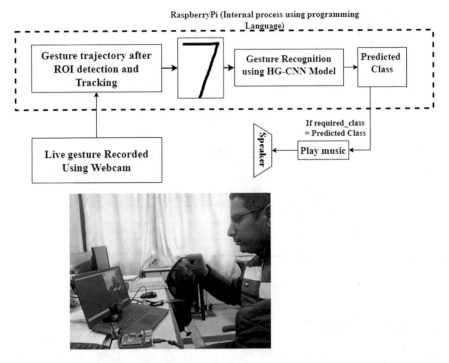

Fig. 9 Hardware implementation in real time

4 Hardware Application

Raspberry Pi [19] 3B+ has been used for hardware development. Logitech Webcam has been used for gesture recording and USB speakers for sound output. First recognition program deployed on to Raspberry Pi. The following schedule is modeled according to the need. For getting weather information, use some weather API that can permit access to the online website to get weather information of any city or country. Similarly, it can also be modeled to get information about current news, medicine reminders (using google calendar API), a simple calculator, etc. The methodology is shown in Fig. 9.

5 Conclusion

In this paper, a modified YOLO detection model with CSRT tracking and new CNN architecture HGCNN has been proposed and described. A broad set of gestures has been recorded for ten numerals in various challenging environments. The accuracy of the detection model is better than the YOLO, and it evenly detects the hand from a far distance. CSRT has been used for tracking. The tracking accuracy is around 96%. The lack of accuracy appears when both hands overlap or cross each other. A novel CNN model HGCNN has been proposed for recognition or prediction. The testing accuracy of 99.34 has been achieved from the proposed CNN model. In the future, some more innovative modifications will employed to the algorithm in order to track hands even when they overlapped or occluded.

Acknowledgements This work is supported by DST (Govt. of India) under the SEED Division [SP/YO/407/2018].

References

1. Dawod AY, Abdullah J, Alam MJ (2010) Hand feature detection from skin color model with complex background, No. November
2. Lamberti L, Camastra F (2011) Real-time hand gesture recognition using a color glove. Lecture notes in computer science (including subseries Lecture Notes in Artificial Intelligence and Lecture Notes in Bioinformatics), vol 6978 LNCS, pp 365–373
3. Saboo S, Singha J (2021) Vision based two-level hand tracking system for dynamic hand gestures in indoor environment. Multimedia Tools Appl 80(13):20579–20598
4. He K, Gkioxari G, Dollár P, Girshick R (2020) Mask r-CNN. IEEE Trans Pattern Anal Mach Intell 42(2):386–397
5. Sridevi S, Kumar Sahu B, Yadav M, Kumar B (2021) Improvised detection of diabetic retinopathy using fast R CNN. 25(5):5153–5161. [Online]. Available: http://annalsofrscb.ro
6. Ren S, He K, Girshick R, Sun J (2017) Faster r-CNN: Towards real-time object detection with region proposal networks. IEEE Trans Pattern Anal Mach Intell 39(6):1137–1149
7. Redmon J, Divvala S, Girshick R, Farhadi A (2016) You only look once: unified, real-time object detection. In. IEEE Conference on computer vision and pattern recognition (CVPR) pp 779–788
8. Zaibi A, Ladgham A, Sakly A (2021) A lightweight model for traffic sign classification based on enhanced LeNet-5 Network. J Sens 2021
9. Kaur T, Gandhi TK (2019) Automated brain image classification based on VGG-16 and transfer learning. In: Proceedings—2019 international conference on information technology, ICIT 2019, pp 94–98
10. Ahmad T, Ma Y, Yahya M, Ahmad B, Nazir S, Haq AU, Ali R (2020) Object detection through Modified YOLO neural network. Sci Program 2020:1–10
11. Srivastava S, Divekar AV, Anilkumar C, Naik I, Kulkarni V, Pattabiraman V (2021) Comparative analysis of deep learning image detection algorithms. J Big Data 8(1). [Online]. Available: https://doi.org/10.1186/s40537-021-00434-w
12. Girshick R (2015) "Fast R-CNN," Proceedings of the IEEE international conference on computer vision, vol 2015 International conference on computer vision, ICCV 2015, pp 1440–1448
13. Sun Y, Xue B, Zhang M, Yen GG, Lv J (2020) Automatically designing CNN architectures using the genetic algorithm for image classification. IEEE Trans Cybernet 50(9):3840–3854
14. Jmour N, Zayen S, Abdelkrim A (2018) Convolutional neural networks for image classification. In: 2018 International conference on advanced systems and electric technologies (IC_ASET), pp 397–402
15. Farhodov X, Kwon O-H, Kang KW, Lee S-H, Kwon K-R (2019) Faster RCNN detection based openCY CSRT tracker using drone data. International conference on information science and communications technologies (ICISCT) 2019:1–3

16. Li S, Wang L, Li J, Yao Y (1813) Image classification algorithm based On Improved AlexNet. J Phys?: Conf Ser 1:2021
17. Szegedy C, Vanhoucke V, Ioffe S, Shlens J, Wojna Z (2016) Rethinking the inception architecture for computer vision. In: Proceedings of the IEEE computer society conference on computer vision and pattern recognition, vol 2016-Decem, pp 2818–2826
18. Lee Y, Im D, Shim J (2019) Data labeling research for deep learning based fire detection system. In: 2019 International conference on systems of collaboration big data, Internet of Things Security (SysCoBIoTS), pp 1–4
19. Nayyar A, Puri V (2015) Raspberry Pi-A Small, powerful, cost effective and efficient form factor computer? A Rev Int J Adv Res Comput Sci Softw Eng 5(12):720–737

Cov-CONNET: A Deep CNN Model for COVID-19 Detection

Swapnil Singh⬤ and **Deepa Krishnan**⬤

1 Introduction

Towards the end of 2019, the world witnessed the onset of an airborne infectious disease called COVID-19 in the city of Wuhan in China, which has later spread to all parts of the world in no time. The World Health Organization in January 2021 announced the SARS-COVID-19 outbreak from Wuhan, a public health emergency of international concern. This has spread widely to all countries of the world very quickly, becoming a massive health crisis crippling big and small industries, financial institutions, government bodies and affecting the lives and livelihood of millions all over the world. There are several techniques for COVID-19 detection, and the most important among them is a type of nucleic acid detection popularly known as the RT-PCR test [1]. Identification of COVID-19 is also undertaken by identifying clinical features from the chest CT scan and chest X-rays. However, detection of COVID-19 and its severity from X-ray and CT scan images require the help of trained medical practitioners. Also, analysis done by researchers in [2] concludes the effectiveness of chest CT scan images over RT-PCR tests in hospitalized patients. Many people worldwide don't have timely access to medical facilities; thus, automated solutions empowered by machine learning and deep learning techniques come in handy.

S. Singh (✉) · D. Krishnan
Computer Engineering Department, Mukesh Patel School of Technology Management and Engineering, NMIMS University, Mumbai, India
e-mail: Swapnil.Singh49@nmims.edu.in

D. Krishnan
e-mail: deepa.krishnan@nmims.edu

The major contributions of the proposed study are as follows:

- We have used transfer learning to harness the advantage of generalization for model learning with minimal parameter tuning which is evident in the significantly high-performance scores for both testing and validation.
- We have compared our model performance with existing literature and is seen to give better performance than other models.

In the proposed work, we developed a custom deep CNN model for detecting and classifying chest CT images into three classes: non-COVID, COVID, and Pneumonia, with significantly higher scores than related works. The subsequent sections have been arranged as follows: Sect. 2 reviews the related work using the machine learning-based approach and deep learning and pre-trained model-based approaches. Section 3 consists of the methodology of the proposed work incorporating the data sets used and the algorithms employed. In Sect. 4, we have presented the results of our study and presented a comparative analysis with existing work. Subsequently, we have extrapolated the work and presented the future work in Sect. 5.

2 Literature Review

Recent research indicates the growing usage of machine learning and deep learning techniques in detecting COVID-19. The literature review that we have undertaken for this research study can be broadly categorized into machine learning-based approaches and deep learning-based approaches.

2.1 Machine Learning-Based Approaches

One of the vital machine learning-based techniques in detecting COVID-19 from chest X-ray images is done by researchers in [3]. The significant highlight of this work is that the authors have used a feature extraction technique based on YCrCb, Kekre-LUV, and CIE-LUV colour spaces. These features thus extracted are fed into ensemble-based machine learning classifiers for training and classification into three classes: COVID-19, Pneumonia and Normal. In this work, the authors have used individual classifiers like Naïve Bayes, Extra Tree, Random Forest, and an ensemble of these classifiers for performance evaluation along with feature extraction techniques mentioned above. Experimental results indicate that YCrCb + Kekre-LUV colour space, when used with Extra Tree, Random Forest, Simple Logistic ensemble, has the highest performance scores. However, the accuracy of the highest performing model is only 84.167%, and sensitivity and F-measure are also found to be 84.2% and 83.8%, respectively.

In research work [4], researchers have used Fuzzy tree transformation and exemplar division to select the features of the chest images. The authors further use iterative neighbourhood feature component feature selector to select the most prominent features, which are further fed to classic machine learning algorithms like Decision Tree, Linear Discriminant, SVM, KNN, and Ensemble methods. SVM classifier with cubic kernel gave the highest accuracy, geometric mean, precision, and recall of 97.01, 97.06, 97.11, and 97.09, respectively. The authors in [5] have also used an ensemble of machine learning algorithms on COVID-19 chest X-ray images. Prior to applying classifiers, the features are extracted using local binary patterns. It is seen that the accuracy of an ensemble of Random Tree, Random Forest, and KNN is the highest. The ensemble technique is seen to be more efficient than individual classifiers, as is evident in [4, 5].

Machine learning classifier in combination with texture feature based is used for COVID-19 detection in [6]. The researchers used 291 images with COVID-19 and 279 images of infectious Pneumonia, and 160 images of Influenza A/B. The texture features are extracted using wavelet transform and matrix computation analysis. Random Forest model built using 500 decision trees obtained optimal performance at image level with AUROC of 0.800, the accuracy of 0.722, sensitivity of 0.770 and specificity of 0.680 and at the individual level with AUROC of 0.858, the accuracy of 0.826, sensitivity of 0.809 and specificity of 0.842. The significance of this work is that model can differentiate COVID-19 from Influenza A/B with high accuracy.

2.2 Deep Learning and Pre-trained Model-Based Approaches

One of the significant works done in deep learning and pre-trained models was proposed by researchers taking advantage of the generalization ability offered by combining powerful StackNet and a combination of deep convolutional neural network (DCNN) [7]. The authors have used the VGG16 pre-trained CNN model for feature extraction. The final features extracted after flattening transformation are fed into StackNet to classify the X-ray images into COVID-19, Pneumonia, and Normal. It is observed that the accuracy and precision in the proposed approach named CovStackNet is 97.9% and 98.8%, respectively. However, the authors have not reported the per-class accuracy, precision scores, and other performance scores.

One of the important works using pre-trained models is a comparative analysis by authors in [8] using Inception V3, Xception, and ResNeXt models. The model performance is obtained using 6432 chest X-ray scan samples, and it is seen that the Xception model outperformed other compared models. The precision, recall, and F1 score obtained for the COVID-19 class using Xception is 0.99%, 0.92%, and 0.95%, respectively. However, the number of samples can be increased, and the performance scores of the model can be validated on other COVID-19 data sets. Another research work in the same direction using transfer learning approaches is done in [9]. The authors used transfer learning networks built on CNN, including ResNet18, ResNet50, SqueezeNet, and DenseNet121, to detect COVID-19 disease in chest X-ray images. The researchers obtained a sensitivity rate of 98% and a specificity rate hovering around 90%. Despite the promising results, the data set used is comparatively small, with 200 COVID images and 5000 non-COVID images. The data set used also did not include chest X-ray images of patients infected with Pneumonia.

Researchers use a deep learning and pre-trained model using the DenseNet201 approach is used by researchers in [10] to detect COVID-19 from chest CT scan images. They have used multiple kernels ELM-based deep neural network with extreme learning machine as a classifier. The final predictions are obtained using majority voting for two classes, COVID and non-COVID, and the accuracy score obtained is 98.36%. However, the data set consists of only 349 images of COVID-19 and 397 images of non-COVID. CoVNet-19, an ensemble deep learning approach [11], is proposed by Priyansh Kedia et al. to detect COVID-19 with the help of chest X-Ray images. In this work, stacked feature extraction is performed using DenseNet21 and VGG16, and further, the extracted features are fed to an SVM classifier for 2-class classification (COVID and non-COVID) and 3-class classification (COVID, non-COVID, and Pneumonia). The proposed work has given accuracy of 98.28% and precision and recall of 98.33%.

CoroNet [12], a deep CNN-based architecture for detecting and classifying COVID, non-COVID cases and Pneumonia, was able to get an accuracy of 95%. The related works reveal the promising capability of deep learning and transfer learning in effectively detecting and classifying COVID, non-OVID, and Pneumonia. The performance measures in most cases range from 95 to 98%, and in our proposed work, we have attempted to improve the performance scores of detections and classification significantly.

3   Methodology

Figure 1 shows the architecture for developing the models. Each section of the architecture has been explained in the subsequent sections.

3.1   Data Set

COVIDx CT-2

Figure 2 portrays a visual representation of the data set; here, class 0 indicates normal class, class 1 indicates the images with Pneumonia, and class 2 indicates the images with COVID-19 positive images. The data set consists of 194,922 CT-scan images of 3745 patients aged 0–93 years. The data set was made by collecting CT-scan images from multiple data sets with patients belonging to more than 15 countries, hence making it a diverse data set. This data set has been labelled by radiologists who have

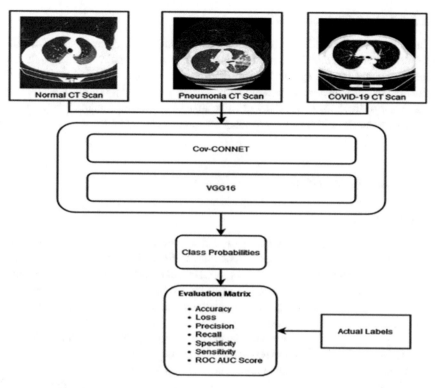

Fig. 1 Model architecture

experience of 30 years and 10 years, respectively [13]. The data set is based on an earlier data set, COVIDx CT-1 [14].

Fig. 2 Classwise
distribution for the COVIDx
CT-2

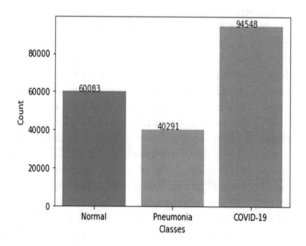

SARS-COV-2 CT-Scan Data Set

The data set consists of 2482 CT scan images consisting of 1252 images for COVID-19 positive images and 1230 images belonging to the Normal class. The data had been collected from hospitals of Sao Paulo in Brazil. The images belong to 120 patients, out of which 62 were male and 58 were female [15]. For our research, we have only utilized the COVID-19 positive images to see our models' performance on unseen images and see how precisely they detect COVID-19.

3.2 Algorithms Used

Cov-CONNET

Convolutional neural network (CNN) is a deep neural network generally used for image classification. The name of CNN comes from the linear operations performed on matrices. There are multiple layers in CNN, like convolutional layers, pooling layers, batch normalization layers, dropout layers, and fully connected layers. The number of filters, filter size, stride, and padding can be tuned for the convolutional layer. Similar to the convolutional layer, the size of filter, stride, and padding can be changed. For the fully connected layer, the number of neurons can be changed. The activation function, optimizer, and loss function can also be tuned based on the application [16]. On a similar note, Table 1 shows the parameters tuned from the Cov-CONNET model proposed in this study. The activation layer of all the layers was set to ReLU, and the activation of the output layer was set to softmax. Figure 3 shows the model of the Cov-CONNET model used for the study.

VGG16

VGG or Visual Geometry Group is an excellent example of transfer learning. For a given domain and learning task and a different domain and learning task, transfer learning helps us improve the learning of the target prediction function for the different domain using the knowledge of the given domain and given learning task [17]. There are various variations of VGG available, the variation being the change in the number of layers. For this study, we use VGG16, a 16-layer convolutional neural network. VGG16 is trained over the ImageNet data set and secured first and second place in the localization and classification tracks, respectively, in the ImageNet Challenge 2014. The data set consists of 1000 classes, hence providing a wide range of classes that helps in transfer learning. The model was tested using the top-1 and top-5 errors [18]. For the VGG16 model, we kept the layers till the flatten layer and added two dense layers, one layer with 1024 neurons and the other as the output layer with three neurons. Figure 4 shows the VGG16 model used in the study. VGG16 was used for the study considering that it has less number of layers compared to most transfer learning models; hence, it is memory efficient and training time efficient. Multiple studies have also shown that VGG16 has proven to be efficient in medical imaging applications.

Table 1 Model parameters for Cov-CONNET

Layers	Parameters
Convolutional layer	3 × 3 size, 32 filters, 1 stride
Pooling	2 × 2 size, 2 stride
Convolutional layer	3 × 3 size, 64 filters, 1 stride
Dropout	0.1
Pooling	2 × 2 size, 2 stride
Convolutional layer	3 × 3 size, 64 filters, 1 stride
Pooling	2 × 2 size, 2 stride
Convolutional layer	3 × 3 size, 128 filters, 1 stride
Dropout	0.2
Pooling	2 × 2 size, 2 stride
Convolutional layer	3 × 3 size, 256 filters, 1 stride
Dropout	0.2
Pooling	2 × 2 size, 2 stride
Fully connected layer	128 neurons
Dropout	0.2
Fully connected layer	3 output classes

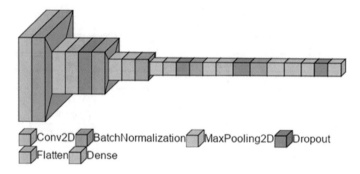

Fig. 3 Proposed model for Cov-CONNET

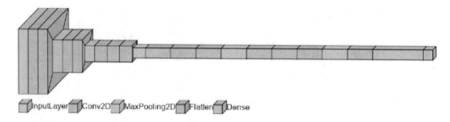

Fig. 4 Proposed model for VGG16

4 Results and Analysis

4.1 Experimental Analysis and Results

The study was performed using Google Colab and Jupyter Notebook environment's GPU environment. For the Google Colab environment, the disc was 78.19 GB, RAM was 12.69 GB, and NVIDIA GPU. The Jupyter Notebook worked on a RAM of 12 GB and a disc space of 256 GB.

Tables 2 and 3 show the testing results and the classwise performance of the Cov-CONNET and the VGG16 model. The models have been trained using the 'adam' optimizer, 'categorical_crossentropy' as the loss function and accuracy as the performance metrics. The batch size was set to 256, and training was done over 20 epochs with a validation split of 10%. The COVIDx CT-2 data set was split into training and testing in an 80:20 ratio, containing 155,937 training images and 38,985 testing images.

Table 2 Testing results for 38,985 samples using model

Models	Cov-CONNET	VGG16
Accuracy	99.3689	99.4229
Loss	0.021297	0.020493
Precision	99.3689	99.4307
Recall	99.3689	99.4229
F1-Score	99.3689	99.4239
Sensitivity	99.3689	99.4229
Specificity	99.3689	99.4229
ROC AUC Score	0.999828	0.999782

Table 3 Classwise results for 38,985 test samples using proposed models

Models	Cov-CONNET			VGG 16		
	Normal	Pneumonia	COVID-19	Normal	Pneumonia	COVID-19
Accuracy	99.6305	99.2604	99.7039	99.9734	99.5557	99.4859
Loss	0.01068	0.02542	0.00919	0.00088	0.01903	0.01465
Precision	99	99	99	98	100	100
Recall	99	99	99	100	99	99
F1-Score	99	99	99	99	100	100
TPR	0.99461	0.98937	0.99498	0.99958	0.99218	0.99175
FPR	0.00277	0.00231	0.00498	0.00738	0.00052	0.00045
TNR	0.99723	0.99769	0.99502	0.99262	0.99948	0.99955
FNR	0.00539	0.01063	0.00502	0.00042	0.00782	0.00825

For Cov-CONNET, the average training accuracy was found to be 98.93%, with an average training loss of 0.0298, whereas the average validation accuracy was found out to be 96.03%. Similarly, for the VGG16 model, the average training accuracy was found to be 98.41%, with an average training loss of 0.0432, whereas the average validation accuracy was found to be 99.09%. Figures 5 and 6 show the trends in training and validation accuracy and loss for Cov-CONNET and VGG16. It can be espied that for Cov-CONNET, the training curves are smooth while there are fluctuations in the validation curve; on the contrary, the validation and training curves both see similar fluctuations for VGG16.

Let's compare the performance of the Cov-CONNET model and VGG16 model based on the parameters given in Table 1. It can be manifested that the performance of VGG16 is faintly better than that of the Cov-CONNET model. Looking at

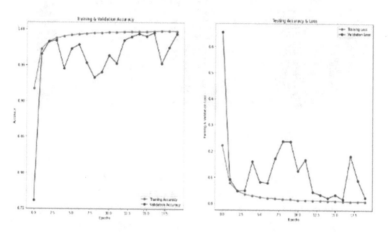

Fig. 5 Trends in training and validation accuracy and loss for Cov-CONNET

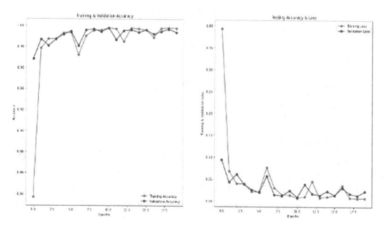

Fig. 6 Trends in training and validation accuracy and loss for VGG16

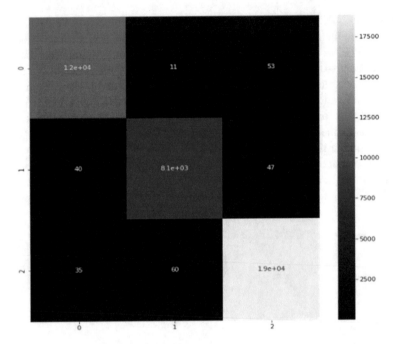

Fig. 7 Confusion matrix for Cov-CONNET

Table 2, we see that the precision of Cov-CONNET is better than that of VGG16 for the Normal class images, therefore, telling us why the ROC AUC score of Cov-CONNET is better than the ROC of VGG16. Figures 7 and 8 show the confusion matrices for the Cov-CONNET and VGG16 models. It can be interposed that VGG16 does not classify COVID-19 positive images as well as for the other two classes. Figures 9 and 10 show the Receiver Operating Characteristic Curve for the Cov-CONNET and VGG16; as seen, the curves are similar, indicating a similar classwise performance of the models on every class.

4.2 Validation Results and Comparative Analysis with Existing Work

The results portrayed in Table 4 show the performance of both the models on the SARS-COV-2 CT-scan data set. As stated earlier, we have only considered the 1252 COVID-19 positive images so that we can test how accurately the model is predicting COVID-19. It was observed that Cov-CONNET correctly classifies 993 images and VGG16 correctly classifies 585 images correctly, hence making our Cov-CONNET the better model.

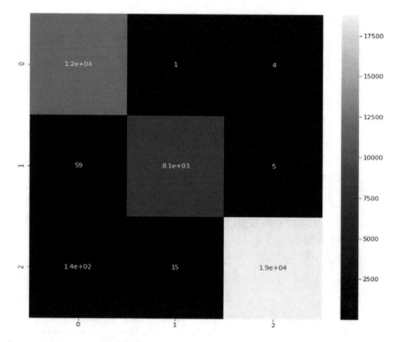

Fig. 8 Confusion matrix for VGG16

In Table 5, we can observe that both of our models have outperformed the COVID-Net CT model, trained on the COVIDx CT-2. The sensitivity for the normal class is more for our VGG16 model. The sensitivity of Pneumonia and COVID-19 given by both of our proposed models are more than that of COVID-Net CT. Table 6 portrays the comparison of our models with previous studies. It can be observed that our models have a better performance than other studies.

5 Conclusion

Considering the growing pandemic and the outflow of growing COVID-19 positive cases, RTPCR becomes a barrier due to its time-consuming nature. CT-scan imaging to the rescue in such a scenario, provide a faster and visual verification of COVID-19. Deep learning models have evidently been proven efficient in COVOD-19 detection. In this study, we have proposed a state-of-the-art convolutional neural network, 'Cov-CONNET' and compared its performance with VGG16. The data sets were trained over the COVIDx CT-2 and achieved an accuracy of 99.3689% using Cov-CONNET and accuracy of 99.4229% using VGG16; we have also presented the classwise performance of the models. We validated the models on the SARS-COV-2 CT-scan data set, and it was observed that the Cov-CONNET performs better in detecting

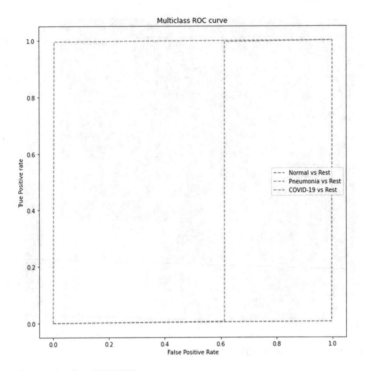

Fig. 9 ROC curve for Cov-CONNET

COVID-19. For future work, we plan to use an ensemble of machine learning and deep learning features and use more transfer learning models for classifying COVID-19. We also aim to segment the infected region in the CT-scan images, which could help identify the severity of the disease.

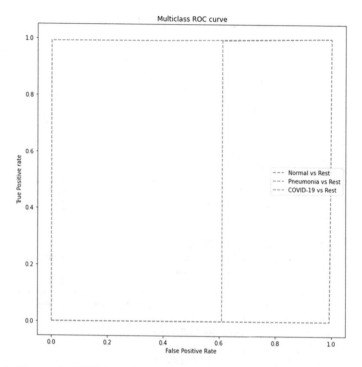

Fig. 10 ROC curve for VGG16

Table 4 Validation results on SARS-COV-2 CT-scan data set

Models	Cov-CONNET	VGG16
Accuracy	78.6846	46.3549
Precision	100	100
Recall	78.6846	46.3549
F1-Score	88.0709	63.3459
Sensitivity	78.6846	46.3549
Specificity	78.6846	46.3549

Table 5 Comparison with existing work

Models		Accuracy	Sensitivity
Cov-CONNET	Normal	99.3689	99
	Pneumonia		99
	COVID-19		99
VGG16	Normal	99.4229	100
	Pneumonia		99
	COVID-19		99
COVID-Net CT [13]	Normal	98.10	99
	Pneumonia		96.20
	COVID-19		98.20

Table 6 Comparison of proposed work with previous studies

Study	Attributes of data set	Feature Extraction and classification methods	Type of Images	Accuracy (%)
[3]	60 COVID-19, 60 Normal, 120 Pneumonia	Extra Tree R Forest, NB, Logistic Regression	X-ray images	90.42
[4]	135 COVID-19, 1250 Normal, 150 Pneumonia	F-transform, MKLBP, and SVM	X-ray images	97.01
[5]	68 COVID-19, 79 Normal, 158 Pneumonia	LBP, SVM, NB, KNN, RTree, RForest	X-ray images	80.328 (average)
[6]	291 COVID-19, 279 Pneumonia, 160 Influenza	Non-subsampled dual-tree complex contourlet transform	CT-scan images	82.60
[7]	2779 Pneumonia, 1583 Normal, 634 COVID-19	CovStackNet	X-ray images	More than 98
[8]	576 COVID-19, 1583 Normal, 1273 Pneumonia	InceptionV3, Xception Net, and ResNeXt	X-ray images	96
[9]	184 COVID-19, 5000 non-COVID	ResNet, SqueezeNet, DenseNet	X-ray images	–

(continued)

Table 6 (continued)

Study	Attributes of data set	Feature Extraction and classification methods	Type of Images	Accuracy (%)
[10]	397 non-COVID-19, 349 COVID-19	AlexNet, GoogleNet, VGG16, MobileNetV2, ResNet18, InceptionV3, MKs-ELM-DNN	CT-scan images	98.36% (highest accuracy for MKs-ELM-DNN)
[11]	798 COVID-19, 2341 Normal, 2345 Pneumonia	DenseNet121, VGG19, SVM	X-ray images	98.28% (ensemble model)
[12]	290 COVID-19, 1203 Normal, 660 Bacterial Pneumonia, 931 Viral Pneumonia	CoroNet	X-ray images	90.21%
[19]	203 COVID-19, 203 Normal	AlexNet, VGG16, GoogleNet, MobileNetV2, SqueezeNet, ResNet34, ResNet50, InceptionV3	X-ray images	98.33% (highest accuracy for AlexNet)
[20]	120 COVID-19, 120 non-COVI-19	GoogleNet and RF	CT-scan images	90.8% (Ensemble Model)
Proposed	94,548 COVID-19, 60,083 Normal, 40,291 Pneumonia	Cov-CONNET, VGG16	CT-scan images	99.4229% (for VGG16)

References

1. Gong Y, Ma T, Xu Y, Yang R, Gao L, Wu S, Li J, Yue M, Liang H, He X, Yun T (2020) Early research on COVID-19: a bibliometric analysis. The Innovation 1:100027. https://doi.org/10.1016/j.xinn.2020.100027
2. Chendrasekhar A (2020) Chest CT versus RT-PCR for diagnostic accuracy of COVID-19 detection: a meta-analysis. J Vascular Med Surg 8:1–4 (2020). https://doi.org/10.35248/2329-6925.20.8.392.Copyright
3. Thepade SD, Chaudhari PR, Dindorkar MR, Bang SV (2020) Covid19 identification using machine learning classifiers with histogram of luminance chroma features of chest x-ray images. In: 2020 IEEE Bombay section signature conference, IBSSC 2020, pp 36–41. https://doi.org/10.1109/IBSSC51096.2020.9332160
4. Tuncer T, Ozyurt F, Dogan S, Subasi A (2021) A novel Covid-19 and pneumonia classification method based on F-transform. Chemom Intell Lab Syst 210:104256. https://doi.org/10.1016/j.chemolab.2021.104256
5. Thepade SD, Jadhav K (2020) Covid19 identification from chest x-ray images using local binary patterns with assorted machine learning classifiers. In: 2020 IEEE Bombay section

signature conference, IBSSC 2020, pp 46–51. https://doi.org/10.1109/IBSSC51096.2020.933 2158

6. Wu Z, Li L, Jin R, Liang L, Hu Z, Tao L, Han Y, Feng W, Zhou D, Li W, Lu Q, Liu W, Fang L, Huang J, Gu Y, Li H, Guo X (2021) Texture feature-based machine learning classifier could assist in the diagnosis of COVID-19. Eur J Radiol 137. https://doi.org/10.1016/j.ejrad.2021. 109602

7. Rabbah J, Ridouani M, Hassouni L (2020) A new classification model based on stacknet and deep learning for fast detection of COVID 19 through X rays images. In: 4th international conference on intelligent computing in data sciences, ICDS 2020. https://doi.org/10.1109/ICD S50568.2020.9268777

8. Jain R, Gupta M, Taneja S, Hemanth DJ (2021) Deep learning based detection and analysis of COVID-19 on chest X-ray images. Appl Intell 51:1690–1700. https://doi.org/10.1007/s10 489-020-01902-1

9. Minaee S, Kafieh R, Sonka M, Yazdani S, Jamalipour Soufi G (2020) Deep-COVID: predicting COVID-19 from chest X-ray images using deep transfer learning. Med Image Anal 65. https:// doi.org/10.1016/j.media.2020.101794

10. Turkoglu M (2021) COVID-19 detection system using chest CT images and multiple Kernels-extreme learning machine based on deep neural network. IRBM 1:1–8. https://doi.org/10.1016/ j.irbm.2021.01.004

11. Kedia P, Anjum, Katarya R (2021) CoVNet-19: a deep learning model for the detection and analysis of COVID-19 patients. Appl Soft Comput 104:107184. https://doi.org/10.1016/j.asoc. 2021.107184

12. Khan AI, Shah JL, Bhat MM (2020) CoroNet: a deep neural network for detection and diagnosis of COVID-19 from chest x-ray images. Comput Methods Programs Biomed 196:105581. https://doi.org/10.1016/j.cmpb.2020.105581

13. Gunraj H, Sabri A, Koff D, Wong A (20021) COVID-Net CT-2: enhanced deep neural networks for detection of COVID-19 from chest CT images through bigger, more diverse learning, pp 1–15

14. Gunraj H, Wang L, Wong A (2020) COVIDNet-CT: a tailored deep convolutional neural network design for detection of COVID-19 cases from chest CT images. Front Med 7:1–11. https://doi.org/10.3389/fmed.2020.608525

15. Soares E, Angelov P, Biaso S, Froes MH, Abe DK (2020) SARS-CoV-2 CT-scan dataset: a large dataset of real patients CT scans for SARS-CoV-2 identification. medRxiv. 2020.04.24.20078584

16. Albawi S, Mohammed TA, Al-Zawi S (2018) Understanding of a convolutional neural network. In: Proceedings of 2017 international conference on engineering and technology, ICET 2017, 1–6 Jan 2018. https://doi.org/10.1109/ICEngTechnol.2017.8308186

17. Pan SJ, Yang Q (2010) A survey on transfer learning. IEEE Trans Knowl Data Eng 22:1345–1359. https://doi.org/10.1007/978-981-15-5971-6_83

18. Simonyan K, Zisserman A (2015) Very deep convolutional networks for large-scale image recognition. In: 3rd international conference on learning representations, ICLR 2015—conference track proceedings, pp 1–14

19. Nayak SR, Nayak DR, Sinha U, Arora V, Pachori RB (2021) Application of deep learning techniques for detection of COVID-19 cases using chest X-ray images: a comprehensive study. Biomed Signal Process Control 64:102365. https://doi.org/10.1016/j.bspc.2020.102365

20. Mayya A, Khozama S (2020) A novel medical support deep learning fusion model for the diagnosis of COVID-19. Proc IEEE Int Conf Advent Trends Multidisc Res Innov (ICATMRI) 2020:2–7. https://doi.org/10.1109/ICATMRI51801.2020.9398317

AI-Based Real Time Object Measurement Using YOLO V5 Algorithm

R. Anand, R. S. Sabeenian, S. Ganesh Kumar, V. Logeshwaran, and B. Soundarrajan

1 Introduction

Open cv is useful for many fields even in medical and for traffic regulations and etc. Measuring the size of an image is equal to computing the distance from camera to an object. If the focal length and the object distance are known then we can be able to find the object's size from an image. First, we need to define a ratio to measure the pixel per metric and we should know the dimension of an object and we need to keep one single left most object as a reference object from an image and sort out the object-contours from left most side to the right most side in an image and it will be used for defining the pixel per metric. The reference object should be placed at corners of the picture(preferably left corner) or based on the object appearance like it's shape/color. Reference object should be uniquely identified compared to other objects in an image.

2 Literature Survey

A novel measuring system using a scan counter method via a CCD camera [1]. It will be easy to find the distance between a CCD camera and to find the object's distance from the camera. For distance measurement two laser projector's are set on the either side of a CCD camera which will produce two parallel rays that will project two bright spots on the object and that object will be identified and it will be noted.

R. Anand (✉) · R. S. Sabeenian · S. Ganesh Kumar · V. Logeshwaran · B. Soundarrajan
Department of Electronics and Communication Engineering, Sona College of Technology, Salem, India
e-mail: anandvimal@gmail.com

R. S. Sabeenian
e-mail: sabeenian@sonatech.ac.in

Then the actual distance from the camera and the object size will be calculated by means of formula. For the area measurement of an object it will count the external clock pulses of the horizontal scan lines covering the area of an object. Then the projected area of an object will be calculated from the algebraic functions.

Types of yolo algorithms play a mojor role in object detection, object size detection and etc. [2]. Yolo v1 algorithm proposes the target through the grid division and finally it will detect the target object through its position to detect the object. Yolo algorithms will detect the tasks directly as regression problem. Compared with yolo v1, yolo v2 has more accuracy and more speed than yolo v1, Because of the multi scale feature maps to detect the objects. After the use of yolo v2, yolo v3 comes into effect, it uses residual network to design Darknet-53 for the feature extraction. It predicts the position of the target object through the method of frame regression prediction. Then after yolo v3 with high accuracy yolo v4 comes into effect, the CSPDarknet-53 of yolo v4 is the core algorithm for the feature extraction. Yolo v4 model is higher than the yolo v3 model in terms of Map values but slightly lower in terms of speed, where yolo v3 uses FPN for un sampling process and yolo v4 uses PaNet network. By using yolo v4 we can enhance the feature through down sampling and finally the feature maps of different layers are used to make predictions. Final yolo algorithm is yolo v5 algorithm, it has higher accuracy and it has better ability to identify and recognize the small objects. When compared to yolo v4, yolo v5 is more flexible and faster. Yolo v5 algorithm is better than both yolo v4 and yolo v3 in terms of both the map values and speed. Yolo v5 uses CSPDarknet as the feature extraction to extract features from the input image. It also uses R-CNN to improve the detection of smaller targets/objects. It proposes to use the Fcos algorithm useful for the calculation in the frame selection area for the object size/object detection and etc.

Estimating the distance and size by means of stereo vision system to measure the absolute size and distance of an object [3]. Many had started to use multi vision sensors for the different types of applications like 3-D image construction, occlusion detection and etc. The proposed system consists of object detection on stereo images and blob extraction for distance, size, and object calculation. It has a advantage that object measurement using stereo vision will be better when compared to the object detection using a single camera.

Estimating the size of an object by the use of open cv libraries, edge canny detection, dilation, and erosion algorithm [3]. In this paper first we need to detect the object by means of edge canny detection then by using morphological operators include dilation and erosion algorithms to close gaps between edges and to find and sort the contours, then finally measuring the dimension of an object. This project was designed by the use of raspberry pi 3 and raspberry pi camera.

Detection of objects in the infrared images by means of yolo v5 algorithm [4]. Yolo Firi for the infrared images used with yolo v5 by optimizing parameters, compressing the channels and etc. It is specially designed for a feature extraction model. CSP module is used to maximize the use of shallow features. Multiscale detection is used to improve the small object detection accuracy. It is experimented with the Kaist and Flir datasets when compared to yolo v4, yolo v5 is increased with 21% on Kaist dataset and the speed is reduced by 62%, parameters are decreased by 84% and the weight size is reduced by more than 94%. Yolo Firi has an improvement of 5–20% in average recall. Image fusion can also be applied to image pre-processing as a data enhancement method by fusing visible and infrared images based on convolutional neural network.

Calculation the size by means of Functional size measurement (FSM) method by quantifying the user requirements of the software [5]. In this paper they present the OO-Method Function points which is a new FSM method for object-oriented system based on a conceptual schemas. It is presented by the following steps of a process model for the software measurement. They present the design of the measurement method and its application study, and to analyze the different types of evaluation to validate the method and to verify its application and results.

3   Flow Diagram

As shown in Fig. 1 first we need to collect some images and store it in a given path and store that image into memory, the path must be specified in a given code. Then perform the grayscale which will be useful to identify the object in a given image and it is better to use gaussian filter to remove unwanted noise from an image. After filtering by means of gaussian filter then the edge detection will be done to detect the edges of an object present in an image. Then by using contours, outline is formed for the edge detected objects and finally we need to sort the object by means of reference object from left to right. Then the looping will happen when if the contour area is less than 100 then neglect that object and if not less than 100 then the distance and size calculation will occur. Distance and size calculation involves width and height of a given image and for finding the pixels per metric ratio we need that two parameters width and height.

4   Methodology

Main aim of our project is to find a optimal solution for measuring the dimension (width and height) of an object present in an image by means of open cv algorithm. To check the dimension we need to take two types of images with and without an object. We need to know about the libraries like simple cv, open cv and etc. to calculate the dimension. In open cv the algorithms used for identification of real time objects and non-real time objects and converting the normal image to grayscale image and Harris corner etc. to be found. In this project 3-D image would be converted to 2-D area image with the invisible background (Fig. 2).

To determine the width of an object from the image we derive a formula to calculate the width of the object.

$$x = y * K + z \tag{1}$$

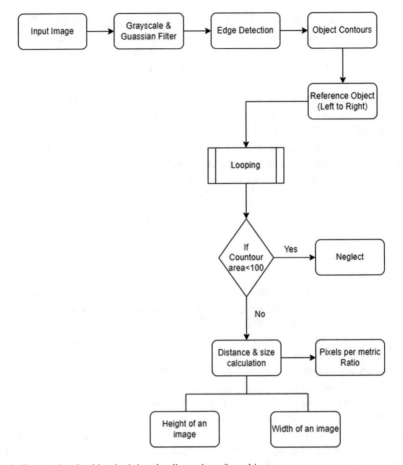

Fig. 1 Process involved in obtaining the dimension of an object

Fig. 2 Proposed methodology

where in the above equation x is defined as the width of an object(measured in pixel), y and z values will depend on the height from camera to an image then y and z value can also be expressed as,

$$y = m1 * \text{height from camera} - b \tag{2}$$

$$z = m2 * \text{height from camera} - b \tag{3}$$

To determine the height of an object from the image we had assumed the reference object as I then the formula to derive the height of an object will be,

$$I = m1 * K + m2 * \text{Height from camera} + b2 \tag{4}$$

In Eq. (4) I is the reference object and K will be denoted as the parameter of the values of distance(measured in pixel). M will referred as the heigth of an object. Distance between the left and right side corner of an image will be referred as the reference object to the physical height. As mentioned above M is the original object's height and N is the reference model height. For using the webcam for real time monitoring of an object if the gamma value will be the 2 times the value of the vertical angle of view. Next our aim is to find the estimated height of an object which will be equal to the value of M. If the I value is greater than the pixel height then the K value will be automatically equals to -1, otherwise it will be equal to 1. Then the estimated height of an object will be derivates as,

$$\text{estimated height} = \text{Height from camera} + k * \tan(\rho) * D \tag{5}$$

In Eq. (5) D is the camera's horizontal distance and it is derivates as

$$D = \cot(y) * \text{height from camera} \tag{6}$$

Next stage is to identify an object by removing the noises from an image and filtering the shadows from an image. CIE lab color model was used to reduce the shadows available in an image. The object can be measured without the help of shadows also, but detection will be difficult when the image has several areas with various colors. Finally, the area with the color will be detected and that object will be taken, and that object will be measured.

4.1 Algorithm—Open CV and Yolo V5

The main goal of open cv is to detect and recognize the image and produces different types of output like recognizing the face from an image and object recognition from an image. In our project we had used open cv and yolo v5 algorithm to detect the object by using the reference object from an image. By using yolo v5 we need to build a model by collecting the dataset images and then training the dataset images and sending that trained images for testing, then the output will be shown. Basically Yolo v5 uses CNN-based object detectors for having high performance. It was found that YOLOv5 performs better when we compare to both YOLOv4 and YOLOv30 in terms of accuracy as shown in Tables 2 and 3. The detection speed of YOLOv3 algorithm was faster when compared to YOLOv4 and YOLOv5 and the detection speed of YOLOv4 and YOLOv5 were same. When compared to overall performance Yolo v5 stands first in position with high accuracy as shown in Table 1 (Figs. 3 and 4).

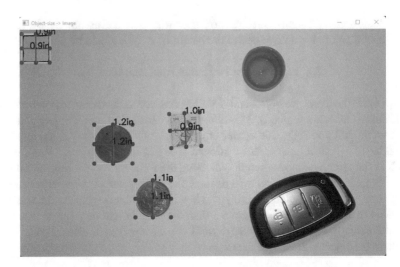

Fig. 3 Output—from original image

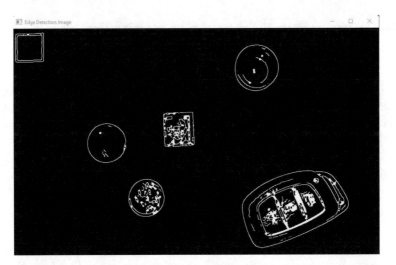

Fig. 4 Output—edge detected image from an original image

Table 1 Yolo v5—object size detection

Camera value	Measurement of an object's width	Measurement of an object's height	Final size of an objects's width	Y position	Final size of an objects's height
352	72	130	103	80	132
127	46	142	71	52	153

Table 2 Yolo v4—object size detection

Camera value	Measurement of an object's width	Measurement of an object's height	Final size of an objects's width	Y position	Final size if an objects's height
352	69	127	103	78	128
127	43	139	71	49	148

5 Result and Discussion

To execute the project, run the python script along with the image(image which is used for object measurement) path. It opens a result window and it shows the measurement of any single object in the picture it shows dimension of a object, our main is to obtain dimension of an object, we need to perform the grayscale which will be useful to identify the object in a given image, and it is better to use gaussian filter to remove unwanted noise from an image.

After filtering by means of gaussian filter then the edge detection will be done to detect the edges of an object present in an image. Then by using contours, outline is formed for the edge detected objects and finally we need to sort the object by means of reference object from left to right.

Then the looping will happen when the contour area criteria is met if not it will neglect that object and if the criteria is met and the condition satisfied then the distance and size calculation will occur.

Distance and size calculation involves the dimension (width and height) of a given image and for finding the pixels per metric ratio we need that two parameters width and height.

While you navigate (right arrow button), it shows measurements of next object, you can navigate to see other object measurements.

The results were also examined by the use of webcam which satisfies the angle of view. The results of the application were tested by means of several images containing several objects with various measurements and set to various heights to check the object measurement. Values obtained from testing are:

By observing Table 1 based on the yolo v5 algorithm, camera value is the value from the heights from the camera. Measurement of an object's width and height is the true dimension of an object from an image. Y position will represent the lower vertical coordinates and the final size of an object's width and height are used to determine the size of an object present in an image.

By observing Table 2 the height and width of an object is less when compared to yolo v5 table. Recall value and the map value is also less when compared to yolo v4. In yolo v4 the recall value is 0.80 and the map value is 0.85 but whereas for yolo v5 the recall value is 0.81 and the map value is 0.87.

Table 3 Yolo v3—Object size detection

Camera value	Measurement of an object's width	Measurement of an object's height	Final size of an objects's width	Y position	Final size if an objects's height
352	66	124	103	73	123
127	41	135	71	44	144

Table 4 Yolo v2—object size detection

Camera value	Measurement of an object's width	Measurement of an object's height	Final size of an objects's width	Y position	Final size if an objects's height
352	62	120	103	68	119
127	38	131	71	40	140

Table 5 Yolo v1—object size detection

Camera value	Measurement of an object's width	Measurement of an object's height	Final size of an objects's width	Y position	Final size if an objects's height
352	58	113	103	64	115
127	34	124	71	36	133

By observing Table 3 for yolo v3 the recall value is 0.69 and the map value is 0.71 when compared to both yolo v4 and yolo v5, yolo v3 is less and the width and height of an object is also less when compared to both yolo v4 and yolo v5.

By observing Table 4 for yolo v2 the recall value is 0.58 and the map value is 0.65 when compared to yolo v3, yolo v4 and yolo v5, yolo v2 is too less and the width and height of an object is also less when compared to yolo v3, yolo v4, and yolo v5.

By observing Table 5 for yolo v1 the recall value is 0.46 and the map value is 0.59 when compared to yolo v2, yolo v3, yolo v4 and yolo v5, yolo v1 is too less and the width and height of an object is also too less when compared to yolo v2, yolo v3, yolo v4, and yolo v5 and the accuracy will be low for yolo v1.

6 Conclusion

In this paper the measurement of an object using open cv with an image and also by using webcam is explained and executed. First we need to take two types of images one with an object and another image with without an object then we can easily find out measuring an object with an image. The output that was found finally will be write back by means of 'imwrite' in python and it will be saved in the folder in the format of image file. By considering the various height and width of an objects we can find the accuracy.

References

1. Sharaiha YM, Marchand-Maillet S (1999) Binary digital image processing. Academic Press, London, UK
2. Kim H, Lin CS, Song J, Chae H (2005) Distance measurement using a single camera with a rotating mirror. Int J Contr Autom Syst 3(4):542–551
3. Liu Y, Lu BH, Peng J et al (2020) Research on the use of YOLOv5 object detection algorithm in mask wearing recognition. World Sci Res J 6(11):276–284
4. Rahman KA, Hossain MS, Bhuiyan MA-A, Tao Z, Hasanuzzaman M, Ueno H (2009) Person to camera distance measurement based on eye-distance. In: 3rd International conference on multimedia and ubiquitous engineering, 2009 (MUE'09), Qingdao, 4-6 June 2009, pp 137–141
5. Muthukrishnan R, Radha M (2011) Edge detection techniques for image segmentation. Int J Comput Sci Inf Tech (IJCSIT) 3(6)
6. Moeslund T (2009) Canny edge detection, AalborgUniversity: Denmark: laboratory of computer vision and media technology
7. Albrecht AJ (1979) Measuring application development productivity. IBM Appl Dev Sympos 83–92
8. Lu M-C, Wang W-Y, Chu C-Y (2006) Image-based distance and area measuring systems. IEEE Sens J 6(2)
9. Pu L, Tian R, Wu H-C (2016) Novel object-size measurements using digital camera. IMCEC Kun Yan
10. Chen S, Fang X, Shen J, Wang L, Shao L (2016) Single-image distance measurement by a smart mobile device. IEEE Trans Cybern, Wang
11. Bochkovskiy A, Wang C Y, Liao H Y M (2020) YOLOv4: optimal speed and accuracy of object detection. ArXiv preprint arXiv:2004.10934
12. Anand R, Shanthi T, Dinesh C, Karthikeyan S, Gowtham M, Veni S (2021) AI based birds sound classification using convolutional neural networks. In: IOP conference series: earth and environmental science, IOP Publishing, vol 785, no 1, p 012015
13. Anand R, Veni S, Aravinth J (2016) An application of image processing techniques for detection of diseases on brinjal leaves using k-means clustering method. In: 2016 international conference on recent trends in information technology (ICRTIT). IEEE, pp 1–6
14. Shanthi T, Sabeenian RS, Anand R (2020) Automatic diagnosis of skin diseases using convolution neural network. Microprocess Microsyst 76:103074
15. Anand R, Shanthi T, Nithish MS, Lakshman S (2020) Face recognition and classification using GoogleNET architecture. Soft computing for problem solving. Springer, Singapore, pp 261–269
16. Anand R, Shanthi T, Sabeenian RS, Veni S (2020) Real time noisy dataset implementation of optical character identification using CNN. Int J Intell Enterp 7(1–3):67–80

Prediction of Heart Abnormality Using Heart Sound Signals

Yashwanth Gadde and T. Kishore Kumar

1 Introduction

Cardiovascular disease (CVD) constitutes the disorders associated with heart and blood vessels. It is one of the significant reasons for the death of millions worldwide [1]. One of the main reasons for CVD is the deposit of fats on the inner walls of the blood vessels. It may be because of hereditary reasons or the lack of proper treatment at the right time. It is essential to detect the disease as early as possible. In economically backward countries, basic healthcare programs for the early detection and treatment of such diseases are not available. If these diseases are not treated early, the mortality rate increases drastically in the future.

Phonocardiogram (PCG), electrocardiogram (EKG/ECG), chest X-ray, CT scan, CAT scan, echocardiogram, exercise stress test, thallium stress test are some of the well-known noninvasive tests to detect CVDs [2]. A phonocardiogram (PCG) is a plot of recordings of the sounds and murmurs made by the heart. Heart sounds are made from blood coursing through the heart chambers as the cardiovascular valves open and close during the heart cycle. Heart sound examination is one of the step-by-step methodologies used during cardiovascular auscultation to analyze heart disease, long-term observing of the patient's wellbeing, and a biometric for training and exploration purposes. Compared with other tests, PCG is less expensive and requires less equipment. Stethoscopes are needed for recognizing heart defects with the help of sounds, yet they have their disadvantages. Only an experienced and trained doctor can precisely analyze the sound produced. Investigating and analyzing PCG signals directly from phonocardiograph (heart sound signal plot) plot is time-consuming and tedious. Machine learning techniques use the available data to analyze the present data and simplify the job.

Y. Gadde · T. K. Kumar (✉)
Department of ECE, National Institute of Technology Warangal, Hanamkonda, Telangana, India
e-mail: kishorefr@gmail.com

© The Author(s), under exclusive license to Springer Nature Singapore Pte Ltd. 2023   669
D. S. Sisodia et al. (eds.), *Machine Intelligence Techniques for Data Analysis and Signal Processing*, Lecture Notes in Electrical Engineering 997,
https://doi.org/10.1007/978-981-99-0085-5_54

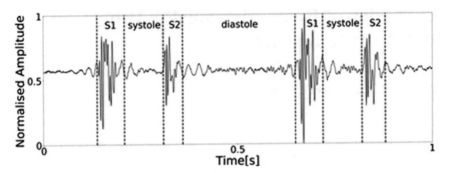

Fig. 1 Phonocardiogram plot

Different steps involved in analyzing a PCG signal through machine learning techniques are preprocessing/denoising, segmentation, and classification. Preprocessing is a method to arrange all the audio samples on the same platform because the recordings in the available PCG data may not necessarily be recorded under the same conditions. The recording equipment, environmental conditions, ambient noise, etc., may differ. Denoising is a method to reduce the noise, which corrupted the audio signal during recording. Adaptive filters, statistical filters, time-frequency techniques, wavelet techniques are various techniques used for PCG denoising.

The cardiac cycle is the behavior of the heart from the beginning of the first beat to the beginning of the next beat. It consists of four different regions, as shown in Fig. 1. $S1$ and $S2$ are fundamental heart sounds. Systole is when the heart contracts and pumps out blood, and diastole is when the heart expands and fills in blood. Segmentation is the method to identify the start and endpoints of each region $S1$, $S2$, period of systole, and diastole of the cardiac cycle. Envelop-based, feature-based, machine learning-based, mel frequency cepstrum coefficient-based, and HMM-based segmentation methods are widely used for segmentation.

Based on the features extracted from the segmented signal, classification is performed to determine whether the signal corresponds to an abnormal or normal heart sound. Support vector machine, decision tree, Naïve Bayes, KNN, logistic regression, convolution neural networks are various machine learning algorithms used for classification.

2   Related Work

In [3], the authors segmented the PCG recording into four stages, and from each stage, time-frequencey features were extracted to identify their characteristics in the T-F domain. The obtained features are applied to the support vector machine to classify the PCG signal. A subset of the PhysioNet dataset was taken with equal no. of normal and abnormal PCG recordings, and accuracy of 86% was obtained. In [4], the authors extracted feature sets of time, frequency, and statistical domains from Kaggle datasets. 8 features from 118 recordings were used and trained using different classifiers such as bagged tree, subspace discriminant, subspace KNN, LDA, quadratic SVM, and fine tree. Out of these 6 classifiers, bagged tree achieved a maximum accuracy of 80.5%. In [5], this paper aims at testing the performance of pre-trained CNN models at the classification of heartbeats. The PASCAL CHSC database was used to train four pre-trained models: VGG16, VGG19, MobileNet, and InceptionV3. The data were processed, and the features were extracted using spectrogram signal representation. The classification accuracies of the models are 80.25%, 85.19%, 72.84%, and 54.32%, respectively. In [6], the authors proposed a method using multidomain features and SVM for classifying. Time interval, the frequency spectrum of states, state amplitude, energy, the frequency spectrum of records, cepstrum, cyclo-stationarity, high-order statistics, and entropy are the 9 features from which 515 features were extracted. Top 400 features were used to classify, and accuracy of 88% was obtained. It is observed that some analyses are made on datasets with fewer samples, which increases the accuracy of prediction, but due to a decrease in the sample space, a wide range of significant samples may be missed. This approach is not wrong in all cases, but if we can bring better results with a full dataset, then the model will be fully efficient.

3 Proposed Work

In our work, the dataset is taken by The 2016 PhysioNet/CinC Challenge [7]. This dataset is a group of 8 different independent databases consisting of heart sound samples recorded in various environmental conditions. These signals are normalized and grouped into a single database; it consists of 3240 heart sound recordings from 1072 patients. All the recordings were annotated as normal and abnormal by physicians/doctors.

In [8], the authors introduced a method to segment PCG by using homomorphic filtering and an HMM. They used PASCAL database for testing. $S1$ sounds showed a high accuracy of 92.4% and while $S2$ sounds showed an accuracy of 93.5%. It was a smaller database, and the performance degraded with noise. In [9], wavelet-based features with HMM are used to segment, but even this combination failed to perform in a noisy environment. We used hidden semi-Markov model for segmenting the audio signal. In [10], the HSMM-based algorithm was first introduced. An initial version of this segmentation algorithm was the hidden Markov model. HSMM and logistic regression for segmentation showed a better performance in the noisy real-world environment. It is a statistical model with unobserved states that yield a sequence of observations, in which the Markovian process is modeled, which means the change to the next state depends only on the present state. An HSMM is similar to HMM model, but the difference is that the semi-Markovian process is modeled here, which means the change to the next state depends on both the present state as well as the time spent on the present state. In [10], the algorithm was trained to detect heart sounds taken from the Michigan heart sound database; however, it did not perform

Fig. 2 Block diagram of total setup

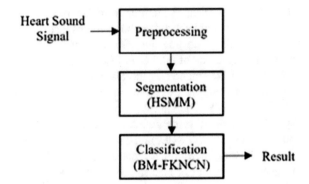

well in a real-time noisy environment. In [11], the authors used HSMM with logistic regression and were trained to detect heart sounds taken from MIT heart sound database. The performance of the algorithm was appreciable even in a real-world noisy environment. KNN is one of the best classification and clustering algorithms, but it is used only on small datasets and became outdated because of some drawbacks. In [12], the authors used DWT features alongside the KNN classifier and showed that DWT features outperformed morphological-based features. In [13], the authors used a combination of simple KNN classifiers with time features, and very high accuracy is obtained on a database of 22 heart sounds. In this work, Bonferroni mean-based fuzzy K-nearest centroid neighbor classifier is used to solve the disadvantages of a simple KNN with outliners. Because of these outliners, the performance deteriorates on datasets with class imbalance. Depending on the nearest local mean vector obtained from NCN concept, Bonferroni mean—FKNCN classifier identifies the class label of an unclassified sample. Imbalanced datasets were taken from the KEEL repository to compute the performance of the classifier. Three datasets were taken from the repository, and these datasets were chosen so that they are imbalanced and have a small set of training samples [14].

4 Methodology

The proposed methodology is shown in Fig. 2. The different stages involved in the classification process are preprocessing/denoising, segmentation, and classification.

4.1 Logistic Regression Hidden Semi Markov Model

The introduction of a probabilistic way of modeling for segmentation of heart sound lead to an increase in accuracy compared to threshold-based methods. HMM is a statistical model used to describe sequential data. It works by preparing guidelines for the possibility of being in particular "hidden states," traveling through those hidden states and watching an observation made by each one of the state. An HMM is defined as the function of transmission matrix (A), observation distribution or emission (B), and initial state distribution(isd). The probability of traveling from one state to the consecutive state at the next time instance ($t+1$) is defined by transmission matrix (A). The probability of the state that creates a particular observation (feature values) at time t is defined by observation distribution (B). The probability of presence in a particular state at the starting point of time is defined by initial state distribution (isd). HMM does not include any details about the expected length of time spent in each state which is its major drawback. The duration of the state is only given by the self-transition probability (A) if we do not include the expected length of time spent in each state. This results in a geometric distribution for the time expected to exist in each state. This distribution monotonically decreases, resulting in the most likely state duration always being a one-time step. To increase the performance of modeling, an extra parameter is needed in the model, $p(d)$. Where $p(d)$ is defined as the probability of existing in a particular state i for a duration d. Now the model is defined as a function of A, B, isd, and p. In [11], it is proved that when including duration densities; the entries for the transition matrix become unity for the transition from present to the next state if it is a permissible transition. This is because the transition between states is now only dependent on the time remaining in each state, and not on the probability of transitioning through the states. This is called HSMM. Since the heart has clear upper and lower limits on the duration of the components of the heart cycle (because of mechanical limitations), it is expected that the inclusion of such details should help to improve segmentation performance. In addition, logistic regression is included in the model, as LR-derived emission or observation probability estimates were used to better discriminate between the states (Figs. 3 and 4).

4.2 Bonferroni Mean-Based Fuzzy K-Nearest Centroid Neighbor Classifier (BM-FKNCN)

In K-nearest neighbor, the number of classified samples (k) in the neighborhood of the unclassified sample decides the class label of the unclassified sample. There are some drawbacks associated with the classic KNN classifier. The classification of a particular sample is decided by its distance from the classified samples; therefore, there is a high chance for the unclassified sample to be classified with a label that has the highest numbers of samples in the dataset. Samples that are already classified are considered equally significant while classifying a new sample, there might be parameters in the dataset which hold higher significance while classifying than others; if all the parameters are treated equally, the least significant parameter may mislead the classification. If the sample space of the datasets is small, then the performance of the classification may get affected by the occurrence of outliers.

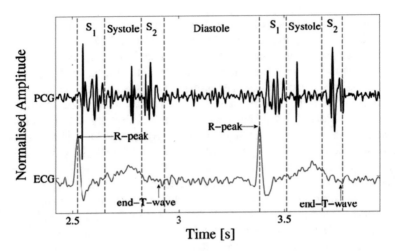

Fig. 3 Example of ECG labeled PCG

Fig. 4 Block diagram of BM-FKNCN classifier

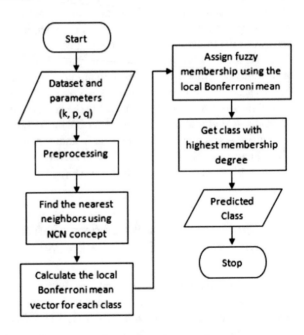

In Bonferroni mean—FKNCN, the theory of fuzzy sets and concept of nearest centroid neighbors are introduced into the KNN algorithm [14]. The classification method based on the Bonferroni mean—FKNCN follows the following steps.

Preprocessing The sample values in the data are normalized in the range [0,1], to keep all the parameter values on the same platform to eliminate the bias due to magnitude.

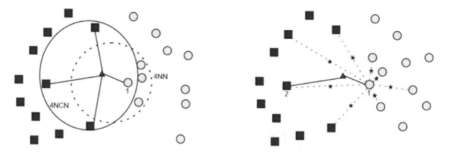

Fig. 5 Nearest centroid neighbors

Finding Nearest Neighbors using NCN concept The NCN method considers both the geometrical placements of the nearest neighbor and also the closeness in distance to the unclassified sample. Here, the first neighbor is the nearest classified sample to the unclassified sample. After that, the centroids are calculated between all the other samples and this first neighbor. The centroid which is closer to the unclassified sample, the corresponding sample other than the first neighbor responsible for that centroid is considered as the second neighbor. The third neighbor is identified based on the distance of centroid of other samples and 1st, 2nd neighbors with the unclassified sample, and so on.

Bonferroni Mean based local mean vectors calculation The Bonferroni mean is a compound function used to present information. It is used in problems with multiple criteria. It will set parameters that depend on the problem. p, q are the parameters used to find the bonferroni mean as shown

$$Br^{p,q}(X) = \left(\frac{1}{n}\sum_{i=1}^{n}x_i^p\left(\frac{1}{n-1}\sum_{i,j=1,j\neq i}^{n}x_i^p\right)\right)^{\frac{q}{(p+q)}}$$

After getting the k nearest centroid neighbors, the set of k nearest centroid neighbors is split and grouped into the class they belong to. Using Bonferroni mean, local mean vector of each class is calculated based on the grouped k nearest centroid neighbors.

Assigning fuzzy membership based on local Bonferroni Mean Vectors Membership value is calculated for an unclassified sample to all the available classes. Now the unclassified sample has a degree associated to each class. A good level of confidence to classify the sample is obtained with the help of the membership degree given to each of the classes. By using membership degrees to each class, the assignment of a specific class to the unclassified sample is never arbitrary; hence, the problem of equal significance to all the parameters given in KNN is solved. The predicted class is the class with the highest membership degree. For example in Fig. 5, for $k = 4$, the classified sample labeled 1 is the first neighbor selected based on the distance from

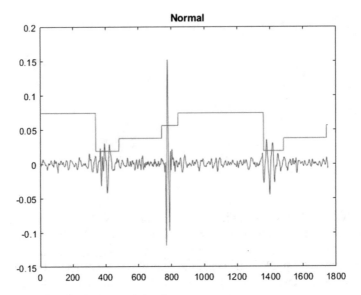

Fig. 6 Segmented normal heart sound signal

the unclassified sample. Then, the centroids are calculated among other classified samples and first neighbor. Based on the centroids calculated, the centroid nearest to the unclassified sample (labeled 2) is the second neighbor, and the third neighbor is identified based on the centroid distance of other samples and 1st, 2nd neighbors with the unclassified sample, and so on till k number of neighbors are identified.

5 Implementation

5.1 Preprocessing and Segmentation

Dataset is taken from PhysioNet Database [7]. The training data consisted of data from six different databases labeled (training a-f). The first training set (training a) consisted of audio files and corresponding ECG files that were recorded simultaneously. It consisted of 409 PCG-ECG pairs. The HSMM algorithm was trained based on the 409 ECG signals that are capable of accurately providing details about the R peak, end of T wave. The $S1$ occurs 0.04–0.06s after the onset of the QRS complex, the $S2$ occurs toward the end of the T wave [15]. After training the algorithm, the transmission matrix (A), observation matrix (B), and initial state distribution (pi) are obtained. With the help of these matrices, the segmentation is performed on the rest of the dataset. For example, the output of the algorithm is shown in Fig. 6 for normal heart sound, Fig. 7 for abnormal heart sound.

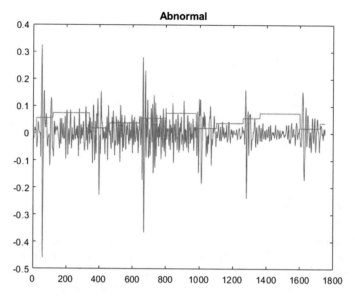

Fig. 7 Segmented abnormal heart sound signal

5.2 Feature Extraction

Features are extracted from the audio signal based on the states information obtained after segmentation. Features extracted are—Amplitude of $S1$, Amplitude of $S2$, Energy in $S1$, Energy in $S2$, Energy in Systole period, Energy in Diastole period, Mean of ratio Systole Duration/Cycle Duration, Mean of ratio Diastole Duration/Cycle Duration, Mean of ratio Systole Duration/Diastole Duration, Mean of the mean absolute amplitude ratios between systole period and $S1$ period in each heartbeat, Mean of the mean absolute amplitude ratios between diastole period and $S2$ period in each heartbeat, Mean Skewness of $S1$, Mean Skewness of Systole, Mean Skewness of $S2$, Mean Skewness of Diastole, Mean Kurtosis of $S1$, Mean Kurtosis of $S2$, and the heart rate. These features are selected based on the observation of the kernal density estimation plot [10]. An example is shown in Fig. 8 for amplitudes $S1$ and $S2$.

Features extracted are saved to a CSV file. This file is further processed to remove outliers. In this work, the outliers are removed based on the z scores. Z score shows how much a given data point value varies from the standard deviation. It is the number of standard deviations of a given data point that lies above or below the mean. Standard deviation is essentially a reflection of the amount of variability inside a given dataset. The data points which lie above 3 or below -3 standard deviations or z score of the data point is above 3 or below -3, the data point is removed and replaced with the average of all the data points corresponding to that parameter/feature. This removal of outliers helps in classifying effectively.

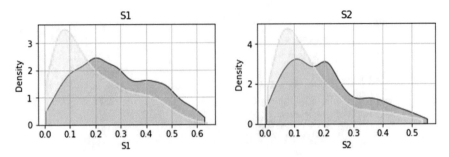

Fig. 8 Kernal density estimation plot of $S1$ and $S2$ amplitudes

5.3 Classification

The CSV file is given as an input to the Python script written on the Jupyter platform 18. The missing values in the CSV file are substituted with the mean value of that attribute for each class label. The dataset is split into 80–20% for training and testing, respectively. The Bonferroni mean-based fuzzy KNCN classifier is implemented. There are 4 input parameters k (no. of nearest neighbors), m (to calculate fuzzy membership degree), p, q (parameters of the Bonferroni Mean). The value of m is given as 2 as recommended in [14], we checked various values for k in [4, 7] and performed simulations with various values of p and q; we found $k = 10, p = 1, q = 1$ are giving better results.

6 Result

We used a fivefold cross-validation statistical method to obtain 5 different training and test sample sets from the same dataset. The training (80% of the samples in the dataset) and testing (20% of the samples in the dataset) data in a CSV file are given as input to various classification algorithms such as support vector machine, decision tree, Naïve Bayes, K-nearest neighbor, logistic regression, and Bonferroni mean—FKNCN. With Bonferroni mean—FKNCN, we obtained a mean accuracy of 86.9% and a maximum accuracy of 88.4%, as shown in Figs. 9 and 10.

7 Conclusion

In this work, phonocardiograms are segmented and classified with the help of a dataset consisting of 3240 heart sound audio samples. If the no. of audio samples increases, training the classifier gets better, resulting in high classification accuracy. The segmentation algorithm in this work is trained with 409 audio samples, to increase this no. we need to obtain more no. of audio samples for which the PCG and ECG are

```
print("SVM Accuracy\t\t\t=\t",np.mean(predictions[:,0]))
print("Decision Tree Accuracy\t\t=\t",np.mean(predictions[:,1]))
print("Naive Bayes Accuracy\t\t=\t",np.mean(predictions[:,2]))
print("KNN Accuracy\t\t\t=\t",np.mean(predictions[:,3]))
print("Logestic Regression Accuracy\t=\t",np.mean(predictions[:,4]))
print('Mean Bonferroni Accuracy\t=\t', np.mean(predictions[:,5]))
print('MAX Bonferroni Accuracy\t\t=\t', np.max(predictions[:,5]))
```

```
SVM Accuracy                   =     0.8237654320987653
Decision Tree Accuracy         =     0.8339506172839506
Naive Bayes Accuracy           =     0.8160493827160493
KNN Accuracy                   =     0.850925925925926
Logestic Regression Accuracy   =     0.8287037037037036
Mean Bonferroni Accuracy       =     0.8694444444444445
MAX Bonferroni Accuracy        =     0.8842592592592593
```

Fig. 9 Comparison between various classification algorithms

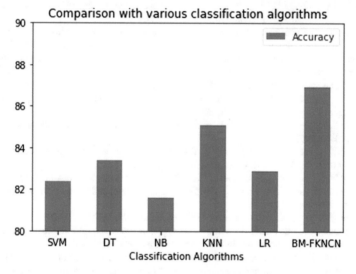

Fig. 10 Plot of accuracies of various classification algorithms

recorded simultaneously if this no. increases, then segmentation accuracy increases significantly. Apart from increasing the no. of samples in the database, carefully selecting the essential samples plays a vital role in classifying accurately. Feature extraction is one of the critical steps in the complete classification problem, finding proper parameters to differentiate between normal and abnormal heart sounds is necessary. This setup can be implemented on an FPGA board for real-time assessment of an audio signal. The time consumed while training the model is huge because of the calculation of neighbors for 3240 samples five times, but it is a one-time process, after which the new unclassified sample is classified simply by calculating ten neighbors and the class is decided.

References

1. Cardiovascular diseases (cvds). https://www.who.int/news-room/fact-sheets/detail/cardiovascular-diseases-(cvds). Accessed on 02 Aug 2022
2. Non-invasive tests and procedures—American heart association. https://www.heart.org/en/health-topics/heart-attack/diagnosing-a-heart-attack/noninvasive-tests-and-procedures. Accessed on 02 Aug 2022
3. Hazeri H, Zarjam P, Azemi G (2021) Classification of normal/abnormal PCG recordings using a time-frequency approach. Analog Integr Circ Sig Proc 109(2):459–465
4. Khan MF, Atteeq M, Qureshi AN (2019) Computer aided detection of normal and abnormal heart sound using PCG. In: Proceedings of the 2019 11th international conference on bioinformatics and biomedical technology. Association for Computing Machinery, New York, NY, USA, pp 94–99
5. Almanifi ORA, Mohd Razman MA, Musa RM, Ab. Nasir AF, Ismail MY, Abdul Majeed PPA (2022) The classification of heartbeat PCG signals via transfer learning. In: Ab. Nasir AF, Ibrahim AN, Ishak I, Mat Yahya N, Zakaria MA, Abdul Majeed PPA (eds) Recent trends in mechatronics towards industry 4.0. Springer Singapore, Singapore, pp 49–59
6. Tang H, Dai Z, Jiang Y, Li T, Liu C (2018) PCG classification using multidomain features and SVM classifier. BioMed Res Int, Hindawi 2018:4205027
7. Classification of heart sound recordings: the physionet/computing in cardiology challenge 2016 v1.0.0. https://physionet.org/content/challenge-2016/1.0.0/. Accessed on 02 Sep 2022
8. Sedighian P, Subudhi A, Scalzo F, Asgari S (2014) Pediatric heart sound segmentation using hidden markov model. In: 2014 36th annual international conference of the IEEE engineering in medicine and biology society, EMBC 2014, pp 5490–5493
9. Castro A, Vinhoza T, Mattos S, Coimbra M (2013) Heart sound segmentation of pediatric auscultations using wavelet analysis 2013:3909–3912
10. seaborn.kdeplot—seaborn 0.11.2 documentation. https://seaborn.pydata.org/generated/seaborn.kdeplot.html. Accessed on 20 Feb 2022
11. Springer DB, Tarassenko L, Clifford GD (2016) Logistic regression-HSMM-based heart sound segmentation. IEEE Trans Biomed Eng 63(4):822–832
12. Bentley PM, Grant PM, McDonnell JT (1998) Time-frequency and time-scale techniques for the classification of native and bioprosthetic heart valve sounds. IEEE Trans Biomed Eng, United States 45(1):125–128
13. Quiceno-Manrique A, godino llorente J, Blanco-Velasco M, Castellanos-Dominguez G (2009) Selection of dynamic features based on time-frequency representations for heart murmur detection from phonocardiographic signals. Annal Biomed Eng 38:118–37
14. Widyadhana A, Putra C, Indraswari R, Arifin AZ (2021) A bonferroni mean based fuzzy k nearest centroid neighbor classifier. Jurnal Ilmu Komputer dan Informasi 14:65–71
15. Balakrishnan M, Kamarulafizam I, Salleh S, Helmi D (2003) Heart sound segmentation algorithm based on instantaneous energy of electrocardiogram 30:327–330

An Intelligent Smart Waste Management System Using Mobile Application and IoT

R. Sajeev Krishna, S. Sridevi, K. Shyam, T. Tamilselvan, R. Ajay Rajan, and C. Deisy

1 Introduction

India is a country with a very high population that generates a large amount of solid urban waste. Smart cities like Madurai generates around 500 metric tons per day. To collect these wastes, dustbins are placed in the roadsides.

Unlike other countries, India has a lot of stray animals that wander around roadsides. The animals like cows, monkeys, and dogs also eat from the garbage transmitting a lot of diseases. The garbage lies scattered all over the dustbin due to the erratic collection of waste by the municipal department. Hence, there is a need to streamline the systems and bring it under a single umbrella by synchronizing the system using an integrated mechanism. The existing problems are

- The dustbin is left unattended, and it overflows often creating an unhygienic environment.
- The waste is eaten by dogs, cows, and monkeys.
- The corporation does not receive real-time status regarding the level of the dustbins.
- There is no proper grievance redressal platform for the citizens.

The solution proposed in this article implements an integrated system to address all the existing problems. The data of the dustbin and other information is stored in the DBMS of the server. The IoT module performs the operation on collecting the data from the dustbin and sending it to the web server. It also repels the animals that comes near the dustbin. When the dustbin is full, then the data is sent to the application that is used by the sanitary workers and officials of the municipal corporation. The dustbin can also repel animals from coming near it, thereby enhancing hygiene.

R. Sajeev Krishna · S. Sridevi (✉) · K. Shyam · T. Tamilselvan · R. Ajay Rajan · C. Deisy
Department of Information Technology, Thiagarajar College of Engineering, Madurai, Tamilnadu, India
e-mail: sridevi@tce.edu

© The Author(s), under exclusive license to Springer Nature Singapore Pte Ltd. 2023
D. S. Sisodia et al. (eds.), *Machine Intelligence Techniques for Data Analysis and Signal Processing*, Lecture Notes in Electrical Engineering 997,
https://doi.org/10.1007/978-981-99-0085-5_55

For the development of smart cities and implementing other development schemes, proper waste management and collection is necessary. This solution is aimed at creating a low cost and an efficient solution to achieve the same.

2 Literature Review

There have been a huge number of researchers who are working in smart waste management system using mobile application or Internet of things (IoT). This section provides some of the literatures from the existing work.

Singhvi et al. [1] developed an IoT device that sent the SMS when the dustbin overflows, and a gas sensor is used to monitor the toxicity levels. The disadvantage in this method is that when the mobile number changes the program in the Arduino must also be changed. It is also not effective to send SMS when the number of recipients is huge. Joshi et al. [2] used a stack-based front-end approach and a wireless sensor network for real-time tracking of the dustbins. Raj et al. [3] have filed a patent for an IoT-based mechanism (201,941,026,559 A). It uses a weight sensing mechanism to detect if the dustbin is full or not.

The mechanism of communicating the smart bin data through a Wi-Fi module was discussed in [4, 5]. Imteaj et al. [6] done research on "Dissipation of waste using dynamic perception and alarming system" by implementing an android app to find nearby dustbin locations in real time. Ankitha et al. [7] proposed "Smart city initiative: Traffic and waste management" in which they are using smart dustbins with different ID numbers and a server to integrate the process. But there is no idea to eliminate the contact of animals from the smart bins [8, 9]. Muruganandham et al. [5] have proposed a system to recycle the uncollected waste directly below the dustbin.

Wang et al. [10] used deep learning convolution neural networks (CNN) to perform the garbage classification task. They also used MobileNetV3 architecture for garbage classification. Out of these two algorithms, MobileNetV3 performs well in terms of accuracy and running time. Krishna et al. [11] proposed the framework to track dustbin status through web servers. When the dustbin reaches its maximum level, a warning message sent to the office via GSM module.

The paper written by Hayati et al. [12] discussed about the generation of waste in MSW landfills and the need to develop it as an alternate energy resource. Smart dustbin [8] was developed to work when there is a people coming near to the dustbin to throw rubbish. The smart dustbin worked based on ultrasonic sensor and motor. The amount of rubbish can be seen by application. Khan et al. [9] developed an efficient framework for waste management that uses machine learning (ML) and the Internet of things (IoT). Rohit et al. [13] introduced the model for smart dust bin that is introduced for smart cities. They have used ultrasonic and Infrared sensors for checking the status of the dustbin. They have used two dustbins and if the dustbin is full, then the message will be sent to the concerned authority. The system worked based on image processing techniques, Arduino UNO microcontroller, moisture sensor, and ultrasonic sensor. In the above literatures, they used either web application or mobile application that is used to detect waste level but not the animal contacts.

To address the above issues and to avoid the contact of animals, this work introduced a GSM/GPRS module to send data about the dustbin to the mobile application via an RDBMS powered server.

3 Proposed Methodology

In this work, we are using a GSM/GPRS module to send data about the dustbin to the mobile application via an RDBMS powered server. Our system provides real-time levels about the status of the dustbin.

It also includes an IR proximity sensor that detects the animals coming near the bin. Upon detection, the ultrasonic sensor emits ultrasound of 40 kHz that repels the animals [11].

- The GSM module sends a HTTP POST request to the server with the dustbin data as the POST parameters.
- The ultrasonic sensor HC-SR04 detects the level of waste in the dustbin.
- The IR proximity sensor detects the presence of any object near the dustbin.
- The mobile application performs a GET request to the server and fetches real-time data from it.

The working can also be understood from the flow (Fig. 1), block diagram (Fig. 2), and the schematic diagram (Fig. 3). The detailed explanation of each diagram is given in sub-modules.

3.1 Microcontroller for Interfacing the Sensors

The microcontroller is Arduino UNO R3 (Ref: https://www.electronicscomp.com/arduino-uno-r3-smd-atmega328p-board-compatible-india) for connecting and integrating the sensors. It contains 14 digital pins. Both the ultrasonic sensor and the proximity sensors are connected to the digital pins. The specifications are given in Table 1 (Fig. 4).

3.2 Sensor for Measurement of Waste Level in Dustbin

The ultrasonic sensor contains sections transmitter and receiver. It works on the principle of echo. The obstacle is the garbage in the dustbin. The transmitter emits ultrasound of 40 kHz which is picked up by the receiver, and the distance is calculated using the below formula (Fig. 5; Table 2).

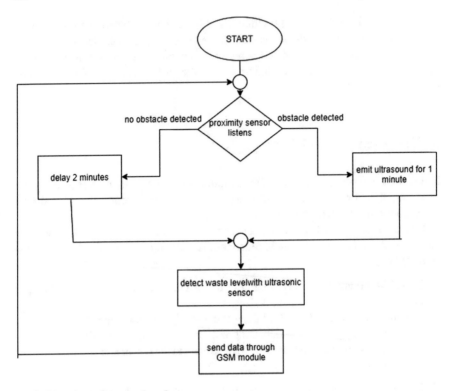

Fig. 1 Flowchart of proposed work

$$\text{distance} = (\text{speed of sound} * \text{time})/2 \qquad (1)$$

$$\text{speed of sound in air} = 334\,\text{m/s} \qquad (2)$$

The effectual angle is the maximum angle at which an object can be placed from the center of the sensor. This placement can be taken in order to deliver accurate reading is ± 11°. Effectual angle is the offset angle in which the sensor works effectively.

3.3 Sensor for Detecting Animals

The proximity sensor (Ref: https://www.engineersgarage.com/proximity-sensors-optical-ultrasonic-inductive-magnetic-capacitive/) contains an IR emitter and detector. When an object approaches the sensor, the bounced back IR light is detected by the sensor. The proximity sensor communicates with the ultrasonic sensor to emit

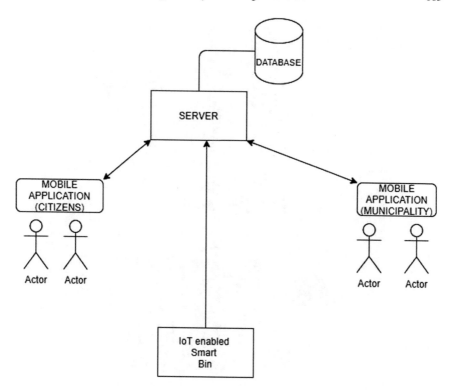

Fig. 2 Block diagram of overall working concepts

ultrasound to repel away the animals. Ultrasound cannot be heard by humans (Fig. 6; Table 3).

3.4 Sensor for Communication

The global system for mobile communication (GSM) module is capable of sending HTTP requests at a data rate of 10 KBps. It works on the 900 MHz band. It can also make voice calls and send messages. GSM (Ref: https://www.electronicwings.com/sensors-modules/sim900a-gsmgprs-module) module is shown in Fig. 7. The specifications are given in Table 4.

3.5 Mobile Application

The mobile application was developed using Android studio using Java programming language. It is polymorphic in nature by virtue of providing different views to

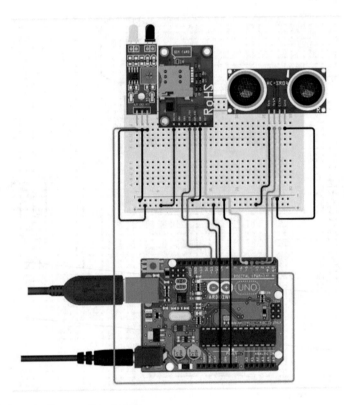

Fig. 3 Schematic diagram of the IoT kit

Table 1 Specification for Arduino UNO R3

Parameters	Specification
Flash memory	32 Kb
Clock speed	16 MHz
Input voltage	7–12 V
Operating voltage	5 V
DC current per I/O pin	20 mA

the citizens and the municipality workers. It contains a login page and other functionalities to send feedback and post grievances for the notice of the municipality and corporation officials. It is possible to locate the next nearest dustbin public taps and public toilets.

Fig. 4 Arduino UNO R3

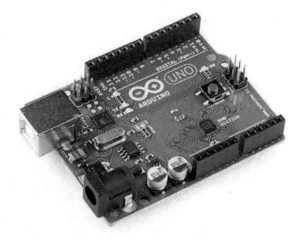

Fig. 5 Ultrasonic sensor HC-SR04

Table 2 Specification for ultrasonic sensor HC-SR04

Parameters	Specification
Operating voltage	5 V
Working current	15 mA
Working frequency	40 kHz range
Effectual angle	< 15°
Operating voltage	5 V

3.6 Server

The server has been developed using Python programming language and the Flask framework. A MYSQL database is used along with it. It receives POST requests from the smart bins and the mobile applications and sends GET requests to the mobile applications to notify about the level of the dustbin. It also stored the details of the geographical locations of the public taps and toilets.

Fig. 6 Proximity sensor

Table 3 Specification for proximity sensor

Parameters	Specification
Operating voltage	5 V
Range	10–20 cm
DC current	20 mA

Fig. 7 GSM sensor

Table 4 Specification of GSM Sensor

Parameters	Specification
Operating voltage	12 V
Data rate	10 KBps
Bandwidth	900 MHz
Working current	2 A
Sim slot	2G

4 Results and Discussion

4.1 Screenshots

These are the screenshots from the mobile application that are used in the proposed work.

Different users like administrator, public, etc., can use the proposed applications. They first signup in the application and then they can login using the application. It is shown in Fig. 8. Home screen dashboard includes grievances, feedback collection, information, and assistant. It is shown in Figs. 9, 10, 11 and 12.

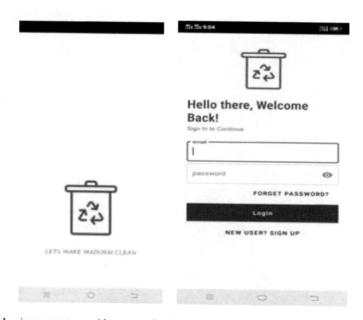

Fig. 8 Login screens to provide separate functionalities for different types of users

Fig. 9 Home screen
dashboard

Fig. 10 Page that displays
the status of the bin in real
time

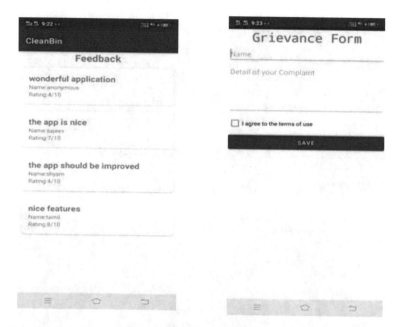

Fig. 11 Grievance redressal forum for users and administrators

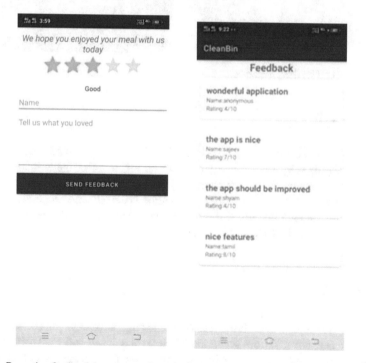

Fig. 12 Reporting feedback by a user and receiving them by an administrator

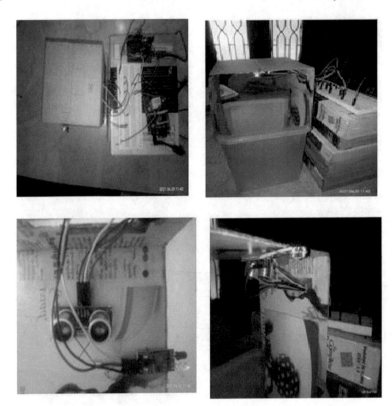

Fig. 13 Low-cost working prototype

4.2 Prototype

The low-cost working prototype is shown in Fig. 13. Here, GSM module is connected to Arduino, proximity sensor, and ultrasonic sensor to detect the animal contacts. When animal contacts found, the alarm will be sent to the concerned authority with the help mobile application. A mobile application is developed to supervise the data and includes extended functionalities to identify the location of the nearest dustbin, public tap, and swachh toilets.

4.3 Usability Testing

The prototype subjects to usability testing. For testing purpose, feedback is collected from 15 users to cover various scenarios. Their feedback was collected using Google Forms. The feedback includes application layout, the performance, and final rating. It is shown in Figs. 14 and 15 (Table 5).

Fig. 14 Feedback about the application layout

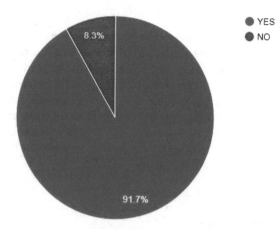

DID YOU LIKE THE APP LAYOUT?

● YES
● NO

8.3%

91.7%

WAS THE PERFORMANCE OF THE APP SATSIFACTORY?

● YES
● NO

22.7%

77.3%

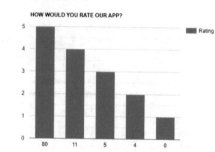

HOW WOULD YOU RATE OUR APP?

■ Rating

Fig. 15 Feedback about the performance and rating of the application Table 5 indicates various types of test cases and their results

Table 5 Test case scenarios and results

Test case No.	Type of testing	Test case	Expected result	Test result
1	Acceptance testing	Dog approaches the dustbin	Ultrasound emitted repels the dog	Passed
2	Acceptance testing	Dustbin becomes full	Sanitary workers are notified	Passed
3	Load testing	The prototype was tested with more number of devices	The prototype works smoothly	Passed
4	Unit testing	The user queries the nearby dustbin	The app locates the nearby dustbin	Passed

5 Conclusion

The proposed system is very much useful for the corporation department for monitoring the smart waste, and further actions can be done. It helps to make the cities smarter and manage the solid waste in an effective manner. Thus, the problem of knowing the waste levels in the dustbins, the overflowing dustbins, finding the animal contacts nearby dustbins are solved. Moreover, the mobile application serves as a platform for citizens to report their issues. This system enables the municipal corporation to work more efficiently. The prototype has been effectively tested against all test cases and scenarios. The product performance has been improved after receiving feedback from the users.

References

1. Singhvi RK, Lohar RL, Kumar A, Sharma R, Sharma LD, Saraswat RK (2019) IoT based smart waste management system: India prospective. In: Paper presented at the proceedings—2019 4th international conference on internet of things: smart innovation and usages, IoT-SIU 2019. https://doi.org/10.1109/IoT-SIU.2019.8777698
2. Joshi J, Reddy J, Reddy P, Agarwal A, Agarwal R, Bagga A, Bhargava A (2017) Cloud computing based smart garbage monitoring system. In: Paper presented at the 2016 3rd international conference on electronic design, ICED 2016, pp 70–75. https://doi.org/10.1109/ICED.2016.7804609
3. Joshua Samuel Raj T, Perumal SR, Dinesh Peter J, Jegatheesan A, Jaya-Chandran (2019) Garbage waste monitoring and collection system with reduced human intervention. Patent application publication India published on 12/07/2019
4. Hayati AP, Emalya N, Munawar E, Schwarzböck T, Lederer J, Fellner J (2018) Analysis the potential gas production of old municipal solid waste landfill as an alternative energy source: preliminary results. In: Paper presented at the IOP conference series: materials science and engineering, vol 334, no 1. https://doi.org/10.1088/1757-899X/334/1/012031
5. Murugaanandam S, Ganapathy V, Balaji R (2018) Efficient IOT based smart bin for clean environment. In: Paper presented at the proceedings of the 2018 IEEE international conference on communication and signal processing, ICCSP 2018, pp 715–720. https://doi.org/10.1109/ICCSP.2018.8524230
6. Imteaj A, Chowdhury M, Mahamud MA (2015) Dissipation of waste using dynamic perception and alarming system: a smart city application. In: Paper presented at the 2nd international conference on electrical engineering and information and communication technology, iCEEiCT 2015. https://doi.org/10.1109/ICEEICT.2015.7307410
7. Ankitha S, Nayana KB, Shravya SR, Jain L (2017) Smart city initiative: traffic and waste management. In: Paper presented at the RTEICT 2017—2nd IEEE international conference on recent trends in electronics, information and communication technology, pro cedings, pp 1227–1231. https://doi.org/10.1109/RTEICT.2017.8256794
8. Alsayaydeh JAJ, Khang AWY, Indra WA, Shkarupylo V, Jayasundar J (2019) Development of smart dustbin by using apps. ARPN J Eng Appl Sci 14(21):3703–3711
9. Khan R, Kumar S, Srivastava AK, Dhingra N, Gupta M, Bhati N, Kumari P (2021) Machine learning and IoT-based waste management model. Comput Intelli Neurosci. https://doi.org/10.1155/2021/5942574
10. Wang C, Qin J, Qu C, Ran X, Liu C, Chen B (2021) A smart municipal waste management system based on deep-learning and internet of things. Waste Manage 135:20–29. https://doi.org/10.1016/j.wasman.2021.08.028

11. Bala Krishna P, Naga Swapna Sri V, Vamsi Krishna T, Satya Sandeep K (2020) Smart city waste management control and analysis system using IoT. Int J Adv Sci Technol 29(6):3178–3186
12. Kolhatkar C, Joshi B, Choudhari P, Bhuva D (2018) Smart E-dustbin. In: Paper presented at the 2018 international conference on smart city and emerging technology, ICSCET 2018.https://doi.org/10.1109/ICSCET.2018.8537245
13. Rohit GS, Chandra MB, Saha S, Das D (2018) Smart dual dustbin model for waste management in smart cities. In: Paper presented at the 2018 3rd international conference for convergence in technology, I2CT 2018. https://doi.org/10.1109/I2CT.2018.8529600

Pan-Sharpening of Multi-spectral Remote Sensing Data Using Multi-resolution Analysis

Aparna S. Menon, J. Aravinth, and S. Veni

1 Introduction

Multi-spectral (MS) images produced by Earth Observation (EO) sensors are widely used for land use-land cover (LULC) applications [1]. Multi-spectral images are a group of greyscale images of a scene captured in different spectral ranges. The typical range of MS varies from 3 to 15. Even though LULC applications demand high spatial resolution, MS images will not be able to deliver images with that resolution due to sharing of energy by all bands. This is where the use of pan-sharpening arises which means merging of MS band with panchromatic band (PAN) [2, 3]. A panchromatic band is a single band greyscale image with higher spatial resolution. Since the MS and PAN band capture the same scene, merging of both would result in an MS image with better spatial resolution. Pan-sharpening methods are broadly classified into 3 sections: Pan-sharpening methods are broadly classified into 3 sections:

1. Component substitution (CS): Here, the components of the MS image are substituted to PAN image. Common techniques used here are intensity-hue-saturation (IHS), Ehlers method, and principal component analysis (PCA) [4].
2. Relative contribution (RC): They provide better spectral information than CS. Commonly used RC techniques are high pass filter (HPF), multiplicative, and Brovey transforms [5].

S. M. Aparna (✉) · J. Aravinth (✉) · S. Veni
Department of Electronics and Communication Engineering, Amrita School of Engineering,
Amrita Vishwa Vidyapeetham, Coimbatore, India
e-mail: sm_aparna@cb.students.amrita.edu

J. Aravinth
e-mail: j_aravinth@cb.amrita.edu

S. Veni
e-mail: s_veni@cb.amrita.edu

© The Author(s), under exclusive license to Springer Nature Singapore Pte Ltd. 2023   697
D. S. Sisodia et al. (eds.), *Machine Intelligence Techniques for Data Analysis and Signal Processing*, Lecture Notes in Electrical Engineering 997,
https://doi.org/10.1007/978-981-99-0085-5_56

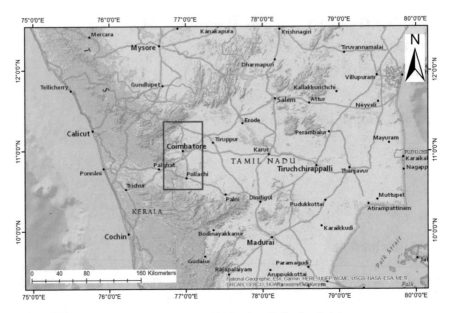

Fig. 1 The study area of Coimbatore, located in Southern India, South Asia

3. Multi-resolution analysis (MRA): Here, the images are decomposed into coefficients and are later merged. In this paper, we apply different MRA techniques to Landsat-8 dataset.

2 Materials and Methods

2.1 Study Area and Dataset

Here, the Landsat-8 MS dataset of coordinates: $1101'06.00''$ N and $7658'2900''$ E of Coimbatore a prominent district in the State of Tamilnadu is considered for dataset Fig. 1. The raw MS and PAN Landsat-8 images are shown in Fig. 2.

2.2 Experimental Setup and Pre-processing

For the Landsat-8 dataset used, ENVI 5.3 has been used for the pre-processing purpose, and the fusion works are done using MATLAB 2020b on a 1.80 GHz CPU and 8GB RAM personal computer.

Fast line-of-sight atmospheric analysis of hypercubes (FLAASH) is applied to the raw Landsat-8 MS datasets (both MS and PAN) using ENVI 5.3 for pre-processing

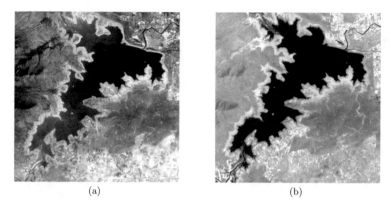

(a)   (b)

Fig. 2 Raw Landsat-8 images of study area **a** RGB band, **b** PAN band

(a)   (b)

Fig. 3 Pre-processed Landsat-8 images of study area **a** RGB band, **b** PAN band

as it helps in atmospheric correction [6]. The pre-processed images are shown in Fig. 3.

2.3 Discrete Wavelet Transform (DWT)

DWT is a tool that helps in hierarchical decomposition. This transform is a time-frequency representation of a signal using wavelets of varying frequencies [8, 9]. In this work, we are using different wavelets such as haar, Daubechies (db2), symlet (sym2), coiflet (coif2), discrete Meyer (dmey), and Fejer-Korovkin (fk4) filters for MRA analysis.

- Haar wavelet: Haar is the simplest wavelet that resembles a step function and is termed as the mother of wavelets [5].

- Daubechies wavelet: These are orthonormal wavelets that made discrete wavelet analysis practically possible [5].
- Symlet wavelet: Symmetric altering of the Daubechies wavelet resulted in symlet, and its properties are same that of Daubechies [5].
- Coiflet wavelet: These are compactly supported wavelets with the highest number of vanishing moments for a given support width [5].
- Discrete Meyer wavelet: The scaling functions of this wavelet are defined in the frequency domain [5].
- Fejer-Korovkin filter: It is a more symmetric but less smooth wavelet than Daubechies [10].

2.4 Steps for Fusion

1. The red–green–blue (RGB) image is converted to hue-saturation-intensity (HSI).
2. The intensity image which is a greyscale and the PAN is applied to the respective MRA, and the approximations and details coefficients are extracted separately.
3. The coefficients are fused taking the maximum of the absolute value of the coefficients for both approximations and details. The inverse transform is applied, and the obtained image is considered as the pan-sharpened intensity image.
4. With the new intensity image, HSI is converted into RGB back which is the pan-sharpened image.

3 Results and Discussion

Figure 4 shows the results of different DWTs applied to the pre-processed image. Six metrices such as correlation coefficients (CC), standard deviation (SD), peak signal to noise ratio (PSNR), root mean square error (RMSE), spectral angular mapping (SAR), and erreur relative globale adimensionnelle de synth'ese (ERGAS) are used for comparison.

Six metrices as mentioned above are used for performance analysis in this paper. The first metric used for comparison is CC. The chart for CC of MS versus fused MS is given in Fig. 5. A higher value is desirable for CC. The figure shows coiflet, symlet, and discrete Meyer having values 0.99, 0.98, and 0.98 for B2, B3, and, B4, respectively, which has the highest CC value.

The second metric used for comparison is SD. The chart for SD of MS versus fused MS is given in Fig. 6. A higher value is desirable for SD. The figure shows all the DWTs having almost the same SD values despite small changes.

The third metric used for comparison is PSNR. The chart for PSNR of MS versus fused MS is given in Fig. 7. A higher value is much desirable for PSNR. Haar gives the highest PSNR value 27.09, 25.57, and 25.6 for B2, B3, and B4, respectively.

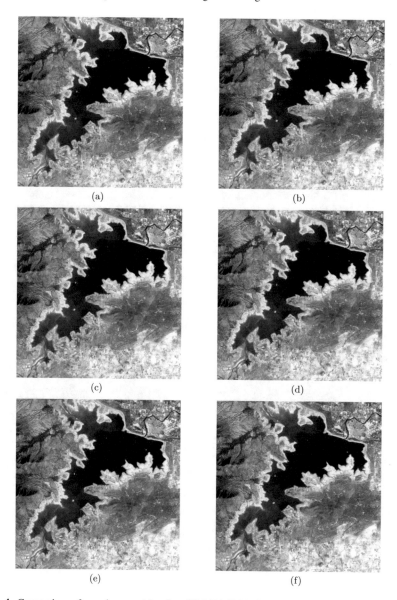

Fig. 4 Comparison of pan-sharpened Landsat-8 B2-B4 (RGB) images derived from different methods **a** DWT (db2), **b** DWT (coif2), **c** DWT (sym2), **d** DWT (dmey), **e** DWT (haar), **f** DWT (fk4)

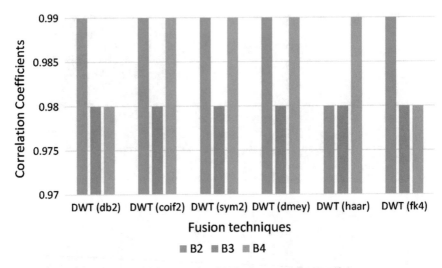

Fig. 5 Chart for evaluation of MS versus fused MS using correlation coefficient

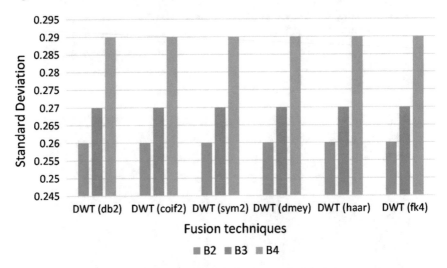

Fig. 6 Chart for evaluation of MS versus fused MS using standard deviation

The fourth metric used for comparison is RMSE. The chart for RMSE of MS versus fused MS is given in Fig. 8. A lower value is desirable for RMSE as the deviation from the reference image will be less. The figure shows all the DWTs having almost the same RMSE values despite small changes.

The fifth metric used for comparison is SAM. The chart for SAM of MS versus fused MS is given in Fig. 9. A lower value is desirable for SAM as spectral angle deviation from the original image would be less. Coiflet gives the lowest SAM value 4.39, 4.79, and 4.78 for B2, B3, and B4, respectively.

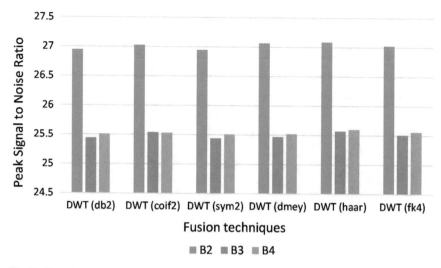

Fig. 7 Chart for evaluation of MS versus fused MS using peak signal to noise ratio

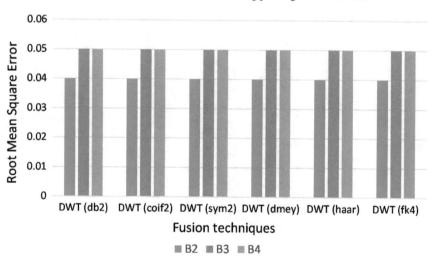

Fig. 8 Chart for evaluation of MS versus fused MS using root mean square error

The sixth metric used for comparison is ERGAS. The chart for ERGAS of MS versus fused MS is given in Fig. 10. A lower value is desirable for ERGAS showing spectral similarity. Coiflet also gives the lowest ERGAS value out of all methods 0.48, 0.57, and 0.62 for B2, B3, and B4, respectively.

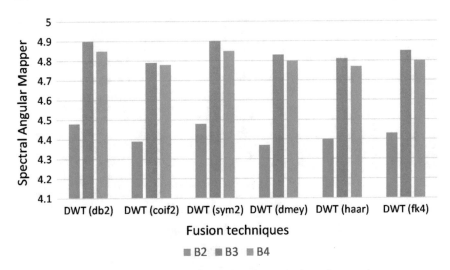

Fig. 9 Chart for evaluation of MS versus fused MS using spectral angular mapping

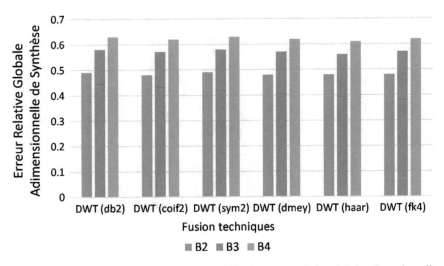

Fig. 10 Chart for evaluation of MS versus fused MS using erreur relative globale adimensionnelle de synthése

4 Conclusion

In this paper, pan-sharpening of Landsat-8 dataset using MRA techniques for enhanced spectral and spatial quality is performed. All the methods have a higher CC ensuring correlation with the original MS dataset. Similar SD and RMSE show lesser variation from the source image. The PSNR value is the highest for Haar wavelet, but Coiflet has the most desirable SAM and ERGAS value making it the most desirable MRA technique out of this which can give a spatially and spectrally rich pan-sharpened image.

References

1. Reshma S, Veni S (2018) Comparative analysis of classification techniques for crop classification using airborne hyperspectral data. In: International conference on wireless communications, signal processing and networking, pp 2272–2276
2. Liu H, Zhang X (2009) Comparison of data fusion techniques for Beijing-1 micro-satellite images. In: Joint urban remote sensing event, pp 20–22
3. Diana AC, Aravinth J (2018) Fusion methods for hyperspectral image and LIDAR data at pixel-level. In: International conference on wireless communications, signal processing and networking, pp 1–3
4. Kpalma K, El-Mezouar C et al (2014) Recent trends in satellite image pan-sharpening techniques. In: 1st international conference on electrical, electronic and computing engineering, pp 2–5
5. Amarsaikhan D, Saandar M (2011) Fusion of multisource images for update of urban GIS in image fusion and its applications. Intech Open Access, pp 127–152
6. Fast line-of-sight atmospheric analysis of hypercubes. https://www.l3harrisgeospatial.com/docs/flaash.html. Last accessed 20/02/2022
7. Goel N, Singh G (2016) Study of wavelet functions of discrete wavelet transformation in image watermarking. Int J Eng Sci 154–160
8. Anand R, Veni S et al (2021) Robust classification technique for hyperspectral images based on 3D-discrete wavelet transform. Rem Sens 13:1255
9. Nidhin PTV, Geetha P Empirical wavelet transform for improved hyperspectral image classification. In: Intelligent systems technologies and applications. Springer International Publishing, pp 393–401
10. Nielsen M (2001) On the construction and frequency localization of finite orthogonal quadrature filters. J Approx Theor 36–52

Detection of Malaria by Using a CNN Model

Dilip Kumar Choubey⊙**, Sandeep Raj, Arju Aman, Rajat Deoli, and Rishabh Hanselia**

1 Introduction

Malaria continues to be a major health concern around the world, affecting around 200 million people and resulting in over 400,000 fatalities each year. It is one of the most frequent infections in the world, affecting children, pregnant women, people with other health problems such as HIV/AIDS, and those who have never been exposed to malaria before. Modern information technology, in addition to the medical and political domain, is playing a vital role in many initiatives to combat this disease. There are over 100 different Plasmodium infections, but only five of them cause malaria in humans.

Plasmodium falciparum (or P. falciparum), Plasmodium malariae (or P. malariae), Plasmodium vivax (or P. vivax), Plasmodium ovale (or P. ovale), Plasmodium knowlesi (or P. knowlesi). The five categories mentioned above are the ones found in humans. Only two of these five are hazardous to humans (P. falciparum. and P. vivax). The malaria parasites are only carried by female Anopheles mosquitos.

D. K. Choubey (✉) · R. Hanselia
Department of Computer Science and Engineering, Indian Institute of Information Technology
Bhagalpur, Bihar, India
e-mail: dkchoubey.cse@iiitbh.ac.in

R. Hanselia
e-mail: rishab.cse.2102020@iiitbh.ac.in

S. Raj · A. Aman · R. Deoli
Department of Electronics and Communication Engineering, Indian Institute of Information
Technology Bhagalpur, Bihar, India
e-mail: sraj.ece@iiitbh.ac.in

A. Aman
e-mail: arju.ece.1709@iiitbh.ac.in

R. Deoli
e-mail: rajat.ece.1735@iiitbh.ac.in

D. S. Sisodia et al. (eds.), *Machine Intelligence Techniques for Data Analysis and Signal
Processing*, Lecture Notes in Electrical Engineering 997,
https://doi.org/10.1007/978-981-99-0085-5_57

When a female mosquito infected with Plasmodium bites a malaria patient, it sucks the patient's blood, which includes Plasmodium parasites. When another person is bitten by the mosquito, the parasites are injected into that person's bloodstream and then transported to the liver. Once the virus migrates from the liver to the red blood cells, which is when the human body begins to display symptoms. This is how the infection spreads.

When a Plasmodium enters the human body, it targets and attempts to damage red blood cells (RBC) and the liver. Each of these mosquito species has a favored habitat. Some species prefer freshwater pools that are small and shallow. Also, where the mosquito life span is greater than expected, transmission is higher because the parasite has more time to thrive and develop inside the mosquito, and mosquitos prefer to bite humans rather than other animals.

It also depends on climatic factors such as rainfall patterns, temperature, and humidity, which can affect mosquito population, life span, and survival. Seasonal transmission is common in many regions. During and after the rainy season, it is at its peak. These symptoms often develop 10–15 days following the infective mosquito bite in humans. Heavy breathing, rapid heart rate, high temperature, shaking chills, sweating, headache, diarrhea, weariness, body aches, yellow skin (jaundice), kidney failure, seizure, bloody feces, and convulsions are some of the symptoms that can occur. Because these symptoms are so similar to those of a normal cold or flu, it can be difficult to discern what a person is suffering from at first. They also don't always arrive within two weeks.

If anybody lives in or travels to a location where malaria is very widespread, they should take precautions to avoid mosquito bites since mosquitoes in that area are so plentiful that catching malaria is almost a certainty. Mosquitoes are most active at dusk and during the night.

It's critical to get medical attention as soon as possible. Malaria has a particularly high risk of causing serious illness in young children, babies, and pregnant women. If diagnosed and treated early malaria can be reduced and deaths can be prevented. Artemisinin-based combination therapy is the best current treatment, especially for P. falciparum malaria (ACT).

If you have a high fever while living in or going to a malaria-prone area, seek medical attention. Even if the symptoms appear weeks, months, or even a year after your trip, you should seek medical attention.

Malaria can be easily detected with a blood test, but a physical exam isn't uncommon, your physician will also perform a physical examination and ask you relevant questions about your recent travel and medical history.

The blood test can tell your doctor:

- Presence of parasite in your blood
- Whether specific treatments would be effective against it
- If your body has ever produced antibodies to combat malaria.

Everything is now done by computer since it provides more accuracy and has the computing power to do any operation in the blink of an eye. One of the most difficult challenges that hospitals face is providing the best treatment to patients at a fair cost. For this, recent breakthroughs in cognitive computing, deep learning, machine learning, and soft computing can be employed to use computer-based information. The prime objective of this research is to encourage the early detection of malaria parasites by using the Convolution Neural Network (CNN) on Malaria Dataset available on Kaggle to train the model. After training, the model can be acquired as a package that can be used as a mobile phone app, with the user receiving the results after uploading a picture. To use the model on a smartphone, we must convert it to a TensorFlow Lite model, as this is the most efficient way to execute the trained model with the limited memory and computational capacity available on smartphones.

Similar data science and machine learning approaches have been employed by Jangir et al. [1], Choubey et al. [2, 3] for the diagnosis and prediction of medical diabetes. The concept was conceived through a review of many published articles, texts, and references such as comparative analysis of classification methods, performance evaluation of classification methods [4], rule-based diagnosis system [5], and classification techniques and diagnosis for diabetes [6, 7] has proven to be extremely useful in the completion of the current work.

The remainder of the paper may be structured as follows: Sect. 2 contains the motivation, Sect. 3 discusses the Related Work, Sect. 4 deals with the proposed methodology, Sect. 5 presents the experimental results and discussion, and at last Sect. 6 commits to the conclusion and future works.

2 Motivation

The most common way to diagnose malaria, which is one of the world's most deadly and life-threatening diseases, is to have a competent technician examine a blood smear under a microscope. This form of diagnosis is based on the expertise and understanding of the individual experimenting. We want to automate this process using modern deep learning tools because manual processes are time-consuming and prone to human error. Our approach should also allow for future training with more data and might be used as a smartphone application.

3 Related Work

Several research publications, including machine learning and deep learning, are studied in this study to gain a broad understanding of how to process a dataset, what algorithms to apply, and how to enhance precision to construct an effective system. The following is a summary of various investigations, including the methodology and methods employed, as well as the findings.

Linder et al. [8, 9] employed morphological factors for feature extraction, and then Principal Component Analysis (PCA) [10] and SVM were applied for classification.

However, when compared to the most up-to-date research methodologies based on deep learning, the accuracy of these models is lower.

Rajaraman et al. [11] implemented a feature extractor for sorting parasite and uninfected blood cells to aid illness detection using a previously trained CNN-based deep learning network. Using basic data, the study uses experimental methods to determine the optimum model layer. Two dense layers which have been fully connected and three convolutional layers make up the CNN model. VGG-16, AlexNet, Xception, DenseNet-121, and ResNet-50 are used to evaluate the performance of extracting characteristics from parasite and uninfected blood cells.

Liang et al. [12] reported a detection accuracy of 97.37% with their 16-layer CNN model, claiming that their approach has achieved a higher performance in comparison to the transfer learning model.

Quinn et al. [13] demonstrated a 3D printable adapter for connecting a smartphone to a microscope, despite the fact that all of the photos in the experiment were captured using a microscope camera, having a greater pixel density than their smartphone camera. They suggested a workflow for automatic analysis of thick blood smears that includes calculating morphological and moment features, as well as training a tree classifier based on these features to distinguish between normal plaques and normal plaques containing parasites. They claim a 97% area under the receiver operating characteristic (ROC) curve for their performance.

Rosado et al. [14] described a supervised classification-based image processing and analysis method for detecting P. falciparum trophozoites and white blood cells in Giemsa-stained thick blood smears. Their automatic detection of trophozoites claimed 80.5% sensitivity and 93.8% specificity using a support vector machine (SVM) and a mix of color, geometric, and texture features, while their white blood cell detection achieved a sensitivity of 98.2 and 72.1% specificity.

Herrera et al. [15] tested the diagnostic performance of the device for automatic RDT interpretation using smartphone technology and image analysis software. The device's diagnostic performance is comparable to that of PDR visual interpretation, and there is no discernible difference between Plasmodium falciparum and Plasmodium vivax. The availability of standardized PDR automatic interpretation services in remote locations, in addition to almost real-time case reports and quality control, will considerably assist the large-scale adoption of PDR-based malaria diagnosis programs.

Cesario et al. [16] discussed the use of mobile support for vector-borne diseases in places where professional medical care is unavailable. They concentrate on the system's picture processing and classification components, with the goal of reducing the possibility of misdiagnosing common diseases like malaria and assisting healthcare providers. Although the progress of the picture categorization and analysis components is mostly described in his article, feedback from healthcare practitioners has been fairly positive.

Dallet et al. [17] described a mobile application platform for Android phones that uses photographs of thin films of blood-stained with Giemsa to detect malaria. The photos are mostly composed of precise morphological processes that may detect white and red blood cells as well as parasites in infected cells. The program also identifies the various stages of parasite life and determines parasitemia levels. The software diagnoses in less than 60 s and has been tested and verified on a variety of Android phone and tablet models.

Skandarajah et al. [18] designed a specialized cell phone microscope that works with a variety of phones. They showed that a quantitative microscope with micron-level spatial resolution can be operated on a variety of mobile phones, with picture color, distortion, and linearity rectified as per requirement. They presented that phones with cameras that are more than 5 megapixels can reach resolutions that are close to the diffraction limit at a wide range of magnifications, including those associated with single-cell pictures. It was also discovered that the autofocus, color gain, and exposure standards on the mobile phone reduced image resolution; if not rectified, they reduced color capture accuracy, and they created algorithms to overcome these roadblocks that make it difficult to produce quantitative photographs.

Pirnstill and Cote [19] introduced a cost-effective spherical polarization microscopy system for imaging malaria pigments known as heavoids, which are a byproduct of the parasite's blood digestion. Even with a specialist microscope engineer, determining the presence of background pigments and other artifacts can be challenging. A polarizing microscope makes it much simpler to see the pigment. Despite the widespread adoption of polarized light microscopy, present commercial equipment has complex designs and requires sophisticated maintenance, making it expensive, and existing microscopic engineers require it for recession. The portable polarizing microscopy design provided by PiRSTILL and COTE suggests that malaria can be detected with high resolution and specificity using easy-to-use, low-cost modular platforms.

Breslauer et al. [20] built an optical microscope placed on a smartphone and demonstrated its effectiveness by imaging Plasmodium falciparum-infected red blood cells and Plasmodium falciparum-infected sputum in the bright field under LED excitation fluorescence. Clinical applications are possible. The resolution is higher than that required to discern the morphology of blood cells and bacteria in all circumstances. They employed digital photos and image analysis software to establish automatic bacterial counts in tuberculosis samples.

Jane and Carpenter [21] proposed the faster R-CNN, an object detection-based model based on CNN. The model is pre-trained on ImageNet [22] before being fine-tuned with their dataset. Another model based on deep relative qualities was proposed by Bibin et al. [23], (DRA) [24].

Razzak and Naz [25] proposed an automated technique for segmenting and classifying malaria parasites that take into account both tasks. Their segmentation network is based on a Deep Aware CNN [26], while their classification network is based on an extreme learning machine (ELM) [27].

Anggraini et al. [28] created an application that separates the backdrop from the blood cells using picture segmentation technology. MOMALA [29] is a smartphone-based application that uses microscopes to quickly detect malaria at a low cost. On blood-stained slides, the MOMALA app can detect the presence of malaria parasites. Connect the phone camera to the microscope eyepiece to photograph and examine the blood smear. Currently, this application is heavily reliant on big, heavy, and difficult-to-transport microscopes.

Researchers have created a mobile app that can instantly identify malaria by taking images [30] of blood samples. We can analyze blood samples without requiring microscope technicians by using a smartphone app. Holding the smartphone in front of the microscope's lens allows the software to evaluate the image of the blood sample and draw a red circle on the malaria parasite. The case was then reviewed by a laboratory worker.

The majority of computer diagnostic tools that use machine learning techniques [31–33] for image analysis rely on decision-making algorithms that are manually constructed. To examine the diversity of the image size, color, background, perspective, and area of interest, the method also necessitates computer vision knowledge.

Deep learning approaches can be used to successfully overcome the common problems in the process of hand-designed feature extraction [34]. The deep learning approach combines a series of sequence layers with hidden nonlinear processing units to reveal hierarchical feature correlations in image data. The abstracted (low level) features contribute to nonlinear decision-making functions and learning complexity, resulting in feature extraction and classification from extreme to extreme [35]. Furthermore, deep learning models outperform kernel-based techniques like support vector machines (SVM) on huge volumes of data and processing resources, making them very scalable [36].

Zhang et al. [37] offer a security system that employs interactive robots for identity verification and access management while restricting access to private cloud data. They created a new form of cognitive IoT paradigm employing cognitive computing technology in later work [38].

4 Proposed Methodology

The proposed methodology entails two stages: Stage first deals with dataset summarization, whereas stage second deals with the proposed algorithm.

4.1 Stage 1: Dataset Summarization

The Malaria Cell Image Dataset [39] is used in this study. There are 27,558 cell pictures in the dataset, with an equal number of parasitized and uninfected cells. All of the photos in a directory cell are included in the dataset set. These photos are divided into two categories: parasitized and uninfected. According to their respective categories, both directories contain 13,780 photos.

We used 11,024 photos from each category to train our model (22,048 training images) and 2255 for testing (5510 testing images). That is, we divided our data set into two sections: training (80%) and testing (20%).

4.2 Stage 2: Convolutional Neural Network

Convolutional neural networks (CNN) are a type of deep learning technique that are commonly used for analyzing visual imagery. CNNs are classified as deep neural networks. It uses a special technique that involves employing a mathematical operation on two functions which in turn outputs a third function expressing how the shape of one is modified by the other.

Convolutional neural networks (CNN) consist of numerous layers of artificial neurons. These artificial neurons try to imitate their biological counterparts and are basically mathematical functions that are used to output activation values by calculating the weighted sum of numerous inputs.

As part of CNN's pseudo-code, the required dataset is trained, and afterward the model is evaluated on the test dataset. The model's output will include accuracy, recall, precision, and the F1-Score.

Algorithm 1 Convolutional Neural Network

Input: X_train & X_valid
1. Procedure CNN MODEL
2. BatchSie=32, epoch=2, filters=128, verbose=1
3. input_length=2160, kernel_size = 5, rate = 0.25
4. algo=Sequential()
   # Dropout Layer
5. algo.add (Dropout(rate = 0.25))
   # Convolutional Layer
6. algo.add (Conv1D(filters = 128, kernel_size = 5, activation = 'relu', input_shape = (27,558,1)))
   # Flatten Layer
7. algo.add (Flatten())
   # Dense Layer
8. algo.add (Dense(1, activation = 'sigmoid'))
   # Compile Function
9. algo.compile (loss = 'binary_crossentropy',optimizer = 'adam',metrics = 'accuracy'])
   # Summary of the model
10. print(algo.summary())
11. For all epochs in (1:NEpoch) **do**
    # Fitting a Model

12. algo.fit (X_train_cnn, y_train, batch_size = 32, epochs= 2, verbose = 1)
   # Model Prediction
13. algo.predict (X_train_cnn,verbose=1)
14. algo.predict (X_valid_cnn,verbose=1)
15.   **End for**
   # Model Evaluation
16. Evaluate thresh=sum (y_train)/length of y_train
17. To print (y_train, y_train_preds_cnn, thresh)
18. To print (y_valid, y_valid_preds_cnn, thresh)
19. **End Procedure**
Output: Accuracy, Recall, Precision, F1-Score

5 Experimental Results and Discussion

In this research article, experimental studies have been performed on Ryzen 7 5800H, max frequency 4.4 GHz, 16 GB RAM, and software used Windows 10, Anaconda Navigator, and Jupyter Notebook.

Here, the malaria image dataset has been divided into 80% and 20%. As we know that classification performance is usually measured in terms of accuracy, recall, precision, and F1-Score. These terms have been briefly explained in [4, 6]. The results of the evaluations of the training and testing dataset are presented below in Table 1.

Table 1 depicts the result of the training and testing set by using CNN Model for Malaria Cell Images Dataset based on some measures.

In the Table 1, it may observe that the CNN model achieved the best performance. As we may observe that accuracy (96%), precision (0.95), recall (0.96), F1-Score (0.954) which are quite good performance for the CNN model.

Figure 1 presents the achieved accuracy of the CNN model for Malaria Cell Images Dataset, whereas Fig. 2 presents the loss graph of the CNN model for Malaria Cell Images Dataset.

Figure 1 represents the prediction capability of CNN model. Figure 1 achieved 96% accuracy which is quiet good.

Figure 2 represents the summation of errors in the CNN model. It showcases how well our model is performing.

Table 1 Training and testing performance of CNN model for malaria cell images dataset

Parameters	Training	Testing
Accuracy (%)	96.39	96
Precision	0.96	0.95
Recall	0.973	0.96
F1-Score	0.96	0.954

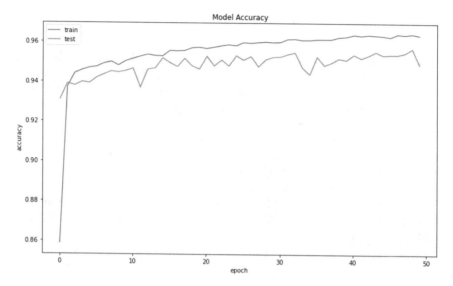

Fig. 1 Achieved accuracy of CNN model for malaria cell images dataset

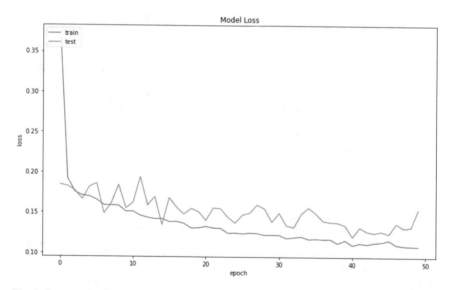

Fig. 2 Loss graph of CNN model for malaria cell images dataset

When comparing the performance of our proposed work for malaria detection to the existing machine learning techniques, we may find that our proposed model outperforms the others.

6   Conclusion and Future Works

In medicine, there are much too many manual works. Even for doctors, determining an illness from raw data is tough, which is why many healthcare businesses are turning to machine learning and deep learning technology to do so. We can provide clinicians with greater information while caring for patients using advanced analytics like these.

The practitioner or internist at the bedside will gain from deep learning. Machine learning can provide objective feedback to boost efficiency, accuracy, and reliability. Medical practitioners will use models like the one proposed in our study to improve current facilities in the malaria detection area, as well as other related domains.

Our model may be further trained on vast amounts of data and better computers with high processing powers as the dataset grows in real time, resulting in increased accuracy and efficiency.

References

1. Jangir SK, Joshi N, Kumar M, Choubey DK, Singh S, Verma M (2021) Functional link convolutional neural network for the classification of diabetes mellitus. Int J Numerical Meth Biomed Eng e3496
2. Choubey DK, Tripathi S, Kumar P, Shukla V, Dhandhania VK (2021) Classification of diabetes by kernel based SVM with PSO. Recent Adv Comp Sci Commun (Formerly: Recent Patents on Computer Science) 14(4):1242–1255
3. Choubey DK, Kumar M, Shukla V, Tripathi S, Dhandhania VK (2020) Comparative analysis of classification methods with PCA and LDA for diabetes. Curr Diabetes Rev 16(8):833–850
4. Choubey DK, Kumar P, Tripathi S, Kumar S (2020) Performance evaluation of classification methods with PCA and PSO for diabetes. Netw Model Anal Health Inf Bioinf 9(1):1–30
5. Choubey DK, Paul S, Dhandhenia VK (2017) Rule based diagnosis system for diabetes. Int J Med Sci 28(12):5196–5208
6. Choubey DK, Paul S (2017) GA_RBF NN: a classification system for diabetes. Int J Biomed Eng Technol 23(1):71–93
7. Choubey DK, Paul S (2016) Classification techniques for diagnosis of diabetes: a review. Int J Biomed Eng Technol 21(1):15–39
8. Linder N, Turkki R, Walliander M, Mårtensson A, Diwan V, Rahtu E, Pietikäinen M, Lundin M, Lundin J (2014) A malaria diagnostic tool based on computer vision screening and visualization of plasmodium falciparum candidate areas in digitized blood smears. PLoS ONE 9:e104855. [PMC free article] [PubMed]
9. Opoku-Ansah J, Eghan MJ, Anderson B, Boampong JN (2014) Wavelength markers for malaria (plasmodium falciparum) infected and uninfected red blood cells for ring and trophozoite stages. Appl Phys Res 6:47
10. Wold S, Esbensen K, Geladi P (1987) Principal component analysis. Chemom Intell Lab Syst 2:37–52
11. Rajaraman S, Antani SK, Poostchi M, et al (2018) Pre-trained convolutional neural networks as feature extractors toward improved malaria parasite detection in thin blood smear images. Peer J 6:e4568
12. Liang Z, Powell A, Ersoy I, Poostchi M, Silamut K, Palaniappan K, Guo P, Hossain M.A, Sameer A, Maude RJ, et al (2016) CNN-based image analysis for malaria diagnosis. In:

Proceedings of the 2016 IEEE international conference on bioinformatics and biomedicine (BIBM). Shenzhen, China, pp 493–496

13. Quinn JA, Andama A, Munabi I, Kiwanuka FN (2014) Automated blood smear analysis for mobile malaria diagnosis mobile point care monitors diagno device design 31:115

14. Rosado L, da Costa JMC, Elias D, Cardoso JS (2016) Automated detection of malaria parasites on thick blood smears via mobile devices. Proc Comput Sci, 90:138–144

15. Herrera S, Vallejo AF, Quintero JP, Arévalo-Herrera M, Cancino M, Ferro S (2014) Field evaluation of an automated RDT reader and data management device for Plasmodium falciparum/Plasmodium vivax malaria in endemic areas of Colombia Malar J 13:87

16. Cesario M, Lundon M, Luz S, Masoodian M, Rogers B (2012) Mobile support for diagnosis of communicable diseases in remote locations. In: Proceedings of the 13th international conference of the NZ chapter of the ACM's special interest group on human-computer interaction. ACM, pp 25–28

17. Dallet C, Kareem S, Kale I (2014) Real time blood image processing application for malaria diagnosis using mobile phones. In: International conference on circuits and systems, IEEE, pp 2405–2408

18. Skandarajah A, Reber CD, Switz NA, Fletcher DA (2014) Quantitative imaging with a mobile phone microscope. PLoS ONE 9:e96906

19. Pirnstill CW, Coté GL (2015) Malaria diagnosis using a mobile phone polarized microscope. Sci Rep 5:13368

20. Breslauer DN, Maamari RN, Switz NA, Lam WA, Fletcher DA (2009) Mobile phone based clinical microscopy for global health applications. PLoS ONE 4:e6320

21. Jane H, Carpenter A (2017) Applying faster R-CNN for object detection on malaria images. In: Proceedings of the IEEE conference on computer vision and pattern recognition workshops. Honolulu, HI, USA, pp 56–61

22. Deng J, Dong W, Socher R, Li L-J, Li K, Fei-Fei L (2009) ImageNet: a large-scale hierarchical image database. In: 2009 IEEE conference on computer vision and pattern recognition. Miami, FL, USA, pp 248–255

23. Bibin D, Nair MS, Punitha P (2017) Malaria parasite detection from peripheral blood smear images using deep belief networks. IEEE Access 5:9099–9108

24. Yang X, Zhang T, Xu C, Yan S, Hossain MS, Ghoneim A (2016) Deep relative attributes. IEEE Trans Multimedia 18(9):1832–1842

25. Razzak MI, Naz S (2017) Microscopic blood smear segmentation and classification using deep contour aware CNN and extreme machine learning. In: 2017 IEEE conference on computer vision and pattern recognition workshops (CVPRW). Honolulu, HI, USA, pp 801–807

26. Kantorov V, Oquab M, Cho M, Laptev I (2016) Contextlocnet: context-aware deep network models for weakly supervised localization. In: Computer vision—ECCV 2016. Springer, pp 350–365

27. Huang GB, Zhu QY, Siew CK (2006) Extreme learning machine: theory and applications. Neurocomputing 70(1–3):489–501

28. Anggraini D, Nugroho AS, Pratama C, Rozi IE, Iskandar AA, Hartono RN (2011) Automated status identification of microscopic images obtained from malaria thin blood smears. In: Proceedings of the 2011 international conference on electrical engineering and informatics, Bandung, Indonesia

29. MOMALA. https://momala.org/malaria-diagnosis/

30. This New App Helps Doctors Diagnose Malaria in Just 2 Minutes. https://www.globalcitizen.org/en/content/app-diagnose-malaria-uganda

31. Ross NE, Pritchard CJ, Rubin DM, Dusé AG (2006) Automated image processing method for the diagnosis and classification of malaria on thin blood smears. Med Biol Eng Compu 44(5):427–436

32. Das DK, Ghosh M, Pal M, Maiti AK, Chakraborty C (2013) Machine learning approach for automated screening of malaria parasite using light microscopic images. Micron 45:97–106

33. Poostchi M, Silamut K, Maude RJ, Jaeger S, Thoma GR (2018) Image analysis and machine learning for detecting Malaria. Transl Res 194:36–55

34. Hossain MS, Al-Hammadi M, Muhammad G (2019) Automatic fruit classification using deep learning for industrial applications. IEEE Trans Industr Inf 15(2):1027–1034
35. Usama M, Ahmad B, Wan J, Hossain MS, Alhamid MF, Hossain MA (2018) Deep feature learning for disease risk assessment based on convolutional neural network with intra-layer recurrent connection by using hospital big data. IEEE Access 6:67927–67939
36. Srivastava N, Hinton G, Krizhevsky A, Sutskever I, Salakhutdinov R (2014) Dropout: a simple way to prevent neural networks from overfitting. J Mach Learn Res 15:1929–1958
37. Zhang Y, Qian Y, Wu D, Shamim Hossain M, Ghoneim A, Chen M (2019) Emotion-aware multimedia systems security. IEEE Trans Multimedia 21(3):617–624
38. Zhang Y, Ma X, Zhang J, Hossain MS, Muhammad G, Amin SU (2019) Edge intelligence in the cognitive internet of things: improving sensitivity and interactivity. IEEE Network 33(3):58–64
39. https://lhncbc.nlm.nih.gov/LHC-downloads/downloads.html#malaria-datasets

Session-Based Song Recommendation Using Recurrent Neural Network

Chhotelal Kumar, Mukesh Kumar, and Amar Jindal

1 Introduction

Session-based item recommendation is a typical recommendation problem that may be seen in various domains such as e-commerce, news, hotel search, classified sites, music, and video recommendation. Because prior user history logs are sometimes unavailable in session-based settings (either because the user is new, not signed in, or not tracked), recommender systems must rely only on the user's activities in current sessions to deliver correct suggestions.

Most of these tasks were previously handled using very simple approaches such as collaborative filtering or content-based methods. Recurrent neural networks (RNNs) have evolved as powerful techniques for modeling sequential data in the deep learning. Speech recognition, machine translation, time series forecasting, and signal processing have all benefited from these models. RNNs have recently been used for the session-based recommendation setting in recommender systems, with improved results [1, 2]. RNNs have the benefit of being able to model the entire session of user interactions, as compared to conventional similarity-based [3] approaches for recommendation (clicks, views, etc.). RNNs may learn the 'theme' of a session by modeling the entire session and hence deliver recommendations with higher accuracy than the previous approaches.

The task of session-based recommendation has been applied to RNNs. Ranking items according to the user preference are one of the main goal of recommendation system. It is less important to rank or score items at the bottom of the list (items that the user will not like), but it is critical to rank correctly the items at the top of the list (items that the user will like) (first 5 positions, 10 positions or 20 positions). To do this using machine learning, one must often employ learning to rank techniques, specifi-

C. Kumar (✉) · M. Kumar · A. Jindal
National Institute of Technology, Patna 800005, India
e-mail: chhotelalk.phd18.cs@nitp.ac.in

© The Author(s), under exclusive license to Springer Nature Singapore Pte Ltd. 2023   719
D. S. Sisodia et al. (eds.), *Machine Intelligence Techniques for Data Analysis and Signal Processing*, Lecture Notes in Electrical Engineering 997,
https://doi.org/10.1007/978-981-99-0085-5_58

cally ranking objectives and loss functions. Ranking loss functions, namely pairwise ranking loss functions, are used in contemporary session-based RNN techniques.

The deep learning algorithms propagate gradient across several layers. So, choosing a suitable ranking loss impacts majorly on the performance. In the case of RNNs 'back in time' over the previous steps, to optimize model parameters, the quality of these gradients originating from the loss function has an impact on the quality of the optimization and the model parameters. The nature of recommendation task involves huge output space due to presence of large number of songs. It presents specific issues that must be considered while developing a good ranking loss function. We shall see that how this enormous output space issue is addressed which is critical in reaching our goals.

The present investigation used ranking loss function in RNNs framework for session-based recommendation system. The BPR-max loss function is one of the popular ranking loss function that has been improve the RNN framework performance. The experiments have been done on 30Music dataset which indicates a considerable boost in the recommendation results as assessed by precision. The proposed framework shows the promising results over traditional recommendation systems.

The rest of the paper is organized as follows. Section 2 discusses about the existing works in this domain. Section 3 describes the proposed methodology of the SBRSs. In Sect. 4, the detailed statistical description of the dataset, evaluation metric, and comparative result analysis has been discussed. Finally, Sect. 5, concludes the work carried out in this paper.

2   Related Work

Most of the approaches proposed in the literature for session-based recommendation involve some form of sequence learning. Early versions of these techniques were based on the idea of frequent sequential patterns and Markov chain. Early techniques were used to predict the online behavior of users. They were also used to predict users' recommendations in various e-commerce and music domain areas [4]. Much of the research in recommendation systems has been on models that operate when a user identification is available, and a clear user profile can be created. As per literature, most of the researches carried out in this field are based on the models such as matrix factorization [5], frequent pattern mining [6], and nearest neighborhood [7]. The item-to-item recommendation technique [3, 7] is one of the most common ways used in the session-based recommender system and a logical solution to the issue of a non-existent user profile. Items that are frequently clicked together in sessions are considered to be similar under this configuration, and an item-to-item similarity matrix is pre-computed from the available session data. During the session, this similarity matrix is simply leveraged to suggest the most similar products to the user that has been recently clicked. Despite its simplicity, it has been proved to be a successful and commonly used technique. While efficient, these approaches only consider the user's most recent click, thereby ignoring information from the previous clicks.

Frequent pattern mining (FPM) [6] is one of the commonly used technique for SBRS, but the computational requirements become challenging. Also, finding suitable algorithm parameters can be challenging. In some cases, frequent item sequences do not lead to better predictions. Mobasher et al. [8] introduced one of the very first session-based methods based on frequent pattern mining for recommending the next Web page to visit.

Markov decision processes(MDP) [9] are well suited for modeling sequential stochastic decision problems. It is defined as a four tuple $\langle S, A, Rwd, tr \rangle$, where S denotes the finite set of states, A denotes a set of actions (actions can be recommendations in case of recommender system), Rwd denotes reward function, and tr denotes the state-transition function. Shani et al. [9] proposed an MDP-based SBRS in the area of e-commerce and demonstrated the value of SBRS from a business perspective. Garcin et al. [10, 11] proposed an MDP-based SBRS that was applied in the news domain to recommend personalized news. Tavakol et al. [12] used a contextual session-based approach and provided a sequential framework for detecting the user's objective (the session's subject) using factored Markov decision processes (fMDPs). Along with MDPs, some other side information has been used on a private e-commerce dataset from a European fashion retailer. Jannach et al. [13] wanted to see how successful it is to use and combine long-term models with short-term adaption techniques. The findings show that preserving visitors' short-term content-based and recency-based profiles can lead to considerable accuracy gains. To make sequential recommendations, Markov chain models assume that the user's future behavior only depends on the very last behavior or last few behaviors. Transition probability between items is evaluated in order to recommend the next item. The transition matrix provides the probability value of recommending an item by the user after consuming a sequence of items. The main drawback of applying Markov chains in the field of session-based recommendation systems is that when trying to capture all the possible sequences of user preferences becomes unmanageable.

Hariri et al. [14] proposed the nearest neighbors-based SBRS in order to generate a context-aware music playlist. More complex approaches based on latent Markov embedding were proposed for playlist generation by Chen et al. [15]. However, the nearest neighbors-based technique and its modified variants have proved to be highly competitive against today's modern neural network methods [16]. The extended general factorization framework (GFF) [17] also has the capability to use session data for recommendations and can be achieved by modeling a session by the sum of its event.

Hidasi et al. [18] presented the possibly first deep learning-based SBRS popularly known as gru4rec for predicting the imminent next item click event. They employed the YOOCHOOSE[1] dataset [19] for assessment and used an item-based nearest neighbors approach as one of their baselines for comparing the results. This model was continually improved later on by applying some different optimization

---

[1] https://www.kaggle.com/chadgostopp/recsys-challenge-2015.

objectives [1, 17]. In terms of prediction accuracy, they discovered that their neural network-based approach outperforms this methodology. Following their method and the increasing interest in neural networks, many other ideas using deep learning approaches to solve session-based recommendations have been deployed. It has been observed that the progress in the field of session-based recommendation seems still limited.

3 Proposed Method

The session-based recommendation system is based upon the session sequences. The present investigation problem uses the sequence of songs. The RNN is well suitable model for the sequential data learning. The RNN model has been widely used in many different domains. In session-based recommendation system, the RNN is capable to learn the sequential patterns of the session sequence. Figure 1 shows the architecture of predicting the next song based on the previous session of songs sequence.

Problem definition: This work focuses on 'next item recommendation' in recommendation system. The term 'item' here is referred as a song. We focus on the prediction problem here, which asks how effectively we can predict the next song the user desires to listen given a series of songs.

Definition of Next Song Recommendation A query q is a collection of songs in a session. Given a query q, the algorithm ranks all candidate songs and recommends the top-k songs most likely to appear following the query q songs. Given a few songs, we predict the best song that may potentially come to the next of them, as seen in Fig. 2 (Figs. 3 and 4).

The listened song log data is a valuable information and considered as an implicit feedback information. This requires no effort on user's side to collect their preferences and privacy concerns much. The proposed method predicts the next song based on the previous sessions listened songs. Since a person state of mind may vary according

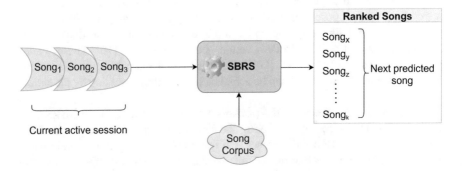

Fig. 1 Proposed framework for song prediction model using RNN

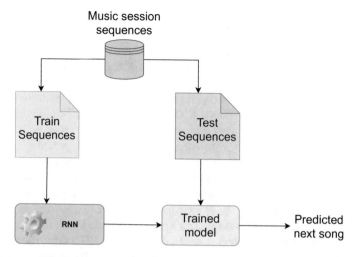

Fig. 2 Next song prediction model

Fig. 3 Next song prediction problem

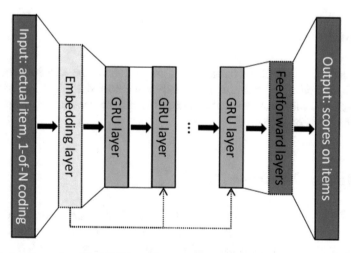

Fig. 4 Architecture of modified RNN used for song recommendation [18]

to the situation, time, activity, and location, the current ongoing session is the most valuable information.

In session-based recommendation, when a user enters a Website and clicked on the first item, it is considered as the initial input of the recurrent neural networks. After first click, each click of the user will produce an output as a recommendation and that output will be dependent on all the previous clicks. In order to encode the categorical data where features do not have any specific order, 1-of-N encoding will be a good solution. In 1-of-N encoding, every song has been mapped as a binary variable. The 1-of-N encoding of the current session has been provided as an input to the RNN model. The term 'N' represents the total number of unique songs present in the song corpus. If the associated song is present in the current session, then the corresponding value will be 1, otherwise 0 in the 1-of-N encoding. The network's input is a single song that is 1-of-N encoded in a vector that represents the entire song space, and the network's output is a vector of similar shape that should yield a ranking distribution for the next song. In the present RNN model, the number of RNN layers are 20. The batch size and learning rate for the training of the RNN model have been kept as 16 and 0.1, respectively. The momentum of the model has been set to 0.1, and to avoid the overfitting of the RNN model, the dropout has been set as 0.1.

In order to efficiently train the proposed model session-parallel mini-batches, sampling technique for getting output and a BPR-max ranking loss function is used.

4 Result and Discussion

In this section, details of the datasets used for evaluating the SBRSs-based model are given as well as the pre-processing steps have been described. After that, the findings of our offline evaluation are presented, and some important conclusion from the results has been drawn.

4.1 Dataset Description

30Music[2] [20] dataset is a real-world publicly available collection of playlists and listening data. LastFM API was used to retrieve the data from Internet radio stations. In this dataset, each sample consists of four attributes, i.e., $session\_id$, $timestamp$, $item\_id$, and $user\_id$. In the pre-processing phase, all the sessions having only one event were removed (Table 1).

---

[2] http://recsys.deib.polimi.it/datasets/.

Table 1 30Music dataset characteristics

Number of sessions	2,764,474
Number of events	31,351,954
Number of users	4,106,341
Number of items	210,633
Minimum session length	1
Maximum session length	198
Average session length	11
Minimum number of session per user	1
Maximum number of session per user	38
Average number of session per users	3.70

4.2 Dataset Splitting Approach

It should be emphasized that just a few researches have addressed their data splitting techniques for model training, validation, and testing. It can be described by two primary types on the basis of whether the user information has been incorporated in the model creation or not. This has been examined on the basis of the public source codes of the relevant authors. Dataset can be split into training and testing in various ways. It depends on whether the user information has been incorporated in the model creation or not. Whenever *user_id* is presented in the dataset, it is better to split the dataset in such a way that every user's last session is kept in testing set, and all the remaining session is kept in the train set. Whenever *user_id* is not present in the dataset, consider the most recent n days of user's sessions as the test set and remaining as the train set. In the present investigation, all the dataset contains *user_id*. So, in order to capture the personalized behavior of the users, this investigation follows the first approach. A user having n sessions which have been split into training and testing set in such a way that first $(n - 1)$ sessions will be in the training set, and the last session (nth session) of every user has been kept in the testing set.

Different ways of data filtering techniques lead to different data scenarios that would result in varying performance. Splitting of the dataset into train set and test set is done in such a way that the last session of every user is taken as a test set, whereas all the remaining sessions are taken as a train set.

4.3 Evaluation Metrics

Recommender systems can recommend a list of items at any point of time, and the user might pick first few items from that list. Therefore, the primary evaluation metrics used in this work are Precision@N, Recall@N, and MRR@N. It is described in terms of recommendation systems as follows: Precision gives the number of items recommended which are relevant out of the total number of items recommended and could be defined by the Eq. 1. Recall gives the number of items recommended which are relevant out of the total number of items that are relevant and could be defined by the Eq. 2. Mean reciprocal rank (MRR) metric finds the rank of the first relevant item in the recommended list and then takes the reciprocal of the rank, the average across all the queries; the measure is known as mean reciprocal rank. It is defined by the Eq. 3.

$$\text{Precision} = \frac{\text{Number of items recommended which are relevant}}{\text{Number of items recommended}} \tag{1}$$

$$\text{Recall} = \frac{\text{Number of items recommended which are relevant}}{\text{Number of relevant items}} \tag{2}$$

$$\text{MRR} = \frac{1}{|Q|} \sum_{i=1}^{|Q|} \frac{1}{ran K_i} \tag{3}$$

4.4 Comparative Result

The result achieved by RNN model and three other models POP, IKNN, and SKNN on the 30Music dataset has been presented in Tables 2, 3, and 4. To compare the

Table 2 Performance comparison at recommendation length of 5

Model	Precision@5	Recall@5	MRR@5
POP	0.003	0.018	0.009
IKNN	0.076	**0.380**	**0.260**
SKNN	0.041	0.205	0.098
V-SKNN	0.195	0.137	0.088
RNN	**0.095**	0.073	0.223

Table 3 Performance comparison at recommendation length of 10

Model	Precision@10	Recall@10	MRR@10
POP	0.003	0.030	0.010
IKNN	0.045	**0.454**	**0.270**
SKNN	0.028	0.285	0.010
V-SKNN	0.159	0.195	0.104
RNN	**0.064**	0.087	0.226

Table 4 Performance comparison at recommendation length of 20

Model	Precision@20	Recall@20	MRR@20
POP	0.002	0.051	0.009
IKNN	0.025	**0.506**	**0.274**
SKNN	0.015	0.307	0.111
V-SKNN	0.112	0.244	0.110
RNN	**0.042**	0.100	0.228

performance of various methods on 30Music dataset, three metrics, such as Precision@N (P@N), Recall@N (R@N), and MRR@N, have been used, where the value of N is set to 5, 10, and 20. For all the metrics mentioned above, greater value implies better performance. The greatest value is written in bold across all approaches.

Tables 2, 3, and 4 show that when the recommendation length is 5, RNN gives better results in terms of precision, whereas in terms of recall and MRR IKNN, it gives better result.

5 Conclusion and Future Work

Being able to predict the short-term interest of a user in an online session is a very relevant problem in practice. In this study, we have compared various computational and algorithmic approaches for session-based song recommendation. In this study, we have used RNN to learn over session sequences in the training sequences set to predict the next song. By observing the above result, it can be seen that the recommendation quality changes when the length of the recommendation list varies.

SBRS generally ignores the user's long-term preference that can be captured by the collaborative filtering-based methods, so in future, we can develop a model which can capture both short-term preferences as well as long-term preferences. Along with session context, features of items of the session can be incorporated to improve the quality of recommendation.

References

1. Hidasi B, Karatzoglou A (2018) Recurrent neural networks with top-k gains for session-based recommendations. In: Proceedings of the 27th ACM international conference on information and knowledge management, pp 843–852
2. Zhang Y, Dai H, Xu C, Feng J, Wang T, Bian J, Wang B, Liu T-Y (2014) Sequential click prediction for sponsored search with recurrent neural networks. In: Proceedings of the AAAI conference on artificial intelligence, vol 28(1)

3. Sarwar B, Karypis G, Konstan J, Riedl J (2001) Item-based collaborative filtering recommendation algorithms. In: Proceedings of the 10th international conference on World Wide Web, pp 285–295
4. Bonnin G, Jannach D (2014) Automated generation of music playlists: survey and experiments. ACM Comput Surv (CSUR) 47(2):1–35
5. Koren Y, Bell R, Volinsky C (2009) Matrix factorization techniques for recommender systems. Computer 42(8):30–37
6. Kamehkhosh I, Jannach D, Ludewig M (2017) A comparison of frequent pattern techniques and a deep learning method for session-based recommendation. In RecTemp@ RecSys, pp 50–56
7. Linden G, Smith B, York J (2003) Amazon. com recommendations: item-to-item collaborative filtering. IEEE Internet Comput 7(1):76–80
8. Mobasher B, Dai H, Luo T, Nakagawa M (2002) Using sequential and non-sequential patterns in predictive web usage mining tasks. In: Proceedings of 2002 IEEE international conference on data mining. IEEE, pp 669–672
9. Shani G, Heckerman D, Brafman RI, Boutilier C (2005) An MDP-based recommender system. J Mach Learn Res 6(9)
10. Garcin F, Zhou K, Faltings B, Schickel V (2012) Personalized news recommendation based on collaborative filtering. In: 2012 IEEE/WIC/ACM international conferences on web intelligence and intelligent agent technology, vol 1. IEEE, pp 437–441
11. Garcin F, Dimitrakakis C, Faltings B (2013) Personalized news recommendation with context trees. In: Proceedings of the 7th ACM conference on recommender systems, pp 105–112
12. Tavakol M, Brefeld U (2014) Factored MDPs for detecting topics of user sessions. In: Proceedings of the 8th ACM conference on recommender systems, pp 33–40
13. Jannach D, Lerche L, Jugovac M (2015) Adaptation and evaluation of recommendations for short-term shopping goals. In: Proceedings of the 9th ACM conference on recommender systems, pp 211–218
14. Hariri N, Mobasher B, Burke R (2012) Context-aware music recommendation based on latent topic sequential patterns. In: Proceedings of the sixth ACM conference on recommender systems, pp 131–138
15. Chen S, Moore JL, Turnbull D, Joachims T (2012) Playlist prediction via metric embedding. In: Proceedings of the 18th ACM SIGKDD international conference on knowledge discovery and data mining, pp 714–722
16. Ludewig M, Jannach D (2018) Evaluation of session-based recommendation algorithms. In: User modeling and user-adapted interaction, vol 28(4–5), pp 331–390
17. Hidasi B, Tikk D (2016) General factorization framework for context-aware recommendations. Data Mining Knowl Discov 30(2):342–371
18. Hidasi B, Karatzoglou A, Baltrunas L, Tikk D (2015) Session-based recommendations with recurrent neural networks. arXiv preprint arXiv:1511.06939
19. Ben-Shimon D, Tsikinovsky A, Friedmann M, Shapira B, Rokach L, Hoerle J (2015) Recsys challenge 2015 and the Yoochoose dataset. In: Proceedings of the 9th ACM conference on recommender systems, pp 357–358
20. Turrin R, Quadrana M, Condorelli A, Pagano R, Cremonesi P (2015) 30Music listening and playlists dataset. In RecSys Posters

The Classification of Breast Cancer Based on Hyper-Tuned AdaBoost Ensemble Model

Devanshu Tiwari, Manish Dixit, and Kamlesh Gupta

1 Introduction

The most common kind of cancer diagnosed in women's is the breast cancer. Breast cancer detection in the preliminary stage proved to be very helpful and eventually enhances the chances of patient's survival [1]. Breast cancer ranks fifth when it comes to the number deaths caused by any cancer worldwide as per Globocan data 2018. The task of breast cancer diagnosis is still having some problems and hard even after there has been a rise in the amount of medical research and technological improvements solely dedicated to the prognosis of cancer. The cancer of breast is the major reason of death among women's after lung cancer [2]. Breast cancer basically originates from the tissues of breast and especially from the interior lining of the lobules or the milk ducts in the breast. Breast cancer cells are the result of the RNA or DNA modification and mutation which eventually transformed a normal cell into a cancer cells. Basically bacteria, chemicals in the air, electromagnetic radiation viruses, fungi, mechanical cell-level injury, parasites, heat, water, food, free radicals, aging of DNA or RNA, and evolution are all can be the reason behind this type of harmful mutation of DNA which leads to breast cancer [3]. Detection of breast cancer at an early stage accompanied by the advance medical treatment will play a major role to prevent deaths from this type of cancer. The mostly employed ways to diagnose breast cancer are by performing regular screening tests.

D. Tiwari (✉)
RGPV, Bhopal, India
e-mail: devanshu.tiwari28@gmail.com

M. Dixit
DoCSE, MITS, Gwalior, India
e-mail: dixitmits@mitsgwalior.in

K. Gupta
DoIT, RJIT, BSF Academy, Tekanpur, Gwalior, India

© The Author(s), under exclusive license to Springer Nature Singapore Pte Ltd. 2023
D. S. Sisodia et al. (eds.), *Machine Intelligence Techniques for Data Analysis and Signal Processing*, Lecture Notes in Electrical Engineering 997,
https://doi.org/10.1007/978-981-99-0085-5_59

The major risk factors associated with breast cancer are the obesity, family background or history, race, genetics, alcohol consumption, avoid doing physical labor for long time, etc. [4]. The health care is one of the most information enrich sectors but still lacking quality knowledge. This mammoth quantity of data can be utilized to unleash as well as to fabricate the relationship with the hidden data in order to perform the early and accurate risk prediction. Data science will play a very important role when it comes to the analysis of this huge amount of healthcare data in order to unleash and learn the hidden patterns for making accurate predictions [5]. The machine learning is an effective data analysis technique that makes a computing device to make predictions in the form of an output by running various algorithms over the datasets [6]. The neural networks, support vector machine, Naive Bayes, k-nearest neighbors, etc. [7] are the examples of mostly used models for many machine learning applications. Utilizing machine learning as a tool to analyze data and then make predictions related to breast cancer diagnosis and risk assessments will play a major role in the early and effective treatment. This type of tool actually gives an opportunity to common people to control risk factors and hence avoid breast cancer in the future.

In this paper, Global dataset, i.e., Wisconsin dataset [8] is taken for the training and evaluation purpose. This dataset is even used to compare the performance of the proposed approach based on AdaBoost ensemble model with the other machine learning and ensemble learning classifiers. The ultimate objective of this paper is to propose an automated approach which can make predictions whether a person is likely to suffer from breast cancer or not after analyzing certain attributes.

2 Literature Review

Machine learning is an evolving field of artificial intelligence and a lot of applications are already developed for the detection and classification of breast cancer in the past decade. Initially, Dhar et al. come up with a brief review paper illustrating major approaches based on machine learning for the prediction of breast cancer [9]. An approach is proposed by Aruna et al. [10] in which decision trees, support vector machine (SVM), and Naive Bayes (NB) machine learning classifiers are used over the Wisconsin dataset. In this study, SVM delivers an accuracy of over 96%. Then, Chaurasia et al. [11] and Asri et al. [12] also compared the performance of various classifiers over the Wisconsin dataset like Naïve Bayes, SVM, neural networks, and decision tree, in which again SVM outperforms other classifiers and achieves an accuracy of over 96%. Then, Delen et al. [13] proposed a breast cancer prediction model utilizing data of over two lakh patients, and these data samples are used to train NB, c4.5 and neural networks. Among these three, c4.5 outperforms other two classifiers in terms of making more accurate predictions. Then, Weng et al. [14] come up with a study to determine an optimal way for the prediction of breast cancer employing SVM, NB, AdaBoost Tree, artificial neural network (ANN) methods along with the Principle Component Analysis (PCA) over Wisconsin Diagnostic

Breast Cancer dataset. Then, a paper illustrating a model for risk prediction on breast cancer utilizing data mining classification techniques was proposed by Williams et al. [15]. In this paper, J48 and NB classifiers are used in order to predict the breast cancer as they require such a system in Nigeria. Then, a computer-aided diagnosis (CAD) system is proposed by the Nithya et al. [16] for the characterization as well as for the detection of breast cancer. This system is based on usage of SVM, NB, and machine-sequential minimal optimization (SVM-SMO) classifiers along with the concept of multiboot and Bagging for the performance enhancement. Then, Oyewola et al. [17] proposed another breast cancer prediction model utilizing mammographic diagnosis. In this approach, linear discriminant analysis (LDA), logistic regression (LR), random forest (FR), SVM, and quadratic discriminant analysis (QDA) machine learning models are used and compared. Then, an accurate and reliable breast cancer prediction model based on multilayer perceptron is proposed by the Agarap et al. [18]. This model is compared with other conventional machine learning classifiers like Gated Recurrent Unit (GRU) SVM, SVM, LR, KNN, and softmax regression. This proposed model achieves an accuracy of 99%.

Another study is conducted by the Westerdijk et al. [19], in which several machine learning classifiers like RF, LR, SVM, neural network, ensemble models, etc., are compared in terms of specificities, accuracies, and sensitivities. An automated system is proposed by the Vard et al. [20] for the prediction of eight different cancer types. These different cancer types are ovarian cancer, lung cancer, breast cancer, etc. This multiclass classification system is based on the Particle Swarm Optimization and statistical feature selection methods for the normalization of datasets and feature reduction, respectively. Along with these two methods SVM, multilayer perceptron neural and decision trees are used for performing the final multiclass classification. Then, Java platform is used by the Pratiwi [21] for the development of highly responsive breast cancer prediction system. Recently, Battan et al. [22] proposed an approach based on Random Forest with Grid search to predict the breast cancer utilizing the Wisconsin Diagnostic Breast Cancer Dataset. Table 1 given simply summarizes some of the major above mentioned machine learning based approaches for the breast cancer prediction.

3 Proposed Approach

The proposed approach consists of major three stages, i.e., data preprocessing, data visualization, and training testing of AdaBoost ensemble model using Grid search and parameter search. The overall proposed approach is illustrated with the help of Fig. 1 given.

Table 1 Summary of major breast cancer prediction machine learning based approaches proposed approach

Author	Classifier used	Result	Dataset
Aruna et al.	SVM, NB and decision tree	SVM RBF offers an accuracy of 96.84%	Wisconsin Diagnostic Breast Cancer Dataset
Chaurasia et al.	Simple Logistic, RepTree, RBF Network	Simple Logistic offers an accuracy of 74%	Local dataset provided by the University Medical Centre at the Institute of Oncology in Yugoslavia
Asri et al.	KNN, NB, SVM, decision tree (c4.5)	SVM offers an accuracy of 97%	Wisconsin Diagnostic Breast Cancer Dataset
Delen et al.	Logistic regression, ANN and Decision tree	Decision tree (C5) offers an accuracy of 93.6%	SEER Cancer Incidence Public-Use Database
Weng et al.	AdaBoost Tree, ANN, SVM, NB along with PCA	SVM offers an accuracy of 97%	Wisconsin Diagnostic Breast Cancer Dataset
Oyewola et al.	LDA, LR, RF, QDA, SVM	SVM offers an accuracy of 95%	Wisconsin Diagnostic Breast Cancer Dataset
Agarap et al.	Linear Regression, Softmax Regression, Multilayer Perceptron (MLP), Nearest Neighbor (NN) search, SVM and GRU-SVM	MLP offers an accuracy of 99%	Wisconsin Diagnostic Breast Cancer Dataset
Westerdijk et al.	RF, LR, SVM, neural network, ensemble models	Ensemble model offers an accuracy of 97%	Wisconsin Diagnostic Breast Cancer Dataset
Battan et al.	RF with grid search method	97–98% accuracy	Wisconsin Diagnostic Breast Cancer Dataset

3.1   Dataset Description

The Wisconsin Diagnostic Breast Cancer (WDBC) dataset [23] is used in paper for the training and validation purpose as it the widely used breast cancer dataset. This dataset basically consist of features values which illustrate the major characteristics of the cell nuclei of the breast image. These features value are derived from the fine needle aspirate (FNA) of a breast mass digitized image. This dataset basically consist of features values which illustrates the major characteristics of the cell nuclei of the benign breast cancer and malignant breast cancer. These features value are derived from the digital images of fine needle biopsies of breast masses of women's either suffering from benign or malignant breast cancer. The fine needle biopsies benign and malignant breast cancer masses [24] are illustrated with the help of Fig. 2. This dataset has thirty two patient attributes in which one attribute is patient ID and one attribute is breast cancer diagnosis result like malignant or benign, whereas

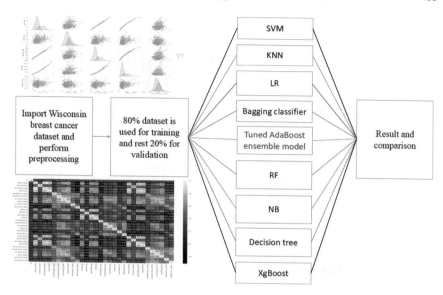

Fig. 1 Overall proposed approach for the breast cancer prediction

rest thirty attributes are associated with breast cancer diagnosis information. These attributes values are taken from around 569 patients in which 38% suffering from malignant breast cancer and rest 62% are diagnosed with benign breast cancer. The thirty attributes of breast cancer diagnosis information are derived from the ten basic attributes in terms of three important measures such as largest value, standard error, and mean. The WDBC dataset is illustrated with the help of Table 2.

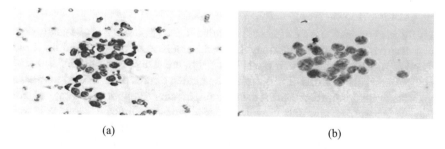

(a)   (b)

Fig. 2 Digitized images of fine needle biopsies **a** benign breast cancer breast mass and **b** malignant breast cancer breast mass

Table 2 Summary of WDBC dataset released in the year 1995

Attribute number	Name of attribute	Worst value	Standard error	Mean
1	Radius	7.93–36.04	0.11–2.87	6.98–28.11
2	Perimeter	50.41–251.20	0.76–21.98	43.79–188.50
3	Texture	12.02–49.54	0.36–4.89	9.71–39.28
4	Smoothness	0.07–0.22	0.00–0.03	0.05–0.16
5	Area	185.20–4254.00	6.80–542.20	143.50–2501.00
6	Concavity	0.00–1.25	0.00–0.40	0.00–0.43 145
7	Compactness	0.03–1.06	0.00–0.14	0.02–0.35
8	Symmetry	0.16–0.66	0.01–0.08	0.11–0.30
9	Concave points	0.00–0.29	0.00–0.05	0.00–0.20
10	Fractal dimension	0.06–0.21	0.00–0.03	0.05–0.10

3.2 Data Preprocessing

The WDBC dataset is preprocessed in order to find out whether it contains any null value or not and as it may affect the accuracy of the overall approach. Then, each attribute values are manually analyze to determine any anomaly, and if there is any unusual values, then those values will be corrected as this dataset is free from such unusual values and error. These preliminary checks are conducted in order to preprocess or clean this dataset.

3.3 Data Visualization

It is the next immediate step after data cleaning. Data visualization is basically done in order to have a better understanding of the numeric data by converting it into visual form with the aid of pair plots, heat maps, etc. This visual form of data play an important role in terms of identifying the hidden trends, patterns, and numerous types of correlations in between the data that might remain unnoticed in large numeric datasets. Apart from this, data visualization also tends to eliminate the noise as well as highlight the important information. As data visualization is very important in terms of identifying outliers, trends and patterns, so it is implemented in this study in the form of pair plots and heat map for the WBCD dataset. Pair plot is simply used to visualize the association in between the different attributes of this WBCD dataset. It displays the data as collection of points. It is a very effective method employed for the identification of trends. Figure 3 simply presents a pair plot named diagnosis which is created among the five attributes like radius mean, texture mean, perimeter mean, area mean, and smoothness mean, whereas the visualization of correlation is carried out with the aid of heat map. A heat map is a two-dimensional illustration of data utilizing different colors. In this study, a diagnosis heat map is presented with

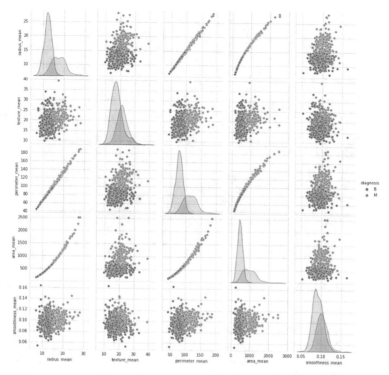

Fig. 3 Pair plot named diagnosis, which is created among the five attributes like radius mean, texture mean, perimeter mean, area mean, and smoothness mean

the help of Fig. 4. The heat map below represents the strength of correlation, in which dark portions represents weak correlation and light portions are represents the high correlation in between the attributes.

3.4 Tuned AdaBoost Ensemble Model

A general approach for improving the learning process performance in machine learning domain is the boosting. AdaBoost classifier simply belongs to the class of ensemble classifiers and is based on the concept of utilization of output of numerous classifiers in order to give the final output. AdaBoost was proposed with a basic thought of making a group of weak learners that tends to learn from the mispredictions and adapt themselves in order to generate accurate results. Normal AdaBoost ensemble classifier tends to give result less accurate so we have perform the hyperparameter tuning of AdaBoost ensemble model with the aid of Grid search. Initially, we have run the AdaBoost classifier with default settings [25], then we create a search grid with the hyperparameters. The two hyperparameters values are the learning rate

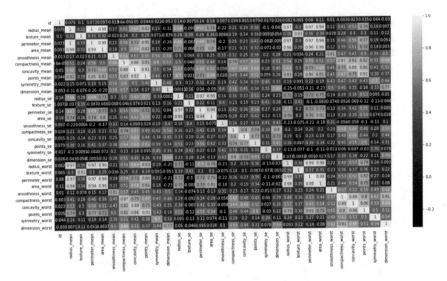

Fig. 4 Heat map of the WBCD dataset

and number of estimators. The number of estimators simply deals with the amount of trees developed and learning rate simply depicts how individual tree contributes to the overall results. As the Grid search tends to find out the best model after trying out all possible combinations. The AdaBoost classifier with learning rate value of 1.0 and the number of estimator's value of 50 tends to deliver the best results in this study.

4 Results and Evaluation

The Google Colaboratory (colab) platform along with the Python 3.6 as an implementation programming language is used for the experimentation and simulation in this study. The performance of the proposed approach is presented with the help of popular classification rates or statistical parameters like Accuracy, Sensitivity, Specificity, Precision, Negative Predictive Value, False Positive Rate, False Discovery Rate, False Negative Rate, Accuracy, and F1 Score. These classification rates are well formulated in Table 3 which also depicts the performance of the proposed approach based on tuned AdaBoost ensemble model. The performance of other conventional machine learning and ensemble learning classifiers are also illustrated with the help of Table 4. The performance comparison graph as well as the ROC curve graph of both the AdaBoost with default settings and tuned AdaBoost are presented with the help of Figs. 5 and 6. These graphs simply showcase that the proposed approach based on tuned AdaBoost is far superior as compare to the other machine learning classifiers.

Table 3 Performance of the proposed tuned AdaBoost ensemble classifier on the WBCD

Classification rates	Their formulas	AdaBoost with default setting	Tuned AdaBoost with Grid search
Accuracy	(TP + TN)/(TP + TN + FP + FN)	94.74	98.25
Sensitivity	TP/(TP + FN)	96.9	98.5
Specificity	TN/(FP + TN)	91.6	97.8
Precision	TP/(TP + FP)	94.1	98.5
Negative Predictive Value	TN/(TN + FN)	95.6	97.8
False Positive Rate	FP/(FP + TN)	8.33	2.13
False Discovery Rate	FP/(FP + TP)	5.88	1.49
False Negative Rate	FN/(FN + TP)	3.03	1.49
F1 Score	2TP/(2TP + FP + FN)	95.5	98.5

Where TP = True positive, TN = True Negative, FP = False Positive, FN = False Negative

Table 4 Performance of the nine machine and ensemble learning classifiers on the WBCD

Classification rates	SVM	KNN	NB	Bagging	XGBoost	Decision tree	LR	RF
Accuracy	95.61	93.04	94.74	96.49	97.37	96.49	90.35	97.37
Sensitivity	98.4	94.12	92.96	98.46	98.48	98.46	92.4	97.37
Specificity	92	91.49	97.67	93.88	95.83	93.88	87.5	97.37
Precision	94	94.12	98.5	95.5	97.01	95.5	91.04	98.67
Negative Predictive Value	97.87	91.49	89.36	97.87	97.87	97.87	89.36	94.87
False Positive Rate	8	8.51	2.33	6.12	4.17	6.12	12.50	2.63
False Discovery Rate	5.97	5.88	1.49	4.48	2.99	4.48	8.96	1.33
False Negative Rate	1.56	5.88	7.04	1.54	1.52	1.54	7.58	2.63
F1 Score	96.18	94.12	95.65	96.97	97.74	96.97	91.73	98.01

5 Conclusion

The proposed approach based on tuned AdaBoost ensemble classifier delivers accuracy of more than 98% which is far better as compare to other conventional machine learning classifiers and ensemble models. The Grid search CV plays an important role when it comes to the hypertuning of parameters and delivers best settings in order to achieve the better results. In the future, this approach could be used to predict the

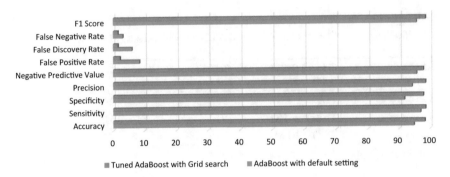

Fig. 5 Performance comparison graph in between AdaBoost with default settings and tuned AdaBoost

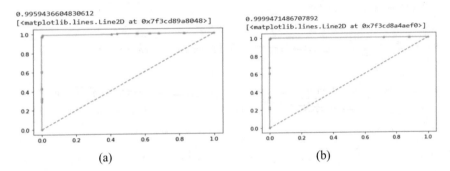

Fig. 6 ROC curve of the proposed approach **a** AdaBoost with default settings **b** tuned AdaBoost ensemble model

brain tumor is benign or malignant once a similar dataset based on biopsies data is available. The future work includes the application of various deep learning and deep transfer learning models over the same dataset in order to analyze the performance.

References

1. Bray F, Ferlay J, Soerjomataram I (2018) Global cancer statistics 2018: GLOBOCAN estimates of incidence and mortality worldwide for 36 cancers in 185 countries. CA Cancer J Clin, pp 394–424
2. Piñeros M, Znaor A, Mery L (2017) A global cancer surveillance framework within non communicable disease surveillance: making the case for population-based cancer registries. Epidemiol Rev, pp 161–169
3. Nechuta SJ, Caan B, Chen WY (2011) The after breast cancer pooling project: Rationale, methodology, and breast cancer survivor characteristics. Cancer Causes Control, pp 1319–1331
4. Manyika J, Chui M, Brown B, Bughin J, Dobbs R, Roxburgh C, Byers A (2011) Big data: the next frontier for innovation, competition, and productivity. Technical Report; McKinsey Global Institute, Washington, DC, USA

5. Alickovic E, Subasi A (2017) Breast cancer diagnosis using GA feature selection and rotation forest. Neural Comput Appl, pp 753–763
6. Mangasarian OL, Street WN, Wolberg WH (1995) Breast cancer diagnosis and prognosis via linear programming. Oper Res, pp 570–577
7. Dubey AK, Gupta U, Jain S (2016) Analysis of k-means clustering approach on the breast cancer Wisconsin dataset. Int J CARS, pp 2033–2047
8. Wolberg WH (1991) Wisconsin breast cancer database. University of Wisconsin Hospitals, Madison, WI, USA
9. Dhar V (2013) Data science and prediction. Commun ACM, pp 64–73
10. Aruna S, Rajagopalan S, Nandakishore L (2011) Knowledge based analysis of various statistical tools in detecting breast cancer. Comput Sci Inf Technol, pp 37–45
11. Chaurasia V, Pal S (2014) Data mining techniques: to predict and resolve breast cancer survivability. Int J Comput Sci Mob Comput, pp 10–22
12. Asri H, Mousannif H, Al Moatassime H, Noel T (2016) Using machine learning algorithms for breast cancer risk prediction and diagnosis. Procedia Comput Sci, pp 1064–1069
13. Delen D, Walker G, Kadam A (2005) Predicting breast cancer survivability: a comparison of three data mining methods. Artif Intell Med, pp 113–127
14. Wang H, Yoon WS (2015) Breast cancer prediction using data mining method. In: Proceedings of the 2015 industrial and systems engineering research conference, Nashville, TN, USA, 30 May–2 June 2015
15. Williams TGS, Cubiella J, Grin SJ (2016) Risk prediction models for colorectal cancer in people with symptoms: a systematic review. BMC Gastroenterol, pp 63
16. Nithya R, Santhi B (2011) Classification of normal and abnormal patterns in digital mammograms for diagnosis of breast cancer. Int J Comput Appl, pp 0975–8887
17. Oyewola D, Hakimi D, Adeboye K, Shehu M (2017) Using five machine learning for breast cancer biopsy predictions based on mammographic diagnosis. Int J Eng Technol IJET, pp 142–145
18. Agarap AFM (2018) On breast cancer detection: an application of machine learning algorithms on the Wisconsin diagnostic dataset. In: Proceedings of the 2nd international conference on machine learning and soft computing, Phuoc Island, Vietnam, pp 5–9, 2–4 Feb 2018
19. Westerdijk L (2018) Predicting malignant tumor cells in breasts. Vrije Universiteit Amsterdam, Netherlands
20. Vard A, Firouzabadi F, Sehhati M, Mohebian M (2018) An optimized framework for cancer prediction using immunosignature. J Med Sig Sens, pp 161
21. Pratiwi PS (2016) Development of intelligent breast cancer prediction using extreme learning machine in Java. Int J Comput Commun Instrum Eng
22. Buttan Y, Chaudhary A, Saxena K (2021) An improved model for breast cancer classification using random forest with grid search method. In: Goyal D, Chaturvedi P, Nagar AK, Purohit S (eds) Proceedings of second international conference on smart energy and communication. Algorithms for Intelligent Systems. Springer, Singapore. https://doi.org/10.1007/978-981-15-6707-0_39
23. Dua D, Graff C (2007) UCI machine learning repository, Irvine, CA: University of California, School of Information and Computer Science
24. Street WN, Wolberg WH, Mangasarian OL (1993) Nuclear feature extraction for breast tumor diagnosis. In: SPIE 1993 International symposium on electronic imaging: science and technology, San Jose, California, pp 861–870
25. Freund Y, Schapire RE (2010) A desicion-theoretic generalization of on-line learning and an application to boosting. Lecture Notes in Computer Science, pp 23–37

Detection of COVID-19-Affected Persons Using Convolutional Neural Network from X-Rays' Images

Bhupendra Singh Kirar, Sarthak Sanjay Tilwankar, Aditya Paliwal, Deepa Sharma, and Dheraj Kumar Agrawal

1 Introduction

1.1 Overview

COVID-19 is a disease that was first detected in China in 2019 and later spread across the entire world. As of September 22, 2021, there have been 228,807,631 cases of COVID-19 reported worldwide. The total number of fatalities has been reported as 4,697,099 by the WHO [1]. This disease is notorious for spreading very easily from one infected person to another. Nearly, every country is struggling to cope up with the speed at which this infection spreads. Slowing down the speed at which the virus spreads depends upon the rate of testing for a particular variant of the COVID-19 virus. In addition to testing, various countries have started immunization programs, but it will take a long time to completely vaccinate such a large population. Therefore, the best course of action is thought to be testing, contact tracing and isolating.

The most important factors that need to be taken care of when deciding a test of COVID-19 are the effectiveness in classifying the cases which are positive, time taken to produce the result and the value of the check. The WHO recommends the RT-PCR test as the most effective method for detecting COVID-19. While the test is effective, it has the drawback of high cost of test kits and time taken to produce the results. Computerized Axial Tomography (CT) scans are also used to detect COVID-19 in certain cases [2]. The advantage of using CT scans is the faster diagnostic rate to produce results, but the accuracy of these tests is less than RT-PCR. Loop-Mediated equal Amplification (LAMP) technique [3] will generate results at

B. S. Kirar · S. S. Tilwankar (✉) · A. Paliwal · D. Sharma
Indian Institute of Information Technology, Bhopal 462003, India
e-mail: 18u02040@iiitbhopal.ac.in

D. K. Agrawal
Maulana Azad National Institute of Technology, Bhopal 462003, India

D. S. Sisodia et al. (eds.), *Machine Intelligence Techniques for Data Analysis and Signal Processing*, Lecture Notes in Electrical Engineering 997,
https://doi.org/10.1007/978-981-99-0085-5_60

intervals associated in Nursing hours; however, it is not widely used owing to its less accuracy.

Protein tests are also used for detection during which the COVID-19 antibodies are used for detection of COVID-19, although these tests have less accuracy. Also, another fact that limits their use is that these tests can detect if a person has previously had COVID but do not detect if the person is currently having the disease [4].

Some of the necessary issues facing current approaches in the detection of COVID-19 are given below:

1. Cavity swab sampling is ordinarily used for testing. During this, a healthcare worker needs to come in contact with an infected person [5]. This collection of samples is to be done by specialist persons. There is a danger of the increase in cross-infection.
2. RT-PCR which is the recommended test by the WHO is quite costly and limited in number; therefore, it is difficult to access in developing countries, since developing countries have fewer kits; there is a chance of large spread of the disease and increase in infections [6].
3. Sensitivity of the Rapid Antigen test cannot alone be considered for initial screening [7].
4. Spread of the COVID infection increases with the delay in results as the COVID-positive person comes in contact with a healthy person before getting traced [8].

2 Literature Survey

This section presents a summary of the existing literature on the topic of detection of COVID-19 using X-rays' images. Table 1 shows the different methods and their accuracy. This table has been formulated from [9]. To overcome the slow testing process of the RT-PCR tests, various other methods were developed including the use of X-ray images. Earlier models utilizing X-ray images were trained on smaller data. Now, more literature has been published to automate the detection process. Sarkar et al. [10] used a transfer learning model on the DenseNet-121 network to classify COVID-affected and normal chest X-ray images and achieved an accuracy of 85.35%. In another experiment, a CNN-based model was also proposed, by Xu et al., based on ResNet-18, and found an accuracy of 86% [11]. Wang and Wong [12] used the transfer model technique on VGG19, MobileNetV2, Xception, Inception and ResNetV2 and reported an accuracy of 92%. DeTraC-ResNet CNN was used by Abbas et al. and they achieved an accuracy of 95.12% [13]. Pachori et al. with their group members also developed various machine and deep learning-based methods for the detection of COVID-19 [14–19].

The purpose of this work is to propose a method for detection of COVID-19 using chest X-rays using machine learning and deep learning approaches.

The organization of the paper is given as: Sect. 3 describes the dataset used and methodology followed. Section 4 discusses the results; Sect. 5 describes the

Table 1 COVID-19 detection approaches

References	Existing approaches	Reported accuracy in %
Sarker et al. [10]	DenseNet-121 network	85.35
Xu et al. [11]	ResNet-18-based CNN network	86
Wang and Wong [12]	COVID-Net for COVID-19 detection	92
Abbas et al. [13]	DeTraC-ResNet-18 CNN	95.12

performance of the model with the help of evaluation methods, and Sect. 6 contains the conclusion of the paper.

3 Methodology

In this paper, COVID-19-affected persons who are detected using convolutional neural network (CNN) from X-rays' images are proposed. It includes the preprocessing, features' extraction followed by CNN for the measurement of various performance parameters. The complete procedures are explained in the subsections of this section. We used 2159 images in the experiment. The machine used to train the model is a 64-bit Laptop with a configuration of 8 GB RAM, 512 GB Storage, Intel i5 10th Generation Processor.

3.1 Dataset

We have used the Kaggle dataset in this proposed method [20]. The dataset distribution is given in Table 2.

The typical chest X-ray images are given in Fig. 1. In Fig. 1, (a) represents the healthy image and (b) represents the COVID-19 image.

Table 2 Dataset distribution

Types of images	Training dataset	Testing dataset
Normal	1266	317
COVID-19	460	116
Total	1726	433

Fig. 1 Typical chest X-ray
images **a** healthy **b**
COVID-19

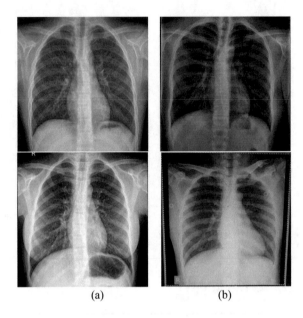

(a)   (b)

3.2 Method

Preprocessing. In this step, all the picture square dimensions are converted to a
dimension of 224*224. Moreover, all the pixels are normalized by multiplying it
with a factor of 1/255 so that the intensity of the pixels is converted to a standard
value in the range 0–1.

Feature extraction. The entropy, geometrical and textural options are extracted
from the preprocessed pictures. Entropy (E) is the estimation of irregularity that
is utilized to explain the feel of the input image. Form options play a vital role to
differentiate the characteristics between traditional and malignant cells. In matter
options, every image is partitioned off into 'n' sub-squares and measured.

Construction CNN model. This model consists of various layers as shown in
Fig. 2. The various layers of CNN Model are given as:

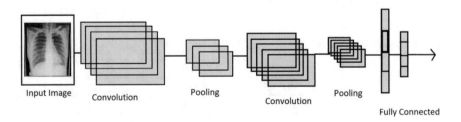

Fig. 2 CNN model [21]

1. Input layer.
2. Convolutional layer.
3. Pooling layer.
4. Dense layer.

Input Layer. The image is resized to 224*224*3, X and Y dimensions are being 224 each and the RGB channel is represented by the 3.

Convolutional Layer. Feature extraction from the input layer is done through the CNN formula with the assistance of a kernel. Kernel is created to go over the whole image and a scalar multiplication was created between the portions of the image lined by the kernel and also the kernel itself. Equation (1) is used to calculate the features' map values.

$$G[M, N] = (f \times h)[m, n] = \sum_{j} \sum_{k} h[j, k] f[m - j, n - k], \tag{1}$$

where f is input image, h is filter and m and n are rows and columns of the result matrix.

Diagram for the convolutional layer indicating input, hidden and output layers is shown in Fig. 3. Figure 3 (1) shows the input layer, from (2) to (5) show hidden layers and (6) shows output layer.

Pooling Layer. This layer is supplemental once the convolutional layer avoids the variance which will lead to the feature map once input is shifted or revolved. Grievous bodily harm pooling is employed in the analysis and the utmost price line by pooling filter was maintained within the feature map. The pooling layer is conjointly used to reduce the computations required to be carried out in the process. A dropout layer is

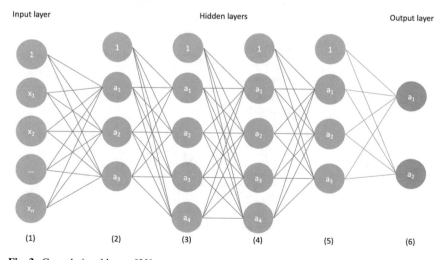

Fig. 3 Convolutional layers [22]

supplemental once pooling for removing overfitting which will happen throughout training.

Dense Layer. Pooling layer output is first flattened into a single-dimensional vector; then, it is applied to the dense layer. The dense layer calculates weights to identify probabilities of the classes. After that, CNN is used with cross-entropy confusion loss with the help of Adam Optimizer. The parameter choice for the proposed model is presented in Table 3.

Table 3 Parameter choice for proposed model

Types of layers	Output	Parameter's count
conv2d (Conv 2D)	(None, 222,222,32)	896
conv2d_1 (Conv 2D)	(None, 220,220,64)	18,496
max_pooling2d (MaxPooling2d)	(None,110,110,64)	0
dropout (Dropout)	(None, 110,110,64)	0
conv2d_2 (Conv2D)	(None, 108,108,64)	36,928
max_pooling2d_1 (MaxPooling2d)	(None, 54,54,64)	0
dropout_1 (Dropout)	(None, 54,54,64)	0
conv2d_3 (Conv2D)	(None, 52,52,128)	73,856
max_pooling2d_2 (MaxPooling2D)	(None, 26,26,128)	0
dropout_2 (Dropout)	(None, 26,26,128)	0
conv2d_4 (Conv2D)	(None, 24,24,128)	147,584
max_pooling2d_3 (MaxPooling2D)	(None, 12,12,128)	0
dropout_3 (Dropout)	(None, 12,12,128)	0
flatten (Flatten)	(None, 18,432)	0
dense (Dense)	(None, 64)	1,179,712
dropout_4 (Dropout)	(None, 64)	0
dense_1 (Dense)	(None, 1)	65

Total params.: 1,457,537. Trainable params.: 1,457,537. Non-trainable params.: 0

3.3 Evaluation of Performances of Proposed Model

The model is now tested on unseen data, and the following metrics are evaluated for the model using the Confusion Matrix [23–25].

$$Accuracy = (TP + TN)/(TP + TN + FP + FN), \tag{2}$$

$$Precision = TP/(TP + FP), \tag{3}$$

$$Recall = TP/(TP + FN), \tag{4}$$

$$F\text{-Measure} = 2 * Precision * Recall/(Precision + Recall), \tag{5}$$

where TP = True Positives, TN = True Negatives, FN = False negatives and FP = False Positives.

4 Results

In this paper, COVID-19-affected persons who are detected using convolutional neural network (CNN) from X-rays' images are proposed. The experiment has been carried in order to categorize X-ray images, as healthy or COVID-19. Preprocessed X-ray images are shown below in Figs. 4a and b for images of Fig. 1 (a) Healthy and (b) COVID-19, respectively. The experiment conducted on a total of 2159 images, out of which 1726 were used for training and 433 were used for testing. The method has been performed on Epochs (Ep) 5, 10, 15, 20. The obtained accuracy, precision, recall, F1-score and support using 20 epochs are 96%, 96%, 94%, 95% and 433, respectively. The evaluation parameters for different numbers of epochs came out as shown in Table 4.

Figure 5 shows the different line graphs of the model accuracy of various epochs. The performance evaluation of various measures for different values of epochs is shown in Fig. 6 for (a) Ep = 5, (b) Ep = 10, (c) Ep = 15 and (d) Ep = 20.

5 Discussion

In this paper, COVID-19 affected persons are detected using convolutional neural network (CNN) from X-rays' images. The obtained accuracy, precision, recall, F1-score and support using 20 epochs are 96%, 96%, 94%, 95% and 433, respectively. Due to a lack of data, many earlier papers had a scope for improvement and for increase in accuracy. In our work, we have trained the model on 1700 images. The

Fig. 4 Preprocessed images for images in Fig. 1, **a** healthy **b** COVID-19

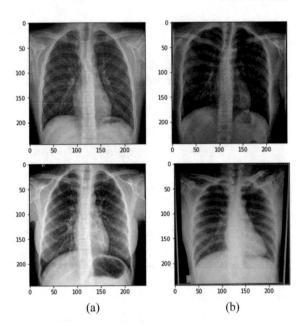

(a)   (b)

Table 4 Evaluation of a constructed model for different number of epochs

No. of Ep	TP	FP	TN	FN	Accuracy	Precision	Recall	F1-score	Support
5	317	116	0	0	0.73	0.37	0.5	0.42	433
10	312	10	106	5	0.96	0.96	0.95	0.96	433
15	311	12	104	6	0.96	0.95	0.94	0.95	433
20	311	11	105	5	0.96	0.96	0.94	0.95	433

comparison of the proposed method with existing methods is shown in Table 6 and Fig. 7.

The comparison of COVID-19 detection methods is given in a graphical way in Fig. 7, where x and y-axis represent the methods and accuracy in percentage, respectively.

From the analysis above, it can be inferred that the proposed methodology performs better than existing methods that are listed in Table 6, and it is also based on a larger dataset. The proposed model gives an accuracy of 96% along with the precision, recall and F1-scores of 96%, 95% and 96%, respectively.

6 Conclusion

The results of the proposed work show that the model created using machine learning along with feature extraction using deep learning CNN performs better. The proposed

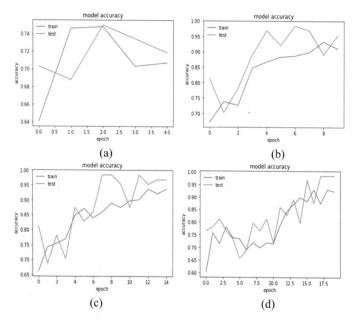

Fig. 5 Development of accuracy for different epochs **a** 5, **b** 10, **c** 15, **d** 20

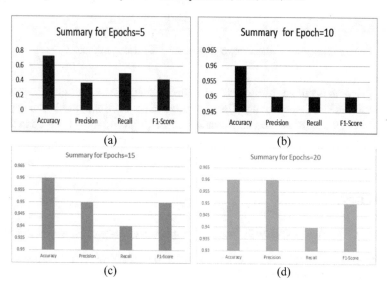

Fig. 6 Development of accuracy for different epochs **a** 5, **b** 10, **c** 15, **d** 20

Table 6 Performance of different existing and proposed works

References	Accuracy (%)
Sarker et al. [10]	85.35
Xu et al. [11]	86.00
Wang and Wong [12]	92.00
Abbas et al. [13]	95.12
Proposed model	96.00

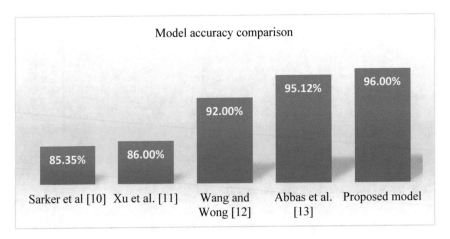

Fig. 7 Model accuracy comparison between proposed and existing methods

model gives an accuracy of 96% along with the precision, recall and F1-scores of 96%, 95% and 96%, respectively. Thus, our method could be used as a better approach for COVID-19 detection. The performance of the proposed model can be increased by handling the data imbalance better and increasing the dataset size.

References

1. WHO Coronavirus (COVID-19) Dashboard. https://covid19.who.int/. Last accessed 21 Nov 2021
2. Li Y, Xia L (2020) Coronavirus disease 2019 (COVID-19): role of chest CT in diagnosis and management. Am J Roentgenol 214(6):1280–1286
3. Notomi T, Okayama H, Masubuchi H, Yonekawa T, Watanabe K, Amino N, Hase T (2000) Loop-mediated isothermal amplification of DNA. Nucleic Acids Res 28(12):e63
4. Rajagopal R (2021) Comparative analysis of covid-19 x-ray images classification using convolutional neural network, transfer learning, and machine learning classifiers using deep features. Pattern Recognit Image Anal 31:313–322
5. Qian Y, Zeng T, Wang H, Xu M, Chen J, Hu N, Chen D, Liu Y (2020) Safety management of nasopharyngeal specimen collection from suspected cases of coronavirus disease 2019. Int J Nurs Sci 7(2):153–156

6. Farfan M, Torres J, O'Ryan M, Olivares M, Gallardo P, Salas C (2020) Optimizing RT-PCR detection of SARS-CoV-2 for developing countries using pool testing. Rev Chil Infectol 37(3):276–280

7. Scohy A, Anantharajah A, Bodéus M, Kabamba-Mukadi B, Verroken A, Rodriguez-Villalobos H (2020) Low performance of rapid antigen detection test as frontline testing for COVID-19 diagnosis. J Clin Virol 129:104455

8. Kretzschmar M, Rozhnova G, Bootsma M, van Boven M, van de Wijgert J, Bonten M (2020) Impact of delays on effectiveness of contact tracing strategies for COVID-19: a modelling study. Lancet Public Health 5(8):E452–E459

9. Albahli S, Ayub N, Shiraz M (2021) Coronavirus disease (COVID-19) detection using X-ray images and enhanced DenseNet. Appl Soft Comput 110:107645

10. Sarker L, Islam MM, Hannan T, Ahmed Z (2021) Covid-DenseNet: a deep learning architecture to detect covid-19 from chest radiology images

11. Xu X et al (2020) A deep learning system to screen novel coronavirus disease 2019 pneumonia. Engineering 6(10):1122–1129

12. Wang L, Wong A (2020) Covid-net: a tailored deep convolutional neural network design for detection of covid-19 cases from chest radiography images. Nature 19549

13. Abbas A, Abdelsamea M, Gaber M (2021) Classification of covid-19 in chest X-ray images using detrac deep convolutional neural network. Appl Intell 51:854–864

14. Chaudhary P, Pachori R (2021) FBSED based automatic diagnosis of COVID-19 using X-ray and CT images. Comput Biol Med 134(104454):1–13

15. Chaudhary P, Pachori R (2020) Automatic diagnosis of COVID-19 and pneumonia using FBD method. In: 1st International workshop on high performance computing methods and interdisciplinary applications for fighting the COVID-19 pandemic (HPC4COVID-19), pp 2257–2263. In: IEEE international conference on bioinformatics & biomedicine, Seoul, S. Korea, 16–19 Dec 2020

16. Bhattacharyya A, Bhaik D, Kumar S, Thakur P, Sharma R, Pachori R (2022) A deep learning based approach for automatic detection of COVID-19 cases using chest X-ray images. Biomed Sig Process Control 71:103182

17. Gaur P, Malaviya V, Gupta A, Bhatia G, Pachori R, Sharma D (2022) COVID-19 disease identification from chest CT images using empirical wavelet transformation and transfer learning. Biomed Sig Process Control, 71:103076

18. Gaur P, Malaviya V, Gupta A, Bhatia G, Mishra B, Pachori R, Sharma D (2021) An optimal model selection for COVID 19 disease classification. In: Biomedical signal and image processing with artificial intelligence. EAI/Springer

19. Nayak S, Nayak D, Sinha U, Arora V, Pachori R (2021) Application of deep learning techniques for detection of COVID-19 cases using chest X-ray images: a comprehensive study. Biomed Sig Process Control 64(102365):1–12

20. KAGGLE Image Dataset Chest X-ray (Covid-19 & Pneumonia). https://www.kaggle.com/prashant268/chest-xray-covid19-pneumonia. Last accessed 21 Oct 2021

21. Convolutional Neural Networks. https://debuggercafe.com/wp-content/uploads/2019/06/conv-1-e1561089320572.png. Last accessed 21 Oct 2021

22. Convolutional Neural Networks. https://www.jeremyjordan.me/convolutional-neural-networks/. Last accessed 21 Oct 2021

23. Kirar B, Agrawal D (2019) Current research on glaucoma detection using compact variational mode decomposition from fundus images. Int J Intell Eng Syst 12(3):1–10

24. Kirar B, Agrawal D (2019) Computer aided diagnosis of glaucoma using discrete and empirical wavelet transform from fundus images. IET Image Process 13(1):73–82

25. Kirar B, Reddy G, Agrawal D (2021) Glaucoma detection using SS-QB-VMD based fine sub band images from fundus images. IETE J Res, pp 1–12. https://doi.org/10.1080/03772063.2021.1959424

Vehicle Theft Identification Using Machine Learning and OCR

Priyam Rai, Eshan Eshwar Kothari, Megha Bhushan⦿, and Arun Negi⦿

1 Introduction

In today's era, it is a necessity to travel from one place to another place in a short amount of time; with the increasing population, the number of vehicles has also increased. The more the number of vehicles on the street, the more is the traffic on the street and up goes the pollution in the environment be it noise pollution or air pollution, vehicles are in a way contributing to increase it. The vehicle includes a unique feature, i.e., the number plate of the vehicle, which determines the owner of the vehicle as well as helps to track down the information regarding the vehicle. The vehicle quantity may have increased in the last decade, but it may have created difficulty for human life.

Vehicular theft happens so often around the world that we rarely think of it. The proposed work will assist the traffic police and officials at checkpoints to ensure that every passing vehicle is safe and legal to be driven on roads. It will help the officials in identifying the theft status of any suspected vehicle based on its attributes such as number plate, vehicular shape, size, color. Further, it will compare the information of the vehicle with the information stored in the database for parameters such as color, number plate, and other important details about the vehicle. It is of utmost importance to check the quality of data [1–8] used for Machine Learning (ML) algorithms. Certain ML algorithms are applied to the dataset to attain some patterns, and further, data are mined and results can be obtained [9–12].

The general users can also make use of it by adding and updating the information related to their stolen vehicle at the nearest police station, checkpoint, or through an application on their smartphone. In this work, the image of the vehicle will be captured

P. Rai · E. E. Kothari · M. Bhushan (✉)
School of Computing, DIT University, Dehradun, India
e-mail: mb.meghabhushan@gmail.com

A. Negi
Deloitte USI, Gurgaon, India

D. S. Sisodia et al. (eds.), *Machine Intelligence Techniques for Data Analysis and Signal Processing*, Lecture Notes in Electrical Engineering 997,
https://doi.org/10.1007/978-981-99-0085-5_61

by a camera [13, 14] and will be used for further processing. When the vehicle gets registered in any RTO office of India, it receives a unique number which is in the format "State code—RTO Code—Number Series—4 Digit number," and therefore, all the registered vehicles in India will have the number in the same format. After receiving the number, the owner of the vehicle must attach the number printed on a plate to the frontside and backside of the vehicle. This plate then becomes the identity of the vehicle. After capturing the number plate, the image of the vehicle is forwarded to the processing engine which helps to preprocess the acquired image to certain conversions that leads to extract the patterns. The patterns are further processed to obtain text from images using various techniques like Character Segmentation and recognition which help to convert an image that contains text to string format. Further, the number of the vehicle is extracted and stored in the server.

Further, the theft status of the vehicle is displayed on the screen whether the vehicle is stolen or not. It will be helpful for the official authorities appointed at Toll Naka and other places to identify whether the vehicle crossing from that checkpoint is a stolen vehicle or not, and it will help in diminishing crime and help people to recover their stolen vehicle [15]. This work deals with theft identification of the vehicles including a management system that helps to update and delete the records in real time.

2   Related Work

In [16], the findings of the work are that this work uses an automatic toll tax collection system and identifies the vehicle during the collision and helps to identify theft detection using RFID. The Telematics Control Unit (TCU) helps to identify the vehicle as a person carrying a vehicle or a goods carrier, and after that, the CSU is used to indicate and allow the vehicle to pass. In [17], OCR is used for high-definition automatic number plate recognition, so the automatic number plate recognition (ANPR) uses three parts NPL, CS, and OCR and also high-definition camera to capture and improve the recognition accuracy. ANPR is built using C++ with OpenCV library which uses edge detection and feature detection techniques [18]. OpenCV library is widely available and presented for all programming languages and hence can be used in any platform either C++ or Python, etc. The video supervision system is used to monitor footage [19].

The finding of the paper is that the system is designed for proper tracking of the vehicles, supervising toll, and arresting actions of those who are not following traffic rules using ANPR expert system in real life. The goal is to design a robust technique for number plate detection using a deep learning model [20].

In [21], the finding of the paper is that the work represents a noble image identification for Indian-based number plate noticing and handling all noisy, low, illuminated, nonstandard, etc., number plates. In [22], the finding of the paper is that they use the datasets of three provinces of Iraq for vehicle images, gathered from the real world in real time by using a device like cameras to form an original raw data of the vehicle's images. Some images capture snowy weather, dusty weather conditions, aiming to produce a practical dataset for automatic number plate detection systems.

In [23], a system was designed for smart service auto. The role of automatic number plate recognition, in the work, is to extract only the characters of a vehicle license number plate from a captured image. Furthermore, preprocessing involves the conversion of an image to a different model and noise reduction. After that, license plate detection involves the use of various edge detection algorithms. In the last two steps, the K-Nearest Neighbor (KNN) classifier is used for Character Recognition and character matching, and finally, the characters are compared with test samples and matched [24]. A model that includes computer graphic processing and neural networks is included. It consists of colored space detection and Convolutional Neural Network (CNN) [25].

In [26], authors have proposed Character Segmentation, Localization of Number Plates based on Computer Vision Algorithm. Also, it proposed a better approach in OCR based on techniques of deep learning. The algorithm of CNN is advanced to identify the license plate after steps of Character Segmentation and NPL.

In [27], it proposed a lot of steps for extracting text from images. Noise is removed from the image using the effective noise removal method. RGB is converted into a grayscale image. Morphological processing is used which helps to detect text more accurately. To detect edges, the edge detection method is used and increases the image intensity level. Gaps in objects are filled as it can affect the overall recognition rate. ANPR systems use OCR to scan the automobile license plates.

In [28], license plate detection (OCR) is used for the recognition of the characters. The cv2.approxPolyDP() counts the number of slides and takes three parameters, namely individual counter, approximation accuracy, and accuracy of the approximation. The system is implemented using OpenCV, and its accuracy is tested on various test subject images.

In [29], some morphological operation is used to detect the number of the vehicle followed by the segmentation approach. After that, bounding box method is applied to each character, and the characters are extracted from the number plate. In [30], Open Computer Vision (OCV) library, the license plate captured in the image is localized and the characters are divided into segments by using KNN algorithm and Python programming language. Table 1 summarizes the existing works related to the vehicle theft identification.

3 Methodology

The proposed work will assist traffic police and officials at checkpoints to ensure that every passing vehicle is safe and legal to be driven on roads. This work will help the officials in identifying the theft status of any suspected vehicle based on its attributes such as number plate, vehicular shape, size, color. This paper will compare the information of the vehicle with the information stored in the database for points such as color, number plate, and other key details about the vehicle. The first step

Table 1 Summary of existing work

Article	Year	Techniques	Description
[16]	2018	• Automatic toll collection • Character Recognition • RFID	• Automatic toll tax collection system and identifies the vehicle during the collision and helps to identify theft detection using RFID
[17]	2018	• NPL • HD camera • OCR • CS	• A high-definition camera is used to improve the accuracy of the results • The recognition rate was found to be 79.84% and the accuracy for the segmentation is found to be 80%
[18]	2018	• OpenCV • Tesseract OCR Engine	• OpenCV is used to do the image processing and Tesseract OCR engine is used to extract all the characters from the image and store it in text
[19]	2018	• NRP • ROI • OCR • CS	• Video supervision system for security monitoring • Sobel edge detection techniques • Morphological applications
[20]	2019	• Neural Network • OCR • LSTM Tesseract	• ANPR setup is used in the real world • A deep neural network is used for license plate detection which makes it robust
[21]	2019	• Python • OpenCV • KNN	• Novel image processing systems • Dealing with noisy low illuminated cross-angled nonstandard front number plate
[22]	2019	• ANPR • GTI DATASETS • The UFPR-ALPR Dataset	• ANPR systems are powerful technology used for detecting and recognizing the number plate • Created a new dataset for a vehicle license plate in North Iraq in different and tough conditions
[23]	2019	• ANPR • Image processing • OpenALPR	• The system allows the real-time organization of a vehicle number plate • ANPR is used for the development, personalizing classics, and increasing prolificacy for the workers
[24]	2020	• OpenCV • KNN • Edge detection • ANPR	• The challenging part of number plate detection is the varying size, shape, and font style of the number plate • To recognize the license plate edge detection and morphological operations, methods are used
[25]	2020	• ANPR • Dark lighting conditions • Color space detection	• A model consisting of computer vision processing and neural network recognition is initiated • Detection of color space and CNN

(continued)

Table 1 (continued)

Article	Year	Techniques	Description
[26]	2020	• ANPR • Character Segmentation • CNN	• A model consisting of computer vision processing and neural network recognition is initiated
[27]	2020	• ANPR • OCR • Edge detection • Morphological processing	• The potential future system used by local authorities and commercial organizations in all aspects of security, surveillance, access control, and traffic management
[28]	2021	• Number Plate Recognition • Character Segmentation • Number Plate Localization	• The system detects and points out the particular vehicle's license plate using Sobel edge detection and morphological operation
[29]	2021	• KNN algorithm • Open CV • Optical Character Recognition	• The license plate is divided into characters and then segregated using the OpenCV library. This has been implemented in the KNN algorithm
[30]	2019	• Character Recognition • ANPR • Character reconstruction	• Morphological transformation, Gaussian smoothing, and Gaussian thresholding have been used in the preprocessing stage

RFID Radio-frequency identification, *TCU* Telematics control unit, *CSU* Central Server Unit, *OpenCV* Open-source computer vision, *ANPR* Automatic number plate recognition, *OCR* Optical Character Reader, *CNN* Convolutional Neural Network, *KNN* K-Nearest Neighbor

involves fetching raw images from various sources like the screen capture from Toll Naka CCTV footage or the parking footage; then, the image is acquired and checked if it has an RGB color profile or BGR color profile. After that, we do the grayscale conversion and find the canny edges of the image; after we successfully obtain the canny edges of the image, we apply a bilateral filter and find the contours of the image. After this process, the image is then passed into the Tesseract OCR engine and the image is converted into a string. Once this string is obtained, we further use the information in the string to find out the number plate of the vehicle from our stored data. Figure 1 represents the whole process.

4 Experiment Details

The environment for the evaluation of proposed method includes an Acer Predator 300 laptop with Windows 10 home of 64 bits, RAM of 16.00 GB 2667 MHz, processor Intel CPU Core i5 at 2.80 GHz, SSD 515 GB, GPU NVIDIA GTX 1060 of 6 GB DDR5 memory using the Visual Studio Code for the editor and Python 3.7 for the software version. Tesseract OCR 5.0.0 was also used to work with the OCR engine.

Fig. 1 Flowchart of
proposed work

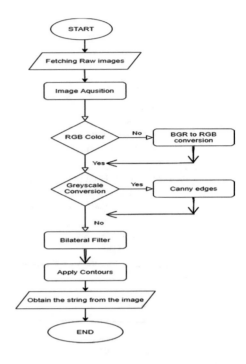

4.1 *Experimental Evaluation*

The experiment results conclude high rates of accuracy, the images used as the sample
were tested for each stage of image preprocessing, and all the stages can be verified
individually. The local images used in the databases were hand captured by the team
members, and the images were acquired using a handheld smartphone. The accuracy
was affected due to some of the images having bad angles and unconventional fonts.
Following are the steps to conduct experiments:

Fig. 2 RGB image

Fig. 3 Grayscale

Fig. 4 **a** Image after applying contours and **b** image after filtering out contours

1. *Image Acquisition*: Image acquisition is the first step in processing an acquired image. The image of the number plate of a vehicle was captured using a 5 megapixels' camera.
2. *Image Undersampling RGB to Grayscale Conversion*: The cv2 image is stored as a NumPy array and cv2.COLOR_RGB2GRAY is a function that converts RGB images into grayscale images. Figures 2 and 3 represent the conversion of RGB to grayscale images. It includes the following steps:

   1. *Bilateral Filtering*: Processing of images having some or high level of noise is not an easy task; Therefore, bilateral filtering is used for removing noise.
   2. *Applying Contours*: The curve joining all the continuous points (along the boundary), having the same color or intensity, is known as a contour. Figure 4a represents the image after applying contours.
   3. *Filter Contours*: It is used for small regions like noise outliers and sharp edges, and contours are filtered from the original images. Figure 4b represents the image after filtering out contours.
   4. *Plate Angle Correction*: A bounding box is applied to each plate in this stage, and an affine transformation is performed, supposing that any of the plates endure from angle distortion. Figure 5a represents the image before plate angle correction, and Fig. 5b represents the image after plate angle correction.
   5. *Cropping the Part with Number Plate*: The image may contain various extra characters in it such as vehicle brand name, logos, or custom texts, which may interrupt the results and add extra unwanted characters. Figure 6a represents the process of identifying the part of the plate with the number plate, and Fig. 6b represents the process of cropping the part with the number plate.

Fig. 5 **a** Image before plate angle correction and **b** image after plate angle correction

(a)   (b)

Fig. 6 **a** Identifying the part with the number plate and **b** cropping the part with the number plate

6. *Building Boxes Over Characters Recognized*: Building boxes involve building imaginary boxes that are around objects that are being checked for a collision like pedestrians on or close to the road, other vehicles, and signs.

4.2 Results and Discussion

The results of the experiments show that the number plate of the vehicle is successfully detected and successfully stored in the database; further, the results of theft detection led to high accuracy as shown in Fig. 7a and b represents the various instances of the steps involved. The problem with the accuracy of the work has been resolved, and better results are obtained as compared to other existing works [23]. Also, the status of the vehicle has been successfully identified (if it is stolen or not). It will further help the officials and higher authorities to detect the passing vehicle and check for the theft status of the vehicle. It can be helpful to the community for reducing the crime in the area and help reducing the efforts of the traffic cops and government officials. Table 2 represents the results attained after the experiments.

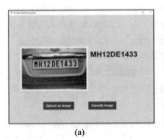 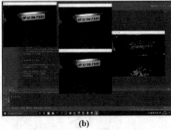

(a)   (b)

Fig. 7 **a** Input window and **b** output for all the processes

Table 2 Results

Original plate	Detected plate	Accuracy (%)
UP32FR5850	UP32FR5850	100
HR26DQ5551	HR26DQ5551	100
KL65K7111	KL65K7111	100
UK14F5981	UK14F5981	100
UP32DM7551	UP32DM7551	100
UK14F9043	UKI4F904E	80
MH13CD0096	WH13(D0096	80
KL07CN8055	KLO7CN8055	90
TN01AS9299	TN01AS9299	100
MH12DE1433	MH12DE1433	100

4.3 Comparative Analysis

This section represents the comparative analysis of the proposed work with the existing work. The work theft identification using ML/OCR for vehicles as compared to the work shown in [26, 27] is a better and advanced version as there is lack of a proper GUI to store and manage the data read; also, there is lack of proper management system for theft status of the vehicles.

5 Conclusion and Future Scope

The proposed method detected and extracted the number plate of vehicles from an image provided by a camera which is then uploaded on a GUI Window using Python. Then, the image is sent to a preprocessing engine and the number of the vehicle is extracted in string format; this string format can be further used to identify the theft status of the vehicles. This model was tested on a test dataset of approximately 100 images and achieved a precision rate of 95%. While considering the remaining 5% of the number plates, 97% out of them were found to be partial matches of at least eight characters recognized correctly in a number plate. It results in a more effective, accurate, and less time-consuming method. It promises to be less vulnerable to errors if implemented in suitable conditions. This method tends to be more cost-efficient than traditionally used methods as they usually require a high-resolution camera and an expensive setup to process and detect number plates from vehicles.

The video feed of traffic footage and real-time processing of the passing vehicles is a limitation of this work. This problem can be handled using deep learning. Moreover, the detection of the plates is crystal clear in the daytime, but the detection in night-time is quite a hassle. Since the lack of a night vision camera and infrared-equipped hardware affects how the image is captured and hence the images lack quality, the processing is hard to do on such images.

References

1. Bhushan M, Kumar A, Samant P, Bansal S, Tiwari S, Negi A (2021) Identifying quality attributes of FODA and DSSA methods in domain analysis using a case study. In: 10th International conference on system modeling & advancement in research trends (SMART), pp 562–567. https://doi.org/10.1109/SMART52563.2021.9676289
2. Bhushan M, Duarte JÁG, Samant P, Kumar A, Negi A (2021) Classifying and resolving software product line redundancies using an ontological first-order logic rule-based method. Expert Syst Appl 168:114167. https://doi.org/10.1016/j.eswa.2020.114167
3. Bhushan M, Goel S, Kumar A (2018) Improving quality of software product line by analysing inconsistencies in feature models using an ontological rule-based approach. Exp Syst 35(3):e12256. https://doi.org/10.1111/exsy.12256
4. Bhushan M, Goel S, Kaur K (2018) Analyzing inconsistencies in software product lines using an ontological rule-based approach. J Syst Softw 137:605–617. https://doi.org/10.1016/j.jss.2017.06.002
5. Bhushan M, Negi A, Samant P, Goel S, Kumar A (2020) A classification and systematic review of product line feature model defects. Softw Qual J 28(4):1507–1550. https://doi.org/10.1007/s11219-020-09522-1
6. Negi A, Kaur K (2017) Method to resolve software product line errors. In: International conference on information, communication and computing technology, pp 258–268. Springer. https://doi.org/10.1007/978-981-10-6544-6_24
7. Bhushan M, Goel S, Kumar A, Negi A (2017) Managing software product line using an ontological rule-based framework. In: International conference on infocom technologies and unmanned systems (trends and future directions) (ICTUS), pp 376–382. IEEE. https://doi.org/10.1109/ICTUS.2017.8286036
8. Bhushan M, Goel S (2016) Improving software product line using an ontological approach. Sādhanā 41(12):1381–1391. https://doi.org/10.1007/s12046-016-0571-y
9. Verma K, Bhardwaj S, Arya R, Islam MSU, Bhushan M, Kumar A, Samant P (2019) Latest tools for data mining and machine learning. Int J Innov Technol Explor Eng 8(9S):18–23. https://doi.org/10.35940/ijitee.I1003.0789S19
10. Pawar S, Bhushan M, Wagh M (2020) The plant leaf disease diagnosis and spectral data analysis using machine learning—a review. Int J Adv Sci Technol 29(9s):3343–3359. http://sersc.org/journals/index.php/IJAST/article/view/15945
11. Suri RS, Dubey V, Kapoor NR, Kumar A, Bhushan M (2021) Optimizing the compressive strength of concrete with altered compositions using hybrid PSO-ANN. In: 4th International conference on information systems and management science (ISMS 2021). Springer, University of Malta, Msida, Malta
12. Kholiya PS, Kapoor A, Rana M, Bhushan M (2021) Intelligent process automation: the future of digital transformation. In: 10th International conference on system modeling & advancement in research trends (SMART), pp 185–190. https://doi.org/10.1109/SMART52563.2021.9676222
13. Sharma S, Nanda M, Goel R, Jain A, Bhushan M, Kumar A (2019) Smart cities using internet of things: recent trends and techniques. Int J Innov Technol Exploring Eng 8(9S):24–28. https://doi.org/10.35940/ijitee.I1004.0789S19
14. Goel R, Jain A, Verma K, Bhushan M, Kumar A (2020) Mushrooming trends and technologies to aid visually impaired people. In: International conference on emerging trends in information technology and engineering (ic-ETITE), pp 1–5. https://doi.org/10.1109/ic-ETITE47903.2020.437
15. Nalavade A, Bai A, Bhushan M (2020) Deep learning techniques and models for improving machine reading comprehension system. Int J Adv Sci Technol 29(04):9692–9710. http://sersc.org/journals/index.php/IJAST/article/view/32996
16. Bhavke A (2018) An automatic toll collection, vehicle identification during collision & theft detection using RFID. Asian J Convergence Technol (AJCT), ISSN-2350-1146

17. Farhat A, Hommos O, Al-Zawqari A, Al-Qahtani A, Bensaali F, Amira A, Zhai X (2018) Optical character recognition on heterogeneous SoC for HD automatic number plate recognition system. EURASIP J Image Video Process 2018(1):1–17
18. Agbemenu AS, Yankey J, Addo EO (2018) An automatic number plate recognition system using OpenCV and Tesseract OCR engine. Int J Comput Appl 180:1–5
19. Patel F, Solanki J, Rajguru V, Saxena A (2018) Recognition of vehicle number plate using image processing technique. Wirel Commun Technol, 2(2)
20. Singh J, Bhushan B (2019) Real time indian license plate detection using deep neural networks and optical character recognition using LSTM Tesseract. In: International conference on computing, communication, and intelligent systems (ICCCIS), pp 347–352
21. Ganta S, Svsrk P (2020) A novel method for Indian vehicle registration number plate detection and recognition using image processing techniques. Procedia Comput Sci 167:2623–2633
22. Yaseen NO, Al-Ali SGS, Sengur A (2019) Development of new Anpr dataset for automatic number plate detection and recognition in North of Iraq. In: 1st International informatics and software engineering conference (UBMYK), pp 1–6
23. Sferle RM, Moisi EV (2019) Automatic number plate recognition for a smart service auto. In: 15th International conference on engineering of modern electric systems (EMES), pp 57–60
24. Chandan R, Veena M (2020) Vehicle number identification using machine learning & OPENCV. Database, 5:6
25. Zhang Y, Qiu M, Ni Y, Wang Q (2020) A novel deep learning based number plate detect algorithm under dark lighting conditions. In: 20th International conference on communication technology (ICCT), pp 1412–1417. https://doi.org/10.1109/ICCT50939.2020.9295720
26. Damak T, Kriaa O, Baccar A, Ayed MB, Masmoudi N (2020) Automatic number plate recognition system based on deep learning. Int J Comput Inf Eng 14(3):86–90
27. Salwan RM, Dhamande PN, Chavhan KP, Waghade AG (2020) Vehicle number plate recognition system
28. Shariff AM, Bhatia R, Kuma R, Jha S (2021) Vehicle number plate detection using python and open CV. In: International conference on advance computing and innovative technologies in engineering (ICACITE), pp 525–529. IEEE
29. Goel T, Tripathi KC, Sharma ML (2020) Single line license plate detection using opencv and tesseract
30. Pavani T, Mohan D (2019) Number plate recognition by using open CV-python. Int Res J Eng Technol (IRJET) 6:4987–4992

Ransomware Attack Detection by Applying Machine Learning Techniques

Siddharth Routray, Debachudamani Prusti, and Santanu Kumar Rath

1 Introduction

With the growth of technology and resources, attackers use numerous feasible and intelligent approaches to create malware that serves their purpose. These computer security threats are categorized as computer viruses, spyware, malware, phishing, and many more as shown in Table 1.

Among these threats, ransomware is a malware that mainly aims to extort individuals or organizations. Ransomware enables extortion by planting denial of service to either a system or resources of that system, resulting in the user not accessing the system. In recent times, ransomware is the most used attack vector as it is irreversible and, unlike other malware, is very difficult to prevent [2]. For example, "SamSam" ransomware in 2018 infected the whole city of Atlanta, the Colorado Department of Transportation, and the Port of San Diego, abruptly terminating services. SamSam ransomware was also used to attack hospitals, municipalities, public institutions, and more than 200 U.S and Canadian companies. SamSam targeted vulnerabilities in file transfer protocol and remote desktop protocol to spread. One such famous attack is "WannaCry" ransomware in 2017, which is considered one of the most devastating ransomware attacks in history.

Ransomware uses asymmetric encryption to encrypt and capture the victim's resources and to decrypt and release the resources, and a certain ransom is demanded from the victim. To implement this, a pair of public–private keys is uniquely generated by the attacker for encryption and decryption of resources, whereas the private key required for decryption is provided by the attacker to the victim only after the ransom

S. Routray (✉) · D. Prusti · S. K. Rath
Department of Computer Science and Engineering, National Institute of Technology Rourkela, Rourkela, Odisha, India
e-mail: siddharth.routray@gmail.com

S. K. Rath
e-mail: skrath@nitrkl.ac.in

© The Author(s), under exclusive license to Springer Nature Singapore Pte Ltd. 2023
D. S. Sisodia et al. (eds.), *Machine Intelligence Techniques for Data Analysis and Signal Processing*, Lecture Notes in Electrical Engineering 997,
https://doi.org/10.1007/978-981-99-0085-5_62

Table 1 Different types of malwares

Type of malware	Description
Sniffers	This software keeps track of the network traffic, analyzes them, and collects information to initiate malware attack [1]
Spyware	As the name suggests, the spyware collects user information such as passwords, pins, and messages without the user knowledge [1]
Adware	This malware tracks users' browsing activity, collects that information, and shows ads on their browsing content and history [1]
Trojan	Trojan disguises itself as a legit application and, upon execution, initiates malicious actions on the user's system [1]
Worms	Upon gaining entry into the user's system, worms replicate themselves exponentially and can cause Distributed Denial of Service or DDoS attacks and even ransomware attacks [1]
Virus	A virus attaches itself to an application and executes itself upon the execution of that application, causing system harm [1]
Rootkits	A rootkit is an application that enables remote access of the user's system to the attacker [1]
Ransomware	Ransomware is an attack through which an attacker disables a user to access his/her system resources and releases those upon payment of ransom [1]

is paid. After the encryption, ransomware prompts the victim for a ransom and provides a specific time to make the payment and release the private key on failing which the captive files are destroyed.

According to Chittooparambil et al., none of the existing methods can afford to detect and prevent ransomware attacks [2]. Among the difficulties, there is a need to come up with different techniques and methods that are to be used to detect ransomware. This study focuses on finding the most efficient techniques using machine learning and concluding with a pipeline of techniques that gives the best accuracy.

In this paper, we will be providing a brief literature review on this field in Sect. 2 followed by the methodology and techniques applied in Sect. 3. The results and observations are recorded in Sect. 4. Finally, we have concluded the work in Sect. 5 followed by references in Sect. 6.

1.1 Motivation

Ransomware is a type of malware that is hard to detect because of its unique attack style and unique behavior, making it a challenge. Ransoms, i.e., the monetary transaction, are involved in ransomware, making it more interesting. Ransomware data are primarily behavioral and exciting to study and analyze.

2 Literature Review

Ransomware attacks date to 1989 and have gained popularity due to their unique and robust attack style. The first-ever documented ransomware attack is "AIDS Trojan," a PC Cyborg virus spread via floppy disks. Moreover, the ransoms were collected via posts. Ransomware attacks slowly gained more popularity since the 2000s. Furthermore, these attacks got more violent and preferred after the introduction of Bitcoins in 2010.

Berrueta et al. [3] focused on detecting cryptographic ransomware, where they analyzed 63 ransomware samples from 52 families to extract the characteristic steps taken by ransomware. They compared the different approaches and classified the algorithms based on the input data from ransomware actions and the decision procedures to distinguish between benign and malign applications.

Zavarsky et al. [4] went on to do an analysis of ransomware on Windows platforms and Android platforms. For this, they used a dataset of 25 significant ransomware families. About 90% of the samples are from Virus Total, 8% are from public malware repositories, and the rest are collected by manually browsing through security forums. Thus, they also found that Windows 10 is pretty effective against ransomware attacks.

Takeuchi et al. [5] detected ransomware using a support vector machine (SVM) classifier. For the study, they used 276 ransomwares and 312 goodware files. And then extracted a specific ransomware feature known as application programming interface (API). The vector representations of the API call logs resulting from ransomware executions are used as training examples for an SVM. And then, they studied the behavior of the API using Cuckoo Sandbox. They achieved an accuracy of 97.48% by using SVM.

Khammas [6] adopted a byte-level static analysis method for detecting ransomware where they achieved high accuracy by using a random forest classifier. Their study has tested different sizes of trees and seeds ranging from 10 to 1000 and 1 to 1000, respectively. They also concluded in their study that a tree size of 100 with a seed size of 1 achieved an accuracy of 97.74% and a high Receiver Operating Characteristic Curve or ROC of about 99.6%. The dataset used by Ban Mohammed Khammas [6] consists of 1680 executable files, including 840 ransomwares executable of different families and 840 goodware files.

Chen et al. [7] developed a ransomware simulation program to demonstrate how to generate malicious I/O operations on the generated feature sequences (GANs). In that case study, they proposed to use GAN to automatically produce dynamic features that exhibit generalized malicious behaviors that can reduce the efficacy of black-box ransomware classifiers.

This section proposes an incremental ransomware detection model using various machine learning techniques. First, we present the architecture of the detection model in Fig. 1. The goal is to train the model by feeding the dataset with no feature selection techniques applied and then further optimizing it by applying feature selection techniques on the dataset and improving the classification's accuracy.

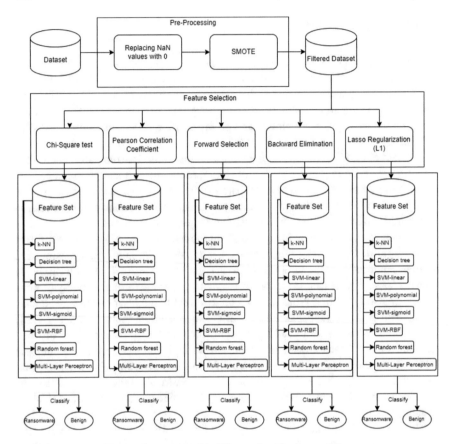

Fig. 1 Proposed architectural approach with different classification techniques

3 Methodology

This section proposes an incremental ransomware detection model using various machine learning techniques. First, we present the architecture of the detection model in Fig. 1. It comprises four stages: Preprocessing the raw dataset, applying the feature selection technique on it, splitting the data into training and testing data, and then finally training the data using some machine learning classifier models. The goal is to train the model by feeding the dataset with no feature selection techniques applied and then further optimizing it by applying feature selection techniques on the dataset and improving the classification's accuracy. For every feature selection technique, we feed the final features to different classifier models, and finally, we achieve a pipeline of technique and model that gives the best accuracy, as shown in Fig. 1.

3.1 Dataset

The dataset consists of historical data records of data breaches and ransomware attacks over 15 years from 2004 to 2020. The dataset is obtained from The University of Queensland repository [8]. The dataset contains 43 features that define the characteristics, behavior, effect, nature, and aftermath. The feature value in this dataset is in a textual form which further needs to be converted to numerical data in further stages to be processed. The dataset consists of 64 ransomware records, and the remaining are benign records. This study has split the dataset into 70% and 30% training and testing sets, respectively.

3.2 Preprocessing

The dataset needs to be preprocessed to be used in further phases, which means that the dataset needs to be cleaned for all null, Not a Number or NAN values, and then the features with most NAN values are also removed from the dataset. The dataset we have is in textual form, which cannot be used, so we need to convert the whole textual dataset into numerical data. For that, we picked every feature and then mapped the unique values of the feature with a unique number. The exact process is repeated, hence making the whole dataset numerical.

3.3 Feature Selection

We take the preprocessed dataset, and before training our model, we have to first check whether we can reduce the features and remove obsolete and unnecessary features [9]. Then, we need to apply some feature selection techniques and determine which technique we get a better result based on the result. Then, we take the most efficient feature selection technique to further stages as discussed below.

Filter method. Filter methods are used to eliminate the feature based on their relevancy [9]. This method calculates the intrinsic property for every feature in the dataset via univariate statistics. Therefore, they are cheaper to implement than wrapper methods while dealing with high-dimensional data. The techniques used from filter methods are discussed below.

Chi-Square test. The Chi-square test is best used for categorical values. In this method, two hypotheses are considered, i.e., null hypothesis and alternative hypothesis [9]. Then, we calculate the Chi-square value of every attribute and check whether the calculated Chi-square distribution is less than 5%. Then, we reject the null hypothesis or accept the null hypothesis. Likewise, we finally get a set of best features at the end to work with.

Correlation Coefficient. This method is used to check the linear relationship of attributes with the target attribute. It picks every attribute from the dataset and checks the correlation between that attribute and the target attribute [9]. If it finds the correlation is very high, we consider that attribute relevant and keep that feature in the dataset.

Wrapper method. It is more complex and expensive than the filter method [9]. In this method, we take every combination of features to create a subset for every combination. Then, we train those features using a model, and likewise, we create models for every such possible subset, and finally, we select the features of that subset with the best model results. As the name suggests, the wrapper method wraps the features' tests and gives an efficient set of features.

Forward Selection. It starts by creating a subset of features by first selecting the best-performing attribute with the target attribute, and then we add the following best-performing attribute until we find the set of best-performing attributes that works best against the target attributes [9].

Backward Elimination. It works exactly the opposite of the forward selection, where we start with training all the attributes against the target attribute and keep on eliminating the best-performing attribute until we get the set of best-performing attributes [9].

Embedded method. It works the same as the wrapper method but with less computational cost [9]. It is also iterative and checks the usefulness of attributes.

LASSO regularization. While computing the best-performing features, Lasso regularization adds a penalty to the model's different attributes that reduce model's freedom and hence helps avoid overfitting [9]. This process helps reduce computational costs. L1 here has the property to reduce coefficients to zero and remove those features. By continuously removing the features, the best subset of features is obtained.

3.4 Synthetic Minority Oversampling Technique

Synthetic minority oversampling technique (SMOTE) is used for imbalanced datasets to generate synthetic data for minority classes by choosing the minority class as one of the k-nearest neighbors (k-NNs) and balancing the dataset [10]. Because we have an unbalanced dataset in our study, we have used SMOTE to balance the dataset.

3.5 Model Training

Different classifiers such as k-NN, SVM (linear, polynomial, sigmoid, radial basis function (RBF)), decision tree, random forest, and multilayer perceptron (MLP) are used to train the model without the feature selection techniques and then with the feature selection techniques.

k-Nearest Neighbor. It is a machine learning algorithm technique. It is a supervised learning technique. k-NN considers various data and categorizes them into specific groups based on their similarities. Upon receiving a new data point, it classifies its group by considering the most k data points near to it [11].

Support vector machine. SVMs are supervised learning algorithms suitable for solving classification problems [12, 13]. SVM generates a hyperplane by choosing the extreme points, which then segregates the high dimensional space into different classes, and upon detecting a new data point, it puts that data point into its desirable class [14]. The kernel functions of SVM are discussed below.

Linear kernel. The bare and fast-performing single-dimensional kernel is suitable for classification problems [15].

Polynomial kernel. It is a directional kernel with one or more dimensions. As a result, it is less efficient and accurate than other kernels [15].

Radial basis function. It is more suitable for nonlinear problems and when there is no prior knowledge of the data. RBF is the preferred kernel in SVM due to its proper data separation and accuracy [15].

Sigmoid kernel. It is similar to a two-layer perceptron model of a neural network. Hence, it is preferred for neural network problems [15].

Decision tree. It is a tree consisting of nodes and branches where each node represents features in an instance that need to be classified, and branches here represent different node values, and any branch taken is a decision made which leads to the leaf node, which is the outcome of the tree [13, 14].

Random forest. It is also known as decision trees which are a collection of many decision trees that overcome the decision tree limit that is overtraining the model and the instability. Random forest uses different training data for different trees, which reduces the overfitting problem. However, each tree in a random forest is generated independently and hence takes more complexity to work with random forest [12].

Multilayer perceptron. It consists of a network of neurons or nodes connected by synapses, and each node and synapse have some weight associated with them. There is one input layer, one output layer, and many hidden layers. MLP works by initially doing an AND operation between the input and weight of the synapses, which generates value at the hidden layers. The hidden layers keep pushing the value after utilizing the activation function at each layer until it generates the output at the output layer. MLP yields outstanding results for classification problems [14].

4   Experimental Results

The above-discussed feature selection techniques and classifiers are used, and the observed results are recorded and contrasted. For example, in the following results, the tangible result is classified into 1 or 0, where 1 represents "ransomware" and 0 means "benign."

Table 2 Contrasting performance of different classifiers in percentage when no feature selection techniques are used

Classifiers	Accuracy	Precision	Recall	f1-score
k-NN	97.56	20.00	25.00	22.22
SVM-linear	96.86	22.22	50.00	30.77
SVM-polynomial	96.52	12.50	25.00	16.67
SVM-sigmoid	90.24	10.00	75.00	17.65
SVM-RBF	97.21	30.00	75.00	42.86
Decision tree	97.91	75.00	37.50	50.00
Random forest	97.56	20.00	25.00	22.22
Multilayer perceptron	97.56	20.00	25.00	22.22

4.1 Without Any Feature Selection Techniques

Initially, no feature selection techniques were used upon the dataset, and different classifiers were used to check the model's performance. The performance of different classifiers is contrasted in Table 2. Without using any feature selection technique, the decision tree performs better than other classifiers.

4.2 Chi-Square Test

To further optimize the results, Chi-square test is used on the dataset, after which we got 31 features to train further, and then the performance from the different models is contrasted as shown in Table 3. After applying the Chi-square test on the dataset and then passing through various classifiers, it has been observed that MLP gave the best performance as compared to other techniques.

Table 3 Contrasting performance of different classifiers in percentage when Chi-square test is used

Classifiers	Accuracy	Precision	Recall	f1-score
k-NN	96.80	25.00	11.11	15.38
SVM-linear	97.09	50.00	20.00	28.57
SVM-polynomial	96.22	25.00	9.09	13.33
SVM-sigmoid	85.47	50.00	4.00	17.41
SVM-RBF	96.51	50.00	16.67	25.00
Decision tree	97.09	12.50	25.00	16.67
Random forest	97.38	14.29	25.00	18.18
Multilayer perceptron	97.67	16.67	25.00	20.00

Table 4 Contrasting performance of different classifiers in percentage when Pearson's correlation coefficient is used

Classifiers	Accuracy	Precision	Recall	f1-score
k-NN	97.97	33.45	30.91	28.82
SVM-linear	85.52	3.21	71.43	6.13
SVM-polynomial	79.34	3.21	71.43	6.13
SVM-sigmoid	84.01	2.17	71.43	4.22
SVM-RBF	81.23	3.18	71.43	6.10
Decision tree	76.58	3.21	71.43	6.13
Random forest	79.81	3.23	71.43	6.17
Multilayer perceptron	75.67	3.23	71.43	6.17

4.3 Pearson's Correlation Coefficient

After being applied to the dataset, the Pearson's correlation coefficient gave six features to be trained further, and the resulting performance of the classifiers is presented in Table 4. Pearson's correlation coefficient gives poor results on the model, and it is found that the k-NN classifier gives the best result here.

4.4 Forward Selection

Forward selection is applied with the desired output of a different range of features, and it is found that the best results are obtained when the "k_features" is set for 15 feature set. The resultant performance of different classifiers is listed in Table 5.

Forward selection technique, when used as feature selection and then trained with different classifiers, we found that random forest gives the best performance result.

Table 5 Contrasting performance of different classifiers in percentage when forward selection is used

Classifiers	Accuracy	Precision	Recall	f1-score
k-NN	97.67	40.00	28.57	33.33
SVM-linear	97.38	37.50	42.86	40.00
SVM-polynomial	97.09	28.57	28.57	28.57
SVM-sigmoid	78.78	7.69	85.71	14.12
SVM-RBF	96.51	27.27	42.86	33.33
Decision tree	97.67	40.00	28.57	33.33
Random forest	97.97	50.00	28.57	36.36
Multilayer perceptron	97.38	33.33	28.57	30.77

Table 6 Contrasting performance of different classifiers in percentage when backward elimination is used

Classifiers	Accuracy	Precision	Recall	f1-score
k-NN	97.38	33.33	28.57	30.77
SVM-linear	96.80	25.00	28.57	26.67
SVM-polynomial	96.80	25.00	28.57	26.67
SVM-sigmoid	87.79	12.77	85.71	22.22
SVM-RBF	96.22	25.00	42.86	31.58
Decision tree	97.38	33.33	28.57	30.77
Random forest	97.97	50.00	28.57	36.36
Multilayer perceptron	97.67	40.00	28.57	33.33

4.5 Backward Elimination

The backward elimination technique is used with an output of 20 feature sets after observing the output of various feature sets. The result of different classifiers is shown in Table 6.

Backward elimination technique, when used as feature selection and then trained with different classifiers, we found that random forest gives the best performance result.

4.6 Lasso Regularization (L1)

Lasso regularization (L1) technique is used with an output of 19 feature sets after observing the performance of various feature sets. The performance measures of different classifiers are shown in Table 7.

Table 7 Contrasting performance of different classifiers in percentage when Lasso regularization (L1) is used

Classifiers	Accuracy	Precision	Recall	f1-score
k-NN	97.38	37.50	42.86	40.00
SVM-linear	97.67	42.86	42.86	42.86
SVM-polynomial	97.38	33.33	28.57	30.77
SVM-sigmoid	74.71	6.52	85.71	12.12
SVM-RBF	95.93	29.41	71.43	41.67
Decision tree	97.67	40.00	28.57	33.33
Random forest	98.26	60.00	42.86	50.00
Multilayer perceptron	96.80	30.00	42.86	35.29

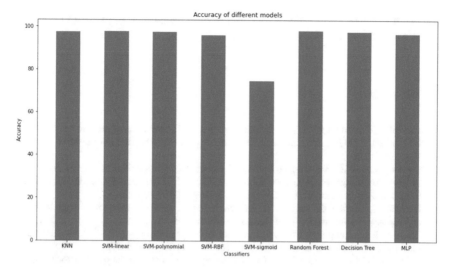

Fig. 2 Contrasting different classifier's accuracy after Lasso regularization (L1) is applied

Lasso regularization (L1) technique, when used as feature selection and then trained with different classifiers, we found that random forest gives the best performance result (Fig. 2).

5 Conclusion

In this study, it is intended to find the best-performing model that classifies an attack as ransomware or benign. The dataset is preprocessed and trained directly with different classification techniques such as k-NN, SVM, decision tree, random forest, and MLP to achieve the research objective. Feature selection techniques have been applied to increase the performance further and optimize the results. The feature selection techniques applied with the classification techniques are Chi-square test, Pearson's correlation coefficient, forward selection, backward elimination, and Lasso regularization (L1). For all the techniques, critical assessment has been carried out to evaluate the performance measures. It has been observed that the incorporation of feature selection techniques has improved the model's performance significantly. Random forest gives the best performance with the Lasso regularization (L1) feature selection technique with a predictive accuracy percentage of 98.26%. Also due to the skewness in the data of the dataset, the important features are considered from the dataset, which improves the predictive accuracy.

This study provides a base model for the classification of ransomware attacks. The model can be further optimized by using different optimization techniques and a combination of different feature selection techniques and classifiers. The future work of this study has a considerable boundary of research to improve and optimize the model and improve the performance.

References

1. Zolkipli MF, Jantan A (2011) An approach for malware behavior identification and classification. In: 2011 3rd International conference on computer research and development, vol 1, pp 191–194. IEEE
2. Chittooparambil HJ, Shanmugam B, Azam S, Kannoorpatti K, Jonkman M, Samy GN (2018) A review of ransomware families and detection methods. In: International conference of reliable information and communication technology. Springer, Cham, pp 588–597
3. Berrueta E, Morato D, Magaña E, Izal M (2019) A survey on detection techniques for cryptographic ransomware. IEEE Access 7:144925–144944
4. Zavarsky P, Lindskog D (2016) Experimental analysis of ransomware on windows and android platforms: evolution and characterization. Proc Comput Sci 94:465–472
5. Takeuchi Y, Sakai K, Fukumoto S (2018) Detecting ransomware using support vector machines. In Proceedings of the 47th international conference on parallel processing companion, pp 1–6
6. Khammas BM (2020) Ransomware detection using random forest technique. ICT Exp, 6:325–331
7. Chen L, Yang CY, Paul A, Sahita R (2018) Towards resilient machine learning for ransomware detection. arXiv: 1812.09400v2 [cs.LG]
8. Ko R, Tsen E, Slapnicar S (2020). Dataset of data breaches and ransomware attacks over 15 years from 2004 to 2020. University of Queensland
9. Visalakshi S, Radha V (2014) A literature review of feature selection techniques and applications: review of feature selection in data mining. In: 2014 IEEE international conference on computational intelligence and computing research, pp 1–6. IEEE
10. Han H, Wang WY, Mao BH (2005) Borderline-SMOTE: a new over-sampling method in imbalanced data sets learning. In: International conference on intelligent computing, pp 878–887. Springer, Berlin, Heidelberg
11. Guo G, Wang H, Bell D, Bi Y, Greer K (2003) KNN model-based approach in classification. In: OTM confederated international conferences on the move to meaningful internet systems, pp 986–996. Springer, Berlin, Heidelberg
12. West J, Bhattacharya M (2016) Intelligent financial fraud detection: a comprehensive review. Comput Secur 57:47–66
13. Prusti D (2015) Efficient intrusion detection model using ensemble methods. Department of Computer Science and Engineering, National Institute of Technology Rourkela, Rourkela, India
14. Kotsiantis SB, Zaharakis I, Pintelas P (2007) Supervised machine learning: a review of classification techniques. Emerg Artif Intell Appl Comput Eng 160(1):3–24
15. Patle A, Chouhan DS (2013) SVM kernel functions for classification. In: 2013 International conference on advances in technology and engineering (ICATE), pp 1–9. IEEE
16. Mercaldo F, Nardone V, Santone A, Visaggio CA (2016) Ransomware steals your phone. formal methods rescue it. In: International conference on formal techniques for distributed objects, components, and systems. Springer, Cham, pp 212–221
17. Kshatri SS, Singh D, Narain B, Bhatia S, Quasim MT, Sinha GR (2021) An empirical analysis of machine learning algorithms for crime prediction using stacked generalization: an ensemble approach. IEEE Access 9:67488–67500. https://doi.org/10.1109/ACCESS.2021.3075140
18. Singh D, Singh S (2020) Realising transfer learning through convolutional neural network and support vector machine for mental task classification. Electron Lett 56(25):1375–1378

Weed Detection in Soybean Crop Using YOLO Algorithm

Dileshwar Dhruw, Ajay Kumar Sori, Santosh Tigga, and Anurag Singh

1 Introduction

The rapid growth of the world's population coupled with climate change puts great pressure on the agricultural sector to increase food production. It is predicted that the world's population will reach nine billion by 2050; therefore, agricultural production should double to meet the growing demand. However, agriculture faces major challenges from the growing risks of plant diseases, pests, and weeds. Increases in weeds, pests, and diseases reduce yields and food quality, fiber yield, and biofuel. Loss is sometimes catastrophic or chronic, but on average accounts for 42% of the production of a few essential food crops. Weeds are unwanted plants that compete with productive plants for light, water, and nutrients and propagate themselves through seeds or the rhizome. They are often toxic, producing thorns and thistles and disrupting crop management by contaminating crop yields. That is why farmers spend billions of dollars on weed control, often without adequate technical support, which leads to poor weed control and reduced crop yields. Therefore, weed control is an important aspect of horticultural crop management, as failure to control weeds effectively leads to reduced yields and product quality. Implementation of chemical and cultural control strategies can lead to negative environmental impacts if not carefully managed. The convolutional neural network (CNN) provides the most efficient extraction efficiency in its detection as it has an excellent ability to detect indirect degenerative material

D. Dhruw (✉) · A. K. Sori · S. Tigga · A. Singh
IIIT Naya Raipur, Chhattisgarh, India
e-mail: dileshwar19102@iiitnr.edu.in

A. K. Sori
e-mail: ajay19102@iiitnr.edu.in

S. Tigga
e-mail: santosh19102@iiitnr.edu.in

A. Singh
e-mail: anurag@iiitnr.edu.in

© The Author(s), under exclusive license to Springer Nature Singapore Pte Ltd. 2023   777
D. S. Sisodia et al. (eds.), *Machine Intelligence Techniques for Data Analysis and Signal Processing*, Lecture Notes in Electrical Engineering 997,
https://doi.org/10.1007/978-981-99-0085-5_63

and represent features. Earlier methods were mostly based on handcrafted feature extraction and machine learning (ML) classifiers. In recent years, there is more push toward applications of deep learning approaches due to their widespread applications in various fields. To the best of our knowledge, however, no previous work has been done so far using the You Only Look Once (YOLO) model in agricultural applications. The proposed work in this paper used the YOLO algorithm to detect weeds in soybean crops as it is the fastest algorithm and detects even the smallest objects with the best performance. Different variants of YOLO, i.e., v3, v4, and v5, have been employed and their comparative performance is analyzed for detecting the soybean weeds. The performance of the proposed framework is tested using a variety of widely used metrics including IoU, mean average precision (mAP), accuracy, memory, and f1 score. Simulation results obtained using YOLO algorithms are robust and accurate with 96% mAP.

2   Related Works

Early weed detection and identification can play a vital role in weed control and can improve the yield of crops and their quality. There are numerous computer-based weed identification methodologies that have been presented in the computer vision field. They can be further classified as handcrafted and deep learning feature extraction. Low-level feature representation is carried out extremely well via handcrafted features. In early stages, techniques were mostly based on handcrafted features and ML approaches [1–4] and now a number of methods were proposed based on deep learning approaches. The works in [1] aimed at developing a stereo vision system for distinguishing between rice plants and weeds and they further propose to classify weeds in two categories using neural networks. Further, a real-time weed/crop detection and classification algorithm was proposed by authors in [2] using the random forest classifier. People have also explored hyperspectral images to detect weeds out of the crops. Authors in [3, 4] have used such images to detect weeds out of the maize crops using a ML approach. A detailed review of the weed detection algorithms based on machine visions may be found in the work [5]. A crop detection methodology using Kalman filtering was proposed in [6] where authors have attempted to classify between different crops. There are also works reported in the literature where computer vision and image processing-based methods were proposed to identify weeds using their imagery features [7–11]. Automated robotics methods are also used in weed control in crops [12, 13]. In recent years, major research work focus is toward deep learning-based approaches due to their efficient detection and classification performances in various engineering fields [14, 15]. In recent years, deep learning-based methods have been used in weed detection too [16, 17]. YOLO is a special type of deep learning architecture popularly used in object detection applications due to its faster and efficient performance [18–20]. This motivated us to use YOLO in weed detection, which delivered good weed detection performance in soybean crops.

3 Methodology

Soybean plants have different types of weeds which impact crops' growth during the growth period of soybean, so it will be better to separate them this time. We have two aspects in this work: (i) identification of weed from crop; (ii) localization of weed and crop. A block diagram of the proposed approach is given in Fig. 1.

3.1 Data Collection and Labeling

The data used to evaluate the proposed weed detection approach is collected from open-source platforms such as Kaggle and a few test pictures are also collected manually from open fields, especially the weed images. Labeling means creating bounding boxes for images with the help of an open-source labeling tool [21]. Creating bounding boxes manually is a very time-consuming task and hence an automated tool is very helpful in labeling the images accurately.

Fig. 1 Block diagram of the proposed approach

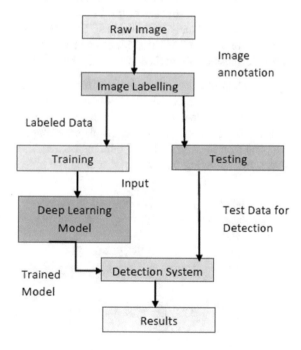

3.2   System Setup

The object identification algorithm in the proposed work aims to identify different weeds and soybean crops. The block diagram of the entire detection system is given in Fig. 1. The proposed model setup employs raw images and then data preprocessing along with image annotation or labeling is done. Then, the dataset is divided into test and training sets. Once the data is segregated, test and train the train set will be used for training the YOLO algorithm. After training, we do testing where we can evaluate our model performance.

3.3   You Only Look Once Algorithm for Weed Detection

For object detection tasks, CNNs have become a popular choice due to their tremendous advancements in computational capability in recent years, which in turn resulted in proposing more efficient deep networks. We explored a few cutting-edge strategies for object identification and characterization in this work for the purposes of weed detection from soybean crops. The proposed work focuses mainly on the three ongoing models namely YOLOv3 [18], YOLOv4 [19], and YOLOv5 [20]. A general architecture of the YOLO algorithm is given in the block diagram depicted in Fig. 2. A brief discussion about the three YOLO models is given below.

YOLOv3: It is an object detection algorithm that is used to identify the area of interest that can be an image or video or live streaming videos. The core idea behind YOLO is to use the complete image as input to the network and generate the output directly with the bounding box and its class. It takes the image and divides it into a prespecified grid, and then these grids form bounding boxes. After going through the number of layers, the bounding box is shown around the object where the confidence for an

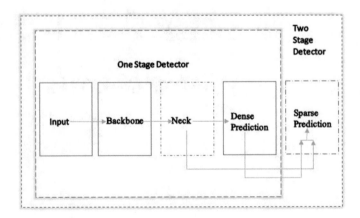

Fig. 2 Block diagram of the architecture of YOLOv4

object is high. It has a simple combination of convolution layer, max pool layer, and YOLO layer.

YOLOv4: YOLOv4 architecture is given in Fig. 2. YOLOv4 is a significant improvement over YOLOv3. The following component of YOLOv4 makes it an attractive package:

- Training of this neural network in a single GPU has become quite comfortable.
- mAP and frames per second have increased to 10% and 12%, respectively, due to new architecture implementation in the backbone and modification in the neck.

YOLOv5: Ultralytics was founded in 2014, spearheading several U.S. Intelligence Community and Department of Defence initiatives in the fields of particle physics, data science, and artificial intelligence. Ultralytics has partnered with several universities, national laboratories, and other start-ups on a variety of successful efforts. Ultralytics has recently created YOLOv5, the world's most advanced Vision AI [22]. With YOLOv5 [14] and its PyTorch implementation, you can quickly get started with object detection.

3.4 Performance Metrics

We have used different widely used performance metrics to evaluate weed detection performance such as mAP, precision, recall, and F1-score [23]. The mAP compares the ground truth bounding boxes and detects bounding boxes, whereas the precision shows the predicted truth values among all truth values and recall shows the truth values among total values.

4 Results and Discussions

The proposed object detection algorithms were implemented in a Python environment on an Intel Xeon Silver 4214 CPU @ 2.20 GHz processor with an Nvidia Quadro P100 GPU. The standard architecture of three models was used. For testing and training, the dataset is divided into a ratio of 20:80. In YOLOv3 and v4, anchors are set to default and we used a learning rate of 0.001 with the batch size 64. The number of iteration for training is decided by the max_batches variable which is set to the maximum value using function max (4000, number of training images, 2000*classes). That means it would run a minimum of 4000 iterations on training. It uses images with a width and height of 416. Whereas YOLOv5 uses images with a width and height of 640 trained for 200 epochs along with an initial learning rate of 0.01 and modifies it by multiplying it by 0.2. We have used 260 images with its text file bounding box for test and 1040 images for training purposes.

4.1 Quantitative Results

In the quantitative analysis, we evaluated the detection performance of the proposed approach. The localization performance of our method will be discussed in the next qualitative result section. The YOLO models were trained on the labeled test images. The training was done up to 1200 iterations, and the optimum set of weights for the network was obtained around 1100 iterations as shown in the model loss versus number of iterations plot shown in Fig. 3 for YOLOv3. Similarly, other YOLO models were also trained. Once trained, the models were tested using the test images and performance metrics.

Each model was calculated in terms of different performance metrics as shown in Table 1. It can be observed that the weed detection performance of YOLOv5 is found to be highest followed by YOLOv4 and v3. The maximum detection accuracy in terms of mAP is 96% with recall of 98%. The mAP variation of the YOLOv5 model

Fig. 3 YOLOv3 model loss variation with respect to number of iteration during training

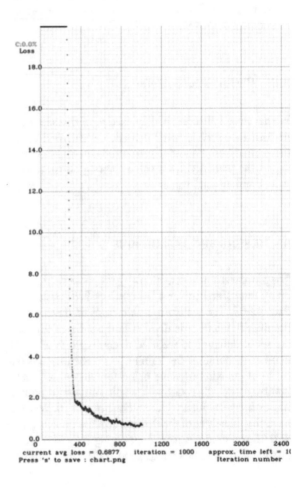

Table 1 Weed detection performance (in %) of the three YOLO models

Algorithm	mAP	Precision	Recall	F-score
YOLOv3	0.72	0.49	0.74	0.61
YOLOv4	0.76	0.51	0.78	0.62
YOLOv5	0.96	0.91	0.98	0.79

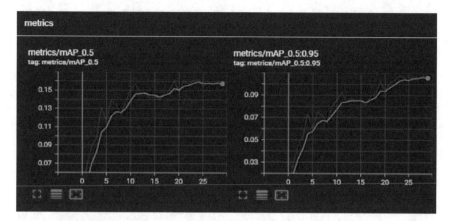

Fig. 4 Variation of mAP of YOLOv5 model

is shown in Fig. 4. The higher value of mAP metric signifies the better average weed detection performance of the YOLOv5 model.

4.2 Qualitative Results

As we discussed the detection performance in the previous section, now we will try to validate the quantitative results through qualitative results. Here, we would demonstrate the localization capability of our approach. As the detection performance of YOLOv3 was much poor, we did not perform location using this; however, we included weed localization results of YOLOv4 and v5 models.

Weed detection and identification results of the YOLOv4 model are shown in Fig. 5. The model creates the bounding box around the plants/weeds along with its identification as plant or weed. As shown in Fig. 5, few plants are identified with bounding boxes around them, but some plants are not inside the bounding box and the height and width of the bounding boxes are not localized properly. Also, some plants are not detected and so the detection accuracy of YOLOv4 is moderate.

The detection and identification performance of YOLOv5 is shown in Fig. 6. Here, the performance on 16 images is shown, and it can be observed that in most of the images, the model is able to accurately detect and identify the plants. The bounding

Fig. 5 Weed detection and localization results by YOLOv4

Fig. 6 Weed detection and localization results by YOLOv5

Table 2 Performance comparison table

Works	Methods	Results (accuracy/mAP)	Precision, recall, F1-score (in %)
Islam et. al. [24]	Image processing using machine learning	0.94 (SVM), 0.63 (k-NN), 0.96 (RF)	0.91, 0.91, 0.89 0.62, 0.62, 0.81 0.95, 0.95, 0.89
Yu et al. [25]	Deep learning (CNN)	0.87 (VGGNet) 0.91 (AlexNet)	0.54, 0.98, 0.70 0.88, 1.00, 0.93
Hu et. al [26]	YOLO v4	0.94	–
Proposed	YOLO v3, v4, v5	0.72 (YOLOv3) 0.76 (YOLOv4) 0.96 (YOLOv5)	0.49, 0.74, 0.61 0.51, 0.78, 0.62 0.91, 0.98, 0.79

boxes are also created accurately around the weeds/plants. The result shows the predicted soybean weed and crop with its probability written with the box. It is also capable of detecting multi plants such as weeds or crops in a single image. So, this can be a better model for detecting weeds in soybean crops and also it can be generalized for other crops too.

As we did not get any similar work on the same database, we found it difficult to compare our results with the existing works. However, just to get an idea about the relative performance of the proposed approach, we summarized the comparison results with other weed detection approaches in Table 2, though it is not a fair comparison. The comparison is done in terms of different metrics used in different works such as detection accuracy, mAP, and F1-score. It is found that the YOLOv5 is performing relatively better in terms of the highest mAP and is much faster. It is more accurate in detecting and identifying even multiple types of weeds in the crop.

5 Conclusion

We explored YOLO-based weed detection and identification approaches in this paper. YOLO algorithms are best known for their accurate detection capabilities, so we have used different versions of this algorithm that include YOLOv3, v4, and v5. It is demonstrated through extensive simulation results that YOLO can be an effective algorithm for detecting and locating weeds. The YOLOv5 algorithm is found to be quite faster in training and detecting weeds and it can detect even small weeds also. Future work in automated agriculture is very promising in India as there is a huge scope of improvement in the performance of the existing methods. The major limitation of the current work is the use of a limited set of annotated datasets. If we can create a large pool of annotated crop–weed test images, we can employ advanced deep learning algorithms to get better detection performance. Scope. This work was found sensitive toward different weed sizes and accordingly, it showed varied detection accuracy. So, proper annotation, especially in test images with multiple weeds, is required to get reliable results.

References

1. Dadashzadeh M, Abbaspour-Gilandeh Y, Mesri-Gundoshmian T, Sabzi S, HernándezHernández JL, Hernández-Hernández M, Arribas JI (2020) Weed Classification for site specific weed management using an automated stereo computer-vision machine-learning system in rice fields. Plants 9:559
2. Alam M, Alam MS, Roman M, Tufail M, Khan MU, Khan MT (2020) Real-time machine-learning based crop/weed detection and classification for variable-rate spraying in precision agriculture. In: Proceedings of the 2020 7th international conference on electrical and electronics engineering (ICEEE), Antalya, Turke, 14–16 April 2020, pp 273–280
3. Gao J, Nuyttens D, Lootens P, He Y, Pieters JG (2018) Recognising weeds in a maize crop using a random forest machine-learning algorithm and near-infrared snapshot mosaic hyperspectral imagery. Biosyst Eng 170:39–50
4. Pantazi X-E, Moshou D, Bravo C (2016) Active learning system for weed species recognition based on hyperspectral sensing. Biosyst Eng 146:193–202
5. Wang A, Zhang W, Wei X (2019) A review on weed detection using ground-based machine vision and image processing techniques. Comput Electron Agric 158:226–240
6. Hamuda E, Mc Ginley B, Glavin M, Jones E (2018) Improved image processing-based crop detection using Kalman filtering and the Hungarian algorithm. Comput Electron Agric 148:37–44
7. Zheng Y, Zhu Q, Huang M, Guo Y, Qin J (2017) Maize and weed classification using color indices with support vector data description in outdoor fields. Comput Electron Agric 141:215–222
8. Choi KH, Han SK, Park K-H, Kim K-S, Kim S (2015) Vision based guidance line extraction for autonomous weed control robot in paddy field. In: Proceedings of the 2015 IEEE international conference on robotics and biomimetics (ROBIO), Zhuhai, China, 6–9 Dec 2015, pp 831–836
9. Tillett N, Hague T, Grundy A, Dedousis A (2008) Mechanical within-row weed control for transplanted crops using computer vision. Biosyst Eng 99:171–178
10. Tang J-L, Chen X-Q, Miao R-H, Wang D (2016) Weed detection using image processing under different illumination for site-specific areas spraying. Comput Electron Agric 122:103–111
11. Berge T, Goldberg S, Kaspersen K, Netland J (2012) Towards machine vision based site-specific weed management in cereals. Comput Electron Agric 81:79–86
12. Nakai S, Yamada Y (2014) Development of a weed suppression robot for rice cultivation: weed suppression and posture control. Int J Electr Comput Electron Commun Eng 8:1736–1740
13. Cordill C, Grift TE (2011) Design and testing of an intra-row mechanical weeding machine for corn. Biosyst Eng 110:247–252
14. Dos Santos CN, Gatti M (2014) Deep convolutional neural networks for sentiment analysis of short texts. In: COLING, pp 69–78
15. Arel I, Rose DC, Karnowski TP (2010) Deep machine learning—a new frontier in artificial intelligence research (Research frontier). IEEE Comput Intell Mag 5(4):13–18. https://doi.org/10.1109/MCI.2010.938364
16. Dos Santos Ferreira A, Freitas DM, da Silva GG, Pistori H, Folhes MT (2017) Weed detection in soybean crops using ConvNets. Comput Electron Agric 143:314–324
17. Yu J, Sharpe SM, Schumann AW, Boyd NS (2019) Deep learning for image-based weed detection in turfgrass. Eur J Agron 104:78–84
18. Redmon J, Farhadi A (2018) YOLOv3: an incremental improvement, Apr 2018 (Online). Available: http://arxiv.org/abs/1804.02767

19. Bochkovskiy A, Wang C-Y, Liao HYM (2020) YOLOv4: optimal speed and accuracy of object detection, Apr 2020 (Online). Available: http://arxiv.org/abs/2004.10934
20. Glenn.jocher. YOLO V5 (Online). Available: https://github.com/ultralytics/yolov5
21. Tzutalin. Labelimg. https://github.com/tzutalin/labelImg
22. Ultralytics.com. https://ultralytics.com/about
23. Deng L, Wang Y, Han Z, Yu R (2018) Research on insect pest image detection and recognition based on bio-inspired methods. Biosyst Eng 169:139–148. https://doi.org/10.1016/j.biosystemseng.2018.02.008
24. Islam N, Rashid MM, Wibowo S, Xu C-Y, Morshed A, Wasimi SA, Moore S, Rahman SM (2021) Early weed detection using image processing and machine learning techniques in an Australian Chilli farm. Agriculture 11(5):387. https://doi.org/10.3390/agriculture11050387
25. Yu J, Schumann AW, Cao Z, Sharpe SM, Boyd NS (2019) Weed detection in perennial ryegrass with deep learning convolutional neural network. Front Plant Sci 10:1422. Published 31 Oct 2019. https://doi.org/10.3389/fpls.2019.01422
26. Hu D, Ma C, Tian Z, Shen G, Li L (2021) Rice weed detection method on YOLOv4 convolutional neural network. In: 2021 international conference on artificial intelligence, big data and algorithms (CAIBDA), pp 41–45. https://doi.org/10.1109/CAIBDA53561.2021.00016

An IoT Machine Learning Model-Based Real-Time Diagnostic and Monitoring System

Joseph Bamidele Awotunde⬤, Sanjay Misra, Sunday Adeola Ajagbe⬤, Femi Emmanuel Ayo⬤, and Ramchandra Gurjar⬤

1 Introduction

The emergence of new technological development gave birth to the Internet of things (IoT), and it is gaining worldwide attention among several researchers. The IoT-based systems have been prominent in the health-care systems for diagnosing, monitoring, and the treatment of patients, thus reducing infectious diseases globally. In the modern health-care system, the IoT-based machine learning (ML) models have been used for the processing of the generated big data in order to generate reliable results that can be used by medical experts for decision-making on various health challenges. The use of interconnected wearable sensors and networks to control diseases utilizing ML and IoT-based systems has proven to be beneficial in the medical field. IoT is a developing field of study in infectious disease epidemiology. The heightened hazards

J. B. Awotunde (✉)
Department of Computer Science, Faculty of Information and Communication Sciences, University of Ilorin, Ilorin 240003, Nigeria
e-mail: awotunde.jb@unilorin.edu.ng

S. Misra
Department of Computer Science and Communication, Ostfold University College, Halden, Norway
e-mail: Sanjay.misra@hiof.no

S. A. Ajagbe
Department of Computer Engineering, Ladoke Akintola University of Technology, Ogbomosho, Nigeria
e-mail: saajagbe@pgschool.lautech.edu.ng

F. E. Ayo
Department of Mathematical Sciences, Olabisi Onabanjo University, Ago-Iwoye, Ogun State, Nigeria
e-mail: ayo.femi@oouagoiwoye.edu.ng

R. Gurjar
Shri Govindram Seksaria Institute of Technology and Science, Indore, India

© The Author(s), under exclusive license to Springer Nature Singapore Pte Ltd. 2023
D. S. Sisodia et al. (eds.), *Machine Intelligence Techniques for Data Analysis and Signal Processing*, Lecture Notes in Electrical Engineering 997,
https://doi.org/10.1007/978-981-99-0085-5_64

of infectious diseases conveyed, on the contrary, have spread throughout the world [1, 2]. Furthermore, the world's integration, combined with the widespread availability of various IoT-based sensors and devices, necessitates the employment of modern technology for diseases diagnosis, forecasting, monitoring, and treatment of various infectious viruses [3].

The huge amount of data generated by the IoT-based system can be processed by using ML models for generating better results that can be used by experts for the treatment of patients [4]. The use of ML algorithms in the processing of big data has significantly transformed the global health-care systems during decision-making, and for real-time diagnosis of patients [4, 5]. These modern technological systems have greatly reduced cost spend on the diagnosis and treatment of various infectious diseases globally.

The capturing of patient data in real time is made possible by the use of IoT-based sensors and devices, thus helping in real-time diagnosis, monitoring, and treatment of various infectious diseases. The IoT-based systems have become one of the most important aspects of people's lives, resulting in the generation of a huge amount of data. Health-care professionals are already using IoT-based sensors in disease diagnosis, monitoring, prediction, and treatment. Individuals and their critical daily tasks are aided by the health-care system that relies on IoT. The most frequent components of conventional supervised ML for illness diagnosis are base traditional algorithms and meta classifiers, and some of these classifiers have been performing wonderfully in health-care systems. The contributions of the paper are:

- This paper proposes an ML-based IoT-based framework for the diagnosis and monitoring of various diseases in real time.
- A dataset generated from heart illness was used as a case study to evaluate the suggested classifier's performance.
- Various performance metrics such as accuracy, precision, recall, F1-score, and ROC were used to evaluate the framework and compare the results with the existing models.

The paper is structured as follows: Sect. 2 presents related works, Sect. 3 presents the proposed framework and discusses the various layers within. Section 4 presents a practical case using the heart disease dataset to test the performance of the proposed system. Section 5 presents the results and discussion of the experiment. Finally, the paper was concluded in Sect. 6 with the presentation of future work.

2   Related Work

Various classifiers were used to evaluate the proposed system and the results were also compared with those of the existing works. The classifiers used were Support Vector Machine (SVM), decision tree, and neural network, and the SVM performed better when compared with the other classifiers. The authors of [6] designed a three-layer IoT-based real-time monitoring and management system. The proposal used an energy-collecting method with piezoelectric instruments coupled to the human body due to the restricted power of sensor batteries.

The authors in [7] used ARIMA with Markov-based models to assess the insulin dosage, by designing an IoT-based prototype for the monitoring and regulations of blood sugar. To increase safety in outdoor activities.

Reference [8] presented a hybrid IoT security and health monitoring system. The proposed model is divided into two stages: (i) data capture and storage, and (ii) the processing of the captured data. Various wearable sensors and devices are used to collect patients' health status from different environments. The authors in [9] constructed an industrial IoT (Health-IoT) system, which is a real-time health monitoring system. This device has a lot of capability for reviewing patient health data, which helps to avoid death situations.

In [10], the authors tended to the setting of IoT alongside its execution from the perspective of u-medical services. The creators proposed a philosophical IoT structure for u-Healthcare. The heterogeneity issue of the information design in the IoT plat-structure was in 2014 by creators in [11], who utilized semantic information model for the handling of the information. Additionally, the asset-based information access architecture aims to deal with IoT data comprehensively. The work additionally introduced an IoT-based design for managing health-related crises to represent the social affair, reconciliation, and interoperability of IoT information.

Continuous time-series readings obtained from wearable sensors have been employed in data-mining activities such as intrusion detection system, forecast, and strategic planning. The methods for developing mHealth-based apps were discussed by the author in [12], including website and application builders for remotely monitoring patients while applying sensor-based medical treatments.

The authors of [13] employed ML to diagnose a brain tumor based on the location of the tumor on (Flair is a simple natural language processing (NLP) library developed and open-sourced by Zalando Research) FLAIR imaging of the brain Magnetic Resonance Imaging (MRI). The dataset used was a coregistered and skull-stripped multimodal brain imaging dataset. The Gabor filter bank was used to construct tecton-map pictures, and the low-level features were recovered when the image was segmented out into superpixels using bilateral filtering. Leave-one-out cross-validation was utilized to locate the entire tumor area, yielding an 88% dice overlap score.

In [14], the authors proposed a cuckoo search-optimized rough set and fuzzy logic to diagnose diabetes and heart diseases using the following steps: Cuckoo search with the rough set was used for feature selection, and fuzzy logic was used for the classification of the diseases. The computational burden was removed and reduced by a rough set and fuzzy set rules, and membership functions are used for the classification of the dataset. The Cleveland and Switzerland heart datasets were used by the authors to test the model, and with a real-time diabetes dataset. The comparison results show that the proposed model performs better when compared with the existing methods. In order to predict the disease, there is a need for a reliable method to diagnose methods.

3 The Machine Learning-Enabled Internet of Medical Things (IoMT)-Based Disease Diagnosis Model

The ML-enabled IoT-based diagnostic system framework proposed in this paper uses a set of interconnected wearables to monitor and diagnose a person's medical condition. The cloud server was used in order to avoid using smartphone memory, and the data were sent straight to the cloud server over wireless networks. The proposed system's framework is shown in Fig. 1.

The first stage is for capturing data from the patients through the devices and sensors based on the IoT that detect and gather physiological body signs and symptoms in order to diagnose various diseases. These devices depend on the types of information from the patient, and the professionals determine which number to be used by any patient [15]. The second stage is the interface and link between the first and third stages. This is the gateway between the two stages using wireless networks to communicate within the IoT-based platform, and the captured data are being stored on a cloud database [15]. Similarly, Apache is used in the third stage to construct the disease's ML-based diagnosis.

The real-time monitoring and diagnosis of these captured data are possible using the ML-based models, which generate and provide real-time information for experts to make final decision and generate useful reports [16]. The limited computing capabilities of IoT-based systems are transformed to useful information using ML-based

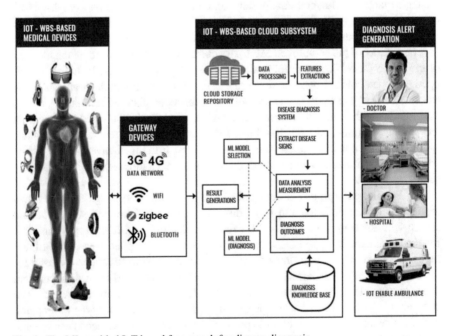

Fig. 1 The ML-enabled IoT-based framework for disease diagnosis

models that can give a high computing capability. Doctors and specialists can quickly make decisions and provide advice using the various ML models to generate a report from captured data with various diseases' mistrustful occurrences. The third stage used the ML models to generate real-time monitoring and diagnosis of various signals and events using the captured data. Heartbeat, blood sugar level, diabetes, hypertension, blood pressure, drug reaction, and dengue are a few examples of such diseases that IoMT-based sensors and devices can be used to capture their symptoms. They help in locating IoT-based devices in 3-dimensional space and calculating the precise number of IoT-based devices. The diagnostics alert-generating step is the final level; the outcome of this layer is a summary of the patients' situation that is delivered to a specialist and an investigator.

3.1 Case Study

In Western states, defibrillation is a medical disease. The term "cardiovascular disease" (CVD) refers to a group of health problems that affect the heart and blood arteries. As an example, plaque formation in the arteries causes hypertension. Corrosion can build and form clots, resulting in a block plate [15, 17]. Various ML-based models have been used to predict patients that are infected with CVD, and several parameters have been employed such as age, smoking habits, and drugs, but many have not yet proven accurate [15, 18]. Growth in the incidence of chronic and infectious diseases can be related to an increase in the world population [1, 2, 19]. The exponential growth in various diseases and health-related data has been used to generate enormous datasets for better and robust health-care systems [20].

3.2 The Machine Learning Algorithms

Lightweight Gradient Boosting Machine

The lightweight gradient boosting machine (LGBM) is a tree-based learning algorithm that uses a gradient boosting system. The tree-based algorithm, under ones' estimation, is the most intuitive because it imitates how a decision is made by humans [21]. LGBM is a fast, distributed, and gradient boosting, and high-performance decision tree based on Microsoft's "Distributed Machine Learning Tools" R&D team. Its distributed architecture has the advantages of speedier acceleration and high performance of training; lower memory usage; greater accuracy; parallel and Graphics Processing Unit (GPU) learning support; and the ability to manage very large data.

These benefits have quickly become common in the area of ML after the release of LGBM, and are commonly used in tasks such as classification, regression, and ranking. The poor learner will develop sequentially using the tree-based learning algorithm, meaning that the first tree we developed will learn how to suit the target variable, then the second tree will learn from the first tree and also learn how to fit the residual, and the next tree will learn how to reduce the residual and fit the residual from the previous tree. Throughout the system, the gradient of the errors is propagated and it is called level-wise tree growth. The growth of the tree is level wise in XGBoost, which distinguishes the LGBM from another gradient boosting algorithm, while CatBoost is more suitable for categorical variables. If you want to create a model with an abundance of data, LGBM is acceptable for use. It is safer to use other ML algorithms if you only have 100 data because your model could cause overfitting.

XGBoost classifier

The XGBoost model is the first of the angle-supporting choice trees, which joins numerous choice trees in a helping way [22]. Leftover portrays the distinctions between the real and anticipated qualities. The example has been prepared till the limit is characterized by the quantity of choice trees. XGBoost follows a similar idea of inclination supporting; to control overfitting and further develop proficiency, it utilizes the number of lifts, learning rate, subsampling proportion, and most extreme tree profundity. All the more explicitly, XGBoost upgrades the capacity focus on the tree size, and the size of the loads directed by standard boundaries of regularization [21, 23]. The XGBoost accomplishes prevalent productivity in a specific pursuit space with various hyperboundaries [24].

Gamma $\gamma \in (\theta, +\infty)$ signifies negligible misfortune decrease, which recalls a split to convey the section for a tree's leaf center point, according to the hyperlimits. The base kid weight $w_{mc} \in (\theta, +\infty)$ is characterized as the base example weight complete, and that truly intends that assuming the tree segment step brings about a leaf hub with the case weight aggregate not exactly w_{mc}, the further segment will be disposed of by the tree. The early stop estimation endeavors to notice the best number of ages that contrast with other hyperlimits given. Finally, sampling techniques and each $r_c \in (0, 1)$ tree also provided the proportion of the section subsample that developed.

Given $X \in \mathbb{R}^{n \times d}$ as a preparation dataset with d elements and n tests, the XGBoost object work in t-th is addressed by

$$\text{Obj}^{(t)} \simeq \sum_{i-1}^{n} \left\{ \ell\left(y_i, \tilde{y}_i^{(t-1)}\right) + g_i f_t(x_i) + \frac{1}{2} h_i f_t^2(x_i) \right\} + \Omega(f_t), \quad (1)$$

$$g_i = \partial_{\tilde{y}(t-1)} \ell\left(y_i, \tilde{y}_i^{(t-1)}\right), h_i = \partial_{\tilde{y}(t-1)}^2 \ell\left(y_i, \tilde{y}_i^{(t-1)}\right), \quad (2)$$

where the misfortune work l is addressed by the main slope g_i, and , h_i with angle of ℓ. The regularization $\Omega(f_t) = \gamma T + \frac{1}{2}\varphi\|\varphi\|^2$ was utilized, where the quantity of leaf hubs is addressed by T. As shown in condition (3), the calculated misfortune l of the preparation misfortune estimates how well the model fits on the preparation information,

Table 1 Metrics for evaluating the performance of the proposed classifiers

Metrics	Definition	Formula
Sensitivity test	Assess your ability to correctly diagnose diabetes illness	$TP/(TP + FN)$
Specificity test	Examine whether you want to stop labeling diabetic illness as commonplace in diagnostic exams	$TN/(TN + FP)$
PPV	Positively predicted value	$TP/(TP + FP)$
NPV	Negatively predicted value	$TN/(TN + FN)$
Accuracy	Evaluate the number of diabetes detectors accurately	$\frac{TN+TP}{TN+FP+FN+TP} \times 100$

where

TN = the healthy people who have been diagnosed with CVD by the mode
TP = denotes a patient with CVD who was positively diagnosed in the dataset
FN = persons who have been diagnosed with CVD but are otherwise healthy
FP = stands for the falsely CVD who was not diagnosed by the mode

$$\ell\left(y_i, \tilde{y}_i^{(t-1)}\right) = y_i ln\left(1 + e^{-\tilde{y}_i}\right) + (1 - y_i)ln\left(1 + e^{\tilde{y}_i}\right) \tag{3}$$

given the t-th training sample $x_i \in \mathbb{R}^d$, and contains K trees, the relating forecast $\tilde{y}_i$ is figured as

$$\tilde{y}_i = \sum_{k=1}^{k} F_k(x_i) \tag{4}$$

$$\text{s.t. } F_k \in \text{XGB, where XGB} = \{F_1, F_2, F_3, \ldots, F_K\}. \tag{5}$$

4 Performance Evaluation Metrics

Performance evaluation metrics is represented in Table 1 and this contains the rationale for the performance ratings.

5 Results and Discussion of the Findings

On the training dataset, two classifier models were utilized, and this section contains the preliminary performance evaluation utilizing all the 13 input parameters. The XGBoost classifier performs better on the results of studies. This indicates that the model's performance is good, with a 95.7% accuracy rate, a precision of 96.0%, a recall of 95.9%, a F1-score of 94.8%, and a ROC of 94.3%. The classifier accurately

identifies around 96% of the test data, and the precision of the two target class groupings is equal. In terms of accuracy, XGBoost classifier has 95.7%, demonstrating its
ability to deliver precise data on a consistent basis. Furthermore, the classifier's ROC-
Area Under the ROC Curve (AUC) score of 94.3% shows a good possibility measure
against LGBM classifier.

5.1 The Proposed Classifier's Performance

Figure 2 shows the results of the two classifiers for the diagnosis of heart disease,
and XGBoost classifier performed reliably better when compared with the LGBM
classifier. The XGBoost gave an accuracy of 95.7%, a precision of 96.0%, a recall
of 95.9%, a F1-score of 94.8%, and a ROC of 94.3%.

The results of the proposed system are presented in Table 2 using various performance metrics for the two classifiers. The findings have shown that the XGBoost
classifier has a better performance when compared with LGBM classifier by an accuracy of 95.7%, a precision of 96.0%, a recall of 95.9%, a F1-score of 94.8%, and a
ROC of 94.3%. The obtained results were based on the dataset by dividing it in 80%
for training and 20% for testing the models. The XGBoost classifier has a maximum
classification result that is above 90% across all metrics used and more consistent
across the two classifiers used for the evaluation.

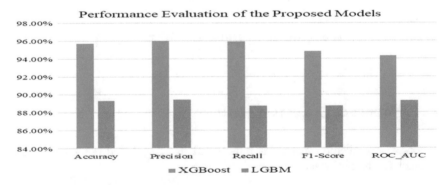

Fig. 2 Performance evaluation of the ML-based classifiers

Table 2 Evaluation of the proposed classifiers

Models	Accuracy (%)	Precision (%)	Recall (%)	F1-score (%)	ROC_AUC (%)
XGBoost	95.7	96.0	95.9	94.8	94.3
LGBM	89.3	89.4	88.7	88.73	89.3

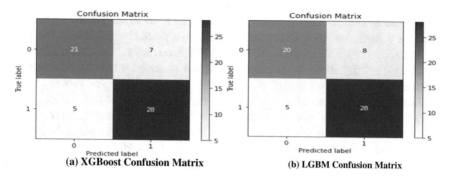

(a) XGBoost Confusion Matrix   **(b) LGBM Confusion Matrix**

Fig. 3 The XGBoost and the LGBM classifier confusion matrix

5.2 Confusion Matrix

This section delves deeper into the evaluation of the models using the confusion matrix for the test set (see Fig. 3). The result shown in Fig. 3 is the confusion matrix for the models that delivered the best recall for each class with the least amount of error. The data were divided into two categories, with "0" representing "no heart disease" and "1" representing "heart disease." For XGBoost, 25 (94%) of the 28 test occurrences of no heart disease were recognized accurately as no heart disease, whereas 3 (6%) were wrongly labeled as having heart disease. Using LGBM classifiers, 5 instances (15.16%) were incorrectly identified as no heart disease, whereas 23 instances (84.85%) were correctly classified as heart illness. As a result of the findings, it was discovered that the majority of the misclassifications were due to the fact that no heart illness was misclassified as heart disease. As a result, we were able to achieve our goal by developing a classification model for detecting cardiac illness.

6 Conclusion and Directions for Future Work

The application of the ML-enabled IoT-based system in recent years is providing various dimensionalities in health care through online services. These have provided a new environment to millions of people in getting new benefits about health-care tips frequently for living a healthy life. The introduction of ML-enabled IoT-based technology and their related devices in medical sectors has strengthened their use in online health-care applications. The paper, therefore, proposed an ML-enabled IoT-based system to diagnose various diseases using heart disease as a case study using captured data from IoT-based sensors/devices. Data were sent to the cloud storage using wirelessly interconnected devices. Such data were generated from various used sensors and devices. The extracted features were processed using ML-based models for the diagnosis of patients from the collected symptoms. From the findings, the XGBoost model performs relatable well when compared with the LGBM model, and other existing models. The XGBoost's performance is better in terms of the metrics used for comparison, with an accuracy of 95.7%, a precision of 96.0%, a recall of 95.9%, a F1-score of 94.8%, and a ROC of 94.3%. The results show that the classifier correctly classifies 95% of the dataset used and provided a balanced sensitivity. Future work will consider the use of features selected to be able to remove reductant parameters from the dataset so as to increase the classifiers of the performance. The use of various security algorithms for the protection of IoT-based systems should also be considered. This will arouse the trust of users to be able to willingly release their data for research use and other important purposes. As data privacy and security are major concerns in the deployment of IoT-based smart health-care devices, this has become critical.

References

1. Awotunde JB, Jimoh RG, AbdulRaheem M, Oladipo ID, Folorunso SO, Ajamu GJ (2022) IoT-based wearable body sensor network for COVID-19 pandemic. Stud Syst Decis Control 2022(378):253–275
2. Awotunde JB, Ajagbe SA, Oladipupo MA, Awokola JA, Afolabi OS, Mathew TO, Oguns YJ (2021) An improved machine learnings diagnosis technique for COVID-19 pandemic using chest X-ray images. Commun Comput Inf Sci 1455 CCIS:319–330
3. Awotunde JB, Ogundokun RO, Misra S (2021) Cloud and IoMT-based big data analytics system during COVID-19 pandemic. Internet Things 2021:181–201
4. Awotunde JB, Bhoi AK, Barsocchi P (2021) Hybrid cloud/fog environment for healthcare: an exploratory study, opportunities, challenges, and future prospects. Intell Syst Reference Lib 2021(209):1–20
5. Marques G, Roque Ferreira C, Pitarma R (2018) A system based on the internet of things for real-time particle monitoring in buildings. Int J Environ Res Public Health 15(4):821
6. Din S, Paul A (2020) Erratum to "Smart health monitoring and management system: toward autonomous wearable sensing for the Internet of Things using big data analytics (Futur Gener Comput Syst 91 (2019) 611–619). Futur Gener Comput Syst 108:1350–1359
7. Otoom M, Alshraideh H, Almasaeid HM, López-de-Ipiña D, Bravo J (2015) Real-time statistical modeling of blood sugar. J Med Syst 39(10):123
8. Wu F, Wu T, Yuce MR (2019) An internet-of-things (IoT) network system for connected safety and health monitoring applications. Sensors 19(1):21
9. Hossain MS, Muhammad G (2016) Cloud-assisted industrial internet of things (IIoT)–enabled framework for health monitoring. Comput Netw 101:192–202
10. Verma P, Sood SK (2018) Cloud-centric IoT-based disease diagnosis healthcare framework. J Parallel Distributed Comput 116:27–38
11. Darwish A, Hassanien AE, Elhoseny M, Sangaiah AK, Muhammad K (2019) The impact of the hybrid platform of the internet of things and cloud computing on healthcare systems: opportunities, challenges, and open problems. J Ambient Intell Humaniz Comput 10(10):4151–4166
12. Awotunde JB, Folorunso SO, Jimoh RG, Adeniyi EA, Abiodun KM, Ajamu GJ (2021) Application of artificial intelligence for COVID-19 epidemic: an exploratory study, opportunities, challenges, and future prospects. Stud Syst Decis Control 2021(358):47–61
13. Rehman ZU, Zia MS, Bojja GR, Yaqub M, Jinchao F, Arshid K (2020) Texture based localization of a brain tumor from MR-images by using a machine learning approach. Med Hypotheses 141:109705

14. Gadekallu TR, Khare N (2017) Cuckoo search optimized reduction and fuzzy logic classifier for heart disease and diabetes prediction. Int J Fuzzy Syst Appl (IJFSA) 6(2):25–42

15. Awotunde JB, Folorunso SO, Bhoi AK, Adebayo PO, Ijaz MF (2021) Disease diagnosis system for IoT-based wearable body sensors with machine learning algorithm. Intell Syst Reference Lib 2021(209):201–222

16. Awotunde JB, Jimoh RG, Oladipo ID, Abdulraheem M (2020) Prediction of malaria fever using long-short-term memory and big data. Commun Comput Inf Sci 1350:41–53

17. Smith JR, Joyner MJ, Curry TB, Borlaug BA, Keller-Ross ML, Van Iterson EH, Olson TP (2020) Locomotor muscle group III/IV afferents constrain stroke volume and contribute to exercise intolerance in human heart failure. J Physiol 598(23):5379–5390

18. Şengül G, Karakaya M, Misra S, Abayomi-Alli OO, Damaševičius R (2022) Deep learning based fall detection using smartwatches for healthcare applications. Biomed Signal Process Control 71:103242

19. Ajagbe SA, Amuda KA, Oladipupo MA, Afe OF, Okesola KI (2021) Multi-classification of Alzheimer disease on magnetic resonance images (MRI) using deep convolution neural network approaches. Int J Adv Comput Res (IJACR) 11(53):51–60

20. Adeniyi EA, Ogundokun RO, Gbadamosi B, Misra S, Kalejaiye O (2022) Classification of swine disease using K-nearest neighbor algorithm on cloud-based framework. In: Artificial intelligence for cloud and edge computing. Springer, Cham, pp 71–90

21. Detrano R, Janosi A, Steinbrunn W, Pfisterer M, Schmid J, Sandhu S, Guppy K, Lee S, Froelicher V (1989) International application of a new probability algorithm for the diagnosis of coronary artery disease. Am J Cardiol 64:304–310

22. Chen T, Guestrin C (2016) Xgboost: a scalable tree boosting system. In: Proceedings of the 22nd ACM sigkdd international conference on knowledge discovery and data mining, pp 785–794

23. Kshatri SS, Singh D, Narain B, Bhatia S, Quasim MT, Sinha GR (2021) An empirical analysis of machine learning algorithms for crime prediction using stacked generalization: an ensemble approach. IEEE Access 9:67488–67500

24. Singh D, Singh S (2020) Realising transfer learning through convolutional neural network and support vector machine for mental task classification. Electron Lett 56(25):1375–1378

Hausa Character Recognition Using Logistic Regression

Akinbowale Nathaniel Babatunde, **Roseline Oluwaseun Ogundokun**,
Ebunayo Rachael Jimoh, Sanjay Misra, and Deepak Singh

1 Introduction

Handwritten character recognition is a critical topic in identifying characters, and it may be used as an exam instance for concepts such as pattern recognition and machine learning's high-efficiency models. Many common elements have evolved to aid machine learning and pattern recognition research [1, 2]. Automatic text extraction from images or videos is a critical and difficult topic in a variety of applications, including document processing, image indexing, video content summary, video retrieval, and video comprehension [3]. Unlike character recognition for scanned documents, recognizing characters in unconstrained images is confounded by a wide scope of variation in backgrounds which consist of irregularities in quality content and detachment of difficult ink from the background [4].

Handwriting recognition (HWR), also known as handwritten text recognition, is a computer's capacity to recognize and understand comprehensible handwritten input from sources such as paper documents, pictures, touch displays, and other

A. N. Babatunde
Department of Computer Science, Kwara State University, Malete, Kwara State, Nigeria
e-mail: akinbowale.babatunde@kwasu.edu.ng

R. O. Ogundokun
Department of Multimedia Engineering, Kaunas University of Technology, Kaunas, Lithuania
e-mail: rosogu@ktu.lt; ogundokun.roseline@lmu.edu.ng

E. R. Jimoh
Department of Computer Science, University of Ilorin, Ilorin, Nigeria

S. Misra
Department of Computer Science and Communication, Ostfold University College, Halden, Norway

D. Singh (✉)
National Institute of Technology Raipur, Raipur, India
e-mail: dsdsingh.cs@nitrr.ac.in

© The Author(s), under exclusive license to Springer Nature Singapore Pte Ltd. 2023   801
D. S. Sisodia et al. (eds.), *Machine Intelligence Techniques for Data Analysis and Signal Processing*, Lecture Notes in Electrical Engineering 997,
https://doi.org/10.1007/978-981-99-0085-5_65

devices [5]. In the field of HWR, there are two major problem areas. The online mode comprises the automated transformation of the collected text as it is created on a particular digitizer or personal digital assistant, where a sensor takes up pen-tip motions including pen-up or pen-down switching. The collected handwritten text is transformed into letter codes, which are then utilized by text-processing software. The dynamic motion while handwriting is recognized by online character recognition software.

In the offline technique, it involves the system to automatically distinguish and change over text in the form of an image into letter codes; the data obtained are the fixed static shape of the character. In offline HWR (OHR), information apprehended is accessible on paper, which is digitized using a scanner. OHR is relatively hard to comprehend as various individuals have diverse handwriting styles [4].

The Hausa language is one of the utmost frequently articulated innate languages in West and Central Africa, with roughly forty to fifty million persons using it as a primary or alternative language. Within the Afro-Asian language phylum, it belongs to the Western branch of the Chadic language superfamily. The Hausa people's homelands are on both sides of the Niger-Niger border, where roughly half of the people voices Hausa as a primary accent, and Nigeria, where around one-fifth of the people voice it as a primary dialectal [5]. The Hausa people are mostly Muslim. Their long-distance trade and pilgrimages to the Islamic Holy Metropolises have spread their language to almost all key metropolitans in West, North, Central, and Northeast Africa. Subject, verb, and object is the fundamental word order. Hausa is a tone language, which means that pitch contrasts matter just as much as consonants and vowels in determining the meaning of a phrase. In Hausa orthography, the tone is not indicated. Accent marks in academic Hausa transcriptions denote tone, which might be high, mediocre, or descending (diacritic).

Despite extensive study on Latin languages, just a few studies on indigenous languages have been published [6]. HWR researchers have looked at recognition in multiple languages as a separate challenge. Each language has its unique collection of characters and properties, making it challenging to create a universal algorithm that works across all languages. Because of its simplicity and low memory requirements, researchers have utilized logistic regression (LR) to extract the properties of letters or numbers. Various aspects of pattern recognition, notably in the area of character recognition, have been successfully used using LR. For character recognition, some researchers have used traditional approaches such as segmentation and recognition, while others have used artificial neural network (ANN)-based methodologies [7]. LR has become an applicable aspect in the field of artificial learning. It's explicit as a statistical analysis model utilized for the prediction of dependent data value by analyzing based on the past perception of a dataset. This methodology empowers an algorithm being used in a machine learning application to classify incoming data based on historical data.

The number of attempts of researchers has improved the efficiency of document image retrieval. Also, research is going on various local languages [5, 8–10]. This work portrays Hausa character recognition; numerous researchers have worked on character recognition of various languages such as English text, Chinese text, and

Yoruba language, but none or very few researchers have worked on Hausa language character recognition, which also prompted taking this research work of Hausa language character recognition. The research tends to introduce the use of deep learning techniques to recognize Hausa text documents.

The remaining part of the article is structured as follows: Sect. 2 presents the methodology and materials used for the implementation of the study. The results and discussion are presented in Sect. 3. The study was concluded in Sect. 4.

2 Materials and Methodology

Before a computer can detect handwritten writing, many processes must be completed. Sample data collection, preprocessing, feature extraction, categorization, and recognition are only a few of them. The flow design for offline handwritten character recognition is shown in Fig. 1. This project's primary focus is on categorization and recognition. The groundwork for implementing a Hausa text character recognition system starts with a collection of diverse Hausa handwritten characters from various people. The character pictures were acquired and recorded in the database using an HP scanner at a resolution of 300 dpi. After preparing the scanned data, a manual categorization method was used to determine each character's intended output. Figure 2 shows the scanned Hausa handwritten characters.

Image preprocessing for optical character recognition assists with setting up an image for subsequent processing (Fig. 1). It helps in expanding the recognition accuracy by eliminating undesirable information that would have been extricated as features. To remove the isolated pixels, the picture is first transformed into a monochrome image, and then the isolated pixels are simply removed from the resulting monochrome image. Preprocessing includes the following stages: Image annotation, grayscale conversion, image binarization, image normalization, and image resizing.

Image segmentation involves separating the text in the image into blocks, lines, words, and ultimately into characters. It is one of the most important stages in this work (Fig. 1). It helps the classifier with extricating highlights from each character. A boundary was set on the character and cropped accordingly such that only the handwritten image was left with the region which contains the necessary features needed for feature extraction. Cropping helps to remove unnecessary features such as the background, thereby leaving just the relevant features.

Feature extraction was used to extract properties that can distinguish a character uniquely and also to remove the properties that can differentiate between similar

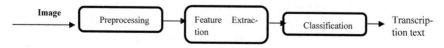

Fig. 1 Block flow of the offline handwritten character recognition

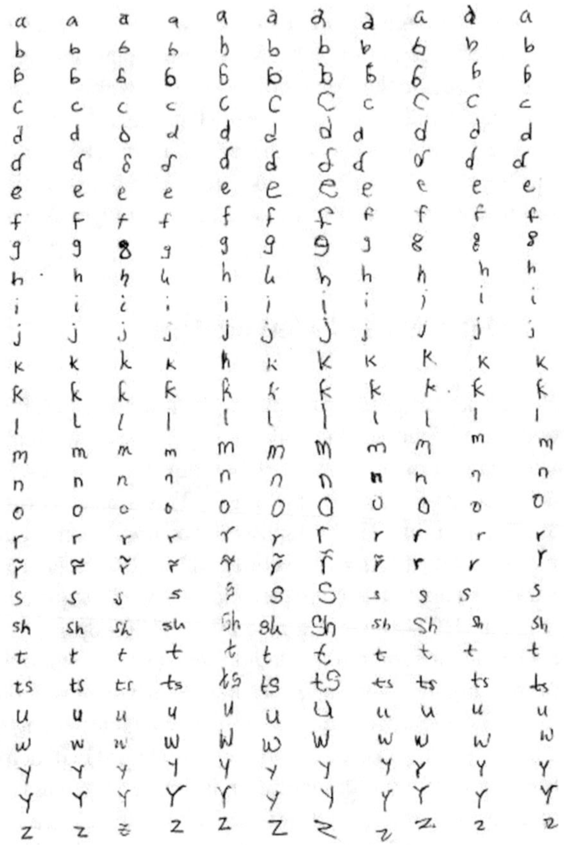

Fig. 2 Samples of captured handwritten of Hausa character

characters (Fig. 1). It distinguishes an area corresponding to a letter from an area corresponding to other letters. It is an essential stage in handwriting character recognition as its effective functioning improves the recognition rate and reduces misclassification. It is utilized to excerpt the features of individual characters which is used to train the system. Statistical feature extraction technique was utilized to excerpt the characters. The major statistical features used in this work are zoning, projections and profiles, and crossings and distances.

The training and recognition stage is the last stage for the developed system. The study used a 10-fold validation before the testing stage was executed. The training and recognition are done using LR. Some of Hausa's uppercase characters will be used for training and testing. For the recognition stage, the feature vector obtained from the feature extraction stage was used for recognition. The feature vectors of the input handwritten image will be compared with the stored feature vectors of the training images to get the result. The unicode of the characters will be used to get the result as the standard keyboard does not have the Hausa alphabet.

Input: Training data

1. For i←1 to k

2. For each training data instance d_i:

3. Set the target value for the regression to
$$z_i \leftarrow \frac{y_j - P(1 \mid d_j)}{[P(1 \mid d_j) . (1 - P(1 \mid d_j))]}$$

4. nitialize the weight of instance d_j to $P(1 \mid d_j) . (1 - P(1 \mid d_j))$

5. finalize a $f(j)$ to the data with class value (z_j) & weights (wj)

Classification Label Decision

6. Assign (class label:1) if $P(1 \mid d_j) > 0.5$, otherwise (class label: 2)

Fig. 3 Logistic regression pseudocode

The method used to train the classification of LR is the back-propagation method, and the pseudocode is shown in Fig. 3.

The flowchart for the proposed Hausa character recognition system is shown in Fig. 4. It displays the sequence steps of how scanned images are being recognized from the point when the application is loaded to sending the images into the application via scanner, to the point at which handwriting images are being recognized.

3 Results and Discussion

The designs obtainable and conferred in the previous segment were developed using Octave programming language using C-sharp (C#) programming language to develop the system's user interface to enable users to use the application without accessing the underline model. The desktop application interface relied on the compiled model that was generated by the LR algorithm written in MATLAB programming language. The program was compiled and tested to avoid errors and an installable execution file was generated so that the package can be installed and run efficiently on computer devices regardless of the version of their operating system as long as it meets the necessary hardware requirements. The application was subjected to various testing, and sample

Fig. 4 Flowchart of the
proposed system

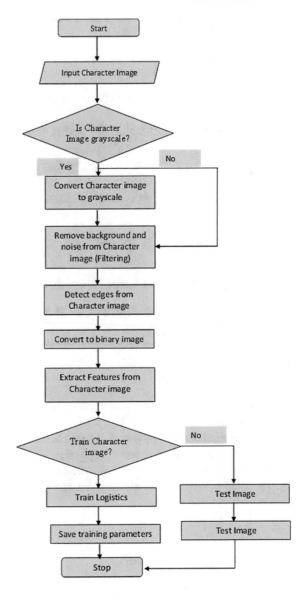

screenshots were presented. The handwritten character pictures were taken from the
constructed Hausa database and tested on the recognition module to verify the built
Hausa handwritten character recognition system. Preprocessing was done on the
test data. The preprocessed pictures' features were extracted and compared to the
learned features vector. If the two feature sets are comparable, it searches the look-up
database for the suitable character and displays the matching digital counterpart of
the handwritten character if it is found, or returns a false negative if it is not.

4 Results and Discussion

The designs obtainable and conferred in the previous segment were developed using Octave programming language using C-sharp (C#) programming language to develop the system's user interface to enable users to use the application without accessing the underline model. The desktop application interface relied on the compiled model that was generated by the LR algorithm written in MATLAB programming language. The program was compiled and tested to avoid errors and an installable execution file was generated so that the package can be installed and run efficiently on computer devices regardless of the version of their operating system as long as it meets the necessary hardware requirements. The application was subjected to various testing, and sample screenshots were presented. The handwritten character pictures were taken from the constructed Hausa database and tested on the recognition module to verify the built Hausa handwritten character recognition system. Preprocessing was done on the test data. The preprocessed pictures' features were extracted and compared to the learned features vector. If the two feature sets are comparable, it searches the look-up database for the suitable character and displays the matching digital counterpart of the handwritten character if it is found, or returns a false negative if it is not.

The extracted pixel value (0–255) in arrays was visualized in 20×20 sized square; 100 characters were generated randomly from the dataset as shown in Fig. 5. All samples for training and testing are converted to portable network graphics (.PNG) to embellish the improved performance of the proposed scheme.

Samples of handwriting Hausa are displayed in Fig. 5. After getting the binarized image, the designed application extracted each handwritten character pixel within the region of interest module and saved its pixel values that range between 0 and 255.

Extracted handwritten characters are shown in Fig. 6. In this research, a total of 1700 data points from 68 individual writers are used in the training of the model. Each Hausa alphabet character was written with a black ball pen on white background paper in a tabulated form.

The performance evaluation metric used is accuracy, and the accuracy result for the training stage indicated 89%. However, the evaluation result shows an accuracy of about 54%. A built-in Octave function, **fminunc** was used as an optimization solver to find the minimum of an unconstrained function. Optimize cost function $J(\circ)$ with parameters $\circ$ in LR was used to optimize the algorithm.

The result of testing the implemented model is shown in Figs. 7 and 8 and 9 shows the simple Hausa character recognition with LR model.

Table 1 shows the comparison of the proposed system with existing systems. Several researchers have done different language recognition using machine learning approaches but only three were used for this study. A comparative analysis is shown in Table 1. It was deduced that Desai [11] had an accuracy of 82% using feed-forward back-propagation neural network method, Khemiri et al. [12] had an accuracy of 79% using Naïve Bayes approach, and lastly Jayech et al. [13] had an accuracy of 82% using dynamic hierarchical Bayesian networks. It was shown that our

Fig. 5 Samples of Hausa
handwriting

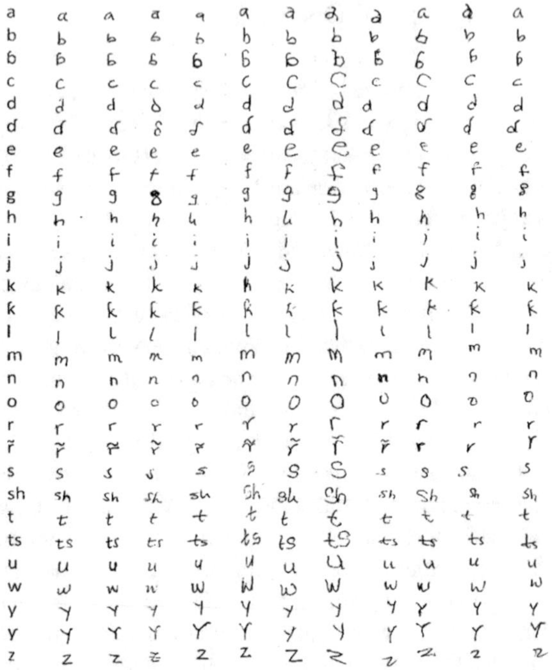

Fig. 6 Extracted
handwritten characters

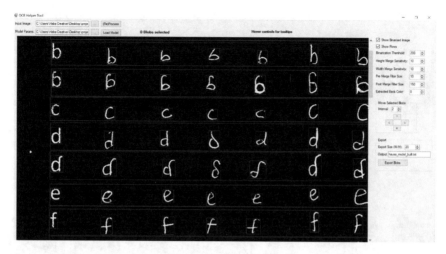

Fig. 7 Character binarization

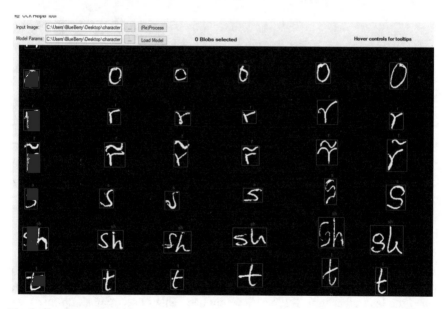

Fig. 8 Character recognition

proposed system is superior by outperforming the three existing systems used for the comparative analysis with an accuracy of 89% (Table 1).

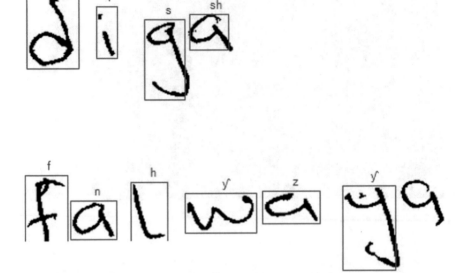

Fig. 9 Simple Hausa character recognition with logistic regression model

Table 1 Comparative analysis with existing systems

Authors	Methods	Accuracy (%)
Desai [11]	Feed-forward back-propagation neural network (FFBPNN)	82
Khémiri et al. [12]	Naïve Bayes (NB)	79
Jayech et al. [13]	Dynamic hierarchical Bayesian networks (DHBN)	82
Proposed system	Logistic regression	89

5 Conclusion

The distinct factor of this work which differentiates this from other related ones is that it has not been implemented using LR. However, a few researchers have tried using the later algorithm, yet they have not been able to accomplish up to 70%of what the project is to accomplish. Some algorithms that have been implemented in accomplishing a similar objective include ANN, K-nearest neighbors, and support vector machine algorithm.

For this research study, the research uses Hausa characters. The previous project has drawn in the utilization of other languages and other classification algorithms. In any case, this research focuses on the use of the Hausa character. The system's user interface was developed with C-sharp (C#) programming language. The desktop application interface relied on the compiled model that was generated by the LR algorithm written in Octave programming language.

The limitation of the proposed system is dataset collection. As the study deals with language recognition and the language is not a universal one, so getting the Hausa language datasets for the implementation of the study was very difficult. Therefore, more dataset generation is suggested for future work and similarly, more machine learning algorithms or deep learning algorithms can be used in future to have a high accuracy and robust system.

References

1. Gupta A, Sarkhel R, Das N, Kundu M (2019) Multiobjective optimization for recognition of isolated handwritten Indic scripts. Pattern Recognit Lett 128:318–325
2. Blessing G, Azeta A, Misra S, Chigozie F, Ahuja R (2019) A machine learning prediction of automatic text-based assessment for open and distance learning: a review. In: International conference on innovations in bio-inspired computing and applications. Springer, Cham, pp 369–380
3. Jain K, Yu B (1998) Automatic text location in images and video frames. Pattern Recogn 31(12):2055–2076
4. Ajao FJ, Olawuyi OD, Odejobi OO (2018) Yoruba handwritten character recognition using freeman chain Code and K-Nearest Classifier. Jurnal Teknologi dan Sistem Komputer 6(2):129–134. https://doi.org/10.14710/jtsiskom.6.4.2018.129-134
5. Ojumah S, Misra S, Adewumi A (2017) A database for handwritten Yoruba characters. In: International conference on recent developments in science, engineering, and technology. Springer, Singapore, pp 107–115
6. Babatunde AN, Abikoye CO, Oloyede AA, Ogundokun RO, Oke AA, Olawuyi HO (2021) English to Yoruba short message service speech and text translator for android phones. Int J Speech Technol 24(4):979–991
7. Haraty R, Ghaddar C (2004) Arabic text recognition. Int Arab J Inf Technol 1:2
8. Akman I, Bayindir H, Ozleme S, Akin Z, Misra S (2011) A lossless text compression technique using syllable-based morphology. Int Arab J Inf Technol 8(1):66–74
9. Sharma I, Anand S, Goyal R, Misra S (2017) Representing contextual relations with Sanskrit word embeddings. In: International conference on computational science and its applications. Springer, Cham, pp 262–273
10. Ajayi LK, Azeta A, Misra S, Odun-Ayo I, Ajayi PT, Azeta V, Agrawal A (2020) Enhancing the low adoption rate of M-commerce in Nigeria through Yorùbá voice technology. In: International conference on hybrid intelligent systems. Springer, Cham, pp 516–524
11. Desai AA (2010) Gujarati handwritten numeral optical character reorganization through neural network. Pattern Recogn 43(7):2582–2589
12. Jayech K, Trimech N, Mahjoub MA, Amara NEB (2013) Dynamic hierarchical Bayesian network for Arabic handwritten word recognition. In: Fourth international conference on information and communication technology and accessibility (ICTA). IEEE, pp 1–6
13. Khémiri A, Echi AK, Belaïd A, Elloumi M (2016) A system for off-line Arabic handwritten word recognition based on Bayesian approach. In: 2016 15th international conference on frontiers in handwriting recognition (ICFHR). IEEE, pp 560–565

Long Short-Term Memory-Driven Recurrent Neural Network for Real-Time Stock Monitoring and Prediction

Shriranjan Patil, Venkatesh Gauri Shankar, Bali Devi, Adwitiya P. Singh, and Nisarg R. Upadhyay

1 Introduction

In simple words, a stock environment is called a 'stock market,' 'equity market,' or 'share market.' People buy and sell stocks, which are 'ownership claims.' A public stock exchange might have some of these, like shares of private companies sold to investors through equity crowdfunding sites. Stock brokerages and online trading platforms are used by most people who invest in the stock market to buy and sell shares of their own company. Most of the time, investments are made with a plan in mind when they are made. The investment strategies are made using Key Performance Indicators (KPIs). The standard KPIs are 'Opening Price,' 'Closing Price,' 'Low,' 'High,' and Volume of the Market [1–3].

Try to figure out the value of company stock or other financial instruments in the future by looking at them on a stock market or other type of exchange. If you know how much a stock will be worth in the future, you could make much money. It says that stock prices are based on all the available information. Any price changes that are not caused by new information will always be slightly different. Those who think they have a lot of ways and techniques that they say can help them figure out what prices will be in the future. For example, stock prices move up and down because there is a lot to buy and sell. Therefore, many people want to buy the stock, so the price increases because there is more demand. If more people want to sell a stock, the price will fall [4]. If you understand supply and demand, it is easy to figure out what makes demand or supply rise or fall, but it is hard to figure out why. They look

S. Patil
Data Science and Engineering, Manipal University Jaipur, Jaipur, Rajasthan, India

V. G. Shankar (✉) · A. P. Singh · N. R. Upadhyay
Information Technology, Manipal University Jaipur, Jaipur, Rajasthan, India
e-mail: venkateshgaurishankar@gmail.com

B. Devi
Computer Science and Engineering, Manipal University Jaipur, Jaipur, Rajasthan, India

© The Author(s), under exclusive license to Springer Nature Singapore Pte Ltd. 2023   813
D. S. Sisodia et al. (eds.), *Machine Intelligence Techniques for Data Analysis and Signal Processing*, Lecture Notes in Electrical Engineering 997,
https://doi.org/10.1007/978-981-99-0085-5_66

at how the market is doing and what is going on in the economy to figure out how they turn [5–7].

It is mostly impossible to predict the stock price of various companies since it involves many factors such as interest rates, inflation, unemployment, current events, natural calamities. Still, it is possible to make an estimated guess of the stock prices [8]. With the help of machine learning and Artificial Intelligence, we will predict the stock value to be beneficial to investors and analyzers instead of going through all the raw data manually. Machine learning is a way to look at data, making it easier to make analytical models [9, 10]. Here, Artificial Intelligence is thought to be based on the idea that machines can learn from data and make their own decisions [2, 6, 11].

When machine learning use Recurrent Neural Networks (RNNs) and a long short-term memory to figure out what the stock market will do, the current study should do this. Two types of neural networks are recurrent and non-recurrent: recurrent and non-recurrent networks. In Recurrent Neural Networks (RNNs), the connections between nodes form a graph, which moves in a certain way over time. Because of this, it can change its behavior over time [12]. People who work with RNNs based on feedforward neural networks use them to process different length sequences of inputs that they get. This way, you can use your memories to process them. Traditional neural networks, music generation, sentiment classification, LSTM, and other types of RNNs are some of the different types of RNNs. To predict the stock price, we will use LSTM (sequential). This is what we will do [5, 8, 13, 14].

There are many ways to learn from a network. Having a long short-term memory is one of them. LSTMs are artificial neural networks called RNNs. They look like a neural network, and they are used to make train data for detection or prediction. These networks, called long short-term memory (LSTMs), are good at classifying and making predictions because the time series can have gaps in the data that make it up that are unknown in length. This makes them good at making these kinds of predictions. To solve the problem of gradients disappearing when training traditional RNNs, LSTMs were made to solve this problem. A significant advantage of the LSTM neural network over RNNs, hidden Markov models, and other sequence learning methods is that the gaps do not matter. It does not care about how long the gaps are. It is not good for long-term dependencies to get in the way of LSTMs [5, 6].

2 Literature Review

Existing research work in predicting the stock market using LSTM, many predictions and other techniques are being applied in the field of the stock market.

Many research papers have been published on this topic over the last two decades. Two exciting approaches are dealing with predictions using autoencoders and from news, tweets, etc. These approaches have been explained below: Mehrnaz et al. [15] have looked at predicting the stock market with autoencoder long short-term memory (AE-LSTM). The autoencoder shows the data and puts them into the LSTM network to figure out the price. Based on data from 1/3/2000 to 11/4/2019, they have looked at their data. Nine technical indicators have been derived from the stock market to predict their movement. The data are split into two groups, and their mean and variance are calculated. The dataset is then cleaned by detecting outliers. The data are then split into the train (80%) and test (20%). The efficiency of their model is demonstrated by comparing the results of GAN with AE-LSTM. AE-LSTM has lower RMSE and MAE than GAN. This proves that their model is accurate. Anshul Mittal et al. [16] predict the stock market by applying sentimental analysis to machine learning algorithms and finding a correlation between 'public sentiment' and 'market sentiment.' The dataset used is 'Dow Jones Industrial.' From June 2009 to December 2009, they have collected 479 million tweets, including name, timestamp, and tweet text. During preprocessing of data, the missing data have been replaced by mean. The four sentiments used here are Happy, Kind, Alert, and Kind. Their methodology to process the data is Word List Generation, Tweet Filtering, Daily Score Computation, and Score Mapping. For this purpose, they have used Granger Causality Analysis to see if mood values from that algorithm can predict how stocks will move. Granger Causality [16] is based on linear regression, but it cannot connect stock prices and moods. Because SOFNNs are used, we can see how moods and stock prices go together. Results from Granger Causality show that Calm and Happy are the main factors that make DJIA values go up or down. They had made seven different algorithms. They found that calm and happiness are better predictors of stock prices for the DJIA. They had even tried SVM and Logistic Regression, which did not work out well. Because they used k-SCV, they had an excellent model.

3 Architecture of LSTM

3.1 Overview of Recurrent Neural Networks (RNN)

RNN is an artificial neural network that can work with time-series data or many sequences. It is usually only used to look at data points that are not linked together [13]. On the other hand, if we have data in a sequence where each data point is linked to the one before it, that means we need to change the neural network to account for how these data points are linked together. 'Memory' is a word that helps RNNs keep track of the states or information of previous inputs, so they can make the subsequent output, which is called the next 'output' (Fig. 1).

At timestamp t, the RNN generates an outcome of ht for any input X_t. Then, at time step t, the RNN generates an outcome of ht for any other input X_t. X_{t+1} and ht are inputs for the RNN in the next step, called $t + 1$. They are used to make the output h_{t+1}. A loop lets information moves from one step to the next to be used again and again. RNNs are not without flaws at all. When the 'context' is from the near past, it works very well to get the right answer. But, when an RNN has to rely on something learned a long time ago to get the right answer, it does not work very well.

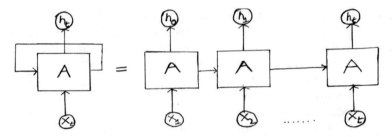

An unrolled recurrent neural network

Fig. 1 Recurrent neural network: sample architecture

Hochreiter [13] and Bengio et al. [17] talked a lot about this limitation of the RNNs, and they talked about it very well. They also looked at the basics to determine why RNNs might not work in long-term situations. The good news is that the LSTMs are made to solve the problem above.

3.2 LSTM Networks

Long short-term memory (LSTM) networks are different from Recurrent Neural Networks because they have colossal memory [17]. When you do this, it makes it easier for you to keep old data in your mind. In this place, there is an end to the disappearing gradient problem of RNN. It can classify and process time lags that are not known in a time series. It can also predict what will happen next. Backpropagation is used to help the model learn how to work. In an LSTM network, three gates connect and work together (Fig. 2).

Input gate—you should look at the input to see which value should change the memory? How does the sigmoid function decide which values to let through, 0 and 1, and which values to keep out? The tanh function gives weight to the values that are passed in. It decides how important they are to you. This ranges from -1 to 1 for each value (Eqs. 1 and 2).

$$i_t = \sigma(W_i[h_{t-1}, x_t] + b_i), \tag{1}$$

$$\check{C}_t = \tanh(W_c[h_{t-1}, x_t] + b_C). \tag{2}$$

Forget gate—Find out which parts of the block will be thrown away. It is up to the sigmoid function. It looks at two things: the previous state (h_{t-1}) and the content input (X_t). Each number in the cell state outputs a number from 0 to 1. C_{t-1} (Eq. 3).

$$f_t = \sigma(W_f[h_{t-1}, x_t] + b_f). \tag{3}$$

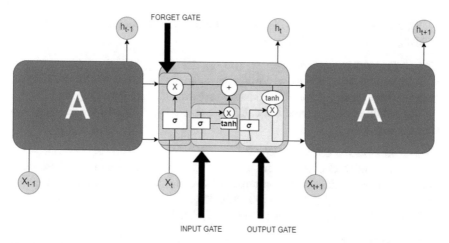

Fig. 2 LSTM network: sample architecture

Output gate—There are two ways to figure out what the output will be: the input and the memory of the block. The sigmoid function must decide which values to let through 0, 1, and so on. A function called tanh gives weight to the values that are passed in. It decides their importance from −1 to 1 and multiplies them with a sigmoid output (Eqs. 4 and 5).

$$o_t = \sigma\left(W_o\left[h_{t-1}, x_t\right] + b_o\right), \tag{4}$$

$$h_t = o_t * \tanh(C_t). \tag{5}$$

4 Material and Methods

As for the dataset, we have collected it from Yahoo Finance [1]. Although any company listed in Yahoo Finance can be entered, the stocks used for this paper are Amazon (AMZN) and Microsoft (MSFT). We have divided the whole methodological part into seven steps, which are given below.

Step 1: *Data extraction*

The data consist of various KPIs, 'Low Price,' 'High Price,' 'Opening Price,' 'Closing Price,' 'Total Volume' extracted from Yahoo Finance [1]. Although any company listed in Yahoo Finance can be entered, the stocks used for this paper are Amazon (AMZN) and Microsoft (MSFT). Our data series covers '2010–01-01'– '2021–12-12.'

Step 2: *Feature extraction*

We must take one of the KPIs to predict our model in this step. So, we will take 'Closing Price' as our key indicator.

Step 3: *Preprocessing the model*

It is preprocessing and transforming the data using scikit-learn (MinMaxScaler) to convert the closing price of the data into an array ranging between 0 and 1.

For creating train data and test data, train data contain 75% of the data, and test data contain 25%.

Step 4: *Building the LSTM model*

To build our model, we are going to use LSTM neural network. We use mean squared error and 'adam' as our optimizer for training. We have used different epochs for our training data (25, 50, 75, 100, 200, 500).

The structure of the model is as follows in Fig. 3.

Step 5: *Output generation*

After the model is trained, we predict the test data and compare it with the actual prices.

Step 6: *Error calculation*

We then calculate Root Mean Squared Error (RMSE), which gives our prediction error concerning actual price.

Step 7: Perform these steps again with different hyperparameters of LSTM to achieve accurate results.

Key contribution or novelty of work:

1. Real-Time Stock Monitoring and Prediction using deep learning model.

```
Model: "sequential"
```

Layer (type)	Output Shape	Param #
lstm (LSTM)	(None, 30, 200)	161600
dropout (Dropout)	(None, 30, 200)	0
lstm_1 (LSTM)	(None, 30, 200)	320800
dropout_1 (Dropout)	(None, 30, 200)	0
lstm_2 (LSTM)	(None, 100)	120400
dropout_2 (Dropout)	(None, 100)	0
dense (Dense)	(None, 1)	101

```
Total params: 602,901
Trainable params: 602,901
Non-trainable params: 0
```

Fig. 3 Proposed customized LSTM model

2. Fine-tuned LSTM model with three layers and activation function of ReLU.
3. Visualized and compared results of two different companies.

5 Result and Discussion

After training the model, our tests have shown different results. As shown, the number of epochs significantly impacted the result. Results are shown from the different iterations of the epoch from 25 to 500. Our model perfectly fits the train data at a specific epoch (i.e., 100 for Amazon and 200 for Microsoft) (Table 1).

Initially, while predicting for Amazon, our model had 'epochs' set to 25, which did not show accurate results as the prices started to show higher values after the year 2020, in which our model had lost the trace of it. As we increased the number of epochs, our model got to know the nature of the data and hence started to show better results. The other dataset, i.e., Microsoft, also had the same initial problem but started to fit in as epochs increased. Our model perfectly fits the train data at a specific epoch (i.e., 100 for Amazon and 200 for Microsoft).

The accuracy of our model is then determined by Root Mean Squared Error (RMSE) in Eqs. 6 and 7. Table 2 presents the resultant RMSE score concerning AMZN and MSFT, whereas Table 3 delivers the accuracy of the AMZN and MSFT datasets.

$$RMSE = \sqrt{\sum_{i=1}^{n} \frac{(\hat{y}_i - y_i)^2}{n}}, \tag{6}$$

$$Accuracy = 1.96 * RMSE \text{ (National Digital Elevation Guidelines [3])}. \tag{7}$$

Here is the actual 'Closing Price' versus 'Predicted Price.' As shown below, Fig. 4 shows that the already real-time predicted prices are nearest to the actual close price. We have compared both Yahoo Finance stocks used for this paper as Amazon (AMZN) and Microsoft (MSFT). Table 4 represents the comparison with existing work.

6 Conclusion and Future Work

This paper uses LSTM to make an RNN for stock prediction. This RNN can be used to predict the value of both companies in the future, it says. Our model results show that it works well, so that we can trust it. As we saw, the prices were close to each other. If you meet the right conditions, it can be very accurate. There is much information you need to understand the stock market and predict the prices. In the future, we will take the mood/sentiments of the market and add them to our model

Table 1 Epoch-based result and visualization of AMZN and MSFT

Epochs	AMZN	MSFT
25 epochs		
50 epochs		
75 epochs		
100 epochs		
200 epochs		
500 epochs		

Table 2 RMSE of AMZN and MSFT: train and test

	Amazon	Microsoft
Train	0.38060957	0.39737493
Test	0.43594788	0.45037993

Table 3 Accuracy of AMZN and MSFT: train and test

Stocks	Accuracy—train (%)	Accuracy—test (%)
Amazon	74.5	84.3
Microsoft	78.4	88.2

AMAZON

	Date	Close	Predicted Price
0	2018-08-09	1898.520020	1868.789185
1	2018-08-10	1886.300049	1870.484741
2	2018-08-13	1896.199951	1873.115479
3	2018-08-14	1919.650024	1876.804321
4	2018-08-15	1882.619995	1881.534546
…	…	…	…
837	2021-12-06	3427.370117	3503.000977
838	2021-12-07	3523.290039	3500.736572
839	2021-12-08	3523.159912	3495.571289
840	2021-12-09	3483.419922	3488.552246
841	2021-12-10	3444.239990	3480.815674

MICROSOFT

	Date	Close	Predicted Price
0	2018-08-09	109.669998	110.414085
1	2018-08-10	109.000000	110.573616
2	2018-08-13	108.209999	110.801529
3	2018-08-14	109.559998	111.048683
4	2018-08-15	107.660004	111.256516
…	…	…	…
837	2021-12-06	326.190002	331.225433
838	2021-12-07	334.920013	330.173004
839	2021-12-08	334.970001	329.019623
840	2021-12-09	333.100006	328.145050
841	2021-12-10	342.540009	327.651337

Fig. 4 Real-time monitoring cum prediction of stocks: AMZN versus MSFT

Table 4 Comparison with existing work

Author	Accuracy	Real-time monitoring	Historical data
Faraz et al. [15]	70%–75% average	No	Yes
Mittal et al. [16]	75.56%	No	Yes
Jagwani et al. [5]	75%–80% average, forecasting based	No	Yes
Li et al. [6]	Review based (approx. 70%–75% average)	No	Yes
Our model	74.5 (real-time monitoring)	Yes	Yes

when we work on this project again. They act in the market based on how the market feels. Economic growth models can help us predict how the stock market will do. We also want to see how this model compares to the linear regression model and how it works with other machine learning algorithms. The limitation of our model is that it works with only huge amount of data not for static limited data.

References

1. Yahoo Finance. https://in.help.yahoo.com/kb/SLN2311.html. Last Accessed 15 Dec 2021
2. Stock Knowledge. https://stockslibrary.com/. Last Accessed 20 Oct 2021
3. Li Y, Pan Y (2021) A novel ensemble deep learning model for stock prediction based on stock prices and news. Int J Data Sci Anal. https://doi.org/10.1007/s41060-021-00279-9
4. Devi B, Shankar VG, Srivastava S, Srivastava DK (2020) AnaBus: a proposed sampling retrieval model for business and historical data analytics. In: Sharma N, Chakrabarti A, Balas V (eds) Data management, analytics, and innovation. advances in intelligent systems and computing, vol 1016. Springer, Singapore. https://doi.org/10.1007/978-981-13-9364-8_14
5. Jagwani MG, Sachdeva H, Singhal A (2018) Stock price forecasting using data from yahoo finance and analysing seasonal and nonseasonal Trend. In: 2018 Second international conference on intelligent computing and control systems (ICICCS), pp 462–467. https://doi.org/10.1109/ICCONS.2018.8663035
6. Li W, Bastos GS (2020) Stock market forecasting using deep learning and technical analysis: a systematic review. IEEE Access 8:185232–185242. https://doi.org/10.1109/ACCESS.2020.3030226
7. Wu JMT, Li Z, Herencsar N et al (2021) A graph-based CNN-LSTM stock price prediction algorithm with leading indicators. Multimedia Syst (2021). https://doi.org/10.1007/s00530-021-00758-w
8. Kumar D, Sarangi P, Verma R (2021) A systematic review of stock market prediction using machine learning and statistical techniques. Mater Today: Proc https://doi.org/10.1016/j.matpr.2020.11.399
9. Devi B, Shankar VG, Srivastava S, Nigam K, Narang L (2021) Racist tweets-based sentiment analysis using individual and ensemble classifiers. In: Micro-electronics and telecommunication engineering. Lecture Notes in Networks and Systems, vol 179. Springer, Singapore. https://doi.org/10.1007/978-981-33-4687-1_52
10. Shankar VG, Devi B, Bhatnagar A, Sharma AK, Srivastava DK (2021) Indian air quality health index analysis using exploratory data analysis. In: Micro-electronics and telecommunication engineering. Lecture Notes in Networks and Systems, vol 179. Springer, Singapore. https://doi.org/10.1007/978-981-33-4687-1_51
11. Devi B, Kumar S, Anuradha, SVG (2019) AnaData: a novel approach for data analytics using random forest tree and SVM. In: Computing, communication and signal processing. Advances in Intelligent Systems and Computing, vol 810. Springer, Singapore. https://doi.org/10.1007/978-981-13-1513-8_53
12. Shankar VG, Devi B, Srivastava S (2019) DataSpeak: data extraction, aggregation, and classification using big data novel algorithm. In: Iyer B, Nalbalwar S, Pathak N (eds) Computing, communication, and signal processing. Advances in intelligent systems and computing, vol 810. Springer, Singapore. https://doi.org/10.1007/978-981-13-1513-8_16
13. Hochreiter S (1997) Jürgen Schmidhuber. Long Short-Term Memory Neural Comput 9(8):1735–1780. https://doi.org/10.1162/neco.1997.9.8.1735
14. Selvamuthu D, Kumar V, Mishra A (2019) Indian stock market prediction using artificial neural networks on tick data. Financ Innov 5:16. https://doi.org/10.1186/s40854-019-0131-7
15. Faraz M, Khaloozadeh H, Abbasi M (2020)Stock market prediction-by-prediction based on autoencoder long short-term memory networks. In: 2020 28th Iranian conference on electrical engineering (ICEE), pp 1–5. https://doi.org/10.1109/ICEE50131.2020.9261055
16. Mittal A (2011) Stock prediction using twitter sentiment analysis

17. Bengio Y, Simard P, Frasconi P (1994) Learning long-term dependencies with gradient descent is difficult. IEEE Trans Neural Netw 5(2):157–166. https://doi.org/10.1109/72.279181
18. NDEP. https://pubs.er.usgs.gov/publication/70182536

Performance Evaluation of TQWT and EMD for Automated Major Depressive Disorder Detection Using EEG Signals

Arti Anuragi, Dilip Singh Sisodia, Ram Bilas Pachori, and Deepak Singh

1 Introduction

Major depressive disorder (MDD), commonly known as depression, is a mood disorder that causes sadness, feeling low, loss of interest, produces suicidal thoughts, which makes the disease life-threatening and must be treated in time. The first treatment recommended or prescribed by the physician for MDD patients is antidepressants. Still, unfortunately, only half of the MDD patients respond to the antidepressants because of the heterogeneous condition. Selecting the appropriate antidepressants can be a time-consuming process, as physicians may rely on their experience or use a trail-and-error approach. Therefore it is essential to predict the response to a specific treatment. Serval investigators have attempted to differentiate the MDD patients who respond to anti-dose by utilizing various methods. Still, the method using electroencephalogram (EEG) signal is preferred more due to feasible, cost-effective, good temporal resolution.

In literature, Hinrikus et al. [1] introduced a novel spectral asymmetry index (SASI) for automated MDD detection evaluated on 36 subjects using 30-min EEG recording. Puthankattil et al. [2] have proposed a methodology using discrete wavelet transform (DWT) to decompose EEG signals into sub-band signals. Then, relative wavelet energy (RWE) features were extracted from decomposed signals and then fed to an artificial feedforward neural network (ANN) for classification which achieved 98.11% accuracy from 11 features. Hosseinifard et al. [3] have presented a new framework where they evaluated both linear and non-linear features for classifying

A. Anuragi (✉) · D. S. Sisodia · D. Singh
Department of Computer Science and Engineering, National Institute of Technology Raipur, G E Road, Raipur, Chhattisgarh 492010, India
e-mail: aanarayandas.phd2018.cs@nitrr.ac.in

R. B. Pachori
Department of Electrical Engineering, Indian Institute of Technology Indore, Simrol, Indore, Madhya Pradesh 453552, India

© The Author(s), under exclusive license to Springer Nature Singapore Pte Ltd. 2023
D. S. Sisodia et al. (eds.), *Machine Intelligence Techniques for Data Analysis and Signal Processing*, Lecture Notes in Electrical Engineering 997,
https://doi.org/10.1007/978-981-99-0085-5_67

healthy and MDD EEG signals. Concluded that using non-linear features such as detrended fluctuation analysis (DFA), Higuchi's fractal dimension (HFD), correlation dimension, and Lyapunov exponent (LE), the proposed framework achieved the classification accuracy of 90%. Faust et al. [4] have proposed a method using the wavelet packet decomposition (WPD) method based on which non-linear entropy features such as approximate entropy, sample entropy, Rényi entropy, and bispectral phase entropy are extracted and classified using various classifiers and achieved the highest classification accuracy of 99.5% using a probabilistic neural network classifier. Acharya et al. [5] proposed a depression diagnosis index for automated detection of MDD and healthy EEG signals using fractal dimension, largest Lyapunov exponent (LLE), sample entropy, DFA, Hurst's exponent, higher-order spectra, and recurrence quantification analysis non-linear features. Mumtaz et al. [6] have derived new features like power at different frequency bands and EEG alpha asymmetry. The researchers employed the receiver operating characteristics (ROC) feature ranking method to determine the most significant features and classified them using various classifiers. They achieved a classification accuracy of 97.6% using logistic regression for automatically discriminating healthy and depressed subjects. Bachmann et al. [7] have presented a new methodology using spectral asymmetry index and non-linear DFA features for signal-channel EEG signals' analysis and achieved the highest accuracy of 91.2%. Acharya et al. [8] have developed a depression severity index based on a one-dimensional (1D) convolutional neural network (CNN), which achieved the highest classification accuracy of 96.0% and 93.5% from the right and left hemispheres, respectively. Fitzgerald et al. [9] have presented a review paper and recommended that gamma rhythm is more capable of discriminating healthy and depressed EEG signals. Azizi et al. [10] have proposed a method using new non-linear geometrical features such as centroid-to-centroid distance, centroid-to-450-line shortest distance, and incenter radius, for discriminating MDD EEG signals. Student t-test was used to obtain statistically significant features, and this research concluded that features extracted from the right hemisphere are more discriminative than that of the left hemisphere.

Various studies have been conducted in the literature to develop MDD detection systems from EEG signals. However, there is still a possibility for improving the system's classification performance. Therefore, a new framework is proposed to develop an automatic MDD detection system based on EEG signals in this study. EEG signals are non-stationary; hence, they cannot be analyzed either using the time- or frequency-domain method. So, for better analysis of EEG signals, wavelet transform methods are used in this study, which is good at localizing both time and frequency simultaneously. In this present work, two wavelet transform methods named empirical mode decomposition (EMD) and tuned-Q wavelet transform (TQWT) are used for classifying MDD EEG signals. The first step is to decompose healthy and MDD EEG signals into sub-band signals. After that, taking motivation from the literature, seven non-linear features are computed from each sub-band signal. To determine statistically significant features, a student t-test is applied. Then, the selected features are passed to the classifiers for discriminating healthy and MDD EEG signals using a single channel. Training and testing of the model are done using the tenfold cross-validation method. First, the model is evaluated on each channel. After that, the highest performing channels are accumulated to improve the performance, which further increases the dimensions of the feature vector. Four feature ranking methods, such as Bhattacharyya space algorithm, entropy, ROC, and Wilcoxon, are employed to reduce feature space in this study and evaluate performance.

The layout of the remaining manuscript is in the following sections: Description of the methods and materials of the proposed framework such as utilized dataset, wavelet methods, extracted features, feature significant test, used classifiers, and their evaluation measures are presented in Sect. 2. In Sect. 3, the results and discussions of the experiment and the comparison of the proposed work with other reported state-of-the-art methods are demonstrated. Finally, the conclusion of the MDD diagnosis is made in Sect. 4.

2 Methods and Materials

The methods and materials of the proposed framework are presented in this section, where Sect. 2.1 provides the studied EEG dataset. Section 2.2 briefly explains the wavelet methods used to decompose EEG signals into IMFs modes. Extracted features are described in Sect. 2.3. Subsequently, Sects. 2.4 and 2.5 explain the method used to determine statistically significant features and classify them using classifiers. Visual illustration of the proposed framework is depicted in Fig. 1. The proposed framework consists of the enlisted steps:

1. EEG signals are segmented into 5 s' epochs.

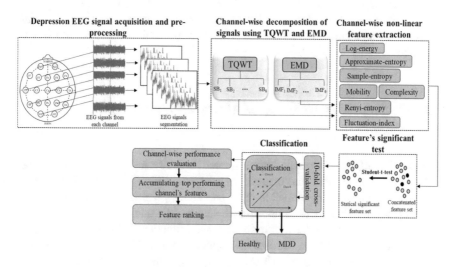

Fig. 1 Illustration of the proposed framework for classifying MDD detection

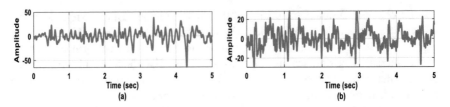

Fig. 2 Epoch of EEG signals of **(a)** MDD, **(b)** healthy patient

2. Decomposition of each EEG signal epoch into sub-band signals by employing two wavelet methods (TQWT and EMD).
3. Determining the ability to distinguish the computed features by student t-test.
4. Evaluating the performance of the proposed framework from individual and accumulated top-performing channel's features.
5. The feature ranking methods are used to obtain the most informative features, corresponding to maximum classification performance.

2.1 Data Acquisition

In this study, we have used the EEG dataset provided by Mumtaz, which was acquired from the Hospital Universiti Sains Malaysia (HUSM) from 34 MDD and 30 healthy patients, of age group ranging 12–77 years [11]. The dataset contains three types of data: eyes closed, eyes opened, and task, having 256 Hz sampling frequency with 20 EEG channels naming A_2A_1, C_3, C_4, C_z, F_3, F_4, F_7, F_8, FP_1, FP_2, F_z, O_1, O_2, P_3, P_4, P_z, T_3, T_4, T_5, and T_6. The present study is carried out using eyes closed dataset. To collect the respective recording, the participants were asked to close their eyes for almost 5 min with minimum eye blinking and head movement, and then, the EEG signals were segmented into 5 s epochs. The epoch of MDD and normal EEG signal is depicted in Fig. 2.

2.2 Decomposing EEG Signal Using TQWT and EMD Methods

- **TQWT method**: The method is considered for analyzing non-stationary signals by employing flexible and discrete wavelet transform (DWT). This wavelet is flexible because of its adaptable input parameters, which are Q-factor signified by Q, the number of decomposition levels as (J), and the oversampling rate denoted as r. By using the non-rational transfer function, TQWT filters are easily implemented [12, 13]. A detailed description of the TQWT method is provided in [14]. In TQWT, the Q value is usually kept high due to the oscillating nature of EEG signals with low frequency. It decomposed each EEG signal into sub-band signals using the input parameters such as (Q, r, J). In this study, we have used the following values 3, 3, and 7 for Q, r, and J, respectively.

- **EMD method**: This method basically breaks the EEG signals into the number of amplitude and frequency modulations termed as IMFs, which have to satisfy the following conditions: (1) in the signal, a number of the extreme and zero-crossing have to be equal or at-most difference of 1 and (2) the mean value of the lower envelope and the upper envelope has to be zero at any point. The IMFs are computed through an algorithm, and the whole algorithm is described in detail in [15]. EMD adaptively finds IMFs of each signal, but we have considered the first six IMFs in this study for feature extraction.

2.3 Feature Extraction

This study extracted seven non-linear features: log energy entropy, approximate entropy, sample entropy, Renyi entropy, mobility, complexity, and fluctuation index for depression EEG signal classification in two different domains, i.e., TQWT and EMD. Each feature is briefly described below:

Let the sub-bands' signals extracted from the wavelet methods containing N sample points be $SB(n) = [SB(1), SB(2), \ldots, SB(N)]$, with n^{th} index.

1. **Log energy entropy (LE)**: It has the capability for better characterizing the non-linear dynamics of non-stationary signals.

$$\text{Log}_{\text{energy}} \text{ (LE)} = -\sum_{n=1}^{N} \left(\log_2 SB(n)\right)^2. \tag{1}$$

2. **Approximate entropy (AE)**: It measures the irregularity of the signals. The greater the AE value, more irregular the signal is. AE has been utilized in many biomedical signal classifications [16, 17] and found that it is robust in short and noisy data. The mathematical expression of AE is provided in [16].

3. **Sample entropy (SE)**: It measures the complexity or irregularity of the signals. The main difference between AE and SE is that it excludes the bias because of self-matches counted in computing AE, and hence, SE improves the performance. Higher the SE value, the more the complexity of the signal. A detailed description with mathematical expression is explained in [18].

4. **Renyi entropy (RE)**: Renyi entropy is computed using the mathematical expression shown in Eq. (2)

$$\text{Renyi-entropy (RE)} = \frac{1}{1-\alpha} \log \sum_{n=1}^{N} (SB(n))^{\alpha}, \text{ where } \alpha > 0 \text{ and } \alpha \neq 1.$$

$$\tag{2}$$

5. **Mobility (M)**: Mobility computes abrupt changes in EEG signals using a mathematical expression presented in Eq. (3).

$$\text{Mobility(M)} = \sqrt{\frac{Var\left(\frac{dSB(n)}{dn}\right)}{Var(SB(n))}}, \text{ where } Var() \text{ is variance operation.} \quad (3)$$

6. **Complexity (C)**: The MDD EEG signals have a more complex pattern than that healthy EEG signals, computed using Eq. (4). The complexity values of depressive features are higher as compared to healthy.

$$\text{Complexity (C)} = \frac{M\left(\frac{dSB(n)}{dn}\right)}{M(SB(n))}. \quad (4)$$

7. **Fluctuation index (FI)**: It measures the intensity of non-stationary EEG signals. The FI of the signal is defined as shown in Eq. (5). It could be found that the FI of MDD EEG signal is more than that of FI of healthy EEG signals.

$$\text{Fluctuation index (FI)} = \frac{1}{N}\sum_{n=1}^{N-1}|SB(n+1) - SB(n)|. \quad (5)$$

2.4 Feature Significant Test

After feature extraction, for wiping off the redundant or non-relevant features and to determine statistical significance features, a student t-test is performed in this study. The statically significant features demonstrate good distinction which is chosen by p-value [19]. The smaller the p-value of the feature, likely to have more significant the difference. In this study, the feature with a p-value less than 0.005 is considered further for classification.

2.5 Classifiers

In the presented study, entropy-based and non-linear features are extracted from sub-band signals obtained from TQWT and EMD wavelet methods and concatenated the obtained features. The significant features are selected and finally passed to three classifiers, named as support vector machine (SVM), k-nearest neighbors (k-NN), and an ensemble classifier. A detailed description of the used classifies is provided in [20–22]. Validation of each classifier is done using a tenfold cross-validation technique which segments the whole dataset into ten equal parts, out of which in each run, one part is considered as a test dataset, whereas the remaining parts are used for training purposes.

Performance evaluation: Classification accuracy is computed as performance assessment in this experiment. The mathematical equation of classification accuracy is expressed as in Eq. (6).

$$\text{Acc}(\%) = \frac{T_{tp} + T_{tn}}{T_{tp} + T_{tn} + T_{fp} + T_{fn}} \times 100, \tag{6}$$

where T_{tp}, T_{tn}, T_{fp}, and T_{fn} in Eq. (6) signify the total number of true positive (tp), true negative (tn), false positive (fp), and false negative (fn) samples. Classification accuracy demonstrates the classifier's ability to discriminate the depression MDD and healthy EEG signals.

3 Results and Discussions

3.1 Results

The proposed framework is implemented and programmed in MATLAB (R2019b) programing platform. EMD, TQWT, feature extraction, and student t-test are the main functions used to extract and select significant features. The decomposed sub-band signals from the TQWT method for 5 s' healthy and MDD EEG signals are presented in Fig. 3. After extracting sub-band signals from the EEG signals using the TQWT and EMD methods, features described in Sect. 2.3 are computed. A total of 56 (8 (sub-band signals) × 7 (features)) features and 42 (6 (first six IMFs) × 7 (features)) are computed from TQWT and EMD methods, respectively, from each channel.

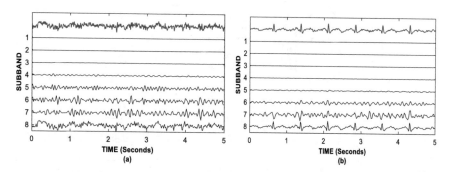

Fig. 3 A_2A_1 channel's decomposed sub-band signals using the TQWT method for **a** healthy and **b** MDD EEG signals

The obtained feature matrix consists of non-significant features, which were eliminated using a statistically significant test known as the student t-test. In this study, the features with a p-value less than 0.005 are selected as significant features for MDD signal detection. Usually, features with a lower p-value signify a better discrimination ability in binary class problems. It is observed from Table 1 that out of 56 features, 50 features are statistically significant with having a p-value less than threshold values (0.05) and the remaining features are discarded as they fail to satisfy the condition. The discarded features are depicted in bold font in Table 1.

After selecting all the significant features, they are fed to an ensemble, k-NN with weighted kernel function, and SVM classifier with the cubic polynomial kernel function for classifying healthy and MDD EEG signals. The classification is done using the MATLAB classification learning application. The tenfold cross-validation technique is used to validate the classification models. The classification performance analysis between two different decomposing methods, EMD and TQWT, is made on a single channel using significant features. The results in terms of classification accuracy of used classifiers and the selected number of statistically significant features are demonstrated in Table 2. The experiment is carried out on twenty channels individually, and the best performing channels are highlighted in bold in Table 2. It is observed from the table that the ensemble tree classifier performs best among other classifiers, and the classification accuracies of the individual channels lie around 94–96% from the EMD method and 97–98% from the TQWT method. To

Table 1 p-values for healthy and MDD EEG signals using the TQWT method from the A_2A_1 channel

Sub-bands	Features						
	LE	AE	SE	M	C	RE	FI
SB_1	1.62×10^{-18}	3.25×10^{-04}	3.75×10^{-06}	1.09×10^{-07}	**0.0387**	3.82×10^{-04}	**0.2852**
SB_2	1.90×10^{-08}	3.49×10^{-08}	8.22×10^{-12}	1.89×10^{-04}	**0.2366**	1.01×10^{-08}	**0.1698**
SB_3	1.90×10^{-63}	7.46×10^{-11}	5.19×10^{-07}	1.70×10^{-51}	2.96×10^{-20}	4.93×10^{-51}	6.72×10^{-27}
SB_4	5.03×10^{-126}	2.87×10^{-05}	8.06×10^{-14}	2.00×10^{-96}	5.79×10^{-45}	5.99×10^{-125}	2.78×10^{-40}
SB_5	2.08×10^{-203}	3.19×10^{-08}	9.56×10^{-06}	1.09×10^{-43}	5.44×10^{-127}	2.21×10^{-220}	6.91×10^{-58}
SB_6	2.86×10^{-227}	1.64×10^{-87}	2.95×10^{-228}	1.09×10^{-43}	1.20×10^{-47}	3.90×10^{-193}	4.95×10^{-237}
SB_7	9.60×10^{-05}	1.35×10^{-21}	1.42×10^{-95}	1.17×10^{-258}	5.96×10^{-97}	**0.8097**	3.77×10^{-04}
SB_8	2.51×10^{-19}	1.08×10^{-08}	1.30×10^{-04}	3.14×10^{-133}	6.39×10^{-68}	**0.0066**	7.02×10^{-13}

Notations: LE—log energy entropy, AE—approximate entropy, SE—sample entropy, M—mobility, C—complexity, RE—Renyi entropy, FI—fluctuation index further improve the proposed framework's performance, the channels are sorted in descending order based on their performance, and then, performance evaluation is made on the top-performing channels. The highest performing individual channels from TQWT methods are P_3, P_z, and A_2A_1. Concatenating features of these channels, we have achieved a classification accuracy of 99.24%. Further, it is noticed while experimenting that increasing the number of channels does not improve the performance and increases the model's complexity. Hence, their results are not reported due to space constraints. While concatenating the channels, the feature dimension also increases. Hence, four feature ranking methods such as entropy, Bhattacharyya space algorithm, ROC, and Wilcoxon are used to optimize the classifier's performance with fewer features. The entropy feature ranking method is based on the divergence between two probability functions [23]. Whereas the Bhattacharya method uses Bhattacharya distance for discriminating the two classes [24], the ROC method depends on the area under the curve of the ROC plot [23], and the Wilcoxon method ranks the features depending on the two-sample unpaired Wilcoxon test [23]. Figure 4 depicts the variation of the ensemble classifier classification accuracies corresponding to the number of the features with the used feature ranking methods. The optimized highest classification accuracy using three channels with 70 features is 99.3%, as shown in Fig. 4. The other parameters such as sensitivity, specificity, precision, Matthews Correlation Coefficient (MCC), and F1-score of the best performing model are evaluated and achieved 99.53%, 98.87%, 98.83%, 98.38%, and 99.18%, respectively. The ROC value of the best performing model is shown in Fig. 5. It is clear from Fig. 5 that the area under the curve of the ensemble classifier from three selected channels using the TQWT method is 1.00.

3.2 Discussions

This study proposed a framework to generate non-linear features based on the TQWT method from three selected channels. We have evaluated this framework on a large dataset that consists of 64 subjects with 20 channels of EEG recording. Comparative analysis between the proposed framework and other state-of-the-art methods for developing an automated MDD detection using EEG signals is presented in Table 3. It can be noticeable from Table 3 that the proposed framework achieved the highest classification accuracy of 99.3% for MDD detection. The proposed framework is a hand-crafted feature extraction model for which EMD and TQWT methods are used. Loh et al. [25] proposed a 2D-CNN model for MDD detection, achieving the highest classification accuracy of 99.25%. They achieved the result using 20 channels with 5-min-long recording. The proposed framework achieved the highest classification accuracy of 99.30% from the three best-performing channels only, which is significantly less than the other reported methods.

The salient features of the proposed framework are as follows:

Table 2 Classifier's performance analysis using EMD and TQWT methods for individual and top-performing channels

	EMD method					TQWT method				
	Channel	No. of features	Ensemble	k-NN	SVM	Channel	No. of features	Ensemble	k-NN	SVM
Performance on individual channel	FP_1	37	94.2	91.4	90.3	FP_1	49	96.5	93.7	92.3
	A_2A_1	34	**96.5**	92.5	93.3	A_2A_1	50	**98.3**	94.9	96.0
	T_4	37	94.2	91.0	90.9	T_4	49	97.7	95.8	95.7
	C_3	37	94.4	91.2	91.6	C_3	49	97.9	96.0	96.2
	P_3	37	94.8	90.1	89.2	P_3	47	**98.5**	96.3	96.6
	F_3	37	94.6	90.8	89.4	F_3	48	97.3	95.3	95.3
	O_1	37	94.0	89.0	89.0	O_1	50	98.0	95.3	95.9
	T_3	36	**95.4**	91.6	92.0	T_3	51	97.9	94.2	95.1
	F_7	33	94.6	91.4	92.1	F_7	52	97.6	94.5	94.5
	T_5	38	94.6	89.7	89.3	T_5	50	98.3	96.0	96.4
	F_z	37	**95.1**	91.6	90.2	F_z	51	97.5	95.8	96.3
	FP_2	30	94.2	91.8	91.5	FP_2	50	97.1	95.5	96.2
	F_4	36	94.6	90.5	90.2	F_4	51	98.0	95.5	95.9
	C_4	38	94.6	90.4	91.0	C_4	51	97.9	95.9	95.9
	P_4	36	93.8	90.1	89.5	P_4	50	98.2	95.6	96.7
	O_2	37	94.1	90.5	90.3	O_2	50	97.6	95.3	94.7
	F_8	37	94.6	90.9	90.5	F_8	49	97.7	94.7	94.9
	T_6	37	93.8	90.7	91.3	T_6	49	97.7	96.0	95.6
	C_z	37	94.7	89.4	89.7	C_z	46	98.2	96.0	95.9
	P_z	36	**95.2**	90.1	90.0	P_z	49	**98.5**	95.9	96.0

(continued)

Table 2 (continued)

	EMD method					TQWT method				
	Channel	No. of features	Ensemble	k-NN	SVM	Channel	No. of features	Ensemble	k-NN	SVM
Performance on top-performing channels	$[A_2A_1, T_3]$	73	97.7	94.4	95.1	$[P_3, P_Z]$	95	98.5	96.5	96.7
	$[A_2A_1, T_3, P_Z]$	106	98.2	95.3	96.8	$[P_3, P_Z, A_2A_1]$	146	**99.24**	97.2	98.4
	$[A_2A_1, T_3, P_Z, F_Z]$	148	**99.1**	95.7	97.2					

Notation: No. of features—number of statistically significant features using the student t-test

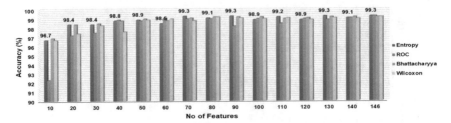

Fig. 4 Feature ranking methods for optimizing the highest performing model

Fig. 5 ROC plot of ensemble classifier from three selected channels using the TQWT method

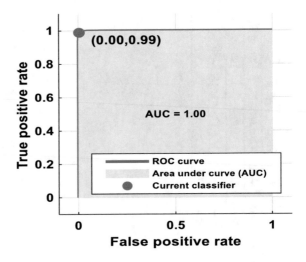

- It is the first time, TQWT wavelet has been used for non-linear feature extraction for automated MDD detection.
- The proposed framework achieved the highest classification accuracy of 99.30% with short EEG signals (5 s), while most of the studies [25–27, 30] use the long duration of EEG signals.
- The model achieved more than 97% classification accuracy from each channel, making the framework robust and better [30].
- This proposed framework is a lightweight method. It does not need to set millions of the parameters like deep networks such as in [25, 28, 29] for achieving the best performance.

4   Conclusion

Detecting MDD automatically using EEG signals is a challenging task. In this proposed framework, hand-crafted features' extraction based on TQWT and EMD methods for automated depression detection systems is developed. The presented

Table 3 Comparison of the proposed framework with other state-of-the-art methods for developing an automated EEG MDD detection model using EEG dataset collected by Mumtaz [11]

Authors/year	Method	Considered EEG recording duration	Average accuracy (%)	No. of channels
Mumtaz et al./2017 [26]	STFT + EMD	1, 2, 3 min	90.5	19
Mahato et al./2019 [27]	Linear and non-linear features based on DWT wavelet	5 min	93.33	19
Mumtaz et al./2019 [28]	1DCNN	1 s	98.32	19
Kang et al./2020 [29]	Asymmetry image, CNN	4 s	98.85	16
Loh et al./2021 [25]	Spectrogram images, CNN	5 min	99.25	20
Aydemir et al./2021 [30]	Melamine pattern	10 s	98.29	1
Proposed method	TQWT wavelet method, non-linear features, ensemble	5 s	99.30	3

method achieved greater than 97% classification accuracy from all twenty EEG channels. This distinctly implies the robustness of the developed model. We obtained 99.1% and 99.27% accuracy from concatenating the highest performing channels using EMD and TQWT methods. For optimizing the performance of the proposed framework, performance is also evaluated on the various number of ranked features using entropy, Bhattacharyya space algorithm, ROC, and Wilcoxon methods. Obtained the highest classification accuracy of 99.30% using three channels (P_3, P_z, and A_2A_1) with 70 features based on the TQWT method from an ensemble classifier. Our findings suggest that three specific channels can distinguish depressive and healthy EEG features. In future, the proposed framework can also be used for detecting other disorder diseases such as sleep apnea, focal, emotion etc.

References

1. Hinrikus H et al (2009) Electroencephalographic spectral asymmetry index for detection of depression. Med Biol Eng Comput 47(12):1291–1299
2. Puthankattil SD, Joseph PK (2012) Classification of EEG signals in normal and depression conditions by ann using rwe and signal entropy. J Mech Med Biol 12(4)
3. Hosseinifard B, Moradi MH, Rostami R (2013) Classifying depression patients and normal

subjects using machine learning techniques and non-linear features from EEG signal. Comput Methods Programs Biomed 109(3):339–345

4. Faust O, Ang PCA, Puthankattil SD, Joseph PK (2014) Depression diagnosis support system based on EEG signal entropies. J Mech Med Biol 14(03):1450035

5. Acharya UR, Sudarshan VK, Adeli H, Santhosh J, Koh JEW, Adeli A (2015) Computer-aided diagnosis of depression using EEG signals. Eur Neurol 73(5–6):329–336

6. Mumtaz W, Xia L, Ali SSA, Yasin MAM, Hussain M, Malik AS (2017) Electroencephalogram (EEG)-based computer-aided technique to diagnose major depressive disorder (MDD). Biomed Signal Process Control 31(2017):108–115

7. Bachmann M, Lass J, Hinrikus H (2017) Single channel EEG analysis for detection of depression. Biomed Signal Process Control 31:391–397

8. Acharya UR, Oh SL, Hagiwara Y, Tan JH, Adeli H, Subha DP (2018) Automated EEG-based screening of depression using deep convolutional neural network. Comput Methods Programs Biomed 161:103–113

9. Fitzgerald PJ, Watson BO (2018) Gamma oscillations as a biomarker for major depression: an emerging topic. Transl Psychiatry 8(1)

10. Azizi A, Alireza, Moridani, Karimi M, Saeedi (2019) A novel geometrical method for depression diagnosis based on EEG signals, vol c

11. Mumtaz W (2021) MDD patients and healthy controls EEG data (new). Figshare. Dataset. https://figshare.com/articles/dataset/EEG_Data_New/4244171. Accessed 28 Dec 2021

12. Bayram I, Selesnick IW (2009) Frequency-domain design of overcomplete rational-dilation wavelet transforms. IEEE Trans Signal Process 57(8):2957–2972

13. Selesnick IW (2011) Wavelet transform with tunable Q-factor. IEEE Trans Signal Process 59(8):3560–3575

14. Anuragi A, Sisodia DS (2017) Alcoholism detection using support vector machines and centered correntropy features of brain EEG signals. In: Proceedings of the international conference on inventive computing and informatics, ICICI, pp 1021–1026

15. Rilling G, Flandrin P, Es P (2003) On empirical mode decomposition and its algorithms. In: IEEE-EURASIP workshop on non-linear signal and image processing, vol 3, pp 8–11

16. Sharmila A, Aman Raj S, Shashank P, Mahalakshmi P (2018) Epileptic seizure detection using DWT-based approximate entropy, Shannon entropy and support vector machine: a case study. J Med Eng Technol 42(1):1–8

17. Guo L, Rivero D, Pazos A (2010) Epileptic seizure detection using multiwavelet transform based approximate entropy and artificial neural networks. J Neurosci Methods 193(1):156–163

18. Kumar M, Pachori RB, Acharya UR (2017) Automated diagnosis of myocardial infarction ECG signals using sample entropy in flexible analytic wavelet transform framework. Entropy 19(9)

19. Acharya UR, Vidya KS, Ghista DN, Lim WJE, Molinari F, Sankaranarayanan M (2015) Computer-aided diagnosis of diabetic subjects by heart rate variability signals using discrete wavelet transform method. Knowl-Based Syst 81:56–64

20. Jakkula V (2006) Tutorial on support vector machine (SVM). Sch EECS Washingt State Univ 37:1–13 (Online). Available: http://www.ccs.neu.edu/course/cs5100f11/resources/jakkula.pdf

21. Sharmila A, Geethanjali P (2016) DWT based detection of epileptic seizure from EEG signals using naive Bayes and k-NN classifiers. IEEE Access 4:7716–7727

22. Chatterjee R, Datta A (2018) Ensemble learning approach to motor imagery EEG signal classification. In: Machine learning in bio-signal analysis and diagnostic imaging. Elsevier, Amsterdam, pp 183–208

23. Sharma R, Pachori RB, Acharya UR (2015) An integrated index for the identification of focal electroencephalogram signals using discrete wavelet transform and entropy measures. Entropy 17(8):5218–5240

24. Kailath T (1967) The divergence and Bhattacharyya distance measures in signal selection. IEEE Trans Commun Technol 15(1):52–60

25. Loh HW, Ooi CP, Aydemir E, Tuncer T, Dogan S, Acharya UR (2021) Decision support system for major depression detection using spectrogram and convolution neural network with EEG signals. Expert Syst

26. Mumtaz W, Xia L, Yasin MAM, Ali SSA, Malik AS (2017) A wavelet-based technique to predict treatment outcome for major depressive disorder. PLoS ONE 12(2):1–30
27. Mahato S, Paul S (2019) Detection of major depressive disorder using linear and non-linear features from EEG signals. Microsyst Technol 25(3):1065–1076
28. Mumtaz W, Qayyum A (2019) A deep learning framework for automatic diagnosis of unipolar depression. Int J Med Inf 132:103983
29. Kang M, Kwon H, Park JH, Kang S, Lee Y (2020) Deep-asymmetry: asymmetry matrix image for deep learning method in pre-screening depression. Sensors (Switzerland) 20(22):1–12
30. Aydemir E, Tuncer T, Dogan S, Gururajan R, Acharya UR (2021) Automated major depressive disorder detection using melamine pattern with EEG signals. Appl Intell 51(9):6449–6466

Smart Crop Recommendation System: A Machine Learning Approach for Precision Agriculture

Preeti Kathiria, Usha Patel, Shriya Madhwani, and C. S. Mansuri

1 Introduction

Agriculture is one of the most crucial sectors worldwide and we all are highly dependent on it. As we know that food is one of the three basic necessities, i.e., food, cloth, and shelter, and survival of living beings is impossible without food. Moreover, agriculture accounts for a countries economy and holds the highest share of the Indian economy. India is an agricultural country that has the second-largest population across the globe; this situation can be problematic as India has more stomachs to feed and there is a lot of diversity in the landmass of the country. Also, in India, more than 70% of people in the rural area are dependent on the production of crops for agriculture [1]. Hence, it is very important to identify which crop will grow on specific land.

Precision Agriculture is the study of increasing crop production, it also helps in taking managerial decisions with the help of various analysis tools and high technological sensors. Numerous researchers have contributed to this noble field, for instance, Mendigoria et al. [2] forecasted morphological features and did the classification of dry bean, Wang et al. [3] evaluated soil organic matter prediction in the urban area of Sanghai.

However, an ML model that can accurately predict the crop which will grow on a particular piece of land can be very helpful and can boost agricultural production.

P. Kathiria · U. Patel (✉) · S. Madhwani · C. S. Mansuri
Nirma University, Sarkhej-Gandhinagar Hwy, Gota, Ahmedabad, Gujarat 382481, India
e-mail: ushapatel@nirmauni.ac.in

P. Kathiria
e-mail: preeti.kathiria@nirmauni.ac.in

S. Madhwani
e-mail: 19mca015@nirmauni.ac.in

C. S. Mansuri
e-mail: 19mca016@nirmauni.ac.in

Therefore, for accomplishing this task and developing the recommendation system we used ML and its various models.

Machine Learning (ML) is a technology that learns from the data which is provided to it. Various ML models are trained on the given data which is then used to predict future outcomes. ML is of 3 types namely: supervised, unsupervised, and reinforcement learning. This work is an example of supervised learning where we gave the models some data to train; as a result ML models learn from the given data and finally recommend the crop. In this work, the problem that we are dealing with is multiclass classification, where each instance can be classified into two or more classes [4].

Smart Crop Recommendation System is a concept that can revolutionize the Agriculture sector. Recommendation systems are generally classified into two categories on the basis of data that is being used for making the inferences, i.e., content-based filtering and collaborative filtering [5]. If farmers know which crop is best suitable for their land, they will utilize the land in the growth of the recommended crop instead of growing the same crop again and again. This system will also be helpful in optimizing crop rotation which is a practice where farmers grow different types of crops in a sequential order to improve the health of the soil and optimize its nutrients. Therefore, Smart Crop Recommendation System can aid and optimize crop rotation and can accurately tell farmers about the crop to grow which is best suitable for their land.

Conclusively, in this paper, we mainly focused on developing a crop recommendation system using various ML models like Support Vector Machine (SVM), Random Forest (RF), Light Gradient Boosting Machine (LGBM), Decision Tree (DT), and K-Nearest Neighbors (KNN). This system will help farmers identify the type of crop they can grow based on the soil properties and other parameters like temperature, rainfall, and humidity in their farms.

1.1 Objective

By the end of 2050, the world population will increase to approximately 9.1 billion, i.e., thirty-four percent increment as of today [6]. India will rise to become the most populated country by 2050; at present, in terms of domestic food production, it is already lagging behind. And there is a huge diversity among lands in different regions of India. Therefore, it is crucial to use these lands efficiently for food production. To ensure food security for the entire world; precision agriculture is one of the solutions [7], even for a diverse country like India.

In this research work, we started with the objective of categorizing the type of crop that can grow on a particular piece of land based on geographical and environmental properties. As we are already aware that there are multiple factors that are responsible for the growth of a crop including but not limited to soil properties like the amount of Nitrogen, Phosphorous, Potassium, Carbon, and pH of soil. Also, environmental properties like temperature, humidity, and rainfall play a vital role in agriculture.

Hence, a system that can recommend the type of crop which will grow based on these properties can be very helpful for farmers and other people in the agriculture sector.

1.2 Research Contribution

The research contributions of this paper are as follows:

- Relative comparison is presented, which can help the researchers understand the work done in this particular field.
- Accuracy of various classifiers are compared and the best performing classifier is suggested on the dataset used.
- Crop prediction based on various environmental and geographical characteristics.

2 Literature Review

Many researchers have worked on precision agriculture previously and have made a huge contribution to this field. There are numerous parameters that were considered during their research like minerals in a soil, temperature, humidity of the land, amount of rainfall a particular land receives, and many more.

Table 1 shows the contribution of various researchers and their work on precision agriculture. Different authors used different approaches and algorithms, for instance, in [15], the authors developed a crop recommendation system that can classify the soil dataset into the type of crop which is recommendable, i.e., Rabi and Kharif. Medar et al.[16] and Namgiri et al. [17] both have explored crop yield Prediction using different ML Techniques. Medar et al.[16] predicted the crop yield using Naive Bayes, KNN, while Namgiri et al. [17] forecasted the same using RF.

Moreover, Abishek et al. [18] worked on rainfall and weather prediction which consists of recreation and refined computer modeling for an accurate prediction as the soil properties are an important factor to consider, as they are related to the climatic and geographical conditions of the land that is being used. The prediction of soil properties primarily contains humidity of soil surface, predicting the nutrients within the soil, and weather conditions during the lifecycle of a crop [7]. The authors [7] analyzed three supervised ML algorithms; KNN, SVM, and DT for the prediction of date plantation; early, normal, and late for wheat crops in Turkey [19]. The author used Deep learning measures to forecast rice crop yield. In [20], the authors have used soil, nutritional properties, and rainfall data recorded for over 31 years. Hence, all the researchers have worked on one thing or the other creating a knowledgebase for future researches.

Table 1 Existing survey on classifiers used for Precision Agriculture

Ref. No.	Year	Objective	Algorithm used	Result
[8]	2021	To analyze ML algorithms and create a hybrid model to predict rainfall	LGBM, SVR, Hybrid	Hybrid model worked better than the other models
[9]	2021	To forecast the crop that is preferable for a particular land, based on soil parameters and seasons	SVM, KNN, RF, ANN	KNN performs the best of all
[2]	2021	Classifying dry beans and forecasting the morphological features.	GPR, SVMR, KNN, LDA, NB	KNN performs best in all
[10]	2021	Creating a model for predicting crop and analyzation of the fundamental method of cultivation	Support Vector Regression, Voting Regression, and RF Regression	SVR provided the best results for predicting the Boro and Wheat yield.
[11]	2021	To predict crop yield	Lemuria algorithm, SVM, DT CLARA, DBSCAN, KNN, K-means and PAM	All models performed good in terms of accuracy
[12]	2021	To develop a robust and reliable machine learning predictive model for TDS simulation using several associated soil physicochemical variables.	SVM, RF, and GBDT	GBDT and SVM performed better than RF
[3]	2021	To evaluate soil organic matter forecasting in the non-rural area (i.e., Shanghai)	PLSR, ANN, SVM	ANN performed better among all
[13]	2022	To forecast soil hydraulic characteristics utilizing Vis-NIR spectral data in the dryland of the Karnataka plateau in the North.	RF, SVM, PLSR	RF and SVM outperformed PLSR model
[14]	2021	Crop yield prediction using remote sensing data	KNN, SVM	Both performed outstanding

3 Implementation

This section contains description of the dataset that we acquired from Kaggle. Various ML models were applied to this dataset to create a crop recommendation system. Further, we explained the evaluation parameters that we considered for measuring the efficiency of the ML models used.

3.1 Dataset Description

Dataset named "Crop Recommendation Dataset" that we used in this research is from Kaggle. This dataset consists of eight features in which seven are independent namely: Nitrogen, Phosphorous, Potassium, temperature (in Celcius), humidity, ph, and rainfall, and one dependent feature namely label(i.e., name of the crop) [21]. Moreover, it consists of 2200 rows of data which leads to various crops suitable for the features. Also, total 22 crops are labeled in this dataset.

As it is observable in Fig. 1, the dataset contains various attributes as previously explained. The value range for each attribute are Nitrogen 0–140, Phosphorous 5–145, Potassium 5–205, temperature 8.83–43.67 °C, humidity 14.26–99.98%, pH 3.50–9.93, and Rainfall 20.21–298.56 mm. Hence, this dataset can be used to create a model for crop recommendation.

While analyzing the dataset, we also decided for finding the features that are important for crop prediction. For this, we did feature selection on the dataset. As per Fig. 2, it is observable that Potassium (K) is one of the most important feature of the dataset with the score of 116710.53, while, pH ranks the last with the feature score of 74.88. The score function used for feature selection in this work is Chi-square.

	N	P	K	temperature	humidity	ph	rainfall	label
0	90	42	43	20.879744	82.002744	6.502985	202.935536	rice
1	85	58	41	21.770462	80.319644	7.038096	226.655537	rice
2	60	55	44	23.004459	82.320763	7.840207	263.964248	rice
3	74	35	40	26.491096	80.158363	6.980401	242.864034	rice

Fig. 1 Dataset visualization

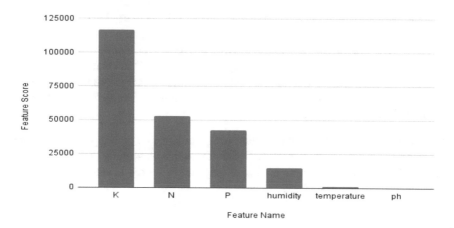

Fig. 2 Feature score

Furthermore, for highlighting the crucial features from the available features, heat map can be utilized, where the features that are more vital are highlighted with a dark shade [22].

3.2 ML Models Used

This research mainly focused on various machine learning models and their implementations. ML models like KNN, LGBM, DT, SVM, and RF were used and all of them performed really well. The dataset was split into 70:30 ratio for training the mentioned ML algorithms. Not only they were able to provide good accuracy but also they were able to predict the suitable crop based on various parameters like soil properties, temperature, humidity, pH, and rainfall requirement.

- **KNN** [19]: KNN is one of the most basic machine learning algorithms which follows the supervised learning approach. It surmises the alikeness betwixt new data and available data, further putting the new data into the category which is akin to available categories [23].
- **LGBM** [8]: It is a decision tree-based gradient boosting framework that occupies less memory and provides more efficiency. In this framework two techniques are used namely: Exclusive Feature Bundling and Gradient-based One Side Sampling which helps it in overcoming the barriers of histogram-based algorithms that are mainly used in all Gradient Boosting Decision Tree frameworks.
- **DT:** DT [19] is a Supervised learning technique that is mostly preferred to solve classification problems. It consists of two types of nodes i.e., Decision node and Leaf node [24]. As the name suggests Decision nodes are utilized to make the decision and can possess multiple branches, while Leaf nodes are the outcomes of a decision and they don't expand to further branches.
- **SVM:** SVM [12] is a very popular supervised learning technique [25], which is used mostly for classification problems just like DT. SVM used in this research is linear SVM because of the dataset, which is in linear form.
- **RF** [26]: It is a popular ML algorithm that is based on ensemble learning. As the name suggests, it is a classifier that contains numerous DTs creating a forest[27, 28]. In this algorithm, it doesn't rely on just a single decision tree instead it takes predictions from numerous DT, and based on the number of votes it further predicts the final outcome.

All the classifiers that we used in this work, performed extremely well, with the minimum accuracy being more than 97% for the worst-performing classifier. Hence, any of these algorithms can be used to identify the best crop to grow based on soil and environmental parameters.

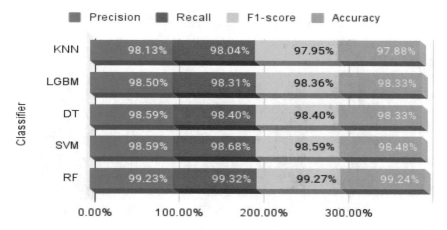

Fig. 3 Modelwise metrics

3.3 Evaluation Parameters

A model cannot be said to be better performing simply because of its accuracy. Therefore, we used various evaluation parameters for the models that we used. Evaluation parameters like Precision, Recall, F1-Score, and Accuracy were used for distinguishing the best one from the used ML models [29].

– **Precision**: It calculates the quantity of positive category forecast that literally belongs to the positive category.
– **Recall**: It calculates the quantity of positive category forecasts created from all positive examples.
– **F1-Score**: It merges recall and precision that is correlative to a particular positive category.
– **Accuracy**: It is a measure that depicts the best model which can recognize patterns and relations on the training or input data provided to the model.

4 Results

In this work, various models such as KNN, LGBM, DT, SVM, and RF. These models were trained on various data related to the crops such as the requirement of minerals, temperature, pH, and rainfall. As per Fig. 3 all the models that we used provided great accuracy and were easily able to predict crops with the provided features.

Figure 3 shows the precision, recall, f1-score, and accuracy of the used models. KNN was the lowest ranking in terms of all the metrics with an accuracy of 97.88%. Moreover, LGBM, DT, and SVM were very close to each other in terms of accuracy,

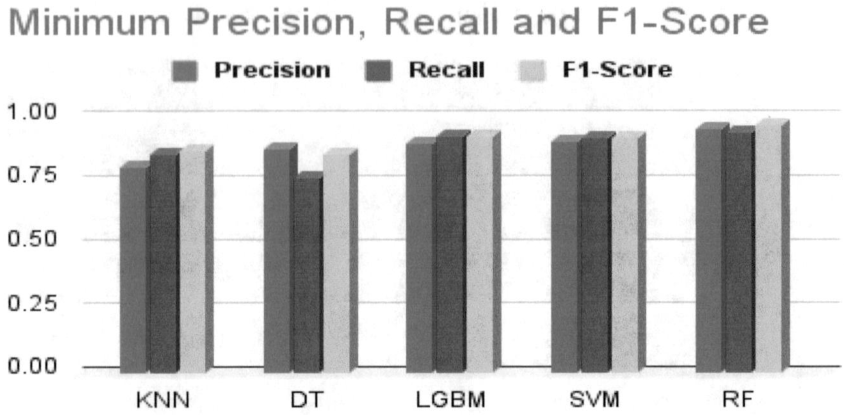

Fig. 4 Modelwise minimum metrics

where LGBM and DT provided an accuracy of 98.33, SVM was leading in terms of accuracy excluding RF. RF was the best model for the dataset we used as it provided an accuracy of 99.24%. However, all the models are able to make predictions, still, we would recommend using RF because of its superiority in terms of all the metrics.

The dataset that we used in this work consists of 22 crops, as a result, different crops produced different metrics. However, the minimum metrics can be observed in Fig. 4 where it is clearly visible that even in terms of minimum precision, KNN is way behind the others providing 81% precision, same is the case with DT in terms of recall, it only provided 77% recall. However, in terms of other metrics excluding the ones previously mentioned, all the classifiers were providing almost similar metrics.

The proposed system can accurately recommend crops when the attributes such as the amount of Nitrogen, Potassium, Phosphorous, Temperature, Humidity, pH, and Rainfall are provided. Therefore, Smart Crop Recommendation System can be very crucial for farmers and the government to take the necessary decision and reduce the wastage of land resources.

5   Research Opportunities and Challenges

– During this research, we faced a challenge finding the proper publicly available dataset, as there is a lack of publicly available datasets like the dataset used in this research.
– The dataset that we used contains only 22 types of crops and it would have been great if a large dataset containing all the major crops was made publicly available.
– Crop Recommendation is the field that needs more exploration and deep study.
– The models we used can be further studied and compared with various neural network models.

6 Conclusion

In this research, we started with the aim to perform multiclass classification, which can help the farmers to know the type of crop that will efficiently grow on their land; based on various parameters. Moreover, the dataset used during this work is publicly available on Kaggle. This dataset, along with various classifiers was used to make a system that can make crop predictions based on various characteristics of a land and its environment; this proposed system was named "Smart Crop Recommendation System". Smart Crop Recommendation System consists of various ML classifiers such as KNN, DT, LGBM, SVM, and RF that are used to make the predictions and all can recommend a crop accurately with the accuracy of 97.88%, 98.33%, 98.33%, 98.48%, and 99.24% respectively. However, RF is recommended based on its performance. Conclusively, Smart Crop Recommendation System can revolutionize how farmers choose to produce a crop and can also benefit other people in the agriculture sector taking vital decisions.

References

1. Agrawal S, Kathiria P, Rawal V, Vyas T. Drone development and embellishing it into crop monitoring and protection along with pesticide spraying mechanism
2. Mendigoria CH, Concepcion R, Dadios E, Aquino H, Alaias OJ, Sybingco E, Bandala A, Vicerra RR, Cuello J (2021) Seed architectural phenes prediction and variety classification of dry beans (phaseolus vulgaris) using machine learning algorithms. In: 2021 IEEE 9th region 10 humanitarian technology conference (R10-HTC), pp 01–06
3. Wang X, Han J, Wang X, Yao H, Zhang L (2021) Estimating soil organic matter content using sentinel-2 imagery by machine learning in shanghai. IEEE Access 9:78215–78225
4. Jesús Fernández-García A, Iribarne L, Corral A, Criado J, Wang JZ. A recommender system for component-based applications using machine learning techniques https://doi.org/10.1016/j.knosys.2018.10.019
5. Banerjee R, Kathiria P, Shukla D (2021) recommendation systems based on collaborative filtering using autoencoders: issues and opportunities. In: Recent innovations in computing. Springer Singapore, Singapore, pp 391–405
6. Alfred R, Obit JH, Chin CPY, Haviluddin H, Lim Y (2021) Towards paddy rice smart farming: a review on big data, machine learning, and rice production tasks. IEEE Access 9:50358–50380
7. Sharma A, Jain A, Gupta P, Chowdary V (2021) Machine learning applications for precision agriculture: a comprehensive review. IEEE Access 9:4843–4873
8. Maliyeckel MB, Chaithanya Sai B, Naveen J (2021) A comparative study of lgbm-svr hybrid machine learning model for rainfall prediction. In: 2021 12th international conference on computing communication and networking technologies (ICCCNT), pp 1–7
9. Patel K, Patel HB (2021) A comparative analysis of supervised machine learning algorithm for agriculture crop prediction. In: 2021 fourth international conference on electrical, computer and communication technologies (ICECCT), pp 1–5
10. Ishak M, Rahaman MS, Mahmud T (2021) Farmeasy: an intelligent platform to empower crops prediction and crops marketing. In: 2021 13th international conference on information communication technology and system (ICTS), pp 224–229
11. Tamil Selvi M, Jaison B (2021) Adaptive lemuria: a progressive future crop prediction algorithm using data mining. Sustain Comput: Inf Syst 31:100577

12. Bokde ND, Ali ZH, Al-Hadidi MT, Farooque AA, Jamei M, Maliki AAA, Beyaztas BH, Faris H, Yaseen ZM (2021) Case study of gypsum soil within Iraq region. Total dissolved salt prediction using neurocomputing models. IEEE Access 9:53617–53635

13. Dharumarajan S, Lalitha M, Gomez C, Vasundhara R, Kalaiselvi B, Hegde R (2022) Prediction of soil hydraulic properties using vis-nir spectral data in semi-arid region of northern Karnataka plateau. Geoderma Regional 28:e00475

14. Kavita, Mathur P (2021) Satellite-based crop yield prediction using machine learning algorithm. In: 2021 Asian conference on innovation in technology (ASIANCON), pp 1–5

15. Kulkarni NH, Srinivasan GN, Sagar BM, Cauvery NK. Improving crop productivity through a crop recommendation system using ensembling technique. https://doi.org/10.1109/CSITSS.2018.8768790

16. Medar R, Rajpurohit VS, Shweta S (2019) Crop yield prediction using machine learning techniques. In: 2019 IEEE 5th international conference for convergence in technology (I2CT), pp 1–5

17. Suresh N, Ramesh NVK, Inthiyaz S, Poorna Priya P, Nagasowmika K, Harish Kumar KVN, Shaik M, Reddy BNK (2021) Crop yield prediction using random forest algorithm. In: 2021 7th international conference on advanced computing and communication systems (ICACCS), vol 1, pp 279–282

18. Abishek B, Priyatharshini R, Akash Eswar M, Deepika P (2017) Prediction of effective rainfall and crop water needs using data mining techniques. In: 2017 IEEE technological innovations in ICT for agriculture and rural development (TIAR), pp 231–235

19. Gümüşçü A, Tenekeci ME, Bilgili AV. Estimation of wheat planting date using machine learning algorithms based on available climate data. https://doi.org/10.1016/j.suscom.2019.01.010

20. Kulkarni S, Mandal SN, Srivatsa Sharma G, Mundada MR, Meeradevi (2018) Predictive analysis to improve crop yield using a neural network model. In: 2018 international conference on advances in computing, communications and informatics (ICACCI), pp 74–79

21. Crop-recommendation (2020)

22. Kumar S, Kumar M, Macintyre J, Iliadis L, Maglogiannis I, Jayne C (2019) Predicting customer Churn using artificial neural network engineering applications of neural networks. Springer International Publishing, Cham, pp 299–306

23. Maya Gopal PS, Bhargavi R (2019) Performance evaluation of best feature subsets for crop yield prediction using machine learning algorithms. Appl Artif Intell 33(7):621–642

24. Koklu M, Ozkan IA (2020) Multiclass classification of dry beans using computer vision and machine learning techniques. Comput Electronics Agric 174:105507

25. Benny A (2021) Prediction of the production of crops with respect to rainfall. Environ Res 202:111624

26. Bhanumathi S, Vineeth M, Rohit N (2019) Crop yield prediction and efficient use of fertilizers. In: 2019 international conference on communication and signal processing (ICCSP), pp 0769–0773

27. Khanal S, Fulton J, Klopfenstein A, Douridas N, Shearer S (2018) Integration of high resolution remotely sensed data and machine learning techniques for spatial prediction of soil properties and corn yield. Comput Electronics Agric 153:213–225

28. Chlingaryan A, Sukkarieh S, Whelan B (2018) Machine learning approaches for crop yield prediction and nitrogen status estimation in precision agriculture: a review. Comput Electronics Agric 151:61–69

29. Sani NS, Rahman AHA, Adam A, Shlash I, Aliff M (2020) Ensemble learning for rainfall prediction. Int J Adv Comput Sci Appl 11(11)

30. Gümüşçü A, Tenekeci ME, Bilgili AV (2020) Estimation of wheat planting date using machine learning algorithms based on available climate data. Sustain Comput: Inf Syst 28:100308

Digital Image Transformation Using Outer Totality Cellular Automata

Sandeep Kumar Sharma and Anil Kumar

1 Introduction

Image is a collection of pixels, which are arranged in the number of rows and columns. The area covered by a pixel is to be considered as cell or grid. The pixels of an image are characterized as brightness, sharpness, color and intensity. The image processing fields, i.e., machine learning, robotics, medical science and artificial intelligence have large number of application, which are to be implemented by analyzing features or characteristics of images [1].

The medical science images, i.e., MRI, CT scan, X-ray, etc., are to be enhanced by processing of image features. The high quality of medical images may help doctors to proper treatment patient disease [2, 3]. The AI, ML and robotics applications, i.e., objects tracking, motion analyzing, automatic monitor a machine, automatic drive the car, pattern recognition, face detection, etc., are to be implemented with the help of high-quality images and accurate methods [4].

The Outer Totality Cellular Automata (OTCA) is a two-dimensional standard approach of cellular automata technology which investigates all the outer neighbor pixels for each pixel of the image [5]. The OTCA method studies features of image pixels and generates a unique function, which can transform the image. The OTCA method is non-deterministic finite automata (NDFA)-based method, which analyzes all the outer neighbors for each pixel of the image. The primary challenge for the OTCA method is to select an appropriate state from the multiple available states. To accomplish the challenge, the OTCA method defines a unique rule as per the number

S. K. Sharma (✉)
Manipal University Jaipur, Jaipur, India
e-mail: sandeepsharma@poornima.org

A. Kumar
Poornima Institute of Engineering and Technology, Jaipur, India
e-mail: anilkumar@poornima.org

© The Author(s), under exclusive license to Springer Nature Singapore Pte Ltd. 2023   851
D. S. Sisodia et al. (eds.), *Machine Intelligence Techniques for Data Analysis and Signal Processing*, Lecture Notes in Electrical Engineering 997,
https://doi.org/10.1007/978-981-99-0085-5_69

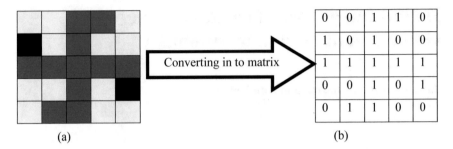

(a)   (b)

Fig. 1 Conversion source data into matrix form

of neighborhoods. The following steps are followed by the OTCA for generating new state of the current state.

Step 1. Analysis neighbors of the image cells.
Step 2. Count live neighbors and create neighbor matrix.
Step 3. Map rule for deciding minima and maxima.
Step 4. Apply rule on the neighbor matrix.
Step 5. Generate new state by examining current state (value of a cell in neighbor matrix); the new state may be 1 or 0.

Consider a window of an input image as depicted in Fig. 1a, and equivalent matrix is depicted in Fig. 1b. The matrix of the source image contains 1 and 0, which represent intensity value of the image pixels [6].

After converting image into matrix form, apply the counter, which can count the neighborhoods with value 1 and create neighbor matrix. Equation 1 could be used to finalize the neighbor matrix. The $Cell_{i,j}$ is the pixel of an input image, whose neighbor matrix $NP_{i,j}$ is creating [7].

$$NM_{i,j} = \bigcup_{\substack{p=i-1 \\ q=j-1}}^{\substack{p=i+1 \\ q=j+1}} Cell_{p,q}. \tag{1}$$

Figure 2 depicts neighbor matrix of the corresponding image. The next step of the OTCA is to map a rule according to the number of neighborhoods. The rule should be applied on neighbor matrix and decide next state, which could be 1 or 0. Figure 2a, b represent the neighbor and output matrices [8]. The OTCA method is able to process any type of color image, i.e., RGB, CMY, gray, etc., and the image should be of any format, i.e., jpeg, bmp, png, etc. This method could be used with the existing methods to increase their efficiency. This method is able to enhance quality of an image and could be used to convert one color space to another color space by calculating a unique threshold value [9].

The next sections of the research paper depict various methods proposed by OTCA methods. The section threshold frequency depicts to calculate a unique threshold

1	1	3	2	1
2	3	3	3	2
3	3	5	3	3
1	3	3	3	2
1	2	3	1	1

0	0	1	0	0
0	1	1	1	0
1	1		1	1
0	1	1	1	0
0	0	1	0	0

(a)   (b)

Fig. 2 Neighbor and output matrices

frequency; section vitality function depicts a method, which could be used to enhance quality of an image; and section luminosity function could be used to convert one color space to another color space.

2 Threshold Frequency

The threshold is a fixed number or a range of numbers which works as transition function of the cellular automata technology. The threshold function is used to change the intensity of a pixel [10]. The threshold function is calculated by counting nearby live neighbors for each cell of a group. Consider a group $G(M, N)$ of M rows and N columns and $Cell_{i,j}$ is the grid or cell of a group G. The threshold frequency of the cell is finalized according to the following ways:

Four-cell threshold frequency (Eq. 2) is finalized by counting left, right, upper and down neighbors.

$$TF_4 = LCell_{p,q-1} + RCell_{p,q+1} + UCell_{p-1,q} + DCell_{p+1,q}. \tag{2}$$

Five-cell threshold frequency (Eq. 3) is finalized by counting left, right, upper, down neighbors and the cell itself.

$$TF_5 = LCell_{p,q-1} + RCell_{p,q+1} + CCell_{p,q} + UCell_{p-1,q} + DCell_{p+1,q}. \tag{3}$$

Eight-cell threshold frequency (Eq. 4) is finalized by counting left, right, upper, down and diagonal neighbors.

$$TF_8 = \bigcup_{\substack{p=i-1\\q=j-1}}^{\substack{p=i+1\\q=j+1}} Cell_{p,q} \text{ if } p \neq q. \tag{4}$$

Nine-cell threshold frequency (Eq. 5) is finalized by counting left, right, upper, down, diagonal neighbors and the cell itself.

$$\text{TF}_9 = \bigcup_{\substack{p=i-1 \\ q=j-1}}^{\substack{p=i+1 \\ q=j+1}} \text{Cell}_{p,q}. \tag{5}$$

The threshold frequency could be finalized by counting neighbors at various radii or levels. The radius is to be considered 1 for all outer adjacent neighbors, radius 2 for next outer neighbors of adjacent neighbors and so on. The radius 1 threshold frequency is equal to eight-cell threshold frequency, which contains maximum eight neighbors, radius 2 threshold frequency contains maximum 16 neighbors, radius 3 threshold frequency contains maximum 24 neighbors and so on. Therefore, the radius r threshold frequency is finalized according to Eq. (6) by taking one extra variable whose initial value is 1.

$$\text{TF}_r = \bigcup_{\substack{p=r-s \\ q=r-s}}^{\substack{p=r+s \\ q=r+s}} \text{Cell}_{p,q} \text{ if } p \neq q, \quad \text{where } s = s + 2. \tag{6}$$

The threshold or transition function of an image is estimated with the help of threshold frequency of the cells. The threshold frequency method could be used with other methods, which can increase efficiency of that method. The experimental results show effect of the threshold frequency method.

3 Luminosity Method

The luminosity method could be used to transform an image from one color space to another color space. The luminosity method is implemented with the help of proposed OTCA method, which takes any type of color image and follows the following steps to convert the image into other type of color image.

Step 1.   Extract intensity of the pixels.
Step 2.   Create a matrix using pixels intensity.
Step 3.   Apply OTCA threshold frequency method.
Step 4.   Generate targeted color image.

Figure 3 depicts an example to convert an RGB color to gray image. The RGB components are extracted from the image, create an intensity matrix and apply OTCA method to convert into gray image.

Equation 7 is used to extract all the three colors from the pixel $\text{PIX}_{i,j}$, and the intensity matrix $\text{IN\_MAT}_{i,j}$ is to be created by applying the equation on all the pixels

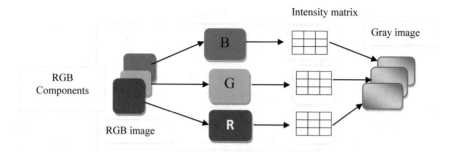

Fig. 3 Luminosity method for RGB to gray conversion

of the image.

$$\begin{aligned} \text{IN}_{\text{MAT}_{i,j}} = \ & \text{PIX}_{i,j}(255, 0, 0) * 0.3 \\ & + \text{PIX}_{i,j}(0, 255, 0) * 0.59 \\ & + \text{PIX}_{i,j}(0, 0, 255) * 0.11. \end{aligned} \tag{7}$$

Here, the values 0 and 255 represent corresponding color value for the 8 bits pixel. The RGB color image components have values from 0 to 255, and the value is known as weight or intensity of the particular component. The RGB color image could be converted to other formats with the help of weight of the colors.

4 Vitality Function

The image processing is playing an important role in recent. The source image of any research may suffer from various types of noise or blurs. So, the source image needs to be smoothed by removing blurs [12, 13]. The vitality function helps for smoothing the source image. The vitality function works on the concepts of low-pass filter, in which corresponding cells' average intensity is measured and all the neighborhood pixels whose intensity is smaller than average intensity are replaced by average intensity value [14]. The step-by-step working of vitality function is as follows:

Step 1. Read an image.
Step 2. Select a group of 3 * 3 pixels and add outer rows and columns with same nearby values to convert into 5 * 5.
Step 3. Find average of each 3 * 3 group using convolution operator and store into another 3 * 3 group.
Step 4. Repeat the step 3 for each group and store value and store values in average group.
Step 5. Compare average group with original group and all the smaller values.

The process of vitality function is explained with the help of example in Fig. 4 [15]. The example takes a region of the size 3 * 3 from source image, and the region is extended by adding virtual row and column the outer of original region with values of region outer cells. The corresponding average region is estimated by multiplying average operator with equivalent region of the extended region. The 3 * 3 region from the extended region should be selected in the way; the cells of original source region should be included one by one.

The below example shows the process to calculate average region by selecting cell similar to first cell (0, 0) marked with yellow color of the source region [16, 17].

$$\begin{aligned}\text{Average Region } (0,0) &= (10+10+10+10+10+10+20+20+20)/9 \\ &= 120/9 = 13\end{aligned}$$

The enhanced region is created by comparing average region and original region. The original region values, which are smaller, are replaced with consequent average region values. The above-defined process is applied on all the cells of source image and produce smoothed image.

10	10	10
20	20	20
30	30	30

a. Source Region

10	10	10	10	10
10	10	10	10	10
20	20	20	20	20
30	30	30	30	30
30	30	30	30	30

Virtual Region

b. Extended Region

$\frac{1}{9}$

1	1	1
1	1	1
1	1	1

c. Average Operator

13	13	13
20	20	20
26	26	26

d. Average Region

13	13	13
20	20	20
30	30	30

e. Enhanced Region

Fig. 4 Working of vitality function

5 Experimental Results

The methods of proposed research are experimented on number of images and applied with number of existing methods. The produced results of the proposed research are compared with the results of existing methods. Various parameters, i.e., Mean Square Error (MSE), Peak Signal Noise Ratio (PSNR), estimated time and accuracy are used to compare corresponding results. Table 1 depicts comparison of MSE values with classical methods, i.e., Sobel, Robert, Prewitt and Canny.

The table depicts results of only five test images, but the proposed method is experimented on 1000 of training and test images. The proposed methods claim that the MSE value of the proposed method is comparatively low; hence, the processed image contains constant edges. Table 2 depicts PSNR comparison of proposed method with existing classical methods, i.e., Robert, Canny, Prewitt and Laplacian. The PSNR value of the corresponding images asserts similarity of the images. The higher value of PSNR depicts that images are closed to each other.

Table 2 depicts results of only five test images, but the proposed method is experimented on 1000 of training and test images. The proposed methods claim that the PSNR value of the proposed method is comparatively high; hence, the processed image contains true edges.

Table 1 MSE analysis of proposed method compared with results of classical methods

Source images	Classical methods' results				Proposed method
	Sobel	Prewitt	Robert	Canny	
SR_IM1	0.18170	0.17454	0.16308	0.18566	0.16089
SR_IM2	0.17160	0.16660	0.15547	0.15715	0.15889
SR_IM3	0.35069	0.35032	0.35132	0.35570	0.32345
SR_IM4	0.16604	0.14276	0.91601	0.94690	0.12343
SR_IM5	0.25594	0.25896	0.27200	0.28200	0.26783

Table 2 PSNR analysis of proposed method compared with results of classical methods

Source images	Classical methods' results				Proposed method
	Robert	Prewitt	Canny	Laplacian	
SR_IM1	5.537	5.711	6.400	5.443	6.3929
SR_IM2	5.785	5.913	6.214	6.167	6.1456
SR_IM3	2.681	2.686	2.673	2.619	2.7453
SR_IM4	5.928	6.584	8.511	8.367	8.5464
SR_IM5	4.040	3.990	3.780	3.620	3.7024

6 Conclusion and Future Scope

The proposed research presented some common methods, which should be used with existing image processing methods. The common methods are directly used to transform the color of an image and enhance quality of the image. The proposed methods are used with popular existing research and produce improved results. The proposed threshold frequency is a unique method, which should be used to perform various image processing applications, i.e., image segmentation, pattern recognition, finger print detection, iris detection, etc., and able to produce efficient results.

The future scope of the proposed research is to implement machine learning applications, which are beneficial to the society.

References

1. Huang J-J, Siu W-C, Liu T-R (2015) Fast image interpolation via random forests. IEEE Trans Image Process 24(10)
2. Agaian SS, Panetta K, Grigoryan AM (2001) Transform-based image enhancement algorithms with performance measure. IEEE Trans Image Process 10(3)
3. Yue H, Sun X, Yang J, Wu F (2015) Image denoising by exploring external and internal correlations. IEEE Trans Image Process 24(6)
4. Sinha K, Sinha GR (2014) Efficient segmentation methods for tumor detection in MRI images. In: IEEE Conference on electrical, electronics and computer science
5. Hamamci A, Kucuk N, Karaman K, Engin K, Unal G (2012) Tumor-cut: segmentation of brain tumors on contrast enhanced MR images for radio surgery applications. IEEE Trans Med Imaging 31(3)
6. Wang T, Cheng I, Basu A (2009) Fluid vector flow and applications in brain tumor segmentation. IEEE Trans Biomed Eng 56(3)
7. Bartunek JS, Nilsson M, Sällberg B, Claesson I (2013) Adaptive fingerprint image enhancement with emphasis on pre-processing of data. IEEE Trans Image Process 22(2)
8. Zhu Q, Mai J, Shao L (2015) A fast single image haze removal algorithm using color attenuation prior. IEEE Trans Image Process 24(11):1057–7149, 3522–3533
9. Jin KH, McCann MT, Froustey E, Unser M (2017) Deep convolution neural network for inverse problems in imaging. IEEE Trans Image Process 26(9):1057–7149, 4509–4522
10. Ancuti CO, Ancuti C, De Vleeschouwer C, Bekaert P (2018) Color balance and fusion for underwater image enhancement. IEEE Trans Image Process 27(1):379–393
11. Zhang Y, Fan Q, Bao F, Liu Y, Zhang C (2018) Single-image super-resolution based on rational fractal interpolation. IEEE Trans Image Process 27(8):3782–3797
12. Talebi H, Milanfar P (2018) NIMA: neural image assessment. IEEE Trans Image Process 27(8):3998–4011
13. Sharma SK, Lamba CS, Rathore VS (2017) Radius based cellular automata approach for image processing applications. IEEE-40222, 8th ICCCNT 2017 IIT Delhi, July 2017
14. Borji A, Cheng M-M, Hou Q, Jiang H, Li J (2019) Salient object detection: a survey 5(2):117–150
15. Sharma SK, Lamba CS, Rathore VS (2017) OTCA approach towards blurred image feature estimation and enrichment. AISC 625:329–338
16. Sharma SK, Lamba CS, Rathore VS (2018) Incessant ridge estimation using RBCA model. AISC 841:203–210
17. Chrysos GG, Favaro P, Zafeiriou S (2019) Motion de-blurring of faces. Int J Comput Vis 127:801–823. https://doi.org/10.1007/s11263-018-1138-7

Community Detection Using Context-Aware Deep Learning Model

Ankur Dahiya and Pradeep Kumar

1 Introduction

In social media websites, users partake in a variety of social activities like sharing their thoughts, likings, knowledge, opinions with people with similar interests. Discovering communities from such a plethora of information is a challenging task. Community is a group of users who have something in common, which can range from belonging to the same geographical location, or people having the same likings or interests. Explicitly mentioned properties are relatively easy to find and can be generally found by using graph theoretical approaches, for example, hierarchical clustering [1], modularity-based approach [2, 3], Spectral Clustering [4], Statistical interference [5]. However, due to the large population using social media, eccentricity in human behavior, it is exceedingly difficult to analyze individual behavior hidden in interactions that people do on these social media platforms; hence, this area is even challenging to interpret but is capable of offering assistance not only to enrich the literature but also find many useful managerial implications. These applications vary across fields, for example, marketing, electoral politics, polarization analytics and fashion. Identifying people with similar interests facilitates conversations, generates content and also keeps social networks attractive. Insights from these communities and user behavior. A group or community can be contemplated as an array of users who "talks about" a subject more often than others.

There can be two ways of identifying these interest-based communities, one of which focusses on the user and focusses on interactions specific to these users, whereas the other method can be item or interest centric, where the interest-based

A. Dahiya (✉) · P. Kumar
Indian Institute of Management, IIM Road, Prabandh Nagar, Lucknow, Uttar Pradesh 226013, India
e-mail: fpm16009@iiml.ac.in

P. Kumar
e-mail: pradeepkumar@iiml.ac.in

© The Author(s), under exclusive license to Springer Nature Singapore Pte Ltd. 2023
D. S. Sisodia et al. (eds.), *Machine Intelligence Techniques for Data Analysis and Signal Processing*, Lecture Notes in Electrical Engineering 997,
https://doi.org/10.1007/978-981-99-0085-5_70

content generated in forms of say tweets, blogs, pictures, etc., can be considered. First approach considers a graph structure approach based on nodes (people) and edges (interactions), whereas the second considers user-generated content in different forms like email, blogs, posts, tweets which characterize their interest analyzed to identify interest-based communities. This can be done by using different content-based supervised or unsupervised learning techniques. This paper uses Convolution Neural Networks with two-word embedding techniques, one of which is famous non-Context-aware word embedding (Word2Vec) and other is state-of-the-art Context-aware technique (ELMo) to model users' interest based on published content.

The contributions of this paper are:

This paper proposes deep learning-based methodology to identify interest-based communities.

We compare and benchmark non-Context-aware (Word2Vec) and Context-aware word embedding (ELMo) techniques with CNN, with respect to interest-based community detection.

The paper is structured in five sections. In the next section, we present literature review related to interest-based community detection; further to this section, we discuss architecture of deep learning model, then we explain datasets and discuss about experimental setup, then we showcase results, and finally, we present conclusion and future scope.

2 Literature Review

A community is a group of people living in the same defined area sharing the same basic values, organization and interests [6]. There can be communities within or outside such spatial and structural communities which are based on other non-place groupings such as profession, religion, interest and some other social bonds [7, 8].

Recent definition being used in research comes from the Computer Science domain which defines community as a group of nodes, in which intra-group connections are much denser than inter-group [8–10]. This definition uses the graph theoretical approach but does not cover semantics in the interactions that these people do on social networks. These interactions also form a community, say for example based on liking, writing blogs, or based on social interactions [11].

There are largely two types of approaches for community detection: graph structure based (topological) or the one which takes the textual information into account (topic-based community detection). Topology-based community detection, researchers visualize them to replicate the real-world using graphs, where vertices denote the real-world unit and edges denote the relationships or interactions between these units or entities. Topic-based community detection researchers believe that similarity of two entities can be measured by the number of words they share in common. There are many different approaches that the researchers have explored, for example, classification, clustering but not many have explored deep learning-based approaches. Deep learning can help overcome limitations of the traditional approaches, proving it to be a superior approach over traditional community detection approaches. Such advantages may range from encrypting feature representations of data which is high-dimensional [12], and graph structural approaches can only utilize properties and information relating to topology of the graph, mostly based matrices like on adjacency matrix and node matrix [5, 10]. Graphs where there is hardly any knowledge on the kind of communities that might exist, where nodes are mostly unlabeled, deep learning performs better since they can learn from different patterns. Deep learning has presented some compelling representations of such properties and structures relating to communities such as [12–14]. Comprehensive survey presented in [15] and [16] on various ways of detection communities suggests many deep learning approaches ranging from statistical methods to the state-of-the-art deep learning methods; however, none of the methods presented so far have used word embedding techniques. Word embedding techniques are known to represent accurate syntactic and semantic relationships [17]. Some word embedding techniques also have started taking sentence information into consideration, hence making them more context aware.

In this paper, we detect interest-based communities using deep learning algorithms. In order to identify interest-based communities, topic-based approaches have been used by researchers, which include usage of topic modeling or word embedding-based approaches. However, Neural Network models have found pretrained word embeddings to be very useful [18], but the usefulness of these word embeddings has not been utilized with Neural Network models. Word embeddings allow words with similar meaning to have similar representations by capturing both syntactic as well as semantic information with reduced dimensionality. These vector representations can be utilized for comparison of similarity. We demonstrate and compare two categories of such word embeddings, i.e., non-Context-aware word embeddings (Word2Vec, glove, FastText, etc.) and Context-aware word embeddings (ELMo and Bert) in the context of detecting interest-based communities using deep learning. Non-Context-aware word embeddings learn from a big corpus of word and then from frequency of occurrence of word and therefore completely ignore the context; for example, meaning of word bank can be a financial institution or a Riverside; non-Context-aware word embeddings do not consider this meaning aspect of words [19]. Context-aware word embeddings, however, take care of semantics of the sequences occurred [20, 21].

In this paper, we detect interest-based communities using deep learning, and we evaluate and compare these Context-aware ELMo and non-Context-aware word embedding Word2Vec.

3 Deep Learning Model Architecture

In order to achieve better discovery of semantic and syntactic structures, we combine Convolution Neural Networks (CNNs) [22] with two-word embeddings, respectively. A Convolutional Neural Network (CNN) is an artificial Neural Network that

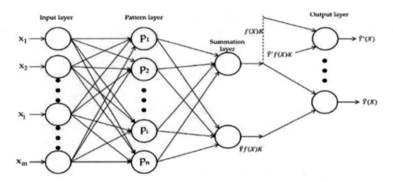

Fig. 1 Deep learning architecture used for evaluation of CNN model using word embedding techniques

is frequently used in various fields such as image classification, face recognition and natural language processing. It uses convolution as a mathematical function instead of matrix multiplication and does not depend on series sequences. CNN has proved to be very important for semantic parsing [23]. Word embedding converts words into numbers and finds some properties related to the words and their neighborhood. There are many ways to do so; hence, many such techniques exist. We analyze non-Context-aware Word2Vec and Context-aware ELMo. Neural Network architecture is given in Fig. 1.

Word2Vec learns from a big corpus of text using the Neural Network model. After training, this can figure out words which are synonymous to each other. Similarity is calculated by cosine similarity between the represented vectors. Word2Vec has successfully been used for NLP-related tasks. Our method of community detection uses a text-based approach; hence, one of the word embedding techniques that have been evaluated is Word2Vec.

ELMo is a Context-aware embedding technique. ELMo uses a bidirectional language model and uses the entire sentence as opposed to the word as explained in Word2Vec. Hence, this method also is capable of capturing the context of the word. Since it uses sentences instead of words, it is also capable of capturing the context of one word in different sentences.

4 Data Set and Experimental Setup

We used two datasets in our experiments. After preprocessing of the data, the dataset was split into train, validation and test sets in a stratified fashion. The training set was assigned 70% data, the validation set 30% and the test set 30%.

The first one has 30,000 tweets from ten categories: Bollywood, Machine Learning, Big Boss, Food, Football, Hollywood, Politics, Cricket, Happy Birthday

Table 1 Number of tweets in each category of the dataset 1

S. No.	Label	Number of tweets
1	Bollywood	5000
2	Machine Learning	217
3	Big boss	4509
4	Food	2797
5	Football	5000
6	Cricket	5000
7	Happy Birthday	1363
8	Hollywood	5000
9	Politics	805
10	Mobiles	309

and Mobiles. This is real-time data collected using Twitter API and the hashtags given.

After preprocessing of the data and removal of duplicate tweets, the data were used for the experimentation (Table 1).

The second dataset contains 317,836 tweets from ten categories, and we mixed tweets from all these categories to form a bigger dataset. All these tweets come from standard datasets with a particular topic which makes it easy to understand the accuracy of the algorithms being tested (Table 2).

a. **Data Preprocessing**

NLTK represents a natural language toolkit used for all the preprocessing tasks such as stemming, POS, tokenization, lemmatizing etc. We checked and removed duplicates before and after the preprocessing. All HTML entities were captured, # and @ characters were replaced with "hashtag" and "mention," respectively. All the stop words, punctuation, Unicode were removed. Pascal case and camel case

Table 2 Number of tweets in each category of the dataset 2

S. No.	Topic	References
1	Python	[24]
2	Cricket	[25]
3	Mothers' Day	[26]
4	Covid19	[27]
5	ISIS	[28]
6	Machine Learning	[29]
7	Health	[30]
8	Bitcoin	[31]
9	Politics	[32]
10	Avengers Endgame	[33]

words were separated using regex (e.g., Pascal Case was converted to Pascal case and Camel Case was converted to Camel Case) and words appearing more than twice consecutively were removed. All the words were converted into lower case.

b. Evaluation Metrics

To be able to evaluate if the tweets have been mapped to respective communities, or in other words to check accuracy of the algorithm, we used widely acceptable matrices like precision, recall and F_1-score. The confusion matrix provides a clearer picture not only of the performance of the prediction model, but also of what categories are predicted correctly and incorrectly, and what kind of errors are made. Metrics for four categories are calculated (TP, FP, FN and TN) using confusion matrix.

Accuracy is a standard evaluation metric for classification. It is the number of appropriate predictions made as a ratio of all predictions made.

Precision is the ratio between the True Positives and all the positives cases. Precision is a good way to decide, when False Positive costs are high.

$$\text{Precision} = \text{True Positives}/(\text{True Positives} + \text{False Positives}) \qquad (1)$$

Recall (Sensitivity) is a metric that measures the number of correct predictions made by all the positive predictions that can be made. Recall is the ratio of True Positives to all the positives in the dataset.

$$\text{Recall} = \text{True Positives}/(\text{True Positives} + \text{False Negative}) \qquad (2)$$

The F_1-score is a harmonic mean of recall and precision, where the best value of F_1-score is 1 and worst is 0. To reach desirable balance between precision and recall, F_1-score is required. The F_1-score might be the best way to use if we need to look for a balance between precision and recall, and there is an unequal distribution of the category.

$$F_1\text{-score} = (2 * \text{Precision} * \text{Recall})/(\text{Precision} + \text{Recall}) \qquad (3)$$

5   Experimental Results

In this section, we provide model and dataset-wise numbers for precision, recall and F_1-score. We compare which model gives better segregation among communities using these measures.

The scores for Convolution Neural Networks with Word2vc and ELMo with respect to detecting interest-based communities are given below. Value of precision, recall and F_1-score with respect to Word2Vec is significantly lower than that of ELMo, when combined with CNN. The results of both the datasets are coherent and indicate that the performance of ELMo is better than Word2Vec (Table 3).

Table 3 Experimental results

Model-CNN	Dataset 1			Dataset 2		
Pretrained embedding	Precision	Recall	F_1-score	Precision	Recall	F_1-score
Word2vec	0.76	0.73	0.74	0.72	0.83	0.77
ELMo	0.84	0.85	0.85	0.94	0.90	0.92

6 Conclusion and Future Scope

Natural language processing is a widely used and high in demand field of Artificial Intelligence. This paper used NLP-based approach to identify interest-based communities. Literature suggests a few supervised and unsupervised learning techniques to detect interest-based communities; however, there is a dearth of literature in community detection approaches using deep learning techniques. This paper uses Convolution Neural Networks (CNNs) to identify such communities. Languages can be very complex to understand and interpret. For example, there can be two meanings to a word, or two words might have a similar meaning. We use word embedding techniques, which are trained on a large corpus of data to suggest similar words. We chose two different type of techniques, non-Context-aware Word2Vec and Context-aware ELMo, and used them with CNN to detect interest-based communities.

This study suggests that the values of precision, recall and F_1-score are higher for ELMo-based CNN as compared to Word2Vec-based CNN ELMo which has performed better on both the datasets.

We have not explored other deep learning algorithms like LSTM, RNN. As a part of further study, which deep learning algorithms performs better in case of detecting interest-based communities, should be explored. We have only compared Word2Vec and ELMo, as a part of this work. Non-Context-aware techniques like Glove, FastText and Context-aware techniques like BERT should also be studied. This study is limited to comparing two deep learning algorithms; however, clustering-based algorithms should be studied in order to conclude which category of algorithms performs better in context of interest-based community detection.

Identifying interest-based influencers can be another important next step to this work, Social Commerce platforms and Word of Mouth campaigns would benefit if these can be mapped.

There can be many useful managerial implications of interest-based community detection; for example, organizations and marketers can use it for targeting marketing, governments can use it to say curb polarization, people with specific interests can be addressed at once and peers can learn from each other, can do upselling and cross-selling, to name a few.

References

1. Aggarwal CC, Wang H (2010) Managing and mining graph data. Adv Database Syst 40:275–301
2. Newman M, Girvan M (2004) Finding and evaluating community structure in networks. Phys Rev E 69(2):26113
3. Zhang P, Moore C (2014) Scalable detection of statistically significant communities and hierarchies, using message passing for modularity. Proc Natl Acad Sci 111(51):18144–18149
4. Fanuel M, Alaíz CM, Suykens JAK (2017) Magnetic eigenmaps for community detection in directed networks. Phys Rev E 95(2):022302
5. He D, Liu D, Jin D, Zhang W (2015) A stochastic model for detecting heterogeneous link communities in complex networks. Proc AAAI, pp 130–136
6. Rifkin SB (1996) Paradigms lost: toward a new understanding of community participation in health programmes. Acta Trop 61(2):79–92
7. Blondel VD, Guillaume JL, Lambiotte R, Lefebvre E (2008) Fast unfolding of communities in large networks. J Stat Mech Theory Exp 2008(10):P10008
8. Dhumal A, Kamde P (2015) Survey on community detection in online social networks. Int J Comput Appl 121(9)
9. Gurini DF, Gasparetti F, Micarelli A, Sansonetti G (2014) iSCUR: interest and sentiment-based community detection for user recommendation on Twitter. In: International conference on user modeling, adaptation, and personalization. Springer, Cham, pp 314–319
10. Yang J, McAuley J, Leskovec J (2013) Community detection in networks with node attributes. In: ICDM
11. Zafarani R, Abbasi MA, Liu H (2014) Social media mining: an introduction. Cambridge University Press
12. Zhang X (2018) Multilayer bootstrap networks. Neural Netw 103:29–43
13. Xie Y, Gong M, Wang S, Yu B (2018) Community discovery in networks with deep sparse filtering. Pattern Recogn 81:50–59
14. Sun F-Y, Qu M, Hoffmann J, Huang C, Tang J (2019) vGraph: a generative model for joint community detection and node representation learning. In: NIPS, pp 512–522
15. Jin D, Yu Z, Jiao P, Pan S, Yu PS, Zhang W (2021) A survey of community detection approaches: from statistical modeling to deep learning. arXiv preprint arXiv:2101.01669
16. Su X, Xue S, Liu F, Wu J, Yang J, Zhou C et al (2021) A comprehensive survey on community detection with deep learning. arXiv preprint arXiv:2105.12584
17. Mikolov T, Yih WT, Zweig G (2013) Linguistic regularities in continuous space word representations. In: Proceedings of the 2013 conference of the North American chapter of the association for computational linguistics: human language technologies, pp 746–751
18. Ma X, Hovy E (2016) End-to-end sequence labeling via bi-directional LSTM-CNNs-CRF. arXiv preprint arXiv:1603.01354
19. Mikolov T, Chen K, Corrado G, Dean J (2013) Efficient estimation of word representations in vector space. arXiv preprint arXiv:13013781
20. Gardner M, Grus J, Neumann M, Tafjord O, Dasigi P, Liu N et al (2018) Allennlp: a deep semantic natural language processing platform. arXiv preprint arXiv:1803.07640
21. Devlin J, Chang MW, Lee K, Toutanova K (2018) Bert: pre-training of deep bidirectional transformers for language understanding. arXiv preprint arXiv:1810.04805
22. LeCun Y, Bengio Y et al (1995) Convolutional networks for images, speech, and time series. In: The handbook of brain theory and neural networks, vol 3361, no 10
23. Grefenstette E, Blunsom P, De Freitas N, Hermann KM (2014) A deep architecture for semantic parsing. arXiv preprint arXiv:1404.7296
24. Dabbas E (2018) Twitter in a DataFrame. Kaggle, 10 Oct 2018. www.kaggle.com/eliasdabbas/twitter_dataframe
25. Sharma K (2020) Tweetss. Kaggle, 10 Feb 2020. www.kaggle.com/kunalsharma01/tweetss
26. Prakash S (2019) Mothers day tweets. Kaggle, 12 May 2019. www.kaggle.com/sprakash08/mothers-day-tweets

27. Preda G (2020) COVID19 tweets. Kaggle, 30 Aug 2020. www.kaggle.com/gpreda/covid19-tweets
28. Aaied A (2020) ISIS Twitter. Kaggle, 24 Dec 2020. www.kaggle.com/aliaaied/isis-twitter
29. Schweder JA (2018) #MachineLearning tweets. Kaggle, 11 Sept 2018. www.kaggle.com/jaschweder/machinelearning-tweets
30. Singh P (2020) Health related tweets. Kaggle, 16 May 2020. www.kaggle.com/prabhavsingh/health-related-tweets
31. Kash (2021) Bitcoin tweets. Kaggle, 30 May 2021. www.kaggle.com/kaushiksuresh147/bitcoin-tweets
32. Wallach D (2018) Financial tweets. Kaggle, 9 Aug 2018. www.kaggle.com/davidwallach/financial-tweets
33. Lolayekar K (2019) Twitter dataset—#AvengersEndgame. Kaggle, 23 Apr 2019. www.kaggle.com/kavita5/twitter-dataset-avengersendgame. The data was taken from Kaggle datasets. After preprocessing of the data and removal of duplicate tweets, 271,584 data points were found to be unique and used for the experiment

Discriminative Feature Construction Using Multi-labeling Approach for Automatic Speech Emotion Recognition

Md. Shah Fahad, Raushan Raj, Ashish Ranjan, and Akshay Deepak

1 Introduction

The classification/identification of a person's emotional state using the speech signal is referred to as "*speech emotion recognition*" (SER). SER offers a wide range of real-world applications. It is useful in applications that demand natural human–computer interaction [1]. The user's sensed emotion aids the system in replying to the user's query in computer tutorial programs [2]. It can be used as a diagnostic tool in psychiatric treatment by medical practitioners [3]. It can aid in the successful communication of emotions between two people in automatic translation systems, according to [4].

Different emotion classes can be depicted using three attributes: arousal (attentiveness), valence (positivity), and dominance (power) on a three-dimensional plane. Hierarchical classification in two-stage and three-stage classifiers is organized using these attributes [5]. Our psychological conduct has prompted us to choose this strategy. In a hierarchical approach, classifiers are constructed like a tree: Emotions are sorted into broad categories at the root node, and the number of candidate classes decreases as you travel deeper down the tree on the basis of these attributes. The

Md. S. Fahad (✉)
Vellore Institute of Technology, Bhopal, India
e-mail: md.shah@vitbhopal.ac.in

R. Raj · A. Ranjan · A. Deepak
National Institute of Technology Patna, Patna, India
e-mail: raushanr.pg20.cs@nitp.ac.in

A. Ranjan
e-mail: ashish.cse16@nitp.ac.in

A. Deepak
e-mail: akshayd@nitp.ac.in

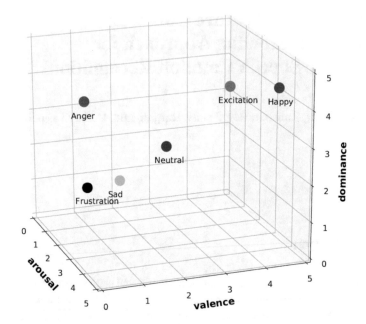

Fig. 1 Three-dimensional plot of anger, excitation, happy, frustration, neutral, and sad emotion using the IEMOCAP database

selection is made for a single class only at the leaf node. While hierarchical systems have higher computational complexity, the increased accuracy is worth it.

Multi-tasking is a technique that enhances generalization by learning shared representation contained in the training signals of related tasks [6]. Such multi-tasking has been applied in many of the problems associated with speech and natural language processing [7, 8]. The main task classifier can be better trained with the secondary-related task. In [9], the authors claimed that various auxiliary tasks, such as phone label, phone, and state context, help to learn the main task (phone-state recognition). They showed better results for phone-state recognition. Facial verification and facial landmark prediction are used as auxiliary task for facial emotion recognition [10]. There are some works in speech emotion recognition that have also employed multi-tasking. In [11], gender and naturalness prediction is used as auxiliary task with the main task of emotion prediction. In [7], each auxiliary task (arousal and valence) is classified into low, mid, and high categories and simultaneously predicted with the main task of emotion prediction. This work used OpenSmile features [12] for training a deep belief neural network.

In Fig. 1, 3D representation (arousal, valence, and dominance) of discrete emotion is plotted. As illustrated in Fig. 1, anger and excitation emotion come under the same arousal and dominance area but different valence area. So, it can be said that anger and excitation have some commonality and some difference. The difference and commonality between anger and excitation emotion can be well established by treating arousal, valence, and dominance as an auxiliary task with the main task of emotion prediction. This work is different from the above works in the following respective: The objective function is changed to multi-label instead of multi-tasking. In this work, discrete emotion classification is used as the main task, and the arousal, valence, and dominance classes are treated as the auxiliary task. The auxiliary task helps to discriminate the primary task and produce discriminative features for emotion prediction.

Following the motivation from the above discussion, the major contributions are as follows:

- The main contribution is to utilize the intrinsic correlations among different emotions with arousal, valance, and dominance attributes using multi-label approach to enhance the performance of speech emotion recognition (SER) system.
- We proposed a deep multi-label framework using convolutional neural network (CNN) to construct feature vectors. These feature vectors are fed as input to the deep neural network for emotion classification. The proposed multi-label framework achieved an improvement of $+3.0\%$ in the unweighted accuracy (UWA) when compared with the single-label framework.

The paper organization is as follows: Sect. 2 introduces the proposed framework. Next, Sect. 3 is a result and discussion. Lastly, Section 4 is a conclusion.

2 Proposed Framework

The proposed framework benefits from the joint learning of discrete categorical and continuous emotion information for the discriminative feature construction. The proposed step involves first utilizing the discrete categorical and continuous information for the formulation of a 13-class problem as discussed in Sect. 2.1. Then, a CNN-based architecture is trained for the 13-class problem, and 128-dimensional features are extracted from an intermediate layer of CNN (discussed in Sect. 2.2). An extra DNN model is constructed using the 128-dimensional features that are discussed in Sect. 2.3.

2.1 Emotion Classification as a Multi-label Classification Problem

Speech emotion datasets, frequently, are composed of emotion labels that denote either the discrete emotional state or two/more fundamental speech emotion dimensions. These popular emotion label schemes are discussed next:

1. **Discrete categorical label**: This is a widely used scheme to denote the emotional state category and denote only the discrete state of person's mood. Common examples include *neutral, anger, sad*, etc.
2. **Continuous dimensional label**: Under this scheme, each emotion state is denoted as a coordinate in a three-dimensional space. These dimensions correspond to three fundamental emotion attributes, namely (i) *valence* (negative vs. positive), (ii) *arousal* (excited vs. calmed), and (iii) *dominance* (strong vs. calmed). For example, anger emotion has following fundamental speech emotion dimensions, *valence*: negative, *arousal*: excited, and *dominance*: strong.

For this work, both discrete and continuous fundamental speech emotion dimensions are used for emotion classification. Each continuous valued fundamental dimensions (*arousal, valence*, and *dominance*) is categorized into discrete values as low, mid, and high. The range of low, mid, and high is taken as [1–2], (2–3.5], and (3.5–5], respectively.

Multi-label emotion classification problem formulation: Let $D_s = \{(X_1, Y_1), (X_2, Y_2), \ldots, (X_N, Y_N)\}$ denote a dataset of N labeled speech utterances, where X_i denotes an utterance and $\{Y_i = y_{i_1}, y_{i_2}, \ldots, y_{i_k}\}$ denotes 13-dimension one-hot vector representing the output label-set for X_i; such that $y_{i_k} = 1$ if X_i exhibits y_{i_k}, else 0. The first four labels are emotions: *anger, excitation, neutral*, and *sad*. The next three labels are *arousal*: A_{low}, A_{mid}, and A_{high}. The subsequent three labels are *valence*: V_{low}, V_{mid}, and V_{high}. The last three labels are *dominance*: D_{low}, D_{mid}, and D_{high}.

2.2 Discriminative Feature Construction

The steps for constructing the discriminative features are as follows:

1. Speech signals are transformed into spectrogram-based representation.
2. Next, generated spectrograms are fed to train the proposed deep CNN-based architecture to solve the 13-class problem (shown in Fig. 2). The proposed architecture is composed of (i) stack of CNN blocks, (ii) global average pooling layer, (iii) dense layer, and (iv) output layer.
3. The speech features are then extracted from the global average pooling layer.

(1) **Log-Mel Spectrogram-based Speech Representation**: For this work, speech signals transformed into *"Mel filter bank spectrograms"* are used as a input. *Mel filter bank spectrogram* is very popular in deep learning framework [13] and has following steps:

1. The pre-emphasized filters are applied on the raw speech signal to boost the components for high frequencies.
2. The speech signal is, next, split to produce frames of size 25 ms in a manner such that two adjacent frames share a common region of 10 ms. Then, the *"Hamming window"* function is applied to each frame.

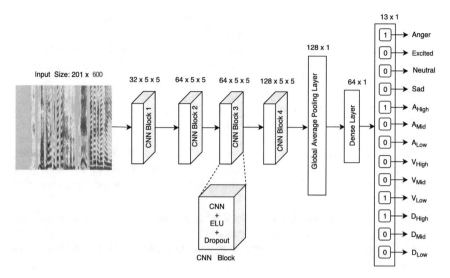

Fig. 2 Deep multi-label architecture for discriminative feature construction

3. The discrete Fourier transform is applied to get the power spectrum with N-point FFT (NFFT) of the length 512.

4. Triangular band-pass filters (40 such filters) on the Mel scale are adopted to get the frequency bands features. Mel scale reflects the "*nonlinear behavior of human auditory system*" that are better at providing higher resolution at lower frequencies. Thus, we get 40 features for each frame.

We have taken 600 frames to fix the input shape to (40 × 600). The utterance having frame less than 600 is padded with zeros. The utterance with more than 600 frames and the frames beyond the 600th frame are removed. Masking layer has been used to normalize the effect of padding.

(2) **Deep CNN architecture**: Deep neural networks, comprising large number of neural network layers (i.e., CNN blocks), are employed to promote multiple levels of feature abstraction. The different layers allow to learn features in a hierarchical manner, where first few layers assist in learning "*low-level*" features (directly utilizing the raw data) and sequent layers performing feature refinement, thus leading to "*high-level*" feature learning down the network. The proposed deep CNN architecture is shown in Fig. 2 that consists of a stack of four CNN blocks for multiple levels of feature abstraction.

CNN block: The CNN block is proposed with following three major neural network layers: (i) a convolution neural network (CNN) layer, (ii) an exponential linear unit (ELU) layer, and (iii) a dropout layer. Figure 2 depicts the CNN block. These are discussed next.

1. Convolution layers with trainable parameters serve as a mapping function to enhance low-level features. Let "$s(a,b)$" denote the input and "w" denote the trainable kernel with convolution layer. The size of kernel is denoted as (x, y) that is initialized randomly. Then, the convolution operation is mathematically defined as

$$O(m, n) = \sum_{a=-x}^{a=+x} \sum_{b=-y}^{b=+y} s(a, b) \cdot w(m - a, n - b) \tag{1}$$

   The size of filters with each CNN block is fixed and taken as (5×5). Further, the number of filters is taken as 32 for first CNN block, 64 for the second and third CNN blocks, and 128 for the final fourth block. Other parameters are also depicted in Fig. 2.

2. The ELU layer is then implemented to improve the network's ability to achieve higher accuracies and convergence rates [14].

3. The dropout layer works as a network regularizer, allowing for generalized learning.

The raw spectrogram is used as the input for the proposed architecture.

Global average pooling: Instead of flattening, the feature maps are reduced by global average pooling [15]. It minimizes the over-fitting by reducing the number of features. It reduces each feature map of $H \times W$ to a single number by taking the average of all $H \times W$ values. After applying global average pooling to each feature map, the size is reduced to $1 \times 1 \times D$, where D is the number of features map. In this work, global average pooling is applied on the last convolution layer.

Dense layer: This layer is composed of 64 neurons with *ReLU* activation function to compress the output from the *global average pooling* layer.

Output layer: The output layer is composed of 13 output neurons, four for discrete emotion classes, and nine for continuous fundamental dimensions as shown in Fig. 2. Each neuron has the *sigmoid* activation function, which is so because, the goal is to predict both the discrete and continuous fundamental dimensions.

Objective function: The multi-label feature construction is formulated as an optimization problem. The goal of this optimization problem is to minimize the average loss of multi-label output. The objective function for the above optimization problem is defined as

$$L(\hat{y}, y) = \frac{-1}{N} \sum_{k \in y} [y_k \log \hat{y}_k + (1 - y_k) \log(1 - y_k)] \tag{2}$$

where

N is the number of classes,
y_k is the actual label of kth class,
$\hat{y}_k$ is a sigmoid function defined as

Input Size : 128 x 1

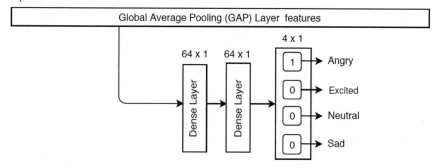

Fig. 3 Global average pooling layer output

$$\hat{y}_k = \frac{1}{1 + e^{-\hat{p}_k}} \tag{3}$$

where $\hat{p}_j$ is the output of last layer of the proposed architecture. The last layer is a deep neural network (DNN) defined as $f(Wd + b)$, where f is the activation function of the layer, W and b are the weights and bias of the layer, respectively, and d is the hidden layer representation of the last layer.

2.3 Speech Emotion Recognition Using Multi-label Features

The 128-dimension features are extracted from the global average pooling layer as shown in Fig. 3. This is done for both train and test datasets. After that, the z-normalization is applied to both the training and testing data. The statistics of z-normalization (mean, variance) are calculated from the neutral emotion of the training data [16]. It normalizes the other factors (speaker and text) and highlights the emotion factor in the feature set. The extracted train features are learned using two-layer DNN as shown in Fig. 3. The number of neurons for both the layers is taken as 64. The output layer now only has four neurons corresponding to four emotions.

3 Result and Discussion

The popular speech emotion dataset, IEMOCAP, comprising five recording sessions —each comprising one male and one female speaker—is used to evaluate the proposed approach. The duration for each individual session is approximately close to five minutes. The dataset contains separate recordings for both scripted and spontaneous cases. In this paper, we have focused on set of four emotion classes, [angry, excited, sad, and neutral] of the improvised, scripted, and all data [17].

For experiments, leave-one-speaker-out (LOSO) cross-validation strategy is used, under which four sessions (i.e., eight speakers) are considered for training, and from the last remaining session, one speaker data is considered for validation and the other speaker data is considered for testing. The IEMOCAP dataset is highly imbalanced, as a result, we calculate both unweighted and weighted accuracy values [18]:

1. **Weighted accuracy (WA)**: It is the number of correctly predicted samples from the total number of samples. This also denotes the *"overall accuracy"*.
2. **Unweighted accuracy (UWA)**: It is the mean of individual class accuracies. The accuracy for a class is computed as the number of correctly predicted samples among the total number of samples in that class. This also denotes *"average class accuracy"* and is same as the *average macro-recall*.

The unweighted accuracy (UWA) is required for an imbalanced dataset, since UWA considers equal weights for each class. In addition, *precision*, *recall*, and F_1-*score* are also used to compare the results.

3.1 Performance Analysis: Single-Label and Multi-label Models

We developed model using both single-label and multi-label.

- **Single-label Model**: To develop the model for single-label, the similar architecture is used that was discussed earlier for multi-label (Fig. 2). However, the single-label model has the output layer with only four neurons for the four discrete emotion classes.
- **Multi-label model**: The multi-label model first used the architecture, shown in Fig. 2, to learn the discrete emotions as well as the continuous emotional attributes, i.e., 13-class problem. The training behavior of the multi-label model is shown in Fig. 4. The blue color indicates the training accuracy, while the orange color indicates the validation accuracy. The maximum validation accuracy was achieved at 63th epoch. At this point, the model was saved and again the spectrogram of each utterance is passed through this model. The output of the global average pooling layer was extracted and used as the feature for the DNN with single-label emotion (Fig. 3).

The t-SNE [19] visualization of features extracted from global average pooling layer using both the single-label and multi-label models is plotted in Fig. 5. Figure 5a shows the features trained using single-label, while Fig. 5b shows the features trained using multi-label. It can be seen from Fig. 5a that the clusters are best formed using multi-label training.

The classification performances of both the single-label model and multi-label model are reported in Table 1. The improvement in UWA using the multi-label features is + 3.0% more than the single-label training. The prediction for *"anger"* class is

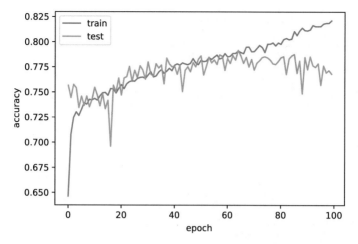

Fig. 4 Training and validation accuracies showing the learning behavior of the multi-label network

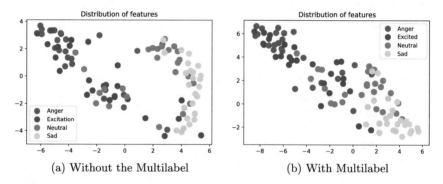

|(a) Without the Multilabel|(b) With Multilabel|

Fig. 5 Distribution of features: the t-SNE plot of the features learned using without and with multi-label is shown in (**a**) and (**b**), respectively

Table 1 Performance comparison: single-label model versus multi-label model

	Single-label model (WA = 65% and UWA = 59%)			Multi-label model (WA = 63% and UWA = 62%)		
Emotion	Precision	Recall	F_1-score	Precision	Recall	F_1-Score
Anger	34	45	39	38	65	48
Excitation	65	48	55	68	43	53
Neutral	65	78	71	63	66	65
Sad	77	64	70	61	74	67

Table 2 Confusion matrix using the multi-label model

Multi-label model (WA = 63% and UWA = 62%)				
Emotion	Anger	Excitation	Neutral	Sad
Anger	**65**	7	22	6
Excitation	10	43	**45**	2
Neutral	5	8	**66**	21
Sad	1	6	19	**74**

Table 3 Comparison with state-of-the-art results on IEMOCAP dataset

Method	Features	WA (%)	UWA (%)
DNN + ELM [21]	MFCC, pitch, and their derivatives	54.34	48.01
LSTM + Attention [20]	Spectrogram, local descriptor	63.50	58.80
DNN-HMM [18]	MFCC	**64.30**	60.86
Multi-label model (proposed)	Spectrogram	63.00	**62.00**

increased in the proposed framework because the auxiliary information (i.e., *valence*) discriminates the *anger* and *excitation* classes in an efficient manner. In contrast, the single-label model has poor prediction for the *anger* class. This is so because the anger emotions are mostly getting classified as the *excitation* emotion. Both *anger* and *excitation* emotions have similar acoustic features.

The confusion matrix for the multi-label model is given in Table 2. As observed, the *excitation* emotion is highly getting classified as the *neutral* emotion. This happens because the *neutral* regions are significantly present in the *excitation* emotion. This is definitely a area of research.

3.2 Comparison with State-of-the-Art Results

Table 3 shows the comparison with state of the result on the IEMOCAP dataset. The best weighted accuracy (WA) is observed with the DNN-HMM-based model that uses MFCC as the input feature [18]. However, the proposed multi-label model produces the best unweighted accuracy (UWA), improving the results by a margin of + 1.14% over the DNN-HMM-based model [18]. The UWA is a vital metric for the imbalance dataset. The proposed method also outperformed other methods that are based on LSTM + Attention [20] and DNN + ELM [21] by a significant margins.

4 Conclusion

This paper used the multi-label framework to generate discriminative emotion features. The *arousal*, *valence*, and *dominance* information are discretized and used as auxiliary task to support the main task (i.e., discrete emotion classification). The CNN network is used to extract the features, and multi-label training is performed. After training, again the training dataset is passed through the model and the predictions of the global average pooling layer are used as features. The features are further used to train a DNN model for discrete emotion class. Due to the auxiliary labels (*arousal*, *valence*, and *dominance*), the features generated using the multi-label framework are more discriminative than the single-label framework and produced better unweighted accuracy. The multi-label framework can be further extended using the gender and speaker information.

References

1. Schuller B, Rigoll G, Lang M (2004) Speech emotion recognition combining acoustic features and linguistic information in a hybrid support vector machine-belief network architecture. In: 2004 IEEE International conference on acoustics, speech, and signal processing, vol 1. IEEE, pp 1–577
2. El Ayadi M, Kamel MS, Karray F (2011) Survey on speech emotion recognition: features, classification schemes, and databases. Pattern Recogn 44(3):572–587
3. France DJ, Shiavi RG, Silverman S, Silverman M, Wilkes M (2000) Acoustical properties of speech as indicators of depression and suicidal risk. IEEE Trans Biomed Eng 47(7):829–837
4. Akagi M, Han X, Elbarougy R, Hamada Y, Li J (2014) Toward affective speech-to-speech translation: strategy for emotional speech recognition and synthesis in multiple languages. In: Signal and information processing association annual summit and conference (APSIPA), 2014 Asia-Pacific. IEEE, pp 1–10
5. Lee C-C, Mower E, Busso C, Lee S, Narayanan S (2011) Emotion recognition using a hierarchical binary decision tree approach. Speech Commun 53(9–10):1162–1171
6. Caruana R (1997) Multitask learning. Mach Learn 28(1):41–75
7. Xia R, Liu Y (2015) A multi-task learning framework for emotion recognition using 2d continuous space. IEEE Trans Affect Comput 8(1):3–14
8. Collobert R, Weston J (2008) A unified architecture for natural language processing: deep neural networks with multitask learning. In: Proceedings of the 25th international conference on Machine learning. ACM, pp 160–167
9. Seltzer ML, Droppo J (2013) Multi-task learning in deep neural networks for improved phoneme recognition. In: 2013 IEEE International conference on acoustics, speech and signal processing. IEEE, pp 6965–6969
10. Devries T, Biswaranjan K, Taylor GW (2014) Multi-task learning of facial landmarks and expression. In: 2014 Canadian conference on computer and robot vision. IEEE, pp 98–103
11. Zhang B, Provost EM, Essl G (2016) Cross-corpus acoustic emotion recognition from singing and speaking: a multi-task learning approach. In: 2016 IEEE International conference on acoustics, speech and signal processing (ICASSP). IEEE, pp 5805–5809
12. Schuller B, Steidl S, Batliner A, Burkhardt F, Devillers L, Müller C, Narayanan SS (2010) The interspeech 2010 paralinguistic challenge. In: Eleventh annual conference of the international speech communication association

13. Chen M, He X, Yang J, Zhang H (2018) 3-d convolutional recurrent neural networks with attention model for speech emotion recognition. IEEE Signal Process Lett 25(10):1440–1444
14. Pedamonti D (2018) Comparison of non-linear activation functions for deep neural networks on MNIST classification task. arXiv preprint arXiv:1804.02763
15. Li P, Song Y, McLoughlin IV, Guo W, Dai L (2018) An attention pooling based representation learning method for speech emotion recognition. In: Interspeech, pp 3087–3091
16. Neumann M, Vu NT (2017) Attentive convolutional neural network based speech emotion recognition: a study on the impact of input features, signal length, and acted speech. arXiv preprint arXiv:1706.00612
17. Busso C, Bulut M, Lee C-C, Kazemzadeh A, Mower E, Kim S, Chang JN, Lee S, Narayanan SS (2008) IEMOCAP: interactive emotional dyadic motion capture database. Lang Resour Eval, 42(4):335
18. Fahad MdS, Deepak A, Pradhan G, Yadav J (2021) DNN-HMM-based speaker-adaptive emotion recognition using MFCC and epoch-based features. Circ Syst Signal Process 40(1):466–489
19. Poličar PG, Stražar M, Zupan B (2019) openTSNE: a modular python library for t-SNE dimensionality reduction and embedding. BioRxiv, p 731877
20. Mirsamadi S, Barsoum E, Zhang C (2017) Automatic speech emotion recognition using recurrent neural networks with local attention. In: 2017 IEEE International conference on acoustics, speech and signal processing (ICASSP). IEEE, pp 2227–2231
21. Han K, Yu D, Tashev I (2014) Speech emotion recognition using deep neural network and extreme learning machine. In: Interspeech 2014
22. Badshah AM, Ahmad J, Rahim N, Baik SW (2017) Speech emotion recognition from spectrograms with deep convolutional neural network. In: 2017 International conference on platform technology and service (PlatCon). IEEE, pp 1–5
23. Mao Q, Dong M, Huang Z, Zhan Y (2014) Learning salient features for speech emotion recognition using convolutional neural networks. IEEE Trans Multimedia 16(8):2203–2213
24. Gupta S, Fahad MdS, Deepak A (2020) Pitch-synchronous single frequency filtering spectrogram for speech emotion recognition. Multimedia Tools Appl 79:23347–23365
25. Fayek HM, Lech M, Cavedon L (2017) Evaluating deep learning architectures for speech emotion recognition. Neural Networks 92:60–68

Intuitionistic Fuzzy Kernel Random Vector Functional Link Classifier

Barenya Bikash Hazarika, Deepak Gupta, and Umesh Gupta

1 Introduction

The support vector machine (SVM) is a popular and powerful machine learning model developed by Cortes and Vapnik [1] which follows the structural risk minimization (SRM) principle. Since it is based on the SRM rule, it minimizes the structural risk and improves the generalization performance. Because of its improved performance, SVM is a commonly used classification model in several fields like pattern classification [2], class imbalance learning [3], and many more [4–7]. However, because the ideal SVM hyperplane is dependent on a limited number of SVs, the SVM is susceptible to noise. Lin and Wang [8] presented a fuzzy SVM (FSVM) to give a new contribution to decision-making by working on a degree of membership (MB) for each training sample. By decreasing the impact of noise and outliers, the FSVM model enhances the performance of SVM [8]. This strategy, on the other hand, has various flaws, for example, the fact that training points located further away from the class center have a greater level of MB than those closer to the class center. To address this difficulty, Atanassov [9] suggested the intuitionistic fuzzy set (IFS), which is more precise in addressing uncertainty concerns and enables precise modeling based on existing data and observations.

B. B. Hazarika
Department of Computer Science and Engineering, Koneru Lakshmaiah Education Foundation, Vaddeswaram, Andhra Pradesh 522502, India

D. Gupta (✉)
Department of Computer Science and Engineering, National Institute of Technology Arunachal Pradesh, Jote, Arunachal Pradesh 791123, India
e-mail: deepakjnu85@gmail.com; deepak@nitap.ac.in

U. Gupta
School of Computer Science Engineering and Technology, Bennett University, Greater Noida, Uttar Pradesh 203206, India

© The Author(s), under exclusive license to Springer Nature Singapore Pte Ltd. 2023   881
D. S. Sisodia et al. (eds.), *Machine Intelligence Techniques for Data Analysis and Signal Processing*, Lecture Notes in Electrical Engineering 997,
https://doi.org/10.1007/978-981-99-0085-5_72

The intuitionistic fuzzy SVM (IFSVM) was proposed by Ha et al. [10, 11]. It does, however, have the ability to lessen the influence of noise to some extent. However, IFSVM needs to solve a quadratic programming problem (QPP) like SVM to get the optimum solution, which increases the computing cost. To resolve this issue, Rezvani et al. [12] came up with an idea of combining the IF number with the popular twin SVM model. IFTSVM solves to small QPPs rather than solving bigger one compared to and thus enhancing the computational efficiency compared to IFSVM. Pao et al. [13] have established that the RVFL is a universal approximator. It provides a fast-training time, a straightforward architecture, and a high degree of generalization. Kernel-based RVFL (K-RVFL) [14], also known as the kernel implementation of RVFL for high generalization and quick learning, was designed to avoid the selection of the number of hidden nodes and hidden mapping function [15]. The K-RVFL is capable to deal with the complex process more effectively due to the complexity embedded in the kernel. A new intuitionistic fuzzy number-based K-RFVL model, IFK-RVFL, is suggested to further boost the generalization performance of K-RVFL.

The prime objectives of this paper are:

1. To propose a novel intuitionistic fuzzy kernel RVFL model for improved classification performance.
2. To analyze the performance of IFK-RVFL with a few related models.

In Sect. 2, the RVFL is discussed. Section 3 elaborates the suggested model. In Sect. 4, numerical experiments are discussed. Lastly, in Sect. 5, the conclusion is explained briefly with future aspects of the work.

2 Related Models

In this section, we discuss a few of the related models. They are the RVFL and the intuitionistic fuzzy number.

2.1 Random Vector Functional Link Network (RVFL)

Assume an input vector of dimension $p \times q$, $x_i \in \Re^q$ where input data matrix is defined as $Z = (x_1, \ldots, x_p)^t$ and $y = (y_1, \ldots, y_p)^t$ considered as the output vector for binary class $+1$ and -1. Here, $\omega_m \in \Re^q$ is the starting seed weight vector to the mth (for $m = 1, \ldots, n$) hidden node, the weight vectors to the hidden layer can be expressed as $W = (\lambda_1, \ldots, \lambda_n)$, where $b \in \Re^n$ is considered as bias with n defines the hidden layer neurons. We have assumed that considered vectors are column vectors. RVFL [13] is also a popular single-layer feedforward network (SLFNs) that arbitrarily allows the weights to the hidden node and sticks them. Here, $\gamma = (\gamma_1, \ldots, \gamma_r)^t$ and $r = n + q$ be the weight vector to the outcome neuron. The activation function is $h_s(x_k) = S(x_s, \lambda, b_k)$ for $s = 1, \ldots, n$ and $k = 1, \ldots, p$ of the sth hidden layer neuron with respect to the kth sample. Z is the $p \times q$ dimensional matrix of training examples. Further, tuning is not required.

The mathematical expression for RVFL [13] is written as,

$$\min \alpha ||\gamma||^2 + || - R\gamma + y||^2, \tag{1}$$

where $R = [H \quad Z]$.

By differentiation (1) with respect to γ, one can find the solution through (2) as follows:

$$\gamma = (R^t R + \alpha I)^{-1} R^t y, \tag{2}$$

where R^t indicates the transpose of R.

At last, for RVFL, the final classifier of any sample is attained as:

$$f(x_s) = \text{sign}([h(x_s) \quad x_s]\gamma), \tag{3}$$

where $H = [h_1(x) \quad \dots \quad h_n(x)]$.

2.2 Intuitionistic Fuzzy Number

An intuitionistic fuzzy (IF) membership degree [12] is having both membership $\Psi(x_i)$ and non-membership $\beta(x_i)$.

2.2.1 The Membership Function ($\Psi(x_i)$)

The IF membership degree ($\Psi(x_i)$) [12] for each sample x_i can be written as:

$$\Psi(x_i) = \begin{cases} 1 - \frac{x_i - N^+}{\xi^+ + c} & \text{if } \psi_i = 1 \\ 1 - \frac{x_i - N^-}{\xi^- + c} & \text{if } \psi_i = -1 \end{cases}, \tag{4}$$

where $c > 0$ considered as a tiny integer to neglect the degree of non-NB in any binary class data points. ψ_i computes the data samples belonging to any of the binary classes. ξ^+ and ξ^- define the radius of the considered binary class data samples. N^+ and N^- signify the centroids of the binary classes, respectively, in such a way:

$$N^+ = \frac{1}{p_i} \sum_{\psi_i = +1} x_i \text{ and } N^- = \frac{1}{p_i} \sum_{\psi_i = -1} x_i. \tag{5}$$

The radius of the binary class data samples is described as:

$$\xi^+ = \max ||x_i - N^+|| \text{ and } \xi^- = \max ||x_i - N^-||.$$

2.2.2 The Non-membership $\beta(x_i)$

The IF non-membership $\beta(x_i)$ [12] is defined as

$$\beta(x_i) = (1 - \Psi(x_i))\Omega(x_i), \tag{6}$$

where $0 \le \Psi(x_i) + \Omega(x_i) \le 1$ and

$$\Omega(x_i) = \frac{\left|\{x_j \ : \ ||x_i - x_j|| \le c, \psi_i \ne \psi_i\}\right|}{\left|\{x_j \ : \ ||x_i - x_j|| \le c\}\right|}. \tag{7}$$

The $c > 0$ is a positive parameter.

2.2.3 The Score Function (SF)

For a new IF MB number, the score function (SF) [12, 16] can be expressed as:

$$\zeta_i = \begin{cases} \Psi_i & \text{if } \beta_i = 0, \\ 0 & \text{if } \Psi_i \le \beta_i, \\ \frac{1-\beta_i}{2-\Psi_i-\beta_i} & \text{otherwise.} \end{cases} \tag{8}$$

3 Intuitionistic Fuzzy Kernel Random Vector Functional Link Network (IFK-RVFL)

As we know RVFL conducts nonlinear mapping of the input data samples to the hidden layer before passing them to the output layer. For attaining the nonlinearity, one can use different popular activation functions. We have to keep in mind that mapping to higher dimensionality from input space before passing to the output layer via direct connection, is one of the ways to attain the nonlinearity. For achieving optimally, we have proposed Intuitionistic fuzzy Kernel Random Vector Functional Link Network (IFK-RVFL) using intuitionistic fuzzy score function, where membership and non-membership functions are defined to give the importance to each data point. Consider the matrix R with the feature mapped input matrix as defined:

$$R = [H \quad \varphi(X)], \tag{9}$$

where $\varphi(X) = (\varphi(x_1), \ldots, \varphi(x_p))^t$ is the feature mapping function.

One can obtain the solution of RVFL (Kernel), from Eq. (9), which can be rewritten with mapped features in this way:

$$\gamma = [H \quad \varphi(X)]^t \left(HH^t + \varphi(X)\varphi(X)^t + \alpha I \right)^{-1} y. \tag{10}$$

Now, implement the kernel trick to the solution for the proposed IFK-RVFL in (10). One can also assume that $k(\cdot, *) = \varphi(\cdot) \cdot \varphi(*) = \varphi(\cdot)^t \varphi(*)$ kernel function, which is selected as random, and the Mercer's kernel matrix is represented in the form:

$$\Pi_G = RR^t : \Pi_{Gi,j} = h(x_i).h(x_j) = K_G(x_i, x_j)$$
$$\Pi_L = X\zeta^t \zeta X^t : \Pi_{Li,j} = x_i x_j = K_L(x_i, x_j). \tag{11}$$

The output function $f(x)$ of IFK-RVFL can be expressed as:

$$f(x) = [h(x) \quad x]R^t \left(RR^t + \frac{I}{\alpha} \right)^{-1} y$$

$$= [h(x) \quad x] \begin{bmatrix} H^t \\ \phi(x)^t \end{bmatrix} \left([H \quad \phi(x)] \begin{bmatrix} H^t \\ \phi(x)^t \end{bmatrix} + \frac{I}{\alpha} \right)^{-1} y$$

$$= \left(h(x)H^t + x\phi(x)^t \right) \left(HH^t + XX^t + \frac{I}{\alpha} \right)$$

$$= \left(\begin{bmatrix} K_G(x, x_l) \\ \vdots \\ K_G(x, x_p) \end{bmatrix} + \begin{bmatrix} K_L(x, x_l) \\ \vdots \\ K_L(x, x_p) \end{bmatrix} \right) \left(\Pi_G + \Pi_L + \frac{I}{\alpha} \right)^{-1} y. \tag{12}$$

At last, the proposed IFK-RVFL model for a new example x can be written in this way:

$$f(x) == \text{sign} \left(\begin{bmatrix} K_G(x, x_l) \\ \vdots \\ K_G(x, x_p) \end{bmatrix} + \begin{bmatrix} K_L(x, x_l) \\ \vdots \\ K_L(x, x_p) \end{bmatrix} \right) \left(\Pi_G + \Pi_L + \frac{I}{\alpha} \right)^{-1} Y. \tag{13}$$

To summarize, in IFK-RVFL, we do not need to choose the number of enhancement nodes or the appropriate mapping function; rather, we must provide the corresponding kernel function.

4 Numerical Experiments

The experiments are performed using MATLAB software on a system with 32 GB RAM, Intel i7 processor embedded with 3.20 GHz processing speed. The Gaussian kernel is used for the experiments which can be shown as: $k(x_p, x_q) = -\exp(-\mu||x_p - x_q||^2)$, where x_p, x_q indicate a data sample. The selection of parameters is very necessary for optimal results for any model. In this work, α and μ are selected from $\{10^{-5}, \ldots, 10^5\}$ and $\{2^{-5}, \ldots, 2^5\}$, respectively, for the reported models along with the proposed IFK-RVFL. The n parameter of RVFL [13] is selected from {20, 50, 100, 200, 500, 1000}. The intuitionistic fuzzy membership parameter ζ for IFSVM [10, 11], IFTSVM [12], and proposed IFK-RVFL is selected from {0.1, 0.2, ..., 1}.

4.1 Simulation on a Few Real-World Datasets

Datasets from different areas are used for experiments. We have collected these datasets from the UCI ML data repository [17]. The description of the datasets is revealed in Table 1.

Classification accuracies with their mean ranks are shown in Table 2. It can be observed from Table 2 that K-RVFL shows comparable or better generalization performance than IFSVM, IFTSVM, RVFL, and K-RVFL. It is further observable from Table 2 that the K-RVFL model shows the lowest average rank and the highest

Table 1 Description of the datasets

Dataset	#Total samples	#Training samples	#Testing samples	#Attributes
Ecoli-0-1_vs_2-3-5	244	147	97	7
Ecoli-0-1_vs_5	240	144	96	6
Ecoli0137vs26	311	187	124	7
Ecoli-0-1-4-6_vs_5	280	168	112	6
Ecoli-0-2-6-7_vs_3-5	224	135	89	7
Ecoli-0-3-4-6_vs_5	205	123	82	7
German credit	1000	600	400	24
Glass-0-4_vs_5	92	56	36	9
Glass2	214	129	85	9
Glass5	214	129	85	9
Haberman	306	184	122	3
LED 7 digit	443	266	177	7
Seeds	210	126	84	7
Yeast3	1484	891	593	8

Table 2 Accuracies and ranks obtained on the real-world datasets (best results are in boldface)

Dataset	IFSVM (rank)	IFTSVM (rank)	RVFL (rank)	K-RVFL (rank)	IFK-RVFL (rank)
Ecoli-0-1_vs_2-3-5	93.75(4)	94.7917(2)	92.7835(5)	93.8144(3)	**95.8763(1)**
Ecoli-0-1_vs_5	95.7895(5)	96.8421(2)	95.8333(3.5)	95.8333(3.5)	**96.875(1)**
Ecoli0137vs26	98.374(2)	95.935(3)	95.1613(4.5)	95.1613(4.5)	**98.3871(1)**
Ecoli-0-1-4-6_vs_5	**100(2)**	**100(2)**	99.6429(4)	**100(2)**	99.1071(5)
Ecoli-0-2-6-7_vs_3-5	**100(2)**	**100(2)**	98.8764(4)	95.5056(5)	**100(2)**
Ecoli-0-3-4-6_vs_5	97.5309(2.5)	97.5309(2.5)	95.122(5)	96.3415(4)	**98.7805(1)**
German credit	75.4386(4)	76.4411(2)	76.15(3)	74.5(5)	**76.5(1)**
Glass-0-4_vs_5	**100(3)**	**100(3)**	**100(3)**	**100(3)**	**100(3)**
Glass2	91.6667(4.5)	91.6667(4.5)	91.7647(3)	**92.9412(1.5)**	**92.9412(1.5)**
Glass5	**98.8095(1)**	97.619(3)	97.4118(4)	96.4706(5)	97.6471(2)
Haberman	72.7273(4)	**76.8595(1)**	64.0984(5)	74.5902(3)	75.4098(2)
Led7digit	92.6136(4)	92.0455(5)	96.8362(2)	94.9153(3)	**97.1751(1)**
Seeds	86.747(4.5)	86.747(4.5)	93.5714(3)	94.0476(2)	**95.2381(1)**
Yeast3	94.0878(2)	93.9189(3)	92.4789(5)	92.5801(4)	**94.7723(1)**
Average	92.6811(3.17)	92.8855(2.82)	92.1236(3.85)	92.6215(3.46)	**94.1935(1.67)**
Win/tie/loss	11/2/2	11/2/2	13/1/1	12/2/1	–

average accuracy. One can see Fig. 1a–b for a clear understanding of the better performance of IFK-RVFL over other reported models in terms of average accuracy and average accuracy ranks. Further, we have shown the win–tie–loss values for each model in the last column of Table 2. The 11/2/2 value of IFSVM indicates that IFK-RVFL wins in 12 cases and ties and losses in 2 cases, respectively. A similar conclusion can be derived for other models.

5 Conclusion and Future Direction

This work proposes a novel RVFL for classification problems that are based on the intuitionistic fuzzy number. In the proposed model, i.e., IFK-RVFL, an intuitionistic fuzzy number is associated with all training samples which consists of both membership and non-membership. Numerical simulations have been undertaken on a few interesting datasets that have been collected from the UCI machine learning repository. Experimental results on a few datasets which are collected from various fields reveal the effectiveness of the proposed IFK-RVFL. No external optimization toolbox was used for solving the optimization problem of IFK-RVFL. Moreover, IFK-RVFL is computationally more efficient compared to IFSVM and IFTSVM. In the future, IFK-RVFL could be remodeled for solving a multiclass classification problem. In addition to that, this model can be effectively used for handling noisy datasets.

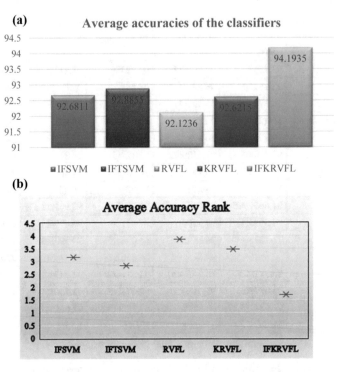

Fig. 1 **a** Performance of the different classifiers in terms of average accuracy obtained on the real-world datasets. **b** Performance of the different classifiers in terms of accuracy ranks obtained on the real-world datasets

References

1. Cortes C, Vapnik V (1995) Support-vector networks. Mach Learn 20(3):273–297
2. Gupta U, Gupta D (2019) Lagrangian twin-bounded support vector machine based on L2-norm. In: Recent developments in machine learning and data analytics. Springer, Singapore, pp 431–444
3. Hazarika BB, Gupta D (2021) Density-weighted support vector machines for binary class imbalance learning. Neural Comput Appl 33(9):4243–4261
4. Borah P, Gupta D (2019) Functional iterative approaches for solving support vector classification problems based on generalized Huber loss. Neural Comput Appl 1–21
5. Hazarika BB, Gupta D, Berlin M (2020) A comparative analysis of artificial neural network and support vector regression for river suspended sediment load prediction. In: First international conference on sustainable technologies for computational intelligence. Springer, Singapore, pp 339–349
6. Gupta U, Gupta D (2021) Kernel-target alignment based fuzzy Lagrangian twin bounded support vector machine. Int J Uncertainty Fuzziness Knowl Based Syst 29(05):677–707
7. Gupta U, Gupta D, Prasad M (2018) Kernel target alignment based fuzzy least square twin bounded support vector machine. In: 2018 IEEE Symposium series on computational intelligence (SSCI). IEEE, pp 228–235
8. Lin C-F, Wang S-D (2002) Fuzzy support vector machines. IEEE Trans Neural Networks 13(2):464–471

9. Atanassov KT (1999) Intuitionistic fuzzy sets. In: Intuitionistic fuzzy sets. Physica, Heidelberg, pp 1–137
10. Ha M, Wang C, Chen J (2013) The support vector machine based on intuitionistic fuzzy number and kernel function. Soft Comput 17(4):635–641
11. Ha MH, Huang S, Wang C, Wang XL (2011) Intuitionistic fuzzy support vector machine. J Hebei Univ (Nat Sci Ed) 3:225–229
12. Rezvani S, Wang X, Pourpanah F (2019) Intuitionistic fuzzy twin support vector machines. IEEE Trans Fuzzy Syst 27(11):2140–2151
13. Pao Y-H, Phillips SM, Sobajic DJ (1992) Neural-net computing and the intelligent control of systems. Int J Control 56(2):263–289
14. Xu KK, Li HX, Yang HD (2017) Kernel-based random vector functional-link network for fast learning of spatiotemporal dynamic processes. IEEE Trans Syst Man Cybern Syst 49(5):1016–1026
15. Chauhan V, Tiwari A, Arya S (2020) Multi-label classifier based on kernel random vector functional link network. In: 2020 International joint conference on neural networks (IJCNN). IEEE, pp 1–7
16. Hazarika BB, Gupta D, Borah P (2021) An intuitionistic fuzzy kernel ridge regression classifier for binary classification. Appl Soft Comput 112:107816
17. Dua D, Graff C (2017) UCI machine learning repository

Mapping of Waterlogged Areas and Silt-Affected Areas After the Flood Using the Random Forest Classifier on the Sentinel-2 Dataset

Shivam Rawat, Rashmi Saini, and Annapurna Singh

1 Introduction

Every year, floods cause huge economic losses, wreaking havoc and resulting in loss of life, infrastructure damage, and loss of livelihood for many people around the world. They may occur for a variety of reasons, such as erratic rainfall, cloudbursts, and climate change [1, 2]. During the monsoon season, which begins in June and lasts until November, Bihar is hit by severe flooding. Waterlogging and silting are two of the major issues that develop after floods submerge an area. Water logging is still a problem until the fall of December. Over 73% of Bihar is at risk of flooding [3].

In the last few years, the remote sensing domain has proven to be extremely useful for mapping the extent of floods by leveraging the potential of very high-quality satellite data and various advanced state-of-the-art modern artificial intelligence algorithms.

This study aims at mapping waterlogged areas and silted areas after the submergence of floods using the pixel-based supervised machine learning approach. The format of this article is as follows: The basic introduction is provided in Sect. 1, and a literature review is provided in Sect. 2. The study area is described in Sect. 3; the dataset preparation and preprocessing are described in Sect. 4; the approach and classifiers used are described in Sect. 5; the results are described in Sects. 6 and 7 summarizes the key conclusions of the study.

S. Rawat (✉) · R. Saini · A. Singh
G.B Pant Institute of Engineering and Technology, Pauri Garhwal 246194, India
e-mail: shivamrawatktd@gmail.com

© The Author(s), under exclusive license to Springer Nature Singapore Pte Ltd. 2023   891
D. S. Sisodia et al. (eds.), *Machine Intelligence Techniques for Data Analysis and Signal Processing*, Lecture Notes in Electrical Engineering 997,
https://doi.org/10.1007/978-981-99-0085-5_73

2 Literature Review

Mapping up of flood entails delineating the water body and flood areas using various geographic information system (GIS) techniques combined with modern, cutting-edge machine learning algorithms. The traditional method for mapping flood extent entails time-consuming and expensive in-person visits [4]. Some of the techniques employed by researchers for mapping floods include masking and thresholding [5], rule-based classification [6], and optimum thresholding based on spectral bands [7].

Satellite-generated land cover maps have become very popular owing to the availability of high-quality data [8]. Two of the most popular satellite datasets are Landsat and Sentinel-2, which provide a spatial resolution of 30 m and 10 m, respectively.

The following are a few of the most widely used indices for flood water delineation: Normalized Difference Water Index (NDWI), modified NDWI (MNDWI), and Automated Water Extraction Index [9–11]. In order to map flood and water zones, the first two indices have been commonly employed. However, these two approaches fall short in their ability to detect floods in metropolitan settings [9, 10]. Using the scikit package, Mobley et al. used random forest (RF) to create a flood map [12]. Due to its huge archive, researchers have used Landsat data extensively to map the extent of the flood. Wang et al. [13] used Landsat-7 TM dataset to determine the flooded areas by separating the water features from non-water features.

Villa and Gianinetto used MNDWI on Landsat data for the flood delineation and mapping of flood damages [14]. Du et al. used MNDWI by pan-sharpening the short-wave infrared band for flood mapping [15].

Land use/land cover (LULC) maps can be produced by using different methods and classification algorithms [16, 17]. The results and the performance of these classification algorithms are influenced by various factors such as the training and testing samples used, the sensor's data employed, and the number of classes [18, 19].

3 Study Area

The study area used in this study is the Khagaria district, located in Bihar, India. The Khagaria district's rectangular extent is given by $25° \ 15' \ 14.1048''$ N to $25° \ 43' \ 54.4512''$ N latitude, and $86° \ 16' \ 44.076''$ E to $86° \ 51' \ 24.696''$ E longitude, with a total area of 1491.84 km^2. Figure 1 describes the study area. Figure 2a, b displays the pre-crisis and post-crisis images.

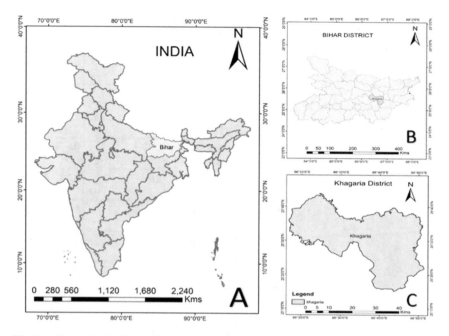

Fig. 1 a The study site located in India. **b** Khagaria district located in the Bihar district. **c** The study area boundary

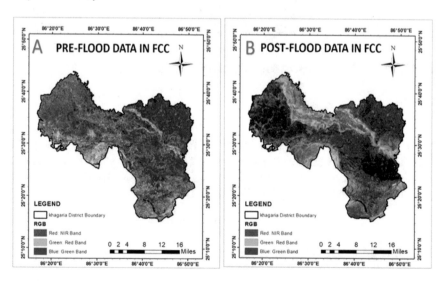

Fig. 2 Khagaria **a** pre-crisis and, **b** post-crisis data in false color composite

Table 1 Band description of Sentinel-2

Band No.	Band name	Resolution (m)	Wavelength (mm)
$(B)^1$	Coastal aerosol band	60	0.433–0.453
$(B)^2$	Blue (B) band	10	0.458–0.523
$(B)^3$	Green (G) band	10	0.543–0.578
$(B)^4$	Red (R) Band	10	0.650–0.680
$(B)^5$	RE vegetation band 1	20	0.698–0.713
$(B)^6$	RE vegetation band 2	20	0.733–0.748
$(B)^7$	RE vegetation band 3	20	0.773–0.793
$(B)^8$	Near infrared (NIR)	10	0.785–0.900
$(B)^{8A}$	Narrow NIR band	20	0.855–0.875
$(B)^9$	Water vapor band	60	0.935–0.955
$(B)^{10}$	SWIR cirrus band	60	1.360–1.390
$(B)^{11}$	$(SWIR)_{Band-1}$	20	1.565–1.655
$(B)^{12}$	$(SWIR)_{Band-2}$	20	2.100–2.280

4 Dataset Preparation and Preprocessing

4.1 Sentinel-2 Data

Sentinel-2A/B satellites were launched by the European Space Agency in the years 2015 and 2017, respectively. It has been extensively used in many fields, such as mapping up of vegetation and LULC, geological/geospatial remote sensing, and ground/surface water mapping [20–24] (Table 1).

4.2 Image Preprocessing

For the given dataset to be free of the effects of scattering and absorption, Sentinel-2 data must undergo atmospheric correction which is done using the sen2cor processor that can be configured using the Sentinel-2 toolbox. It converts the images into their true reflectance form that can be used for further processing [25].

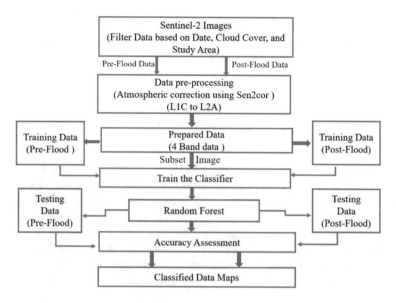

Fig. 3 Classification process used in the study

5 The Methodology and Classifiers Used

5.1 Methodology

In this study, the machine learning-based supervised classification method is employed which uses pixel as an individual unit. Figure 3 describes the overall classification process methodology incorporated in the study. The high-resolution Google Earth imagery is used to create two sets of training data, i.e., one before the crisis and one after the crisis using the prepared dataset. After the classification, the post-flood damages that include waterlogged and silted areas have been calculated using the RF classifier.

5.2 Classifiers Used

Random Forest

RF is a very popular and highly effective machine learning algorithm that uses bootstrapping techniques to build a group of decision trees [26, 27]. It combines the techniques of bootstrapping and random feature selection. Some of the samples in bootstrapping will be chosen frequently, while some of the samples will not be picked at all. Each tree participates in the voting process and based on the votes from the trees, the most popular tree is selected as the output. There are two important parameters in the RF algorithm that need to be tuned, which are mtry (a measure of how many independent variables were sampled at each split) and ntree (the total count of trees) [26].

6 Results and Discussion

This study aims to map the flood damages that consist of waterlogged areas and silted areas after the submergence of floods on the Sentinel-2 dataset using the RF classifier. In this study, a four-band dataset consisting of the red, blue, green, and near-infrared bands is used to classify the pre-crisis and post-crisis images. The training as well as testing data is generated using Google Earth Imagery, and the number of data samples taken is shown in Table 2. The classification is done using the caret package in the R framework using the RF classifier with default values of all the tuning parameters. For the study area, six LULC classes were identified after the careful analysis of the study area, namely (1) water body, (2) vegetation, (3) fallow land, (4) built-up, (5) waterlogged areas, and (6) silted areas, covering a total area of 1491.84 km^2 of the Khagaria district, Bihar (Table 3).

For the accuracy assessment of the classifier, a few accuracy measures are used (overall accuracy (OA), Kappa score (k), precision (P), and recall (R)). For further details, see Tables 4 and 5, which give the confusion matrix for the individual classes for the pre-crisis and post-crisis classified data, respectively. It can be observed from the confusion matrix (Table 5) that there is a misclassification between the waterlogged areas and the water body. Also, a few of the silted areas have been misclassified as built-up areas. Table 6 gives an area change analysis of pre-crisis and post-crisis classified images using the RF. The impacted class consisted of waterlogged areas and silted areas. The quantitative analysis of post-crisis classified images shows that waterlogged areas and silted areas have shown great expansion, which can also be realized by the obtained maps as shown in Fig. 4. The results of the post-crisis dataset

Table 2 Number of testing samples used in the study

Number of samples	Water body	Vegetation	Fallow land	Built-up areas	Waterlogged areas	Silted areas
Pre-crisis	800	900	800	500	500	500
Post-crisis	800	900	600	500	600	600

Table 3 Performance metrics used

Classifier used	Pre-crisis dataset		Post-crisis dataset	
	Overall accuracy	Kappa score	Overall accuracy	Kappa score
RF	84.95	0.817	83.325	0.798

indicate that waterlogged areas have expanded from 22.40 to 245.60 km^2. Whereas the silted areas have increased from 7.22 to 81.53 km^2. The classifier shows good producer and user accuracy for waterlogged areas and silted areas. The resulting recall and precision values achieved using RF for waterlogged areas are 81% and 100% for the pre-crisis classified dataset, and 67.5% and 92.2% the for post-crisis classified dataset, respectively. Whereas for silted areas, the resulting recall and precision values are 60.4% and 86.2% for the pre-crisis classified dataset, and 82.1% and 87.2% for the post-crisis classified data, respectively. The maps show that the majority of the impact is in the district's northwest and southeast regions.

7 Conclusion

The major objective of this study was to use the Sentinel-2 dataset to map the waterlogged and silted areas that had been affected by floods. The OA along with kappa score (k) for pre-crisis classified dataset obtained using RF are 84.95% and 0.817, respectively. Whereas for post-crisis dataset, the accuracy metrics obtained are 83.325% and 0.798, respectively. The results of post-crisis classified data show that waterlogged areas and silted areas have expanded from 22.40 km^2, 7.22 km^2 to 245.60 km^2, 81.53 km^2, respectively. Also, the classifier shows fair producer's and user's accuracy for the affected class that consists of waterlogged areas and silted areas. Waterlogging and siltation make areas inaccessible and unusable for agricultural activity, causing severe damage to the soil and the ecosystem. The majority of the impact is in the district's northwest and southeast regions, according to post-crisis data maps. In future, more sophisticated methods such as deep learning and convolutional neural networks may be used to achieve better results.

Table 4 Confusion matrix for pre-crisis RF classified data

Class	Water body	Vegetation	Fallow land	Built-up areas	Waterlogged areas	Silted areas	Classification overall	User's accuracy (precision)
Water body	784	0	0	0	95	0	879	0.891
Vegetation	0	815	0	0	0	0	815	1.0
Fallow land	0	85	640	0	0	0	725	0.882
Built-up areas	16	0	160	452	0	198	826	0.547
Waterlogged areas	0	0	0	0	405	0	405	1.0
Silt-affected areas	0	0	0	48	0	302	350	0.862
Truth overall	800	900	800	500	500	500	4000	
Producer's accuracy (recall)	0.98	0.905	0.8	0.904	0.81	0.604		

Table 5 Confusion matrix for post-crisis RF classified data

Class	Water body	Vegetation	Fallow land	Built-up areas	Waterlogged areas	Silted areas	Classification overall	User's accuracy (precision)
Water body	735	0	0	0	175	3	913	0.805
Vegetation	2	811	6	0	4	0	823	0.985
Fallow land	0	79	453	36	16	0	584	0.775
Built-up areas	0	2	134	436	0	104	676	0.644
Waterlogged areas	26	8	0	0	405	0	439	0.922
Silt-affected areas	37	0	7	28	0	493	565	0.872
Truth overall	800	900	600	500	600	600	4000	
Producer's accuracy (recall)	0.918	0.901	0.755	0.872	0.675	0.821		

Table 6 Pre/post-crisis classified dataset change analysis

Class description	Pre-crisis data (area in km^2)	Post-crisis data (area in km^2)
Water body	40.18	108.62
Vegetation	708.11	575.31
Fallow land	573.70	354.22
Built-up areas	140.23	126.56
Waterlogged areas	22.40	245.60
Silted areas	7.22	81.53

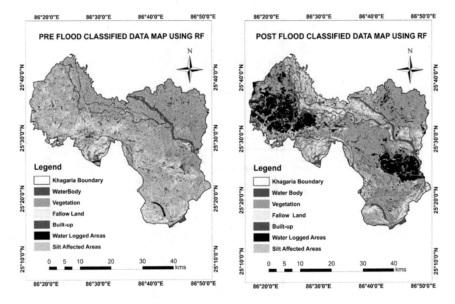

Fig. 4 RF pre-crisis and post-crisis classified data maps

References

1. Bezak N, Mikos M (2019) Investigation of trends, temporal changes in intensity-duration-frequency (IDF) curves and extreme rainfall events clustering at regional scale using 5 min rainfall data. Water 11:2167
2. Schumann GJP, Moller DK (2015) Microwave remote sensing of flood inundation. Phys Chem Earth 83–84:84–95
3. Kansal ML, Kishore KA, Kumar P (2017) Impacts of floods and its management—a case study of Bihar. Int J Adv Res 5:1695–1706
4. Brackenridge R, Anderson E (2006) MODIS-based flood detection, mapping, and measurement: the potential for operational hydrological applications. In NATO Science series IV: earth and environmental sciences
5. Voormansik K, Praks J, Antropov O, Jagomägi J, Zalite K (2013) Flood mapping with TerraSAR-X in forested regions in Estonia. IEEE J Sel Top Appl Earth Obs Remote Sens 562–577

6. Carroll M, Townshend JR, DiMiceli CM, Noojipady P, Sohlberg R (2009) A new global raster water mask at 250 m resolution. Int J Digit Earth 2(4):291–308
7. Wang Z, Liu J, Li J, Zhang D (2018) Multi-spectral water index (MuWI): a native 10-m multi-spectral water index for accurate water mapping on Sentinel-2. Remote Sens 10(10):1643
8. Tanguy M, Chokmani K, Bernier M, Poulin J, Raymond S (2017) River flood mapping in urban areas combining Radarsat-2 data and flood return period data. Remote Sens Environ 198:442–459
9. McFeeters SK (1996) The use of the normalized difference water index (NDWI) in the delineation of open water features. Int J Remote Sens 17(7):1425–1432
10. Xu H (2006) Modification of normalised difference water index (NDWI) to enhance open water features in remotely sensed imagery. Int J Remote Sens 27(14):3025–3033
11. Feyisa GL, Meilby H, Fensholt R, Proud SR (2014) Automated water extraction index: a new technique for surface water mapping using Landsat imagery. Remote Sens Environ 140:23–35
12. Mobley W, Sebastian A, Blessing R, Highfield WE, Stearns L, Brody SD (2021) Quantification of continuous flood hazard using random forest classification and flood insurance claims at large spatial scales: a pilot study in southeast Texas. Nat Hazards Earth Syst Sci 21:807–822
13. Wang Y, Colby JD, Mulcahy KA (2002) An efficient method for mapping flood extent in a coastal floodplain using Landsat TM and DEM data. Int J Remote Sens 23(18):3681–3696
14. Villa P, Gianinetto M (2006) Monsoon flooding response: a multi-scale approach to water-extent change detection
15. Du Y, Zhang Y, Ling F, Wang Q, Li W, Li X (2016) Water bodies' mapping from Sentinel-2 imagery with modified normalized difference water index at 10-m spatial resolution produced by sharpening the SWIR band. Remote Sens 354:1–19
16. Waske B, Braun M (2009) Classifier ensembles for land cover mapping using multitemporal SAR imagery. ISPRS J Photogramm Remote Sens 64:450–457
17. Li C, Wang J, Wang L, Hu L, Gong P (2014) Comparison of classification algorithms and training sample sizes in urban land classification with Landsat Thematic Mapper imagery. Remote Sens 6:964–983
18. Heydari SS, Mountrakis G (2018) Effect of classifier selection, reference sample size, reference class distribution and scene heterogeneity in per- pixel classification accuracy using 26 Landsat sites. Remote Sens Environ 204:648–658
19. Hamad R (2020) An assessment of artificial neural networks, support vector machines and decision trees for land cover classification using Sentinel-2A data. Appl Ecol Environ Sci 8:459–464
20. Saini R, Ghosh SK (2018) Exploring capabilities of Sentinel-2 for vegetation mapping using random forest. In: International archives of the photogrammetry, remote sensing and spatial information sciences, vol 42, pp 1499–1502
21. Saini R, Ghosh SK (2019) Crop classification in a heterogeneous agricultural environment using ensemble classifiers and single-date Sentinel-2A imagery. In: Geocarto international
22. Luo X, Tong X, Pan H (2020) Integrating multiresolution and multitemporal sentinel-2 imagery for land-cover mapping in the Xiongan New Area, China. IEEE Trans Geosci Remote Sens 59:1029–1040
23. Werff H, Meer F (2016) Sentinel-2A MSI and Landsat 8 OLI provide data continuity for geological remote sensing. Remote Sens 8
24. Du Y, Zhang Y, Ling F, Wang Q, Li W, Li X (2016) Water bodies' mapping from sentinel-2 imagery with modified normalized difference water index at 10-m spatial resolution produced by sharpening the SWIR band. Remote Sens 8(4):354
25. Main-Knorn M, Pflug B, Louis J, Debaecker V, Müller-Wilm U, Gascon F (2017) Sen2Cor for Sentinel-2. In: Conference-proceedings-of-SPIE, image and signal processing for remote sensing, vol 10427
26. Belgiu M, Drăguţ L (2016) Random forest in remote sensing: a review of applications and future directions. ISPRS J Photogramm Remote Sens 114:24–31
27. Breiman L (2001) Random forests. Mach Learn 45(1):5–32

The book series *Lecture Notes in Electrical Engineering* (LNEE) publishes the latest developments in Electrical Engineering—quickly, informally and in high quality. While original research reported in proceedings and monographs has traditionally formed the core of LNEE, we also encourage authors to submit books devoted to supporting student education and professional training in the various fields and applications areas of electrical engineering. The series cover classical and emerging topics concerning:

- Communication Engineering, Information Theory and Networks
- Electronics Engineering and Microelectronics
- Signal, Image and Speech Processing
- Wireless and Mobile Communication
- Circuits and Systems
- Energy Systems, Power Electronics and Electrical Machines
- Electro-optical Engineering
- Instrumentation Engineering
- Avionics Engineering
- Control Systems
- Internet-of-Things and Cybersecurity
- Biomedical Devices, MEMS and NEMS

For general information about this book series, comments or suggestions, please contact leontina.dicecco@springer.com.

To submit a proposal or request further information, please contact the Publishing Editor in your country:

China

Jasmine Dou, Editor (jasmine.dou@springer.com)

India, Japan, Rest of Asia

Swati Meherishi, Editorial Director (Swati.Meherishi@springer.com)

Southeast Asia, Australia, New Zealand

Ramesh Nath Premnath, Editor (ramesh.premnath@springernature.com)

USA, Canada

Michael Luby, Senior Editor (michael.luby@springer.com)

All other Countries

Leontina Di Cecco, Senior Editor (leontina.dicecco@springer.com)

** This series is indexed by EI Compendex and Scopus databases. **

Lecture Notes in Electrical Engineering

Volume 997